Perspectives on European Earthquake Engineering and Seismology

GEOTECHNICAL, GEOLOGICAL AND EARTHQUAKE ENGINEERING

Volume 34

Series Editor

Atilla Ansal, *School of Engineering, Özyeğin University, Istanbul, Turkey*

Editorial Advisory Board

Julian Bommer, *Imperial College London, U.K.*
Jonathan D. Bray, *University of California, Berkeley, U.S.A.*
Kyriazis Pitilakis, *Aristotle University of Thessaloniki, Greece*
Susumu Yasuda, *Tokyo Denki University, Japan*

For further volumes:
http://www.springer.com/series/6011

Atilla Ansal
Editor

Perspectives on European Earthquake Engineering and Seismology

Volume 1

Editor
Atilla Ansal
School of Engineering
Özyeğin University
Istanbul, Turkey

ISSN 1573-6059 ISSN 1872-4671 (electronic)
ISBN 978-3-319-07117-6 ISBN 978-3-319-07118-3 (eBook)
DOI 10.1007/978-3-319-07118-3
Springer Cham Heidelberg New York Dordrecht London

Library of Congress Control Number: 2014946618

© The Editor(s) (if applicable) and the Author(s) 2014. The book is published with open access at SpringerLink.com.

Open Access This book is distributed under the terms of the Creative Commons Attribution Noncommercial License which permits any noncommercial use, distribution, and reproduction in any medium, provided the original author(s) and source are credited.

All commercial rights are reserved by the Publisher, whether the whole or part of the material is concerned, specifically the rights of translation, reprinting, re-use of illustrations, recitation, broadcasting, reproduction on microfilms or in any other way, and storage in data banks. Duplication of this publication or parts thereof is permitted only under the provisions of the Copyright Law of the Publisher's location, in its current version, and permission for commercial use must always be obtained from Springer. Permissions for commercial use may be obtained through RightsLink at the Copyright Clearance Center. Violations are liable to prosecution under the respective Copyright Law.

The use of general descriptive names, registered names, trademarks, service marks, etc. in this publication does not imply, even in the absence of a specific statement, that such names are exempt from the relevant protective laws and regulations and therefore free for general use.

While the advice and information in this book are believed to be true and accurate at the date of publication, neither the authors nor the editors nor the publisher can accept any legal responsibility for any errors or omissions that may be made. The publisher makes no warranty, express or implied, with respect to the material contained herein.

Printed on acid-free paper

Springer is part of Springer Science+Business Media (www.springer.com)

Preface

The collection of articles compiled in this first volume of the series entitled as *Perspectives on European Earthquake Engineering and Seismology* is composed of some of the keynote and theme lectures presented during the Second European Conference on Earthquake Engineering and Seismology (2ECEES) held in Istanbul. The remaining keynote and theme lectures will be compiled in the second volume of the series that will be published after the Conference. Since the Conference is a joint event of European Association of Earthquake Engineering (EAEE) and European Seismological Commission (ESC), the lectures thus articles cover the major topics of earthquake engineering and seismology along with priority issues of global importance.

On the occasion of the 50th anniversary of the establishment of the European Association of Earthquake Engineering, and for the first time in the book series on *Geotechnical, Geological, and Earthquake Engineering*, we will be publishing an open access book that can be downloaded by anybody interested in these topics. We believe that this option adopted by the Advisory Committee of 2ECEES will enable wide distribution and readability of the contributions presented by very prominent researchers in Europe.

The articles in this first volume are composed of five keynote lectures, first of which given by Robin Spence, the recipient of the third Prof. Nicholas Ambraseys Lecture Award. His lecture is titled *"The full-scale laboratory: the practice of post-earthquake reconnaissance missions and their contribution to earthquake engineering"*. The other four keynote lectures are by Mustafa Erdik on *"Rapid earthquake loss assessment after damaging earthquakes"*, Paolo E. Pinto on *"Existing buildings: the new Italian provisions for probabilistic seismic assessment"*, Matej Fischinger on *"Seismic response of precast industrial buildings"*, and Marco Mucciarelli on *"The role of site effects at the boundary between seismology and engineering: lessons from recent earthquakes"*.

The remaining 15 chapters are the EAEE Theme Lectures that are presented by: Tatjana Isakovic on *"Seismic analysis and design of bridges with an emphasis to Eurocode standards"*, Michael N. Fardis on *"From performance- and displacement-based assessment of existing buildings per EN1998-3 to design of new concrete structures in fib MC2010"*, Elizabeth Vintzileou on *"Testing of historic masonry structural elements and/or building models"*, Carlos Sousa Oliveira on *"Earthquake risk reduction: from scenario simulators including systemic interdependency to impact indicators"*, Roberto Paolucci on *"Physics-based earthquake ground shaking scenarios in large urban areas"*, Gian Michele Calvi on *"A seismic performance classification framework to provide increased seismic resilience"*, Katrin Beyer on *"Towards displacement-based seismic design of modern unreinforced masonry structures"*, Mario De Stefano on *"Pushover analysis for plan irregular building structures"*, Alessandro Martelli on *"Recent development and application of seismic isolation and energy dissipation and conditions for their correct use"*, Dina D'Ayala on *"Conservation principles and performance-based strengthening of heritage buildings in post-event reconstruction"*, Helen Crowley on *"Earthquake risk assessment: present shortcomings and future directions"*, George Mylonakis on *"The role of pile diameter on earthquake-induced bending"*, Amir Kaynia on *"Predictive models for earthquake response of clay and quick clay slopes"*, Kemal Önder Çetin on *"Recent advances in seismic soil liquefaction engineering"*, and Martin Wieland on *"Seismic hazard and seismic design and safety aspects of large dam projects"*.

    The Editor and the Advisory Committee of the Second European Conference on Earthquake Engineering and Seismology appreciate the support given by the Istanbul Governorship, Istanbul Project Coordination Unit for the publication of the Perspectives on European Earthquake Engineering and Seismology volumes as Open Access books.

Istanbul, Turkey   Atilla Ansal

Contents

1 **The Full-Scale Laboratory: The Practice of Post-Earthquake Reconnaissance Missions and Their Contribution to Earthquake Engineering**... 1
Robin Spence

2 **Rapid Earthquake Loss Assessment After Damaging Earthquakes**.. 53
Mustafa Erdik, K. Şeşetyan, M.B. Demircioğlu, C. Zülfikar, U. Hancılar, C. Tüzün, and E. Harmandar

3 **Existing Buildings: The New Italian Provisions for Probabilistic Seismic Assessment**...................................... 97
Paolo Emilio Pinto and Paolo Franchin

4 **Seismic Response of Precast Industrial Buildings**................. 131
Matej Fischinger, Blaž Zoubek, and Tatjana Isaković

5 **The Role of Site Effects at the Boundary Between Seismology and Engineering: Lessons from Recent Earthquakes**........... 179
Marco Mucciarelli

6 **Seismic Analysis and Design of Bridges with an Emphasis to Eurocode Standards**..................................... 195
Tatjana Isakovic and Matej Fischinger

7 **From Performance- and Displacement-Based Assessment of Existing Buildings per EN1998-3 to Design of New Concrete Structures in *fib* MC2010**.................................. 227
Michael N. Fardis

8 **Testing Historic Masonry Elements and/or Building Models**...... 267
Elizabeth Vintzileou

9 Earthquake Risk Reduction: From Scenario Simulators
   Including Systemic Interdependency to Impact Indicators....... 309
   Carlos Sousa Oliveira, Mónica A. Ferreira, and F. Mota Sá

10 Physics-Based Earthquake Ground Shaking Scenarios in Large
   Urban Areas.. 331
   Roberto Paolucci, Ilario Mazzieri, Chiara Smerzini, and
   Marco Stupazzini

11 A Seismic Performance Classification Framework to Provide
   Increased Seismic Resilience.............................. 361
   Gian Michele Calvi, T.J. Sullivan, and D.P. Welch

12 Towards Displacement-Based Seismic Design of Modern
   Unreinforced Masonry Structures.......................... 401
   Katrin Beyer, S. Petry, M. Tondelli, and A. Paparo

13 Pushover Analysis for Plan Irregular Building Structures....... 429
   Mario De Stefano and Valentina Mariani

14 Recent Development and Application of Seismic Isolation
   and Energy Dissipation and Conditions for Their Correct Use.... 449
   Alessandro Martelli, Paolo Clemente, Alessandro De Stefano,
   Massimo Forni, and Antonello Salvatori

15 Conservation Principles and Performance Based Strengthening
   of Heritage Buildings in Post-event Reconstruction............ 489
   Dina D'Ayala

16 Earthquake Risk Assessment: Present Shortcomings and Future
   Directions.. 515
   Helen Crowley

17 The Role of Pile Diameter on Earthquake-Induced Bending...... 533
   George Mylonakis, Raffaele Di Laora, and Alessandro Mandolini

18 Predictive Models for Earthquake Response of Clay
   and Sensitive Clay Slopes................................ 557
   Amir M. Kaynia and Gökhan Saygili

19 Recent Advances in Seismic Soil Liquefaction Engineering...... 585
   K. Önder Çetin and H. Tolga Bilge

20 Seismic Hazard and Seismic Design and Safety Aspects
   of Large Dam Projects................................... 627
   Martin Wieland

Chapter 1
The Full-Scale Laboratory: The Practice of Post-Earthquake Reconnaissance Missions and Their Contribution to Earthquake Engineering

The Third Nicholas Ambraseys Lecture

Robin Spence

Abstract This paper aims to review the nature and practice of earthquake reconnaissance missions since the earliest examples to today's practice, and to try to show some of the ways in which the practice of earthquake engineering today has benefitted from field observations. To give some historical background, the nature of some of the earliest recorded field missions are reviewed, notably that of Mallet following the 1857 Neapolitan earthquake; the achievements of the UNESCO-supported missions of the period 1963–1980 are considered; and the nature and contributions made by several national earthquake reconnaissance teams (EERI based in the United States, EEFIT based in the UK, and more briefly the Japanese Society for Civil Engineering, the German Earthquake Task Force, and AFPS based in France) are reviewed. The paper then attempts to summarise what have been the most important contributions from the field observations to several aspects of earthquake engineering, particularly to understanding the performance of buildings, both engineered and non-engineered, including historical structures, to geotechnical effects, to gaining understanding of the social and economic consequences of earthquakes, and to loss estimation from future scenario events. The uses and limitations of remote sensing technologies to assess damage caused by an earthquake are considered. Finally, possible changes in earthquake field missions to meet anticipated future challenges and opportunities are discussed.

R. Spence (✉)
Emeritus Professor of Architectural Engineering, Cambridge University, Cambridge, UK

Cambridge Architectural Research Ltd, Cambridge, UK
e-mail: robin.spence@carltd.com

1.1 Introduction

Engineering progresses through innovation, through the development of theories to explain observed phenomena, and through testing of those theories in the laboratory and in the field. In the case of earthquake engineering, field observation assumes a particular importance, because the science which needs to be applied, both in estimating the ground motions to be designed for, and in predicting the performance of structures under these ground motions is still relatively poorly understood, and also because earthquakes occur in any one location so infrequently.

A decade ago, in his keynote address to the 12th European Conference on Earthquake Engineering (Ambraseys 2002), Nicholas Ambraseys quotes a colleague's definition of the earthquake engineer as the professional who "designs structures whose shapes he cannot analyse, to resist forces he cannot predict, using materials the properties of which he does not understand, but in such a way that the client is not aware of it". Ambraseys was pointing to the alarming fact that for all our scientific and technological achievements, earthquake losses keep increasing with time, stretching the credibility of the earthquake engineering profession: and over many years he strongly argued the need for more systematic learning of the lessons from past earthquakes to improve performance.

The title of this talk is taken from the concluding remarks of Ambraseys' Mallet-Milne Lecture (Ambraseys 1988), which emphasises the importance of field observation through post-earthquake reconnaissance missions, and identifies some of the most important roles of such missions:

> It is increasingly apparent that the site of a damaging earthquake is undoubtedly a full-scale laboratory, in which significant discoveries can be made by keen observers - seismologists, geologists, engineers, sociologists and economists. As our knowledge of the complexity of earthquakes has increased we have become more and more aware of the limitations which nature has imposed on our capacity to predict, on purely theoretical grounds, the performance of engineering structures, of the ground itself or of a community. It is the long-term study of earthquakes and fieldwork that offers the unique opportunity to develop a knowledge of the actual situation created by an earthquake disaster... It is field observations and measurement that allow the interaction of ideas and the testing of theories....Much computer effort has been devoted to solving problems based on guessed parameters ... more data from field observation and measurement are now required.

The major disasters which have occurred since those words were written have only served to demonstrate their validity, and there has, in the last 25 years, been a steady growth in the number and quality of field reconnaissance missions, and in the understanding gained from them of the essential aspects of earthquake actions, the behaviour of different types of structures, and the response of communities in different societies to large earthquakes. But many barriers to the achievement of effective post-event reconnaissance still exist, from organisational and funding difficulties to long delays in the implementation of field observations into design practice.

This paper aims to review the nature and practice of earthquake reconnaissance missions since the earliest examples to today's practice, and to try to point out some of the ways in which the practice of earthquake engineering today has benefitted from field observations. To give some historical background, the nature of some of the earliest recorded field missions will be reviewed; the achievements of the UNESCO-supported missions of the period 1963–1980 will be considered; and the nature and contributions made by several national earthquake reconnaissance teams (EERI in the US, EEFIT in UK, the Japanese Society for Civil Engineering and others) will be reviewed. The paper will finally try to summarise what have been the most important contributions from the field observations to several aspects of earthquake engineering, particularly to understanding the performance of buildings, to geotechnical effects, to gaining understanding of the social and economic consequences of earthquakes, and to loss estimation from future scenario events. The future of earthquake field missions will be discussed.

The UNESCO field missions were interdisciplinary field missions in which engineers studied alongside geologists and seismologists, sciences which depend to a large degree on field observation and measurement, and much was gained from this collaboration. Since about 1980, such interdisciplinary missions have become less common, since the style and timing, as well as the funding of post-earthquake seismological investigations has become very different from that of earthquake engineering missions. A limitation of this paper is that it concentrates on lessons for earthquake engineering rather than seismology, which is a topic for another author.

1.2 Early Field Investigations

Perhaps the earliest field investigation with a scientific purpose was that of De Poardi following the 1627 M = 6.8 earthquake in the Gargano Region on the Adriatic Coast of Southeastern Italy. The earthquake was destructive, with a maximum intensity Imax = X (MCS), and liquefaction along the coast; there was also a strong tsunami that inundated the low-lying coastland (De Martini et al. 2003). De Poardi's map shows the towns and villages affected with different symbols to indicate the different levels of damage (Fig. 1.1). Fish are depicted being thrown out of the coastal Lesina Lake which was seriously affected by the tsunami, corresponding to contemporary eyewitness accounts which reported that the lake completely dried out for many hours after the shock and many fish were stranded. Thus Poardi's map may claim to be the first macroseismic intensity map (De Martini et al. 2003, Musson, pers comm).

The 1755 Lisbon earthquake of course was the occasion for important studies of earthquake and tsunami effects, though since Lisbon, the primary focus of the disaster, was also the capital city these cannot properly be said to be the result of a reconnaissance mission. The Marques de Pombal, Prime Minister at the time, was given charge of the emergency management (as it would today be called), and reconstruction planning. One of his notable moves was the systematic collection of quantitative information on the degree of shaking and the effects it produced. His questionnaire, sent out to local officials and the clergy, included questions such as:

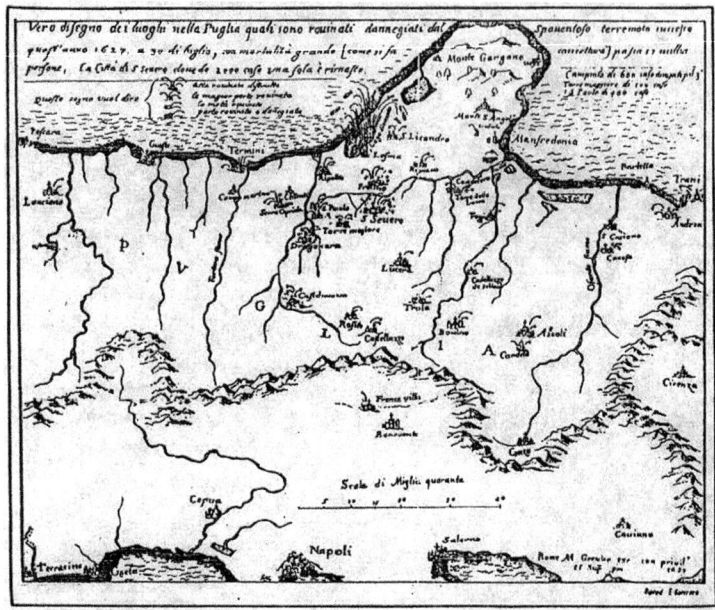

Fig. 1.1 De Poardi's map of the damage caused by the 1627 Gargano earthquake (Based on De Martini et al. 2003, a forerunner of modern isoseismal maps)

How long did the earthquake last? How many shocks were felt? What damage was caused? Did animals behave strangely?, and was thus arguably the forerunner of today's online Did You Feel It? questionnaires (Dewey et al. 2000). Another of Pombal's actions was to order the reconstruction of the Baixa District, close to the Tagus, not in the closely-packed heavy masonry construction which had proved so vulnerable to the ground shaking, but with broad avenues and use of a braced timber frame construction (the gaiola system), which is still the main form of construction in that area today (Cardoso et al. 2013).

1.3 Mallet's Investigation of the 1857 Neapolitan Earthquake

The most significant earthquake reconnaissance mission prior to the twentieth century was undoubtedly that of Robert Mallet, who investigated the effects of the 1857 Great Neapolitan Earthquake, and who in his subsequent report (Fig. 1.2) justifiably laid claim to have established the first principles of observational seismology (a term which Mallet was the first to use).

Mallet, from Ireland, was by profession an engineer, having taken over his father's Dublin foundry at the age of 21. Through involvement with the learned

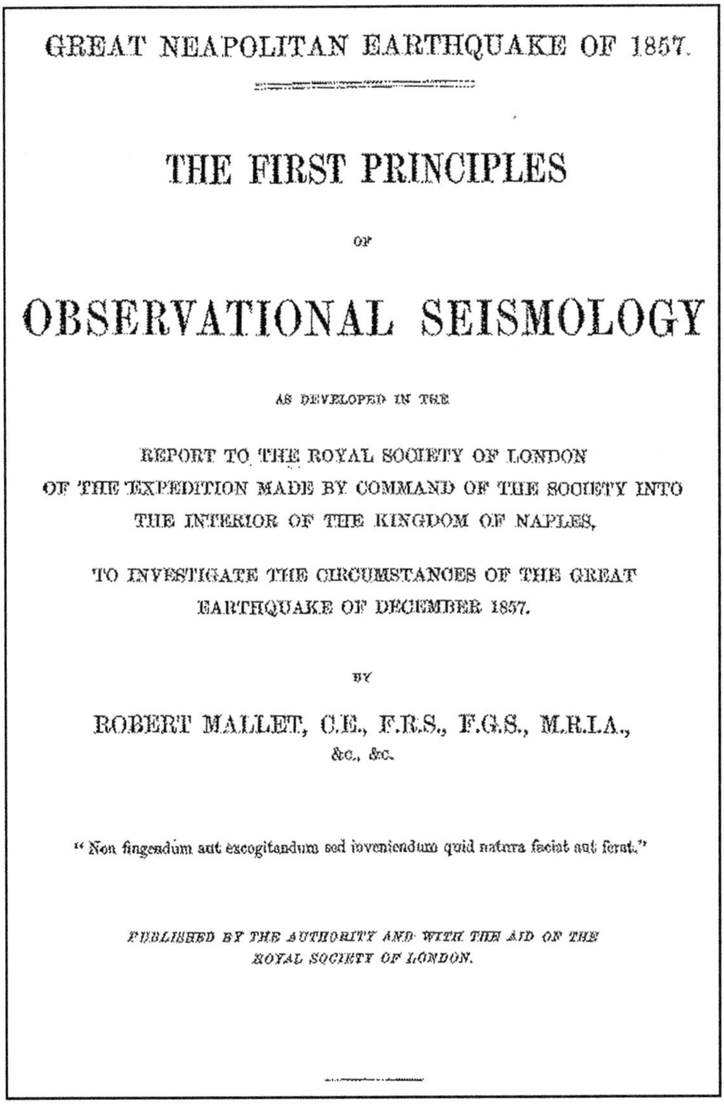

GREAT NEAPOLITAN EARTHQUAKE OF 1857.

THE FIRST PRINCIPLES

OF

OBSERVATIONAL SEISMOLOGY

AS DEVELOPED IN THE

REPORT TO THE ROYAL SOCIETY OF LONDON
OF THE EXPEDITION MADE BY COMMAND OF THE SOCIETY INTO
THE INTERIOR OF THE KINGDOM OF NAPLES,

TO INVESTIGATE THE CIRCUMSTANCES OF THE GREAT
EARTHQUAKE OF DECEMBER 1857.

BY

ROBERT MALLET, C.E., F.R.S., F.G.S., M.R.I.A.,
&c., &c.

"Non fingendum aut excogitandum sed inveniendum quid natura faciat aut ferat."

PUBLISHED BY THE AUTHORITY AND WITH THE AID OF THE ROYAL SOCIETY OF LONDON.

Fig. 1.2 Cover of Mallet's report on the 1857 Neapolitan earthquake

societies of the time, first the Royal Irish Academy and later the British Association, he became interested in earthquake mechanics, and wrote a paper in 1847 in which he set out a view (not in fact a new one, Musson 2013) that an earthquake consists in the transmission through the solid crust of the earth of a wave of elastic compression, and that this could explain the previously observed rotation of monuments in earthquakes. He was convinced that this theory could be used to locate the focus of an earthquake using the effects on buildings and objects at the surface, but he needed a large earthquake to test his hypothesis. This earthquake was to be the Neapolitan earthquake of 1857, a decade later; but before he undertook this field mission, he had made two other important contributions to seismology. The first of these was a large catalogue of over 7,000 historical earthquakes from 1606 BC to 1842, developed from a variety of sources, and accompanied by a map of global seismicity remarkable for its accuracy in identifying most of the earthquake belts known today (notably not the mid-ocean ridges). The second was a design for a seismograph; this was never built but may have influenced the design of Palmieri's later working seismograph.

Mallet explains his purpose in undertaking the mission in the first chapter of his report (Mallet 1862), so elegantly expressed it is worth quoting at length:

> An earthquake, like every other operation of natural forces, must be investigated by means of its phenomena or effects. Some of these are transient and momentary and leave no trace after the shock, and such must either be observed at the time, or had from testimony. But others are more or less permanent and from the terrible handwriting of overturned towns and buildings, may be deciphered, more or less clearly, the conditions under which the forces that overthrew them acted, the velocity with which the ground underneath moved, the extent of its oscillations, and ultimately the point can be found, in position and depth beneath the earth's surface, from which the original blow was delivered, which, propagated through the elastic materials of the mass above and around, constituted the shock. . . . . .
>
> (There are) two distinct orders of seismic enquiry. By the first we seek to obtain information as to the depth beneath the surface of the earth at which those forces are in action whose throbbings are made known to us by the earthquake and thus to make one great and reliable step towards a knowledge of the nature of these forces themselves; and this is the great and hopeful aspect in which seismology must be chiefly viewed and valued. By the second order of enquiry we seek to determine the modifying and moulding power of earthquake on the surface of our world as we now find it; to trace its effects and estimate its power upon man's habitation and upon himself.

Thus Mallet's goals were both seismological and engineering; and the paragraph quoted can indeed be taken, as a statement of the general aims which have guided post-earthquake reconnaissance missions to the present day.

The arrangements made by Mallet for the field mission are instructive, and are set out clearly in the introductory Chapter of his report (Mallet 1862). The earthquake occurred on 16th December 1857, and began to be reported in England about 24th December. On 28th December Mallet wrote to the President of the Royal Society suggesting the importance to science of sending "a competent observer" and offering to undertake this himself, estimating the cost at £50. He received (with the support of Charles Lyell) approval on 21st January, spent the next 5 days getting letters of approval from the Royal Society, the Minister for Foreign Affairs and "some noble or eminent scientific persons" to assist his travel into the earthquake affected area, and departed on 27th January. He travelled overland through Paris and Dijon where he consulted with eminent geologists; arrived in Naples on 5th February, and had to wait for a further 5 days for approval from the King, setting off on 10th February, accompanied by "a trustworthy staff of persons", including an interpreter, who he had recruited while waiting for permission.

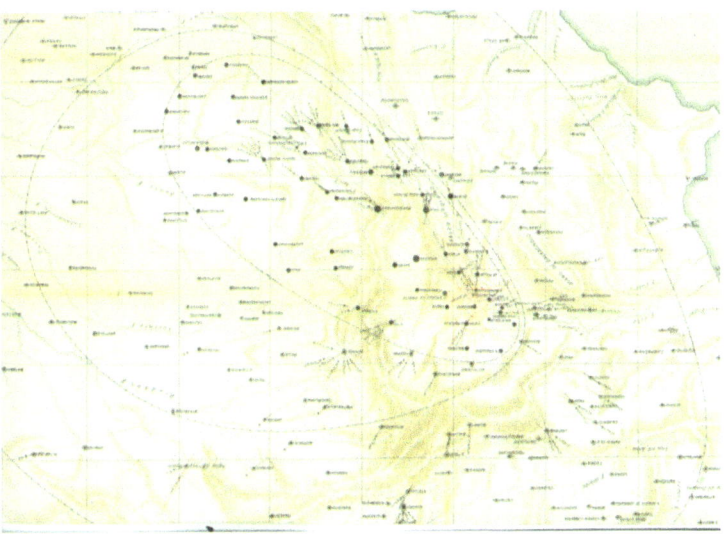

Fig. 1.3 Mallet's isoseismal map of the 1857 Neapolitan earthquake (Mallet 1862)

Once in the field, his method of working was to make use of detailed observations of the effects of the earthquake: cracks in masonry walls, fallen and overturned objects, the size, orientation and displacement of which he used to estimate the direction of the earthquake wave and also its angle of emergence, and even the velocity of the ground shaking. For this purpose he used a series of mechanical equations governing the movement of objects given an initial impulse, and some hypotheses about the position, size and direction of cracking in masonry walls under an emerging earthquake shock. By his own admission it was in many places extremely difficult to make any sense of the chaotic damage visible, but he learnt to make use of a subset of buildings which were typical, suitably oriented, and standing away from adjacent buildings. By plotting the direction and strength of shock in a total of 78 locations, he found a strong convergence and was able to determine a focus (at Caggiano), and plot a series of isoseismals (his own term) (Fig. 1.3) showing areas in four categories, essentially: those destroyed, those heavily damaged with fatalities, those slightly damaged, and those where the earthquake was felt (Musson 2013). He also estimated the focal depth from his estimates of the angle of emergence which had a mean value of 10.6 km.

All of these deductions look reasonable today, but given what we now know about the complexity of ground motion and its effects on buildings, the method of deducing not only direction but also angle of emergence of the earthquake waves is questionable. The chronology of the journey and what was observed at each location is exhaustively recorded in the report, which when finally produced had more than 700 pages. Mallet was also able to commission a photographer, Alphonse Bernoud, to travel the same route later, taking the first earthquake damage photos.

Fig. 1.4 Drawing, based on photograph, of damage in Polla from Mallet's report on the 1857 Neapolitan earthquake (Mallet 1862)

Figure 1.4 is a drawn reproduction of one of several hundred also published with his Report, many of them designed to be viewed stereoscopically.

While the contribution to seismology, and the development of an approach which could be used by others, was the main aim of Mallet's investigation, the report is full of important insights about the local construction techniques of the time and their failings. He makes the observation several times that where buildings are well-built, they were very little if at all damaged by the earthquake. The sketches and photos clearly demonstrate the principal mechanisms of failure of masonry structures, and the attempts to describe these in mathematical equations of equilibrium anticipate later important lines of enquiry about vulnerability and strengthening measures. So does his assembly of the available statistics on

Fig. 1.5 The damage to Polla, in Irpinia, in the 1857 earthquake (from frontispiece of Mallet 1862)

fatalities, which numbered more than 10,000. The concluding remarks in the report are striking:

> All human difficulties to be dealt with must be understood: were understanding and skill applied to the future construction of houses and cities in Southern Italy, few if any human lives need ever again be lost by earthquakes; which must there recur in their times and seasons.

Unfortunately the reconstruction efforts following the 1857 earthquake substantially rebuilt the towns and villages of this area in the same manner as before; and when another major earthquake struck the same region in 1980, the destruction was just as severe and extremely similar in nature to that of 1857, and a further 3,000 deaths occurred. The town of Polla was affected by both earthquakes, and Figs. 1.5 and 1.6 show identical views of Polla following the two events, demonstrating the similarity of the damage, the former from the Mallet report, the later one taken by the author during a field reconnaissance there in 1981 (Spence et al. 1982).

The methods proposed by Mallet did not find immediate scientific application, and his report (perhaps because of its severe criticism of Italian seismologists of the day) was little noticed in Italy until some 20 years after its publication (Ferrari 1987). Then first an Englishman (Johnson-Lavis), and subsequently the great seismologist Giuseppe Mercalli applied Mallet's methods to the 1883 and 1885 earthquakes on the island of Ischia, then to the 1884 Andalusian earthquake and finally to the Ligurian earthquake of 1887, and in the process elaborated and extended them. The method was also taken up in India (Melville and Muir Wood 1987). However, within another 10 years instrumental seismology had arrived, and epicentres were in future to be located by instrumental means, a surer and less time-consuming approach. From the 1890s onwards, field investigations were concerned

Fig. 1.6 Polla from the same location as Fig. 1.6 in 1981 after the Irpinia earthquake. Note similarity of building form and construction (Photo by author)

more with the determination of intensity, using the newly devised macroseismic intensity scale of Rossi and Forel (Melville and Muir Wood 1987). However, Omori (1908), after the 1908 Messina earthquake, used observations of overturned bodies to locate the point of origin of the event.

Fig. 1.7 S.V. Medvedev, S. Bubnov and N.N. Ambraseys, founding members of EAEE, in Skopje in 1964

1.4 UNESCO Field Missions 1962–1980

Over period of nearly 20 years from 1962, UNESCO supported at least 23 post-earthquake reconnaissance missions. Nicholas Ambraseys was the leading figure in this programme: according to Michael Fournier d'Albe, then Head of the UNESCO Natural Hazards Programme, it was Ambraseys who was largely instrumental in persuading the UNESCO Secretariat in the early 1960s that "a useful purpose might be served by UNESCO sending international multidisciplinary teams to conduct field studies of damaging earthquakes as soon as possible after their occurrence", and Ambraseys himself carried out the first of such studies of the Buyin-Zara earthquake in Iran in 1962. He subsequently participated in a further 12 of these studies; he gave a shape and a cohesion to the programme, and he made sure that the findings of the studies were properly recorded and made available to the governments of the countries concerned and to the wider research community.

An important element of the missions was their multi-disciplinarity: they all included seismologists, geologists and engineers. Many distinguished engineers and scientists participated in one or more of the missions, including J. Despeyroux, A Zatopek, A.A. Moinfar, S. Bubnov, T.P. Tassios and J.S Tchalenko. Indeed the 1964 Skopje Conference, at which the European Association for Earthquake Engineering was founded, took place as a direct result of the 1963 UNESCO mission to Skopje (Fig. 1.7).

A summary account of the programme was given by Ambraseys at the Intergovernmental Conference on the Assessment and Mitigation of Earthquake Risk organised by UNESCO in Paris in 1976. The general objective of the missions, simply stated, was "to investigate the cause and effects of such events for the purpose of adding to scientific and practical knowledge for the mitigation of their disastrous consequences" (Fournier-D'Albe 1986). More specifically Ambraseys (1976a) states that:

> It is only through properly-run field studies that ground deformation or faulting associated with an earthquake can be discovered and studied and the bearing on local risk assessed. Existing building codes and regulations as well as the efficacy of their enforcement and implementation, can only be tested after an earthquake. It is only through well-designed and efficient field studies that the economic and social repercussions of an earthquake disaster can be identified so as to avoid undesirable results in future events.

The composition of the missions was dictated by the circumstances, whether the affected area was urban and small, rural and large, or not easily accessible. But a key aspect of the missions was that they were based on a small number of international experts, and drew in expertise from local organisations as far as possible. One further aim was to bring to the country and install a portable network of seismic stations, or at least a strong-motion accelerograph, although that proved possible in only a few cases. There was also a target that the mission should aim to arrive within 72 h of the earthquake's occurrence, but this was never achieved, and the typical delay, mainly due to the waiting for permission from the host Government, was typically 3 weeks. However, once in place, the field studies typically lasted 3 or 4 weeks or more, much longer than is typical of many reconnaissance missions today.

Table 1.1 identifies the earthquakes for which the UNESCO Missions which took place between 1962 and 1980, and Figs. 1.10, 1.11 and 1.12 show the locations of the earthquakes studied. Reports on all these events were published by UNESCO. General features of all these reports are:

- Information on the regional and local seismicity, including usually a detailed listing of all historical and instrumentally recorded damaging earthquakes.
- An account of the actual earthquake and its overall effects, including foreshocks and aftershocks.
- Details and analysis of any strong motion recordings available.
- Detailed description of any surface faulting, and other geological or geotechnical features observed, with maps and photographs.
- Description of typical forms of building construction found, and description, place by place of the extent and types of damage, with maps and photographs.
- Description of notable civil engineering structures and any damage sustained.
- Assessments of macroseismic intensity at the different locations visited, and where possible the preparation of preliminary intensity maps.
- Recommendations for reconstruction.

Overall, this is an immense record of earthquake effects in more than 20 - earthquake-prone locations, involving a huge individual research effort. Ambraseys

Table 1.1 Summary of field missions, 1962–2013, undertaken by UNESCO, EERI, EEFIT, GTF, AFPS and JSCE

Year/Date	Earthquake	Country	Magnitude	EERI	EEFIT	German Taskforce	UNESCO	AFPS	JSCE
01/09/1962	Buyin-Zara	Iran	7.2				x		
21/02/1963	Barce	Libya	5.6				x		
26/07/1963	Skopje	Yugoslavia	6.0				x		
20/03/1966	Toro	Uganda	6.4				x		
19/08/1966	Varto	Turkey	6.8				x		
17/10/1966	Lima	Peru	7.6				x		
22/07/1967	Mudurnu	Turkey	7.1				x		
29/07/1967	Caracas	Venezuela	6.5				x		
10/12/1967	Koyna	India	6.5				x		
01/08/1968	Luzon	Philippines	7.5				x		
31/08/1968	Dasht-e-Bayaz	Iran	7.1				x		
26/10/1969	Banja Luka	Yugoslavia	6.3				x		
08/03/1970	Gediz	Turkey	7.1				x		
07/04/1970	Luzon	Philippines	7.2				x		
31/05/1970	Ancash-Chimbote	Peru	7.5				x		
30/07/1970	Karnaveh	Iran	6.7				x		
09/02/1971	San Fernando, California	USA	6.5	x					
10/04/1972	Ghir	Iran	6.2				x		
23/12/1972	Managua	Nicaragua	6.2	x			x		
28/08/1973	Veracruz	Mexico	7.0	x					
03/10/1974	Lima	Peru	7.2	x					
28/12/1974	Pattan	Pakistan	6.4						
01/08/1975	Oroville, California	USA	5.8	x					
04/02/1976	Motaqua Fault	Guatemala	7.5	x			x		
06/05/1976	Friuli	Italy	6.5	x					
06/05/1976	Gemona	Italy	6.5				x		
28/07/1976	Tangshan	China	7.8	x					

(continued)

Table 1.1 (continued)

Year/Date	Earthquake	Country	Magnitude	EERI	EEFIT	German Taskforce	UNESCO	AFPS	JSCE
17/08/1976	Mindanao	Philippines	7.9	x					
04/03/1977	Vrancea	Romania	7.2	x			x		
23/11/1977	Caucete and San Juan	Argentina	7.4	x					
19/12/1977	Gisk	Iran	5.8				x		
12/06/1978	Miyagi	Japan	7.5	x					
20/06/1978	Salonica	Greece	6.4	x					
29/11/1978	Oaxaca	Mexico	7.9	x					
14/03/1979	Guerrero	Mexico	7.7	x					
15/04/1979	Montenegro	Montenegro	7.2	x					
15/10/1979	Imperial County, California	USA	6.4	x					
24/01/1980	Greenville, California	USA	6.4	x					
09/06/1980	Mexicali, Baja	Mexico	6.2	x					
10/10/1980	El-Asnam	Algeria	7.3	x			x		
08/11/1980	Offshore Trinidad, California	USA	7.0	x					
23/11/1980	Campania-Basilicata	Italy	7.2	x					
24/02/1981	Alcionides Isles	Greece	6.6, 6.3	x					
09/01/1982	Miramichi, New Brunswick	Canada	5.7	x					
19/06/1982	San Salvador	El Salvador	7.0	x					
31/03/1983	Popayan	Colombia	5.5	x					
02/05/1983	Coalinga, California	USA	6.5	x					
26/05/1983	Nihan-Kai-Chubu	Japan	7.7	x					
18/10/1983	Borah Peak, Idaho	USA	7.0	x					
30/10/1983	Northeastern Turkey	Turkey	7.1	x					
08/11/1983	Liege, Belgium	Belgium	5.0			x			
16/11/1983	Kaoiki, Hawaii	USA	6.6	x					
24/04/1984	Morgan Hill, California	USA	6.2	x					
14/09/1984	Nagano-Ken Seibu	Japan	6.9	x					

1 The Full-Scale Laboratory: The Practice of Post-Earthquake Reconnaissance...

Date	Location	Country	Magnitude				
03/03/1985	Llolleo	Chile	7.5	x	x		
19/09/1985	Guerrero-Michoacan (Mexico City)	Mexico	8.0	x	x		
13/09/1986	Kalamata	Greece	6.2	x			
10/10/1986	San Salvador	El Salvador	5.5	x	x		
02/03/1987	Edgcumbe	New Zealand	6.3	x			
01/10/1987	Whittier Narrows, California	USA	5.9	x			
06/11/1988	Yunnan	China	7.6, 7.2	x			
25/11/1988	Saguenay, Quebec	Canada	6.0	x			x
07/12/1988	Spitak	Armenia	6.8	x			x
17/10/1989	Loma Prieta, California	USA	6.9	x	x		
27/12/1989	Newcastle, Australia	Australia	5.4		x		
30/05/1990	Vrancea, Romania	Romania	7.0		x		
16/07/1990	Luzon	Philippines	7.8	x	x		
21/06/1990	Manjil	Iran	7.7	x	x		x
13/12/1990	Augusta, Sicily	Italy	5.6		x		
22/04/1991	Valle de la Estrella	Costa Rica	7.7	x			
20/10/1991	Garhwal	India	6.2	x			
13/03/1992	Erzincan	Turkey	6.9	x	x	x	
13/04/1992	Roermund	Netherlands	5.3	x			x
22/04/1992	Joshua Tree, California	USA	6.1	x			
28/06/1992	Landers and Big Bear, California	USA	6.6	x			
12/10/1992	Cairo	Egypt	5.9	x			
25/03/1993	Scott Mills, Oregon	USA	5.6	x			
12/07/1993	Hokkaido-Nansei-Oki	Japan	7.8	x			
08/08/1993	South End of Island	Guam	8.1	x			
20/09/1993	Klamath Falls, Oregon	USA	5.7	x			
29/09/1993	Killari, Latur District, Maharashtra	India	6.2	x			x
17/01/1994	Northridge, California	USA	6.7	x	x		x

(continued)

Table 1.1 (continued)

Year/Date	Earthquake	Country	Magnitude	EERI	EEFIT	German Taskforce	UNESCO	AFPS	JSCE
04/10/1994	Kuril Islands/Hokkaido Toho-oki	Japan	8.2	x					
15/11/1994	Mindoro Island	Philippines	7.0	x					
17/01/1995	Kobe	Japan	6.9	x	x				
13/05/1995	Grevena (Central-North)	Greece	6.6	x					
30/07/1995	Antofagasta, Chile	Chile	8.0			x			
14/09/1995	Ometepec	Mexico	7.2	x					
01/10/1995	Dinar	Turkey	6.0	x					
07/10/1995	Sungaipenuh	Indonesia	7.0	x					
09/10/1995	Manzanillo	Mexico	7.6	x					
22/11/1995	Aqaba	Egypt	7.1	x					
03/02/1996	Lijiang	China	6.2	x					
17/02/1996	Irian Jaya Region	Indonesia	8.1	x					
18/02/1996	St-Paul-de-Fenouillet	France	5.2					x	
21/02/1996	Chimbote	Peru	7.4	x					
12/11/1996	Nazca	Peru	7.5	x					
10/05/1997	Ardekul	Iran	7.3	x					
22/05/1997	Jabalpur	India	5.8	x					
09/07/1997	Cariaco	Venezuela	6.9	x				x	
26/09/1997	1997 Italy Series (Umbria-Marche)	Italy	5.5, 5.9, 5.3, 5.5, 5.7	x	x				
27/06/1998	Adana-Ceyhan	Turkey	6.2	x				x	x
17/07/1998	Near North Coast	Papua New Guinea	7.0	x					
25/01/1999	El Quindio	Colombia	6.2	x	x			x	x
29/03/1999	Chamoli	India	6.6	x					
08/06/1999	Martinique	Martinique	5.6					x	
15/06/1999	Tehuacan	Mexico	6.5	x					
17/08/1999	Kocaeli	Turkey	7.6	x	x				x

Date	Location	Country	Magnitude				
31/08/1999	Izmit, Turkey	Turkey	5.1				x
07/09/1999	Athens	Greece	5.9	x			
21/09/1999	Chi-Chi	Taiwan	7.7	x	x		x
12/11/1999	Duzce, Turkey	Turkey	7.2		x		
13/01/2001	El Salvador	El Salvador	7.7				x
26/01/2001	Bhuj	India	7.7	x			x
28/02/2001	Nisqually, Washington	USA	6.8	x	x		
23/06/2001	Southern Peru	Peru	8.4	x			x
03/02/2002	Sultandagi	Turkey	6.2	x			x
31/03/2002	Northeast Taiwan	Taiwan	7.1	x			
22/06/2002	Ab garm-abhar-avaj-shirin su	Iran	6.5				x
31/10/2002	Molise	Italy	5.9	x			
02/11/2002	Sumatra	Indonesia	7.4	x			
03/11/2002	Denali, Alaska	USA	7.9	x			
21/01/2003	Colima, Mexico	Mexico	7.6	x			
01/05/2003	Bingöl	Turkey	6.4	x			x
21/05/2003	Boumerdes (Northern Algeria)	Algeria	6.8	x	x		x
26/05/2003	Minami Sanriku	Japan	7.0				x
26/07/2003	Northern Miyagi	Japan	5.1				x
22/09/2003	Puerto Plata	Dominican Republic	6.5	x			
25/09/2003	Hokkaido	Japan	8.3	x			x
22/12/2003	San Simeon, California	USA	6.5	x			
26/12/2003	Bam (Southeastern Iran)	Iran	6.6	x			x
24/02/2004	North Coast	Morocco	6.4	x			x
28/09/2004	Parkfield (Central California)	USA	6.0	x			
23/10/2004	Niigata Ken Chuetsu	Japan	6.6	x			x
21/11/2004	Les Saintes	Guadeloupe	5.1				x
26/12/2004	Sumatra-Andaman Islands and Indian Ocean Tsunami	India, Indonesia, Maldives, Singapore, Sri Lanka, Thailand	9.0	x	x		x
20/03/2005	Kyushu	Japan	6.6				x

(continued)

Table 1.1 (continued)

Year/Date	Earthquake	Country	Magnitude	EERI	EEFIT	German Taskforce	UNESCO	AFPS	JSCE
28/03/2005	Northern Sumatra	Indonesia	8.7	x					x
13/06/2005	Tarapaca	Chile	7.8	x					
08/10/2005	Kashmir	India, Pakistan	7.6	x	x	x			x
08/01/2006	Kythira Island (Southwestern Greece)	Greece	6.9	x					
26/05/2006	Java	Indonesia	6.3	x		x			x
17/07/2006	Java	Indonesia	7.7						x
26/12/2006	Taiwan	Taiwan	7.1	x					
06/03/2007	Western Sumatra, Indonesia	Indonesia	6.4	x					x
25/03/2007	Noto Peninsula (offshore)	Japan	6.7	x					x
16/07/2007	Honshu (offshore)	Japan	6.6	x				x	x
15/08/2007	Near Central Coast, Peru	Peru	8.0	x	x				x
12/09/2007	Southern Sumatra	Indonesia	8.5	x					x
13/09/2007	Sumatra	Indonesia	7.9	x					
14/11/2007	Antofagasta-Tocopilla, Chile	Chile	7.7	x		x			
01/01/2008	Nura, Kygryzsta	Kygryzsta	5.6			x			
12/05/2008	Wenchuan	China	7.9	x	x				x
08/06/2008	Offshore Greece	Greece	6.3	x					
14/06/2008	Honshu	Japan	6.9						x
06/04/2009	L'Aquila, Italy	Italy	6.3	x	x	x			x
28/05/2009	Honduras (offshore)	Honduras	7.3	x					
11/08/2009	Suraga Bay	Japan	6.5						x
02/09/2009	Java, Indonesia	Indonesia	7.0	x					
29/09/2009	Samoan Islands	Samoa	8.0	x	x				
30/09/2009	Padang, Indonesia	Indonesia	7.6	x	x				x
10/01/2010	Eureka	USA	6.5	x					
12/01/2010	Haiti	Haiti	7.0	x	x				

Date	Location	Country	Magnitude				
27/02/2010	Maule, Chile (offshore)	Chile	8.8			x	x
27/02/2010	Tori Shima	Japan	5.2				x
04/04/2010	Baja California, Mexico	Mexico, USA	7.2	x			
04/09/2010	Canterbury, New Zealand	New Zealand	7.1	x			
20/12/2010	Hosseinabad	Iran	6.5	x			
22/02/2011	Christchurch, New Zealand	New Zealand	6.3	x	x		
11/03/2011	Tohoku Japan	Japan	9.0	x	x		
23/08/2011	Virginia	USA	5.8	x			
18/09/2011	Sikkim	Bhutan, India, Nepal	6.9	x			
23/10/2011	Eastern Turkey	Turkey	7.1	x		x	
20/03/2012	Ometepec, Mexico	Mexico	7.4	x			
11/08/2012	Varzaghan-Ahar, Iran	Iran	6.3, 6.4	x			
07/11/2012	Champerico, Guatemala	Guatemala	7.4	x			
11/11/2012	Shwebo, Myanmar	Myanmar	6.8	x			
09/04/2013	Bushehr, Iran	Iran	6.4	x			
20/04/2013	Lushan County, China	China	6.6	x			

said that he had himself spent, in total, more than 5 years of his life on such field investigations. A few of the more notable findings of specific missions are worth summarising.

1.4.1 The $M = 6.1$ Skopje Earthquake of 26 July 1963

The earthquake, though not of great magnitude, was of shallow depth, and had its epicentre close to or within the city. The report concentrated on damage within Skopje itself, a city which had grown very rapidly from a population of 47,000 in 1947 to 220,000 in 1962. Damage was in some areas very severe, but much of the city's infrastructure was left intact or repairable; the spatial damage distribution was difficult to understand. Varying soil conditions, marked variations in the standards of construction, particularly in reinforced concrete structures, and the effect of the 1962 Vardar floods on basements and subsoil conditions were all thought to have played a part. Flexible structures were found to have behaved far better than rigid ones (UNESCO 1963).

1.4.2 The $M = 6.8$ Varto-Üstükran Earthquake of 19 August 1966

Damage was over a wide, largely rural, area of Eastern Turkey, and many houses of traditional adobe or stone masonry construction collapsed. Some houses used reinforced concrete, but construction standards were very poor. It was impossible to assess macroseismic intensity above MMI VII + in rural areas, because in many places all buildings collapsed at this intensity; damage from a series of foreshocks in the months before the August earthquake probably contributed to this. The report concluded that, for this reason, past assessments of intensity in developing countries may have been systematically overestimated (Ambraseys and Zatopek 1967).

1.4.3 The $M = 7.1$ Mudurnu Valley Earthquake of 22 July 1967

This earthquake, on a section of the North Anatolian Fault with many previous recorded events, caused more than 80 km of surface rupture. The fault displacement was traced along the whole of this rupture length, with a maximum right lateral displacement of 1.9 and 1.2 m vertical; observations on power lines suggested that there was considerable additional displacement away from the immediate surface rupture. Damage was very severe over a wide area, but damage in the immediate vicinity of the fault break was no higher than that at distances as much as 10 km from the fault. As for the Varto earthquake, it was impossible to assess intensities above MMI VII because almost all adobe construction collapsed. There was a very large difference between the performance of adobe and timber-frame buildings, which survived well. There were significant ground displacements and associated liquefaction in and around Sapanca Lake (an observation which was to be repeated in the 1999 Kocaeli and Duzce earthquakes, which also affected this part of the fault zone) (Ambraseys et al. 1968).

1.4.4 The $M = 6.4$ Pattan Earthquake of 28 December 1974

The earthquake affected a mountainous region of Northern Pakistan characterised by steep slopes and deep valleys, with a relatively small seasonally migrant agricultural population. The focal depth, as deduced from the seismic array at the Tarbela Dam 130 km south, was relatively shallow, about 5 km, and the directionality of movement was in accordance with expected movement on the Himalayan thrust; however there was no observed surface faulting. Widespread rockfalls damaged roads; and the earthquake occurred in winter, making access to many of the affected places difficult. Nevertheless the UNESCO team were able to visit most of the worst damaged settlements, often on foot, and record the damage distribution. Stone masonry is a common material of construction in the area, and marked differences in level of damage were noted according to the form of construction. In many cases the roofs (flat packed earth on timber rafters), were supported independently of the timber-laced rubble-filled walls on separate timber columns (Fig. 1.8); in other cases the roofs were directly supported on the walls. The houses which had bearing walls were found to have suffered severely from the earthquake, but those with independent columns much less. (This observation was to be followed up in the 1980 International Karakoram Project, Spence et al. 1983). There were very few modern structures in the area. Brick masonry buildings with good quality mortar were little damaged, but others were damaged severely. Bridges generally survived intact, but the Karakoram Highway was seriously affected by rockfalls in many places (Ambraseys et al. 1975).

1.4.5 The $M = 6.3$ Gemona di Friuli Earthquake of 6 May 1976

The earthquake was the first visited by a UNESCO team to occur in an area with a large number of buildings of historical importance. The main objective of this mission was to study damage to structures, rather than investigate the geological and seismological aspects. The team accordingly consisted of two architects and an

Fig. 1.8 Stone masonry buildings with roofs supported on timber columns independently of the walls, from UNESCO mission to 1974 Pattan earthquake (Ambraseys et al. 1975)

engineer (Ambraseys), and the report divides into three separate parts. The report notes the unusually large number of large aftershocks, associated with earthquakes in this region. The damage caused by this repeated activity compounded that resulting from the age and poor quality of much construction. Construction methods typical of the Friuli region of Italy are described in detail, and the many weaknesses in the stone masonry leading to damage and collapse are described; these were further compounded by improper repair, war damage and previous earthquakes; it is noted, though, that many houses were saved from collapse by the use of tie-rods in masonry walls which held walls together. A detailed listing of the historic churches and palazzi damaged by the earthquake is given, covering a very wide area but with a particular concentration in the historic towns of Gemona and Venzone. A section discusses the loss of life and injuries, and its demographic distribution, and analyses possible reasons for higher casualties among the young adult population in the older town centres, probably the first time this issue was considered in a field mission report (Ambraseys 1976b).

1.4.6 The $M = 7.2$ Romania Earthquake of 4 March 1977

This report, like that of Friuli, is also a compilation of separate reports, that of Ambraseys dealing with the earthquake and its principal effects, and that of Despeyroux dealing with the behaviour of buildings. The earthquake was deep (110 km); it occurred in the same area, with a very similar magnitude and depth as a previous one in 1940 (and, indeed a later one in 1990). Both earthquakes caused moderate damage over a wide area (around 80,000 km^2), with a particular concentration of damage in Bucharest about 200 km away from the focus. Much of the damage was sustained by older reinforced concrete frame buildings which had either been damaged in 1940 or built without provision for earthquake loading (Fig. 1.9). By contrast, small brick bearing wall structures suffered relatively minor damage. The recording of a strong motion accelerograph from the Building

Fig. 1.9 Damage to older reinforced concrete buildings in Bucharest in the 1977 Romania earthquake, recorded by the UNESCO mission (Ambraseys and Despeyroux 1978)

Research Institute in Bucharest was analysed, and the response spectrum approximately extracted, showing a peak between 1.5 and 2 s. The concentration of damage in 6–12 storey RC frame buildings (with fundamental periods of 0.7–1.6 s (the ascending branch of the response spectrum) is thus explained. Over the whole affected area, intensity assessment was made very difficult because of the lack of damage caused by high-frequency ground motion. Earlier attempts to provide a microzonation of Romania and Bucharest are shown to have been ineffective for this event: the report notes that there was "not the slightest similarity in pattern between the predicted and observed damage pattern". The importance of reconsidering the design codes to be able to deal with both long–period motions from distant earthquakes and local, shallow earthquakes is emphasised (Ambraseys and Despeyroux 1978).

The 1980 El Asnam Mission was the last such UNESCO mission. While it lasted, the UNESCO programme made vital contributions to the understanding of earthquake effects across a wide area of the world. Fault systems were mapped, ground motion and response spectra and their distribution was reported and analysed where possible, the distribution of damage across the affected zone was explored, the effects of subsoil conditions investigated, and the performance of a variety of types of building, including historical structures in several cases, was also investigated. One particular aspect of this was demonstrating the relative performance of different traditional building types in a way which is today less common, as field missions nowadays concentrate more on engineered and modern structures. The style and contents of the UNESCO Reports was to become a template for those of later field missions.

But the inherent inter-disciplinarity of the UNESCO Missions was perhaps difficult to keep going as the field investigation techniques of the different disciplines matured and it also became more common to involve research students in the data collection. And as Fournier-D'Albe (1986) states in the Foreword to the compilation of field reports, the administrative obstacles that such UN-sponsored international missions had to overcome were steadily increasing. From 1980 onwards earthquake engineering reconnaissance missions organised by national societies, and supported by research councils and by industry began to become more common, while earth scientists have tended to conduct separate studies with different itineraries and timescales.

1.5 EERI Learning from Earthquakes Programme (1972–2014)

The Earthquake Engineering Research Institute, based in Oakland California, was founded in 1949, and has conducted post-earthquake field investigations, both of US and non-US earthquakes from its inception. However, until 1971 these missions were ad-hoc responses to the events, largely focussed on investigating damage to buildings. The 1971 San Fernando earthquake was the stimulus to establishing EERI's Learning from Earthquakes (LFE) programme; it became clear from that event that advance planning and coordination would have been beneficial to achieve the maximum benefit in understanding the damage, and ensuring that all aspects of the event were examined, and avoiding the tendency of individual surveys to duplicate each others' investigations. The LFE programme was formalised in 1973, with three principal activities: conducting field investigations; developing guidelines for conducting post-earthquake investigations that enable consistent data to be collected; and disseminating the lessons learned (EERI 1986, 1995a). For many years funding for the LFE programme has been provided by the US National Science Foundation.

Today, after mounting investigations of nearly 300 events, EERI has developed a highly professional approach to the mounting and management of field missions, and can claim to be the world's leading earthquake field investigation organisation. With a large worldwide individual membership, EERI is in many respects an international organisation with a global outreach. As well as documenting each separate mission, EERI has also documented the overall learning from its field missions in a number of different publications (EERI 1986, 2004).

EERI is notified on a 24-h basis of all global earthquakes likely to have been damaging by the National Earthquake Information Service of USGS; the Executive Committee then has responsibility for deciding which earthquakes EERI will investigate. The level of response is determined by the location and extent of damage. In general terms for small earthquakes in the USA or for moderate earthquake abroad, EERI identifies members in the area who can be asked to conduct a short investigation and produce a brief report for the EERI newsletter. For earthquakes outside the USA which "hold potentially significant lessons for US practice", a multidisciplinary reconnaissance team of 4–8 members is sent into the field for typically 1 week or more; EERI members from the affected country are often members, sometimes the leaders, of such reconnaissance teams. The aims of such reconnaissance teams (EERI 1995a) are:

1. To collect the available perishable data in an effort to learn as much as possible about the nature and extent of damage and identify possible gaps in existing research or in the practical application of scientific, engineering and policy knowledge, and
2. To make recommendations regarding the need for further research and suggest possible foci.

For significant earthquakes in the USA, similar reconnaissance teams may be mounted, but for US events EERI also works closely with local universities or companies which are mounting their own investigations to ensure that all available observations are assembled and reported.

In either case the findings of each reconnaissance mission are recorded in a Reconnaissance Team Report, sometimes, for major events, in a special issue of the EERI journal Earthquake Spectra (www.earthquakespectra.org), and more recently by an online report. All reports are available through a web portal at https://www.eeri.org/projects/learning-from-earthquakes-lfe/lfe-reconnaissance-archive/.

Aspects normally investigated by EERI post-earthquake reconnaissance missions are very broad, but generally include the following:

- Geosciences
- Geotechnical engineering
- Engineered buildings
- Industrial facilities
- Lifelines and transportation structures
- Architectural and non-structural elements
- Emergency management and response
- Societal impacts
- Urban planning and public policy implications

Each of these topics normally constitutes a chapter of the final report. Where appropriate a chapter on tsunami impacts may also be included. The level of geoscience investigation varies: but is usually primarily associated with the level of ground shaking and its distribution, with less attention to the investigation of the underlying faulting as was attempted by the earlier UNESCO mission teams.

An important feature of EERI's programme are the detailed procedures laid down for the recruitment, briefing, activity in the field, and post-event debriefing of the reconnaissance team members, all of whom are volunteers. A balanced team

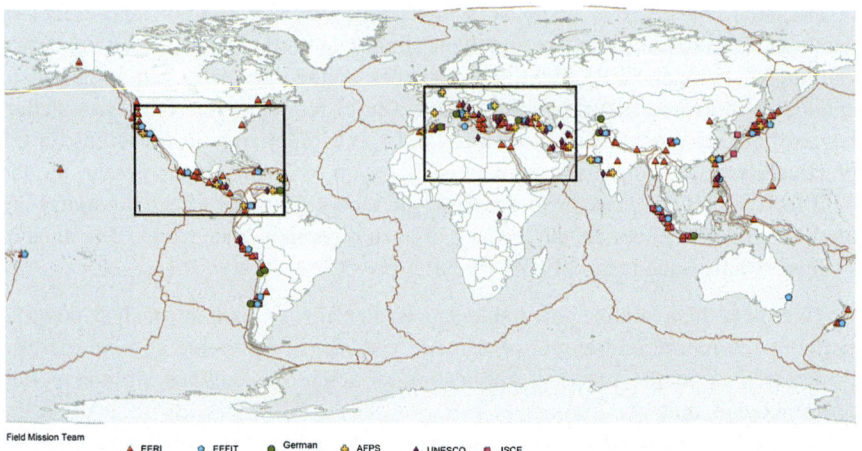

Fig. 1.10 Locations of all field investigations by different reconnaissance teams 1962–2013

membership whose members are experienced and capable to deal with all the above aspects is selected. For non-US earthquakes a team leader and team members able to speak the local language are sought. Advice is also given on dealing with the media, and the responsibility of all team members for contributing to the final reports is emphasised (EERI 1995a).

The total number of field missions of all types conducted by EERI since the 1971 San Fernando earthquake is 290, of which 138 have led to Reconnaissance Team Reports or Earthquake Spectra articles. Of these only 34 were in the USA, Canada or Mexico, the remaining 104 were elsewhere in the world. On average there have been about four such missions per year since 1990. Table 1.1 lists all of the 138 events reported in detail, and their locations are shown on the maps, Figs. 1.10, 1.11 and 1.12.

The cumulative learning from all of these field missions is immense. An early review was made in the publication *Reducing Earthquake Hazards* (EERI 1986), and learning was more briefly reviewed in *Learning from Earthquakes* (EERI 2004). A selection of some of the most important contributions noted by these publications, many of which are now widely accepted generalisations, includes the following.

1.5.1 Contributions to Structural Engineering

It has consistently been found that well-designed, well detailed and well-constructed buildings resist earthquake-induced forces without excessive damage, though designing to code does not necessarily protect against severe damage; damage and collapse of buildings can often be attributed to poor construction

1 The Full-Scale Laboratory: The Practice of Post-Earthquake Reconnaissance...   27

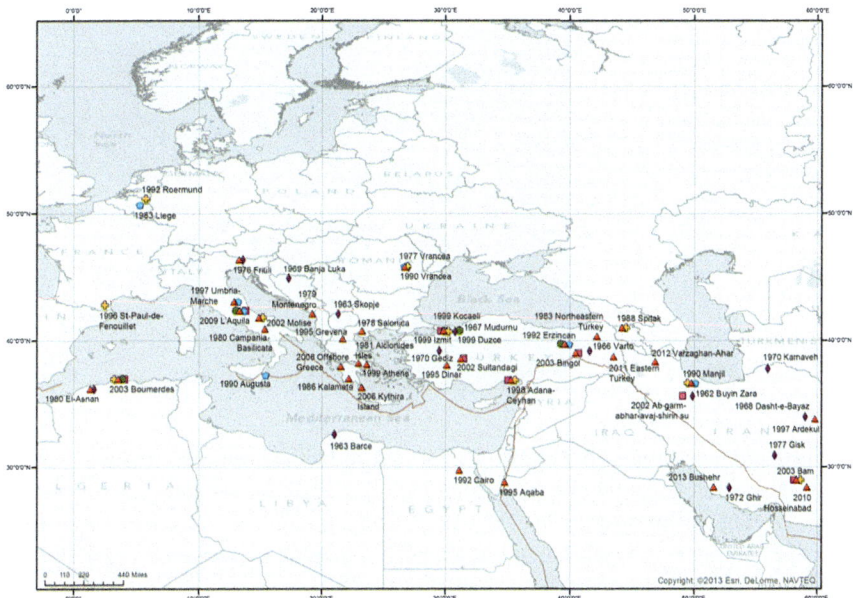

Fig. 1.11 Detail of Fig. 1.10 for European region

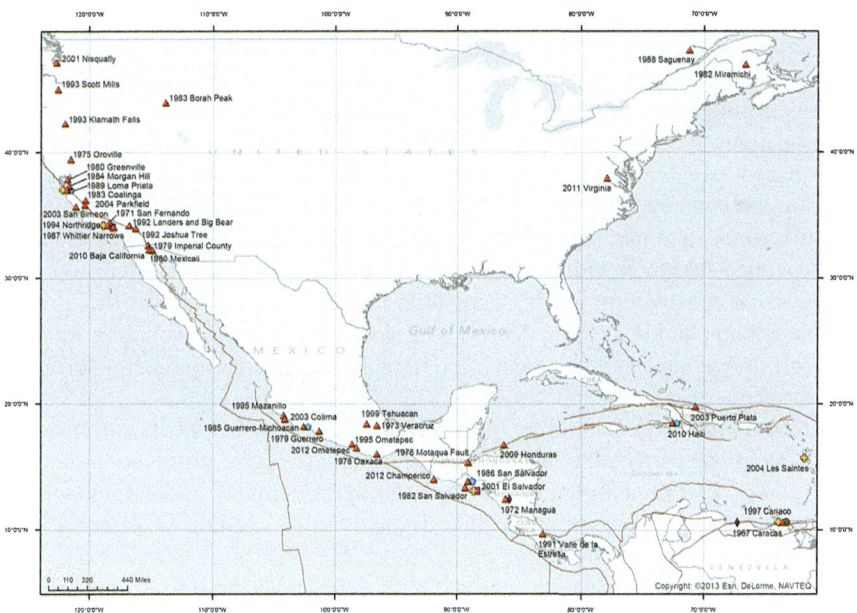

Fig. 1.12 Detail of Fig. 1.10 for USA and Central America region

Fig. 1.13 Damage to precast concrete garage structure in the 1994 Northridge earthquake from the EERI photographic dataset for that earthquake (EERI 1995a)

practice and lack of quality control. Detailing for ductility and redundancy provide safety against collapse: a complete load path designed for seismic forces must be provided. The stiffness of the lateral load resisting system has a major effect on both structural and non-structural damage. Properly designed horizontal diaphragms are essential. Irregularities in both plan and elevation can have a very significant effect on earthquake performance, especially soft stories. Inadequate distance between buildings can result in pounding damage. Stiff elements not considered in the design can strongly affect the seismic response of a building (Fig. 1.13).

The relative performance of structures with different load-resisting systems has shown that unreinforced masonry buildings have performed poorly, though better if strengthened with steel ties; by contrast reinforced and confined masonry buildings have performed well. Steel frame buildings have generally performed well, though investigations following the 1994 Northridge and 1995 Kobe earthquakes found unexpected levels of damage to welded connections. Performance of precast and pre-stressed concrete buildings depend critically on the connection of the elements; exterior panels and parapets need strong anchoring to protect life safety. Though timber frame structures often perform comparatively well, various recent forms of wood frame construction has been found to have serious weaknesses. Reinforced concrete frame buildings often demonstrate similar weaknesses, including the roles of a soft storey, nonductile elements, and irregularities in contributing to damage or collapse.

The importance of such observations consists not only in their occurrence and reporting in one earthquake, but in the repetition of the same observation in many earthquakes in different regions with differing patterns of ground motion, in building stocks designed to different codes and built according to differing local practices.

These and other observations derived from field studies have led, often through subsequent research programmes (such as that of Arnold and Reitherman 1982), to the progressive development of the building codes for earthquake-resistant construction in the USA, from ATC3-06 (ATC 1978) through to the current version of the International Building Code (International Code Council 2012). The US codes, in turn, have influenced earthquake construction codes in other countries of the world. Thus a direct link can be traced between the structural engineering findings of these EERI Field Reconnaissance Missions and today's best practice in the structural design of buildings worldwide. Field mission experience has also led to the definition of a small number of Model Building Types (FEMA 2003; Jaiswal et al. 2011) used in loss estimation studies, and to the development of standards for the evaluation of existing buildings to assess whether they should be strengthened (ASCE 2003). Field investigations have also helped gain acceptance for new technologies such as seismic isolation and semi-active control (Booth, pers comm).

1.5.2 Contributions to Site Effects and Geotechnical Engineering

Field investigations of the distribution of damage, coupled with the increasing availability of strong ground motion recordings of the main shock and aftershocks, has led to a better understanding of the role played by site conditions on the amplification of ground motion and the types and distribution of damage to structures. Prior to 1999 there were only eight strong ground motion recordings worldwide within 20 km of the fault for earthquakes greater than M = 7 (EERI 2004). In the last 15 years this situation has been transformed by the much wider availability of such records which, coupled with field observation of damage, has enabled a much better understanding to be gained of the role played by soil amplification, topographical effects, location in relation to the fault, and the nature of the ground motion, on the damage to structures caused by earthquakes.

As a result of this, it is now widely recognised that no single parameter of ground motion can be used to define the damage capability of strong ground motion, and that features such as fault-rupture type, duration, frequency content, and the ratio of vertical to horizontal ground motion amplitudes have to be considered in different ways for different classes of structures. In some especially well-instrumented events such as 1994 Northridge, effects of ground motion directivity and of high vertical acceleration on damage distributions have been observed. For different regions, ground motion prediction equations have been developed through which it is possible to estimate the ground motion for locations where it has not been measured directly.

Liquefaction effects have been observed in reconnaissance missions following a number of earthquakes, notably 1989 Loma Prieta, 1995 Kobe, 1999 Kocaeli, 2004 Niigata 2010 Haiti and 2011 Christchurch events, which have enabled an extensive database of liquefaction effects to buildings, bridges, port structures and pipelines to be assembled, enabling improvements in the design of such structures in soils with a liquefaction potential. Field missions have enabled similar advances in understanding of the deformations caused by the displacement at surface fault ruptures and by landslides.

1.5.3 Contributions to Lifeline Engineering

Investigations into the performance of lifelines have been a crucial aspect of EERI reconnaissance missions. Bridges and highway structures, gas and water pipelines, and electrical power generation and transmission systems all suffered damage in recent earthquakes. The data assembled by field missions has included damage, lost service and needed repair. This has identified both systems that have performed well and those that failed; and has resulted in numerous changes to design practices including better characterisation of ground motion, better specification of materials, anchorage details and welding practices. Damage to the power supply system in the 1999 Taiwan earthquake demonstrated the importance of building redundancy into lifeline systems.

1.5.4 Contributions to Social Science (and Urban Planning)

Since 1977 social scientists have regularly contributed to field reconnaissance missions, studying aspects of mitigation, response and preparedness, and more recently post-earthquake recovery. These observations have been used in the design of disaster plans for areas of the US which have not experienced an earthquake. From such studies, conducted in many different societies, certain general conclusions have been reached. It is now widely understood that that the most effective search and rescue activity is neighbourhood-based, involving informal groups of individuals who are on the scene because they live or work there; this has been used in the US to develop training programmes for neighbourhood groups. It is also understood that self-protective practices applicable for well-designed structures do not work in poorly built or weak masonry structures. Observations of emergency response procedures adopted in different situations have demonstrated a need for a more integrated approach to building design, land-use planning and emergency response in many seismic hazard areas. Experience in communities affected by tsunamis has provided important lessons in the best way to manage the distribution of warnings to potentially affected communities. Strategies for providing temporary shelter in different societies have been observed and their effectiveness reviewed. More recent field missions have revisited areas affected by earthquakes after a lapse of some months or years, and a database is being assembled of longer-term recovery experiences, which will provide data on the relative success of, for example, centralised or decentralised approaches to recovery. In recent events, the availability of rapid post-event damage estimation (e.g. using the USGS PAGER, or QLARM approach, Jaiswal et al. 2011; Trendafilowski et al. 2011) has enabled an early assessment of recovery needs. The impact of such early warnings has been assessed in recent events.

1.5.5 Use of Information Technology

EERI has been involved in pioneering the use and development of new information technology tools for post-earthquake reconnaissance. High-resolution satellite imagery has now been available for more than 10 years, and was first used to examine damage after the 2001 Gujarat earthquake in India (Saito et al. 2004). More recently, the satellite image providers have been able to rapidly make available before-event and after-event images of the most badly affected areas at less than 1 m resolution, and these have been used to guide reconnaissance missions in the field. Field investigations (2003 Bam, 2010 Haiti) have experimented with the use of VIEWS, a satellite linked video camera for recording damage, enabling a large increase in the speed of capturing building-by-building damage data in ground surveys. In recent earthquakes EERI has, in conjunction with ImageCat, deployed the GEOCAN network, a method of obtaining a rapid building-by-building damage assessment directly from satellite imagery using crowd sourcing (this technology is further discussed in Sect. 1.7). After recent events EERI has established a web-based data assembly and dissemination tool, called the Virtual Clearinghouse, on the EERI website. This enables the field team, the researcher community and EERI to upload data and communicate rapidly. The Virtual Clearinghouse has been mounted for 12 events since 2009.

1.6 EEFIT (1982–2014)

The UK-based Earthquake Engineering Field Investigation Team (EEFIT) was founded in 1982. Its direct origin was a field investigation of the 1980 Irpinia Earthquake in Southern Italy by the author with several UK colleagues (Spence et al. 1982). Because of logistical difficulties, this field investigation did not take place until four months after the earthquake, and it was realised that for field missions to be most effective they should occur earlier; for this to be possible, a team should be ready to mobilise at short notice, with procedures and funding sources in place beforehand. In 1982 EEFIT was formed as "a UK-based group of engineers, architects and scientists who seek to collaborate with colleagues in earthquake-prone countries in the task of improving the seismic resistance of both traditional and engineered structures". It was supported by both the Institution of Civil Engineers though SECED (the British national section of IAEE) and the Institution of Structural Engineers (IStructE). From the outset EEFIT was envisaged as a collaboration between academic institutions and the practising engineering profession.

EEFIT exists solely to facilitate the formation of investigation teams which are able to undertake, at short notice, field studies following major damaging earthquakes and to disseminate the findings to engineers, academics, researchers and extent the general public. The objectives are to collect data and make observations leading to improvements in design methods and techniques for strengthening and retrofit, and where appropriate to initiate longer-term studies. Field training for engineers involved in earthquake-resistant design practice and research is also one of its key objectives. Recently EEFIT has extended its activities by conducting two longitudinal studies, one to L'Aquila (Rossetto et al. 2014) and one to Tohoku, Japan; the objectives of these were to better understand the recovery process and how engineers can contribute to this. The observations and findings from these missions are published in detailed reports and usually include sections on:

- Mission methodology
- The earthquake affected region
- Seismological aspects
- Types of damage, including distribution and extent, on both engineered and non-engineered structures
- Social and economic effects of the earthquakes.

EEFIT reports can be freely downloaded from http://www.istructe.org/resources-centre/technical-topic-areas/eefit/eefit-reports and contain many valuable descriptions of failure and detailed photographs.

For any major reported earthquake, the EEFIT management committee decides whether the event might merit an investigation; if so, EEFIT members are invited to express an interest in joining a mission; the management committee then decides whether a mission is justified, who should be invited to participate and who should be the team leader. The team leader, a person with experience of previous missions and if possible also with knowledge of the country affected, organises the logistics of the mission, including making local contacts and obtaining any permissions needed. Team members are briefed by the team leader including any necessary risk assessments, and asked to sign a form committing them, among other things, to contribute to the final report. Since the late 1980s IStructE has provided the secretarial support for EEFIT. The relatively small recurrent central office costs of running EEFIT are met by IStructE, as well as membership subscriptions and corporate sponsorship. The time and mission expenses of practising engineers are provided by their employers, while the expenses of academic participants is met by specific grants from the UK Engineering and Physical Sciences Research Council, using an accelerated application procedure. Since 2010 EPSRC has provided funding for a 5-year programme of work, which has ensured that reconnaissance missions can continue to be supported, and has enabled follow-up missions to take place (Rossetto et al. 2014; Booth et al. 2011a).

Between 1982 and 2014 EEFIT reconnaissance team have visited and produced reports on 29 separate earthquakes, including most of the significant events of the period, with two of these (2009 L'Aquila and 2011 Tohoku) having had follow-up missions. A list of these events is shown in Table 1.1, and the locations are shown in Figs. 1.10, 1.11 and 1.12. Eight of these events have been in the wider European area (in countries with EAEE membership, Fig. 1.11). Collaboration with other national teams has been an important feature of these missions where possible, and

1 The Full-Scale Laboratory: The Practice of Post-Earthquake Reconnaissance...

Fig. 1.14 Damage to the Basilica of S. Francisco at Assisi in the 1997 Umbria-Marche Earthquake: investigation of performance of historical structures has been a regular feature of EEFIT missions (Spence 1998)

EEFIT has collaborated with teams from France, Italy, Turkey, USA, Chile, Peru and New Zealand.

The findings of EEFIT reports echo, in many respects, those of the EERI missions listed earlier. An important aspect of EEFIT's mission is in the training of younger engineers and scientists, and this has been achieved by the participation of over 100 engineers and scientists in EEFIT missions, more or less equally divided between industry and academia. EEFIT members have been involved in the development of Eurocode 8, now governing the design of structures in most EU countries, helping to bring field observations into new code provisions. As in the USA, field mission findings have been the basis for a number of important subsequent research programmes (Booth et al. 2011a) including:

- Development of guidelines for the post-earthquake investigation of historical structures and non-engineered buildings Fig. 1.14, and approaches for the repair and strengthening of masonry structures (Hughes and Lubkowski 1999; Patel et al. 2001).
- Development of vulnerability functions for masonry structures and historic centres (D'Ayala 2013) and the need for code provisions for vernacular structures (D'Ayala and Benzoni 2012); these are further discussed in Sect. 1.8.
- The development of databases of earthquake damage data: in recent years these have been web-based searchable databases, which enable cross-event comparisons to be made, such as CEQID (Spence et al. 2011) and GEMECD (So et al. 2012); these are further discussed in Sect. 1.8.

- Soil amplification and other effects following the Mexico earthquake of 1985 (Steedman et al. 1986; Heidebrecht et al. 1990).
- Seismic hazard and risk in areas of low seismicity (Chandler et al. 1991; Pappin et al. 1994).
- Modelling of tsunami impacts on structures (Allsop et al. 2008).
- Mitigation of liquefaction effects on foundations (Brennan and Madabhushi 2002).
- Performance of earth dams in earthquakes (Madabhushi and Haigh 2005).
- Understanding human casualties associated with building damage in earthquakes (So et al. 2008); this is further discussed in Sect. 1.8.
- Assessment and validation of damage estimates from satellite and aerial images (Booth et al. 2011b; Foulser-Piggott et al. 2014); this work is further discussed in Sect. 1.8.
- Relationships between ground motions and observed damage (Goda et al. 2013)

These research programmes have in their turn, affected both engineering practice and design regulations in the country affected and elsewhere. Of equal importance, perhaps, have been the establishment of lasting collaborations with colleagues and research teams in the affected countries, which, particularly in the EU countries, have led to UK involvement in long-term funded collaborations such as RiskUE (2001–2004), LessLoss (2004–2008) and PERPETUATE (2009–2012).

1.7 Other Post-Earthquake Field Reconnaissance Teams

This discussion has emphasised the UNESCO, EERI and EEFIT missions primarily because these were deliberately set up to be international in scope, and also because these are the best documented archives of earthquake damage descriptions available in the English language. But post-earthquake reconnaissance missions and associated reports on damage have been made by many other organisations and by individual efforts; there are national teams in many countries set up to undertake post-earthquake reconnaissance, notably in Japan, France, Germany, Italy, Greece, Turkey and China. Many university groups have fielded reconnaissance missions to study particular aspects of earthquakes; consultancy, insurance and modelling companies have fielded their own reconnaissance missions to obtain data for their own purposes, some of which has been published; and the literature can yield many thousands of individual observations of earthquake damage, which can be of great value, particularly eyewitness accounts by acute observers such as that of Rev Charles Davy documenting his experiences of the 1755 Lisbon earthquake (Davy 1755), Swedish doctor Axel Munthe describing his experiences in the ruins of Messina in 1908 (Munthe 1929), or writer Jack London's account of the 1906 San Francisco (London 1906). To conclude this section, the aims and achievements of three further teams with international scope will be briefly summarised.

1.7.1 Japanese Society for Civil Engineering (JSCE)

Since 1993 JSCE has had a programme of sending field investigation teams to all major events both in Japan and overseas. Multidisciplinary teams have investigated strong motion, engineering and post-disaster response aspects of the events, and reports from 1996 to 2010 are available on the JSCE website (www.jsce.or.jp/library/eq_repo/index.html). The 38 reports covering this period are listed in Table 1.1, and their locations are shown in Figs. 1.10, 1.11 and 1.12. Ten of these were in Japan, 12 of the others elsewhere in Asia. The joint JSCE team investigation with the Architectural Institute of Japan and the Japan Geotechnical Society after the 1999 Kocaeli earthquake in Turkey, involving a joint team of Japanese and Turkish scientists, was perhaps the most intensive investigation of that event, including a detailed building by building survey of more than 2000 buildings in the heavily damaged town of Gölcük (AIJ 2000).

1.7.2 German Task Force (GTF)

The German Task Force for Earthquakes is a multidisciplinary response team which was founded in 1993; it consists of scientists from geosciences, structural engineering, sociologists and rescue specialists. It has three subsections: geology and geophysics (the main core of the taskforce), building and underground studies, and economic and societal affairs (Eggert et al. 2014). An important aspect of GTF missions is the deployment of a network of strong motion instruments in the affected area, sometimes in collaboration with other scientific teams. Since 1993 GTF participated in 22 national and international rapid response actions after earthquakes. Eleven of these are listed in Eggert et al. (2014) of which seven had structural engineering participation in the team. Dates and locations of these are listed in Table 1.1 and shown in Figs. 1.10, 1.11 and 1.12. The seismological data acquired is stored within the GEOFON data archive at GFZ Potsdam (http://geofon.gfz-postadam.de/waveform/). The building-related reconnaissance mission reports are available online at http://www.edac.biz/field_missions/german_taskforce_for_earthquakes.html?L=1

1.7.3 AFPS (Association Francaise du Genie Parasismique)

AFPS is a French society set up in 1983 on the initiative of Jean Despeyroux to promote the study of earthquakes and their consequences, and to promote measures to mitigate their effects and to protect human life. One of its central activities has been to send field missions to areas affected by earthquakes, especially, but not exclusively in French speaking countries. The first of these field missions was to the

1988 Spitak Armenia earthquake, and the AFPS website lists reports on 22 earthquakes since that time which have been visited by AFPS teams. These are listed in Table 1.1, and shown in Figs. 1.10, 1.11 and 1.12. Reports on all these events are available through the AFPS website (www.afps-seisme.org). The 92-page Report on the 2003 Boumerdes Algeria earthquake (Mouroux 2003) is probably the most detailed available record of that event.

1.8 Some Contributions of Post-Earthquake Field Missions to Earthquake Engineering

1.8.1 Understanding Performance of Non-engineered Structures

From Mallet onwards, field reconnaissance missions have frequently found that a large proportion of the damage has been suffered by so-called "non-engineered" structures, mostly ordinary domestic buildings built according to the local vernacular, but also larger public buildings, churches, mosques etc which may be of historical importance. Sections discussing the performance of non-engineered or vernacular structures often form a part of the field reconnaissance reports, especially those of UNESCO and EEFIT, both of which organisations specifically set out to record such damage.

Performance of non-engineered and/or historical buildings are discussed in detail for example in the UNESCO reports on the 1966 Varto, 1967 Mudurnu (Ambraseys et al. 1968), 1974 Pattan (Ambraseys et al. 1975) and 1976 Friuli earthquake and in the EEFIT reports on the 1990 Romania, (Pomonis 1990), 1992 Erzincan (Williams 1992), 1997 Umbria-Marche (Spence 1998) and 2010 Maule, Chile (Lubkowski 2010) earthquakes. Additionally historical structures formed an important part also of the EEFIT report on the 2009 L'Aquila earthquake (Rossetto 2009). Other field investigators, notably Langenbach (2000), have focused exclusively on investigation of vernacular structures. In the 1997 Umbria-Marche and 2010 Maule earthquake it was possible to observe the performance of buildings which had been strengthened by relatively recent interventions specifically to improve their earthquake resistance (Fig. 1.15).

The conclusions of such investigations reveal much of interest about the comparative performance of different forms of traditional construction, and also about the performance of traditional structures by comparison with more recent engineered ones. In a variety of field reports, it has been observed that lightweight structures, using timber frames, have had a surprisingly good performance. Local traditions such as *quincha* and *bahareque* in Central and South America, *himis* and *baghdadi* in Turkey, and also masonry-infilled timber frame construction *dhajji diwari* in Kashmir performed comparatively well (Spence 2007) (Fig. 1.16). In Pakistan, as noted earlier, the UNESCO mission following the 1974 Pattan

Fig. 1.15 Investigation of the performance of strengthened historical structures formed part of the EEFIT reconnaissance following the 2010 Maule Chile earthquake (Lubkowski 2010)

Fig. 1.16 Dhajji Diwari construction in Kashmir, found to have performed much better than more recent forms of construction in Indian Kashmir following the 2005 Kashmir earthquake (Langenbach 2000)

earthquake observed much better performance in stone masonry buildings in which the flat roof was independently supported on timber columns than in those buildings in which the roof was directly supported by the walls (Ambraseys et al. 1974) (Fig. 1.8). However, conversely, many local traditional building types, especially those using field stone masonry or earthen construction, performed very poorly, and uniformly collapsed at relatively low levels of ground motion. Buildings with heavy mud roofs, or vaulted roofs, have been found to perform very poorly. But also certain forms of timber-frame structure, such as the traditional heavy-roof construction in Kobe, often performed badly (Chandler and Pomonis 1995).

For historical structures, several studies have concentrated on identifying the particular mechanisms of damage using methods proposed by Lagomarsino et al. 1997. Common mechanisms of damage found in the 1997 Umbria-Marche, 2009 L'Aquila and 2010 Maule earthquakes include shear cracks in walls, separation of walls at corners, overturning of facades, collapse of masonry arches and vaults, and separation of roof trusses from supporting walls. Strengthening interventions intended to improve performance seem in some cases to have contributed to the failure, as for example in the case of the Basilica of S Francesco at Assisi in 1997 (Spence 1998), or more recent evidence of failure of several churches in L'Aquila (Cimellaro et al. 2011) and Maule Chile (D'Ayala and Benzoni 2012).

It is worth considering what have been the benefits of such field investigations for earthquake engineering, given that these are structure types which are not designed by engineers. One benefit is in loss modelling: the accumulation of data on damage enables us to model the performance of these building types, some of which continue to be built in large numbers, and to estimate, for future events, what damage and attendant casualties will occur given any particular ground motion scenario. A second, more positive benefit is that the observation of relative damage enables good practice to be identified. Many "building for safety" programmes have been set up, in recent years (ASAG 1996; Schilderman 2004), which have had the aim of bringing good earthquake resistant design practice to the construction of small buildings in rural areas through builder training, for example in the application of timber or reinforced concrete ring-beams to masonry structures, improving masonry bonding, promoting improved *quincha* construction etc., and nowadays using grouting or reinforced masonry (NSET 2005). There have been to date still relatively few such programmes and most have been confined to areas which are in the process of reconstruction following an earthquake; but they will be important as long as housing in earthquake risk areas continues to be owner-built rather than engineered. And this will continue to be an important role, currently rather overlooked, for the engineering profession.

A further benefit is in the application to the protection of historical monuments. In countries such as Italy and Greece, protection of the national heritage of historical monuments has a high priority, and a huge number of valuable monuments are at risk from earthquakes and other hazards. The observation of damage from past earthquakes has enabled a number of common mechanisms of damage to be classified (Lagomarsino et al. 1997; D'Ayala 2013); and this enabled not only modelling of expected damage from future earthquakes, but also has led to development of techniques for improving the earthquake-resistance of such structures with minimal impact on the integrity of the ancient fabric of the monument. Such work has been the core of two recently completed EU-funded research programmes PERPETUATE (www.perpetuate.eu) and NIKER (www.niker.eu) (D'Ayala and Paganoni 2014). Thus earthquake field reconnaissance missions have fed directly and indirectly into important earthquake engineering work in the protection of Europe's historic monuments.

1.8.2 Understanding Human Casualties

Understanding of the direct and indirect causes of casualties (deaths and injuries) in earthquakes is of importance to help formulate appropriate mitigation strategies, to develop public advice for self-protection, for the planning of search and rescue, and also to enable loss modelling to include estimates of potential numbers of people killed and injured in future earthquake scenarios. Most of what is currently understood about human casualties is derived from post-earthquake field investigations: although immediate post-earthquake reconnaissance missions have contributed important data on the most vulnerable locations and building types, much of the detailed understanding has come from a relatively small number of detailed surveys of earthquake survivors which have taken place in the months following earthquakes. The factors influencing the likely numbers of casualties in any future event are numerous. An epidemiological summary of the available studies by Petal (2011) has identified 5 classes of variables affecting casualty rates:

- Individual (demographics, location, individual behavior)
- Hazard (nature of the ground motion)
- Building (construction type, level and type of damage)
- Mitigation (household preparedness and first response skills)
- Response (speed and effectiveness of search and rescue)

Alexander examined the casualty data following the 2009 L'Aquila earthquake, in which 308 people were killed, and related this to demographic factors and also to the nature of the damage and collapse of the local building stock (Alexander 2011), with a view to proposing better self-protective behavior.

Koyama et al. (2011) carried out an extensive questionnaire survey in Ojiya City following the 2004 Niigata earthquake in Japan to understand the relationship between location, types and severity of injuries and the arrangement of the building and its furniture, and the activity of the occupants at the time of the earthquake. The aim was to help in loss modelling and to develop strategies for a life-loss reduction strategy. So et al. (2008), with the help of local co-workers, carried out investigations using a survivor questionnaire following the 2005 Pakistan, 2006 Yogjakarta and 2007 Pisco earthquakes to identify the most important causal pathways of injuries and deaths, including examination of types level and causes of injuries, the form of construction and level of damage of the building occupied, and the extent of rescue and post-event treatment available. Figure 1.17 shows the interconnected set of factors found to affect the occurrence of deaths and injuries.

From such investigations it is clear that it is the level and type of building damage that is the predominant variable affecting death and injury rates, the bulk of casualties occurring when the building not only suffers catastrophic damage, but collapses with significant volume loss. However, many other variables such as time of day, the nature of the ground motion, and the behavior of the occupants can have an important modifying influence on these casualty rates. Working with the USGS PAGER, So (2014) has developed estimates for the likely range of fatality rates which will be associated with building collapse for different classes of building taking account of their likely collapse patterns, to improve casualty estimates provided in the PAGER early post-earthquake alerts, which are now widely used by humanitarian agencies in the planning of emergency response (Jaiswal et al. 2011).

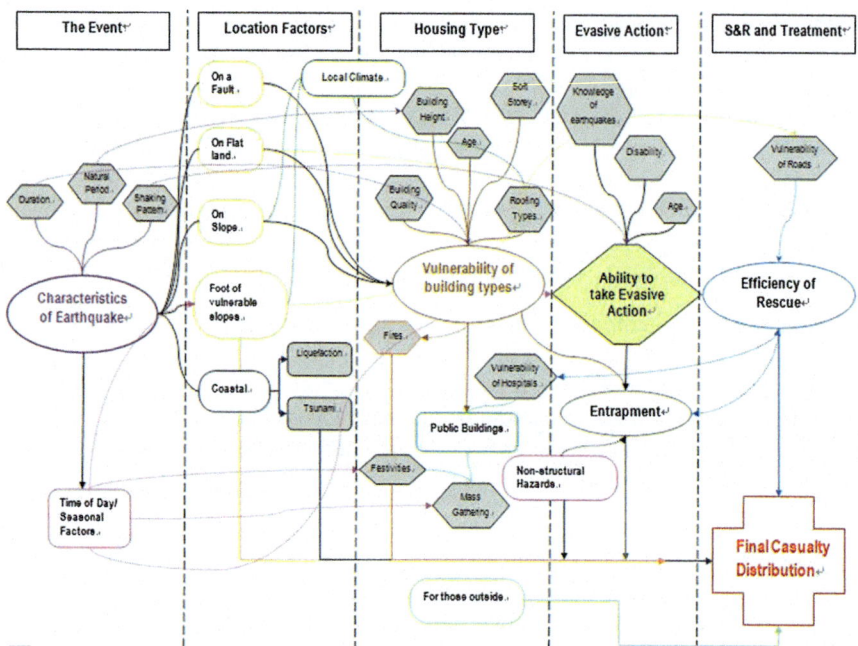

Fig. 1.17 Causes of human casualties in earthquakes: derived from post-earthquake reconnaissance studies by So et al. (2008)

1.8.3 Assembly of Data on Earthquake Consequences

A number of the post-earthquake field missions considered have acquired damage data in a statistical form, either from field surveys or compiled from local reports. This has indeed been a main aim of several EEFIT missions. In the past, the data were made available through the mission-specific publication reports and through the research articles that discuss the observed vulnerability of selected building classes or cross-event summaries (Coburn and Spence 2002). However with the advent of new tools that allow the creation and design of web-accessible data architecture, a much wider accessibility of the data is now possible. Moreover, the publication in 2009 of the USGS ShakeMap archive (http://earthquake.usgs.gov/shakemap), provides an estimate of the ground shaking at any location in any past event. This enables cross-event analyses against a consistent set of estimated ground motions and their variable impacts for the first time. The Cambridge Earthquake Impacts Database (CEQID) (Spence et al. 2011) has been designed and assembled to take advantage of these new tools.

CEQID (www.ceqid.org) is based on earthquake damage data assembled since the 1960s, complemented by other more recently published and some unpublished data. The database assembles the data into a single, organised, expandable and web-accessible format, with a direct access to event-specific shaking hazard maps.

Analytical tools are available which enable cross-event relationships between casualty rates, building classes and ground motion parameters to be determined. The Database is freely accessible to all users, and uses a simple xml format suitable for data mining. Location maps and images of damage are provided for each earthquake event. The Database links to the USGS ShakeMap archive to add data on local intensities and on measured ground shaking.

Currently the database contains data on the performance of more than 1.3 million individual buildings, in over 600 surveys following 51 separate earthquakes, and the total is continuously increasing. The database also has a casualty element, which gives total recorded casualties (deaths, seriously and moderately injured), and casualty rates as a proportion of population with definitions of injury levels used, and information on dominant types of injury, age groups affected etc. Of the 51 events currently in the database, 23 were in Asia and the Pacific (12 of which were in Japan), 17 in Europe, Turkey and North Africa, and 11 in North or South America. Most of the surveys have been done in events since 1990; among these 51 events, 18 were prior to 1990, 21 between 1990 and 2000, and 14 since 2000. Of the 1.3 million buildings in the database, 0.45 million do not have a well-defined building or structural typology given; of the remainder, 78 % are of timber frame, 14 % masonry, 5 % reinforced concrete, and 3 % are of other structural types. Thus, in spite of its size, CEQID in its current state is patchy in global coverage, and in terms of building typologies.

The cross event analysis tools of CEQID allow the construction of charts of empirical damage data related to consistent measures of ground motion derived from the USGS Shakemap archive to be used to show the relationship between damage and any chosen measure of ground motion. Thus post-earthquake damage data can be used directly to enable empirical vulnerability relationships to be developed for any given building type, making an important contribution to loss modelling capability.

1.8.4 GEM Earthquake Consequences Database

A more substantial assembly of earthquake consequence data has, over the last 3 years, been taking place within the framework of GEM (the Global Earthquake Model), to complement a series of other hazard and risk components of the model (www.globalquakemodel.org). Like CEQID, GEMECD is also open-access, GIS-based and related to ground motion parameters derived from the USGS shakemap archive, but its scope and the number of events for which data are assembled is wider (So et al. 2012).

GEMECD assembles consequence data of five different categories as follows:

(a) Ground shaking damage to standard buildings (67 events)
(b) Human casualty studies and statistics (26 events)

Fig. 1.18 Types of earthquake consequences considered in the GEM Earthquake Consequences database (So et al. 2012)

(c) Ground shaking consequences on non-standard buildings, critical facilities, important infrastructure and lifelines (22 events),
(d) Consequences due to secondary, induced hazards (landslides, liquefaction, tsunami and fire following) to all types of inventory classes (24 events, 13 of which are related to landslides)
(e) Socio-economic consequence and recovery data (18 events)

GEMECD has been designed in such a way as to be able to capture the full spectrum of earthquake consequences which can be visualised as a matrix of the interaction between the various inventory assets and the earthquake-related damage agents, as shown in Fig. 1.18. Like CEQID, GEMECD also has cross-event analysis tools which can be used to enable cross-event analyses to be derived for given inventory classes, and levels of ground motion, leading to more robust empirical vulnerability relationships. GEMECD can be accessed at http://www.globalquakemodel.org/what/physical-integrated-risk/consequences-database/

1.8.5 Post-Earthquake Image Archives

Photographic images of geological impacts, damaged buildings and facilities have formed an important element of the record of field investigations from the earliest days, from Mallet's field investigation onwards. Photographs of damage accompany all UNESCO Mission reports though they were not separately archived. Both

EERI and EEFIT have compiled photographic datasets from all recent missions, including many images which were not included in mission reports, and these are now available in digital form. Since 2008, ImageCat, MCEER and UCL and several other collaborators have developed the Virtual Disaster Viewer (VDV) (www.virtualdisasterviewer.com) which links geolocated photos and other images with MS Virtual earth maps to provide an online tool for viewing damage and other earthquake effects from a particular event. Data from six earthquakes as well as several windstorm and tsunami events can be viewed.

EEPImap is a new tool, currently under development at Cambridge Architectural Research which forms the first searchable photographic archive of earthquake damage photographs (http://www.eepimap.com). It is based on a georeferenced photographic database containing attributes of individual buildings and other structures and the level of damage sustained. It can be searched online to provide cross-event datasets corresponding to a range of possible facility types and damage attributes. Currently it contains over 15,000 photographs from 40 events including most of those visited by EEFIT, and has facilities for easy uploading of additional data, so it is continually being expanded (Foulser-Piggott 2013). EEPImap is designed to be compatible with risk components of the Global Earthquake Model (GEM).

1.8.6 Use and Limitations of Remote Sensing

Aerial imagery for the identification of areas of serious damage in earthquakes has been used for some years (Saito et al. 2004), and an international consortium of research teams to promote this use has existed since 1994 (Eguchi and Massouri 2005). Since their first availability around 2000, high-resolution optical satellite images as well as aerial images have been increasingly employed for early post-earthquake damage assessments at a building-by-building and local level. The potential benefits of such deployments are considerable: large damaged areas can be surveyed rapidly without being hampered by the emergency operation on the ground; rescue services can be directed to areas or buildings of greatest need; and the extent of damage can be assessed, leading to a valuable early estimate of reconstruction costs or insurance payouts, of value to international aid organizations, bi-lateral/multi-lateral donors and to the insurance industry. Early work established that the human eye is better able to distinguish features of damage than computerised image analysis (Saito et al. 2004), and this has been the basis of much application since then. The Bam earthquake gave a strong spur to such work: 13 separate papers on aspects of remote sensing were submitted to the Earthquake Spectra special issue on that event (Eguchi and Massouri 2005).

The development of web-based crowd-sourcing techniques in recent years has created a further boost to the potential of such methods, enabling a large team of experienced people to share the task of building-by-building assessment over a large damaged area, so that an overall assessment can be produced very rapidly.

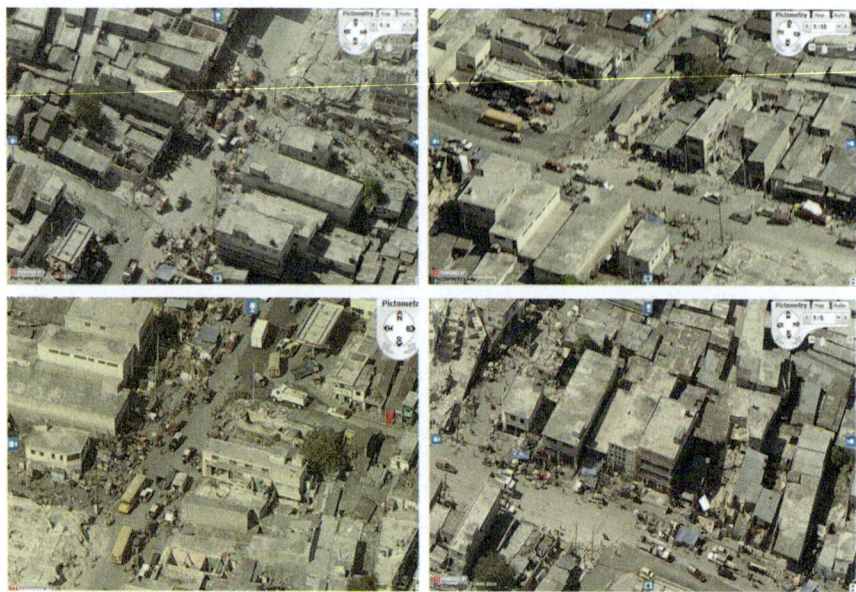

Fig. 1.19 Pictometry images of damage in the 2010 Haiti earthquake (Booth et al. 2011a)

After the 2010 Haiti earthquake, a team of more than 600 people, the GEOCAN network, was assembled by EERI within a few days of the earthquake, and produced a first damage map of the urban area of Port-au-Prince within a week of the occurrence of the event; and within 3 weeks a second more extensive and detailed study was prepared by the same team, involving damage assessments of 107,000 buildings. The result of this was used for the validation of rapid sample ground-based assessment results carried out for the World Bank/UN/EU Post-Disaster Needs Assessment (Corbane et al. 2011). There are thus considerable financial implications for the accuracy of such estimates.

Following the 2010 Haiti earthquake GEOCAN deployment, an independent on the ground validation exercise took place. The EEFIT reconnaissance mission looked closely on the ground at a very small sample of 142 buildings in the GEOCAN dataset. A new aerial imagery technique, Pictometry, which involves multi-angle images of each location with a horizontal resolution of better than 25 cm, was also used to obtain a further damage dataset of 1241 buildings (Fig. 1.19) (Booth et al. 2011b). After the 2011 Christchurch earthquake, a further GEOCAN deployment took place, identifying damage levels for some 5000 buildings in affected area, and this was able to be assessed against Building Safety Evaluations for these same buildings conducted by the Christchurch City Council (Foulser-Piggott et al. 2014).

These two studies, though complicated by many methodological difficulties, were able to establish that, although most of the buildings identified by interpretation of the remotely sensed image as being seriously damaged were in reality seriously damaged, much of the heavy damage on the ground, including building collapses, were missed in the remote assessments. Heavy damage and collapse was obscured by vegetation, by proximity to other buildings, because the lower floor of a building collapsed, leaving upper stories and roofs intact, or because major damage ultimately leading to demolition was simply invisible from outside the building. Typically no more than around 40 % of the buildings which ground surveys identified as heavily damaged or collapsed were identified as such in the aerial imagery. The extent of underestimate depended on the resolution of the image, the level of experience of the image analyst, the construction typology of the building, and the type of damage. Damage to masonry buildings was easier to identify than that to either timber frame or reinforced concrete buildings; damage caused by foundation failure or subsoil liquefaction (a very important class of damage in the Christchurch earthquake) proved particularly difficult to identify (Foulser-Piggott et al. 2014).

Many recommendations were made as a result of these studies to improve the results of future remotely-sensed damage assessments; and improvements in the quality of the available imagery will certainly continue to be made. Indeed it is probable that photography from low-level pilotless aircraft will in the near future be able to augment substantially the remotely sensed data available. But remote sensing cannot in the near future be expected to become a substitute for post-earthquake field reconnaissance. Assessments from remote sensing can be very useful to identify areas where damage is concentrated; to identify blocked roads and collapsed bridges; to identify areas of liquefaction (especially where these are associated with sand boils), and major landslides. They can also be used to make an approximate assessment of overall damage if enough is known about the likely omission errors in such assessments. But the detail of damage, the performance of different construction typologies, and the relationship of damage to quality of construction will continue to need investigations by experienced observers on the ground, at close quarters to, and where possible inside, the damaged buildings. Future remote sensing assessments should be planned to be coupled with field deployments to validate the results and to provide more of the detail which remote sensing cannot supply.

1.9 The Future of Earthquake Field Missions

Over the last 30 years there has been a huge change in the technology available to support earthquake field missions. Digital photography, GPS positioning, the internet, mobile phone networks, high resolution satellite reconnaissance, social media have all arrived and made their mark on the way earthquake reconnaissance missions are conducted. This is in contrast to the construction technologies whose performance is being investigated, which have changed comparatively little in that time. Technology will continue to evolve at a rapid pace in both predictable and unpredicted ways, allowing improvements in speed of operation, in communication between team members and base, and in the capturing of detail: through photographic communication, some people will be able to contribute to the work of field missions without travelling to the affected area. For example, developments such as EEPI Map will allow for the crowdsourcing of photographs from general members of the public which can be assessed remotely and can help produce a rapid damage assessment of an area.

As discussed above, the development of higher-resolution and other forms of remote sensing is not likely to eliminate the need for investigators in the field to view damage from close range. But it will enable teams to organise their field operations with support from continuously updated and pre-analysed remote sensing images. The development of databases of the building stock inventory (already in development through the GEM project) will enable teams to have access to pre-event data and images of each damaged object. As a response to such changes field teams may in future be smaller, more focussed on special aspects and deployed at different times.

The collection of building-by-building data on damage has been an important feature of the work of some reconnaissance missions, and it is largely through such damage surveys that empirical fragility relationships for loss estimation have been developed. It is often assumed by reconnaissance teams that detailed building damage surveying will be done, over time, by national authorities and made available. But such official damage data often turns out to be inadequate for use in loss estimation, with damage levels and construction typologies poorly defined, and undamaged buildings often omitted. Assembling damage data through well-chosen local building-by-building sample datasets will continue to be of vital importance, and field surveys can now be supported through remote sensing to locate appropriate samples across a range of areas, not just those most heavily damaged.

There is still a need to improve the level of international collaboration between field mission teams. Table 1.1 shows that the sites of a number of the most important earthquakes in recent years have been visited by multiple teams, which usually work independently of each other. In many of the affected countries significant expertise in earthquake engineering now exists, and it is vital for visiting reconnaissance teams to work with local experts, to learn from them, and share their own knowledge. This already happens, but should be extended in future.

Recent events have shown that in many parts of the world, especially in poorer countries, there is an urgent need to improve the earthquake resistance of much of the existing building stock, as well as improve the standards of new buildings for the future. Thus future post-earthquake field missions are likely to be as much concerned with helping with developing resilience as recording damage: this will give rise to a need for a series of missions at different stages of the recovery cycle, and the involvement of more expertise from complementary disciplines such as sociology and urban planning. EEFIT and EERI already have funding in place permitting such operations. Given the probability of large urban disasters in the future it is important that field mission organisations make plans to be able to mount field missions in potentially challenging situations (such as that in Haiti in 2010). It may also be that established field mission teams, now already familiar with studying tsunami impacts, should consider mounting, or supporting, field investigations following non-earthquake disasters such as volcanic eruptions or major typhoons where there is a similar need for rapid deployment to assemble perishable data.

1.10 Conclusions

- This paper set out to review the nature and practice of earthquake reconnaissance missions since the earliest examples to today's practice, and to point out some of the ways in which the practice of earthquake engineering today has benefitted from field observations.
- After a brief historical background the paper has concentrated on the missions of 5 separate groups, active in the last 50 years, those of UNESCO, EERI, EEFIT and more briefly the Japanese Society for Earthquake Engineering, AFPS in France and the German Task force, all of whom have been involved in multiple international missions in that time.
- Between these teams, 258 post-earthquake reconnaissance missions have been mounted, and they have investigated, and have reported on, 178 separate events. Of these 37 were in the European area, 64 in Asia, 64 in the Americas, 7 in Africa, and 6 in Australasia and the Pacific. The style of mission has varied considerably, from the small expert interdisciplinary scientist/engineer teams of UNESCO spending several weeks in the field to today's larger, more multidisciplinary teams with many specialists, but often on shorter initial missions sometimes backed by follow-up studies.
- Reports on each mission have been prepared and those of current teams are available on their websites which have been referenced; often these have been accompanied by published papers.
- The cumulative contribution of these field teams to earthquake engineering, seismology and to understanding the social and economic consequences of earthquakes has been considerable, leading to improved design codes and design practices, to better understanding of human behaviour and guidance to inhabitants of earthquake zones, and the accumulation of data on earthquake consequences enabling estimation of possible losses in future events to be made.
- An important benefit to recent field studies has been the increasing availability of strong motion records of earthquakes, making it possible to link damage observations to the level and characteristics of the causative ground motion.
- For engineered buildings, repeated observations of the same types of damage in many earthquakes has driven the development of the current generation of design codes; buildings designed and built to these codes have largely performed well in subsequent earthquakes.
- Field investigations of the distribution of damage coupled with the increasing availability of strong ground motion recordings of the main shock and aftershocks, has led to a better understanding of the role played by site conditions on

the amplification of ground motion and the types and distribution of damage to structures.
- The data on performance of lifelines assembled by field missions has identified both systems that have performed well and those that failed; and has resulted in numerous changes to design practices.
- Studies of the behaviour of people and communities has made numerous contributions to preparedness planning, to organisation of search and rescue and to the improved planning of longer-term recovery.
- The differences in the performance of domestic scale non-engineered structures of different forms of construction has become better understood, enabling guidelines to be developed for safer reconstruction in especially rural areas, and leading to effective building for safety programmes in reconstruction.
- The likely mechanisms of collapse of historical masonry buildings have been identified, and some inappropriate earlier attempts at strengthening measures identified, leading to the development of appropriate techniques for strengthening and protecting historical monuments.
- The causes of human casualties resulting from building damage in earthquakes have become better understood, enabling better early estimation of likely losses, better design of effective measures for self-protection of the population, and better planning for early search and rescue activity.
- The data acquired from past field missions has in recent years become more systematically documented and archived using web-based database technology, so that data can easily be accessed and retrieved, and so that cross-event analysis of damage and other impacts to particular components of the built environment , social and economic activities can be conducted.
- Remote sensing technology has begun to make a contribution to the recording of earthquake damage, making possible early assessments of likely impacts. Much remains to be done to realise the full potential of these technologies, but their application will enhance rather than replace field investigations.
- Future field missions will make use of rapidly developing technology for viewing, recording and communicating mission activities. They will be more interdisciplinary, carry out repeat missions, and concerned increasingly with developing resilience. They should not abandon collection of building-by-building damage data through local surveys.

Acknowledgements The author is greatly indebted to contributions from a numbers of colleagues who have reviewed parts of this paper and provided additional background material, notably Roger Musson, James Jackson, Edmund Booth, Marjorie Greene, Antonios Pomonis, Dina D'Ayala, John Douglas, Vicki Kouskouna, Roxane Foulser-Piggott, Emily So, Sean Wilkinson, Göcke Tonuk and Sebastian Hainzl. The maps and tables of mission locations were prepared by Hannah Baker.

Open Access This chapter is distributed under the terms of the Creative Commons Attribution Noncommercial License, which permits any noncommercial use, distribution, and reproduction in any medium, provided the original author(s) and source are credited.

References

Alexander DE (2011) Mortality and morbidity risk in the L'Aquila Italy earthquake of 6 April 2009 and lessons to be learned, Chapter 13. In: Spence R, So E, Scawthorn C (eds) Human casualties in earthquakes: progress in modelling and mitigation. Springer, Berlin

Allsop W, Rossetto T, Robinson D, Charvet I, Bazin PH (2008) A unique tsunami generator for physical modeling of violent flows and their impact. In: Proceedings of the 14th world conference on earthquake engineering, Beijing. CD rom, Paper no. 15-00

Ambraseys NN (1976a) Field studies of earthquakes. Discussion paper for intergovernmental conference on the assessment and mitigation of earthquake risk, UNESCO, Paris

Ambraseys NN (1976b) Italy: the Gemona di Friuli Earthquake of 6 May 1976, Part 2, Serial FMR/CC/SC/ED/76/169, UNESCO Paris

Ambraseys NN (1988) Engineering seismology. Earthquake Eng Struct Dyn 17:1–105

Ambraseys NN (2002) Engineering seismology in Europe. Keynote talk, 12th European conference on earthquake engineering, London

Ambraseys NN, Despeyroux J (1978) Roumania: Le Tremblement de Terre du 4 Mar, 1977, Part 1 The earthquake and its principal effects, Serial No FMR/SC/GEO/78/102, UNESCO Paris

Ambraseys NN, Zatopek A (1967) Turkey: the Varto Ustukran earthquake of 19 August 1966: earthquake reconnaissance mission, Serial WS/0267.81-AVS, UNESCO, Paris

Ambraseys NN, Zatopek A, Tasdemiroglu M, Aytun A (1968) Turkey: the Mudurnu Valley (West Anatolia Earthquake of 22 July 1967), Serial 622/BMS.RD/AVS, UNESCO, Paris

Ambraseys NN, Lensen G, Moinfar A (1975) Pakistan: The Pattan Earthquake of 28 December 1974, Serial no FMR/SC/GEO/75/134, UNESCO, Paris

Architectural Institute of Japan (2000) Report of the damage investigation of the 1999 Kocaeli earthquake in Turkey, AIJ, JSCE, Japanese Geotechnical Society

Arnold C, Reitherman R (1982) Building configuration and seismic design. Wiley, New York

ASAG (1996) ASAG's intervention through technology upgrading for seismic safety. Ahmedabad Study and Action Group, Ahmedabad

ASCE (2003) Seismic evaluation of existing buildings (ASCE/SEI 31-03), American Society of Civil Engineers

ATC (1978) Amended tentative provisions for the development of seismic regulations for buildings ATC-3-06, Applied Technology Council, California http://www.atcouncil.org/pdfs/atc306.pdf

Booth E, Wilkinson S, Spence R, Free M, Rossetto T (2011a) EEFIT: the UK earthquake engineering field investigation team. Forensic Eng 164(3):117–121

Booth ED, Saito K, Spence R, Madabhushi G, Eguchi R (2011b) Validating assessments of seismic damage made from remote sensing. Earthq Spectra Special edition on Haiti earthquake 20:S1, S157–S178

Brennan AJ, Madabhushi SPG (2002) Effectiveness of vertical drains in mitigation of liquefaction. Int J Soil Dynam Earthq Eng 22(9–12):1059–1065

Cardoso R, Lopes M, Bento R, D'Ayala D (2013) Historic, braced frame timber buildings with masonry infill ('Pombalino' buildings)" Report # 92 World Housing Encyclopedia. http://www.world-housing.net

Chandler A, Pomonis A (eds) (1995) The Hyogo-ken Nanbu (Kobe) Earthquake of 17 January 1995: a field report by EEFIT. Institution of Structural Engineers, London

Chandler AM, Pappin JW, Coburn AW (1991) Vulnerability and seismic risk assessment of buildings following the 1989 Newcastle, Australia earthquake. Bull N Z Soc Earthq Eng 24 (2):116–138

Cimellaro GP, Reinhorn AM, De Stefano A (2011) Introspection on improper seismic retrofit of Basilica Santa Maria di Collemaggio after 2009 Italian earthquake. Earthq Eng Eng Vib 10:153–161. doi:10.1007/s11803-011-0054-4

Coburn A, Spence R (2002) Earthquake protection. Wiley, Chichester

Corbane C, Carrion D, Lemoine G, Broglia M (2011) Comparison of damage assessment maps derived from very high spatial resolution satellite and aerial imagery produced for the 2010 Haiti earthquake. Earthq Spectra 27(S1):S199–S218

D'Ayala D (2013) Assessing the seismic vulnerability of masonry buildings. Handbook of seismic risk analysis and management of civil infrastructure systems. Woodhead Publishing, Elsevier, USA, pp 334–365

D'Ayala D, Benzoni G (2012) Historic and traditional structures during the 2010 Chile earthquake: observations, codes, and conservation strategies. Earthq Spectra 28(S1):425–451. doi:10.1193/1.4000030

D'Ayala DF, Paganoni S (2014) Testing and design protocol of dissipative devices for out-of-plane damage. Proc ICE-Struct Build 167(1):26–40

Davy C (1755) The earthquake at Lisbon, 1755. In: Tappan EM (ed) The world's story: a history of the world in story, song and art, 14 vols. (Boston: Houghton Mifflin, 1914), vol. V: Italy, France, Spain, and Portugal, pp 618–628

De Martini P, Burrato P, Pantosti D, Maramai A, Graziani L, Abramson H (2003) Identification of tsunami deposits and liquefaction features in the Gargano area (Italy): paleoseismological implication. Ann Geophys 46(5):883–902

Dewey J, Wald D, Dengler L (2000) Relating conventional USGS modified Mercalli intensities to intensities assigned with data collected via the Internet. Seismol Res Lett 71:264

EERI (1986) Reducing earthquake hazards: lessons learned from earthquakes, Publication 86-02. EERI, Oakland

EERI (1995a) The EERI Northridge Earthquake of January 1994, Collection (CD). EERI, Oakland

EERI (1995b) Post earthquake investigation field guide. www.eeri.org. Accessed 6 Jan 2014

EERI (2004) Learning from earthquakes: the EERI learning from earthquakes program: a brief synopsis of major contributions. www.eeri.org. Accessed 6 Jan 2014

Eggert S, Walter TR, Luhr B-G, Woith H, Hainzl S, Milkereit H, Grosser H, Sobiesiak M, Tillman F, Schwarz J, Dahm T, Zscau J (2014) 20 years of German Taskforce for Earthquakes: Motivation, Strategy and Data Access (under review for Geochemistry, Geophysics, Geosystems)

Eguchi R, Massouri B (2005) Use of remote sensing Technologies for Building Damage Assessment after the 2003 Bam Iran Earthquake. Earthq Spectra 21(S1):S207–S212

FEMA (2003) Hazus technical manual. Federal Emergency Management Agency, Washington, DC

Ferrari G (1987) Mallet's method after Mallet in Italy. In: Guidoboni E, Ferrari G (eds) Mallet's Macroseismic survey on the Neapolitan Earthquake of 16th December 1857. SGA Storia Geofisica Ambiente, Bologna

Foulser-Piggott R (2013) Earthquake engineering photographic investigation map. SECED Newsl 24(3):8–10. ISSN 0967-859X

Foulser-Piggott R, Spence R, Eguchi R, King A (2014) Using remote sensing for building damage assessment: the geocan study and validation for the 2011 Christchurch earthquake. submitted to Earthquake Spectra (under review)

Fournier-D'Albe EM (1986) Foreword. In: Ambraseys NN, Moinfar AA, Tchalenko JS (eds) UNESCO earthquake reconnaissance mission reports 1963–1981. UNESCO, Paris

Goda K, Pomonis A, Chian SC, Offord M, Saito K, Sammonds P, Fraser S, Raby A, Macabuag J (2013) Ground motion characteristics and shaking damage of the 11th March 2011 Mw9.0 Great East Japan earthquake. Bull Earthq Eng 11(1):141–170

Heidebrecht AC, Henderson P, Naumoski N, Pappin JW (1990) Seismic response and design for structures located on soft clay sites. Can Geotechn J 27(3):330–341

Hughes R, Lubkowski Z (1999) The survey of earthquake damaged non-engineered structures. http://www.istructe.org/resources-centre/technical-topic-areas/eefit/eefit-reports

International Code Council (2012) 2012 international building code. International Code Council

Jaiswal KS, Wald DJ, Earle PS, Porter K, Hearne M (2011) Earthquake casualty models in the USGS prompt assessment of global earthquakes for response (PAGER) system. Chapter 6 In:

Spence R, So E, Scawthorn C (eds) Human casualties in earthquakes: progress in modelling and mitigation. Springer, Dordrecht

Koyama M, Okada S, Ohta Y (2011) Major factors controlling earthquake casualties as revealed via a diversified questionnaire survey in Ojiya City for the 2004 Mid-Niigata Earthquake. Chapter 14 In: Spence R, So E, Scawthorn C (eds) Human casualties in earthquakes: progress in modelling and mitigation. Springer, Dordrecht

Lagomarsino S, Brencich A, Bussolino F, Moretti A, Pagnini LC, Podesth S (1997) Una nuova metodologia per il rilievo del danno alle chiese: prime considerazioni sui meccanismi attivati dal sisma. Ingegneria Sismica 14(3):70–82

Langenbach R (2000) What can we learn from traditional construction in seismic areas? Conservationtech. www.conservationtech.com/IstanCon/keynote. Accessed 20 Jan 2014

London J (1906) The story of an eyewitness. Colliers Magazine, May 5th 1906

Lubkowski Z (ed) (2010) The Mw8.8 Maule Chile Eathquake of 27 February 2010: a preliminary field report by EEFIT. Institution of Structural Engineers, London

Madabhushi SPG, Haigh SK (2005) Performance of earth dams during the Bhuj Earthquake of 26th January 2001. Dams Reserv 15(3):14–24

Mallet R (1862) Great Neapolitan earthquake of 1957. The first principles of observational seismology, Facsimile Edition, SGA Storia Geofisica Ambiente, Bologna

Melville C, Muir Wood R (1987) Robert Mallet, first modern seismologist. In: Guidoboni E, Ferrari G (eds) Mallet's macroseismic survey on the Neapolitan Earthquake of 16th December 1857. SGA Storia Geofisica Ambiente, Bologna

Mouroux P (ed) (2003) Le Seisme du 21 Mai, 2003 en Algerie, Rapport préliminaire de la mission AFPS. AFPS, Paris

Munthe A (1929) The story of San Michele. Murray, London, pp 398–408

Musson RMW (2013) A history of British seismology: the 14th Mallet-Milne lecture. Bull Earthq Eng 11:715–786

NSET (2005) Earthquake resistant construction of buildings, curriculum for Mason training, NSET. http://www.nset.org.np

Omori F (1908) Preliminary report on the Messina-Reggio Earthquake of Dec 28, 1908

Pappin JW, Coburn AW, Pratt CR (1994) Observations of damage ratios to buildings in the epicentral region of the 1992 Roermond earthquake, the Netherlands. Geologie en Mijnbouw 73:299–302

Patel DB, Patel DB, Pindoria K (2001) Repair and strengthening guide for earthquake damaged lowrise domestic buildings in Gujarat, India. http://www.arup.com/_assets/_download/download197.pdf. Accessed 01 June 2011

Petal M (2011) Earthquake casualties research and public education. Chapter 3 In: Spence R, So E, Scawthorn C (eds) Human casualties in earthquakes: progress in modelling and mitigation. Springer, Dordrecht

Pomonis A (ed) (1990) The Vrancea, Romania earthquakes of 30–31 May 1990: a field report by EEFIT. Institution of Structural Engineers, London

Rossetto T (ed) (2009) The L'Aquila Italy Earthquake of 6 April 2009: a preliminary field report by EEFIT. Institution of Structural Engineers, London

Rossetto T, D'Ayala D, Gori F, Persio R, Han J, Novelli V, Wilkinson SM, Alexander D, Hill M, Stephens S, Kontoe S, Elia G, Verrucci E, Vicini A, Shelley W, Foulser-Piggott R (2014) The value of multiple earthquake missions: the EEFIT L'Aquila Earthquake experience. Bull Earthq Eng. doi:10.1007/s10518-014-9588-y

Saito K, Spence RJS, Going C, Markus M (2004) Using high-resolution satellite images for post-earthquake building damage assessment: a study following the 21 January 2001 Gujarat Earthquake. Earthq Spectra 20(1):145–169

Schilderman T (2004) Adapting traditional shelter for disaster mitigation and reconstruction: experiences with community-based approaches. Build Res Inf 32(5):414–426

So E (2014) Derivation of fatality rates for use in a semi-empirical earthquake loss estimation model (draft) USGS Open File Report

So E, Spence R, Khan A, Lindawati T (2008) Building damage and casualties in recent earthquakes and tsunamis in Asia: a cross-event survey of survivors. In: Proceedings of 14th world conference on earthquake engineering, Beijing. CD rom

So EKM, Pomonis A, Below R, Cardona O, King A, Zulfikar C, Koyama M, Scawthorn C, Ruffle S, Garcia D (2012) An introduction to the global earthquake consequences database (GEMECD). Paper 1617 in 15th world conference on earthquake engineering, Lisbon

Spence R (ed) (1998) The Umbria March Earthquakes of 26 September 1997: a field report by EEFIT. Institution of Structural Engineers, London

Spence R (2007) Saving lives in earthquakes: successes and failures in seismic protection since 1960. Bull Earthq Eng 5:139–251

Spence RJS, Nash DFT, Hughes RE, Taylor CA, Coburn AW (1982) Damage assessment and ground motion in the Italian earthquake of 23.11.1980. In: Proceedings of 7th European conference on earthquake engineering, Athens

Spence R, So E, Jenkins S, Coburn A, Ruffle S (2011) A global earthquake building damage and casualty database. Chapter 5 In: Spence R, So E, Scawthorn C (eds) Human casualties in earthquakes: progress in modelling and mitigation. Springer, Dordrecht

Steedman RS, Booth ED, Pappin JW, Mills JH (1986) The Mexico earthquake of 19th September 1985, some lessons for the engineer. In: Proceedings of 8th European conference on earthquake engineering, Lisbon, pp 83–89

Trendafilowski G, Wyss M, Rosset P (2011) Loss estimation module in the second generation software QLARM, Chapter 7. In: Spence R, So E, Scawthorn C (eds) Human casualties in earthquakes: progress in modelling and mitigation. Springer, Dordrecht

UNESCO (1963) The Skopje earthquake of 26 July 1963: report of the UNESCO technical mission. UNESCO, Paris

Williams MS (ed) (1992) The Erzincan, Turkey Earthquake of 13 March 1992: a field report by EEFIT. Institution of Structural Engineers, London

Chapter 2
Rapid Earthquake Loss Assessment After Damaging Earthquakes

Mustafa Erdik, K. Şeşetyan, M.B. Demircioğlu, C. Zülfikar, U. Hancılar, C. Tüzün, and E. Harmandar

Abstract This article summarizes the work done over last decades regarding the development of new approaches and setting up of new applications for earthquake rapid response systems that function to estimate earthquake losses in quasi real time after an earthquake. After a critical discussion of relevant earthquake loss estimation methodologies, the essential features and the characteristics of the available loss estimation software are summarized. Currently operating near real time loss estimation tools can be classified under two main categories depending on the size of area they cover: Global and Local Systems. For the global or regional near real time loss estimation systems: GDACS, WAPMERR, PAGER, ELER and SELENA methodologies are. Examples are provided for the local rapid earthquake loss estimation systems including: Taiwan Earthquake Rapid Reporting System, Real-time Earthquake Assessment Disaster System in Yokohama, Real Time Earthquake Disaster Mitigation System of the Tokyo Gas Co., IGDAS Earthquake Protection System and Istanbul Earthquake Rapid Response System.

2.1 Introduction

As illustrated in Fig. 2.1 (after Böse 2006), management of earthquake risks is a process that involves pre-, co- and post-seismic phases. Earthquake Early Warning (EEW) systems are involved in the co-seismic phase. These involve the generation of real time ground motion estimation maps as products of real-time seismology

M. Erdik (✉) • K. Şeşetyan • M.B. Demircioğlu • C. Zülfikar • U. Hancılar • C. Tüzün • E. Harmandar
Kandilli Observatory and Earthquake Research Institute, Department of Earthquake Engineering, Boğaziçi University, İstanbul, Turkey
e-mail: erdikm@gmail.com

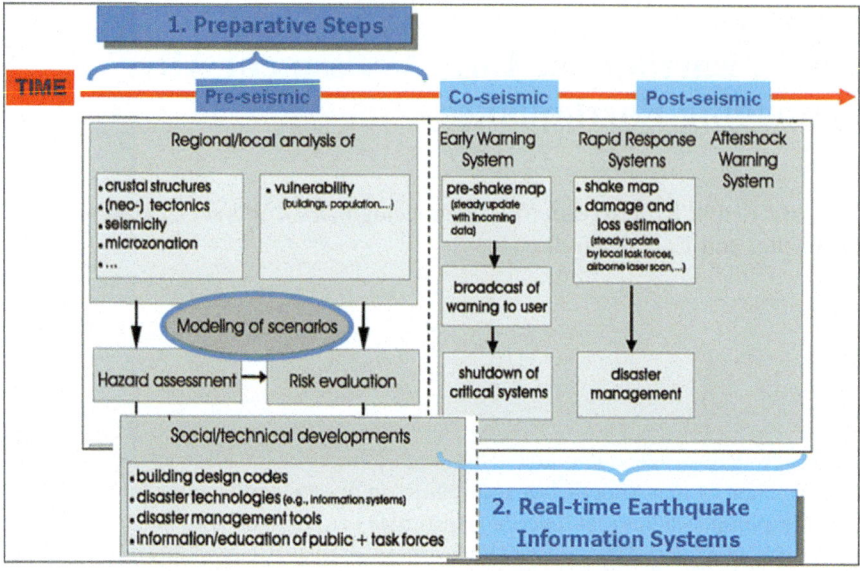

Fig. 2.1 Pre- co- and post-earthquake risk management activities (After Böse 2006)

and/or generation of alarm signals directly from on-line instrumental data. The Rapid Response Systems take part immediately after the earthquake and provide assessment of the distribution of ground shaking intensity (so-called ShakeMaps) and information on the physical (buildings) damage, casualties (fatalities) and economic losses. This rapid information on the consequences of the earthquake can serve to direct the search and rescue teams to the areas most needed and assist civil protection authorities in the emergency action. As such, the need for a rapid loss estimate after an earthquake has been recognized and requested by governments and international agencies.

This study will critically review the existing earthquake rapid response systems and methodologies that serve to produce earthquake loss information (building damages, casualties and economic losses) immediately after an earthquake.

Potential impact of large earthquakes on urban societies can be reduced by timely and correct action after a disastrous earthquake. Modern technology permits measurements of strong ground shaking in near real-time for urban areas exposed to earthquake risk. The assessments of the distribution of strong ground motion, building damage and casualties can be made within few minutes after an earthquake. The ground motion measurement and data processing systems designed to provide this information are called Earthquake Rapid Response Systems.

The reduction of casualties in urban areas immediately following an earthquake can be improved if the location and severity of damages can be rapidly assessed by the information from Rapid Response Systems. Emergency management centers of both public and private sector with functions in the immediate post-earthquake period (i.e. SAR, fire and emergency medical deployments) can allocate and prioritize resources to minimize the loss of life. The emergency response capabilities can be significantly improved to reduce casualties and facilitate evacuations by permitting rapid and effective deployment of emergency operations. To increase its effectiveness, the Rapid Response data should possibly be linked with the incident command and emergency management systems.

Ground motion data related with power transmission facilities, gas and oil lines and transportation systems (especially fast trains) allow for rapid assessment of possible damages to avoid secondary risks. Water, wastewater and gas utilities can locate the sites of possible leakage of hazardous materials and broken pipes. The prevention of gas-related damage in the event of an earthquake requires understanding of damage to pipeline networks and prompt shut-off of gas supply in regions of serious damage.

Available near real time loss estimation tools can be classified under two main categories depending on the size of area they cover: (1) Global/Regional Systems and (2) Local Systems.

For the global or regional near real time loss estimation efforts, Global Disaster Alert and Coordination System (GDACS, http://www.gdacs.org), World Agency of Planetary Monitoring Earthquake Risk Reduction (QLARM, http://qlarm.ethz.ch), Prompt Assessment of Global Earthquakes for Response (PAGER, http://earthquake.usgs.gov/earthquakes/pager) and NERIES-ELER (http://www.koeri.boun.edu.tr/Haberler/NERIES%20ELER%20V3.1_6_176.depmuh) can be listed.

Several local systems capable of computing damage and casualties in near real time already exist in several cities of the world such as Yokohama, Tokyo, Istanbul, Taiwan, Bucharest and Naples (Erdik et al. 2011).

2.2 Earthquake Loss Estimation Methodology

An extensive body of research, tools and applications exists that deals with all aspects of loss estimation methodologies. The components of rapid earthquake loss estimation will be addressed following the structures of the HAZUS-MH (2003) and the OpenQuake (Silva et al. 2013) earthquake loss assessment model. Both of these developments use comprehensive and rigorous loss assessment methodologies that can only be adapted to rapid earthquake loss assessment after intelligent simplifications.

The HAZUS-MH Earthquake Model (HAZUS-MH 2003) is developed to provide a nationally applicable methodology for loss estimates of damage and loss to buildings, essential facilities, transportation and utility lifelines, and population based on scenario or probabilistic earthquakes. HAZUS first discusses the inventory data including the Collection and Classification schemes of different systems, attributes required to perform damage and loss estimation, and the data supplied with the methodology. The loss assessment methodology that HAZUS uses consists of the main components of: Potential Earth Science Hazard, Direct Physical

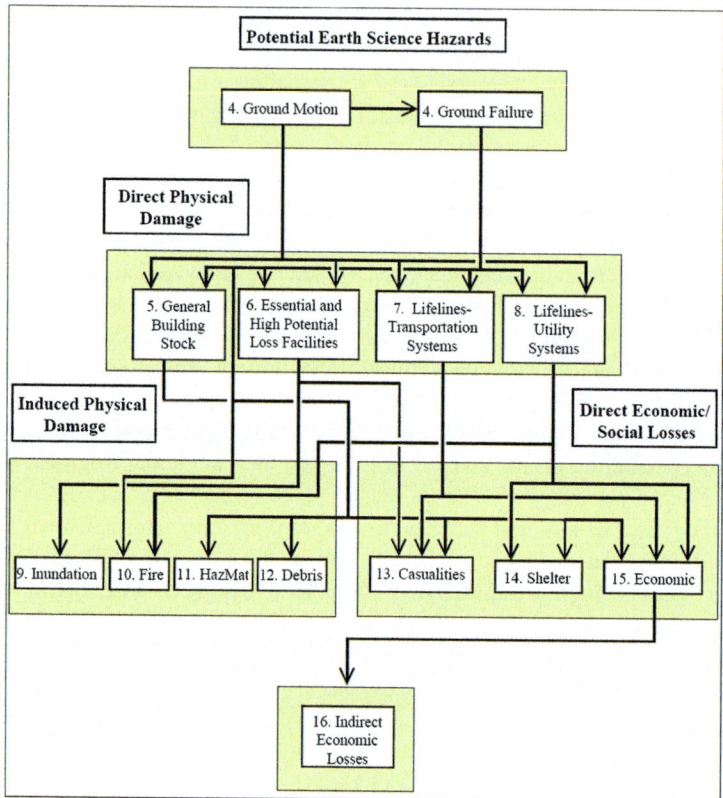

Fig. 2.2 Flowchart of the HAZUS earthquake loss estimation methodology

Damage, Induced Physical Damage and Direct Economic/Social Loss, as illustrated in the flowchart provided in Fig. 2.2. As indicated by arrows on the flowchart, modules are interdependent with output of some modules acting as input to others.

The main ingredients of the HAZUS loss assessment methodology are as follows.

- Potential Earth Science Hazards: Potential earth science hazards include ground motion, ground failure (i.e., liquefaction, landslide and surface fault rupture) and tsunami/seiche.
- Direct Physical Damage: Encompasses the modules for General Building Stock, Essential and High Potential Loss Facilities, Lifelines – Transportation and Utility Systems. The General Building Stock module determines the probability of Slight, Moderate, Extensive and Complete damage to general building stock through the use of fragility curves, that describe the probability of reaching or exceeding different states of damage given peak building response, and the building capacity (push-over) curves, that are used (with damping-modified demand spectra) to determine peak building response

- Induced Physical Damage: This module models the damage caused by Inundation, Fires Following Earthquakes, Hazardous Materials Release and Debris.
- Direct Economic/Social Losses: Casualties, Shelter Needs and Economic Loss models are encompassed under this component. The Casualty module describes and develops the methodology for the estimation of casualties, describes the form of output, and defines the required input. The methodology is based on the assumption that there is a strong correlation between building damage (both structural and nonstructural) and the number and severity of casualties. The module for Direct Economic Losses describes the conversion of damage state information into estimates of economic loss. The methodology provides estimates of the structural and nonstructural repair costs caused by building damage and the associated loss of building contents and business inventory. The Indirect Economic Losses are also treated as an extension of this module.

A recent development on earthquake loss estimation based on comprehensive methodologies is the OpenQuake project (http://www.globalquakemodel.org/openquake/) which has been initiated as part of the global collaborative effort entitled Global Earthquake Model (GEM) (http://www.globalquakemodel.org). OpenQuake is a web-based risk assessment platform, which offers an integrated environment for modeling, viewing, exploring and managing earthquake risk (Silva et al. 2013). The engine behind the platform currently has five main calculators (Scenario Risk, Scenario Damage Assessment, Probabilistic Event Based Risk, Classical PSHA-based Risk and Benefit-Cost Ratio). The Scenario Damage Assessment calculator uses a rigorous methodology in estimating damage distribution due to a single, scenario earthquake, for a spatially distributed building portfolio, which can be used for post-earthquake loss assessment. Workflow of the Scenario Damage Assessment is provided in Fig. 2.3, after Silva et al. (2013).

In this methodology, a finite rupture definition of the earthquake needs to be provided, along with the selected GMPE. A set of ground-motion fields is computed, with the possibility of considering the spatial correlation of the ground-motion residuals. Then, the percentage of buildings in each damage state is calculated for each asset the fraction of buildings in each damage state using the fragility models. By repeating this process for each ground-motion field, a list of fractions (one per damage state) for each asset is obtained to yield the mean and standard deviation of this list of fractions for each asset. The absolute building damage distribution is obtained by multiplying the number or area of buildings by the respective fractions with confidence intervals (Crowley and Silva 2013).

The key ingredients of the OpenQuake scenario risk assessment methodology are as follows.

- Rupture model (Finite Rupture Definition): The definition of the finite rupture model, specified by a magnitude and a rupture surface geometry, is a key input for scenario risk and damage analysis. The rupture surface geometry can be as simple as the hypocenter point or complex, described by the rake angle and other fault geometrical surface attributes, depending on the level of knowledge.
- Fragility model: Fragility is defined as the probability of exceeding a set of limit states, given a range of intensity measure levels. A fragility model can either be

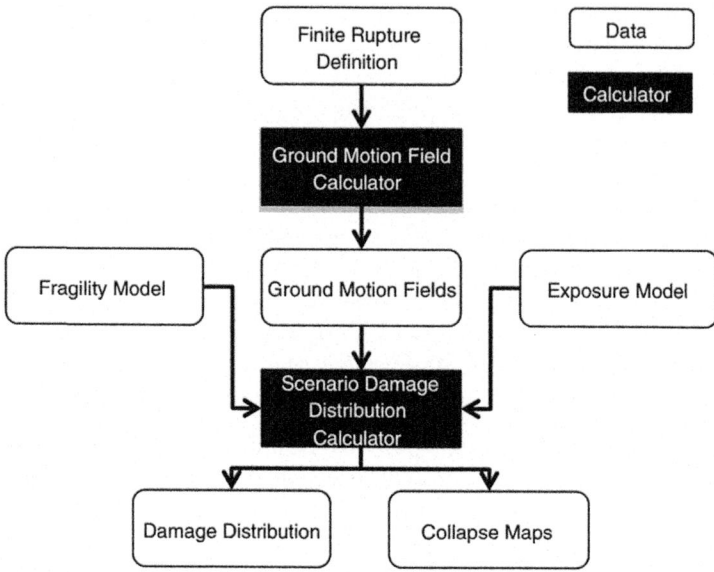

Fig. 2.3 Workflow of scenario risk assessment

defined as: discrete fragility models, where a list of probabilities of exceedance per limit state are provided for a set of intensity measure levels, or as continuous fragility models, where each limit state curve is modeled as a cumulative lognormal function, represented by a mean and standard deviation.
- Exposure Model: The exposure model contains the information regarding on the assets (physical elements of value) exposed to the earthquake hazard within the region of interest. A number of attributes (such as: construction type/material, height, age and value) are required to define the characteristics of each asset. Building taxonomy (classification scheme) and the geographic location respectively allows for the association of the asset with the appropriate fragility function and the site-specific seismic hazard.

The important ingredients of both of these earthquake loss estimation methodologies, in consideration of the "rapid" assessment of earthquake losses, are Ground Motion, Direct Physical Damage to General Building Stock and as Direct Economic/Social (Casualties) Losses.

2.2.1 Ground Motion

Bird and Bommer (2004) has shown that 88 % of damage in recent earthquakes has been caused by ground shaking, rather than secondary effects (e.g. ground failures, tsunamis). As such the quantification of the vibratory effects of the earthquakes is of prime importance in rapid loss assessments.

Almost all deterministic earthquake loss assessment schemes rely on the quantification of the earthquake shaking as intensity measure parameters in geographic gridded formats. The earthquake shaking can be determined theoretically for assumed (scenario) earthquake source parameters through ground motion prediction relationships GMPE's (i.e. attenuation relationships) or using a hybrid methodology that corrects the analytical data with empirical observations, after an earthquake. Either procedure yields the so-called, maps that display the spatial variation of the peak ground motion parameters or intensity measures. We owe this "ShakeMap" term to the USGS program that provides near-real-time maps of ground motion and shaking intensity following significant earthquakes in the United States as well as around the Globe (http://earthquake.usgs.gov/eqcenter/shakemap/). ShakeMap uses instrumental recordings of ground motions, kriging techniques and empirical ground motion functions to generate an approximately continuous representation of the shaking intensity shortly after the occurrence of an earthquake (Wald et al. 2005). In this connection Harmandar et al. (2012) has developed a novel method for spatial estimation of peak ground acceleration in dense arrays. The presented methodology estimates PGA at an arbitrary set of closely spaced points, in a way that is statistically compatible with known or prescribed PGA at other locations. The observed data recorded by strong motion stations of Istanbul Earthquake Rapid Response System are used for the development and validation of the new numerical method.

The data that are generated via ShakeMap can be used as inputs for the casualty and damage assessment routines for rapid earthquake loss estimation. In USA, and increasingly in other countries, these maps are used for post-earthquake response and recovery, public and scientific information, as well as for loss assessment and disaster planning.

Needless to say, for rapid loss assessment after an earthquake the fast and reliable information on the source location and magnitude is essential. Most rapid loss basements (e.g. PAGER and QLARM) rely on teleseismic determinations of epicenters. This reliance can create error in loss estimations, especially in populated areas, since the mean errors in real-time teleseismic epicenter solutions, provided by U.S. Geological Survey (USGS, the PDE) and/or the European Mediterranean Seismological Center (EMSC), can be as large as 25–35 km (Wyss et al. 2011).

Real-time seismology has made significant improvements in recent years, with source parameters now available within short time after an earthquake. In this context, together with the development of new ground motion predictive equations (GMPEs) that are able to account for source complexity, the generation of strong ground motion shaking maps in quasi-real time has become ever more feasible after the occurrence of a damaging earthquake (Spagnuolo et al. 2013).

The increased availability of seismic intensity data (such as those from "Did You Feel It-DYFI" type programs) immediately following significant earthquakes offers the opportunity to supplement instrumental data for the rapid generation of ShakeMaps. With minor filtering and with sufficient numbers, the intensity data reported through DYFI were found to be a remarkably consistent and reliable measure of earthquake effects (e.g., Atkinson and Wald 2007).

2.2.2 Direct Physical Damage to Building Stock

For the assessment of direct physical damages, general building stock inventory data and the associated fragility relationships are needed.

2.2.2.1 Inventory

To perform a seismic loss assessment, an inventory of the elements at risk should be defined. The classification systems used to define the inventories, the necessary inputs for each level of analysis and the default databases should be compatible with the fragility relationships. The definition of a classification system for the characterization of the exposed building stock and the description of its damage is an essential step in a risk analysis in order to ensure a uniform interpretation of data and results. For a general building stock the following parameters affect the damage and loss characteristics: structural (system, height, and building practices), nonstructural elements and occupancy (such as residential, commercial, and governmental). Building taxonomies define structure categories by various combinations of use, time of construction, construction material, lateral force-resisting system, height, applicable building code, and quality (HAZUS-MH 2003; EMS-98-Grünthal 1998; RISK-UE 2001–2004). The inter-regional difference in building architecture and construction practices should be reflected in building classifications for the development of inventories and fragility information. Only limited number of countries and cities has well developed building inventories. Several efforts are underway, such as PAGER and Global Earthquake Model-GEM (www.globalquakemodel.org) projects, to develop global building inventory databases.

Publicly available data includes: UN-Housing database, UN-HABITAT, UN Statistical database on Global Housing (1993) housing censuses, Population and Housing Censuses of individual countries (United Nations 2005), the World Housing Encyclopedia (WHE) database developed by EERI (2007).

In order to quantify earthquake risk of any selected region or a country of the world within the Global Earthquake Model (GEM) framework (www.globalquakemodel.org/), a systematic compilation of building inventory and population exposure is indispensable. Through the consortium of leading institutions and by engaging the domain-experts from multiple countries, the GED4GEM project has been working towards the development of a first comprehensive publicly available Global Exposure Database (Gamba et al. 2012).

ELER software (Sect. 2.4.4 of this chapter) uses a proxy procedure that relies on land use cover and population distributions to develop regional scale building inventories (Demircioglu et al. 2009).

2.2.2.2 Fragility Functions

A seismic fragility function defines loss (here, probability of buildings in various damage states as a result of direct physical damage) as a function of shaking intensity measure. The fragility functions can be classified under three main groups: Empirical (damage probability matrices or vulnerability functions based on field surveys, typology or expert judgement), Analytical (using capacity spectrum or other non-linear static procedures, collapse mechanism-based or displacement-based methods) or Hybrid.

The statistical method for the development of structural fragility functions is empirical that is, it employs loss data from historical earthquakes. The observed damage at various locations can be correlated to instrumental ground motion, intensity or some measure of intensity (Spence et al. 1992). Statistically derived building damage probability matrices (DPM) where first proposed by Whitman et al. (1973). The DPMs developed in the ATC-13 (1985) use expert opinion. He essentially partitioned the observed damage data from the 1971 San Fernando earthquake using various structural classes (taxonomy) and damage state categories as a function of the ground motion intensity (MMI). The statistical (or observed) methods are of greater relevance with non-engineered buildings where substantial damage data is available. The statistical approach offers conceptual simplicity and confidence since it is based on empirical loss data. However, the averaging effect of the definition of the intensity between different building types and damage states sets a limit to their applications. Using the EMS'98 (Grünthal 1998) intensity definitions, Giovinazzi and Lagomarsino (2004) developed a method on the basis of beta damage distribution and fuzzy set theory to produce DPM's. This method has been incorporated into the ESCENARIS and ELER earthquake loss assessment tools (Sect. 2.3). Empirical vulnerability curves (Rossetto and Elnashai 2003) and PSI-via-MSK (Spence et al. 1991) and are developed to give a continuous function of intensity versus damage.

Analytical (or predicted) fragility refers to the assessment of expected performance of buildings based on calculation and building characteristics, or on judgment based on the "expert's" experience. The fragility relationships refer to the structural damage states defined (essentially on the basis of displacement drifts) as Slight, Moderate, Extensive and Complete. Each fragility curve is associated with a standard deviation that encompasses the uncertainties stemming from damage threshold, capacity spectrum and the seismic demand.

An analytical method for estimating seismic fragility that uses nonlinear pseudo-static structural analysis is described by Kircher et al. (1997), where the lateral force versus the lateral displacement curve of the building structure, idealized as an equivalent nonlinear, single degree of freedom (SDOF) system, is obtained. This curve is transformed to the spectral displacement-spectral acceleration space to obtain the so-called capacity spectrum. Building capacity spectra vary between different buildings reflecting structural types, local construction practices and building code regulations.

The analytical fragility procedure, commonly called the Capacity Spectrum Method, essentially involves the comparison of the capacity of a structure, represented by the capacity spectrum, with the seismic demand represented by an acceleration displacement response spectrum (ADRS – Mahaney et al. 1993). The "performance point" of a model building type is obtained from the intersection of the capacity spectrum and the demand spectrum and this is then input into fragility curves which allow the probability of exceeding a number of damage states, given this performance point.

The capacity spectrum method, originally derived by Freeman (1998), is first implemented within the HAZUS procedure (FEMA 1999, 2003) as well as in many other earthquake loss estimation analyses: HAZ-Taiwan (Yeh et al. 2000, 2006), Risk-UE (Mouroux et al. 2004; Mouroux and Le Brun 2006), EQRM (Robinson et al. 2005), SELENA (Molina and Lindholm 2005 and ELER (Erdik et al. 2008, 2010; Hancılar et al. 2010).

DBELA (Displacement-Based Earthquake Loss Assessment) method (Crowley et al. 2004; Bal et al. 2008a) relies on the principles of direct displacement-based design method of Priestley (1997, 2003). DBELA method compares the displacement capacities of the substitute SDOF models of the buildings are compared with the seismic demand at their effective periods of vibration at different levels of damage. Buildings are classified on the basis of their response mechanisms: beam-sway or column-sway and the displacement capacities and periods of vibration for each damage state computed. Structural displacements are used to define the limit states of damage.

2.2.3 Casualties as Direct Social Losses

One of main reasons for rapid earthquake loss estimation is to estimate the spatial distribution of casualties, such that the search and rescue (SAR) and other emergency response activities can be prioritized and rationally coordinated. Casualty estimations encompass significant uncertainties since the casualty numbers vary greatly from one earthquake to another and they are poorly documented.

Apart from simple correlations with intensity or magnitude and population density, the casualty numbers are generally estimated via a correlation with the damage state experienced by a structure, the time of day, the structural use, and other factors. ATC-13 (1985) casualty estimation model consists of tabulated injury and death rates related to a building's level of damage, or damage state, providing a 4:1 ratio of serious injuries to deaths, and 30:1 ratio of minor injuries to deaths. The model does not provide any differentiation of structural types, suggesting only taking 10 % of the rates for light steel and wood-frame structures.

The casualty estimation model of Coburn and Spence (2002) is based on the distribution of buildings in the complete damage state (D5) as defined in EMS'98. The number of deaths is obtained by multiplication of D5, average people in each collapsed building, percentage of occupants at time of shaking, expected trapped occupants, mortality at collapse and mortality post-collapse. However, it is not in event tree format and does not account for non collapse (damage) related casualties, nor does it account for the population not indoors at the time of earthquake. Coburn and Spence (2002) notes that especially for cases of moderate levels of damage, i.e. those where fewer than 5000 buildings were damaged, the casualty estimations could be highly inaccurate. Irrespective of the methodology chosen, casualty numbers are computed for three different day time scenarios (night time, day time, and commuting time). This methodology was then improved through the LessLoss methodology of Spence (2007a) with other damage states also taken into account in terms of fatalities. In addition, updated casualty and injury ratios were produced based on a greater set of earthquakes. So and Spence (2009) explored further the relationship of building.

HAZUS-MH (2003) model estimates casualties directly caused by structural or nonstructural damage under four severity levels to categorize injuries, ranging from light injuries (Severity Level 1) to death (Severity Level 4). The model provides casualty rates for different structural types and damage states. Relevant issues in casualty estimation such as occupancy potential, collapse and non-collapse vulnerability of the building stock, time of the earthquake occurrence, and spatial distribution of the damage, are included in the methodology. Casualties caused by a postulated earthquake can be modeled by developing a tree of events leading to their occurrence.

Recent empirical methods of Porter et al. (2008a, b), Jaiswal et al. (2009) and Jaiswal and Wald (2010c) have concentrated on the key parameters of intensity as the hazard metric versus fatality to population ratios or the death rate in collapsed buildings, using expert opinion related collapse ratios and historical data. The earthquake fatality rate is defined as total killed divided by total population exposed at specific shaking intensity level. The total fatalities for a given earthquake are estimated by multiplying the number of people exposed at each shaking intensity level by the fatality rates for that level and then summing them at all relevant shaking intensities. The fatality rate is expressed in terms of a two-parameter lognormal cumulative distribution function of shaking intensity. The parameters are obtained for each country or a region by minimizing the residual error in hindcasting the total shaking-related deaths from earthquakes recorded between 1973 and 2007. A new global regionalization scheme is used to combine the fatality data across different countries with similar vulnerability traits.

The study of the socio-economic losses associated with past earthquakes has gained a new dimension with the development of the worldwide catalogue of damaging earthquakes and secondary effects database (CATDAT) (Daniell et al. 2011c, 2012b). CATDAT has been created using over 20,000 information sources to present loss data from 12000+ historical damaging earthquakes since 1900, with 7000+ examined and validated before insertion into the database. In addition to seismological information, each earthquake includes parameters on building damage data and socio-economic losses. CATDAT have facilitated the study of socio-economic earthquake losses and the derivation of associated fragility/vulnerability relationships. Daniell (2014) has developed an approach to rapidly calculate fatalities and economic losses from earthquakes using the input of intensity based map and historical earthquakes as a proxy over multiple temporal and spatial scales. The population and its social and economic status for each earthquake were compared to the detailed socio-economic data in CATDAT to produce the functions. Temporal relationships of socio-economic losses were explored in order to calibrate loss functions.

2.2.4 Estimation of Economic Losses

Financial loss is, essentially, the translation of physical damage into total monetary loss using local estimates of repair and reconstruction costs. Studies on economic impacts of earthquakes have been usually examined in two categories: (a) loss caused by damage to built environment (direct loss), and (b) loss caused by interruption of economic activities (indirect loss). Simple economic loss models are based on direct calculation of property values multiplied by some form of damage metric.

HAZUS-MH (2003) estimates losses at three levels of accuracy: Levels 1, 2, and 3.

Level 1: A rough estimate based solely on data from national databases (demographic data, building stock estimates, national transportation and infrastructure data) included in the HAZUS-MH software distribution.

Level 2: A more accurate estimate based on professional judgment and detailed information on demographic data, buildings and other infrastructure at the local level.

Level 3: The most accurate estimate based on detailed engineering input that develops into a customized methodology designed to the specific conditions of a community.

The level of accuracy encompassed in "Level 1" can be suitable for post-earthquake rapid economic loss assessment.

Through use of statistical regression techniques, data from past earthquakes can be used to develop relationships (Loss Functions) for predicting economic losses. However the existing economic loss data are scarce, biased for heavy damage and could also be proprietary. Loss functions can be estimated by using analytical procedures in connection with a Monte Carlo simulation technique. However, such procedures are not intended for rapid loss estimation type applications.

Losses are generally calibrated to damage states in order to determine direct losses. The definition of the slight, moderate and heavy damage classes in terms of losses has a large variation in terms of potential loss estimates. Let alone the rapid assessment, even the formal quantification of economic losses is a very challenging issue. The technical manual of HAZUS-MH states that the total uncertainty (including that of the ground shaking) is "possibly at best a factor of two or more".

Chan et al. (1998) have proposed a quick and approximate estimation of earthquake loss using with detailed local GDP and population data, instead of the detailed building inventory required in traditional loss estimation methodologies. This method has been used for numerous case studies. Their method combines seismic hazard, GDP, population data, published earthquake loss data, and the relationship between GDP and known seismic loss, to estimate earthquake loss from the following relationship:

$$L = \Sigma \, P(I) \times F(I, GDP) \times GDP \tag{2.1}$$

where L is the economic loss, P(I) is the probability of an earthquake of intensity I, and F(I,GDP) is a measure of the area's fragility to earthquake damage for the given GDP value and the earthquake of intensity I. The GDP is used as a macroeconomic indicator to represent the total exposure of an area in the earthquake loss estimation. In this study F(I,GDP) is determined from the relationship between reported losses from earthquakes to the computed GDP of the affected area. Since GDP is usually provided for a country, it must be apportioned over the nation to the affected area. For this purpose Chan et al. (1998) relies on the correlation between GDP and population density.

The estimates of the direct economic losses due to building damage, which consist of capital stock loss, are relatively easier to be included in rapid loss assessments. These losses are generally quantified as Loss Ratios (LR) – the loss as a percentage of the building replacement value. The economic losses to other elements of the built environment and indirect economic losses, representing the losses due to various forms of post-earthquake socioeconomic disruptions (such as employment and income, insurance and financial aids, construction, production and import-export of goods and services) cannot be rationally included in rapid earthquake loss assessment estimations.

Jaiswal and Wald (2011, 2013) have developed a model of economic losses based on economic exposure versus intensity correlations to rapidly estimate economic losses after significant earthquakes worldwide. The requisite model inputs are shaking intensity estimates made by the ShakeMap system, the spatial distribution of population available from the LandScan database, modern and historic country or sub-country population and Gross Domestic Product (GDP) data, and economic loss data from Munich Re's historical earthquakes catalog. Earthquakes from 1980 to 2007 were examined using economic loss estimates from past events from the MunichRe NatCat Service database. The methodology uses a wealth index as a proxy for exposure, multiplying this in much the same way as a multiplier-output ratio has been applied in Chen et al. (1997a). The process consists of using a country specific multiplicative factor to accommodate the disparity between economic exposure and the annual per capita GDP, and it has proven successful in hindcasting past losses. Although loss, population, shaking estimates, and economic data used in the calibration process are uncertain, approximate ranges of losses can be estimated for the primary purpose of gauging the overall scope of the disaster and coordinating response. The proposed methodology is both indirect and approximate and is thus best suited as a rapid loss estimation model for applications.

Daniell et al. (2012a) has analysed the trends in economic losses (direct, indirect and insured) in earthquakes since 1900 using CATDAT Damaging Earthquakes Database and developed methodologies for the rapid assessment of economic losses (Daniell 2014). In order to compare the economic losses of the historic earthquakes, the losses were converted into today's dollars.

2.2.5 Uncertainties in Loss Estimation

Uncertainties are inherent in any loss estimation methodology. They arise in part from incomplete scientific knowledge concerning earthquakes, earthquake ground motion and their effects upon buildings and facilities. They also result from the approximations and simplifications that are necessary for comprehensive analyses. Incomplete or inaccurate inventories of the built environment, demographics and economic parameters add to the uncertainty. These factors can result in a range of uncertainty in loss estimates produced by the HAZUS-MH Earthquake Model, possibly, at best, a factor of two or more. HAZUS-MH (2003).

The earthquake loss estimations should consider the uncertainties in seismic hazard analyses, and in the fragility relationship. There exits considerable amount of epistemic uncertainty and aleatory variability in ShakeMaps. Accuracy of the ShakeMap is mainly related to two factors: (1) the proximity of a ground motion observation location, i.e. the density of the strong ground motion network in the affected area, and (2) the uncertainty of estimating ground motions from the GMPE, most notably, elevated uncertainty due to initial, and unconstrained source rupture geometry. The epistemic uncertainties become highest for larger magnitude events when rupture parameters are not yet well constrained (Wald et al. 2008). Aleatory uncertainties may be reduced if the bias correction with recorded amplitudes is performed directly on the ground surface rather than at bedrock level which the case in the current ShakeMap application (USGS, ShakeMap).

The reliability of the fragility relationships is related to the conformity of the ground motion intensity measure with the earthquake performance (damage) of the building inventory. Estimates of human casualties are derived by uncertain relationships from already uncertain building loss estimates, so the uncertainties in these estimates are compounded (Coburn and Spence 2002).

It is possible to examine the effect of cumulative uncertainties in loss estimates using discrete event simulation (or Monte-Carlo) techniques if the hazard and that the probability distribution of each of the constituent relationships is known. The general finding of the studies on the uncertainties in earthquake loss estimation is that the uncertainties are large and at least as equal to uncertainties in hazard analyses (Stafford et al. 2007).

2.3 Earthquake Loss Estimation Software Tools

For known inventories of buildings and under conditions where the earthquake hazard in terms of ground shaking distribution can be assessed rapidly after an earthquake, these tools can be adapted for rapid loss estimation. Daniell (2009, 2011b) has provided a comprehensive comparison between different earthquake loss estimation software packages, in terms of their applicability regions, exposure resolution (district, city, regional, country), hazard (deterministic predicted, deterministic observed, probabilistic), vulnerability type (analytical, empirical, socioeconomic). Strasser et al. (2008) has provided a comparison of five selected European earthquake loss estimation software packages (KOERILOSS-ELER, SELENA, ESCENARIS, SIGE-DPC and DBELA), using Istanbul as a test bed. The packages considered common inputs in terms of ground motions, building inventory and population; however the fragility functions and modelling assumptions differed in each package. The overall estimates of building damage were close to each other. However, the results often substantially differed at grid cell level. In terms of social losses, the predictions from the various approaches show a large degree of scatter, mostly driven by differences in the casualty rates assumed.

A brief description and references for the selected earthquake loss assessment software can be given as follows:

2.3.1 HAZUS

HAZUS-MH (FEMA and NIBS 2003) is developed by the United States Federal Emergency Management Agency (FEMA) for the prediction and mitigation of losses due to earthquakes (HAZUS), hurricanes and floods (Whitman et al. 1997; Kircher et al. 2006). The package is intended for U.S. applications only and includes federally collected data as default. The inventory is classed based on 36 different types of building based on construction standards and material as well as size and building use. HAZUS-MH MR2 version, released in 2006, includes the capability for rapid post-event loss assessment.

2.3.2 EPEDAT

The EPEDAT (Early Post-Earthquake Damage Assessment Tool) is designed by EQE International, Inc. for post-earthquake loss estimation (Eguchi et al. 1997). The output encompasses damage (building and lifelines) and casualty for California based on county specific housing and demographic data. It is Windows-based and uses Modified Mercalli Intensity to quantify the hazard.

2.3.3 SIGE

SIGE, developed by Italian National Seismic Service of the Civil Protection Department, is used for rapid approximate estimate of the damage (Di Pasquale et al. 2004). The first update of the program (FACES) considers linear sources, directivity effects, and the influence of focal depth. The most recent modification of the codes has been implemented in a new model called ESPAS (Earthquake Scenario Probabilistic Assessment).

2.3.4 KOERILOSS

A scenario-based building loss and casualty estimation model developed by Bogazici University (Erdik and Aydinoglu 2002; Erdik et al. 2003a, b; Erdik and Fahjan 2006) for estimating earthquake losses in Istanbul, Izmir, Bishkek and Tashkent. Derivatives of the model were used in the EU FP5 LessLoss project as well as for the assessment of scenario earthquake losses in Amman. The methodology considers both deterministic (scenario) and probabilistic forecasting approaches. The fragility calculations can be based on empirical results (EMS intensity-based) or on a response-spectrum-based method similar to HAZUS. It is used for rapid loss assessment in connection with the Istanbul Earthquake Rapid Response System, described in Sect. 2.5.3 of this chapter.

2.3.5 ESCENARIS

ESCENARIS (Roca et al. 2006) is the software tool developed for Catalonia. The methodology relies on the use of scenario-based earthquake hazards and intensity-based empirical fragility functions of Giovinazzi (2005). The losses are based on the building stock and classes of social impact.

2.3.6 CAPRA

CAPRA (Central American Probabilistic Risk Assessment – www.ecapra.org) Project has developed a region-specific Earthquake Loss Estimation model using a Web 2.0 format. It is currently under construction (Anderson 2008).

2.3.7 LNECLOSS

LNECLOSS is a software package developed by the Laboratorio Nacional de Engenharia Civil (LNEC) in Lisbon, Portugal (Sousa et al. 2004). LNECloss is an earthquake loss assessment tool, integrated on a Geographic Information System (GIS), which comprises modules to compute seismic scenario bedrock input, local soil effects, fragility and fragility analysis, human and economic losses. LNECloss was applied to Metropolitan Area of Lisbon (Zonno et al. 2009).

2.3.8 SELENA

SELENA (Seismic Loss Estimation Using a Logic Tree Approach) is a software package developed at NORSAR for earthquake building damage assessment (Molina and Lindholm 2005). SELENA uses the capacity-spectrum method (HAZUS methodology, ATC-55-ATC 2005) with a logic tree-based weighting of input parameters that reportedly allows for the computation of confidence intervals. GIS software can be utilized at multiple levels of resolution to display predicted losses graphically. Detailed information on SELENA is provided in Sect. 2.4 of this chapter.

2.3.9 DBELA

DBELA (Displacement-Based Earthquake Loss Assessment) is an earthquake loss estimation tool currently being developed at the ROSE School/EU-Centre in Pavia (Crowley et al. 2004; Calvi et al. 2006; Bal et al. 2008a). The methodology is essentially based on comparison of the displacement capacity of the building stock (grouped by structural type and failure mechanism) and the imposed displacement demand from a given earthquake scenario. The methodology aims to allow a good correlation with damage, ease of calibration to varying building stock characteristics and systematic treatment of all sources of uncertainty. It takes into account the uncertainties associated through the process for demand and capacity. Applications of the methodology were carried out for loss assessment in the Marmara Region (Bommer et al. 2006).

2.3.10 EQSIM

EQSIM (EarthQuake damage SIMulation) is the rapid earthquake damage estimation component of the Disaster Management Tool (DMT) currently being developed at the University of Karlsruhe (Baur et al. 2001; Markus et al. 2004). The loss estimation methodology is based on the adaptation capacity spectrum method used in HAZUS to reflect the European building practice. EQSIM has been used to assess earthquake losses in Bucharest on the basis of scenario earthquakes (Wenzel and Marmuraenu 2007).

2.3.11 QUAKELOSS

QUAKELOSS is a computer tool for estimating human loss and building damage due to Earthquakes developed by the staff of the Extreme Situations Research Center in Moscow. An earlier version of this program and data set is called EXTREMUM (Larionov et al. 2000). QUAKELOSS software is used by the World Agency of Planetary Monitoring and Earthquake Risk Reduction (WAPMERR) to provide near-real-time estimates of deaths and injuries caused by earthquakes anywhere in the world. The building inventory reportedly incorporates data from about two million settlements throughout the world.

2.3.12 NHEMATIS

NHEMATIS (Natural Hazards Electronic Map and Assessment Tools Information System) has been developed Emergency Preparedness Canada (Couture et al. 2002). It is a national-scale automated facility for the collection and analysis of natural hazard information combined with characterizations of population and infrastructure to allow analyses of risks. Similar to HAZUS, NHEMATIS integrates an expert system rule base, geographic information system (GIS), relational databases, and quantitative models to permit assessment of the hazard impact.

2.3.13 EQRM

EarthQuake Risk Management (EQRM), developed by Geoscience Australia, is an event-based tool for earthquake scenario ground motion and scenario loss modeling as well as probabilistic seismic hazard and risk modeling (Robinson et al. 2005, 2006). The risk assessment methodology is based on the HAZUS methodology with some modifications to adapt it to Australian conditions. It has the potential to be used with earthquake monitoring programs to provide automatic loss estimates.

2.3.14 OSRE

The Open Source Risk Engine (OSRE), developed in Kyoto University – Graduate School of Engineering – Department of Urban Management, is multi-hazards open-source software that can estimate the risk (damage) of a particular site (object) given a hazard and the fragility with their associate probability distributions (AGORA-Alliance for Open Risk Analysis, http://www.risk-agora.org). The catalogue fragility data for different facility classes was obtained from ATC-13.

2.3.15 ELER

The Joint Research Activity 3 (JRA3) of the EU Project NERIES has developed a methodology and software "Earthquake Loss Estimation Routine – ELER" (ELER V3.1 2010; Erdik et al. 2008, 2010) for rapid estimation of earthquake damages and casualties throughout the Euro-Med Region. ELER is designed as open source software to allow for community based maintenance and further development of the database and earthquake loss estimating procedures. The software provides for the estimation of losses in three levels of analysis. These levels of analysis are designed to commensurate with the quality of the available building inventory and demographic data. Detailed information on ELER is provided in Sect. 2.4 of this chapter.

2.3.16 MAEVIZ

MAEviz, developed in the Mid-America Earthquake Center in University of Illinois, integrates spatial information, data, and visual information to perform seismic risk assessment and analysis (http://mae.ce.uiuc.edu/software_and_tools/maeviz.html). It can perform earthquake risk assessment for buildings (structural and non-structural damage), bridges and gas networks with a built-in library of fragility relationships. In addition to applications in USA and important application of the software has been conducted for the Zeytinburnu District of Istanbul (Elnashai et al. 2007).

2.4 Earthquake Rapid Loss Assessment Systems

Available near real time loss estimation tools can be classified under two main categories depending on the size of area they cover: (1) Global or Regional Systems and (2) Local Systems. For the global or regional near real time loss estimation efforts the following developments will be considered:

- Global Disaster Alert and Coordination System – GDACS (http://www.gdacs.org),
- World Agency of Planetary Monitoring Earthquake Risk Reduction – WAPMERR (http://www.wapmerr.org),
- Prompt Assessment of Global Earthquakes for Response – PAGER (http://earthquake.usgs.gov/eqcenter/pager/),
- Earthquake Loss Estimation Routine – ELER (http://www.koeri.boun.edu.tr/Haberler/NERIES%20ELER%20V3.1_6_176.depmuh)
- Seismic Loss Estimation using a Logic Tree Approach – SELENA (http://selena.sourceforge.net/selena.shtml)

A description of the important rapid earthquake loss assessment systems with global or regional coverage will be provided in the following sub-sections.

2.4.1 PAGER (Prompt Assessment of Global Earthquakes for Response)

PAGER (Prompt Assessment of Global Earthquakes for Response) is an automated system that produces content concerning the impact of significant earthquakes around the world, informing emergency responders, government and aid agencies, and the media of the scope of the potential disaster. PAGER has three separate methodologies for earthquake loss estimation as part of their package (empirical, semi-empirical and analytical). PAGER rapidly assesses earthquake impacts by comparing the population exposed to each level of shaking intensity with models of economic and fatality losses based on past earthquakes in each country or region of the world (Earle et al. 2009a, b). PAGER products are generated for all earthquakes of magnitude 5.5 and greater globally and for lower magnitudes of about 3.5–4.0 within the US. PAGER's results are posted on the USGS Earthquake Program Web site (http://earthquake.usgs.gov/) and sent in near real-time to emergency responders, government agencies, and the media. In the hours following significant earthquakes, as more information becomes available, PAGER's content is modified.

2.4.1.1 Process

The following steps are used in the PAGER methodology:

1. After the magnitude and hypocenter of an earthquake are determined. The PAGER process begins for each new event with the determination of the earthquake source parameters, macroseismic data and the resulting ShakeMap. For large earthquakes ShakeMaps are further constrained (if available, within several hours) by finite-fault waveform inversions (Wald et al. 2008). The

ShakeMaps are constrained, if available, by measurements from strong-motion seismometers in the region surrounding the ruptured fault. In case ground motion recordings are insufficient, ShakeMaps are constrained using empirical ground motion prediction equations based on magnitude, site amplification, and distance to the fault. Observations reported by people in the shaken region using the USGS "Did You Feel It" system (Wald et al. 1999) are converted to estimates of shaking intensity and also used to constrain the ground motion distribution. ShakeMap generates a soil/rock site-specific ground-motion amplification map based on topographic slope and then converts the estimated ground motions to a map of seismic intensities.

2. Following the determination of the shaking distribution, PAGER takes the grid shaking parameter values produced by ShakeMap and determines the settlements (Geonames, http://www.geonames.org) and the population (LandScan) database in each grid cell (accounting for time of day, Jaiswal and Wald 2008a) exposed to each level of Intensity (MMI).

3. Based on the population exposed to each shaking intensity level, the PAGER system estimates total shaking-related losses based on country-specific models developed from economic and casualty data collected from past earthquakes.

4. PAGER's output is distributed by e-mail and is available on the USGS Earthquake Program webpage (http://earthquake.usgs.gov/pager/). The maps and tables in this output provide a quick assessment of the estimated impact of the earthquake. The maps provide an indication of the geographic extent of the shaking and distribution of the affected population. The Earthquake Impact Scale provides alert levels for fatalities and economic losses. These alert levels are based on the range of most likely losses due to earthquake shaking and the uncertainty in the alert level can be gauged by the histogram, depicting the percent likelihood that adjacent alert levels (or fatality/loss ranges) occur. The table included provides information on the impact of an earthquake by providing the total number of people within the map boundary estimated to have experienced each MMI level from I (not felt) to X (extreme) and information on possible building damage at different MMI levels for resistant and vulnerable structures.

2.4.1.2 Building and Population Inventories and Fragilities

EXPO-CAT (http://earthquake.usgs.gov/research/data/pager/expocat) provides first-order estimates of the number of people exposed to significant global earthquakes since 1973 using current PAGER methodology (Allen et al. 2009a, b). It combines earthquakes in the Atlas of ShakeMaps (Allen et al. 2008) with a gridded global population database to estimate population exposure at discrete levels of macroseismic intensity. Present-day population exposure is estimated using the Landscan global population database. Combining this population exposure dataset with historical earthquake loss data provided for the calibration loss methodologies against the set of ShakeMap hazard outputs.

Currently a first-order building inventory database compiled from: the housing data of the United Nations (UN 1993) and UN Habitat (2007); data compiled by Population and Housing Censuses of individual countries (UN 2005) and; the World Housing Encyclopedia (WHE) database developed by the Earthquake Engineering Research Institute (EERI 2007) is available (Jaiswal and Wald 2008a, b; Wald et al. 2009a, b). At the country level, the inventory database contains estimates of building types categorized by material, lateral force-resisting system, use, and occupancy characteristics.

In a collaborative effort between the US Geological Survey, the Earthquake Engineering Research Institute, and the World Housing Encyclopedia (http://www.world-housing.net/), experts from around the world have estimated the distribution of predominant buildings types in each of 26 countries, and provided by judgment or statistical survey collapse fragility functions for the predominant structure types in each country (Jaiswal and Wald 2008b; Porter et al. 2008a, b). Operationally, the current PAGER system relies on the empirically-based loss approach (Wald et al. 2008).

The collapse fragility functions developed for global building types using the procedure described in Jaiswal et al. (2011) is expected to form a starting building damage estimation model within the PAGER semi-empirical vulnerability model.

PAGER's fatality loss models (Wald et al. 2008; Jaiswal and Wald 2010) stems from the wide, global variability in the built environment and uncertainty associated with inventory and structural vulnerability data, as well as the knowledge about past casualties in different countries. The empirical model relies on country-specific earthquake loss data from past earthquakes and makes use of calibrated casualty rates for future prediction. For this purpose, a three tiered approach is adopted for fatality estimation. In the empirical approach, a fatality rate is proposed as a proportion of the population exposed at each intensity level, and depends on the shaking intensity according to a lognormal function, with values of the two separate parameters defining the function, and an uncertainty factor, each for different countries or regions of the world. This empirical approach is mostly adaptable for the developing regions of the world, where the available data does not permit for an analytical analysis to be conducted. The PAGER semi-empirical approach aims to develop a better casualty estimate by using, for the area affected at each intensity level, the number of buildings and their vulnerability to collapse at the estimated ground shaking, combined with an estimate of the fatality (or lethality) rate as a proportion of total occupants, given collapse.

2.4.1.3 Economic Loss Estimation

In order to estimate economic losses an assessment of the economic exposure at various levels of shaking intensity is used. Since the economic value of all the physical assets exposed at different locations in a given area is generally not known and extremely difficult to compile at a global scale, In the absence of such a dataset, the total Gross Domestic Product (GDP) exposed at each shaking intensity is estimated by multiplying the per-capita GDP of the country by the total population exposed at that shaking intensity level. The total GDP thus estimated at each intensity is then scaled by an exposure correction factor, which represents a multiplying factor to account for the disparity between wealth and/or economic assets to the annual GDP (Jaiswal and Wald 2011).

For this development at least four damaging earthquakes that occurred within a country or region during the observation period between 1973 and 1980. Since only a few countries experienced large, damaging earthquakes for which loss values are available during the observation period, it was necessary to aggregate some countries into regions using the "Human Development Index" (HDI) to estimate the parameters of the economic loss ratio function. The economic exposure obtained using this procedure is a proxy estimate for the economic value of the actual inventory that is exposed to the earthquake.

2.4.2 GDACS: The Global Disaster Alert and Coordination System

The Global Disaster Alert and Coordination System – GDACS (http://www.gdacs.org/) provides near real-time alerts about natural disasters around the world and tools to facilitate response. GDACS is a joint initiative of the United Nations Office for the Coordination of Humanitarian Affairs (OCHA) and the European Commission that serves to consolidate and improve the dissemination of disaster-related information, in order to improve the coordination of international relief efforts. It started as GDAS, but was later coupled with the coordination information system of the UN Office of Coordination of Humanitarian Affairs-Virtual On-site Operations Coordination Center (the OCHA Virtual OSOCC, http://vosocc.unocha.org, http://vosocc.gdacs.org). GDACS collects near real-time hazard information and combines this with demographic and socio-economic data to perform a mathematical analysis of the expected impact. This is based on the magnitude of the event and possible risk for the population. The result of this risk analysis is distributed by the GDACS website and alerts are sent via email, fax, and SMS to subscribers in the disaster relief community, and all other persons that are interested in this information.

GDACS collects earthquake information from: United States Geological Survey National Earthquake Information Center (NEIC), European-Mediterranean Seismological Centre (EMSC), GEOFON Program of the GFZ Potsdam and Japan Meteorological Agency (JMA).

Using the reported earthquake parameters, a three level alert based on the LandScan population dataset and the population fragility (European Commission Humanitarian Aid Department Global Needs Assessment Indicator) in the region of interest. Currently, the evaluation of the potential humanitarian impact of earthquakes considers (1) earthquake magnitude, (2) earthquake depth, (3) population within 100 km of epicenter, and (4) national population fragility. The last two elements are automatically calculated by GIS based on the earthquake epicenter, the LandScan population dataset and ECHO's Global Needs Assessment indicator. The alerts are considered on the basis of the so-called alert score which combines the earthquake magnitude and depth, size of the exposed population and the country-specific fragility index. The alert score is transformed into three alert levels: red, orange and green.

2.4.3 WAPMERR-QLARM World Agency of Planetary Monitoring and Earthquake Risk Reduction

QLARM (http://qlarm.ethz.ch) provides loss estimates for earthquakes in global scale after the event. The post-earthquake alerts issued include number of fatalities and injured, as well as average damage to buildings in the affected settlements. This service is being carried out in partnership between WAPMERR (World Agency of Planetary Monitoring and Earthquake Risk Reduction) and the Swiss Seismological Service (SED-ETH, Zurich). The estimates in the current version include: (1) The expected percentage of buildings in each of five damage states in each settlement, (2) the mean damage state in each settlement, (3) the numbers of fatalities and injured, with error estimates, in each settlement (Trendafiloski et al. 2009b). The loss estimates are reportedly provided in about 30 min after the earthquake.

QLARM is an outgrowth of the former QUAKELOSS software, the computer tool used to estimate the building damage and casualties (Trendafiloski et al. 2009a). Loss estimations are done for the QLARM worldwide database constructed of: (1) point city models for the cases where only summary data for the entire city are available; and, (2) discrete city models where data regarding city sub-divisions (districts) are available (Trendafiloski et al. 2009b). The ground shaking for the settlements is computed based on the magnitude, epicenter and depth of the event using global and regional ground motion prediction models. Soil amplification is estimated using either local data to derive an amplification factor for each discrete city model or global data based on Vs30 values derived from topographic slopes from Allen and Wald (2007).

QLARM calculates the expected building damage using intensity based fragility models, calibrated using about 1,000 earthquakes for which losses are known. Distribution of building stock and population are attributed to these city models. In the data base of QLARM, the population of about two million settlements is known and each settlement has a profile of building fragility. Fragility classes are assigned to different building types considering the fragility table given by the European Macroseismic Scale EMS-98 (Grünthal 1998). Building and population distributions are constructed using the percentage of the number of buildings and population belonging to a particular fragility class. QLARM population database is constructed using national census data and the online sources World Gazetteer and

Geonames. Opinion of local experts, World Housing Encyclopedia and PAGER database are additional sources used to improve the population database. Population distribution by time of the event is taken into account using the model proposed by Coburn and Spence (2002).

The European Macroseismic Method of Giovinazzi (2005) is used to calculate building damages. The fragility models are pertinent to EMS-98 fragility classes and correlate the mean damage grade µD ($0 \leq µD \leq 5$) with the seismic intensity and the fragility index.

Human losses are estimated using the casualty event-tree model proposed by Stojanovski and Dong (1994). The probability of occurrence of casualty state for a given seismic intensity is calculated as a product of the damage probabilities for seismic intensity and the casualty probabilities for damage grades of EMS-98. It is claimed that the human losses are estimated within a factor of 2 for past earthquakes.

2.4.4 ELER: Earthquake Loss Estimation

The Joint Research Activity JRA-3 of the EU Project NERIES aims at establishing rapid estimation of earthquake damages, casualties, shelters and food requirements throughout the Euro-Med Region. Within the scope of this activity, a rapid loss estimation tool (ELER, http://www.koeri.boun.edu.tr/Haberler/NERIES%20ELER%20V3.1_6_176.depmuh) is developed by researchers from KOERI, Imperial College, NORSAR and ETH-Zurich. The loss estimation is conducted under three levels of sophistication as elaborated in Fig. 2.4.

The ground motion estimation methodology is common in all levels of analysis. The shake mapping methodology is similar to the USGS Shake Map (Wald et al. 2005). Based on the event parameters the distribution of PGA, PGV, SA ($T = 0.2$ s) and SA ($T = 1$ s) are estimated based on a choice of ground motion prediction models. Local site effects are incorporated either with the Borcherdt (1994) methodology or, if available, with the use of Vs30 based amplification functions within the ground motion prediction equations (GMPE). If strong ground motion recordings are also available, the prediction distributions are bias corrected using the peak values obtained from these recordings. Geo-spatial analysis can be also employed in this step, through the Modified Kriging Method. EMS-98 Intensity distributions are obtained based on computed PGA and PGV values using the procedure proposed by Wald et al. (1999). For site-specific analysis, Vs30 values (average shear wave propagation velocity in upper 30 m of the soil medium) are obtained from regional geology (Quaternary, Tertiary, Mesozoic (QTM) maps) or slope-based Vs30 maps (Allen and Wald 2007).

After the estimation of the spatial distribution of selected ground motion parameters, earthquake losses (damage and casualty) can be estimated at different levels of sophistication, namely Level 0, 1 and 2. The differentiation of these levels of

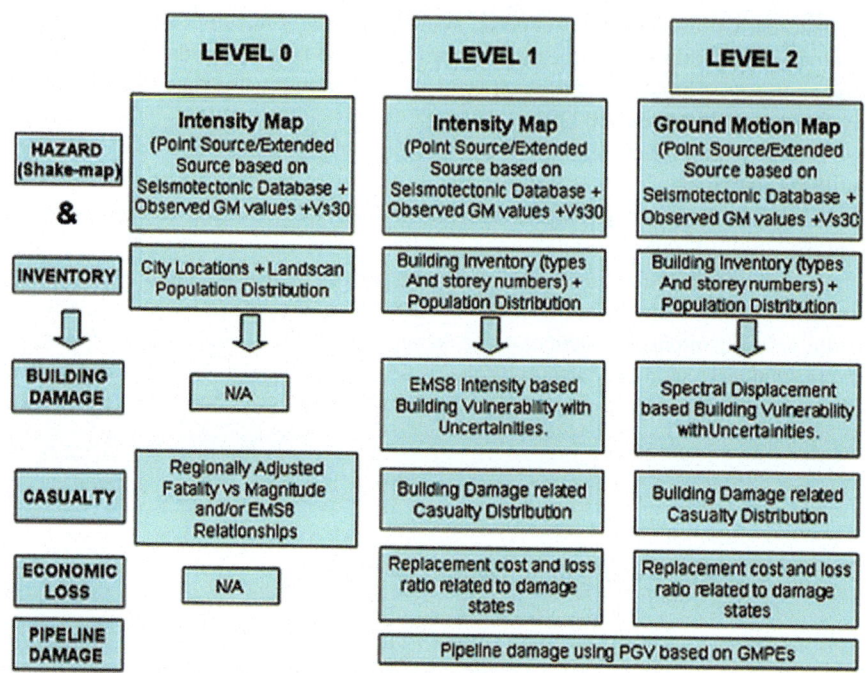

Fig. 2.4 The levels of analysis incorporated in the ELER software

analysis is essentially controlled by the availability of building inventory and demographic data (Demircioglu et al. 2009; ELER v3.1 2010; Erdik et al. 2010).

Both Level 0 (quite similar to PAGER system of USGS) and Level 1 analyses of ELER software are based on obtaining intensity distributions analytically and estimating total number of casualties either using regionally adjusted intensity-casualty or magnitude-casualty correlations (Level 0) or using regional building inventory databases (Level 1). Level 1 type analysis uses EMS98 (Grünthal 1998) based building fragility relationships of Lagomarsino and Giovinazzi (2006) to estimate building damage and casualty distributions.

Level 2 type analysis corresponds to the highest sophistication level in the loss estimation methodology developed. The building damage and casualty distributions are obtained using analytical fragility relationships and building damage related casualty fragility models, respectively. The Level 2 module of ELER aims at assessing the building damage and the consequential casualties using methodologies similar to HAZUS-MH (2003).

2.4.4.1 Demographic and Building Inventory

For all levels of analysis the 30 arc sec (about 1 km) grid based LandScan (Oak Ridge National Laboratory 2011) population data are used. For both the Level 1 and Level 2 analyses options exist for the use of local demographic data for casualty estimation.

ELER is structured in such a way that a building inventory can be classified in terms of any classification system as long as the empirical and/or mechanical fragility relationships associated with each building type is defined by the user. The HAZUS (FEMA 2003), EMS-98 (Grünthal 1998), and RISK-UE (2001–2004) building taxonomies are used as the default main classification systems in the development of ELER. The user has the capability of defining custom fragility curves by "Building Database Creator" tool.

The regional scale building inventory used in Level 1 analysis corresponds to an approximated (proxy) European database consisting of the number of buildings and their geographic distribution. This approximated building database is obtained from CORINE Land Cover (European Environment Agency 1999), LandScan population database and Google Earth (http://earth.google.com) and is provided within ELER as the default data for Level 1 analysis. Following the determination governing land cover classes for each country, the basic methodology used in obtaining the country basis proxy distribution of the number of buildings (per unit area in each building class) is as follows (Demircioglu et al. 2009; ELER v3.1 2010; Erdik et al. 2010):

1. Select suitable sample areas from Google Earth for each Corine Land Cover class in all countries
2. Obtain the actual number of buildings in each sample area, automatically using image processing techniques.
3. Approximate the total number of buildings in each country by spreading the sample area building counts to the country
4. Verify (and adjust) the number of buildings thus obtained by computing the population per building for each Corine Land Cover class, and by also checking with the actual number of buildings in a country if such information has been obtained from the corresponding country's statistical office.

The corresponding RISK-UE building taxonomy classes were identified and the associated percentages have been used to convert the grid based number of buildings to an inventory of differentiated structural types in each country. The grid based distribution of the number of buildings and population thus obtained is aggregated to 30 and 150 s arc grids to form the default data for Level 1 analysis.

2.4.4.2 Building Damage Estimation

Different fragility relationships and building damage assessment methodologies are used under the different levels of analysis.

The Level 0 analysis does not include any building damage assessment. The physical damage in cities and other populated areas can be inferred through the intensities given by the Shakemaps.

For Level 1 damage assessment analysis, the intensity based empirical fragility relationships developed by Lagomarsino and Giovinazzi (2006) are used. ELER software allows for the incorporation of a regional variability factor in these relationships.

Level 2 analysis is essentially intended for earthquake risk assessment (building damage and consequential human casualties) in urban areas (Hancılar et al 2010). As such, the building inventory data for the Level 2 analysis will consist of grid (geo-cell) based urban building (HAZUS or user-defined similar typology) and demographic inventories. The building damage assessment is based on the analytical fragility relationships based on the Capacity Spectrum Method (so-called HAZUS methodology).

For the representation of seismic demand the 5 %-damped elastic response spectrum provided EC8 Spectrum (Eurocode 8, CEN 2003) or IBC 2006 Spectrum (International Building Council 2006) is used. For the estimation of the so-called "Performance Point", the intersection pint of the capacity and the demand curves, ELER uses the procedures based on: the Capacity Spectrum Method specified in ATC-40 (1996), its recently modified and improved version Modified Acceleration-Displacement Response Spectrum Method (FEMA-440) and the Coefficient Method originally incorporated in FEMA-356 (2000). ELER also incorporates another nonlinear static procedure, the so-called "N2 – Reduction Factor Method" method (Fajfar 2000) where the inelastic demand spectra is modified using ductility factor based reduction factors.

2.4.4.3 Casualty Estimation

The casualty estimation is done by using regionally adjusted intensity casualty or magnitude-casualty correlations based on the Landscan population distribution inventory. The module can use previously calculated intensity grid (with the Hazard Module) or a custom intensity grid. There are three possible algorithms for computing the casualty estimation: (a) Samardjieva and Badal (2002), (b) RGELFE (1992), and (c) Vacareanu et al. (2005). The uncertainty regarding the results of this module is substantial, however, it can be a very fast way of providing casualty estimates, based on minimum data that can be easily available.

Casualties in Level 1 analysis is assessed on the basis of the simple correlations with fatalities and the number of buildings damaged beyond repair. The rates of severe injuries were obtained by revising those suggested in ATC-13 (1985) using regional post-earthquake casualty data. The casualty estimation methodology of Coburn and Spence (2002) based on the number of buildings in D5 damage state of EMS98 is also coded in ELER.

The estimation of casualties in Level 2 analysis is the one used in HAZUS based on the number of buildings of a given type at different damaged states and the associated casualty rates. The casualty rates corresponding to reinforced concrete and masonry structures given in HAZUS-MH (FEMA 2003) are adopted in ELER. The module computes, after obtaining probabilities for buildings in different damage states (five damage states: slight, moderate, extensive, complete and total collapse), estimates for human casualties, based on HAZUS-MH rates. The output from the module consists of a casualty breakdown by injury severity level, defined by a four level injury severity scale.

2.4.5 SELENA: Seismic Loss Computation Engine

SELENA (Seismic Loss Estimation using a Logic Tree Approach) is a software tool for seismic risk and loss assessment (http://selena.sourceforge.net/selena.shtml). It relies on the principles of capacity spectrum methods (CSM) and follows the same approach as the loss estimation tool for the United States HAZUS-MH (2003). A logic tree-computation scheme has been implemented in SELENA to account for epistemic uncertainties in the input data. The user has to supply a number of input files that contain the necessary input data (e.g., building inventory data, demographic data, definition of seismic scenario etc.) in a simple pre-defined ASCII format. SELENA computes ground shaking maps for various spectral periods (PGA, Sa(0.3 s) and Sa(1.0 s), damage probabilities, absolute damage estimates (including Mean Damage Ratios MDR) as well as economic losses and numbers of casualties. Flowchart of a deterministic analysis using SELENA is provided in Fig. 2.5.

In SELENA the provision of seismic demand can be done by assigning PGA or spectral accelerations at 0.3 and 1 s, obtained from seismic hazard assessment, to the geographical units. SELENA can compute the ground motion parameters by built-in GMPRs for deterministic scenario earthquakes. For real time analysis, data from strong motion stations (at least PGA values) can also be used with certain limitations. Based on these ground motion parameters SELENA generates site-specific response spectra based on IBC-2006 (International Code Council 2006), Eurocode 8 (CEN 2003) and Indian seismic building code IS 1893.

SELENA uses analytical approach for obtaining building damage with different user-selectable methodologies: (1) the traditional capacity spectrum method (CSM) as proposed by ATC-40 (ATC 1996), (2) the Modified Acceleration Displacement Response Spectra (MADRS) method according to FEMA 440 (FEMA 2005) and (3) the Improved Displacement Coefficient Method (I-DCM) as given by FEMA 440 (FEMA 2005). Damage probabilities and absolute estimates of structural building damage are computed for the five damage states no, slight, moderate, extensive and complete. The associated economic losses and casualties are estimated on the basis of available building stock inventory, replacement values and demographic information, by adopting the methodology described by HAZUS-MH (2003).

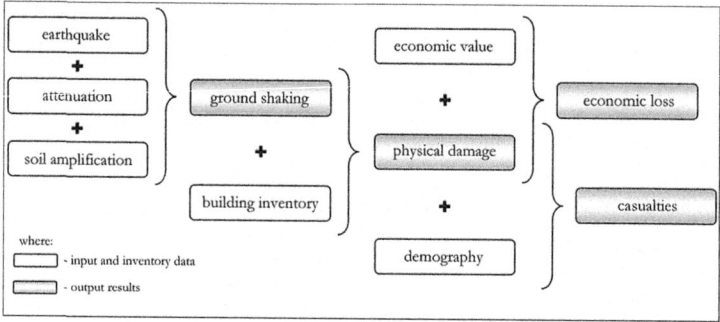

Fig. 2.5 Principle flowchart of a deterministic analysis using SELENA

The methodology applied in order to estimate the number of human casualties follows basically the HAZUS-MH (2003) approach or the basic approach following Coburn and Spence (2002). For the estimation of economic losses HAZUS-MH (2003) approach is adopted with the possibility to modify the replacement cost percentage.

2.5 Local Earthquake Rapid Loss Assessment Systems

Several local systems (country-, city- or, facility-specific) capable of computing damage and casualties in near real time already exist in several regions of the world. For example the Taiwan Earthquake Rapid Reporting System, the Real-time Earthquake Assessment Disaster System in Yokohama (READY), The Real Time Earthquake Disaster Mitigation System of the Tokyo Gas Co. (SUPREME) and the Istanbul Earthquake Rapid Response System (IERRS) provide near-real time damage estimation after major earthquakes (Erdik and Fahjan 2006). Almost all of these systems are based on the assessment of demand in real time from dense strong motion instrument arrays and the estimation of damage on the basis of known inventory of elements exposed to hazard and the related fragility relationships. After an earthquake the shaking and damage distribution maps are automatically generated on the basis of the ground motion intensity measure data received from the field stations, building inventory and the fragility relationships.

2.5.1 Earthquake Rapid Reporting System in Taiwan

Earthquake Rapid Reporting and Early Warning Systems in Taiwan, operated by Taiwan Central Weather Bureau, uses a real-time strong-motion accelerograph network that currently consists of 82 telemetered strong-motion stations distributed across Taiwan, an area of 100 km × 300 km. The rapid reporting system can offer information about 1 min after an earthquake occurrence, that includes earthquake location, magnitude and shaking maps (Tsai and Wu 1997; Teng et al. 1997; Wu et al. 1998, 1999, 2004; Shin and Teng 2001; Wu and Teng 2002).

Central Weather Bureau of Taiwan operates two dense digital strong-motion networks: (1) The Taiwan Rapid Earthquake Information Release System (TREIRS), and (2) The Taiwan Strong Motion Instrumentation Program (TSMIP).

TREIRS can obtain earthquake magnitude, epicenter location and focal depth within 90 s after occurrence of earthquakes. The TSMIP system consist of more than 650 stations spaced approximately every 5 km in populated areas in Taiwan. The Early Seismic Loss Estimation (ESLE) module has been developed and integrated with the application software "Taiwan Earthquake Loss Estimation System (TELES) provides decision support soon after occurrence of strong earthquakes for emergency providers (Yeh et al. 2003). TELES software, essentially modeled after HAZUS, acts as a decision support tool in emergency responses. The ESLE module is automatically triggered after receiving earthquake alerts. The estimated damages and casualties are then provided in the form of maps and tables automatically. Currently, the time span to complete the hazard analysis and damage assessment needs 3–5 min depending on the earthquake magnitude, epicenter location and focal depth.

2.5.2 Istanbul Earthquake Rapid Response System

To assist in the reduction of losses in a disastrous earthquake in Istanbul a dense strong motion network has been implemented. All together this network and its functions is called Istanbul Earthquake Rapid Response and Early Warning System (IERREWS). The system is designed and operated by Bogazici University with the logistical support of the Governorate of Istanbul, First Army Headquarters and Istanbul Metropolitan Municipality (Erdik et al. 2003a, b; Erdik and Fahjan 2006; Şeşetyan et al. 2011). Currently 230 strong motion recorders (including those from the IGDAŞ network) are stationed in dense settlements in the Metropolitan area of Istanbul in on-line mode for Rapid Response information generation. Post-earthquake rapid response information is achieved through fast acquisition, analysis and elaboration of data obtained from these stations.

The Rapid Response part of the IERREWS System satisfies the COSMOS (The Consortium of Organizations for Strong-Motion Observation Systems) Urban Strong-Motion Reference Station Guidelines (www.cosmos-eq.org) for the location of instruments, instrument specifications and housing specifications. The relative instrument spacing is about 2–3 km which corresponds to about 3 wavelengths in firm ground conditions and more than 10 wavelengths for soft soils for horizontally propagating 1 s shear waves. For communication of data from the rapid response stations to the data processing center and for instrument monitoring a reliable and redundant GSM 3G communication system (backed up by dedicated landlines and a microwave system) is used.

After an earthquake, the ground motion parameters, spectral displacements at selected periods, are calculated at each station location, are interpolated to determine the spectral displacement values at the center the geo-cells. The earthquake demand at the center of each geo-cell ($0.005°$ units) is computed through interpolation of these spectral displacements using two-dimensional splines. For the generation of Rapid Response information (Loss Maps) the ELER software is used (Şeşetyan et al. 2011).

The loss estimation relies on the building inventory database, fragility curves and the direct physical damage and casualty assessment techniques. The building inventories (in 24 groups) for each geo-cell together with their spectral displacement curves are incorporated in the software. The casualties are estimated on the basis of the number of occupancies and degree of damage suffered by buildings. The resulting rapid response (i.e. LossMap) information is communicated to the concerned emergency response centers (currently Istanbul Governorate, Istanbul Municipality and First Army Headquarters).

Another application called "SOSEWIN-Self Organizing Seismic Early Warning Information Network", based on the innovative technology of self-organizing networks, has been set up in the Atakoy region of Istanbul as a prototype (Picozzi et al. 2008). In contrast to centralized conventional Early Warning approach, the SOSEWIN system uses new, low-cost wireless sensing units, specifically designed to form a dense decentralized wireless mesh network (Fleming et al. 2009). The sensors allow the performance of onsite, independent analysis of the ground motion and the real-time communication of estimated parameters. The dedicated algorithms in the system provide the decision to issue warning within the wireless mesh network itself and reduces the lead-time for early warning activities. As a long-term aim of the SOSEWIN system, the use of low-cost sensing nodes by a range of end users including the general public will provide valuable input for higher resolution ShakeMaps with neighborhood-scale loss assessments. In this regard, the increase of SOSEWIN sensing units will complement existing earthquake early warning and rapid response systems.

2.5.3 IGDAS: Istanbul Natural Gas Earthquake Response System

Istanbul Gas Distribution (IGDAS) is the primary natural gas provider in Istanbul to 5 Million subscribers, and operates an extensive system of 9,867 km of gas lines, with 704 district regulators and 474,000 service boxes.

A real time risk mitigation system, currently encompassing 110 strong motion accelerometers located at critical district regulators, became operational in 2013 (Bıyıkoğlu et al. 2012 The real-time ground motion data is transmitted to the

IGDAS server at SCADA center and KOERI through 3G. The system works integrated with IERREWS with the total of 230 strong motion stations. The real-time ground shaking maps and grid based pipeline damage maps including pipeline components such as bends, tees, district regulators, isolation joints, valve rooms and service boxes are obtained.

The IGDAS Earthquake Response System follows four stages as below:

1. Real-time ground motion data is transmitted from IGDAS and KOERI stations to the IGDAS Scada Center and KOERI.
2. During an event EW information is sent from IGDAS Scada Center to the IGDAS stations at district regulators.
3. Automatic Shut-Off depending on the treshold level of certain parameters at each IGDAS district regulator is applied, and calculated parameters are sent from stations to the IGDAS Scada Center and KOERI.
4. Integrated ground shaking and damage maps are prepared immediately after the earthquake event.

2.5.4 REaltime Assessment of Earthquake Disaster in Yokohama (READY)

In 1997 the city of Yokohama installed a dense strong-motion array for earthquake disaster management. The array (called, REal-time Assessment of earthquake Disaster in Yokohama -READY System) consists of 150 strong motion accelerographs at a spacing of about 2 km. In addition borehole strong motion systems at installed at nine different locations for liquefaction monitoring. It is currently used for strong motion monitoring, real-time seismic hazard and risk assessment and damage gathering systems (Midorikawa 2005). These stations are connected to three observation centers, the disaster preparedness office of the city hall, the fire department office of the city and Yokohama City University, by the high-speed and higher-priority telephone lines.

When the accelerograph is triggered by an earthquake, the station computes ground-motion parameters such as the instrumental seismic intensity, peak amplitudes, predominant frequency, total power, duration and response spectral amplitudes. The seismic intensity data is conveyed to the city officials by the pager, and the intensity map of the city is drawn within a few minutes after the earthquake. The map is immediately open to the public through the Internet and local cable TV.

Rapid assessment of the damage to the timber houses is computed and mapped on the basis of their dynamic characteristics and the response spectrum of ground motion. The damage map is displayed with other information such as locations of hospitals, refuges and major roads for emergency transportation (Midorikawa 2004; Ariki et al. 2004).

2.5.5 Tokyo Gas: Supreme System

To cope with earthquake related secondary disasters, the new real-time disaster mitigation system for a city gas network has been developed by Tokyo Gas Company. since 1998 for the purpose of realization of dense real-time seismic motion monitoring, quick gas supply shut-off, prompt emergency response and efficient restoration work. In 2001, Tokyo Gas successfully started the operation of SUPREME, which employs 3,800 SI sensors and remote control devices at all the district regulator stations in its service area (3,100 km^2). In order to avoid earthquake risks due to leakage of gas from breakage of buried pipes, Tokyo Gas Co. Ltd. has developed and put into use a real-time safety control system, called SUPREME (http://www.tokyo-gas.co.jp/techno/stp3/97c1_e.html). The system monitors the earthquake motion at 3,800 district regulators using spectrum intensity sensors, interprets the data, and assesses gas pipe damage in order to decide whether or not the gas supply should be interrupted (Yamazaki et al. 1995; Shimizu et al. 2004 and 2006; Inomata and Norito 2012). Spectrum intensity sensors computes the Housner Intensity (Housner 1961) based on the integral of the 5 % damped response spectra between the periods of 0.1 and 2.5 s.

SUPREME interpolates SI values for 50 m meshes to calculate the number of damaged locations in each mesh in real time, based on SI values observed after disasters and data of geotechnical investigations (local site effects on ground motion) obtained in advance. SUPREME is also equipped with logic to simultaneously estimate the risk of liquefaction and to calculate damaged locations (Inomata and Norito 2012).

2.6 Comments and Conclusions

Impact of large earthquakes in urban and critical facilities and infrastructure can be reduced by timely and correct action after a disastrous earthquake. Today's technology permits for the assessments of the distribution of strong ground motion and estimation of building damage and casualties within few minutes after an earthquake.

The reduction of casualties in urban areas immediately following an earthquake can be improved if the location and severity of damages can be rapidly assessed by the information from Rapid Response Systems. The emergency response capabilities can be significantly improved to reduce casualties and facilitate evacuations by permitting rapid, selective and effective deployment of emergency operations.

The ground motion measurement hardware, data transmission systems and the loss assessment methodologies and software needed for the implementation of such Earthquake Rapid Response Systems have reached to a degree of development that can ensure the feasible application of such systems and services throughout the world.

Recent earthquakes provided opportunities for evaluation of the operational rapid loss assessment systems. The Center for Disaster Management and Risk Reduction Technology (CEDIM, www.cedim.de) has critically evaluated rapid loss assessments done after the M7.2 Van Earthquake (Eastern Turkey) of 23 October 2011 in connection with their comprehensive forensic investigations (Wenzel et al. 2012). In Van earthquake event, alerts of major earthquake activity came first from from KOERI, SARBIS, EMSC and USGS. There was much difference in initial hypocenter information from different agencies and the estimates from ELER, PAGER, WAPMERR, CATDAT-EQLIPSE showed a large range of losses. The ELER based rapid loss assessment provided by KOERI proved to be very close to the final losses doe to correct location of the earthquake source used (Wenzel et al. 2012; Erdik et al. 2012; CEDIM 2011).

The 2011 Tohoku earthquake is an example that illustrates the importance of post-event analysis. Fifteen alerts were issued by PAGER/ShakeMap in time periods ranging from within 23 min to 6 months after the earthquake. Rapid loss estimations loss estimation for the Tohoku earthquake of 11 March 2011 is compared in Daniell et al. (2011a). It is shown that a number of rapid earthquake loss estimation software packages (PAGER, QLARM, EXTREMUM) have created reasonable estimates of loss in quick time after a disaster. However, the earthquake data alone was not sufficient to produce reliable loss estimates because of the associated tsunami.

Uncertainties in real-time estimates of human losses are a factor of two, at best. And the size of the most serious errors can be an order of magnitude. They can be generated by hypocenter errors, incorrect data on building stock, and magnitude errors, especially for large earthquakes. Several studies have shown that casualty models currently used for rapid post-event casualty estimation involve a high degree of uncertainty. This is essentially due to uncertainty in the earthquake's source parameters and also our lack of knowledge on built environment, its fragility characteristics, and of the survival rates in an earthquake. For example, Spence and So (2011) have compared the performance of WAPMEER and PAGER in the estimation of casualties in several recent earthquakes. They found significant underestimations and overestimations depending on the earthquake. The reduction of the uncertainties inherent in the basic ingredients of earthquake loss assessment is an important issue that needs to be tackled in the future for viability and reliability of rapid loss assessments. Improvement in the speed and quality of moment tensor information, including estimates of rupture direction and fault finiteness, will be needed for refining loss estimates especially in regions without dense local seismograph networks.

Much remains to be done to produce more reliable rapid loss estimates after earthquakes. It is believed that the increasing number of scientific studies, outcomes of the relevant EU projects (such as NERIES, SAFER, NERA and REAKT), ongoing refinements in PAGER methodologies, as well as the expected achievements of the Global Earthquake Model (www.globalquakemodel.org) project will provide the correct directions and developments in this regard.

Open Access This chapter is distributed under the terms of the Creative Commons Attribution Noncommercial License, which permits any noncommercial use, distribution, and reproduction in any medium, provided the original author(s) and source are credited.

References

AGORA-Alliance for Open Risk Analysis. http://www.risk-agora.org
Allen TI, Wald DJ (2007) Topographic slope as a proxy for seismic site-conditions (VS30) and amplification around the globe. U.S. Geological Survey Open-File Report 2007-1357, 69 p
Allen TI, Wald DJ, Hotovec AJ, Lin K, Earle PS, Marano KD (2008) An Atlas of ShakeMaps for selected global earthquakes. U.S. Geological Survey Open-File Report 2008-1236, 47 p
Allen TI, Marano KD, Earle PS, Wald DJ (2009a) PAGER-CAT: a composite earthquake catalog for calibrating global fatality models. Seism Res Lett 80(1):57–62. doi:10.1785/gssrl.80.1.57
Allen TI, Wald DJ, Earle PS, Marano KD, Hotovec AJ, Lin K, Hearne M (2009b) An Atlas of ShakeMaps and population exposure catalog for earthquake loss modeling. Bull Earthq Eng 7. doi:10.1007/s10518-009-9120-y
Anderson E (2008) Central American Probabilistic Risk Assessment (CAPRA): objectives, applications and potential benefits of an open access architecture, global risk forum, GRF Davos, August 2008
Ariki F, Shima S, Midorikawa S (2004) Earthquake disaster prevention of Yokohama City. J Jpn Assoc Earthq Eng 4(3) (Special Issue):148–153
ATC-13 (1985) Earthquake damage evaluation data for California, Report ATC-13. Applied Technology Council, Redwood City
ATC-40 (Applied Technology Council) (1996) Seismic evaluation and retrofit of concrete buildings, vol 1. Applied Technology Council, Redwood City
ATC-55 (Applied Technology Council) (2005) Improvement of nonlinear static seismic analysis procedures, FEMA 440. Applied Technology Council, Redwood City, p 392
Atkinson GM, Wald DJ (2007) "Did you feel it?" intensity data: a surprisingly good measure of earthquake ground motion. Seismol Res Lett 78:362–368
Bal IE, Crowley H, Pinho R (2008) Displacement-based earthquake loss assessment for an earthquake scenario in Istanbul. J Earthq Eng 11(2):12–22
Baur M, Bayraktarli Y, Fiedrich F, Lungu D, Markus M (2001) EQSIM – a GIS-based damage estimation tool for Bucharest. Earthquake Hazard and Countermeasures for Existing Fragile Buildings, Bucharest
Bird JF, Bommer JJ (2004) Earthquake losses due to ground failure. Eng Geol 75(2):147–179
Bıyıkoğlu H, Türkel V, Seralıoğlu MS (2012) Istanbul natural gas network risk mitigation system. In: Proceeding of the world gas conference, Kuala Lumpur, 2012
Bommer J, Pinho R, Crowley H (2006) Using a displacement-based approach for earthquake loss estimation. In: Wasti ST, Ozcebe G (eds) Advances in earthquake engineering for urban risk reduction. Springer, Dordrecht
Borcherdt RD (1994) Estimates of site-dependent response spectra for design (method and justification). Earthq Spectra 10:617–654
Böse M (2006) Earthquake early warning for Istanbul using artificial neural networks, PhD thesis, University of Karlsruhe, Germany
Calvi GM, Pinho R, Magenes G, Bommer JJ, Restrepo-Vélez LF, Crowley H (2006) The development of seismic vulnerability assessment methodologies over the past 30 years. ISET J Earthq Tech 43(4):75–104
CEDIM Forensic Earthquake Analysis Group (2011) Reports 1-4 - Van Earthquake 2011, Karlsruhe, Germany

Chan LS, Chen Y, Chen Q, Chen L, Liu J, Dong W, Shah H (1998) Assessment of global seismic loss based on macroeconomic indicators. Nat Hazards 17:269–283

Chen Q-F, Chen Y, Chen L (1997a) Earthquake loss estimation with GDP and population data. Acta Seismol Sin 10(4):791—800

Coburn A, Spence R (2002) Earthquake protection, 2nd edn. Wiley, Chichester

CORINE Land Cover (1999) European Environment Agency, Brussels, Belgium

Couture R, Evans SG, Locat J (2002) Introduction. Nat Hazards 26(1–6):2002

Crowley H, Silva V (2013) OpenQuake engine book: risk. GEM Foundation, Pavia, July 2013

Crowley H, Pinho R, Bommer JJ (2004) A probabilistic displacement-based fragility assessment procedure for earthquake loss estimation. Bull Earthq Eng 2(2):173–219

Daniell JE (2009) Open source procedure for assessment of loss using global earthquake modelling (OPAL-GEM Project): CEDIM Research Report 2009-01 (CEDIM Loss Estimation Series). Karlsruhe, Germany

Daniell JE (2011b) Open Source Procedure for Assessment of Loss using Global Earthquake Modelling software (OPAL). Nat Hazards Earth Syst Sci 7(7):1885–1899

Daniell J (2014) The development of socio-economic fragility functions for use in worldwide rapid earthquake loss estimation procedures, PhD thesis, Karlsruhe Institute of Technology, Germany

Daniell E, Vervaeck A, Wenzel F (2011a) A timeline of the socio-economic effects of the 2011 Tohoku Earthquake with emphasis on the development of a new worldwide rapid earthquake loss estimation procedure. In: Australian Earthquake Engineering Society 2011 conference, Barossa Valley, Australia

Daniell JE, Khazai B, Wenzel F, Venvaeck A (2011c) The CATDAT damaging earthquakes database. Nat Hazards Earth Syst Sci 11(8):2235–2251

Daniell JE, Khazai B, Wenzel F, Venvaeck A (2012a) The worldwide economic impact of earthquakes, Paper No. 2038. In: Proceedings of the 15th world conference of earthquake engineering, Lisbon, Portugal

Daniell JE, Venvaeck A, Khazai B, Wenzel F (2012b) Worldwide CATDAT damaging earthquakes database in conjunction with Earthquake-report.com – presenting past and present socio-economic earthquake data, Paper No. 2025. In: Proceedings of the 15th world conference of earthquake engineering, Lisbon, Portugal

Demircioglu MB, Erdik M, Hancilar U, Sesetyan K, Tuzun C, Yenidogan, Zulfikar AC (2009) Technical Manual – Earthquake Loss Estimation Routine ELER-v1.0, Bogazici University, Department of Earthquake Engineering, Istanbul, March 2009

Di Pasquale G, Ferlito R, Orsini G, Papa F, Pizza AG, Van Dyck J, Veneziano D (2004) Seismic scenarios tools for emergency planning and management. In: Proceedings of the XXIX general assembly of the European seismological commission, Potsdam, Germany

Earle PS, Wald DJ, Jaiswal KS, Allen TI, Marano KD, Hotovec AJ, Hearne MG, Fee JM (2009a) Prompt Assessment of Global Earthquakes for Response (PAGER): a system for rapidly determining the impact of global earthquakes worldwide. U.S. Geological Survey Open-File Report 2009-1131

Earle PS, Wald DJ, Allen TI, Jaiswal KS, Porter KA, Hearne MG (2009b) Rapid exposure and loss estimates for the May 12, 2008 MW 7.9 Wenchuan earthquake provided by the U.S. Geological Survey's PAGER system. In: Proceedings of the 14th world conference earthquake engineering Beijing, China, Paper, 8 pp

EERI (Earthquake Engineering Research Institute) (2007) www.eeri.org. World Housing Encyclopedia online database http://www.world-housing.net

Eguchi RT, Goltz JD, Seligson HA, Flores PJ, Blais NC, Heaton TH, Bortugno E (1997) Real-time loss estimation as an emergency response decision support system: the Early Post-Earthquake Damage Assessment Tool (EPEDAT). Earthq Spectra 13(4):815–832

ELER v3.1 (2010) Earthquake loss estimation routine, technical manual and users guide, Bogazici University, Department of Earthquake Engineering, Istanbul. http://www.koeri.boun.edu.tr/Haberler/NERIES%20ELER%20V3.1_6_176.depmuh

Elnashai, AS, Hampton S, Karaman H, Lee JS, Mclaren T, Myers J, Navarro C, Sahin M, Spencer B, Tolbert N (2007) Overview and applications of Maeviz-Hazturk 2007. J Earthq Eng 12(S2),:100–108

EMS-98, European Seismic Commission Working Group on Macroseismic Scales, European Macroseismic Scale, Luxembourg. http://www.gfz-potsdam.de/pb5/pb53/projekt/ems/eng/index_eng.html

Erdik M, Aydinoglu N (2002) Earthquake performance and fragility of buildings in Turkey: report prepared for World Bank disaster management facility, Washington, DC

Erdik M, Fahjan Y (2006) Damage scenarios and damage evaluation. In: Oliveira CS, Roca A, Goula X (eds) Assessing and managing earthquake risk. Springer, Dordrecht, pp 213–237

Erdik M, Aydinoglu N, Fahjan Y, Sesetyan K, Demircioglu M, Siyahi B, Durukal E, Ozbey C, Biro Y, Akman H, Yuzugullu O (2003a) Earthquake risk assessment for Istanbul metropolitan area. Earthq Eng Eng Vib 2(1):1–25

Erdik M, Fahjan Y, Özel O, Alçık H, Mert A, Gül M (2003b) Istanbul earthquake rapid response and early warning system. Bull Earthq Eng 1:157–163, Kluwer

Erdik M, Cagnan Z, Zulfikar C, Sesetyan K, Demircioglu MB, Durukal E, Kariptas C (2008) Development of rapid earthquake loss assessment methodologies for Euro-MED region. In: Proceedings of the, 14 world conference on earthquake engineering, Paper ID: S04-004

Erdik M, Sesetyan K, Demircioglu MB, Hancilar U, Zulfikar C, Cakti E, Kamer Y, Yenidogan C, Tuzun C, Cagnan Z, Harmandar E (2010) Rapid earthquake hazard and loss assessment for Euro-Mediterranean Region. Acta Geophys 58(5):855–892

Erdik M, Şeşetyan K, Demircioğlu MB, Hancılar U, Zülfikar C (2011) Rapid earthquake loss assessment after damaging earthquakes. Soil Dyn Earthq Eng 31(2):247–266

Erdik M, Kamer Y, Demircioğlu M, Şeşetyan K (2012) 23 October 2011 Van (Turkey) earthquake. Nat Hazards 64:651–665

Eurocode 8, CEN (2003) Eurocode 8: design of structures for earthquake resistance – Part 1: general rules, seismic actions and rules for buildings, prEN 1998-1, Doc CEN/TC250/SC8/N335, Comité Européen de Normalisation, Brussels, Belgium

Fajfar P (2000) A nonlinear analysis method for performance based seismic design. Earthq Spectra 16(3):573–592

FEMA (1999) HAZUS earthquake loss estimation methodology, technical manual, Prepared by the National Institute of Building Sciences for the Federal Emergency Management Agency. Federal Emergency Management Agency, Washington, DC

FEMA (2003) HAZUS-MH technical manual. Federal Emergency Management Agency, Washington, DC

FEMA (Federal Emergency Management Agency) (2006) HAZUS-MH MR2 technical manual. Federal Emergency Management Agency, Washington, DC

FEMA&NIBS (1999) Earthquake loss estimation methodology – HAZUS 1999. Federal Emergency Management Agency/National Institute of Building Science, Washington, DC

FEMA-356 (2000) Prestandard and commentary for the seismic rehabilitation of building. Federal Emergency Management Agency and National Institute of Building Science, Washington, DC

FEMA-440 (2005) Improvement of nonlinear static seismic analysis procedure. Federal Emergency Management Agency, Washington, DC

Fleming K, Picozzi M, Milkereit C, Kuhnlenz F, Lichtblau B, Fischer J, Zulfikar C, Ozel O (2009) The Self-organizing Seismic Early Warning Information Network (SOSEWIN). Seismol Res Lett 80(5):755–771

Freeman SA (1998) Development and use of capacity spectrum method. In: Proceedings of the 6th U.S. National conference on earthquake engineering, Oakland, EERI

Gamba P, Cavalca D, Jaiswal KS, Huyck C, Crowley H (2012) The GED4GEM project: development of a global exposure database for the global earthquake model initiative. In: Proceedings of the 15th WCEE, Lisbon, 2012

GDACS, Global Disaster Alert and Coordination System, http://www.gdacs.org

GEM, Global Earthquake Model, (www.globalquakemodel.org)

Geonames, http://www.geonames.org

Giovinazzi S (2005) Fragility assessment and the damage scenario in seismic risk analysis. PhD thesis, Department of Civil Engineering of the Technical University Carolo-Wilhelmina at Braunschweig and Department of Civil Engineering, University of Florence, Italy

Giovinazzi S, Lagomarsino S (2004) A Macroseismic model for the fragility assessment of buildings. 13th World conference on earthquake engineering. Vancouver, Canada

Grünthal G (ed) (1998) European Macroseismic Scale 1998 (EMS-98). Cahiers du Centre Européen de Géodynamique et de Séismologie 15, Centre Européen de Géodynamique et de Séismologie, Luxembourg, 99 pp

Hancılar U, Tuzun C, Yenidogan C, Erdik M (2010) ELER software – a new tool for urban earthquake loss assessment. Nat Hazards Earth Syst Sci 10:2677–2696

Harmandar E, Çaktı E, Erdik M (2012) A method for spatial estimation of peak ground acceleration in dense arrays. Geophys J Int 191:1272–1284. doi:10.1111/j.1365-246X.2012.05671.x

HAZUS-MH (2003) Multi-hazard loss estimation methodology: earthquake model – technical manual (www.fema.gov/plan/prevent/hazus)

Housner GW (1961) Vibration of structures induced by seismic waves Part I. Earthquakes. In: Harris CM, Crede CE (eds) Shock and vibration handbook. McGraw-Hill, New York, 50-1-50-32

IBC (International Building Code) (2006) International Code Council, USA

Inomata W, Norito Y (2012) Result of SUPREME (Super-dense Real time monitoring Earthquake system for city gas supply) in The Great East Japan Earthquake. Proceedings of the international symposium on engineering lessons learned from the 2011 Great East Japan Earthquake, 1–4 March 2012, Tokyo, Japan

Jaiswal KS, Wald DJ (2008a) Developing a global building inventory for earthquake loss assessment and risk management. In: Proceedings of the 14th world conference on earthquake engineering, Beijing, China, 8 p

Jaiswal KS, Wald DJ (2008b) Creating a global building inventory for earthquake loss assessment and risk management. U.S. Geological Survey Open-File report 2008–1160, 103 p

Jaiswal K, Wald D (2010) An empirical model for global earthquake fatality estimation. Earthq Spectra 26(4):1017–1037

Jaiswal KS, Wald DJ (2011) Rapid estimation of the economic consequences of global earthquakes (Open-File Report No. 2011-1116). Retrieved from http://pubs.usgs.gov/of/2011/1116

Jaiswal KS, Wald DJ (2013) Estimating economic losses from earthquakes using an empirical approach. Earthq Spectra 29(1):309–324

Jaiswal KS, Wald DJ, Hearne M (2009) Estimating casualties for large earthquakes worldwide using an empirical approach: U.S. Geological Survey Open-File Report OF 2009–1136. Retrieved from http://pubs.usgs.gOv/of/2009/1136/pdf/OF09-1136.pdf

Jaiswal K, Wald D, D'Ayala D (2011) Developing empirical collapse fragility functions for global building types. Earthq Spectra 27(3):775–795

Kircher CA, Nassar AA, Kustu O, Holmes WT (1997) Development of building damage functions for earthquake loss estimation. Earthq Spectra 13:663–682

Kircher CA, Whitman RV, Holmes WT (2006) HAZUS earthquake loss estimation methods. Nat Hazards Rev 7(2):45–59

Lagomarsino S, Giovinazzi S (2006) Macroseismic and mechanical models for the fragility and damage assessment of current buildings. Bull Earthquake Eng 4:445–463,

LandScan, http://www.ornl.gov/sci/landscan/

Larionov V, Frolova N, Ugarov A (2000) Approaches to fragility evaluation and their application for operative forecast of earthquake consequences. In: Ragozin A (ed) All-Russian conference "Risk- 2000". ANKIL, Moscow, pp 132–135

Maeviz, Mid-America Earthquake Center in University of Illinois, http://mae.ce.uiuc.edu/software_and_tools/maeviz.html

Mahaney JA, Freeman SA, Paret TF, Kehoe BE (1993) The capacity spectrum method for evaluating structural response during the Loma Prieta earthquake. In: Proceedings of the 1993 national earthquake conference, Memphis

Markus M, Fiedrich F, Leebmann J, Schweier C, Steinle E (2004) Concept for an integrated disaster management tool. In: Proceedings of the 13th world conference on earthquake engineering, Vancouver, BC, Canada

Midorikawa S (2004) Dense strong-motion array in Yokohama, Japan, and its use for disaster management. NATO Meeting, Kusadasi, Turkey

Midorikawa S (2005) Dense strong-motion array in Yokohama, Japan, and its use for disaster management. In: Gulkan P, Anderson JG (eds) Directions in strong motion instrumentation. Springer, Dordrecht, pp 197–208

Molina S, Lindholm C (2005) A logic tree extension of the capacity spectrum method developed to estimate seismic risk in Oslo, Norway. J Earthq Eng 9(6):877–897

Mouroux P, Le Brun B (2006) Presentation of RISK-UE project. Bull Earthq Eng 4:323–339

Mouroux P, Bertrand E, Bour M, Le Brun B, Depinois S, Masure P, RISK-UE Team (2004) The European risk-UE project: an advanced approach to earthquake risk scenarios. Proceedings of the 13th world conference on earthquake engineering, Vancouver, BC, Canada

NERIES Project-http://www.neries-eu.org

Oak Ridge National Laboratory (2011) LandScan™: Digital raster data. http://web.ornl.gov/sci/landscan/index.shtml

Picozzi M, the SAFER and EDIM Work Groups (2008) Seismological and early warning activities of the SOSEWIN. Geophys Res Abstr 10, EGU2008-A-07001, 2008 SRef-ID:1607-7962/gra/EGU2008-A-07001

Porter K, Jaiswal K, Wald D, Earle P, Hearne M (2008a) Fatality models for the U.S. Geological Survey's Prompt Assessment of Global Earthquakes for Response (PAGER) system. In: Proceedings of 14th world conference on earthquake engineering, Beijing, China, 8 pp

Porter KA, Jaiswal KS, Wald DJ, Green M, Comartin C (2008b) WHE-PAGER project: a new initiative in estimating global building inventory and its seismic vulnerability. In: Proceedings of 14th world conference on earthquake engineering, Beijing, China, 8 pp

Priestley MJN (1997) Displacement-based seismic assessment of reinforced concrete buildings. J Earthq Eng 1(1):157–192

Priestley MJN (2003) Myths and fallacies in earthquake engineering – revisited. In: The Mallet-Milne Lecture. IUSS Press, Pavia, Italy

QLARM, (earthQuake Loss Assessment for Response and Mitigation – http://qlarm.ethz.ch)

RGELFE (1992) Estimating losses from earthquakes in China in the forthcoming 50 years, China Seismology Bureau, Seismology Publications, Beijing

RISK-UE (2004) The European Risk-UE Project: an advanced approach to earthquake risk scenarios. (2001–2004) www.risk-ue.net

Robinson D, Fulford G, Dhu T (2005) EQRM: Geoscience Australia's earthquake risk model. Geoscience Australia Record 2005/01. Geoscience Australia, Canberra, p 151

Robinson D, Fulford G, Dhu T (2006) EQRM: Geoscience Australia's earthquake risk model: technical manual version 3.0, Book Bib ID 3794291, Geoscience Australia

Roca A, Goula X, Susagna T, Chavez J, Gonzalez M, Reinoso E (2006) A simplified method for fragility assessment of dwelling buildings and estimation of damage scenarios in Spain. Bull Earthq Eng 4(2):141–158

Rossetto T, Elnashai A (2003) Derivation of vulnerability functions for European-type RC structures based on observational data. Eng Struct 23(10):1241–1263

Samardjieva E, Badal J (2002) Estimation of the expected number of casualties caused by strong earthquakes. Bull Seismol Soc Am 92(6):2310–2322

Şeşetyan K, Zulfikar C, Demircioglu M, Hancilar U, Kamer Y, Erdik M (2011) Istanbul earthquake rapid response system: methods and practices. Soil Dyn Earthq Eng. doi:10.1016/j.soildyn.2010.02.012

Shimizu Y, Yamazaki F, Isoyama R, Ishida E, Koganemaru K, Nakayama W (2004) Development of realtime disaster mitigation system for urban gas supply network. In: Proceedings of the 13th WCEE, Vancouver, BC, Canada, 1–6 August 2004, Paper No. 157

Shimizu Y, Yamazaki F, Yasuda S, Towhata I, Suzuki T, Isoyama R, Ishida E, Suetomi I, Koganemaru K, Nakayama W (2006) Development of real-time safety control system for urban gas supply network. J Geotech Geoenviron Eng, ASCE 132(2):237–249

Shin TC, Teng TL (2001) An overview of the 1999 Chi-Chi, Taiwan, earthquake. Bull Seism Soc Am 91:895–913

Silva V, Crowley H, Pagani M, Monelli D, Pinho R (2013) Development of the OpenQuake engine, the Global Earthquake Model's open-source software for seismic risk assessment. Nat Hazards 2014, 72:1409–1427

So E, Spence R (2009) Estimating shaking-induced casualties and building damage for global earthquake events (Final Technical Report, NEHRP Grant. No.08HQGR0102)

Sousa ML, Campos Costa A, Carvalho A, Coelho E (2004) An automatic seismic scenario loss methodology integrated on a geographic information system. In: Proceedings of the 13th world conference on earthquake engineering, Vancouver, Canada

Spagnuolo E, Faenza L, Cultrera G, Herrero A, Michelini A (2013) Accounting for source effects in the ShakeMap procedure: the 2000 Tottori and the 2008 Miyagi earthquakes. Geophys J Int 194:1836–1848

Spence RJS (2007) Earthquake disaster scenario prediction and loss modeling for urban areas: LESSLOSS report. IUSS Press, Pavia

Spence RJS, So EKM (2011) Human casualties in earthquakes: modelling and mitigation. In: Proceedings of the ninth pacific conference on earthquake engineering building an earthquake-resilient society, 14–16 April 2011, Auckland, New Zealand

Spence RJS, Coburn AW, Sakai S, Pomonis A (1991) A parameterless scale of seismic intensity for use in seismic risk analysis and vulnerability assessment. In: Society for References Earthquake and Civil Engineering Dynamics (ed) Earthquake blast and impact: measurement and effects of vibration. Elsevier Applied Science, Amsterdam

Spence R, Coburn RW, Pomonis A (1992) Correlation of ground motion with building damage: the definition of a new damage-based seismic intensity scale. In: Proceedings of the 10th world conference on earthquake engineering, Rotterdam, The Netherlands

Stafford,PJ, Strasser FO, Bommer JJ (2007) Preliminary report on the evaluation of existing loss estimation methodologies, Report prepared for EU FP6 NERIES Project, Department of Civil & Environmental Engineering, Imperial College, London

Stojanovski P, Dong W (1994) Simulation model for earthquake casualty estimation. In: Proceedings of fifth US national conference on earthquake engineering, Paper No. 00592, Chicago

Strasser FO, Stafford PJ, Bommer JJ, Erdik M (2008) State-of-the-Art of European earthquake loss estimation software. In: Proceedings of the 14th WCEE, Beijing, China

Teng TL, Wu YM, Shin TC, Tsai YB, Lee WHK (1997) One minute after: strong-motion map, effective epicenter, and effective magnitude. Bull Seism Soc Am 87:1209–1219

Trendafiloski G, Wyss M, Rosset PH, Marmureanu G (2009a) Constructing city models to estimate losses due to earthquakes worldwide: application to Bucharest, Romania. Earthq Spectra 25(3):665–685. http://www.wapmerr.org/City_model_Spectr.pdf

Trendafiloski G, Wyss M, Rosset PH (2009b) Loss estimation module in the new generation software QLARM. In: Proceedings of the second international workshop on disaster casualties, Cambridge, June 2009

Tsai YB, Wu YM (1997) Quick determination of magnitude and intensity for seismic early warning. 29th IASPEI meeting, Thessaloniki, Greece

United Nations (UN) (1993) Housing in the world-graphical presentation of statistical data. United Nations, New York

United Nations (2005) Population and Housing Censuses of individual countries

UN-HABITAT (2007) Housing in the world –demographic and health survey (personal communication)

USGS, PAGER, Prompt Assessment of Global Earthquakes for Response (http://earthquake.usgs.gov/eqcenter/pager/)

USGS, ShakeMap, http://earthquake.usgs.gov/eqcenter/shakemap/

USGS, ShakeCast (http://earthquake.usgs.gov/shakecast/)

Vacareanu R, Lungu D, Aldea A, Arion C (2005) WP7 handbook – seismic risk scenarios, Bulletin of the Technical University of Civil Engineering Bucharest, 2/2003, pp 97–117

Wald DJ, Quitoriano V, Heaton TH, Kanamori H (1999) Relationships between peak ground acceleration, peak ground velocity, and Modified Mercalli Intensity in California. Earthq Spectra 15(3):557–564

Wald DJ, Worden BC, Quitoriano V, Pankow KL (2005) ShakeMap manual: technical manual, user's guide, and software guide. U.S. Geological Survey Techniques and Methods, book 12, section A, chap. 1. Reston, Virginia, U.S. Geological Survey: 132

Wald DJ, Earle PS, Allen TI, Jaiswal K, Porter K, Hearne M (2008) Development of the U.S. Geological survey's pager system (prompt assessment of global earthquakes for response). In: Proceedings of the 14th world conference on earthquake engineering, 12–17 October 2008, Beijing, China

Wald DJ, Jaiswal KS, Marano KD, Bausch D (2009a) Developing casualty and impact alert protocols based on the USGS Prompt Assessment of Global Earthquakes for Response (PAGER) system. In: Proceedings of the 1st international disaster casualty workshop, Cambridge, England, 10 pp

Wald DJ, Jaiswal K, Marano K, Earle P, Allen TI (2009b) Advancements in casualty modeling facilitated by the USGS Prompt Assessment of Global Earthquakes for Response (PAGER) system. In: Proceedings of the 2nd international disaster casualty workshop, Cambridge, England, 8 pp

Wenzel F, Marmuraenu G (2007) Rapid earthquake information for Bucharest. J Pure Appl Geophy 164(5):929–939

Wenzel F, Daniell JE, Khazai B, Kunz-Plapp T (2012) The CEDIM Forensic Earthquake Analysis Group and the test case of the 2011 Van earthquakes. In: Proceedings of the 15thWCEE, Lisbon

WHE (World Housing Encyclopedia) (http://www.world-housing.net/)

Whitman RV, Reed JW, Hong ST (1973) Earthquake damage probability matrices. In: Proceedings of the 5th world conference on earthquake engineering, Rome, Italy

Whitman RV, Anagnos T, Kircher CA, Lagorio HJ, Lawson RS, Schneider P (1997) Development of a national earthquake loss estimation methodology. Earthq Spectra 13(4):643–661

World Agency of Planetary Monitoring Earthquake Risk Reduction (WAPMERR, http://www.wapmerr.org)

Wu YM, Teng TL (2002) A virtual sub-network approach to earthquake early warning. Bull Seism Soc Am 92:2008–2018

Wu YM, Shin TC, Tsai YB (1998) Quick and reliable determination of magnitude for seismic early warning. Bull Seism Soc Am 88:1254–1259

Wu YM, Chung JK, Shin TC, Hsiao NC, Tsai YB, Lee WHK, Teng TL (1999) Development of an integrated seismic early warning system in Taiwan – case for the Hualien area earthquakes. TAO 10:719–736

Wu, Y-M, Teng T-L, Hsiao N-C, Shin T-C, Lee WHK, Tsai Y-B (2004) Progress on earthquake rapid reporting and early warning systems in Taiwan. Department of Geosciences, National Taiwan University, Taipei, Taiwan

Wyss M, Elashvili M, Jorjiashvili N, Javakhishvili Z (2011) Uncertainties in teleseismic epicenter estimates: implications for real-time loss estimate. Bull Seismol Soci Am 101:482–494

Yamazaki F, Katayama T, Noda S, Yoshikawa Y, Ohtani Y (1995) Development of large-scale city-gas network alert system based on monitored earthquake ground motion. Proc JSCE 525 (I-33):331–340 (in Japanese)

Yeh CH, Jean WY, Loh CH (2000) Damage building assessment for earthquake loss estimation in Taiwan. In: Proceedings of the 12th world conference on earthquake engineering, Auckland, New Zealand, Paper No. 1500

Yeh CH, Loh CH, Tsai KC (2003) Development of earthquake assessment methodology in NCREE, Joint NCREE/JRC workshop international collaboration on earthquake disaster mitigation research, Report No. NCREE-03-029, pp 83–92

Yeh CH, Loh CH, Tsai KC (2006) Overview of Taiwan earthquake loss estimation system. Nat Hazards 37(1–2):23–37

Zonno,G, Carvalho A, Franceschina G, Akinci A, Campos Costa A, Coelho E, Cultrera G, Pacor F, Pessina V, Cocco M (2009) Simulating earthquake scenarios in the European Project LESSLOSS: the Case of Lisbon.,In: Mendes-Victor LA, Oliveira CS, Azevedo J, Ribeiro A (eds) The 1755 Lisbon Earthquake: Revisited. Springer, Dordrecht

Chapter 3
Existing Buildings: The New Italian Provisions for Probabilistic Seismic Assessment

Paolo Emilio Pinto and Paolo Franchin

Abstract In Europe, the reference document for the seismic assessment of buildings is the Eurocode 8-Part3, whose first draft goes back to 1996 and, for what concerns its safety format, has strong similarities with FEMA 276. Extended use of this document, especially in Italy after the 2009 L'Aquila earthquake has shown its inadequacy to provide consistent and univocal results. This situation has motivated the National Research Council of Italy to produce a document of a level higher than the one in force, characterized by a fully probabilistic structure allowing to account for all types of uncertainties and providing measures of performance in terms of mean rates of exceedance for a selected number of Limit States (LS). The document, which covers both reinforced concrete and masonry buildings, offers three alternative approaches to risk assessment, all of them belonging to the present consolidated state of knowledge in the area. These approaches include, in decreasing order of accuracy: (a) Incremental dynamic analysis on the complete structural model, (b) Incremental dynamic analysis on equivalent SDOF oscillator(s), (c) Non-linear static analysis. In all three approaches relevant uncertainties are distinguished in two classes: those amenable of description as continuous random variables and those requiring the set-up of different structural models. The first ones are taken into account by sampling a number of realizations from their respective distributions and by associating each realization with one of the records used for evaluating the structural response, the latter by having recourse to a logic tree. Exceedance of each of the three considered Limit States: Light or Severe damage and Collapse, is signaled by a scalar indicator Y, expressing the global state of the structure as a function of that of its members, taking a value of one when the Limit State is reached. For the first two LS's, which relate to functionality and to economic considerations, the formulation of Y is such as to leave to the owner the choice of the acceptable level of damage, while

---

P.E. Pinto (✉) • P. Franchin
Department of Structural and Geotechnical Engineering, University of Rome La Sapienza, Via Antonio Gramsci 53, 00197 Rome, Italy
e-mail: profpaolopinto@gmail.com; paolo.pinto@uniroma1.it; paolo.franchin@uniroma1.it

for the Collapse LS the formulation is obviously unique. An application to a real school building completes the paper.

3.1 Preamble

In spite of the availability (officially since 2005, but with preliminary versions since 1996) of Eurocode 8 Part3 (EC8/3) dealing with seismic assessment and retrofitting of buildings, the relevance for Italy of a document of this type had escaped the attention of both the authorities and the profession until a small earthquake occurred in 2002 caused the complete collapse of a school and the death of all the young students inside. This fact produced a national scandal and the awakening in the general public of the consciousness of the seismic risk potentially affecting all types of constructions, the old as well as the recent ones.

The situation prompted the Department of Civil Protection to take action in two directions: preparing a technical document dealing with the analytical seismic assessment of buildings, and emanating an ordnance requiring that all important public facilities be subjected to assessment within 5 years time. The technical document can be regarded essentially as the translation of the EC8/3: it has been made official in 2008 by the competent Ministry (NTC2008) Ministero Infrastrutture (2008) and its use mandatory in July 2009, right after the April 6th 2009 L'Aquila earthquake.

3.1.1 The Present Normative State and the Purpose of the New Document Issued by the National Research Council

It will be understood that due to the ordnance of 2003 a very large number of buildings has been by now subjected to seismic assessment using basically EC8/3, so that experience on its merits and limitations rests on solid statistical bases. Critical aspects have emerged from the use of EC8/3, not only in Italy, but in a number of other European Countries as well, and plans for an improved version are under way. The consensus existing on major critical aspects allows for just a brief mention to be made here.

(a) Performance must be checked with reference to three Limit States. These are formulated in terms of system performance, but then the verifications, for reinforced concrete (RC) buildings, must be carried out in terms of member behavior, independently of the number and the importance of non-complying members. This inconsistency is a major cause of dispersion of the results obtained by different analysts.

(b) The uncertainties regarding the structure are grouped into three types, namely: those related to geometry, to the properties of the materials and to the details of reinforcement (for RC structures.) Three levels of knowledge are considered, each one characterized by a combination of the knowledge acquired on the three types of uncertainty, and a so-called "confidence factor (CF)" is associated to each level. In many cases in practice, however, the achievable state of knowledge does not fit in any of the levels above, due to non-uniform quality/quantity of information on the three aspects, with the consequent uncertainty on the value of CF to be adopted.
(c) The CF factors are to be applied to the material properties, which are only one of the many sources of uncertainties, and in the majority of cases are of comparatively much lesser relevance on the outcome of the assessment.
(d) Little if any guidance is given on the modeling of the structure, e.g. on the use of classical fiber elements or of stiffness/strength degrading models. Yet different choices on these aspects are rather consequential on the definition of the attainment of the LS's, especially for that of collapse.

In consideration of the above mentioned limits, the National Research Council (CNR) decided to prepare a document of a level higher than the one in force, in which the performance-based concept, which is claimed to be at the base of most of the modern design codes, is implemented in explicit probabilistic terms, allowing thus uncertainties of all nature to be taken into consideration and introduced into the assessment process, with their relevance on the final outcome properly reflected.

For what concerns the probabilistic procedures adopted the choice has been to adhere to the now well consolidated state-of-the-art, avoiding refinements deemed as inessential, in order to make the document accessible to a larger audience.

The CNR documents, denominated "Instructions", do not have the status of "state laws", as it is the case for the Ministerial norms, so they cannot replace or contrast with the latter, but they enjoy a high scientific reputation, and recourse to them is frequent in case of dubious or absent indications in the norms. It is auspicable and plausible that the future revision of the norms will take profit of both the format as well of the content of the new document.

3.1.2 The Content of the CNR Instructions

The main content of the document is subdivided into the following chapters.

1. Introduction
2. Methodological aspects common to all typologies:

   - Limit States
   - Target performances
   - Seismic action
   - Knowledge acquisition

- Uncertainty modeling
- Structural analysis
- Identification of LS exceedance
- Assessment methods.

3. Specific provisions for masonry buildings

- Response modeling
- Probabilistic capacity models

4. Specific provisions for reinforced concrete buildings

- Response modeling
- Probabilistic capacity modeling

5. Commentary to the text
6. Example application to a masonry building
7. Example application to a reinforced concrete building

The present paper illustrates all material devoted to reinforced concrete buildings.

3.2 Methodological Aspects Common to All Typologies

3.2.1 Limit States

The Limit States are defined with reference to the performance of the building in its entirety including, in addition to the structural part, also non-structural ones like partitions, electrical and hydraulic systems, etc.

The following three Limit States are considered:

- Damage Limit State (SLD): negligible damages (no repair necessary) to the structural parts, and light, economically repairable damages to the non-structural ones.
- Severe Damage (also called life safeguard) Limit State (SLS): loss of use of non-structural systems and a residual capacity to resist horizontal actions. State of damage uneconomic to repair.
- Collapse prevention Limit State (SLC): the building is still standing but would not survive an aftershock.

Check against the attainment of the SLC is mandatory, in consideration of the general lack of reserve ductility of non-seismically designed buildings (contrary to the proven large reserve possessed by buildings designed according to present seismic codes).

3 Existing Buildings: The New Italian Provisions for Probabilistic Seismic... 101

Table 3.1 Minimum levels of protection in terms of maximum tolerated λ_{LS} (values in the table are multiplied by 10^3) as a function of building class

Limit state	Class I	Class II	Class III	Class IV
SLD	64.0	45.0	30.0	22.0
SLS	6.8	4.7	3.2	2.4
SLC	3.3	2.3	1.5	1.2

3.2.2 Target Performances

A distinction is made among buildings depending on the socio-economic consequences of their LS exceedance, and four Classes of importance are identified.

The required level of protection for each Class and each Limit State is formulated in terms of the mean annual frequency of exceedance (MAF): λ_{LS}.

The proposed maximum values of λ_{LS} are such as to ensure approximately the same level of protection as currently required by the national seismic code for the different Classes and LS's for new buildings. They are reported in Table 3.1.

The values in the table have been calculated using the approximate expression due to Cornell et al. (2002):

$$\lambda_{LS} = \lambda_S\left(S_{\hat{D}=\hat{C}}\right)\exp\left|\frac{1}{2}\frac{k_1^2}{b^2}\left(\beta_D^2 + \beta_C^2\right)\right| \tag{3.1}$$

expressing the MAF of the LS as the MAF of the seismic intensity inducing a median demand equal to the median capacity, times an amplification factor accounting for the uncertainty in demand β_D and capacity β_D, as well as the slopes of the hazard curve k_1 and of the intensity – demand relation b. If the common values $k_1 = 3$, $b = 1$, $\beta_D = \beta_D = 0.3$ are introduced, the exponential factor takes the value ~2.25. Taking for $\lambda_S\left(S_{\hat{D}=\hat{C}}\right)$ the inverse of the mean return period T_R of the seismic action to be considered for each Class and LS in the current deterministic code, leads to $\lambda_{LS}^* = 2.25/T_R$, which corresponds e.g. for Class II buildings (ordinary) and the severe damage LS to: $2.25/475 = 0.0047$.

3.2.3 Seismic Action

In line with the adopted IM-based approach, the seismic action is characterized in terms of:

- the mean hazard curve for the site
- a set of time histories of the seismic motion, used for the calculation of the fragility $p_{LS}(s)$

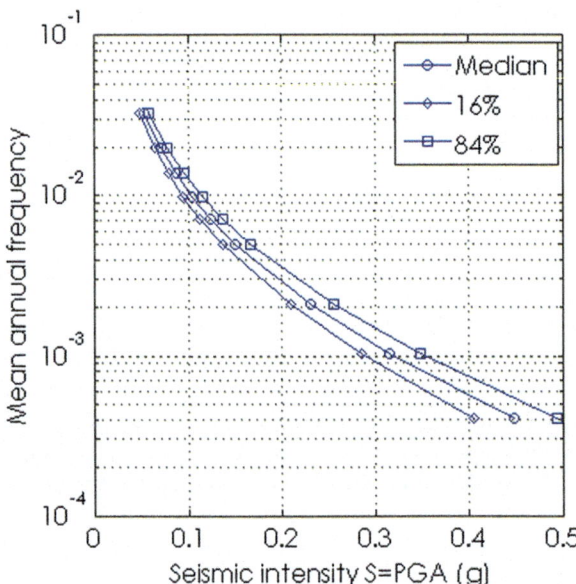

Fig. 3.1 Median and 16 %/ 84 % fractile hazard curves

A discrete hazard curve (Fig. 3.1) for any site in Italy can be obtained from the *median* uniform hazard spectra (UHS) provided in the national code for nine values of the mean return periods, ranging from 30 to 2,475 years. The UHS are provided at the nodes of a square grid with sides of about 5 km. The hazard in a point inside a grid is obtained by interpolation of the values at its four corners.

For any given value T of the structural period the nine values of $S_a(T)$ provide a point-wise median hazard curve to which, for the purpose of the evaluation of λ_{LS}, a quadratic interpolation function is applied.

As suggested in the SAC-FEMA procedure (Cornell et al. 2002), the epistemic uncertainty on the hazard curve is accounted for by using its mean value, instead of the median, which is obtained by multiplying the latter by an amplification factor:

$$\bar{\lambda}(s) = \lambda_{S,50\%}(S)\exp\left(\frac{1}{2}\beta_H^2\right) \quad (3.2)$$

where the uncertainty on the hazard is:

$$\beta_H = \sigma_{\ln S} = \frac{\ln S_{84\%} - \ln S_{16\%}}{2} \quad (3.3)$$

The above expression is obtained assuming a lognormal distribution for the intensity S at any given λ_S, and the uncertainty should be evaluated at the intensity with MAF close to λ_{LS} (an iteration is therefore required).

The time histories to be used for response analysis can be either natural records or artificially generated motions, provided these latter are able to reproduce the same mean, variance and correlation of the spectral ordinates of the natural motions.

The selection of the natural records can be made, according to the state of the practice, using the technique of disaggregation of the hazard in terms of magnitude M, distance R and epsilon: it is suggested that the above data are obtained for values of the IM characterized by a MAF in the interval from 1/500 to 1/2,000. The use of more refined techniques for record selection is also allowed (Bradley 2013; Lin et al. 2013).

The minimum number of motions is 30.

The selection of the records should be made among those recorded on rock or stiff soil. If the site is characterized by soft soil (e.g. V_{s30} in the interval 180–360 m/s, or less) a site response analysis is mandatory. Equivalent linear methods can be used for this purpose if significant inelastic response at the higher intensities is not expected, otherwise fully non-linear methods must be employed.

Uncertainties regarding soil profile and geotechnical parameters should be treated in the same way as those related to the structure above soil, see 3.2.5).

For sites in proximity of known active faults the probability of occurrence of pulse-like motions must be evaluated and the selection of records should proportionately reflect it.

3.2.4 Knowledge Acquisition

Given that a fully exhaustive (i.e. deterministic) knowledge of an existing building in terms of geometry, detailing and properties of the materials is realistically impossible to achieve, it is required that every type of incomplete information be explicitly recognized and quantified, for introduction in the assessment process in the form of additional random variables or of alternative assumptions. Since the number and the relevance of the considered uncertainties has an obvious bearing on the final evaluation of the risk, and consequently on the cost of the upgrading intervention, the search for a balance between the cost for additional information and the potential saving in the intervention should be a guiding criterion in the knowledge acquisition process.

Based on the above consideration the provisions do not prescribe quantitative minima for the number of elements to be inspected, the number of samples to be taken, etc. They ask instead for a sensitivity analysis to be carried out on one or more preliminary models of the building (variations on a first approximation of the final model). For RC structures this analysis is of the linear dynamic type (modal with full elastic response spectrum), which is adequate to expose global modes of response (regular or less regular) and to provide an estimate of the member chord rotations demands to be compared with yield chord rotation capacities. The latter, being quite insensitive to the amount of reinforcement, can be obtained based on gross concrete dimensions and nominal steel properties. The results of these analyses would then provide guidance on where to concentrate tests and inspections.

The extension of these tests depends on the initial amount of information. If original construction drawings are available, only limited verification of the actual reinforcement details is required, through concrete removal over an area sufficient to expose longitudinal and transverse reinforcement (and estimate spacing). When drawings are incomplete or missing, the extension of test/inspections must increase to understand the "designer's modus operandi" in view of replicating it (this is regarded as more effective than blindly applying the ruling provisions at the time in a simulated design).

3.2.5 Uncertainty Modeling

All types of uncertainties are assumed to belong to either one of the following two classes:

- those describing variations of parameters within a single model, amenable to a description in terms of random variables, with their associated distribution function
- those whose description requires consideration of multiple models, to each of which a subjective mass probability function is associated.

The uncertainties belonging to the first class include: the seismic intensity at the site, governed by the hazard function, the record-to record variability, described by a set of records, all material properties, related both to the soil and to the structure, normally described as lognormal variables, and the model error terms of the capacity models, also usually described as lognormal variables.

The uncertainties belonging to the second class include, among others, the geometry of the structure (e.g. presence and dimension of certain elements whose precise identification would be too invasive), the reinforcement details in important places, alternative models for the capacity of the elements, alternative models for the behavior of the components (e.g. degrading or non degrading).

Uncertainties of this class are treated with the logic tree technique, where mass probabilities are assigned to the alternative assumptions for each of uncertain factor. The concept is illustrated in Fig. 3.2 in which the alternative assumptions are made at each node, and the result obtained with any particular sequence of assumptions (the branches of the tree) is weighted by the product of the mass probabilities assigned to the each of them, due to the assumed independence of the factors (X, Y and Z in the figure).

3.2.6 Structural Analysis and Modeling

Exclusive recourse to non-linear methods of analysis, accounting for material and geometric non-linear phenomena, is considered in the provisions. The analysis can

3 Existing Buildings: The New Italian Provisions for Probabilistic Seismic...

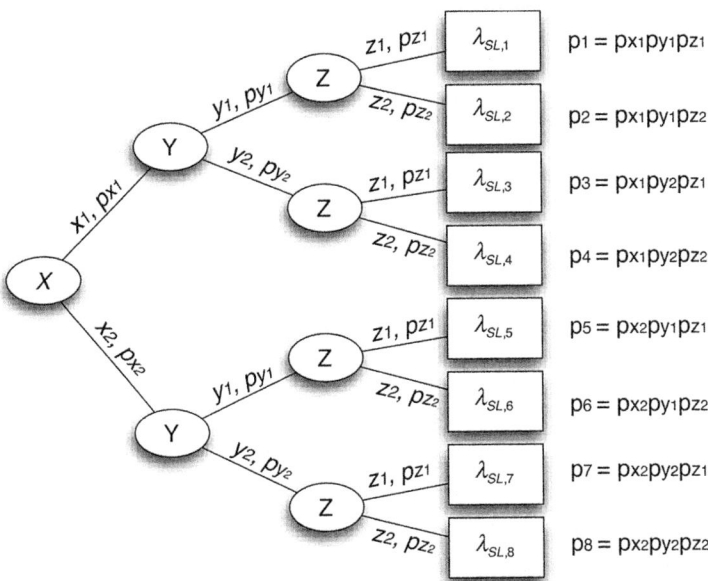

Fig. 3.2 Logic tree

be either static or dynamic, and guidance is given for the application, as it will be illustrated in the following (recall that for reason of space this paper covers only the part relative to RC buildings, the part devoted to masonry buildings is at least equivalent in terms of extension and detail).

The structural model must be tri-dimensional, with simultaneous excitation applied along two orthogonal directions.

Regarding the behavior of the structural members (beams and columns) under cyclic loading of increasing amplitude two modeling approaches are considered, as shown in Fig. 3.3.

- Non-degrading, i.e. stable hysteretic behavior without degradation of strength but overall degradation of stiffness (Takeda-type models)
- Degrading, where both stiffness and strength degrade with increasing cyclic amplitude down to negligible values.

The document provides in Chap. 4 an overview of the state of the art on this latter type of models for RC structures.

It is important to note from now that the use of the two different types of models has important reflexes in the identification of the collapse limit state of the structure.

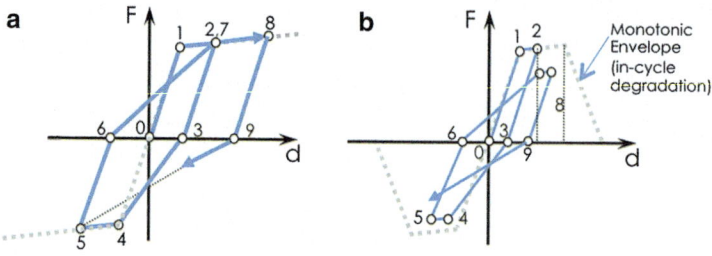

Fig. 3.3 Non-degrading (**a**) vs degrading (**b**) nonlinear modeling

3.2.7 Identification of LS Exceedance

Exceedance of each LS is signaled by a scalar indicator Y, expressing the global state of the structure as a function of that of its members, taking a value equal or larger than unity. Its definition depends on the considered LS. For the first two LS's, of light and severe damage, which pertain functionality and economic feasibility of repair actions, the choice of an appropriate threshold is left to the analyst in accordance to the owner/stakeholder requirements. The formulation of Y for the collapse limit state, related to safety, is stricter and does not leave space for subjective choices on the analyst side.

3.2.7.1 Light Damage

For the purpose of the identification of the light damage LS, the building is considered as composed by N_{st} structural members and N_{nst} non-structural components:

$$Y_{SLD} = \frac{1}{\tau_{SLD}} \max \left[\sum_{i=1}^{N_{st}} w_i I\left(\frac{D_i}{C_{i,SLD}}\right); \sum_{j=1}^{N_{nst}} w_j I\left(\frac{D_j}{C_{j,SLD}}\right) \right] \quad (3.4)$$

In the above expression, D and C indicate the appropriate demand and capacity values, I is an indicator function taking the value of one when $D \geq C$ and zero otherwise, and the w's are weights summing up to one, accounting for the importance of different members/components. The indicator Y attains unity when the max function equals τ_{SLD}, a user-defined tolerable maximum cumulative damage. (e.g. something in the range 3–5 %).

3.2.7.2 Severe Damage

For the purpose of the identification of the severe damage LS, the indicator Y is formulated in terms of a conventional total cost of damage to structural and non-structural elements as:

$$Y_{SLS} = \begin{cases} \dfrac{1}{\tau_{SLS}} \left[\alpha_{st} \sum_{i=1}^{N_{st}} w_i c \left(\dfrac{D_i}{C_{i,SLS}} \right) + (1-\alpha_{st}) \sum_{j=1}^{N_{nst}} w_j c \left(\dfrac{D_i}{C_{j,SLS}} \right) \right] & \text{if } Y_{SLC} < 1 \\ 1 & \text{if } Y_{SLC} \geq 1 \end{cases}$$

(3.5)

where α_{st} is the economic "weight" of the structural part (i.e. about 20 % in a low- to mid-rise residential building); $c(D/C)$ is a conventional cost function which starts from zero for $D=0$ and reaches unity, i.e. the replacement cost for the element, for $D=C_{SLS}$ (with C_{SLS} usually a fraction of the ultimate capacity of the element); as for the light damage LS, the indicator function attains unity when the quantity within square brackets equals τ_{SLS}, a user-defined fraction of the total building value over which repair is considered economically not competitive with demolition and replacement. Obviously if collapse occurs Y_{SLS} is set to 1.

3.2.7.3 Collapse

As anticipated, the identification of this LS depends on the modeling choices (see §2.6).

If non-degrading elements are adopted, the system is described as a serial arrangement of a number of elements in parallel, so that the Y variable takes the expression (Jalayer et al. 2007):

$$Y_{SLC} = \max_{i=1,N_s} \min_{j \in I_i} \dfrac{D_j}{C_{j,SLC}} \qquad (3.6)$$

where N_S is the number of parallel sub-systems (cut-sets) in series, and I_i is the sets of indices identifying the members in the i-th sub-system. This formulation requires the a priori identification of the cut-sets. Carrying out this task is in general not immediate, since the critical cut-set depends on the dynamic response and changes from record to record.

If all elements are of the "degrading" type, i.e. they are able to simulate all types of failure, accounting for the interaction of bending and shear, the collapse state $Y=1$ is identified with the occurrence of the so-called "dynamic instability", that is, when the curve intensity-response becomes almost flat. In order to identify the point on the curve corresponding to $Y=1$ one can use the expression:

$$Y_{SLC} = (1+\Delta) - \dfrac{S'}{S'_0} \text{ with } 0 < S' < S'_0 \qquad (3.7)$$

with values for Δ in the interval 0.05–0.10, corresponding to a small residual positive stiffness, in order to avoid numerical problems.

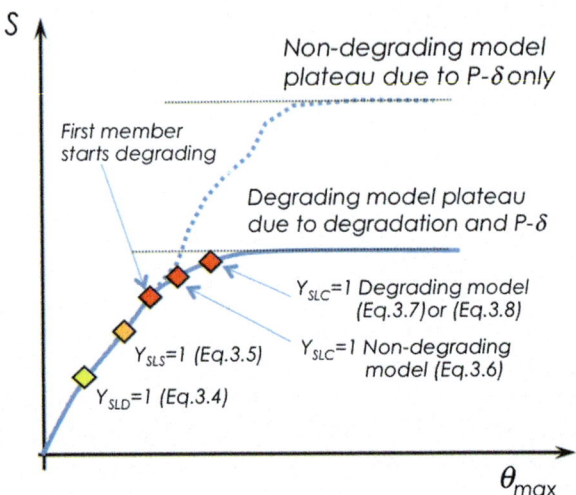

Fig. 3.4 Intensity vs response curve (also known as IDA curve, see 3.2.8.1), as a function of modeling choices

Finally, if the elements are of the degrading type but the adopted formulation cannot account for all possible collapse modes, the indicator variable can be expressed as:

$$Y_{SLC} = \max\left[(1+\Delta) - \frac{S'}{S'_0}; \max_{nsm}\left(\frac{D}{C}\right)\right] \quad (3.8)$$

which simply indicates that the collapse condition is attained for the most unfavorable between dynamic instability and the series of the "non simulated (collapse) modes". Typically, this set includes the axial failure of columns. Care should be taken in selecting the columns to be included in the evaluation of (3.8), limiting it only to those that can really be associated with a partial/global collapse.

The Fig. 3.4 shows an idealized intensity-response relation S vs θ_{max} (maximum interstorey drift ratio), with marks on the points corresponding to LS's according to the above definitions.

3.2.8 Assessment Methods

As already indicated in 3.2.2, the outcome of the assessment is expressed in terms of the mean annual frequency of exceeding any of the proposed three Limit States: λ_{LS}. Differently formulated or additional Limit States could be considered without any modification of the procedure.

The mean annual frequency is obtained using the Total Probability Theorem, as the integral of the product of the probability of exceedance of the LS conditional to the value $S = s$ of the seismic intensity (denominated as "fragility"), times the probability of the intensity being in the neighborhood of s. This latter is given by the absolute value of the differential of the hazard function at $S = s$:

$$\lambda_{LS} = \int_0^\infty p_{LS}(S)|d\lambda_S(S)| \tag{3.9}$$

The integral can be evaluated numerically. However, if the hazard is approximated with a quadratic fit in the log-log plane ($\ln\lambda_S = \ln k_0 + k_1 \ln s + k_2 \ln^2 s$), and the fragility function is assumed to have a lognormal shape, closed forms for the evaluation of the integral are available.

The lognormal assumption is indeed adopted in the provisions based on the international general consensus. The fragility thus takes the form:

$$p_{LS}(S) = p(Y_{LS} \geq 1 | S = s) = p(S_{Y_{LS}=1} \leq s) = \Phi\left(\frac{\ln s - \mu_{\ln S_{Y=1}}}{\sigma_{\ln S_{Y=1}}}\right) \tag{3.10}$$

requiring evaluation of two parameters only: the mean and the standard deviation of the logarithm of the seismic intensity inducing the unit value of the Limit State indicator: $Y = 1$.

The document provides three alternative methods, indicated in the following as A, B and C, for the evaluation of the fragility. All methods require a 3D model of the structure.

3.2.8.1 Method A: Incremental Dynamic Analysis on the Complete Model

Recourse is made to the well known technique usually referred to as Incremental Dynamic Analysis (IDA) (Vamvatsikos and Cornell 2002): it consists in subjecting the complete 3D model of the structure to a suite of n time-histories (each with two orthogonal horizontal components, the vertical component being normally omitted in case of ordinary buildings), each time-history being scaled at increasing intensity levels. At each level of S the value of Y is calculated, and the set of (S,Y) points are plotted to obtain a curve in the intensity-response plane, denoted as "IDA" curve.

A sample of values of S leading to $Y = 1$ is obtained from the set of n IDA curves, as shown in Fig. 3.5, left: this sample is used to evaluate the parameters $\mu_{\ln S_{Y=1}}$ and $\sigma_{\ln S_{Y=1}}$.

The effect of the uncertainties that can be modeled as continuous can be approximately determined by associating to each ground motion a sample of the uncertainties taken from their distributions (the approach is acceptable if the number of time-histories is adequate to describe at least approximately the distribution of the r.v.'s). The effect of the introduction of the uncertainties is visible on the IDA curves by their larger spread (Fig. 3.5, right).

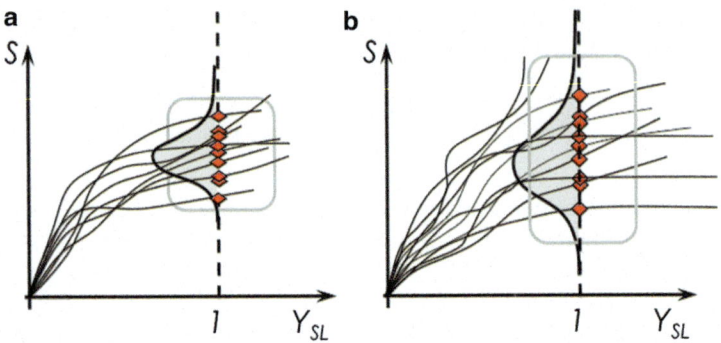

Fig. 3.5 IDA curves and samples of the SY = 1 intensity values: (**a**) including record-to-record variability only, (**b**) with structural uncertainty

3.2.8.2 Method B: Incremental Dynamic Analysis on an Equivalent Single Degree-of-Freedom Oscillator

This method differs from the previous one for the fact that the incremental dynamic analyses are carried out on a (number of) "equivalent" single degree–of–freedom (SDOF) oscillators, obtained through nonlinear static (NLS) analysis on the 3D model. Any of the available types of NLS analysis can be adopted, as appropriate for the case at hand.

The global curve relating base shear to the top displacement obtained from the pushover becomes the force-displacement relationship of a simple oscillator, which for the purpose of the response analysis is approximated with a multi-linear relationship.

The number of the needed SDOF oscillators equals the number of modes contributing significantly to the total 3D response. On each SDOF an IDA analysis is performed for all of selected time-histories: for any time-history, modal responses, obtained translating the maximum dynamic response of each SDOF in the response of the 3D structure, at the same intensity level are combined by an appropriate rule (SRSS or CQC) to yield the total response. The latter is used to compute the indicator variable for each LS. Then collection of $S_{Y=1}$ values and evaluation of the fragility parameters $\mu_{\ln S_{Y=1}}$ and $\sigma_{\ln S_{Y=1}}$ proceeds as per method A.

The effect of the uncertainties that can be modeled as continuous can be treated in the same approximate way as in Method A. In this case the pushover analyses must be repeated on different structures each one characterized by a different realization of the uncertainties, and associated one-to-one with the selected motions.

3.2.8.3 Method C: Non-linear Static Analysis and Response Surface

This method is again based on nonlinear static analysis. The main differences with respect to method B are two: demand on the SDOF oscillators is determined using the response spectra of the selected time-histories (the actual response can be obtained using any of the available methods for obtaining the inelastic displacement response from an elastic spectrum), and the effect of the system-related uncertainties that can be modeled as continuous is determined through the use of the Response Surface technique (Pinto et al. 2004).

The two parameters of the fragility function are determined as follows.

The log-mean is obtained from the median response spectrum of the selected time-histories (scaled to the same S=s), whose intensity is scaled upwards until $Y=1$ is obtained:

$$\mu_{\ln S_{Y=1}} = \ln S_{Y=1}|_{S_{a,50\%}(T)} \tag{3.11}$$

The logarithmic standard deviation is assumed as independently contributed by two factors: the variability of the response due to the variability of the ground motions (given $S=s$), and the variability due to the randomness of the material properties:

$$\sigma_{\ln S_{Y=1}} = \sqrt{\sigma^2_{\ln S_{Y=1},S} + \sigma^2_{\ln S_{Y=1},C}} \tag{3.12}$$

The first of the two terms is evaluated from the response spectra fractiles at 16 and 84 % from the selected time-histories (scaled to the same S=s) according to:

$$\sigma_{\ln S_{Y=1},S} = \frac{\ln S_{Y=1|16\%} - \ln S_{Y=1|84\%}}{2} \tag{3.13}$$

The influence on $S_{Y=1}$ of the continuous random variables, denoted by X_k, is studied by expressing $\ln S_{Y=1}$ as a linear response surface, in the space of the normalized variables $x_k = (X_k - \mu_{X_k})/\sigma_{X_k}$:

$$\ln S_{Y=1} = \alpha_0 + \sum_k \alpha_k x_k + \varepsilon \tag{3.14}$$

The normalized variables are assigned the values ± 1 in correspondence of their fractile values of 16 % and 84 %. The N parameters α_k are obtained through a complete factorial combination of the variables at two levels $(+1,-1)$. For each of the $M = 2^N$ combinations the median spectrum is increased up to the value producing $Y = 1$. The values attributed to the normalized variables ($+1$ or -1) for each of the combinations are the rows of a so-called "matrix of experiments" **Z**, and the corresponding values of $\ln S_{Y=1}$ form a vector of "response" denoted as **y**.

The parameters α_k are then obtained from the expression:

$$\alpha = \left(\mathbf{z}^T\mathbf{z}\right)^{-1}\mathbf{z}^T\mathbf{y} \qquad (3.15)$$

from which the component of $\sigma_{\ln S_{Y=1}}$ related to the uncertainty in the structure ("capacity") follows as:

$$\sigma_{\ln S_{Y=1},C} = \sqrt{\sum_k \sum_j \alpha_k \alpha_j \rho_{x_i x_j} + \sigma_\varepsilon^2} \qquad (3.16)$$

where σ_ε^2 is the variance of the residuals, and the facts that ε and $\mathbf{x}$ are independent, and the latter are correlated standard variables with correlation coefficient ρ has been used.

3.3 RC Specific Provisions

This chapter complements the general Chap. 2, by providing detailed indications on modeling of response and capacity for RC structures. As mentioned before the document is based exclusively on nonlinear analysis and prescribes a mandatory verification of the collapse LS. Inelastic models that describe response up to collapse, however, are still not in the average technical background of engineers, and, also, they are still evolving toward a more mature and consolidated state. In recognition of this, the document introduces formulations for the identification of the collapse LS that allow a correct use of the mainstream non-degrading models (3.6), but leaves the door open to the use of more advanced degrading models (3.7). Further, in order to guide the user in the selection of the latter, it provides a brief reasoned classification of inelastic response models.

3.3.1 Response Models

Models for beam-columns, joints and masonry infills are presented, though the former are obviously given the major attention. In particular, collapse modes of RC columns are described, as schematically shown in Fig. 3.6. The figure illustrates the possible modes of collapse in a monotonic loading condition, in terms of shear force-chord rotation of the member. In all three cases the plot shows with dashed grey lines the monotonic response in a pure flexural mode, with the usual I, II and III stages up to ultimate/peak strength, followed by a fourth descending branch to actual collapse, and the shear strength envelope. The latter starts with $V_{R,0}$ and decreases as a function of deformation, measured in terms of ductility μ. Depending on whether the two curves cross before flexural yield, after, or do not cross at all, the member fails in brittle shear, ductile shear or flexure. In all cases, collapse occurs

Fig. 3.6 Collapse modes of RC columns (chord rotations at peak strength, usually denoted as ultimate values θ_u, are here differentiated as either shear θ_V or flexural θ_f)

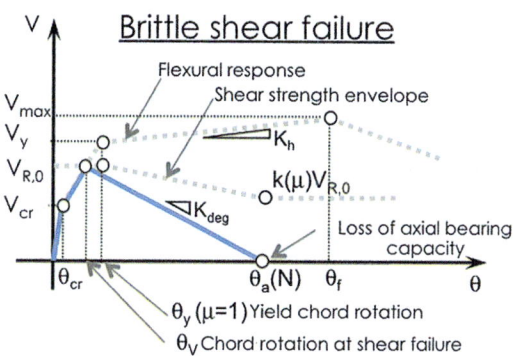
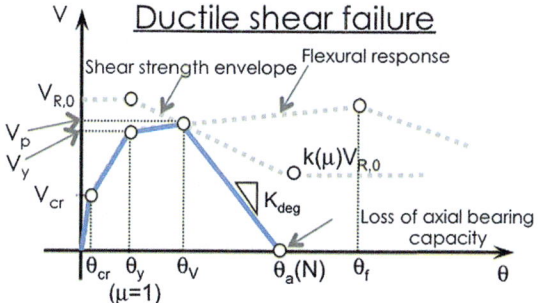
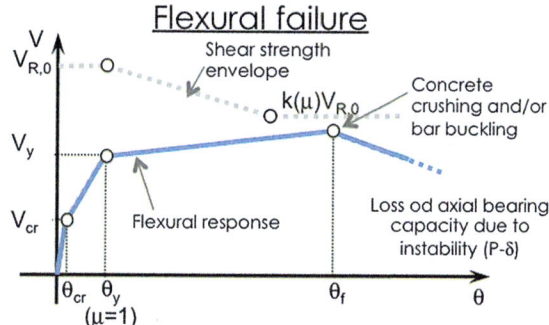

due to loss of vertical load-bearing capacity ($V_R = N_R = 0$) at the end of the degrading branch.

In cyclic loading at large amplitude the response presents a second contribution to degradation, which is cyclic degradation, as shown in Fig. 3.7.

Available models can be classified in mechanical and phenomenological. The state of the art of purely mechanical models is not yet capable of describing the full range of behaviour of RC members illustrated in Figs. 3.6 and 3.7 (especially for brittle and ductile shear collapse). Currently, if the analyst wishes to incorporate degrading models, the only viable option is to use phenomenological (e.g. Ibarra et al. 2005) or hybrid models (Elwood 2004). These models, however, also have

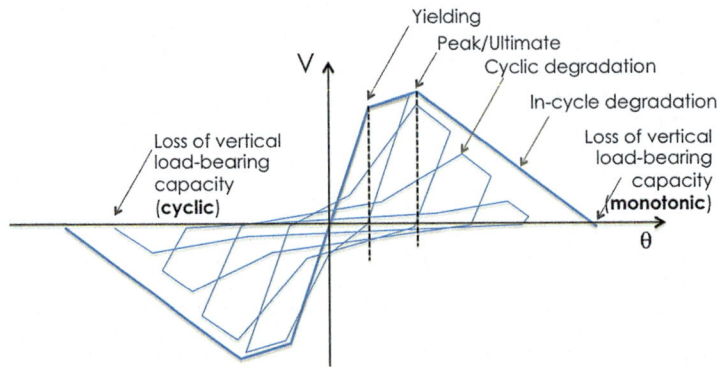

Fig. 3.7 Cyclic and in-cycle components of degradation (Response shown is from Ibarra et al. model)

their limitations and, for instance, rely heavily on the experimental base used to develop them, which is often not large enough (e.g. for the Ibarra et al. model, the proportion of ductile shear and flexural failures dominate the experimental base, resulting in limited confidence on the model capability to describe brittle failures). Further, computational robustness is an issue with all these models.

Figure 3.7 shows the monotonic backbone (e.g. for the ductile shear collapse mode) and the cyclic response. It is important to note that the deformation thresholds corresponding to state transitions and ultimately to collapse are different for monotonic and cyclic loading. This fact is highlighted in Fig. 3.8, where the peak/ultimate and axial failure rotations are clearly identified as different in the monotonic and cyclic loading.

The user is advised that consistency is essential in the choices of response, capacity and LS identification formulas.

If non-degrading models are chosen, one should use (3.6) for collapse identification, with peak deformation thresholds $\theta_{u,cyclic}$ that account on the capacity side for the degradation disregarded on the response side.

If degrading models are used, (3.7) or (3.8) are employed, and the monotonic deformation thresholds, $\theta_{u,mono}$, $\theta_{a,mono}$, etc are used as input parameters for the response model (together with degradation parameters).

3.3.2 Capacity Models

A survey of probabilistic models for the deformation thresholds shown before, grouped by LS, is presented in the document. Requirements for an ideal set of models are stated explicitly: consistency of derivation of thresholds of increasing amplitude (i.e. yield, peak and axial deformation models derived based on the same experimental tests, accounting also for correlations), and an experimental base

3 Existing Buildings: The New Italian Provisions for Probabilistic Seismic... 115

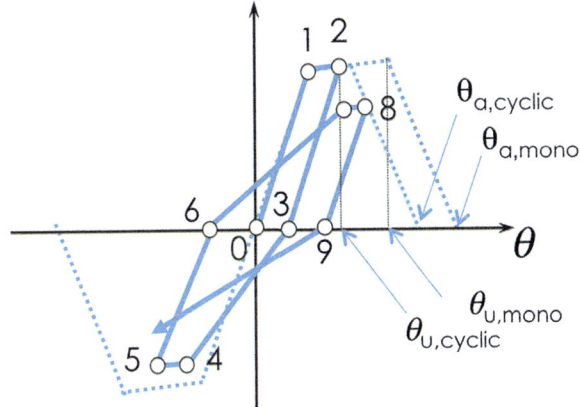

Fig. 3.8 Deformation limits for monotonic and cyclic loading

covering the full range of behaviours (different types of collapse, different reinforcement layouts, etc) in a balanced manner. Such a set of models is currently not available.

One set of models that comes closer to the above requirements, and is used in the application illustrated in the next section, is that by Haselton et al. (2008), which consists of predictive equations for the parameters of the Ibarra et al. (2005) degrading hysteretic model. Haselton et al, however, provide only mean and standard deviation of the logarithm of each parameter, disregarding pair-wise correlation, in spite of the fact that the equations were established on the same experimental basis. Also, as already anticipated, brittle shear failures are not represented.

Figure 3.9 shows the tri-linear moment-rotation monotonic envelope according to the Ibarra model, with qualitative (marginal) probability density functions (PDFs) for its parameters, as supplied by Haselton et al. (2008). Not all the parameters can be independently predicted at the same time, to maintain physical consistency of the moment-rotation law. In the application the stiffness at 40 % and 100 % of yield, and the rotation increment $\Delta\theta_f$ and $\Delta\theta_a$ have been used (darker PDFs in the figure). Use of the latter two in place of θ_f and θ_a ensures that situations with $\theta_f > \theta_a$ cannot occur. The equation for θ_y is redundant since θ_y is obtained from M_y and K_y. As described in the application, care has been taken in ensuring that K_y is always larger than $K_{40\%}$. The latter is used as an intermediate value between I and II stage stiffness, since the model is tri-linear. Finally, Haselton et al. (2008) provide also a marginal model for the parameter regulating cyclic degradation in the Ibarra model, i.e. the normalized total hysteretic energy $E_t/(M_y\theta_y)$.

The document provides also equations by Biskinis and Fardis (2010a, b), adopted since 2005 in earlier form in Eurocode 8 Part 3 (CEN 2005) and in latest fib Model Code (fib 2010), as well as by Zhu et al. (2007). These equations, however, are calibrated to provide cyclic values of the deformation thresholds, and their use is thus appropriate for LS identification when non-degrading models are used.

Fig. 3.9 Deformation limits for monotonic loading with schematic indication of the marginal PDF of each parameter

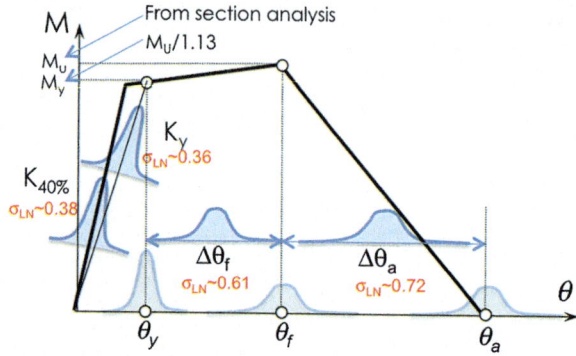

3.3.2.1 Biaxial Verification

Most response and all available capacity models are applicable for deformation in a single plane of flexure, while the document requires mandatory use of tri-dimensional models. While this does not represent a limitation for beams and for joints, with the exception of corner ones, columns are always subjected to biaxial deformation.

If degrading models are employed, currently the only option is to use the same model independently in the two orthogonal planes of flexure, disregarding interaction.

When non-degrading models are employed, interaction can be accounted for on the response side e.g. by use of fibre-discretized sections, and on the capacity side through the use of an "elliptical" rule for the evaluation of the local, member-level capacity-to-demand ratio (Biskinis and Fardis 2010a, b):

$$y = \sqrt{\left(\frac{\theta_2}{\theta_{2,LS}}\right)^2 + \left(\frac{\theta_3}{\theta_{3,LS}}\right)^2} \qquad (3.17)$$

where θ_2 and θ_3 are the rotation demands in the two orthogonal planes, and $\theta_{2,LS}$ and $\theta_{3,LS}$ are the corresponding (cyclic) capacities for the LS under consideration.

3.4 Example Application to an RC Building

3.4.1 Premise

The document contains example applications to two real buildings, one in unreinforced masonry and the other in reinforced concrete. Together, the two examples illustrate the application of the three assessment methods presented in the provisions: methods A and C with reference to the masonry building and method B with reference to the RC building.

The seismic risk assessment of the RC building has been carried out twice, using both non-degrading and degrading models, denoted as A and B, respectively. This has been done to provide users with an order of magnitude of the expected differences between the two approaches. Actually, the document provides results also for a third analysis with masonry infills, not reported here.

3.4.2 Description of the Structure

The building, shown in Fig. 3.10, is one of three blocks making up a school complex in southern Italy, built in the early 1960s. The structure consists of an RC space frame with extradosed beams and one-way hollow-core slabs, developing for three storeys over a sloping site. The lower storey is constrained since it is under-ground on three sides.

3.4.3 Seismic Action

For the purpose of the evaluation the building has been located at a site in the Basilicata region. Seismic hazard from the current design code, in terms of uniform hazard spectra at nine return periods, has been used to reconstruct median and fractile hazard curves at the first mode period of the structure (see later). The median curve has been interpolated with a quadratic polynomial in log-log space ($k_0 = 8.134 \times 10^{-5}$, $k_1 = 3.254$, $k_2 = 0.303$). Fractile curves have been used to compute a value of the hazard dispersion $\beta_H = 0.3$ (at a frequency between 1/500 and 1/1,000 years, close to the value of collapse MAF).

Thirty ground motion records have been selected from an aggregated database obtained merging the European Strong Motion database, and the Italian ITACA and SIMBAD databases. Records have been selected in the $M_w = [5.6;6.5]$ and $d_{epi} = [10 \text{ km};30 \text{ km}]$ ranges (Fig. 3.11), centred around the values obtained from PSHA deaggregation in the same 1/500 and 1/1,000 years frequency range.

3.4.4 Preliminary Analysis and Test Results

No construction or design drawings were available. Based on an existing architectural survey, a structural survey was conducted to reconstruct the gross concrete frame dimensions. Based on these and on values for material properties, loads and reinforcement assumed based on the ruling design code at the time of construction a preliminary model was set–up (Fig. 3.12, where loads are shown in red, with height

Fig. 3.10 North-east view of the building

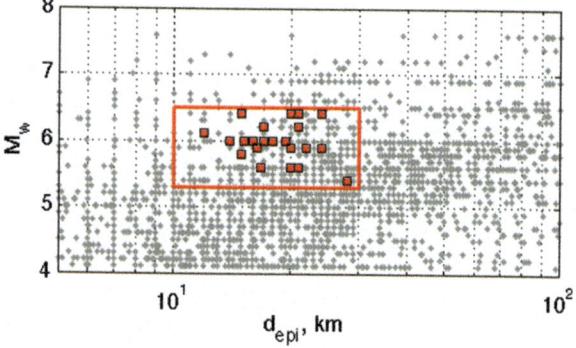

Fig. 3.11 Magnitude and distance bin used in the selection of recorded motions. Red dots indicate selected records

proportional to intensity). Modal analysis with full elastic response spectrum has provided the location where the largest inelastic deformation demand is expected. The most stressed columns are framed in red in Fig. 3.13, where actual members chosen for inspection and material sampling (at ground floor) are circled in blue. The results are reported in Table 3.2.

3.4.5 Structural Modeling

Structural analysis has been carried out using the general-purpose FE package OpenSees (McKenna et al. 2010). The behaviour of RC beam-column joints has not been modelled. Beams and columns have been modelled by means of elastic frame elements with zero length at the two ends, with independent uniaxial

Fig. 3.12 View of preliminary analysis model with loads distributed to beams (one-way slabs)

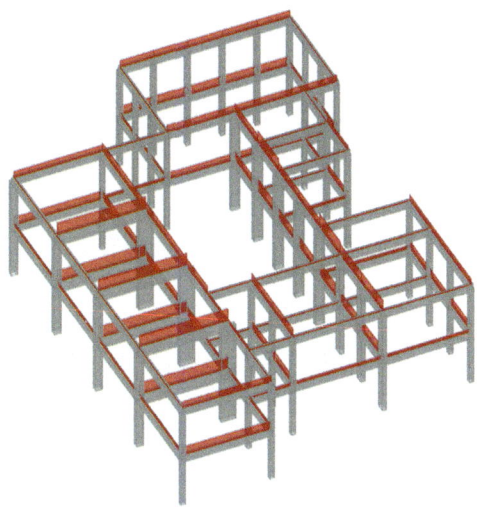

constitutive laws on each degree of freedom.[1] The adopted moment-rotation law is the tri-linear one by Ibarra et al. (2005), in the implementation by (Lignos and Krawinkler 2012), and shown in Fig. 3.14 for the two orthogonal planes of flexure of one of the columns. Axial force-bending moment interaction is not included in the model, therefore a constant axial force needs to be assigned at the beginning of the analysis for determination of the model parameters. A single gravity load analysis on the median model has been used to determined axial forces in all columns, and these have been used for all random realizations of the structure (see next section).

Parameters for the Ibarra model have been predicted with the set of equations calibrated by Haselton et al. (2008). These equations include one that provides the degradation parameter:

$$\gamma = \frac{E_t}{M_y \theta_y} \qquad (3.18)$$

Actually, the Opensees implementation of the Ibarra model requires in input the degradation parameter in the form:

$$\Lambda = \lambda \theta_p = \frac{E_t}{M_y \theta_p} \theta_p = \gamma \theta_y \qquad (3.19)$$

Since method B has been used for the assessment (see later), a unique value of the degradation parameter needs to be assigned to the equivalent oscillator of each mode. The average value of Λ over the columns has been used.

---

[1] This option is easy to implement with a simple script in Tcl/Tk and is more robust than using a lumped plasticity element formulation, since it leaves complete control to the analyst through the global solution algorithm.

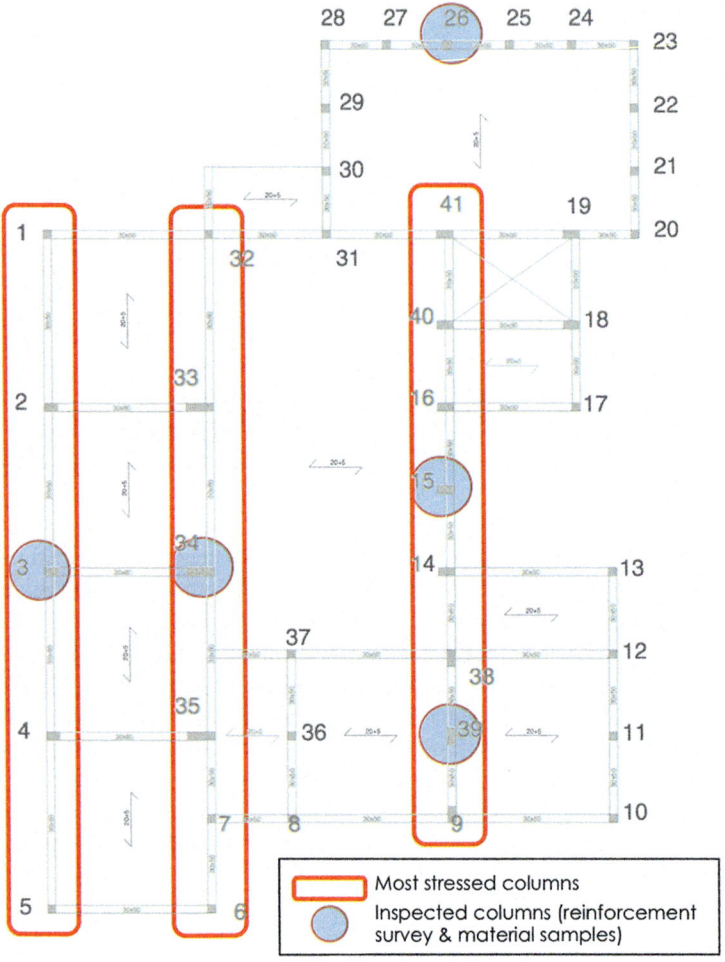

Fig. 3.13 Plan of inspections

As anticipated, the risk analysis has been performed twice, for both degrading and non-degrading models. In the latter case, for the sake of simplicity, the same Ibarra model has been used, but with zero, rather than negative, post-peak stiffness (e.g. M-θ curves in Fig. 3.14 go flat after 3.1 % and 3.2 %, respectively). Equation (3.6) has been used to check the collapse LS, and cyclic thresholds by Zhu et al. (2007) have been used for the ductile shear (θ_V) or flexural (θ_f) peak deformation. Each member has been attributed a ductile shear or flexural threshold based on the classification criterion proposed in Zhu et al., i.e. shear if geometric transverse reinforcement percentage lower or equal to 0.002, or shear span ratio lower than 2 (squat member), or plastic shear $V_p = 2M_u/L$ larger than 1.05 the shear strength (according to Sezen and Mohele 2002). Zhu et al. model for cyclic axial failure threshold θ_a has also been used for the non-degrading model.

Table 3.2 Results of tests on columns at ground floor

Member	B (mm)	H (mm)	Long. Reinf.	Transv. Reinf.	f_c (MPa)	f_y (MPa)
P3	300	500	6ɸ20	2ɸ6/200	16.7	–
P15	300	600	6ɸ20	2ɸ6/200	15.4	–
P26	300	300	4ɸ12	2ɸ6/200	17.8	–
P34	300	1,000	8ɸ20	2ɸ6/200	11.9	337
P39	300	500	6ɸ12	2ɸ6/200	11.6	370

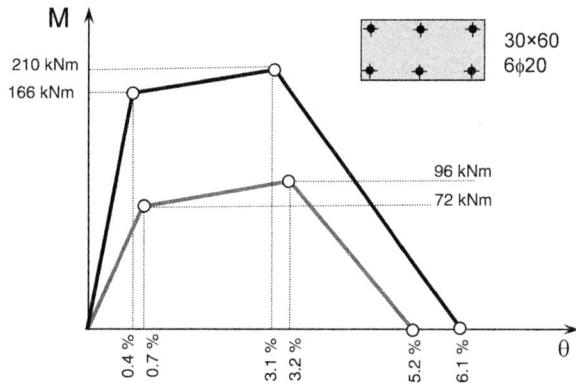

Fig. 3.14 Moment-rotation in two orthogonal planes

3.4.6 Uncertainty Modeling

In this application uncertainties that require analysis of alternative models, to be treated with the logic tree technique, have not been considered.

The uncertainties included in the assessment are:

- Material strengths: f_c and f_y, and ultimate concrete deformation ε_{cu}, which determine the constitutive law of the materials and enter into: (a) the stiffness of the elastic members, (b) section analysis leading to M_u, c) predictive formulas for deformation thresholds;
- Monotonic incremental deformation $\Delta\theta_f = \theta_f - \theta_y$ and $\Delta\theta_a = \theta_a - \theta_f$, and the cyclic degradation parameter γ, the latter two only for the degrading model;
- Cyclic deformation thresholds θ_f, θ_V and θ_a, for the non-degrading model;

All variables have been modelled as lognormal. As anticipated, statistical dependence of parameters within the same member or between same-parameter across different members has been modelled through assumed correlation coefficients.

In particular, in order to ensure that within each member $K_{40} > K_y$, perfect correlation has been assumed, a single standard normal random variable $\varepsilon_i \sim N(0,1)$ has been sampled in each member, and then amplified by the corresponding

Table 3.3 Distribution parameters for the random variables

RV	Median	Log-std	Correlation
f_c (MPa)	14.0	0.20	0.7
ε_{cu}	0.006	0.20	0.7
f_y (MPa)	338.0	0.10	0.8
K_{40}	Haselton et al.	0.38	0.8
K_y	Haselton et al.	0.36	0.8
$\Delta\theta_f$	Haselton et al.	0.61	0.8
$\Delta\theta_a$	Haselton et al.	0.72	0.8
θ_f	Zhu et al.	0.35	0.8
θ_V	Zhu et al.	0.27	0.8
θ_a	Zhu et al.	0.35	0.8
γ	Haselton et al.	0.50	0.8

logarithmic standard deviation to yield the factors $\exp(\varepsilon_i\,\sigma_{\ln K40})$ and $\exp(\varepsilon_i\,\sigma_{\ln Ky})$ that multiply the corresponding medians.

Similarly, in order to avoid situations where a very ductile element loses axial bearing capacity prematurely, the variables $\Delta\theta_f$ and $\Delta\theta_a$ have been considered perfectly correlated and a single normal variable has been sampled as done for the stiffnesses.

Finally, in a way of simplicity, same-variables across different members (stiffnesses, deformation thresholds and material properties) have been considered equicorrelated, independently of distance (one could have used a distance-dependent correlation coefficient, with an exponential or squared exponential model, differentiating correlation lengths in the vertical and horizontal directions), with values reported in Table 3.3.

Figure 3.15 shows the moment-rotation law of a member for median values and one of the 30 samples of the random variables. The figure reports also in dashed line the non-degrading branch of the M-θ law, and the corresponding cyclic thresholds used for LS checking.

3.4.7 Method B and Response Analysis via Modal Pushover

The assessment has been carried out with method B, which uses IDA on equivalent oscillators obtained through nonlinear static analysis to characterize response. Several proposals are available in the literature for the determination of an approximate IDA curve starting from nonlinear static analysis, e.g. (Vamvatsikos and Cornell 2005; Dolsek and Fajfar 2005; Han and Chopra 2006). The latter, based on the modal pushover analysis (MPA) technique (Chopra and Goel 2002), has been chosen here due to its easy implementation with commercial analysis packages, since it uses invariant force patterns, and its applicability to general spatial geometries (Reyes and Chopra 2011). Differently from (Reyes and Chopra 2011), however, herein a single excitation that accounts for both orthogonal components of ground motion has been used. This excitation is derived as follows.

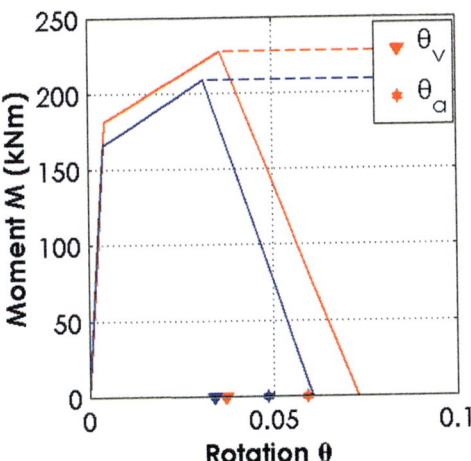

Fig. 3.15 Moment-rotation law for median values (*blue*) and one sample (*red*) of the random variables

The equations of motion for a discrete nonlinear MDOF system subjected to two components of ground motion along the X and Y axes are:

$$\mathbf{M\ddot{u} + C\dot{u} + F(u)} = -\mathbf{M}(\mathbf{t}_X a_X + \mathbf{t}_Y a_Y) \quad (3.20)$$

with usual meaning of symbols and omitting the time dependence of input acceleration and response quantities. Adopting, according to the MPA method, the modal decomposition also in presence of nonlinear resisting forces, one gets:

$$M_i \ddot{q}_i + C_i \dot{q}_i + F_i = -(L_{iX} a_X + L_{iY} a_Y) \quad i = 1, \ldots, n \quad (3.21)$$

where $M_i = \boldsymbol{\phi}_i^T \mathbf{M} \boldsymbol{\phi}_i$, $C_i = \boldsymbol{\phi}_i^T \mathbf{C} \boldsymbol{\phi}_i$, $F_i = \boldsymbol{\phi}_i^T \mathbf{F}$ is the projection of **F** on the i-th mode, and $L_{iX,Y} = \boldsymbol{\phi}_i^T \mathbf{M} \mathbf{t}_{iX,Y}$. Upon dividing (3.16) by the modal mass one gets:

$$\ddot{q}_i + 2\xi_i \omega_i \dot{q}_i + \frac{F_i}{M_i} = -(\Gamma_{iX} a_X + \Gamma_{iY} a_Y) \quad i = 1, \ldots, n \quad (3.22)$$

Finally, by further dividing (3.17) by the largest (dominant) of the two load participation factors L, e.g. that associated with the X component, one arrives at the equation of motion of a nonlinear oscillator having F/L-D force-displacement law, excited by an excitation which combines the two orthogonal input motions:

$$\ddot{D}_i + 2\xi_i \omega_i \dot{D}_i + \frac{F_i}{L_i} = -\left(a_X + \frac{\Gamma_{iY}}{\Gamma_{iX}} a_Y\right) \quad i = 1, \ldots, n \quad (3.23)$$

The assessment starts with modal analysis. For each significant vibration mode two nonlinear static analyses are carried out, one for each sign of the forces. The result of each nonlinear static analysis will consist of a database of local responses, i.e. matrices of nodal displacements, of size (n_{steps} x n_{nodes} x n_{dofs}), or of member deformations, of size (n_{steps} × $n_{members}$ × $n_{deformations}$), plus a curve, usually called capacity curve, linking the base shear V_b to the displacement of a control degree of freedom u_c, usually taken to be that with the largest modal coordinate. The capacity curves are approximated by trilinear laws and transformed into F/L-D format. Each trilinear equivalent oscillator is then subjected to IDA with the 30 selected motions and local responses are obtained by interpolation of the corresponding database at the maximum displacement of the oscillator (for each motion and intensity level). Total responses are obtained from modal ones, at the same intensity $S = s$, by a suitable combination rule (SRSS or CQC). Based on total response, LS indicator functions Y are evaluated.

3.4.8 Results

Modal analysis of the median model (i.e. a model with median values assigned to all random variables) shows that the first three modes cumulatively account for more than 80 % of the total mass in both plan directions (Fig. 3.16). These mode shapes are the same for models A and B, since they have the same elastic properties.

These three modes are chosen for nonlinear static analysis. Figure 3.17, left, shows the corresponding results in terms of capacity curves with reference to model A. The figure shows also the tri-linear approximations of the curves used as monotonic backbone for the equivalent oscillators. The post-peak negative stiffness for this non-degrading model is entirely due to geometric effects (P-δ). Figure 3.17, right, compares the capacity curves for the three considered modes obtained with model A (red) and B (black), respectively. The curves depart from each other only after some excursion in the inelastic range, when the first local failure (exceedance of the ultimate/capping deformation) occurs. The total number of pushover analysis amounts to 2 signs × 3 modes × 30 models = 270, as shown in Fig. 3.18.

Figure 3.19 shows further details of the nonlinear static analysis, with the capacity curve of one of the 30 random realizations of Model B, subjected to modal forces according to its 3rd mode, in the positive sign, and the deformed shapes (same scale) at three steps corresponding to increasing levels of inelastic demand. The first and second step (S1 and S2 in the figure) correspond to the yield and peak displacement in the tri-linear approximation of the capacity curve, the last step S3 is midway between the peak and the last point. The deformed shapes report also the level of inelastic demand in plastic hinges, according to the convention already used in (Haselton and Deierlein 2007): hollow circles denote potential plastic hinge zones, blue and red circles denote inelastic demands lower and higher of the peak rotation, respectively. The diameter, for blue and red circles, is proportional to the D/C ratio. The blue circle fills completely the hollow black circle when $y = 1$ (Eq. 3.12), with $\theta_{LS} = \theta_f$ or θ_V. It can be observed that along the descending branch increases at some locations to more than three times the diameter of the black circle. This situation is numerically possible since the loss of axial load-bearing capacity is not modelled, and the analysis proceeds with redistribution

3 Existing Buildings: The New Italian Provisions for Probabilistic Seismic... 125

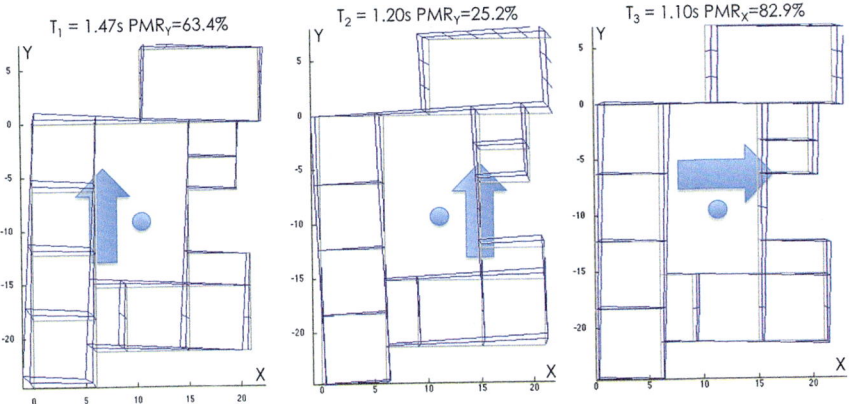

Fig. 3.16 Plan view of the first three mode shapes, with participating mass ratios in the dominant direction of each mode ("median" model)

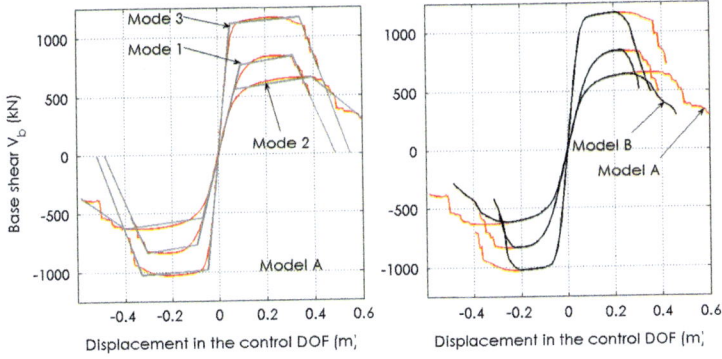

Fig. 3.17 Pushover curves for model A and B

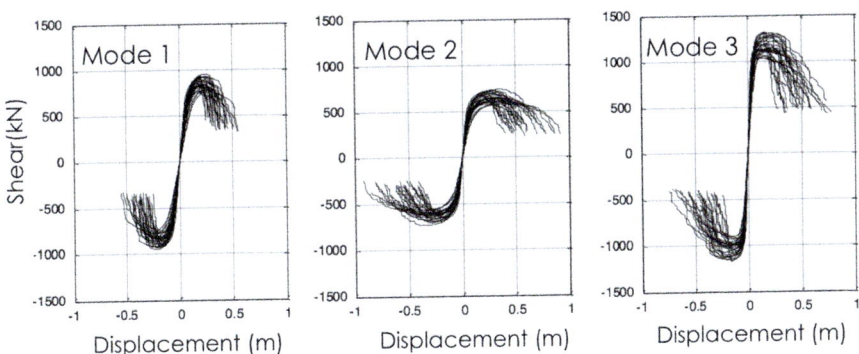

Fig. 3.18 Pushover curves of 30 random samples of model A

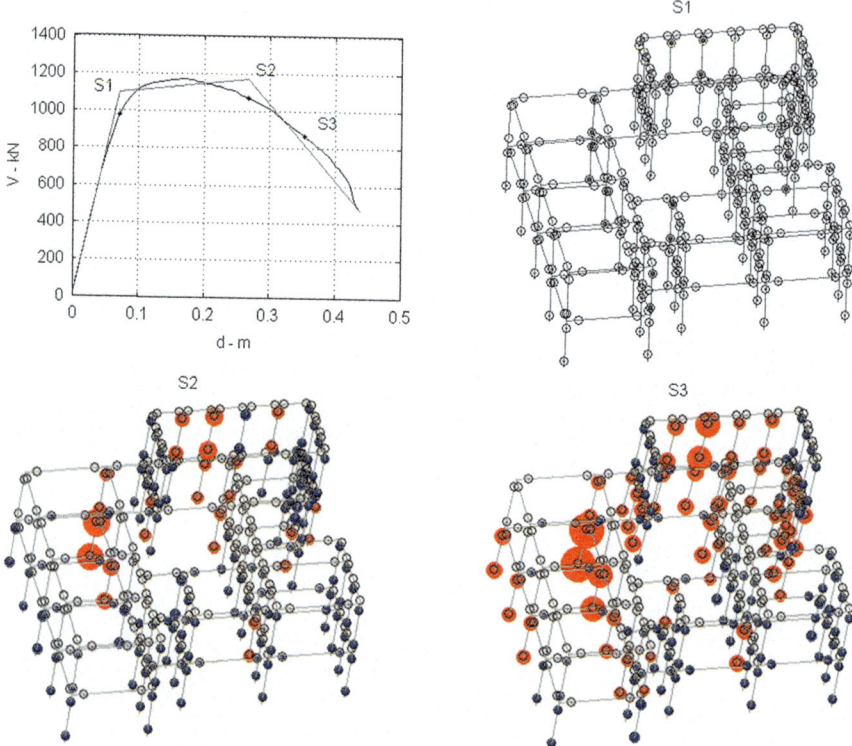

Fig. 3.19 Model B, Mode 3, pushover curve and deformed shapes at three different displacement levels, with indication of plastic hinge deformations (*hollow circles*, *blue circles* and *red circles* denote potential plastic hinges, active hinges before peak/ultimate deformation and hinges in the degrading post-peak branch, respectively)

of shear demand on the adjacent members. This fact, however, does not compromise the analysis since the axial collapse mode is actually checked a posteriori, using the θ_a model from Zhu et al. (2007) in conjunction with (3.8).

Figure 3.20 shows the response time-series for the equivalent oscillator (Model B, Mode 3, first random sample and associated motion) at three increasing intensity levels, shown below in terms of force-displacement loops. Depending on whether the largest response displacement has a positive or negative sign, the local responses at node/member level are interpolated from the database relative to the positive or negative pushover.

Finally, Figs. 3.21 and 3.22 show the IDA and the fragility curves for model A (left) and B (right), respectively. Green, blue and red dots on the IDA curves mark the attainment ($Y=1$) of the damage, severe damage and collapse LS. Each cloud of points is used to determine the log-mean and log-standard deviation of the intensity leading to the corresponding LS: $S_{Y=1}$, parameters of the fragility curves reported below.

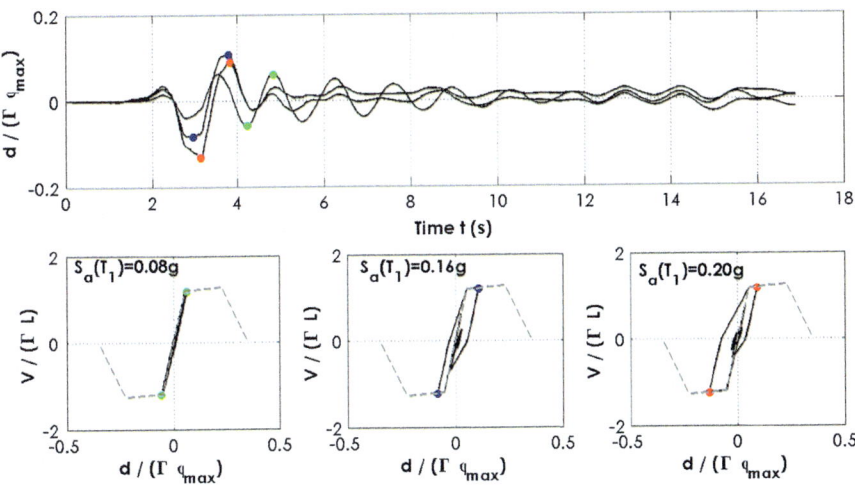

Fig. 3.20 Model B, Mode 3, response of the equivalent oscillator to the same motion at three increasing intensity levels (*top*) and corresponding force-deformation loops (*bottom*)

Convolution of the fragility curves in Fig. 3.22 with the hazard curve for the corresponding intensity measure, $S = S_a(T_1)$, yields the values of the mean annual rate of exceedance of the three LS's reported in Table 3.4. The table reports also the MAF thresholds for this school building (Class III structure, Table 3.1). As it can be seen, for the considered example the MAFs from the two modelling approaches are practically coincident for all LSs.

In conclusion, the example shows that the method is of relatively lengthy but rather straightforward application to real buildings, requiring in sequence a modal analysis, random sampling of model realizations, pushover analyses with invariant modal patterns, tri-linear approximation of capacity curves, expeditious IDA on equivalent SDOF oscillators, interpolation in the local response databases and CQC/SRSS combination, fragility parameters evaluation by simple statistical operations on the $S_{Y=1}$ intensity values. As long as MPA can provide a reasonable approximation of the dynamic response, Method B is a computationally effective alternative to Method A (IDA on complete model), since it requires determination and handling of much smaller response databases: where Method A requires determination of $n_{responses} \times n_{steps} \times n_{IM\text{-}levels}$ quantities per each record/model pair (with e.g. $n_{steps} = 2,000$ steps and $n_{IM\text{-}levels} = 10$), Method B requires determination of $n_{responses} \times n_{steps} \times n_{modes}$ quantities only (with e.g. $n_{steps} = 100$ steps and $n_{modes} = 3 \div 5$), since the IDA is carried out on a SDOF oscillator.

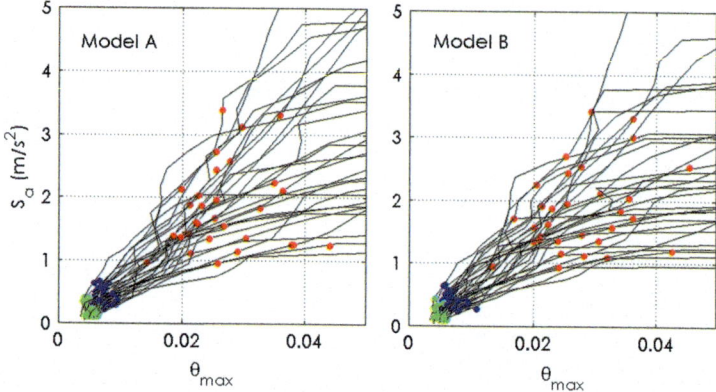

Fig. 3.21 IDA curves with indication of intensity leading to each LS for all records

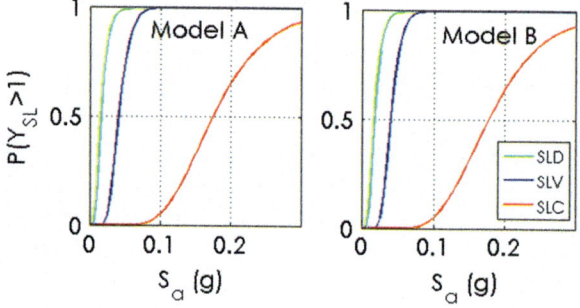

Fig. 3.22 Fragility curves

Table 3.4 Mean annual frequencies of LS exceedance for the two models and corresponding thresholds

Model	A	B	Threshold
λ_{SLD}	0.03150	0.03040	0.0300
λ_{SLS}	0.01270	0.01310	0.0032
λ_{SLC}	0.00119	0.00117	0.0015

3.5 Conclusions

The paper illustrates the latest Italian provisions issued by the National Research Council as Technical document 212/2013, for the probabilistic seismic assessment of existing RC and masonry buildings. These provisions are thought to overcome the limitations of the current normative, though they are not intended to replace them but, rather, to provide higher-level methods and tools for special applications and to inform possible revisions of the code in the future. The main merits of the document are:

(a) The systematic treatment of the problem of identification of global LS exceedance, in a manner consistent with their verbal description, with the introduction of LS indicator variables differentiated as a function of LS and modelling option.
(b) The explicit probabilistic treatment of all uncertainties, related to ground motion, material properties, modelling, geometry, detailing. In particular, the distinction of uncertainties that can be described within a single structural model via random variables and uncertainties that require the use of multiple models (logic tree) is introduced.
(c) The mandatory use of nonlinear analysis methods for response determination, and of ground motion time-series (preferably natural recorded) for the description of the seismic motion variability, irrespectively of the analysis method (dynamic or static).

Acknowledgements This paper focuses on the general part and on RC-specific provisions of the CNR-DT212 document, and includes an example application to a real RC building. Prof. Sergio Lagormarsino and Dr. Serena Cattari from University of Genoa have contributed to the drafting of the document, in particular for the masonry-specific provisions and application. The production of the document has been funded by Department of Civil Protection under the grant "DPC-Reluis 2010–2013" to the Italian Network of University Laboratories in Earthquake Engineering (ReLUIS).

Open Access This chapter is distributed under the terms of the Creative Commons Attribution Noncommercial License, which permits any noncommercial use, distribution, and reproduction in any medium, provided the original author(s) and source are credited.

References

Biskinis D, Fardis MN (2010a) Flexure-controlled ultimate deformations of members with continuous or lap-spliced bars. Struct Concr 11(2):93–108
Biskinis D, Fardis MN (2010b) Deformations at flexural yielding of members with continuous or lap-spliced bars. Struct Concr 11(3):127–138
Bradley B (2013) Ground motion selection for seismic risk analysis of civil infrastructure. In: Tesfamariam S, Goda K (eds) Handbook of seismic risk analysis and management of civil infrastructure systems. Woodhead Publishing Ltd, Cambridge. ISBN 978-0-85709-268-7
Chopra AK, Goel RK (2002) A modal pushover analysis procedure for estimating seismic demands for buildings. Earthq Eng Struct Dyn 31(3):561–582
Cornell C et al (2002) Probabilistic basis for 2000 SAC Federal Emergency Management Agency Steel Moment Frame Guidelines. J Struct Eng 128(4):526–533
Dolsek M, Fajfar P (2005) Simplified non-linear seismic analysis of infilled concrete frames. Earthq Eng Struct Dyn 34(1):49–66
Elwood K (2004) Modelling failures in existing reinforced concrete columns. Can J Civ Eng 31:846–859
fib, International Federation of Structural Concrete (2013) Model code for concrete structures 2010. Ernst & Sohn, Berlin. ISBN 978-3-433-03061-5
Han S, Chopra A (2006) Approximate incremental dynamic analysis using the modal pushover analysis procedure. Earthq Eng Struct Dyn 35:1853–1873

Haselton CB, Deierlein GG (2007) Assessing seismic collapse safety of modern reinforced concrete moment frame buildings, Blume center Technical report 156

Haselton CB, Liel AB, Taylor Lange S, Deierlein GG (2008) Beam-column element model calibrated for predicting flexural response leading to global collapse of RC frame buildings, PEER report 2007/03

Ibarra LF, Medina RA, Krawinkler H (2005) Hysteretic models that incorporate strength and stiffness deterioration. Earthq Eng Struct Dyn 34(12):1489–1511

Jalayer F, Franchin P, Pinto PE (2007) A scalar damage measure for seismic reliability analysis of RC frames. Earthq Eng Struct Dyn 36:2059–2079

Lignos D, Krawinkler H (2012) Sidesway collapse of deteriorating structural systems under seismic excitations. Blume Center technical report 177

Lin T, Haselton CB, Baker JW (2013) Conditional spectrum-based ground motion selection. Part I: hazard consistency for risk-based assessments. Earthq Eng Struct Dyn 42(12):1847–1865

McKenna F, Scott MH, Fenves GL (2010) Nonlinear finite-element analysis software architecture using object composition. ASCE J Comput Civ Eng 24(1):95–107

Pinto PE, Giannini R, Franchin P (2004) Seismic reliability analysis of structures. IUSS Press, Pavia. ISBN 88-7358-017-3

Reyes J, Chopra A (2011) Three dimensional modal pushover analysis of buildings subjected to two components of ground motion, including its evaluation for tall buildings. Earthq Eng Struct Dyn 40:789–806

Sezen H, Mohele JP (2004) Shear strength model for lightly reinforced concrete columns. ASCE J Struct Eng 130(11):1692–1703

Vamvatsikos D, Cornell CA (2002) Incremental dynamic analysis. Earthq Eng Struct Dyn 31(3):491–514

Vamvatsikos D, Cornell C (2005) Direct estimation of seismic demand and capacity of multidegree-of-freedom systems through incremental dynamic analysis of single degree of freedom. J Struct Eng 131(4):589–599

Zhu L, Elwood K, Haukaas T (2007) Classification and seismic safety evaluation of existing reinforced concrete columns. J Struct Eng 133(9):1316–1330

(NTC2008) Ministero Infrastrutture, 2008. D.M.14/1/2008 "Norme Tecniche per le Costruzioni" (Testo integrato con la Circolare n°617/C.S.LL.PP. del 2/2/2009)

(EC8-3) Comité Européen de Normalisation (2005) Eurocode 8: design of structures for earthquake resistance – Part 3: assessment and retrofitting of buildings

Chapter 4
Seismic Response of Precast Industrial Buildings

Matej Fischinger, Blaž Zoubek, and Tatjana Isaković

Abstract The most common structural system of the precast industrial buildings in Europe consists of an assemblage of cantilever columns tied together with beams. Typical beam-to-column connection in these structures is constructed with steel dowels. Although this system has been used for decades, its seismic response was poorly understood, which reflected in ambiguous code requirements and conservative approach. Therefore, along with innovative precast structural solutions (not discussed in this paper), this system was the main focus of the continuous European research in the past two decades. The key results of this vast research effort (including unprecedented cyclic, PSD and shake table experiments on large-scale structures) led by the associations of the precast producers in Europe and the Politecnico di Milano are presented and discussed in this paper. The details are provided for the work done at the University of Ljubljana. The results of these research projects led to some major modifications and improvements of the relevant chapter in Eurocode 8, when this was evolving from the initial informative annex to the final code provision. Refined FEM models for the complex behaviour of the dowel beam-to-column connections as well as macro models for the post-critical analysis of the complete structures were proposed. Single-storey and multi-storey structures were investigated and the design formulas to estimate high shear and storey-force amplification due to higher-modes effect in multi-storey structures were derived. The design guidelines for connections of precast structures under seismic actions were prepared. Systematic risk studies were done indicating that this structural system can be safe in seismic regions if all Eurocode 8 provisions as well as the recommendations based on the presented research are considered. These include the capacity design of the connections. Behaviour factor for such precast systems was studied and the values initially proposed in preEC8 were modified

M. Fischinger (✉) • B. Zoubek • T. Isaković
Faculty of Civil and Geodetic Engineering, University of Ljubljana, Ljubljana, Slovenia
e-mail: mfischin@ikpir.fgg.uni-lj.si; matej.fischinger@fgg.uni-lj.si; blaz.zoubek@fgg.uni-lj.si; tatjana.isakovic@fgg.uni-lj.si

(increased). However, it was shown that drift limitations typically govern the design and that the nominal value of the behaviour factor is not so decisive. The key factors contributing to the good seismic behaviour of this system (assuming that the connections are properly designed) is the low value of the compressive axial force in the columns confined with adequate hoops and the overstrength caused by drift limitation requirements. Cladding-to-structure interaction has been one of the most poorly understood components of the system, which is now the topic of the on-going research.

4.1 Introduction

Seismic behaviour and seismic safety of precast structures has been frequently discussed. However, when such discussion refers to precast structures in general, it is pretty much displaced and meaningless. Precast buildings are defined as structures made of pre-fabricated elements assembled into the structural system on the construction site. Obviously the behaviour of such systems depends predominantly on the details of the connections, which may differ essentially from one to the other precast system. So, empirical evidence from the past earthquakes shows everything from good structural response (Fig. 4.1) to complete disasters (Fig. 4.2).

While the tragedy of the Spitak 1988 earthquake in Armenia (EERI 1989) imposed large distrust onto precast structures in general, it should be noticed that at the same time large panel precast structures behaved quite well in spite of the poor construction practice. Therefore any generalized and superficial conclusions that precast structures are bad or good are non-professional and unacceptable. We should be fully and constantly aware that even a single life, which might be lost in the structures designed by ourselves or by the codes developed by us imposes a huge moral obligation onto us.

For these reasons the specific precast system, discussed in this paper, was extensively studied for two decades. Based on these results the relevant sections of Eurocode 8 were substantially modified and hopefully improved. The overview of the main research results is given in the following sections. Although, in general the observations are positive, one should be aware of the strict design requirements that are needed to ensure good performance. It is hoped that this presentation will give better insight into the seismic response and behaviour of this frequently used precast system, which is required for the objective evaluation of its performance.

Simply speaking, the analysed system consists of an assemblage of cantilever columns tied together with beams (Figs. 4.3 and 4.4).

The key element in this system is the beam-to-column connection. Among many different solutions the dowel beam-to-column connection is most frequently used (Fig. 4.5). This connection is practically hinged and the system indeed behaves as an assemblage of cantilever columns tied by beams.

4 Seismic Response of Precast Industrial Buildings

Fig. 4.1 Undamaged structural system of the precast industrial building after the L'Aquila 2009 earthquake

Fig. 4.2 Large panel precast structure standing among the rouble of the precast frames, which caused a terrible tragedy during the Spitak 1988 earthquake

This precast system has been used in Europe to construct about 50 millions of square meters of buildings per year. Such buildings house a predominant share of industrial facilities in many European countries. Recently they are also used for multi-storey office buildings and shopping centres housing thousands of people (Fig. 4.6). Therefore the potential seismic risk is high. However, due to the complex and complicated seismic behaviour of these buildings our knowledge is still limited and the design practice and codes need further improvements.

The paper is built predominantly on the research results gained within several large EU projects organized during the past two decades. The authors have been actively participating in these projects and cooperating with large consortia

Fig. 4.3 Schematic representation of the structural system of an industrial building

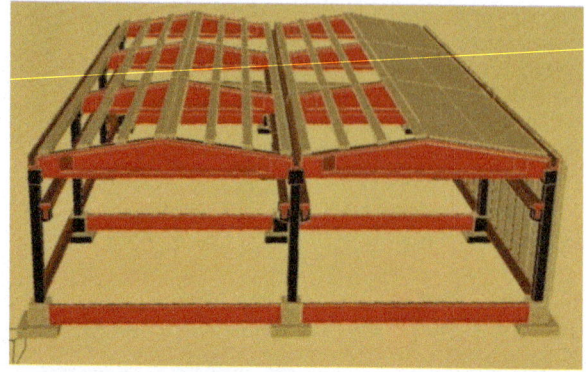

Fig. 4.4 Structural system consists of an assemblage of cantilever columns tied together with beams and floor structures

Fig. 4.5 Beam-to-column dowel connection is clearly seen in the upper floor. In the lower floor the beam has been already installed and the steel sockets will be grouted

Fig. 4.6 The analysed precast system is frequently used for large multi-storey buildings. The picture shows a huge shopping centre in Ljubljana to be visited by several 10,000 visitors a day

(Fischinger et al. 2011b) of European Associations of precast producers, enterprises and research institutions. While the results were always discussed within the consortia and the conclusions were typically agreed by all participants, the opinions and conclusions presented in this paper are those of the authors and do not always reflect the views of all the partners.

Most important general results of these projects are only briefly summarized and they are used as the framework of the paper. The details are then provided for the work done at the University of Ljubljana with the particular emphasis on the response of beam-to-column dowel connections and cladding-to-structure connections, inelastic response analysis of precast industrial buildings, behaviour factors and higher modes effects in multi-storey buildings. The most important result of this research has been the improvement of the design practice governed by the modified provisions in the relevant chapter of Eurocode 8 (CEN 2004), which has been immediately applied by sponsoring associations and companies.

4.2 Post-Earthquake Inspections

In spite of the frequent use of the analysed precast system, the information about its behaviour during the earthquakes has been sparse and sometimes controversial (see also the Introduction). Although good structural behaviour prevails (Fig. 4.7), it sometimes goes hand in hand with collapses. Again one should pay attention on seemingly small but important details. During Friuli earthquake, good behaviour was observed (Fajfar et al. 1978; EERI 1979). However, in Friuli quite long period structures were exposed to short, high-frequency ground motion with relatively weak energy content and low displacements in the range of predominant structural periods. During the recent Emilia earthquakes, which occurred near-by, a lot of damage was reported (i.e. Bournas et al. 2013a). But here, most of this damage

Fig. 4.7 Structural system of most precast industrial buildings designed for seismic action remained undamaged during recent Italian earthquakes

Fig. 4.8 Many cladding panels fell off the structures during recent Italian earthquakes (Toniolo and Colombo 2012)

should be attributed to the fact that the majority of the buildings were not designed for the earthquake action (this region has been considered as seismic only from the year 2003 on). However, there were also some collapses in new buildings. A strong low frequency content of the N-S component of the May 29th earthquake might contribute to the increased damage. Similar reason might increase damage in the case of Vrancea earthquake (Tzenov et al. 1978) and during some Turkish earthquakes.

After the other recent Italian earthquake – L'Aquila, good structural behaviour of the precast industrial buildings was reported (Figs. 4.1 and 4.7; Toniolo and Colombo 2012). But in both, Emilia and L'Aquila earthquakes heavy damage to claddings was observed (Fig. 4.8; Toniolo and Colombo 2012). The problem of claddings will be discussed in a separate chapter.

During the Montenegro earthquake (Fajfar et al. 1981) damage to precast structures was small and it was predominantly due to the soil effects and the rotation of foundations (Fig. 4.9). After the recent earthquakes in Turkey (Saatcioglu et al. 2001; EERI 2000; Arslan et al. 2006) statistics show small, but nevertheless considerable number (3 % of the total inventory) of collapses and heavy damage (Fig. 4.10).

Fig. 4.9 Collapse of the roof structure in Montenegro earthquake due to soil effects and poor connections

Fig. 4.10 This collapsed precast structure illustrates the importance of ties in precast buildings

Fig. 4.11 Collapse of the beam during Emilia earthquake due to the loss of the seating; general view – *left* and the detail of the support – *right* (Bournas et al. 2013a)

Fig. 4.12 Collapse of the roof system during Montenegro earthquake due to soil effects and poor connections

Regardless all the differences in observations, the main causes of the damage to the investigated precast system were similar in all cases and all countries:

- Failure of the connections, as the main cause of damage and collapse (Figs. 4.2, 4.9, 4.11 and 4.12);
- Lack of mechanical connections between the columns and roof girders in old buildings and in supposedly aseismic regions (Bournas et al. 2013a);
- Lack of ties (Fig. 4.10);
- Insufficient in-plain stiffness of the roof/floor structures;

Fig. 4.13 Displacements are typically very large

- Torsional response due to asymmetric stiffness distribution;
- Poor detailing of hoop reinforcement in columns (Figs. 4.14 and 4.15);
- Unpredicted large displacements (Fig. 4.13) associated with too-short seating and poor connections;
- Poor foundation in soft soil (Fig. 4.12);
- Detachment of claddings (Fig. 4.8).

4.3 Past Research – General Overview

Compared to cast-in-place structures, all types of precast structures have received relatively little attention which has reflected in slow development of codes. In particular, precast industrial buildings, which are discussed in this paper, are not used in some countries (USA, Japan, New Zealand) that lead in earthquake engineering. Research there has predominantly addressed systems with flexural-resistant and prestressed connections (i.e. PRESS – PREcast Seismic Structural System; Priestley 1996; Shiohara and Watanabe 2000). Consequently, there has been very little information about the precast industrial buildings in the state-of the-art (at the time of publication) reports as it was the ATC-8 action – Design of prefabricated buildings for earthquake loads (ATC-08 1981). More recent report (FIB 2003) of the fib-Task group 7.3 (the first author of this paper was a member of

Fig. 4.14 Collapse of a precast column due to poor confinement

Fig. 4.15 The mistake in the construction of the hoops (and the resulting impaired confinement) led to heavy damage of the precast column in the PRECAST full-scale test

the group) contains some, but still very limited information. Surprisingly so, in the past even the major Balkan research project (UNDP/UNIDO 1985) addressed predominantly the large panel precast systems with very little attention to precast industrial buildings.

Due to the poor knowledge the only possible and right solution in the early developments of Eurocode 8 was to adopt quite conservative approach for seismic design of precast (industrial) buildings. Simply speaking, practically elastic response was required for the analysed system (see also the discussion of the behaviour factors in Sect. 4.10). This was a great shock for the industry, used to the same or at least similar seismic forces and structural details in precast and cast-in-situ structures. The authors of this paper fully support the idea that energy dissipation capacity of any precast system to be used in mass production should be first experimentally and analytically verified. For the precast system addressed in this document, systematic verification has been done within five large research projects (Toniolo 2012):

- Cyclic and PSD tests of precast columns in socket foundations (ASSOBETON; 1994/96)
- Comparison of the seismic response of the precast and cast-in-situ portal frame (ECOLEADER project; 2002/03)

- PRECAST – Precast Structures EC8; 2003/06
- SAFECAST – Performance of Innovative mechanical Connections in Precast Building Structures under Seismic Conditions; 2009/12
- SAFECLADDING – Improved Fastening systems of Cladding Wall Panels of Precast Buildings in Seismic Zones; on-going project; 2013/15

All these projects were sponsored by the associations of precast producers and SMEs in Italy, Greece, Turkey, Portugal, and Spain, which demonstrates the interest of the industry in this topic. The research providers (European Laboratory for Structural Assessment, JRC-ELSA; Istanbul Technical University – ITU; LABOR; National Laboratory of Civil Engineering in Lisbon, LNRC; National Technical University of Athens, NTUA; Politecnico di Milano., POLIMI and University of Ljubljana; UL) were coordinated by The Politecnico di Milano under the scientific leadership of Professor Giandomenico Toniolo. The key activities and results of these projects are very briefly introduced in the next section, followed by more detailed description of the selected results contributed by the research group in Ljubljana.

4.4 European Research in Support of the Eurocode-8 Developments

4.4.1 Cyclic and PSD Tests of Precast Columns in Socket Foundations (ASSOBETON)

The aim of the research (Saisi and Toniolo 1998) was to investigate the ductility and energy dissipation capacity of precast columns at realistic foundation conditions (Fig. 4.16). Substantial ductility (3.5–4.5) was demonstrated. This is to be expected for an element with relatively low compressive axial force (typical for the columns in one-storey industrial buildings), symmetric reinforcement and considerable confinement. Good behaviour was further enhanced due to the absence of the splice in the critical region and construction in controlled environment. It should be noted that the larger ductility displacement value was achieved only if the spacing of the hoops in the critical region was 3.5 times of the longitudinal bar diameter (about 7.5–10 cm). It is interesting to note that this complies with the practice in Slovenia (former Yugoslavia) after the Montenegro earthquake. On the other hand the valid Eurocode requirements can be less stringent.

Fig. 4.16 ASSOBETON tests on precast columns with pocket foundation performed at ELSA

4.4.2 Comparison of the Seismic Response of the Precast and Cast-In-Situ Portal Frame (ECOLEADER)

The project (Ferrara et al. 2004; Biondini and Toniolo 2002; Biondini and Toniolo 2004), approved for ECOLEADER (European Consortium of Laboratories for earthquake and Dynamic Experimental research) funding, was aimed at demonstrating the practical equivalence between the behaviour factor of precast and cast-in-situ single-storey industrial buildings (Fig. 4.17).

Both structures were designed to have the same fundamental period. Quite similar behaviour of both structures was observed – Fig. 4.18

This supports the supposition that the same behaviour factor can be used for the precast and cast-in-situ structure of this type (Biondini and Toniolo 2002, 2004). But this result by itself does not mean, in any case, that either of the structures had

4 Seismic Response of Precast Industrial Buildings

Fig. 4.17 Precast (*left*) and cast-in-situ (*right*) ECOLEADER prototypes during testing

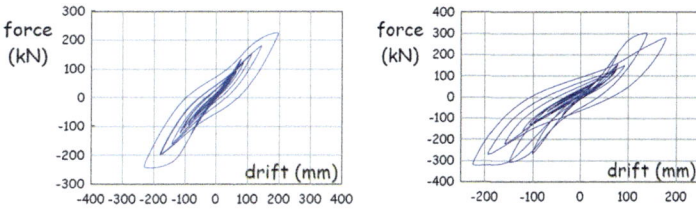

Fig. 4.18 The PSD response of the precast (*left*) and cast-in-situ (*right*) ECOLEADER prototypes

the same energy dissipation capacity as the multi-storey, multi-bay frame designed with the weak-beam-strong-column concept (see discussion in Sect. 4.10). Therefore further research of this important issue was needed.

4.4.3 PRECAST – Seismic Behaviour of Precast Concrete Structure with Respect to EC8

The goal of the PRECAST project (Toniolo 2007) was to assess (experimentally and numerically) and to calibrate the design rules for (industrial) precast concrete structures in Eurocode 8. It was a logical continuation of the ECOLEADER project. Similar to ECOLEADER a full scale one-storey precast structure (Figs. 4.19a, b) was tested with PSD and cyclic experiments. However, this structure, supported by six 5 m high columns, had two bays and realistic floor/roof structure (the one in ECOLEADER was rigid slab) composed with slab panels, once oriented in the direction of the loading (Fig. 4.19b), and the other time perpendicular to this direction (Fig. 4.19a). In initial – elastic tests, cladding panels were added (Fig. 4.19a), which were then removed at higher levels of loading. Tolmezzo record modified to fit EC8 (soil B) spectrum with peak ground acceleration 0.05, 0.14, 0.35 g (design acceleration) and 0.525 g was used in tests. PRECAST project

Fig. 4.19 (**a**) PRECAST EC8 building with cladding panel; load perpendicular to the slab panels. (**b**) PRECAST EC8 building; claddings removed; load in the direction of panels

provided valuable information about the seismic response, which was subsequently used in numerical analyses (see Sect. 4.5) and systematic risk studies (see Sect. 4.9).

Most important results of the project were:

– The structure had large overstrength. Yielding in the columns was not observed until the last PSD test with maximum ground acceleration 0.525 g. Only much stronger cyclic loading, applied at the end of testing, imposed near collapse mechanism. It should be noted, however, that even this very large structure had still smaller spans (mass) compared to those in real structures.
– Therefore a systematic numerical study was done showing good response and acceptable risk for a whole set of realistic one-storey structures used in practice (Kramar et al. 2010a).
– Extremely large top displacements (8 % drift or 40 cm) were recorded at the ultimate stage. As a surprise yield drift was over 2 %. See more details in the following Sect. 4.5, discussing the inelastic model for the columns.
– Seemingly quite flexible floor structure worked pretty much as a rigid diaphragm, regardless of the orientation of the floor panels
– Cladding panels changed the response significantly

4.4.4 SAFECAST – Performance of Innovative Mechanical Connections in Precast Building Structures Under Seismic Conditions

As discussed in the previous section, PRECAST project demonstrated good seismic performance of one storey precast industrial buildings. However, this result was still far for being conclusive. First of all, it has been obvious that realistic behaviour of connections determines the response of any precast structure. And even the capacity of most commonly used connections was not known. Furthermore, the

Fig. 4.20 The SAFECAST structure tested at ELSA

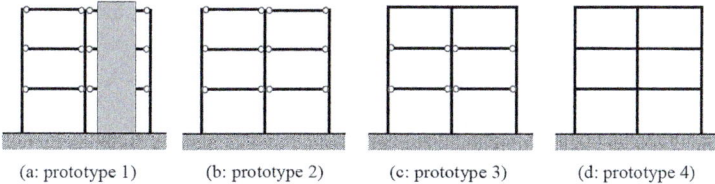

Fig. 4.21 Structural layouts of the four prototypes (Bournas et al. 2013b)

inelastic response and the behaviour of multi-storey structures were far from being understood. To fill these gaps in the knowledge project SAFECAST was initiated (Toniolo 2012). Together with many parallel tests on individual existing and innovative connections, the main experimental research consisted of PSD tests on a full-scale 3-storey precast concrete building, performed at ELSA (Fig. 4.20; for the details of the structure see Negro et al. 2013 and Bournas et al. 2013b).

To use the expensive specimen as efficiently as possible, for different structural solutions were tested one after another (Fig. 4.21). The level of damage in the columns was limited to enable to carry out this sequence of tests.

Structural system of the prototype 2 complied with the definition given in the introduction of this paper (an assemblage of multi-storey columns with all hinged beam-to-column joints). In prototype 1 two symmetrically placed precast structural walls were added to stiffen the system. These walls were disconnected after the test of the prototype 1. In prototype 3 it was attempted to reduce the flexibility of the system by making the joints in the top story moment-resistant. Innovative dry connections were installed and activated for this purpose. In prototype 4 all joints were subsequently made moment-resistant. Different floor diaphragms were used in each floor. Box type elements were used in the first floor. Pre-topped double tee diaphragm was used in the second floor. Separated slab elements were installed in the third floor to simulate openings in the roofs used for architectural reasons. The same ground motion as in the case of PRECAST was used. Prototypes 1 and 2 were exposed to maximum ground accelerations 0.15 and 0.30 g, prototype 3 to $a_{gmax} = 0.3$ g and prototype 4 to $a_{gmax} = 0.3$ g and 0.45 g.

Only brief overview of the key observations and conclusions is given below:

- Overall, the response of the structure was good up to the ultimate limit state design levels.
- However, extremely high influence of higher mode effects was observed in prototype 2 (see Sect. 4.8 for more detailed analysis). This imposes very high seismic storey forces and the demand on joints. This demand would not be identified by traditional design. The structure was very flexible with inter-story drifts up to 2.4 %. It is believed that such multi-storey precast structures are difficult to be designed without some kind of stiffening measures.
- The use of precast structural walls in prototype 1 reduced the maximum inter-story drift to 0.7 %. At the same time the rigid diaphragm action was not impaired (with a certain exception of the top story with separated floor panels). But it should be noted that the walls (with the same stiffness) were placed symmetrically in the floor plan. Asymmetry in real design may impose significant torsional effects and large forces to transfer through the floor structures into the walls.
- Making moment-resistant connections only at the top floor in the prototype 3 (which could be a practical solution in real life) reduced the fundamental period for only 23 % in comparison with the structure with hinge joints.
- The solution in prototype 4 was more efficient. However, the innovative dry joints were only semi-rigid (large slips were observed due to the problems in technology of construction)
- The tests provide valuable data for numerical modelling

SAFECAST project provided important knowledge about the strength and deformation capacity of the most common types of connections used in the design practice (in particular beam-to-column connection, which will be discussed in more detail in Sect. 4.6). Additionally many innovative connections were proposed and tested (these results cannot be published here).

The most important outcome of the SAFECAST project, based on the mutual effort of all the partners in the consortium, is a set of design guidelines for connections of precast structures under seismic actions (Negro and Toniolo 2012). It is hoped that this document (or at least parts of it) will be subsequently incorporated into Eurocode 8 provisions.

These guidelines are based on the experience obtained by testing a large number of different typical connections. However, it is obvious that there are many different variations and even completely different types of connections used in the construction practice. Therefore, one should be extremely careful when extrapolating the design guidelines to other types of connections (more detailed discussion is given in Sect. 4.10) (Fig. 4.22).

Fig. 4.22 Most important result of the SAFECAST project

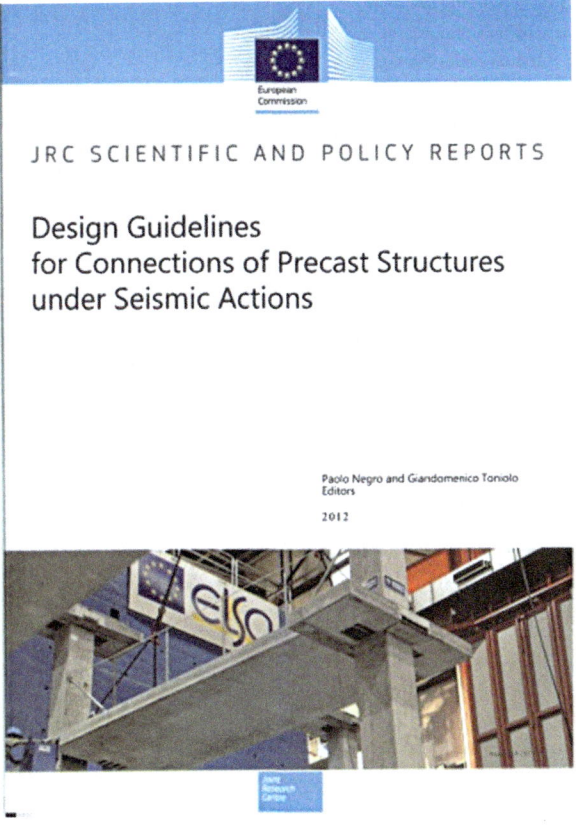

4.4.5 SAFECLADDING – Improved Fastening Systems of Cladding Wall Panels of Precast Buildings in Seismic Zones

SAFECAST project indicated that the most poorly (in fact wrongly) understood connection in the precast industrial buildings is the cladding-to-structure connection. It was traditionally supposed that the existing connections separate the cladding panel from the structure. Panels were usually considered only as added mass in the structural model. Therefore these connections were designed for the inertia forces contributed by the mass of the panel only as well as only in the direction perpendicular to the panel. However, in many cases traditional connections cannot accommodate the very large relative displacements between the structure and the panels. In such a case the panel and the columns begin to act together as a single structure. The connections are then loaded by inertia forces contributed by the total mass of the structure, which act in the plane of the panel. This observation was drastically confirmed during the recent L'Aquila and Emilia earthquakes (Figs. 4.8 and 4.23).

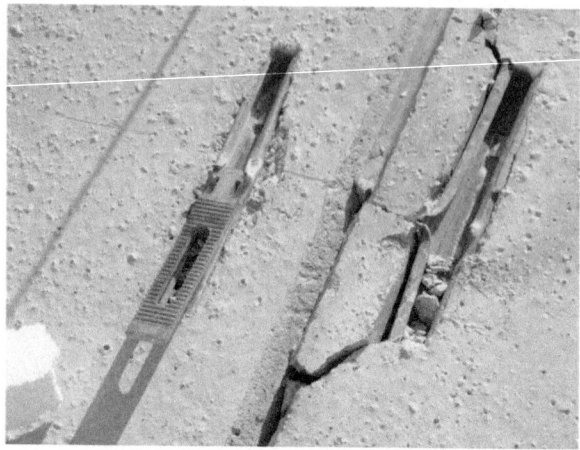

Fig. 4.23 Failure of the traditional hammer-head connection during the Emilia earthquake

In SAFECLADDING project four different solutions to this problem are being proposed and analysed:

- Additional research is under way to understand better the capacity and the demand in the case of the existing connections. This will improve the design practice. However, it is expected that in cases of strong earthquakes the collapse of existing types of connections cannot be always prevented. In such cases second line of defence measures will be proposed. This research will be presented in more detail in Sect. 4.7.
- New connections allowing for larger relative displacements will be proposed.
- Integrated (dual) systems are studied. In these studies panels are designed as a part of the lateral resisting (dual) system
- Dissipative connections seemed to be very promising solution.

In addition to a large number of tests on the individual types of connections the key experiment will be again performed at ELSA. A sequence of 22 tests are planned to be performed on a single-storey two-bay full-scale structure. In each test the arrangement of panels and the type of the connections will be varied.

4.5 Modelling of the Inelastic Seismic Response of Slender Cantilever Columns

A slender cantilever column may represent a class of one-storey industrial buildings with strong connections. Therefore, we start more detailed discussions of the research done with the presentation of the inelastic model for slender columns. Whatever, this problem might appear trivial and several extensive data bases (PEER 2007; Panagiotakos and Fardis 2001) related to the cyclic behaviour of RC columns exist,

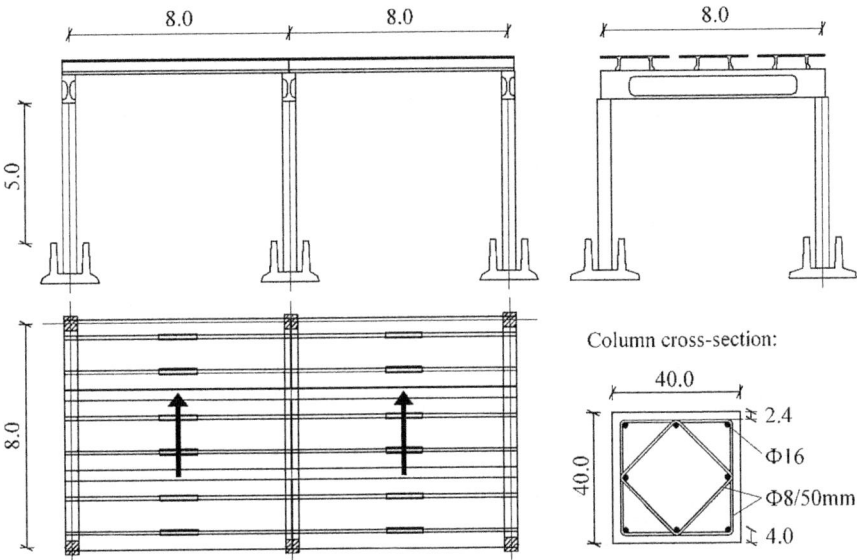

Fig. 4.24 Plan of the analyzed PRECAST structure, showing the typical column cross-section

practically no information was available about the behaviour of very slender cantilever columns having shear span ratios of more than ten. This is especially true for the post-peak behaviour at large drift ratios, which should be understood and clearly defined when using up-to-date performance-based procedures and seismic risk studies. Therefore, a numerical model based on the full-scale PSD and cyclic tests done at ELSA (see Sect. 4.4.3) within the PRECAST project (Toniolo 2007) was proposed and verified by the research team at UL (Kramar 2008; Fischinger et al. 2008).

The plan of the tested structure (Fig. 4.19; walls were disconnected) is given in Fig. 4.24. The shear-to-span ratio of the columns was 12.5. They were designed according to the EC8 standard. The study was later extended to the lightly reinforced columns, not observing the minimum requirements of EC8.

Very specific behaviour of the columns with high shear-span ratio was observed (Fischinger et al. 2008). The deformability and the deformation capacity of the columns were large (Fig. 4.25). The yield drift was 2.8 % (much more than the values reported for columns with smaller shear-spans). In the final cyclic test, the columns exhibited quite stable response up to a large drift close to 7 %. Buckling of the longitudinal reinforcement bars then led to subsequent tension failure of the bars in the first column. The strength of the structure dropped considerably, but it was stabilized by the other five columns. A 20 % drop in maximum strength was observed at about 8 % drift, following considerable in-cycle strength degradation and the flexural failure of several columns. Very short height of the plastic hinge (only half of the cross-section dimension of the column) was observed.

The beam-column model with lumped plasticity was chosen. However, most existing hysteretic models had problems to describe the observed behaviour. The best results were obtained using Ibarra hysteretic model (Ibarra et al. 2005) that

Fig. 4.25 The ultimate drift of 8 % (top displacement equal to 40 cm) was observed in PRECAST full-scale test

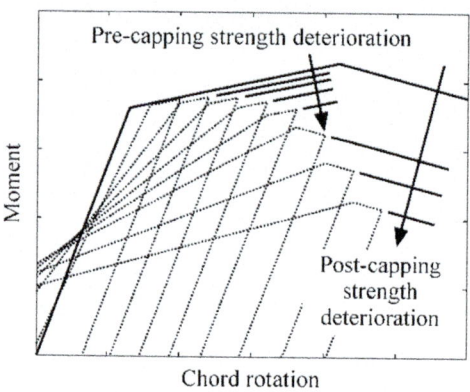

Fig. 4.26 Strength deterioration in the Ibarra's model

accounts for history-dependent strength and stiffness deterioration. The behaviour is first described by a monotonic backbone curve. Pre-capping and post-capping cyclic strength deterioration, based on the energy dissipation criterion, is then considered (Fig. 4.26). Haselton (2006) has calibrated Ibarra hysteretic model for a large number of column tests. If Haselton expressions, except for the yield drift (which was determined analytically taking into account empirical corrections for pull-out and shear-slip), were used, the match of the analytical and experimental results was very good (Figs. 4.4 and 4.27).

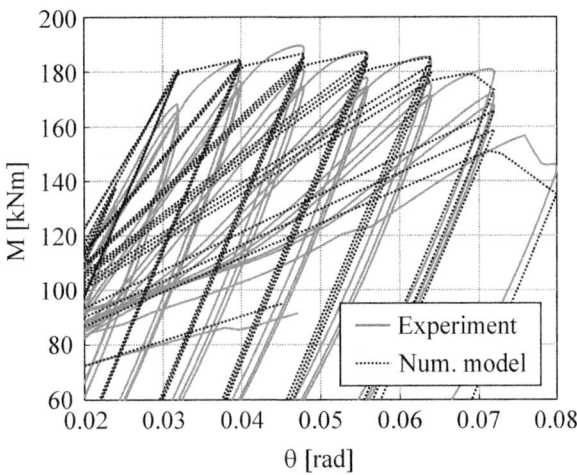

Fig. 4.27 Numerical versus experimental results

4.6 Cyclic Response of Beam-to-Column Dowel Connections

Beam-to-column connections are extremely important for the integrity and safety of the precast industrial structure. The majority of collapses during earthquakes are due to the fall of the beam. Nowadays it is obvious that the connection should not rely only on the friction and that some kind of mechanical connection should be provided. The most common solution is the use of steel dowels (Fig. 4.28). This option has been used for decades. Nevertheless, the design (if done at all) was predominantly based on engineering feeling and the requirements of non-seismic loading. However, the correct approach would be the use of the capacity design, which is in fact required by Eurocode 8. For this we obviously need to know capacity of the dowel connection and the demand imposed during seismic action (the latter will be discussed in Sect. 4.8). Neither of them was understood enough. Therefore a good deal of the experimental and numerical research effort within SAFECAST was devoted to this connection. Static and cyclic tests at large relative rotations between the beam and column were done at UL in Ljubljana (Kramar et al. 2010b; Fischinger et al. 2012a, 2013) (Figs. 4.29 and 4.28), static, cyclic and shake table test were performed at NTUA in Athens (Fig. 4.30) and shake table tests were done AT LNEC in Lisbon (Fig. 4.31).

In this paper we present mainly the research performed in Ljubljana. Three types of connections were tested (Fig. 4.32): (1) single centric dowel (typical for roof beam to column connection), (2) single eccentric dowel (for comparison) and (3) two eccentric dowels (typical for floor beam to column connection).

While several experiments were done in the past to estimate the dowel strength (i.e. Vintzeleou and Tassios 1987) they were restricted to pure shear and specimen without hoop reinforcement. Special purpose of the tests at UL was to study the behaviour of the connections at very large relative rotation between the beam and the column (Fig. 4.25) observed in the previous PRECAST project. It should be

Fig. 4.28 Typical eccentric beam-to-column dowel connection, which was tested at UL

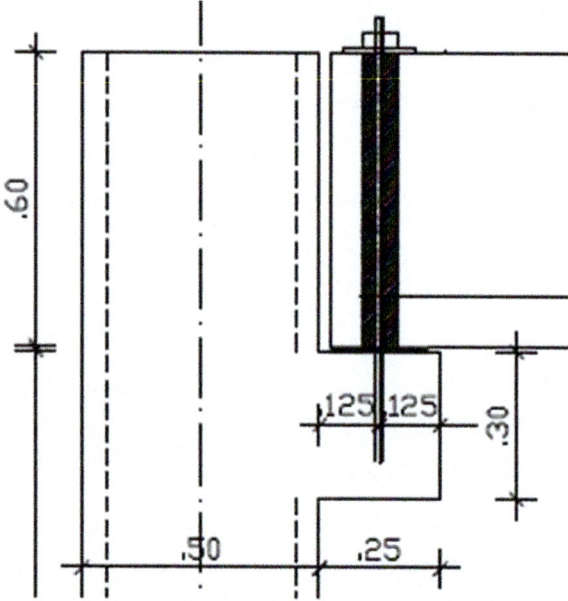

Fig. 4.29 Test of the dowel connection at large relative rotation between the beam and column

Fig. 4.30 Shake table test of the beam-to-column dowel connection at NTUA

Fig. 4.31 Shake table test of the beam-to-column dowel connection at LNEC

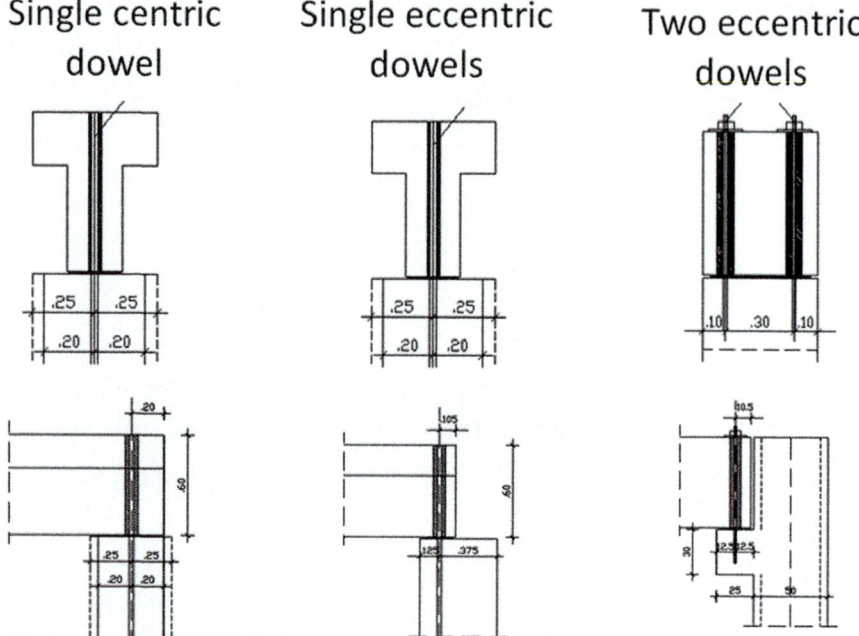

Fig. 4.32 Typical beam-to-column connections tested at UL

Fig. 4.33 Rupture of the dowel in the case of large concrete cover

noted that this large drifts are needed to justify the energy dissipation capacity (behaviour factor) assumed in the design.

Two types of failure of the investigated connections were identified: (a) the rupture of the dowel and crushing of the surrounding concrete (Fig. 4.33) and (b) the failure of the beam to column joint due to the insufficient tension strength of concrete and stirrups surrounding the dowel (Fig. 4.34).

Fig. 4.34 Failure in the case of small concrete cover and in the case of the insufficient tension strength of concrete and stirrups surrounding the dowel

The type of the failure and strength of the connections strongly depended on: (a) the distance of the dowel from the edge of the column and beam, (b) the amount of the provided stirrups in beam and column, and (c) the amount of relative rotations between the column and the beam. Due to the large relative rotations, the 20 % reduction of the strength of the connections was identified. In asymmetric connections the strength was also influenced by the direction of the loading, since the distance of the dowel from the edges of column and beam was different. It has been confirmed that the cyclic strength of the connections was 50–60 % of the strength measured during the monotonic tests (as it was noticed in the previous research). In the majority of cases, the formulas existing at the time of the experiment, which can be used to estimate the strength of the dowel itself, underestimated the actual strength. The difference between the predicted and actual strength was even larger in the case of other types of failure.

To understand the mechanism of the response better and to propose the design formulas and procedures, extensive numerical studies were done. FEM models were developed and used (Zoubek et al. 2013b) The models and the results are presented in the following subsections for (a) dowels embedded deep into the column's concrete core – large concrete cover ($c \geq 6d_d$; c is the dimension of the concrete cover and d_d is the diameter of the dowel) and (b) dowels placed close to the edge of the column – small concrete cover ($c \leq 6d_d$).

4.6.1 Capacity of the Beam-Column Connection with Dowels Embedded Deep in the Concrete Core

Behaviour of dowels embedded deep in the concrete core is mainly characterized by the dowel action mechanism for which numerical models have already been developed and experimentally tested in some previous studies (Dulascska 1972; Højlund-Rasmussen 1963; Engström 1990; Vintzeleou and Tassios 1986; Zoubek et al. 2013a, b). The simple models assume that the strength of the dowel is reached

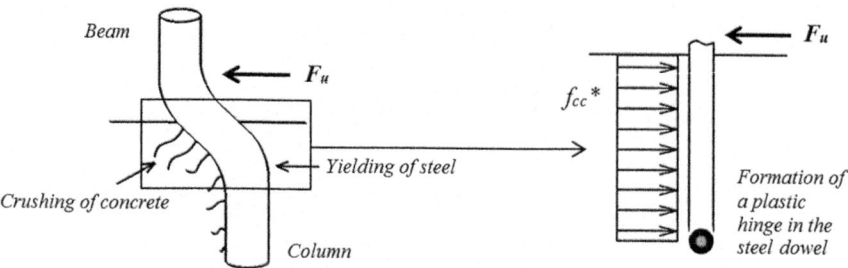

Fig. 4.35 Failure mode of the dowel mechanism

at simultaneous yielding of the dowel and crushing of the surrounding concrete (see Fig. 4.35).

Assuming the failure mechanism presented in Fig. 4.35, the following formula can be used to analytically evaluate the ultimate resistance of the dowel connection at monotonic loading:

$$R_{u,m} = F_{u,m} = \alpha_0 \cdot d_b^2 \cdot \sqrt{f_{cm} \cdot f_{ym}}$$

f_{cm} [MPa] ... mean uniaxial compressive strength of concrete

f_{ym} [MPa] ... mean yield strength of steel

d_b [mm] ... diameter of the dowel

Coefficient α_0 varies among different authors from 1.0 to 1.3 and mainly depends on the increase of the concrete compressive strength due to tri-axial state of stresses in front of the dowel (f_{cc}^* in Fig. 4.35).

In the case of cyclic loading the strength should be reduced because of the cyclic degradation of concrete and steel. Vintzeleou and Tassios (1986) suggested a reduction factor of 0.7 (0.5 for design purposes):

$$R_{u,c} = 0.7\, R_{u,m} = 0.95 \cdot d_b^2 \cdot \sqrt{f_{cm} \cdot f_{ym}}$$

Based on the results of the experiments performed in the frame of the SAFECAST project (Psycharis and Mouzakis 2012) proposed a modified formula, which accounts for cyclic behaviour of the realistic beam-to-column dowel connections:

$$R_{u,c} = C_0/\gamma_R \cdot d_b^2 \cdot \sqrt{f_{ck} \cdot f_{yk}},$$

where C_0 is a factor ranging between 0.9 and 1.1 and takes into account the influence of relative rotations between the beam and the column . Based on the

4 Seismic Response of Precast Industrial Buildings

tests performed at the University of Ljubljana (see Sect. 4.6) value of 0.9 should be used to account for the strength reduction due to the large relative rotations. γ_R is a general safety factor to account for the uncertainties in the experimental procedure and the limited number of experimental data used in the derivation of this formula. Value of 1.3 for γ_R was suggested in Psycharis and Mouzakis (2012). This formula was adopted in the Design recommendations (Negro and Toniolo 2012), but γ_R was not included.

This expression is predominantly empirical and no detailed analysis of the failure mechanism leading to this result was done within the SAFECAST project. Therefore the understanding of the behaviour was incomplete and consequently the generalization of the formula to the cases not tested within the project was difficult. To get a generally applicable tool to estimate the capacity of the dowel connections some sophisticated finite element analyses were performed to understand the behaviour in detail and to support the formula (Zoubek et al. 2013a, b). Good correlation between the numerical results and the values given by the formula has demonstrated the ability of the proposed numerical tool.

4.6.2 Capacity of the Beam-Column Connections with Dowels Placed Close to the Edge of the Column

In the case of dowel connections with dowels placed close to the edge, premature splitting of concrete can occur before the dowel mechanism, described above, can develop (Fig. 4.36).

This brittle failure was thoroughly investigated in Fuchs et al. (1995; Fig. 4.36a). Based on the extended experimental study, the following formula was proposed to predict the capacity of the eccentric anchor:

$$R_{no} = 1.0(d_d f_{cc})^{0.5}(l/d_d)^{0.2} c_1^{0.5},$$

where $1 < 8 \, d_d$ is the effective embedment depth and c_1 is the distance from the centre of the dowel to the free edge of the concrete element in the direction of loading. To take into account the dimensions of the concrete element and the eccentricity of loading in the case of a group of anchors (coefficients ψ_4 and ψ_5), the following correction of the resistance R_{no} is needed:

$$R_n = (A_v/A_{v0})\psi_4\psi_5 R_{n0},$$

where A_v is the actual projected area at side of concrete member, idealizing the shape of the fracture area of individual anchor as a half-pyramid with side lengths $1.5 \, c_1$ and $3c_1$ (Fig. 4.1), while A_{vo} is the projected area of one fastener unlimited by

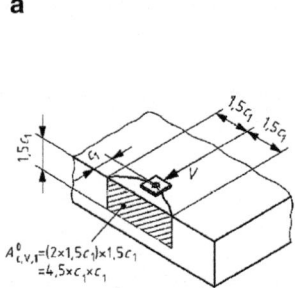

Fig. 4.36 (a) Simplified design model of the concrete failure zone for fasteners in a thick concrete member as proposed in CEN/TS (1992-4-2 2005) and Fuchs et al. (1995) and (b) failure of the eccentric beam-column dowel connection at the end of the cyclic test performed at the University of Ljubljana

corner influences, spacing or member thickness, idealizing the shape of the fracture area as a half-pyramid with side length 1.5 c_1 and $3c_1$. Similar formulas are also included in CEN/TS 1992-4-2 and *Design Guidelines for Connections of Precast Structures under Seismic Actions* (Negro and Toniolo 2012).

The presented closed expressions usually underestimate the capacity of the actual eccentric beam-to-column dowel connections due to the inadequate evaluation of the contribution of the confining reinforcement, which definitely helps to improve the integrity of the connection and prevent the brittle failure. In CEN/TS 1992-4-2 the resistance of the eccentric anchor is allowed to be increased by factor 1.4 if closely spaced stirrups are provided in the region around the connection. Even though the standard recognizes the importance of the confinement, the approach seems to be too simplified. The authors therefore suggest an alternative method. The capacity of the eccentric dowel connection should be estimated by appropriate usage of the Strut and tie model (Fig. 4.3). The compressive stresses in concrete are equilibrated with the tension stresses in the confining reinforcement. The assumed directions of the compression diagonals for the connection with one or two dowels were supported with the finite element model presented in Zoubek et al. (2013a, b, last column in Fig. 4.37).

The procedure was tested against the experimental results obtained within the SAFECAST procedure (Zoubek et al. 2014a, b). Very good match with the experimental results was demonstrated. It was also shown that the formulas proposed in CEN/TS (2005) greatly underestimate the capacity of the connections in the case of spalling of concrete edge.

4 Seismic Response of Precast Industrial Buildings

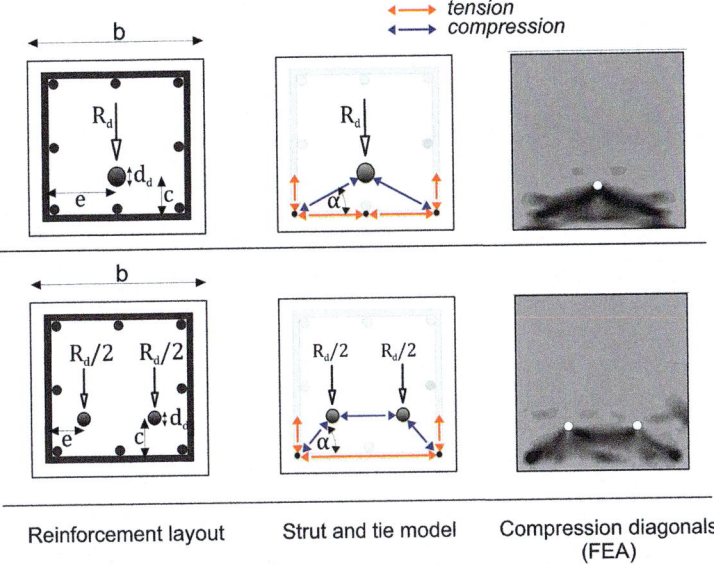

Fig. 4.37 Proposal for the calculation of the resistance of the eccentric dowel connection with one or two dowels using truss and tie model

4.7 Cyclic Response of Typical Cladding-to-Structure Connections

Cladding-to-structure connections have been among the less understood connections in industrial buildings. Actually the problem was typically avoided by assuming that claddings are separated from the structure. During stronger earthquakes the relative displacements are so large that this is definitely not true and very complex interaction is imposed. To analyse this interaction one must know the imposed demand as well as the capacity of the connection. Up to now the extensive study of the capacity of the most typical connections used in Europe was already completed within the SAFECLADDING project.

Typical mechanical connections, which are used to attach the cladding panels to the structural system of precast buildings depend on the orientation of the panels. Vertical as well as horizontal panels are widely used. Therefore, some typical representatives of both groups of connections were included in the plan of the experiments. Three types of mechanical connections, presented in Fig. 4.38 were tested.

In order to optimize the experiments as much as possible, the same setup (see Fig. 4.39) was used for all investigated connections (Fig. 4.39).

Altogether 30 tests were performed. In general three types of tests were accomplished:

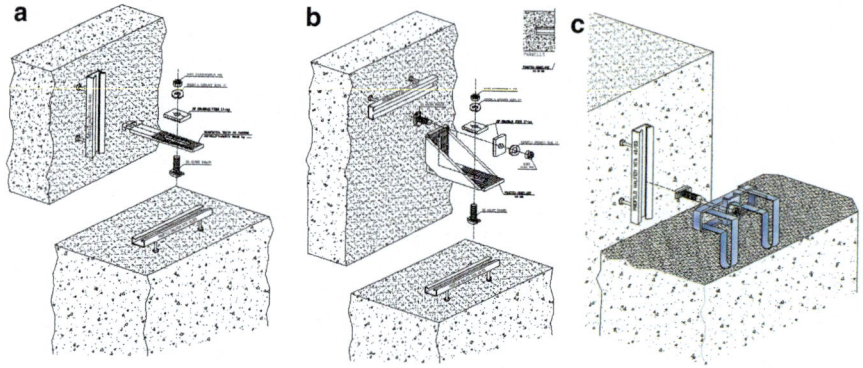

Fig. 4.38 Tested cladding-to-structure connections (**a**) Typical connection of the vertical panel and the beam. (**b**) Typical angle connection. (**c**) The connection, which is used to attach the horizontal panels to the columns

Fig. 4.39 Basic configuration of the setup

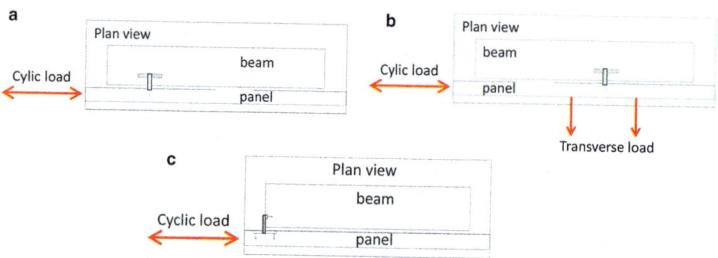

Fig. 4.40 Schemes of the tests. (**a**) Uniaxial shear test. (**b**) Biaxial shear test. (**c**) Uniaxial sliding test

– Uniaxial shear tests (see Fig. 4.40a): The load was applied in the horizontal direction in parallel to the longitudinal axis of the panel. The direction of the load was perpendicular to the channel mounted in the panel and perpendicular to the hammer-head strap.

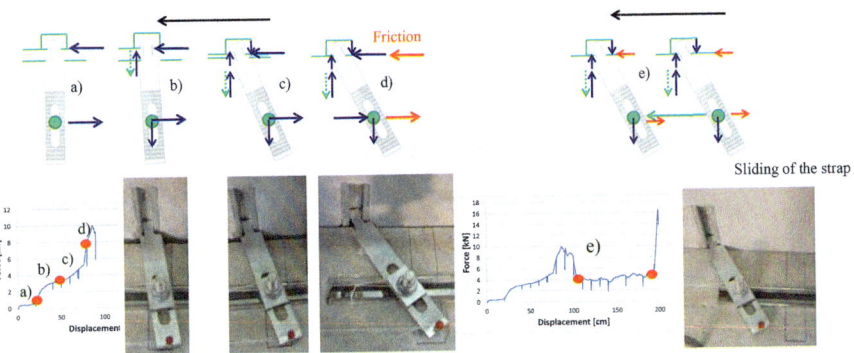

Fig. 4.41 The main steps of the response of connections presented in Fig. 4.38a

- Biaxial shear tests (see Fig. 4.40b): The specimens were loaded in two horizontal directions perpendicularly and in parallel with the longitudinal axis of the panel. The hammer-head strap was loaded in shear and tension simultaneously. The hammer head strap was loaded in its strong direction.
- Uniaxial sliding tests (see Fig. 4.40c): The load was applied in the horizontal direction in parallel to the longitudinal axis of the panel. The channel mounted in the panel was loaded in parallel to its longitudinal axis. The hammer-head strap was loaded perpendicularly to its weak direction. These tests gave information of the response of the sliding connections in the vertical direction.

The hammer-head connection presented in Fig. 4.38a is very common in the construction practice, yet hits behaviour at cyclic loading in the plane of the panel was never tested before. The main phases of the response are summarized in Fig. 4.41. In order to make this presentation clearer, the main steps are explained on the example of the connection loaded only in one direction. This mechanism is activated when the connections are loaded perpendicularly to the strong axis of the strap.

In the beginning the strap can rotate without restrictions (a). The displacements of the panel and the rotations of the strap increase simultaneously. When the displacements of the panel are large enough the head of the strap is stacked into the channel. Consequently, the force in the connection is increased (b). Plastic deformations of the head of the strap increase (c). When the displacements are large enough, the gap between the panel and the beam is closed (d). The force almost instantly considerably increases due to the activated friction between the panel and the beam. All these phases are visible in the force-displacements diagram, presented in Fig. 4.4. They are marked by red spots. The strength of connections subjected to cycling loading was considerably smaller than that observed in the monotonic tests.

In connections presented in Fig. 4.38b the failure of the channel mounted in the panel was typically observed. The screw was pulled out from the channel. The same type of the failure was observed in the special connections of the horizontal panels and columns (Fig. 4.38c). In both cases the response was considerably different compared to the connections presented in Fig. 4.38a. The strength was larger, particularly in the connections of the horizontal panels and columns. More details about the response of the tested connections can be found in Isakovic et al. (2013).

4.8 Higher Modes Effects in Multi-Storey Precast Industrial Buildings

Initial research was mostly devoted to single-storey buildings, which are indeed most frequently used. But nowadays, there has been more and more demand for complex multi-storey buildings (Fig. 4.6). The question arises, to what extend the research findings for single-storey buildings can be extended to multi-storey structures? It was found that there are several issues specific to multi-storey buildings. Obviously the columns are higher and loaded with higher compressive axial force. Consequently the margin of overstrength may be lower. The assumption of perfectly hinged connections between the beams and columns leads to models with very slender cantilevers, which might be unrealistic. However, the most specific and important problem is related to the higher modes effect. This can increase the shear forces in columns and first of all the demand on the connections for several times, compared to the values indicated by classical design procedures. If we did not consider this effect properly, the capacity design cannot be done. This problem was identified already in the PRECAST project (Fischinger et al. 2007). Later it was experimentally demonstrated and analytically studied in detail within the SAFECAST project.

Blind predictions of the response of the SAFECAST full-scale structure indicated very important higher mode effect. This was particularly obvious in the case of the prototype 2 (Fig. 4.21b) with hinged beam-to-column connections. The actual test proved that the prediction was correct. The good match of the predicted and experimental results (Fig. 4.42) also proved that the analytical models were efficient.

Shear magnification factors were systematically studied by inelastic response analyses on five realistic three–storey cantilevered structures, typical for the construction practice in Europe (Fischinger et al. 2011a). The same height of the stories (3.3, 3.2 and 3.2 m) as in the case of the full-scale SAFECAST structure (Fig. 4.20) were assumed. Buildings were modelled as single multi-storey columns. To each of the five buildings/columns different value of the normalized axial force ν_d ($0.05 \leq \nu_d \leq 0.20$) was assigned to reflect actual spans and loads used in practice. The buildings were designed according to Eurocode 8, using standard design procedures based on the results of the equivalent elastic spectrum modal analysis ($a_{g,max} = 0.25$ g and Soil Type B) considering one half of the inertia characteristics of the un-cracked sections. The same reduction as for DCH cast-in-situ frames $q = 4.5$ was assumed. The response history analyses were performed using OpenSees with a set of accelerograms, matching the EC8 spectrum.

4 Seismic Response of Precast Industrial Buildings

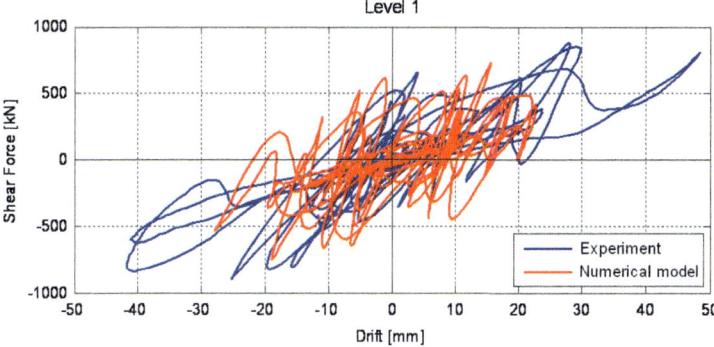

Fig. 4.42 PSD response of the SAFECAST prototype 2 (Fig. 4.21b) confirmed very large effect of higher modes, which was numerically predicted

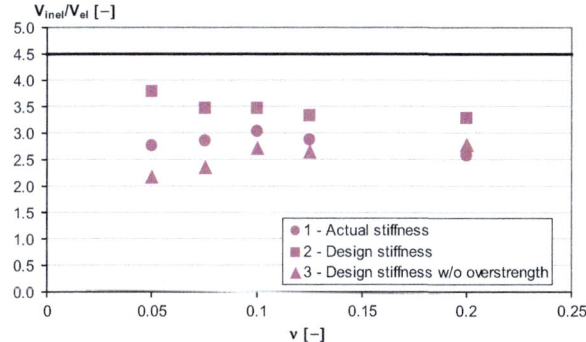

Fig. 4.43 Shear magnification ratios evaluation for the five analysed structures using different stiffness/overstrength models

Figure 4.43 shows the shear magnification factor (the ratio between the shear forces obtained by the inelastic analyses and those obtained by the equivalent elastic spectrum modal analysis) for the five investigated structures, identified by their normalized axial force value. For each structure, three different assumptions regarding stiffness of the columns and overstrength were considered in the inelastic response analyses. In the Fig. 4.43 the circles denote results of the model based on the actual stiffness during response (model 1). Squares indicate the results obtained with the inelastic analysis using the bilinear model having the same initial stiffness as it had been used in design (one half of the inertia characteristics based on the un-cracked section were used) – model 2. Model 3 (triangles) is basically the same as the model 2, except for the overstrength, which is not considered.

The results show that, as expected, shear forces are strongly influenced by the overstrength originating from different sources (including the usual assumptions about initial stiffness). In any case, the actual shear forces in multi storey cantilevered structures are considerably higher than the forces foreseen by the equivalent linear-elastic lateral force analysis, or by the modal response spectrum analysis specified in the codes. Simply said, this magnification occurs due to flexural overstrength and the amplified effect of the higher modes in the inelastic range.

It has been demonstrated that the similar shear magnification factor as proposed in Eurocode 8 for ductile (DCH) RC structural walls can be used also in the case of multi-storey cantilever columns in precast buildings (see Rejec et al. 2012 for definitions and derivation of the formula):

$$\varepsilon = q \cdot \sqrt{\left(\frac{\gamma_{Rd}}{q} \cdot \frac{M_{Rd}}{M_{Ed}}\right)^2 + 0.1 \cdot \left(\frac{S_e(T_c)}{S_e(T_1)}\right)^2} \quad \begin{Bmatrix} \leq q \\ \geq 1.5 \end{Bmatrix}.$$

It is important to note that the shear magnification factors for shear forces as large as the behaviour factor q are possible. Shear forces are directly further related to the seismic (storey) forces, which are the inertial forces acting on the floors of a structure and can be calculated as the difference between the total shear force above and below each floor. In precast structures these forces are particularly important as they determine the design of the floor system as well as beam-to-column connections. Therefore the study of the amplifications of the shear forces was extended to seismic storey forces and similar (modified) amplification factors ε were proposed.

4.9 Seismic Collapse Risk of Precast Industrial Buildings

The research, which is described in the previous sections, has provided the models and tools needed for a robust and reliable assessment of seismic risk of the precast industrial buildings. The result of these risk studies have been then of great importance for the calibration of the design requirements proposed for Eurocode 8. The study was done in two phases. First a systematic study of single-storey buildings with strong connections (assuming that the proper capacity design procedure was applied) was done (Kramar et al. 2010a). Then the study was extended to multi-storey structures with strong and weak connections (Fischinger et al. 2012b).

The limit state of the structure was defined as the inability of a system (column) to support gravity loads because of excessive lateral displacement. The collapse capacity of the structure (column) was predicted with the deteriorating numerical model (see Sect. 4.5) considering P-delta effects. The Intensity Measure (*IM*)-based variation of the recently popular PEER methodology (Fajfar and Krawinkler 2004) was used to estimate the probability of exceeding a structural limit state. The methodology is illustrated in Fig. 4.44.

It is based on the Incremental Dynamic Analysis (IDA). IDA involves a series of dynamic analyses performed under several values of the intensity. The result is an IDA curve which is a plot of response values (i.e. damage measure – *DM*) versus the intensity levels (i.e. intensity measure – *IM*). The collapse of the structure occurs

Fig. 4.44 Schematic of the IM-based approach

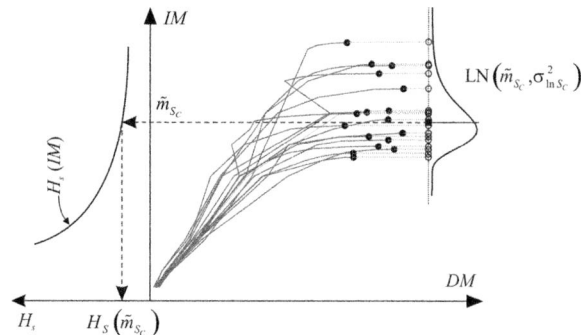

when the *DM*s increase in an unlimited manner for exceedingly small increments in the *IM* (collapse is indicated as the black dot on the IDA curve in Fig. 4.44). Considering the record-to-record variability and the uncertainty in the numerical modeling, large number of IDA curves corresponds to the same structure, thus resulting in large number of limit state intensities (S_c). Separate analysis is involved in order to determine the seismic hazard function (H_s). The hazard function is defined as the probability that the intensity of the future earthquake will be greater than or equal to the specific value. Finally, limit state probability is calculated as the hazard function multiplied by the probability density function (PDF) of the limit state intensity and integrated over all values of the intensity. Presuming the lognormal distribution of the limit state intensity and exponential form of the seismic hazard function, limit state probability of the structure can be derived analytically.

The appropriate limit value for the probability of collapse has been proposed based on the recommendations suggested by the Joint Committee on Structural Safety (JCSS 2001). It is important to note that only regular buildings were analysed.

Whereas the uncertainties in the parameters used in the PEER methodology have often been only roughly estimated, a rigorous analysis of the effect of uncertainty in the model parameters on the dispersion of the collapse capacity of the analyzed precast system (columns) was made. The dispersion due to uncertainty in the model parameters was large (conservative estimates vary from 0.18 to 0.33 depending on the column) and of similar size as the usual value of record-to-record variability (0.4 according to ATC). Both methods, the more rigorous Monte Carlo method and the simpler first order method, yielded comparable results.

4.9.1 Seismic Collapse Risk of Single-Storey Precast Industrial Buildings with Strong Connections

The mass of the structure tested within the PRECAST project (the total mass of the prototypes was 57.9 t, which resulted in the average mass of 9.6 t per individual

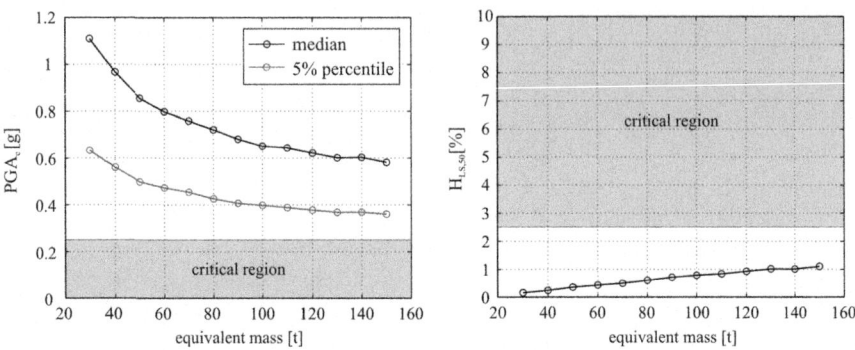

Fig. 4.45 Seismic risk (EC8 detailing requirements are considered)

column) was low compared to the mass in the real structures. Therefore a systematic parametric study was done (Kramar et al. 2008) for a whole range of possible average masses in the practical applications (10–150 t, which corresponded to the tributary roof area of 230 m^2). The results for the column with the cross-section dimensions 60/60 cm are presented here. Record-to-record variability was considered by means of 50 accelerograms generated to simulate the seismic action according to EC8. The hazard function was derived from the design acceleration values for return periods of 475 (0.25 g), 1,000 (0.3 g) and 10,000 years (0.55 g) for the area of Ljubljana (Kramar 2008; Kramar et al. 2010a).

Two different cases were analysed. In one case EC8 detailing requirements (in particular 1 % minimum longitudinal reinforcement and the minimum code required confinement) for DCH structures were considered. In the other case only the calculated (statically required) reinforcement was taken into account without considering detailing requirements. In this case the resulting amount of the reinforcement was much lower and similar to the reinforcement observed in some existing structures (although seismic force reduction factor 4.5 was used in both cases).

Seismic risk was estimated based on the following criteria. Capacity of the structure was expressed in terms of PGA (PGA$_C$). Reference value (5 % percentile of PGA$_C$) was compared to the design acceleration of 0.25 g. Probability of collapse in 50 years for the area of Ljubljana (H$_{LS,50}$) was considered. While details are given in Kramar et al. (2010a), the results are summarised in Figs. 4.45 and 4.46.

Minimum EC8 detailing requirements provide the analysed structures with sufficient overstrength so that the seismic risk is acceptably low (the probability of collapse is 0.1–1.2 % in 50 years). However, if only the calculated reinforcement is considered (disregarding the minimum detailing requirements), the conservative estimate of seismic risk is very high (the probability of collapse is 1.0–8.5 % in 50 years). The results have been used to obtain a quantitative evaluation of the force reduction factor used in the Eurocode 8 standard.

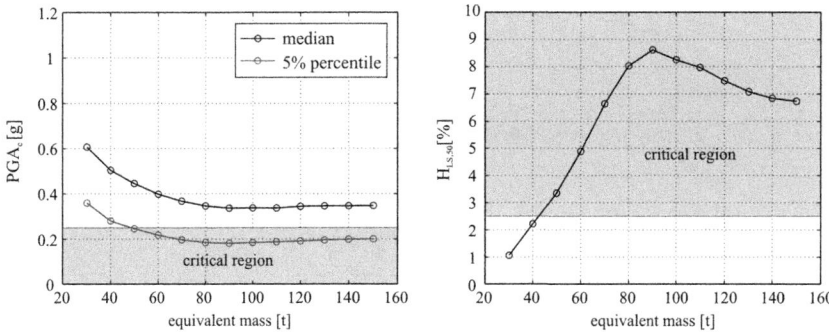

Fig. 4.46 Seismic risk (EC8 detailing requirements are not considered)

4.9.2 Seismic Collapse Risk of Multi-Storey Precast Industrial buildings with Strong and Weak Connections

The analysis described in the previous sub-section was extended to a set of realistic regular multi-storey precast buildings, which are commonly used in the Slovenian/European practice (Fischinger et al. 2012b). The investigated structural system (Fig. 4.47) consisted of 3 multi-storey cantilevers connected with the hinged beams. In accordance with the common practice, the structure had either 2 or 3 floors. The height of the first storey was assumed equal to 7 m, while the height of the subsequent stories was taken equal to 5 m. The amount of mass (i.e. vertical loading) and thus the size of the column cross-sections were varied within the range determined by the Eurocode standards. The structures vary depending on the column cross-section (bxh), and maximum normalized axial force measured at the base of a middle column (ν_d).

Realising that major seismic risk associated with many existing prefabricated systems is related to the inferior behaviour of connections, realistic strength of the beam-to-column connections as measured during experiments (weak connections) was considered and compared with the results obtained with the assumption of the strong connections.

Some typical results are shown in Figs. 4.48 and 4.49 discussed in the following text.

The design of multi-storey cantilever columns in precast structures is governed by drift and slenderness limitations. This study re-confirmed that the resulting cross-sections of the columns are large – in most realistic cases between 60 × 60 and 80 × 80. Taking into account the minimum longitudinal reinforcement requirement (1 %), this results in a considerable overstrength. So the peak ground acceleration capacity for structures with strong connections was frequently (for v_d between 0.1 and 0.15) several times higher than the design ground acceleration.

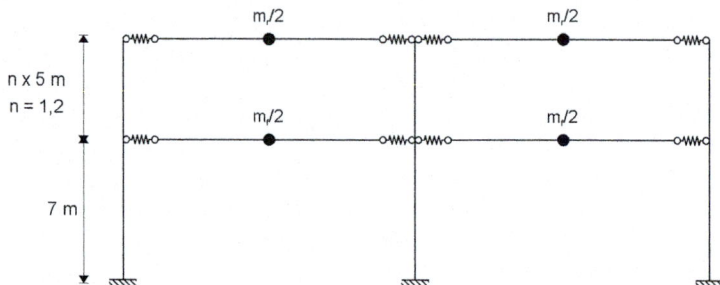

Fig. 4.47 Numerical model of multi-storey precast structures

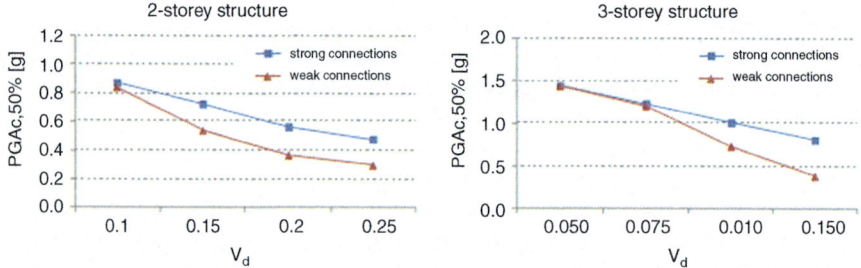

Fig. 4.48 Median PGA capacity of the structures

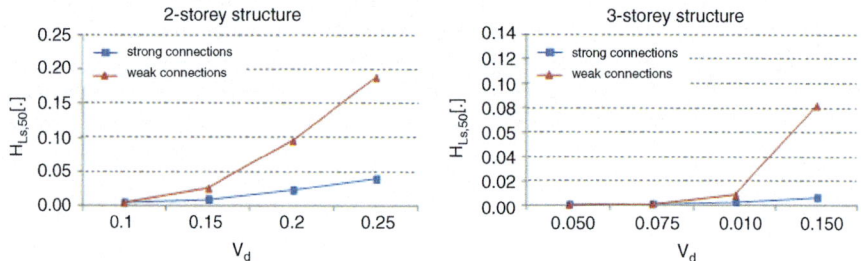

Fig. 4.49 Frequency of exceeding the limit state in 50 years

Accordingly, the probability of collapse in 50 years was sufficiently low in comparison with the recommended values. In the analysed structures value $v_d = 0.15$ corresponds approximately to the vertical load of 10 kN/m² acting on a tributary area of 100 m². Larger loads than this could be considered as rather exceptional. However, for structures carrying such large masses, overstrength is not so pronounced. In these cases, the stiffening of the system by concrete walls/cores or the use of dissipative elements is needed. The same applies, if the designer wants to reduce the large cross-sections of the columns.

Note that the cross-section of the columns in 3-storey structures were larger than in the case of the 2-storey structures (this was mainly due to the drift limitations). Therefore the peak <u>normalized</u> axial force of the 3-storey structures was smaller compared to the 2-storey structures. When comparing the 2-storey and 3-storey structures, columns with the same normalized axial force should be considered. If such comparison is made, it can be concluded that the results are similar for both structures.

It was particularly important to evaluate the influence of the realistic (weak) connections on the seismic risk. For structures with lower masses (v_d from 0.1 to 0.15) the risk did not increase compared to the risk assessed in the case of structures with strong connections (indicated that the strength of standard connections was sufficient). However, in structures with larger masses the connections were damaged and the risk drastically increased. This confirms the conclusion that the capacity design of connections is strictly needed. According to these results, beam-column connection with median capacity of 165 kN (this have been experimentally determined value for the connections typically used in the design practice) should only be used for the structures where v_d does not exceed approximately 0.15. In other cases stronger connections should be used or structural walls should be used to strengthen the system. In general, this results demonstrate that beam-column connections cannot dissipate a large amount of the energy introduced by the seismic loading. Soon after the yielding occurs, the failure of the connections follows, resulting in high seismic risk.

4.10 Eurocode 8 Implications

The key result of the presented projects has been a set of proposed improvements (either proposed or already incorporated) of the relevant requirements in Eurocode 8. Among many contributions, the most important are:

- The calibration of the behaviour factor;
- The proposed designed methodology for the design of the typical connections in precast industrial buildings (Negro and Toniolo 2012) based on the experimentally verified capacity of the connections (see Sect. 4.5)
- The proposed methodology for the realistic evaluation of the demand in the multi-storey columns in precast industrial buildings (see Sect. 4.8)
- The proposal of many innovative solutions in precast construction (not discussed in this contribution)
- Systematic risk evaluation supporting the design recommendations (see Sect. 4.9)
- The evaluation of the capacity of the cladding-to-structure connections and the on-going research on the methodology of the design, which would explicitly account for the cladding contribution (see Sect. 4.7)

While most of these results were already discussed in the previous sections, additional and more detailed comments on the choice of the behaviour factor are given in this section.

Behaviour factor is a semi-empirical parameter, which reflects many partial factors. Some of them can be experimentally or analytically calibrated, yet another, like those reflecting local construction practices, are almost impossible to consider with rigorous approach and require a good deal of engineering feeling. In particular in a complex structural system composed of many details and components of extremely different ductility, it is an illusion to give a precise value for the behaviour factor valid for all different systems. Nevertheless, the results of the presented research projects have contributed a lot towards better understanding of the energy dissipation capacity the precast industrial buildings and the definition of the behaviour factor.

Considering the above mentioned ambiguities and the lack of relevant knowledge as well as mixed field observations, one can understand that the value of the behaviour factor was changing dramatically during the evolution of the Eurocode 8.

Before Eurocode standards were introduced, most designers used the same value of seismic forces for cast-in-situ frames and precast structures. Therefore a specific note in the paragraph B1.2(2) of (CEN 1995) "Single storey industrial buildings with doubly hinged beams should be distinguished from the normal frame system" came as a shock. Strictly applying the standard one-storey precast industrial structures should be designed as inverted pendulum structures (the structural system is a set of cantilevers and more than 50 % of the mass is concentrated at the top of the cantilevers). This requirement practically meant that precast industrial structures should be designed for elastic response. However, (CEN 1995) explicitly allowed in paragraph B3.2(3) that $q_0 = 3$ can be used for precast columns in single-storey industrial buildings, which are not integrated into frames under the following conditions: (a) the top of the columns are connected with ties along both principal directions of the building and (b) the number of columns is at least six. This value was predominantly based on the engineering judgment and compromise. The ambiguity of the topic was further stressed by the fact that Annex B, which covered seismic design of precast structures, was only informative. In any case, there was not explicit reference to multi-storey structures.

The authors believe that considering the limited level of the information available at that time and the risk of the catastrophes with most damaging consequences, the proposals in CEN (1995) were fully justified. However, it was also clear that in many cases they were very conservative and they were jeopardizing (without proper research evidence) the competitiveness of a large sector of construction industry. In addition field evidence showed quite good behaviour of precast industrial buildings in spite of the fact that they were typically designed with the same behaviour factor as the cast-in-situ structure would be. Moreover, making the columns unnecessary strong would increase the demand on the most vulnerable components of the structural system – connections, and it might have a contraproductive effect.

Therefore the extensive research, presented in this paper, was initiated (Sect. 4.4). Based on the results of the ECOLEADER project a very important change was incorporated into the final pre-standard (CEN 2003 – prEN 1998-1-2003) and subsequently into the standard valid today (CEN 2004). A note was added to the definition of the "inverted pendulum structure", saying that "One-storey frames with column tops connected along both main directions of the building and with the value of the column normalized axial load v_d nowhere exceeding 0.3, do not belong to this category". This note was included (as explained by the author – Professor Toniolo) with the purpose to define precast industrial buildings as "frames" and consequently allowing to use the same behaviour factor for precast industrial buildings and cast-in-situ frame systems. This change might not be identified and understood at the first sight, since it is not a part of the chapter 5.11 (Precast concrete structures), but appears only as a note within the definitions of the structural systems and there is no explicit statement that such systems are frame systems.

In fact, in spite of the experimental results of the ECOLEADER project and the extensive supporting analytical studies, the proof for this very important change was not conclusive. First (as discussed in Sect. 4.4.2) ECOLEADER proved only that one-storey precast industrial building behaved similar (even better) than the one-storey cast-in-situ frame with strong beam. This is not to say that such cast-in-situ frame has the same energy dissipation as the multi-storey, multi-bay frame designed by a weak beam – strong column concept. Moreover, this is certainly not the case. Additionally, it should be considered that some important simplifications were used in the experiment and analyses (strong beam-to-column connections, rigid diaphragm, regular building, and construction in the controlled environment). Therefore it was obvious that further research effort is urgently needed to verify this important decision better.

PRECAST structure demonstrated (see Sect. 4.9 on risk analyses) that it is feasible to use such high behaviour factors, but with the condition that the drift limitations and minimum reinforcement requirements are fully respected. There were several factors contributing to the demonstrated good behaviour: (a) typical low compressive axial force in the columns of the single-storey buildings; (b) inherent overstrength due to the drift and slenderness limitations as well as minimum reinforcement; (c) confinement at the base of the columns (however, ASSOBETON tests indicated that the maximum spacing of the transverse should be even shorter than those required at the present by EC8). And first of all, the beam-to-column connections used in the existing practice should be designed by using capacity design rules.

Finally, after the careful study and analyses the authors are now convinced that all the debate about the behaviour factor has not been that important. What typically determines the response of the structure, is the inherent overstrength imposed by drift and slenderness limitations and not the strength determined by the behaviour factor. Of course practically elastic design (as required in the earliest stages of the pre-standard) was demonstrated as overly conservative and in some cases even contra productive. But on the other hand, there is practically no need to insist on the use of the same behaviour factor for industrial buildings and cast-in-situ frames (based on weak beam – strong column concept). One might argue that the columns themselves exhibit the ductility of more than four. But it should be considered that the drifts (up to 10 %) and top displacements (up to 1 m) needed to exploit this energy dissipation capacity of the column are impractical to achieve.

SAFECAST project brought additional research evidence for the multi-storey buildings (they were never studied before and there has been no explicit requirements for these structures in EC8) and for structures with realistic connections. The experiment at ELSA showed good behaviour of the 3-storey structure designed by $q = 4.5$. However, the mass was, in spite of the large specimen, still small compared to realistic structures. The systematic risk study (Sect. 4.9.2) showed that the use of the behaviour factor 4.5 for the DCH structure was fine. But again, the stiffness and strength were dictated by the drift and slenderness limitations rather than by the behaviour factor.

4.11 Conclusions

Not to repeat again all the specific and detailed conclusions given in the individual sections of this report, only the overall understanding of the seismic response of the precast industrial buildings, which the authors obtained during many years of the study, is presented and summarized in the conclusions.

1. When we refer to a precast system, we shall clearly and carefully determine all the details (in particular the connections and ties) of the system. Generalization can be incorrect and dangerous since even seemingly minor differences can change the behaviour considerably. Therefore since 1981 the Slovenian (former Yugoslav) code (...) required (Article 39 and 44) that the prototypes of prefabricated buildings or structures which are produced industrially in large series (except for wooden structures) and which are designed in zones of seismic activity VIII or IX, shall be checked experimentally and by inelastic dynamic analyses. While at the present, there is no such explicit requirement in the Eurocode 8, it was sensibly considered in the Design guidelines for connections of precast structures under seismic actions (Negro and Toniolo 2012).
2. Such verification was fully accomplished for the structural system of the precast industrial buildings with dowel beam-to-column connections, which is discussed in this paper and which is very frequently used in Europe. Unprecedented experimental, numerical and risk studies were done.
3. The authors are convinced that such structural system can be designed as safe in the seismic regions if all Eurocode requirements and research recommendations described in this paper are considered. This in particular includes drift limitations and capacity design of the key connections in the system.

4. The document Design guidelines for connections of precast structures under seismic actions (Negro and Toniolo 2012), produced within the SAFECAST project, provides a valuable tool for this purpose.
5. Among many different connections in the system, the beam-to-column dowel connection was particularly well studied. The capacity of the connections at large relative rotations between the column and beam were investigated. The behaviour of this connection is now fully understood and the design formulas and the design methodology are provided. Note that the design procedure for the connection with the dowels close to the edge of the column was recently improved (Zoubek et al. 2014a, b) and considerably modified in comparison with the formulas given in the Design guidelines.
6. Innovative (i.e. dissipative) connections and new structural solutions were studied within the presented research projects. However, this paper is restricted to the traditional existing systems (due to the rights and patents of the industrial partners as well as due to the limitation of the length of the paper).
7. Multi-storey structures were extensively studied in addition to the single-storey structure. Several additional problems were identified. Most important is the problem of higher modes effect, which highly increases the demand for the connections and for the shear resistance of columns. Magnification factors for shear and seismic storey forces were proposed.
8. Drift limitations require very large dimensions of the columns in the multi-storey system using dowel (hinged) beam-to-column connection. While also multi-storey building can be safely designed in seismic regions, it is a general impression that multi-storey structures need some kind of stiffening, either in the form of additional cores (the connections of the core to the precast system should be carefully designed!) or (semi) rigid beam-to-column connections. Other promising solution is the use of energy-dissipating devices.
9. Effective numerical models were proposed, including the refined FEM models to describe the complex response of the dowel connection and macro models of the post-critical behaviour of the slender columns with very high shear-span ratio.
10. Cladding-to-structure connections were very poorly understood in the past. The authors realized that for decades we have been using in design the model, which is not correct. Using existing connections, cladding cannot be fully separated from the structure during strong earthquakes. The interaction between the cladding panels and the columns should preclude large displacement, which are needed to justify the energy dissipation capacity (behaviour factors) assumed in the design. Complex realistic interaction is still under investigation within the SAFECLADDING project.
11. Finally, it should be noted that all presented research was restricted to regular structures.

Acknowledgements The presented research was mainly supported by the three large EU projects: (a) PRECAST project (Grant agreement no. G6RD-CT-2002-00857) – "Seismic Behaviour of Precast Concrete Structure with respect to EC8" in the Fifth Framework Programme (FP5) of the European Commission, (b) SAFECAST project "Performance of Innovative Mechanical

Connections in Precast Building Structures under Seismic Conditions" (Grant agreement no. 218417–2), and (c) SAFECLADDING project – "Improved Fastening systems of Cladding Wall Panels of Precast Buildings in Seismic Zones" (Grant agreement no. 314122), both in the Seventh Framework Programme (FP7) of the European Commission. All projects were led by the associations of the precast producers in Europe and the Politecnico di Milano. The leading persons in this research were in particular Professor Giandomenico Toniolo and Dr. Antonella Colombo. The large scale experiments were performed at the ELSA European Laboratory for Structural Assessment of the Joint Research Centre of the European Commission at Ispra (Italy) along with numerous experiments and numerical studies done at the Istanbul Technical University, National Laboratory for Civil Engineering in Lisbon, National Technical University of Athens Polytechnic University of Milan, the Slovenian National Building and Civil Engineering Institute (ZAG), and University of Ljubljana. The work was further supported by the Ministry of Higher Education, Science and Technology of Slovenia. During the time of this long-going research many post-graduate students and post-doc researchers contributed to the results of the research group in Ljubljana. The authors are in particular grateful to Dr. Miha Kramar for his outstanding contributions. The authors are also obliged to Professor Giandomenico Toniolo and Dr. Paolo Negro, who contributed valuable material, which was used in the introductory sections of this paper.

The research results were always discussed between the members of the consortia of the above-mentioned research projects and general agreement was found in most cases. Nevertheless, it is obvious that the conclusions of such complex research cannot be exactly defined and they cannot be always supported by all the partners. Therefore the views and the conclusions in this paper are solely those of the authors, although the paper was in general approved by all the partners.

Open Access This chapter is distributed under the terms of the Creative Commons Attribution Noncommercial License, which permits any noncommercial use, distribution, and reproduction in any medium, provided the original author(s) and source are credited.

References

Arslan MH, Korkmaz HH, Gulay DG (2006) Damage and failure pattern of prefabricated structures after major earthquakes in Turkey and shortfalls of the Turkish Earthquake code. Eng Fail Anal 13(4):537–557
ATC-08 (1981) Proceedings of a workshop on design of prefabricated concrete buildings for earthquake loads. ATC, NSF, Redwood City, California, USA
Biondini F, Toniolo G (2002) Probabilistic calibration of behaviour factors of EC8 for cast-in-situ and precast frames. In: 17th BIBM Congress, Istanbul
Biondini F, Toniolo G (2004) Validation of seismic design criteria for concrete frames based on Monte Carlo simulation and full scale pseudodynamic tests. In: 13th WCEE, Vancouver
Bournas A, Negro P, Taucer FT (2013a) Performance of industrial buildings during the Emilia earthquakes in Northern Italy and recommendations for their strengthening. Bull Earthq Eng, published on-line, June 2013
Bournas A, Negro P, Molina FJ (2013b) Pseudodynamic tests on a full-scale 3-storey precast concrete building: behavior of the mechanical connections and floor diaphragms. Eng Struct 57:609–627
CEN (1995) Eurocode 8 – Design provisions of earthquake resistance of structures, part 1–3: specific rules for various materials and elements, ENV 1998-1-3:1995. European Committee for Standardization, Brussels
CEN (2003) Eurocode 8: design of structures for earthquake resistance – part 1: general rules, seismic actions and rules for buildings, prEN 1998–1, Draft No. 6, version for translation (Stage 49). European Committee for Standardization, Brussels

CEN (2004) Eurocode 8: design of structures for earthquake resistance – part 1: general rules, seismic actions and rules for buildings, EN 1998–1. European Committee for Standardization, Brussels

CEN/TS 1992-1-1 (2005) Design of fastenings for use in concrete – part 4–2: headed fasteners, 89/106/EEC, European Comittee for Standardization, Brussels

Dulascska H (1972) Dowel action of reinforcement crossing cracks in concrete. JACI 69–70:754–757

EERI (1979) Friuly, Italy earthquakes of 1976. Earthquake Engineering Research Institute, Oakland, California, USA

EERI (1989) Armenia earthquake reconnaissance report. Earthq Spectra supplement, Oakland, California, USA

EERI (2000) Kocaeli, Turkey, Earthquake of August 17, 1999. Earthq Spectra supplement, Oakland, California, USA

Engström B (1990) Combined effects of dowel action and friction in bolted connections. Nordic Concrete Research, The Nordic Concrete Federation, Publication no. 9, Oslo 1990, pp 14–33

Fajfar P, Krawinkler H (eds) (2004) Performance-based seismic design concepts and implementation. In: Proceedings of an International Workshop, PEER, University of California, Berkeley

Fajfar P, Banovec J, Saje F (1978) Behaviour of prefabricated industrial building in Breginj during the Friuli earthquake. In: 6th ECEE, Dubrovnik, vol 2. pp 493–500

Fajfar P, Duhovnik J, Reflak J, Fischinger M, Breška Z (1981) The behaviour of buildings and other structures during the earthquakes of 1979 in Montenegro, IKPIR publication 19A, University of Ljubljana, Ljubljana

Ferrara L, Colombo A, Negro P, Toniolo G (2004) Precast vs. cast-in-situ reinforced concrete industrial buildings under earthquake loading: an assessment via pseudodynamic tests. In: 13th WCEE, Vancouver, Canada

FIB (2003) Seismic design of precast concrete building structures, Federation International du Beton, Lausanne, Bulletin 27

Fischinger M, Kramar M, Isaković T, Kante P (2007) Seismic behaviour of precast concrete structures with respect to EC8, final report on the contribution of the University of Ljublajana, Chapter 5: two-storey precast industrial building, University of Ljubljana, Ljubljana, pp 51–78

Fischinger M, Kramar M, Isaković T (2008) Cyclic response of slender RC columns typical of precast industrial buildings. Bull Earthq Eng 6(3):519–534, http://www.springerlink.com/content/m282220243851270/fulltext.pdf

Fischinger M, Ercolino M, Kramar M, Petrone C, Isaković T (2011a) Inelastic sesmic shear in multi-storey cantilever columns. COMPDYN 2011, Corfu, Greece, 10 p

Fischinger M, Kramar M, Isaković T (2011b) Inelastic seismic response of precast industrial buildings – research in support of EC8. MASE 14. Maced Assoc Struct Eng 1:39–50

Fischinger M, Zoubek B, Kramar M, Isaković T (2012a) Cyclic response of dowel connections in precast structures. In: 15th WCEE, Lisbon, 10 p

Fischinger M, Kramar M, Isaković T (2012b) Seismic risk assessment of multi-storey precast structures. In: 15th WCEE, Lisbon, 10 p

Fischinger M, Zoubek B, Isaković T (2013) Seismic behaviour of the beam-to-column dowel connections – macro modelling. COMPDYN 2013, Kos Island, Greece, 8 p

Fuchs W, Eligehausen R, Breen JE (1995) Concrete capacity design (CCD) approach for fastening to concrete. ACI Struct J 92(1):73–94

Haselton CB (2006) Assessing seismic collapse safety of modern reinforced concrete moment frame buildings. Dissertation, Stanford University, Stanford, California

Hojlund-Rasmussen B (1963) Betoninstobe tvaerbelastade boltes og dornesbaereevne (Resistance of embedded bolts and dowels loaded in shear). Byngninsstatiske Meddelser, Copenhagen, p 34

Ibarra LF, Medina RA, Krawinkler H (2005) Hysteretic models that incorporate strength and stiffness deterioration. Earthq Eng Struct Dyn 34(12):1489–1511

Isaković T, Zoubek B, Lopatoč J, Urbas M, Fischinger M (2013) Report and card files on the tests performed on existing connections. Deliverable 1.2, SAFECLADDING, University of Ljubljana, Ljubljana

JCSS (2001) Probabilistic model code, part 1: basis of design, 12th draft. Joint Committee on Structural Safety, Lyngby, Denmark

Kramar M (2008) Seismic vulnerability of the precast reinforced concrete structures. Ph.D. thesis (in Slovenian) University of Ljubljana, Ljubljana

Kramar M, Isaković T, Fischinger M (2008) Seismic collapse safety of RC columns in precast industrial buildings. In: 14th WCEE, Beijing, 8 p

Kramar M, Isaković T, Fischinger M (2010a) Seismic collapse risk of precast industrial buildings with strong connections. Earthq Eng Struct Dyn 39(8):847–868, http://onlinelibrary.wiley.com/doi/10.1002/eqe.970/pdf

Kramar M, Isaković T, Fischinger M (2010b) Experimental investigation of "pinned" beam-to-column connections in precast industrial buildings. In: 14th ECEE, Ohrid, 8 p

Negro P, Toniolo G (2012) Design guidelines for connections of precast structures under seismic actions. Publication Office of the European Union, Luxemburg

Negro P, Bournas D, Molina FJ (2013) Pseudodynamic tests on a full-scale 3-storey precast concrete building: global response. Elsevier Eng Struct J 57:594–608

Panagiotakos TB, Fardis MN (2001) Deformations of reinforced concrete members at yielding and ultimate. ACI Struct J 98(2):135–148

PEER Structural Performance Database (2007) Pacific earthquake engineering research center. University of California, Berkeley, http://nisee.berkeley.edu/spd

Priestley MJN (1996) The PRESS program: current status and proposed plans for phase III. PCI J 41:22–40

Psycharis IN, Mouzakis HP (2012) Shear resistance of pinned connections of precast members to monotonic and cyclic loading. Eng Struct 41:413–427

Rejec K, Isaković T, Fischinger M (2012) Seismic shear force magnification in RC cantilever structural walls, designed according to Eurocode 8. Bull Earthq Eng 10(2):567–586

Saatcioglu M, Mitchell D, Tinawi R, Gardner NJ, Gillies AG, Ghobarah A, Anderson DL, Lau D (2001) TheAugust 17, 1999 Kocaeli (Turkey) earthquake-damage to structures. Can J Civ Eng 28(8):715–773

Saisi A, Toniolo G (1998) Precast RC columns under cyclic loading: an experimental programme oriented to EC8. Studi e ricerche, Scuola di specializzazione. Politecnico di Milano, vol. 19, pp 373–414

Shiohara H, Watanabe F (2000) The Japan PRESS precast concrete connection design. In: 12th WCEE, Aukland

Toniolo G (2012) European research on seismic behaviour of precast structures. In: 2012 NZSEE conference, Lausanne, Switzerland

Toniolo G (coordinator) (2007) Final report of the EU Research Project: seismic behaviour of precast concrete structures with respect to EC8 (Contract No. G6RD-CT-2002-00857), Brussels

Toniolo G, Colombo A (2012) Precast concrete structures: the lessons learned from the L'Aquila earthquake. Struct Concr 13:73–83

Tzenov L, Sotirov L, Boncheva P (1978) Study of some damaged industrial buildings due to Vrancea earthquake. In: 6th ECEE Dubrovnik, vol 6. pp 59–65

UNDP/UNIDO (1985) Building construction under seismic conditions in the Balkan region. UNDP/UNIDO Project RER/79/015, vol 2: design and construction of prefabricated reinforced concrete building systems, Vienna Austria; Skopje Macedonia

Vintzeleou EN, Tassios TP (1986) Mathematical model for dowel action under monotonic and cyclic conditions. Mag Concr Res 38:13–22

Vintzeleou EN, Tassios TP (1987) Behaviour of dowels under cyclic deformations. ACI Struct J 84(1):18–30

Zoubek B, Fischinger M, Isaković T (2013a) Seismic behaviour of the beam-to-column dowel connections – FEM analysis. COMPDYN 2013, Kos Island, Greece, 8 p

Zoubek B, Isaković T, Fahjan Y, Fischinger M (2013b) Cyclic failure analysis of the beam-to-column dowel connections in precast industrial buildings. Eng Struct 52:179–191

Zoubek B, Fischinger M, Isaković T (2014a) Estimation of the cyclic capacity of beam-to-column dowel connections in precast industrial buildings. Submitted for possible publication in Bulletin of Earthquake Engineering, Springer, Netherlands

Zoubek B, Fischinger M, Isaković T (2014b) Seismic response of dowel connections in precast industrial buildings, Second European conference on earthquake engineering and seismology, Istanbul, 25–29 August 2014

Chapter 5
The Role of Site Effects at the Boundary Between Seismology and Engineering: Lessons from Recent Earthquakes

Marco Mucciarelli

Abstract This paper summarises the experience gathered on the field following four recent earthquakes: in 2009 at L'Aquila, Italy; in 2010 at Lorca, Spain; in 2011 at Christchurch, New Zealand; in 2012 at Emilia, Italy. These quakes provided useful lessons at the boundary between seismology and engineering, about the difference between what we expected to happen, thanks to more or less simplified models, and what happened in reality. The topics dealt with are: (1) the reliability of "free-field" strong motion recordings, discussing the role of accelerometer housing, spurious transient, city-soil effect, and the possible over-correction of displacements; (2) the mismatch between code provision and observed spectral acceleration due to the role of velocity inversions, the influence of topography, the softening and hardening non-linearity, (3) the importance of vertical component considering the time distribution of phases arrivals and the presence of amplification due to P-velocity contrasts.

5.1 Introduction

In the past 5 years, four moderate magnitude earthquakes caused substantial economic damage and a death toll from dozens to hundreds of casualties each. Namely, they are the 2009 L'Aquila earthquake, Italy; the 2010 Lorca earthquake, Spain; the 2011 Christchurch earthquake, New Zealand; the 2012 Emilia

M. Mucciarelli (✉)
Seismological Research Centre, National Institute of Oceanography and Experimental Geophysics, Trieste, Italy

Engineering School of Basilicata University, Potenza, Italy
e-mail: marco.mucciarelli@unibas.it

earthquake, Italy. All of them happened in densely populated, industrialised area previously subjected to seismic classification.

There were debates following each of those events about the reliability of seismic hazard studies, the implementation of site effects in seismic codes and about the limit of damage that is acceptable by designers but unacceptable (or misunderstood) by population. I had the opportunity, with colleagues of different research groups, to perform field studies in all these areas, noting similarity and differences. This paper tries to summarise the role of the difference between what we expected to happen, thanks to more or less simplified models, and what happened in reality. We all accept that models are a need to simplify theories and make them useful to practitioners, but there is a threshold of disagreement between models and reality that must not be trespassed.

5.2 How Reliable Are "Free-Field" Strong Motion Recordings?

In recent years, it was acknowledged the importance of ground-truthing microzonation maps or Vs30 studies by summarising some lessons learned from large earthquakes and recent earthquake site response studies that utilise earthquake recordings from dense seismic networks or ambient noise measurements (Cassidy and Mucciarelli 2010).

But if we want to considered the instrumental recordings as the truth against which our model should be tested, we must be sure of the reliability of such data. Recent earthquakes have shown that in some cases particular care should be taken before using recorded data. In some cases the owners of an accelerometric network provided to pre-check the strong motion recordings and decided not to disseminate corrupted data. This was the case of the 2009, L'Aquila earthquake when the Italian department of Civil Protection did not distribute the recording of main shock at AQM station. The accelerometer, set to 1 g full-scale, saturated due to a partial detachment of the instrument from the pillar (Zambonelli et al. 2011); In other cases, problems with the recordings were encountered as listed in the following.

5.2.1 Housing and City-Soil Effects

The influence of buildings on free-field ground motion recordings has been postulated for the first time more than 30 years ago (Jennings 1970), and confirmed both by experiments and numerical simulations. Ditommaso et al. (2009a, b) showed that the peak and spectral parameters are the most affected, while the integral ones are not so disturbed. This is due to the fact that the presence of the structure has both the effect of a damper (thus reducing the total energy) and of a filter, focusing energy in the band of building eigenfrequencies.

5 The Role of Site Effects at the Boundary Between Seismology and...

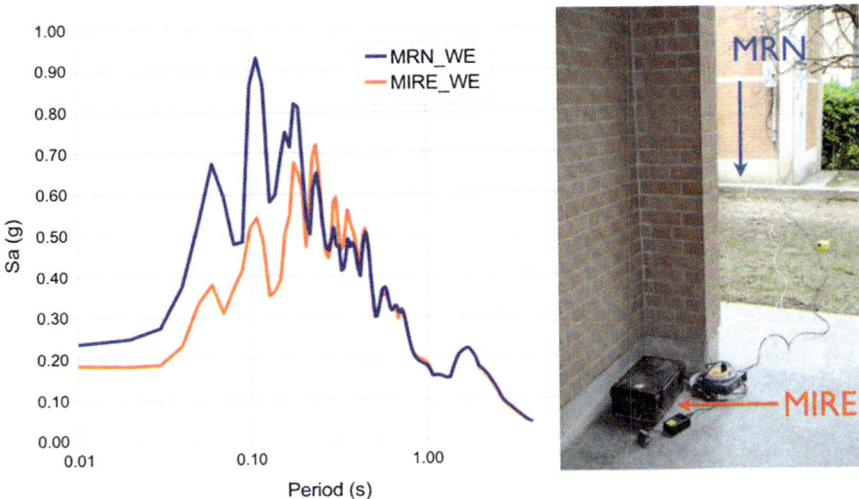

Fig. 5.1 Comparison of ground response spectra inside and outside a building for the 29.05.12, M = 5.8 shock in Mirandola, Emilia

During the Emilia sequence, an accelerometer (MIRE) was installed in free-field at 5 m from the existing RAN station, located inside a small electrical substation (MRN). The response spectra of the second strongest shock of the sequence (Ml = 5.8, 29.05.12) showed a noticeable agreement at the two locations (see Fig. 5.1), except that for the short period range, where the recording inside the substation showed peaks much higher than in the free-field station. It is possible that several strong motion recorded in urban areas depend on housing or on the vicinity to oscillating buildings.

5.2.2 Over-Correction of Displacements

The Emilia second strongest shock provided a lot of strong motion data very close to the epicentre. This posed the problem of correction of accelerometric recordings. Figure 5.2 shows the comparison between the uncorrected and corrected time histories from the vertical component of station MIRE (see Fig. 5.1). The uncorrected data shows a permanent displacement of about 30 cm. INSAR data and modelling from different authors shows that this location suffered a 15 cm static coseismic displacement.

The standard de-trending and filtering procedure could introduce spurious frequencies due to the presence of a real permanent displacement that does not allow for having zero-mean corrected recordings. In the future the availability of high-frequency GPS data co-located with seismic and accelerometric station will provide an unbiased estimate of real ground motion.

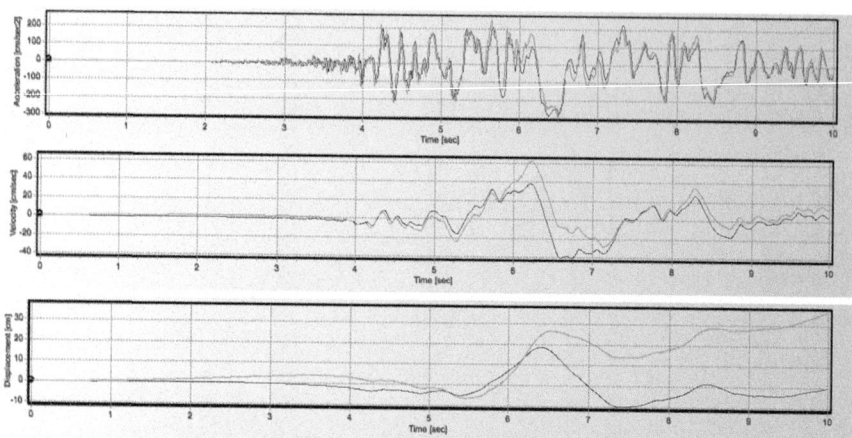

Fig. 5.2 Corrected (*blue*) and uncorrected (*grey*) strong motion recording at MIRE. From *top* to *bottom*: acceleration, velocity and displacements

Fig. 5.3 The accelerometric station in the basement of the old jailhouse, Lorca

5.2.3 *Spurious Transient in Strong Motion Recordings*

During the 2010 Lorca earthquake, a valuable strong motion recording was available thanks to a station of Red Sismica Nacional located in the historical city centre, very close to the epicentre. The station was installed in the basement of the old jailhouse (see Fig. 5.3).

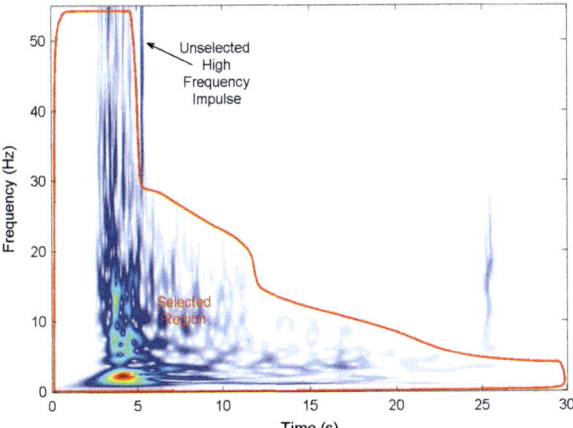

Fig. 5.4 Application of a band-variable filter (Ditommaso et al. 2012) to the recording of the mainshock in Lorca

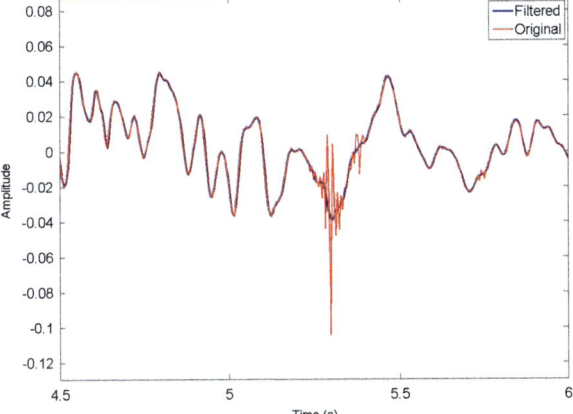

Fig. 5.5 Enlargement of the accelerometric recording of the mainshok in Lorca

During the mainshock, some heavy objects close to the accelerometer fell on the reinforced concrete pillar of the station. This caused a strong, high-frequency acceleration transient in the recording. Using a band variable filter based on Stockwell transform (Ditommaso et al. 2012) it was possible to carefully remove this spurious peak.

Figure 5.4 shows the area selected for filtering in the time frequency domain, while Fig. 5.5 compares the time histories before and after the filtering, showing the accuracy of the band variable filter in preserving the signal outside the area selected for removal.

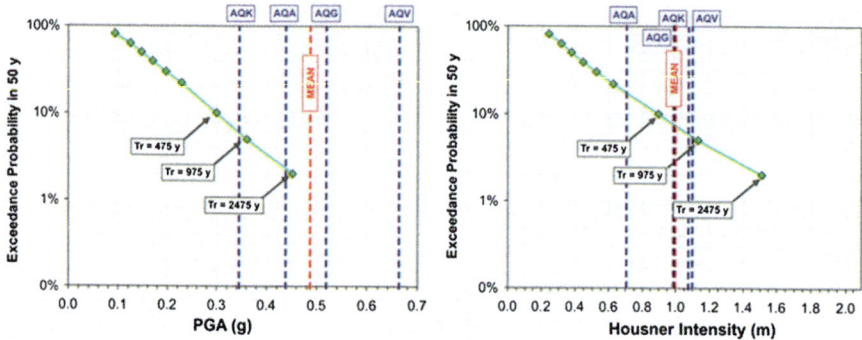

Fig. 5.6 From Masi et al. (2011) Exceedance probability in 50 years at L'Aquila provided by the NTC2008 code for soil classB in terms of PGA (on the *left*) and Housner Intensity (on the *right*); the maximum values of the horizontal components recorded at four stations are also displayed (*blue dashed lines*) together with their mean (*red dashed lines*)

5.3 Comparison Between Code Spectra and Observed Strong Motion

A careful evaluation of site effects is crucial for the activity of validation of PSHA estimates. Procedures like the one proposed by Albarello and D'Amico (2008) requires to know if the set of recordings to be compared with estimates are obtained on rock or if they have to be deconvolved to a rock-equivalent condition.

The L'Aquila and Emilia earthquakes provided contrasting evidences. For l'Aquila event, the difference between the observed recordings and code provision was mainly due to the choice of parameters used rather than in a bias in base hazard estimates or insufficient description of site effects . After correcting for soil class according with Vs30, Masi et al. (2011) showed that Housner Intensity provided much better results than PGA (Fig. 5.6), and was well correlated with site seismic hazard obtained from the long series of macroseismic data available.

On the other hand, in Emilia it was observed (Gallipoli et al. 2014) that while code provision largely underestimated the recorded values, the convolution of expected motion at a rock site with a 1-d velocity profile down to 120 m instead of Vs30 soil class greatly improved the agreement. This difference it is probably due to the fact that the sediment in the Aterno valley (L'Aquila) are coarse and less than 40 m thick, while in the Po valley (Emilia) the soil is very soft and bedrock is hundreds of meters deep, the condition where Vs30 gives its poorest performances as a proxy of site amplification (Gallipoli and Mucciarelli 2009).

5.4 When Reality Is Far from Models

5.4.1 Need for Nanozonation?

During L'Aquila earthquake the variation of damage due to site effects was shown to vary abruptly over a very short distance. The most striking example was observed in the village of San Gregorio. After the microzonation performed following the ICMS08 (Indirizzi e Criteri per la Microzonazione Sismica, Guidelines For Seismic Microzonation) for the basic level, including a new, detailed geological mapping at 1:5000 scale, it was no possible to explain a peculiar damage observed: a three-story, reinforced concrete (RC) building had the first floor collapsed. The remaining two stories fell with a displacement in the horizontal projection of about 70 cm. Buildings located at a short distance had little or no damage reported.

Mucciarelli et al. (2011a) performed a geophysical and geologic survey at the site. The acceleration and ambient noise recordings showed a high amplification in the slope direction. Geo-electrical tomography showed a strong discontinuity just below the building. A very soft material (possibly fault cataclasites) was found in a borehole down to 17 m from ground level, showing a shear wave velocity that starts at 250 m/s, increases with depth and has an abrupt transition in calcarenites at 1,150 m/s. The surface geophysical measurements in the vicinity of the site have not shown similar situations, with flat HVSR curves as expected for a rock outcrop, except for a lateral extension of the soft zone (these results are summarised in Fig. 5.7). The analysis on the quality of the building materials has yielded values higher than average for the age and type of construction, and no special design or construction deficiencies have been observed. A strong, peculiar site effect thus appears to be the most likely cause of the damage observed, extending at a very limited scale, in an area slightly wider than building foundations. This sound like a warning for anyone that may think to use microzonation studies as input data for design of a specific structure and not for the urban planning aim they are designed for.

5.4.2 Velocity Inversions

The EuroCode 8 soil classification in Vs30 classes, adopted following the scheme of NEHRP recommendations, considers a soil-over-bedrock scheme, with mechanical properties improving with depth. The possibility of velocity inversions is not taken into account. The L'Aquila earthquake showed that this kind of geo-lithogical situation was more common than previously thought. In some instances, a stratum of well-cemented breccia (conglomerates), even 30 m thick, was overlying softer soil deposits, giving amplification in a situation that could be easily mistaken for a bedrock site. An example of this kind of velocity inversion is given in Gallipoli et al. (2011) for the Poggio Picenze village (see Fig. 5.8).

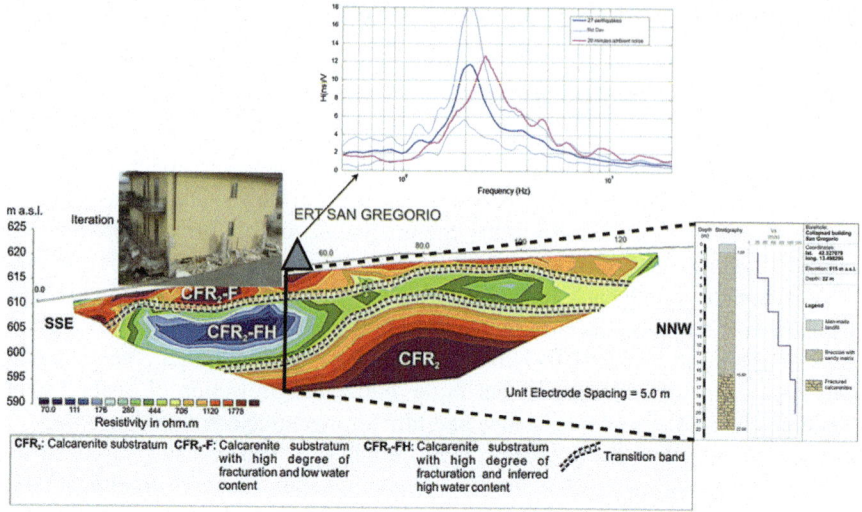

Fig. 5.7 Summary of surveys in San Gregorio from Mucciarelli et al. (2011a), see text for details

In other instances, a further soft stratum was present at the top of the sequence, giving rise to a more complex amplification pattern, that is visible since HVSR measurements have a double peak. This results in amplification of seismic motion over a wider range of frequencies, and was related to damage enhancement as clearly shown for the L'Aquila historical centre (Fig. 5.9) by Del Monaco et al. (2013).

5.4.3 The Role of Topographic Amplification

During the L'Aquila, 2009 seismic sequence, the temporary installation of accelerometric networks provided a test of the Italian anti-seismic provisions about topographic amplifications. Two morphological situations were particularly suitable for the test: Castelnuovo, where two accelerometers located on the same lithology at the hill top and halfway along the slope provided the ideal case to test the proposed rule of linear increment of amplification along the slope, and Navelli, where the combination of code topographic and stratigraphic amplification factors was similar, given a station on a rocky slope and one on a flat alluvial valley. Gallipoli et al. (2013). showed that *"in neither case the observation matches code provisions. For Castelnuovo, there is a frequency dependence that shows as the code is over-conservative for short periods but fails to predict amplification in the intermediate range. For Navelli, the code provision is verified for long periods, but*

5 The Role of Site Effects at the Boundary Between Seismology and... 187

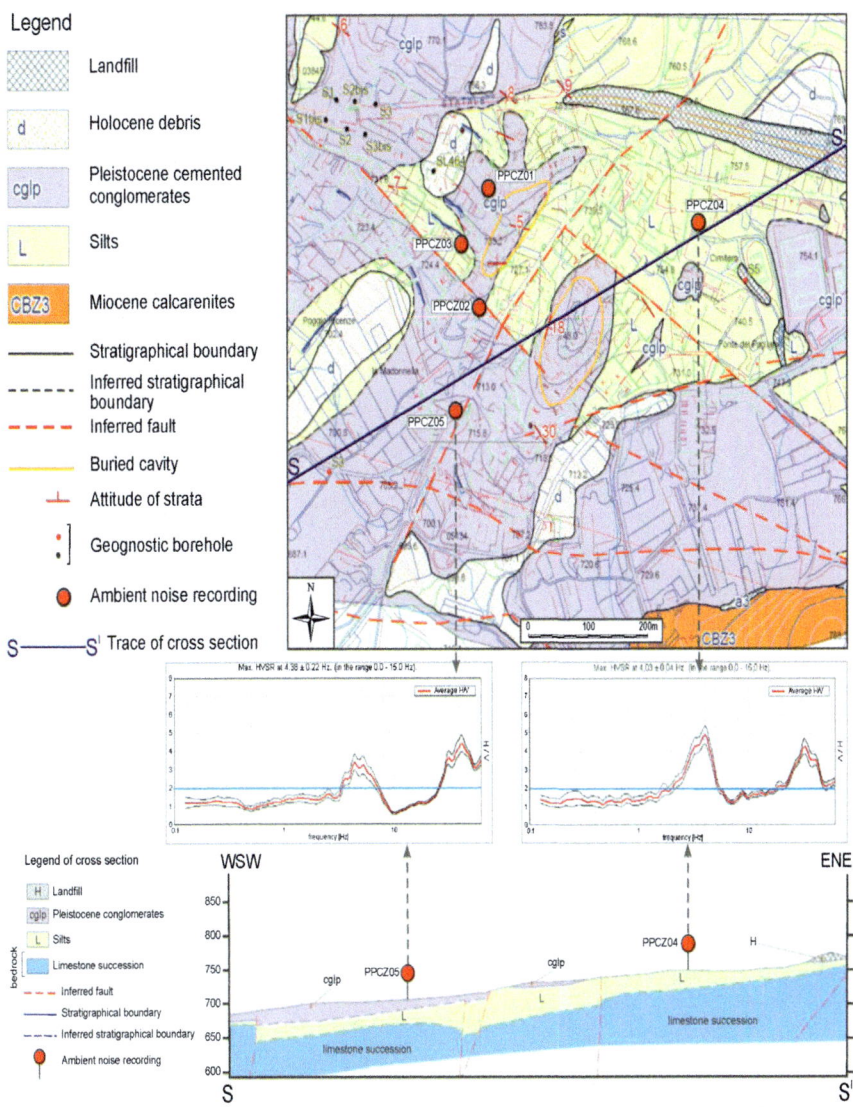

Fig. 5.8 The geological map and geological section with HVNSR (PPCZ04 and PPCZ05) of Poggio Picenze, from Gallipoli et al. (2011)

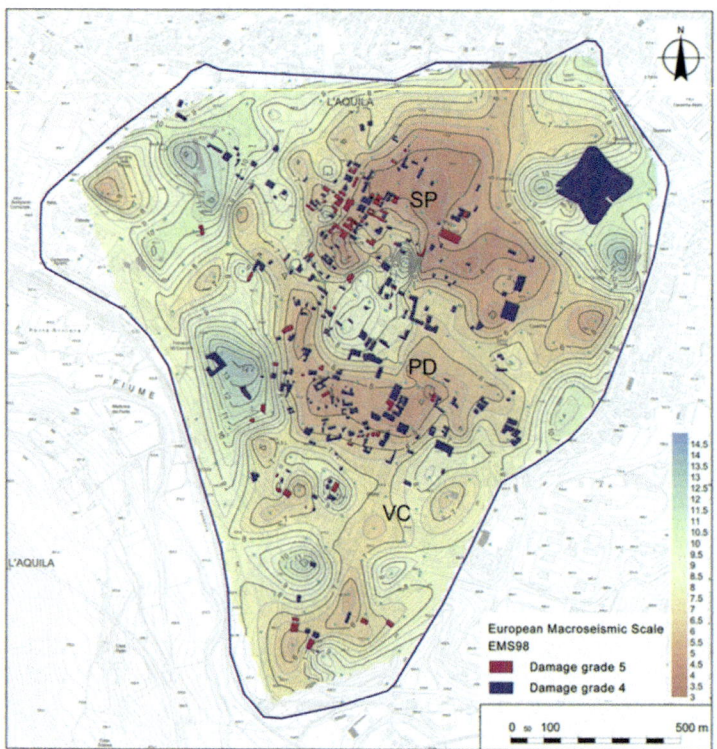

Fig. 5.9 From del Monaco et al. (2013); location of the severely damaged buildings (DG5 and 4 in EMS'98 damage grade, Tertulliani et al. 2011) and contouring of the second resonance frequency peak from HVNSR analysis in L'Aquila historical centre

in the range around the site resonance frequency the stratigraphic amplification proves to be three times more important than the topographic one."

Figure 5.10 reports the Navelli case.

5.4.4 The Role of Non-linearity

The L'Aquila and the Christchurch earthquake provided interesting evidence about the role of non linearity in seismic response.

The analysis of two arrays in the Upper (L'Aquila) and Lower Aterno valley (Navelli) showed that softening soil non-linearity played a role only of soft, fine and well graded basins like in Navelli. Mucciarelli et al. (2011b) found a few percent decrease in fundamental frequency and amplification between the largest ($M > 4$) aftershocks and lesser aftershocks and noise. On the contrary, Puglia et al. (2011) did not find any evidence on non-linearity in the response of the coarser, inter-digited soils of the Upper Aterno valley.

5 The Role of Site Effects at the Boundary Between Seismology and... 189

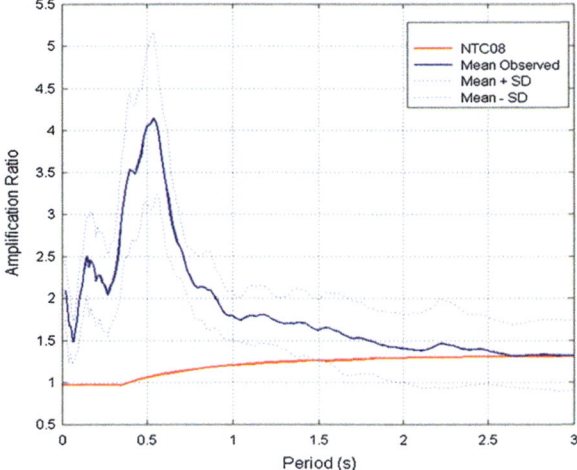

Fig. 5.10 Comparison between code provisions (*red*) and observed amplification ratio (*blue*) in Navelli between closely spaced stations, one on a rocky slope and one on a flat alluvial valley

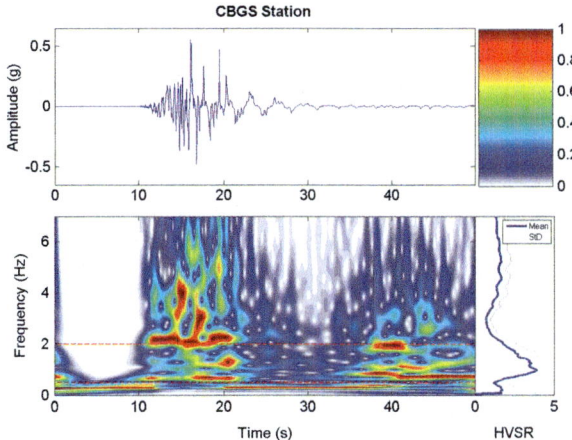

Fig. 5.11 Comparison between normalized S-transform and HVSR at GeoNet CBGS accelerometric station

In Christchurch it was possible to observe hardening non-linearity in action. Mucciarelli (2011) analysed jointly noise and accelerometric recordings, using the S-transform. The result (Fig. 5.11) shows that the energy of the largest horizontal component for coda waves is at frequencies lower than the fundamental one determined by HVSR, but in an earlier phase, the time-domain trace and the S-transform show high-frequency acceleration peaks, the evidence of the hardening non-linearity first described by Bonilla et al. (2005), due to hysteretic dilatant behaviour of non-cohesive, partially saturated soils.

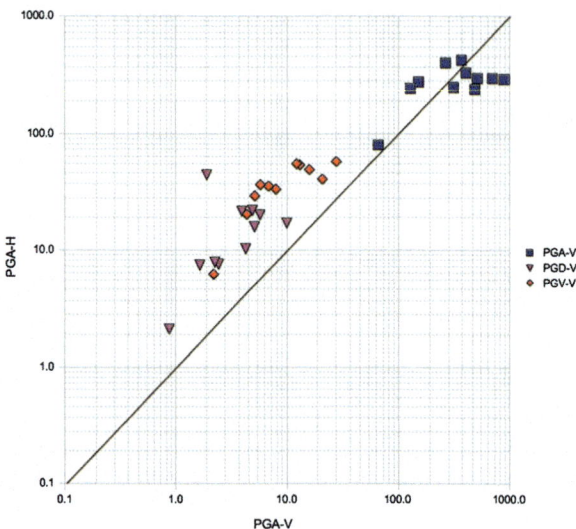

Fig. 5.12 comparison between horizontal and vertical component of the peak parameters of ground motion PGA, PGV, PGD in the near field of Emilia, 2012 earthquake. The units are respectively cm/s^2, cm/s and cm

5.4.5 Vertical Component and P-Wave Amplification

The Emilia sequence had two similar magnitude main events separated by 9 days. While there was only an accelerometric station active during the first shock, several organisation (INGV; CNR-IMAA, OGS, RAN) installed temporary network in the epicentral area. When the second shock occurred it was thus possible to have a large number of near field recordings. Figure 5.12 summarises the relationship between horizontal and vertical component of the three peak parameters of ground motion (PGA, PGV, PGD). It is possible to see that while for velocity and displacement the horizontal peak is always larger, for acceleration the majority of near-field peaks is larger in the vertical component. These large vertical accelerations are overlooked by present day Italian seismic code.

5.4.6 Time Distribution of Seismic Actions

Some important lessons from these recent earthquakes came from the time-domain representation of data.

Analysing the previously described data from the Christchurch earthquake using the cumulative Housner intensity, calculated from $T = 0$ for incrementing time intervals, it possible to evaluate the importance of the transition from linear behaviour in the beginning to hardening non linearity in the middle and softening non-linearity at the end (Fig. 5.13).

It is possible to see that during the hardening non-linearity phase the Housner intensity recorded is enough to cause damages corresponding to the VIII EMS

Fig. 5.13 Cumulative Housner Intensity at GeoNet CBGS accelerometric station

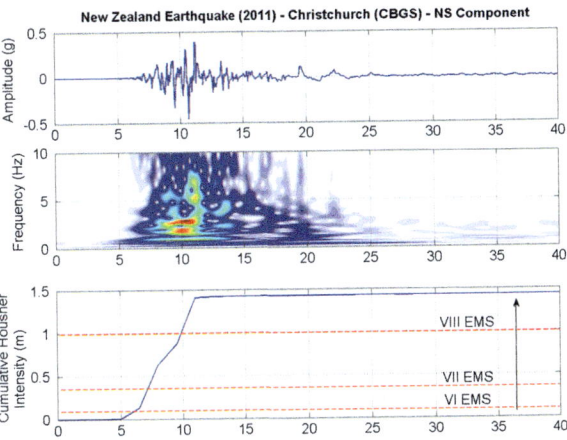

Fig. 5.14 S-transform of the 29 May earthquake at MIRE station

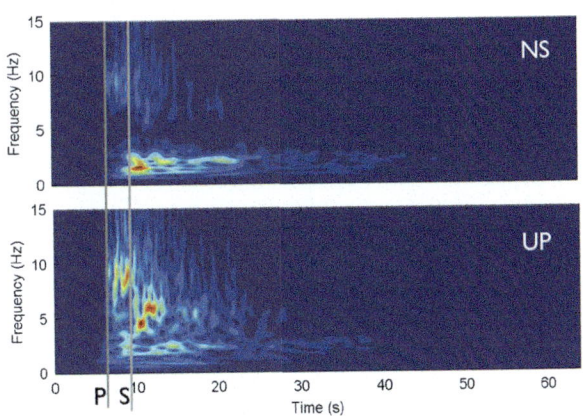

degree. When finally there is the onset of softening non-linearity, the Cumulative Housner intensity is already more than 90 % of the total. This should induce care when using simplified 1-d linear-equivalent models for site seismic response that do not take into account hardening non linearity and are not able to reproduce correctly in time-domain the onset of softening non-linearity.

Another lesson learned from frequency-time domain during the Emilia earthquake is the role of the combination of vertical and horizontal strongest phases. A peculiar kind of damage of this earthquake was the failure of several pre-fab industrial facility. Most of damage was caused by the fact that the beam were not connected to pillars, but the contact was pure friction. A loss of vertical load could have caused the reduction of friction and subsequently the collapse of the beams.

A look to the frequency domain representation of the recordings at MIRE stations (Fig. 5.14) shows that there is, as expected, a strong phase of vertical motion connected to the arrival of the P waves, when the horizontal motion is minimal. Unexpectedly there is also a strong pulse in the vertical component practically synchronous with the arrival of S-waves. This could have been the cause of many observed collapse of industrial facilities.

5.5 A Look to the Future

Three main field of activity are envisaged for the future.

1. A federation of accelerometric borehole arrays in Italy. The motivation of this project arises from the need of improving existing installations, provide uniform site characterisation of sites (Fo, velocity profiles, etc.), bring together the owners in order to share good practices and finally to provide a web portal for the public dissemination of results. The availability of well characterised sites where the absolute site amplification is known, beside improving GMPEs could also be a resource for hands-on training of practitioners that could test their skills and their equipment against the available knowledge.
2. The consideration of building soil-resonance. The importance of resonance was highlighted for the Emilia quake by the striking case of two twin buildings whose different damage was caused by the different fundamental frequency of foundation soil even at close distance (Castellaro et al. 2014). During the L'Aquila earthquake it was possible to determine the frequency decay due to different level of damage on a large set of buildings (Ditommaso et al. 2013). The availability of these data made possible the study of the relationship between height, damage and fundamental frequency. Since the microzonation studies will provide in few years iso-frequency maps of the most hazardous municipality, it will then be possible to map the resonance-prone buildings, both for elastic and post-yield frequency.
3. A move toward a two-parameters soil classification. As in other parts of Europe (see, e.g. Pitilakis et al. 2013) also in Italy similar studies are carried on (Luzi et al. 2011). It is now time to implement these study into seismic code abandoning the Vs30 classification scheme.

Acknowledgements I would like to acknowledge the fruitful co-operation with several colleagues of CNR-IMAA, INGV, OGS, Basilicata University, GFZ, CSIC-UCM and Canterbury University.

Above all my co-authors, I would like to give special thanks to Maria Rosaria Gallipoli of CNR-IMAA that made possible most of the papers reviewed here.

Open Access This chapter is distributed under the terms of the Creative Commons Attribution Noncommercial License, which permits any noncommercial use, distribution, and reproduction in any medium, provided the original author(s) and source are credited.

References

Albarello D, D'Amico V (2008) Testing probabilistic seismic hazard estimates by comparison with observations: an example in Italy. Geophys J Int 175(3):1088–1094

Bonilla LF, Archuleta RJ, Lavallée D (2005) Histeretic and dilatant behavior of cohesionless soils and their effects on nonlinear site response: field data observation and modeling. Bull Seismol Soc Am 95:2,373–2,395

Cassidy JF, Mucciarelli M (2010) The importance of ground-truthing for earthquake site response. In: Proceedings of the 9th U.S. National and 10th Canadian conference on earthquake engineering, Toronto, Ontario, Canada , 25–29 July 2010. Paper no. 758

Castellaro S, Padrón LA, Mulargia F (2014) The different response of apparently identical structures: a far-field lesson from the Mirandola 20th May 2012 earthquake. Bull Earthq Eng. doi:10.1007/s10518-013-9505-9

Del Monaco F, Tallini M, De Rose C, Durante F (2013) HVNSR survey in historical downtown L'Aquila (central Italy): site resonance properties vs. subsoil model. Eng Geol 158:34–47

Ditommaso R, Parolai S, Mucciarelli M, Eggert S, Sobiesiak M, Zschau J (2009a) Monitoring the response and the back-radiated energy of a building subjected to ambient vibration and impulsive action: the Falkenhof Tower (Potsdam Germany). Bull Earthq Eng. doi:10.1007/s10518-009-9151-4

Ditommaso R, Mucciarelli M, Gallipoli MR, Ponzo FC (2009b) Effect of a single vibrating building on free-field ground motion: numerical and experimental evidences. Bull Earthq Eng. doi:10.1007/s10518-009-9134-5

Ditommaso R, Mucciarelli M, Ponzo FC (2012) Analysis of non-stationary structural systems by using a band variable filter. Bull Earthq Eng 10(3):895–911. doi:10.1007/s10518-012-9338-y

Ditommaso R, Vona M, Gallipoli MR, Mucciarelli M (2013) Evaluation and considerations about fundamental periods of damaged reinforced concrete buildings. Nat Hazards Earth Syst Sci 13:1903–1912

Gallipoli MR, Mucciarelli M (2009) Comparison of site classification from Vs30, Vs10, and HVSR in Italy. Bull Seismol Soc Am 99:340–351

Gallipoli MR, Albarello D, Mucciarelli M, Bianca M (2011) Ambient noise measurements to support emergency seismic microzonation: the Abruzzo 2009 earthquake experience. Boll di Geofisica Teorica ed Appl 52(3):539–559. doi:10.4430/bgta0031

Gallipoli MR, Bianca M, Mucciarelli M, Parolai S, Picozzi M (2013) Topographic versus stratigraphic amplification: mismatch between code provisions and observations during the L'Aquila (Italy, 2009) sequence. Bull Earthq Eng 11(5):1325–1336

Gallipoli MR, Chiauzzi L, Stabile TA, Mucciarelli M, Masi A, Lizza C, Vignola L (2014) The role of site effects in the comparison between code provisions and the near field strong motion of the Emilia 2012 earthquakes. Submitted to Bull Earthq Eng. doi:10.1007/s10518-014-9628-7

Jennings PC (1970) Distant motion from a building vibration test. Bull Seismol Soc Am 60:2037–2043

Luzi L, Puglia R, Pacor F, Gallipoli MR, Bindi D, Mucciarelli M (2011) Proposal for a soil classification based on parameters alternative or complementary to Vs,30. Bull Earthq Eng 9:1877–1898. doi:10.1007/s10518-011-9274-2

Masi A, Chiauzzi L, Braga F, Mucciarelli M, Vona M, Ditommaso R (2011) Peak and integral seismic parameters of L'Aquila 2009 ground motion: observed vs. code provision values. Bull Earthq Eng 9:139–156

Mucciarelli M (2011) Ambient noise measurements following the 2011 Christchurch earthquake: relationships with previous microzonation studies, liquefaction, and nonlinearity. Seismol Res Lett 82(6):919–926

Mucciarelli M, Bianca M, Ditommaso R, Vona M, Gallipoli MR, Giocoli A, Piscitelli S, Rizzo E, Picozzi M (2011a) Peculiar earthquake damage on a reinforced concrete building in San Gregorio (L'Aquila, Italy): site effects or building defects? Bull Earthq Eng 9:825–840. doi:10.1007/s10518-011-9257-3

Mucciarelli M, Bianca M, Ditommaso R, Gallipoli MR, Masi A, Milkereit C, Parolai S, Picozzi M, Vona M (2011b) Far field damage on RC buildings: the case study of Navelli during the L'Aquila (Italy) seismic sequence, 2009. Bull Earthq Eng 9:263–283. doi:10.1007/s10518-010-9201-y, ISSN: 1570–761X

Pitilakis K, Riga E, Anastasiadis A (2013) New code site classification, amplification factors and normalized response spectra based on a worldwide ground-motion database. Bull Earthq Eng 11:925–996. doi:10.1007/s10518-013-9429-4

Puglia R, Ditommaso R, Pacor F, Mucciarelli M, Luzi L, Bianca M (2011) Frequency variation in site response as observed from strong motion data of the L'Aquila, 2009 seismic sequence. Bull Earthq Eng 9(3):869–892. doi:10.1007/s10518-011-9266-2

Tertulliani A, Arcoraci L, Berardi M, Bernardini F, Camassi R, Castellano C, Del Mese S, Ercolani E, Graziani L, Leschiutta I, Rossi A, Vecchi M (2011) An application of EMS98 in a medium-sized city: the case of L'Aquila (Central Italy) after the April 6, 2009 Mw 6.3 earthquake. Bull Earthq Eng 9:67–80

Zambonelli E, de Nardis R, Filippi L, Nicoletti M, Dolce M (2011) Performance of the Italian strong motion network during the 2009, L'Aquila seismic sequence (central Italy). Bull Earthq Eng 9(1):39–65

Chapter 6
Seismic Analysis and Design of Bridges with an Emphasis to Eurocode Standards

Tatjana Isakovic and Matej Fischinger

Abstract Bridges are quite different from buildings regarding their dimensions, structural systems and in general regarding their seismic response. Thus the specialized standards for their seismic design are needed. One of them is Eurocode 8/2 standard (EC8/2), which considerably improved the design practice. It is well organized, practically oriented and designer friendly.

In Slovenia it has been used for years. Some experiences, obtained during its application in practice are presented. Four issues are addressed: (1) the correlation between pre-yielding stiffness and strength of structures as well as the reduction of the seismic forces and equal displacement rule, (2) the application of the nonlinear static (pushover) methods of analysis, (3) the estimation of the shear strength of RC columns, and (4) the protection of the longitudinal reinforcement in RC columns against buckling.

It was concluded that pre-yielding stiffness and strength of structures are strongly correlated. The pre-yielding stiffness is different for different levels of selected strength. This does not negate the equal displacement rule. The EC8/2 is one of the rare standards that explicitly recognized the quite important correlation between chosen strength of structures and corresponding pre-yielding effective stiffness. Accordingly, the equal displacement rule is presented in a modified way. Different interpretations of this rule are discussed in the paper.

The EC8/2 introduced the nonlinear static pushover methods into the design practice. The way of their use is examined in the paper. Specifics in the application of the single mode pushover methods and the scope of their applicability are discussed. Some of the alternative methods are briefly overviewed.

It was found that EC8/2 provisions related to the estimation of the shear strength of some typical bridge columns can be quite conservative. Some of the alternative

T. Isakovic (✉) • M. Fischinger
Faculty of Civil and Geodetic Engineering, University of Ljubljana, Jamova 2, 1000, Ljubljana, Slovenia
e-mail: tisak@ikpir.fgg.uni-lj.si; tatjana.isakovic@fgg.uni-lj.si; mfischin@ikpir.fgg.uni-lj.si; matej.fischinger@fgg.uni-lj.si

methods are presented and discussed on the example of the experimentally investigated columns. It is concluded that the estimation of the shear strength, in general, is far from being solved and it demands further investigations.

It was also found that some requirements of EC8/2 related to the prevention of buckling of the longitudinal reinforcement in RC columns are not interpreted in an appropriate way; thus their corrections are needed.

6.1 Introduction

Bridges are specific structures whose structural concept is mostly related to functionality. They give the impression of being rather simple structures whose seismic response could be easily predicted. Therefore, in the past, a little attention was paid to their seismic design. Usually, the design methodologies, developed primarily for the analysis and design of buildings were also uncritically applied to bridges. In many cases this approach was/is inappropriate, since the structural system of bridges, dimensions, and their seismic response, in general, is considerably different from buildings.

The need for special consideration, which is adjusted to specific properties of bridges, has been recognized and the practice has been changed. An example of this good practice is the Eurocode standard, which comprises a part Eurocode 8/2 (CEN 2005a) – EC8/2 that regulates the seismic design of bridges.

This standard includes many modern design principles of the seismic engineering, which were usually not taken into account in the design practice in the past, and very often they are not taken into account even in the nowadays practice. In some countries, e.g. in Slovenia, it has been used for many years. In the beginning, the pre-standard version of EC8/2 (CEN 1994) was applied. Although the early applications were unofficial, most of the bridges built on the main highways in Slovenia were designed taking into account its requirements. For the last 6 years it has been used as an official standard in Slovenia.

Based on the experiences obtained during its application, it can be concluded that EC8/2 definitely considerably improved the seismic design of bridges. It is well organized, practically oriented and designer friendly.

In this paper some of the experiences, obtained when applying the standard in the practice and a critical overview of some of its requirements are presented. They are listed in the next paragraphs.

1. The reduction of the seismic forces and equal displacement rule are well known and they are regularly used in the design practice. Nevertheless, sometimes they are applied, neglecting the correlation between the strength of the structure and corresponding pre-yielding effective stiffness (initial effective stiffness). As a consequence some researchers and designers expressed their doubts about this basic principle of the seismic engineering. Following the previous discussion

about these issues, and solutions that are proposed in EC8/2, the problem of the correlated strength (reduction of forces) and equivalent initial stiffness is examined in Sect. 6.2.
2. Seismic load is the strongest load that threatened the bridge in the seismically prone areas. Accordingly, many structures can be exposed to significant plastic deformations and its response can be significantly nonlinear. Nevertheless, the elastic linear methods are usually used for their analysis.

    In bridges, which are supported by piers having very different stiffness and strength, a considerable redistribution of the effects of the seismic load in the transverse direction of the bridge is usually observed comparing to the results of the linear analysis. Consequently, the nonlinear methods are needed in such cases, since the linear methods cannot estimate the response realistically. This was recognized by the EC8/2 standard as well. In bridges, where the significant redistribution of the seismic effects is expected, the nonlinear analysis is suggested as an option to estimate their seismic response more realistically.

    As an alternative to the nonlinear dynamic analysis, which is still too demanding for the everyday design, a simplified nonlinear pushover method, N2 method (Fajfar and Fischinger 1987) is included to the EC8/2. This method was primarily developed for the analysis of buildings. Therefore some important modifications are needed when it is applied to bridges. They are discussed in Sect. 6.3. Moreover, the limitations of the method are analyzed and possible alternatives are briefly presented.
3. It has been observed that EC8/2 requirements related to the estimation of the shear strength can be quite conservative for some typical types of bridge columns (e.g. hollow box columns). Namely, the contribution of the concrete to the shear strength should be quite often neglected even if the displacement demand is relatively low. Since this contribution can be as large as the half of the total shear strength of a column, quite a large shear reinforcement could be required if this contribution is not taken into account.

    It should be noted that according to the organization of the Eurocode standards, this subject is primarily related to Eurocode 2 standard, EC2 (CEN 2004a), where the procedure for estimation of the shear strength is defined. However, these already conservative requirements of EC2 are in some cases additionally tightened by EC8/2, which sometimes additionally reduces the already low level of the shear strength defined in EC2. This issue is discussed in Sect. 6.4.
4. The ductility capacity of the column (bridge) strongly depends on the ability of its lateral reinforcement to sustain the buckling of the longitudinal flexural reinforcement and to ensure the adequate confinement of the concrete core. These two functions of lateral reinforcement were in the past designs in many cases neglected, and are not considered even in some new designs. This can lead to undesirable brittle types of failure and irreparable types of damage. In EC8/2 a special attention is devoted to these problems. However, some provisions require certain modifications, which are discussed in Sect. 6.5.

6.2 The Strength and the Effective Stiffness – The Equal Displacement Rule

According to EC8/2 bridges can be designed so that their behaviour under the design seismic action is either ductile or limited ductile/essentially elastic. The type of the response depends on the chosen global behaviour factor. It defines the global level of the reduction of forces that would be obtained in the structure, which responds to the seismic load elastically and have the same effective pre-yielding stiffness as the analysed structure. In EC8/2 the limited ductile and ductile response corresponds to the behaviour factor of 1.5 and 3, respectively.

When the larger reduction of forces (larger behaviour factor) is chosen, the larger global ductility capacity of the analysed structure is required, since the displacement demand in a structure, which respond to the seismic load elastically and the corresponding structure with the reduced strength and the same pre-yielding stiffness and mass are in general approximately the same. This is so called equal displacement rule, which is more strictly speaking, applicable to structures with medium and long periods of vibrations.

This basic principle of the seismic engineering is usually illustrated with the idealized force-displacement diagram presented in Fig. 6.1a. The larger reduction of the force means that the smaller strength and the larger ductility capacity of the structure should be provided. In this presentation the pre-yielding stiffness is independent of the level of the force reduction (strength). For the reasons explained in the next paragraphs, this presentation is applicable only to *different* structures with the *same effective pre-yielding stiffness* and *different strengths*.

In general it cannot be applied to *one structure* with the same geometry of structural components and different levels of the provided strength. For this case, the equal displacement rule can be presented in a different way, as it is illustrated in Fig. 6.1b for medium and long period structures. It is assumed that the yield displacements are approximately the same; regardless of the strength (explanation is provided later in this section). For the sake of simplicity the rule is presented for the case of a simple cantilever. For more complex structure it is discussed later in this section.

In Fig. 6.1b three types of response (three levels of force reduction) are examined: (1) The essentially elastic response (presented with black line), (2) limited ductile (presented with red line) and (3) ductile response (presented with blue line). The F_{e1}, F_{e2} and F_{e3} represent the elastic forces, which correspond to certain effective pre-yielding stiffness that is correlated with the chosen strength (reduction of forces or chosen behaviour factor). Forces F_{y2} and F_{y3} are reduced forces. They are defined as it is proposed in EC8/2 reducing the force F_{e1} by factors 1.5 and 3. Thus, F_{y2} is 1.5 and F_{y3} is 3 times smaller than F_{e1}, respectively. Seismic displacements corresponding to three examined types of response are denoted as D_1, D_2 and D_3 respectively. Corresponding yielding displacements are denoted as D_{y1}, D_{y2} and D_{y3}.

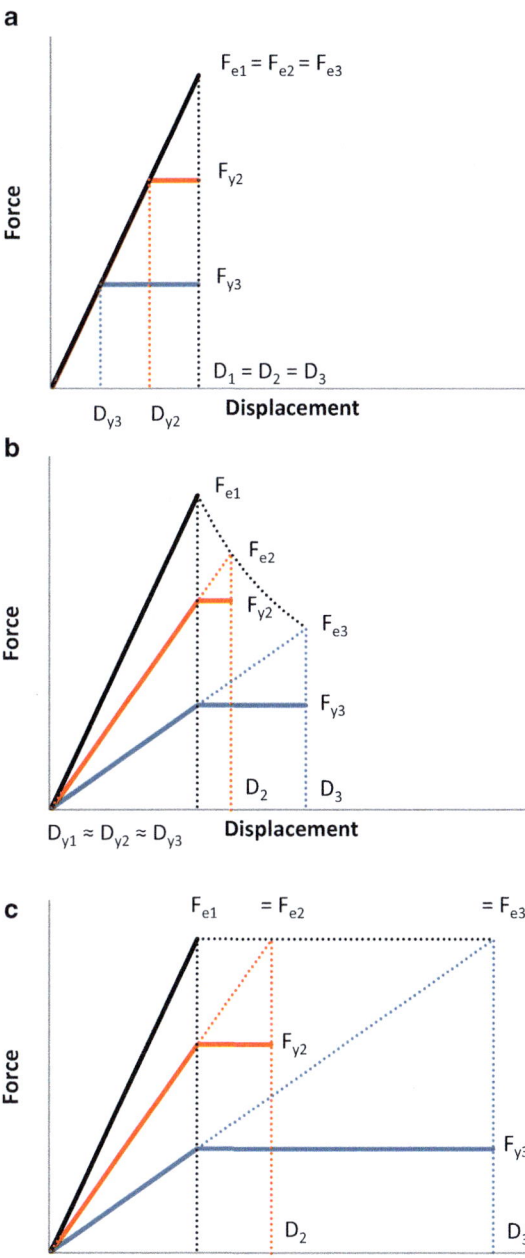

Fig. 6.1 Different interpretations of the equal displacement rule. (**a**) Traditional interpretation of the equal displacement rule. (**b**) The equal displacement rule, where the correlation between the strength and the stiffness is taken into account. (**c**) Interpretation of the equal displacement rule in EC8/2

Contrary to the interpretation in Fig. 6.1a, where the effective pre-yielding stiffness is independent of the level of the force reduction, in the interpretation, presented in Fig. 6.1b, this stiffness varies based on the chosen level of strength or the chosen level of the force reduction. Moreover, the seismic displacements D_1–D_3 as well as the elastic forces F_{e1}–F_{e3} are not the same (as in Fig. 6.1a) and are, in general, also dependent on the chosen reduction of forces.

A superficial analysis of the diagrams, presented in Fig. 6.1b, can lead to a conclusion that equal displacement rule is invalid. This opinion is recently often expressed by different researchers. However, the precise examination of the presented diagrams confirms that equal displacement rule is not doubtful. The seismic displacements D_1–D_3 are still the same as those that characterize the *corresponding* elastic response, calculated taking into account the adequate (corresponding) effective pre-yielding stiffness. The ratio of the seismic displacements and yield displacements are still approximately the same as the corresponding level of the force reduction. Note that actual global reduction of forces is somewhat smaller than 1.5 and 3, since the corresponding elastic forces F_{e2} and F_{e3} are also smaller than F_{e1}, which was used to select the reduced strength F_{y2} and F_{y3}.

In other words, the equal displacements rule is valid, but it should be adequately interpreted, taking into account the correlation between the strength of the structure and the corresponding pre-yielding stiffness as well as the corresponding reduced demand. It is applicable for each level of the chosen strength individually. This is illustrated in Fig. 6.1b.

The strong correlation between the strength and effective pre-yielding stiffness is crucial for the proper interpretation of Fig. 6.1b. Therefore it is analysed in more details in the next paragraphs. For the sake of clarity, this relationship is analysed on the example of simple cantilever column (presented in Fig. 6.2a). It is assumed that the strength of the column is chosen and that it is expressed in term of the force F_y.

The selected level of force can be resisted providing an appropriate bending moment resistance at the bottom of the cantilever $M_y = F_y \times h$. In this expression M_y is the bending moment corresponding to yielding of the cantilever, h represents its height and F_y the force that should be resisted (chosen strength).

For the sake of simplicity, let's assume that the response of the analysed structure is perfectly elasto-plastic (there is no strain hardening after yielding). This means that the moment M_y represents also the bending moment capacity that corresponds to the chosen level of force reduction.

Moment M_y can be further correlated with the yield curvature Φ_y using the simple expression:

$$\Phi_y = \frac{M_y}{EI_{eff}} \qquad (6.1)$$

where E is the modulus of elasticity and I_{eff} the effective moment of inertia of the bottom most critical cross-section. The yield curvature depends first of all on the yield strain of the reinforcement and the effective depth of the cross-section. The

Fig. 6.2 The cantilever column

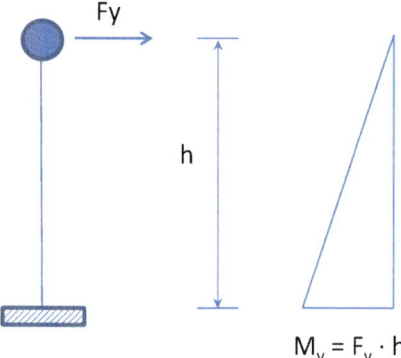

Fig. 6.3 Moment – curvature relationship of one cross section for different levels of axial forces

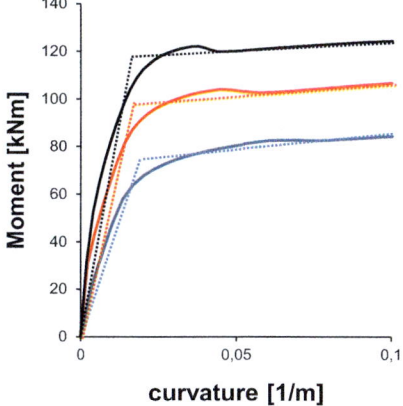

variations of the axial force and the corresponding variation of the yield moment have only slight influence to the value of the yield curvature. This is documented on the example, presented in Fig. 6.3. More examples can be found elsewhere (e.g. Priestley et al. 2007).

Considering a small variation of the yield curvature, it is evident from Eq. (6.1) that the variations of the yield moment (bending moment capacity) has considerable influence only to the effective moment of the inertia I_{eff}. Consequently it has also considerable influence to the effective pre-yielding stiffness. Since the curvature Φ_y is almost independent of the level of the yield moment, the effective pre-yielding stiffness and yield moment are explicitly correlated. In other words, the effective stiffness cannot be randomly chosen, when the yield moment (strength) is selected and vice versa. In general, the effective stiffness varies proportionally to the strength. This is illustrated in Fig. 6.1b.

As it was mentioned before, the seismic displacements corresponding to different strength levels are not equal (displacements D_1–D_3 in Fig. 6.1b). Instead, the yield displacements (displacements D_{y1}–D_{y3} in Fig. 6.1b) are quite similar and almost independent of the strength (taking into account that the yield curvature is not strongly correlated with the strength).

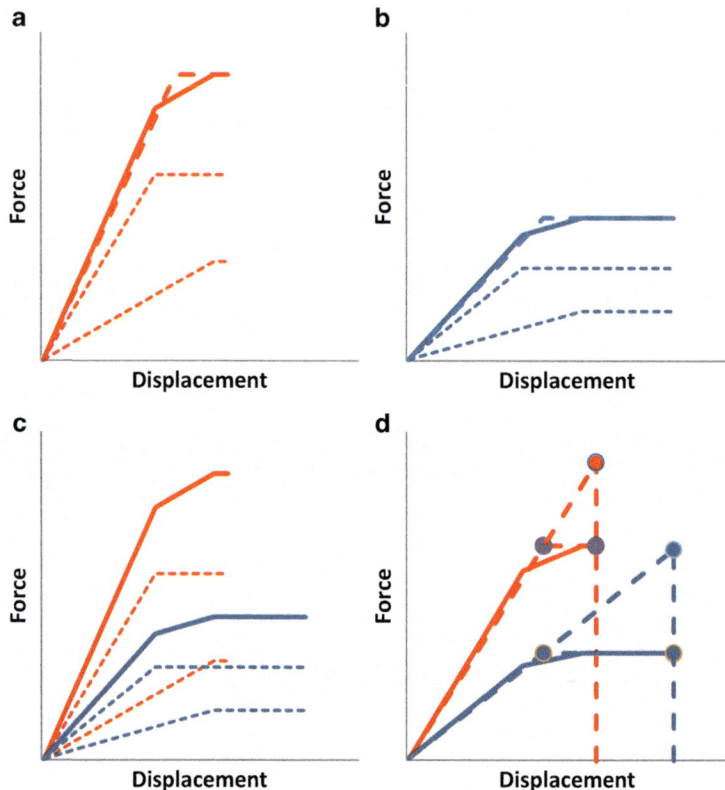

Fig. 6.4 Equal displacement rule in the case of bridge, supported by two columns

In more complex structures the relationship between global effective pre-yielding stiffness and the strength is not so straightforward. In general, iterations are needed, particularly when the bridge is analysed in the transverse direction and when the analysed structure is irregular, supported by columns of different heights and strengths. However, the conclusions, presented above are in general essentially the same. The effective stiffness and strength are correlated, and the effective stiffness varies proportionally to the variations of the strength.

This is illustrated in Fig. 6.4, where the response of a bridge, supported by two columns of different strength and stiffness is analysed in its longitudinal direction. For each column a force-displacement relationship is defined (red dashed lines in Fig. 6.4a). The total stiffness of the structure can be obtained summing the stiffness of both columns. Thus, the total force-displacement diagram can be determined summing the forces in both structural components (bold solid red line in Fig. 6.4a). The effective pre-yielding stiffness of the whole structure can be defined taking into account equal energy rule (bold dashed red line in Fig. 6.4a). This stiffness defines the equivalent period of the structure, which further influences the seismic displacements (see red lines in Fig. 6.4d).

6 Seismic Analysis and Design of Bridges with an Emphasis to Eurocode Standards 203

When the strength of both components is decreased, the effective stiffness is also decreased. This is illustrated in Fig. 6.4b, where the equivalent stiffness (bold blue dashed line) is defined in the same way as it was explained on the example, presented in Fig. 6.4a. Both cases are compared in Fig. 6.4c, where it is evident that reduction of strength also means the reduction of the effective stiffness. It can be concluded that the strength and the stiffness are strongly correlated also in more complex structures. In other words, if the strength of the structure is chosen, the stiffness of single components and the global stiffness cannot be randomly selected and vice versa.

The seismic displacements of the analysed bridge can be estimated using the equal displacement rule presented in Fig. 6.4d. The presentation of this rule is essentially the same as in the simple cantilever structure. The yield displacements are almost independent of the strength. Contrary, the seismic displacements significantly vary depending on the pre-yielding stiffness and the chosen strength.

As it was mentioned earlier, the correlation between stiffness, strength, and seismic displacement demand is more complex than in the simple cantilever beam. The equivalent pre-yielding stiffness is not a simple sum of the pre-yielding stiffness of single components (as it is illustrated in Fig. 6.4a, b). In general iterations are needed.

The correlation between the effective per-yielding stiffness and the strength is recognized in the standard EC8/2 (see Fig. 6.1c). The interpretation of the equal displacement rule is similar to that presented in Fig. 6.1b, with an important difference. The strength of all structures exhibiting the elastic response is presented to be the same (forces F_{e1}, F_{e2} and F_{e3}).

Taking into account the EC8 acceleration spectrum, it can be concluded that in many medium and long period structures, the elastic forces determined in this way are overestimated. Consequently the seismic displacements are also overestimated. This means that an additional safety is introduced to the design. Taking into account the complexity of the response (e.g. the redistribution of the seismic effects in the nonlinear range) and considering that standard EC8/2 does not require explicit examination of the available displacement ductility capacity (it is ensured by special detailing rules) this additional safety is feasible. It should be noted that in the case of highly irregular structures, where in the nonlinear range the considerable redistribution of seismic effects between the single components can occur, the examination of the seismic response using the nonlinear methods (see next Section) is highly recommended.

The elastic forces F_{e1}, F_{e2} and F_{e3} could be the same for certain short period structure with periods suited to the resonant region of spectrum. However, in this region the seismic displacement defined using the equal displacement rule should be modified (increased).

6.3 The Nonlinear Static Pushover Analysis

The EC8/2 standard recognized the need for more reliable estimation of the highly nonlinear seismic response of bridges. It introduced the nonlinear methods into the design practice: (a) the most refined nonlinear response-history analysis (NRHA), as well as (b) simplified nonlinear pushover based method – N2 method.

In most of the cases, the most refined NRHA is still quite complex to be used in the everyday design. It requires a lot of experiences regarding the modelling of the dynamic response of structures and an appropriate modelling of the seismic loading as well. The specialized software is needed. Thus, to simplify the nonlinear analysis and to make it more regulated, different simplified nonlinear methods can be used. There are many variations of different simplified nonlinear methods proposed, mostly for the analysis of buildings. They can be divided regarding the influence of the higher modes and variability of the important mode shapes based on the different levels of the seismic load.

The simplest methods assume that the response is governed by one predominant mode, which does not essentially change when the seismic load is changed. These methods can be characterized as the single-mode non-adaptive methods.

The next more complex group of methods takes into account the influence of the higher modes, but still suppose that these modes are essentially independent of the seismic intensity. These are so called multimode non-adaptive pushover methods.

The more complex methods take into account the influence of the higher modes as well as their changes based on the seismic intensity. These are so called multimode adaptive methods.

The accuracy of these methods depends on many parameters. A comprehensive analysis of these parameters as well as the list of different pushover methods can be found in FEMA-440 (2005). This document is related mostly to buildings. More specialized information about the application of different pushover methods for the analysis of bridges can be found in Kappos et al. (2012).

In this paper the single-mode non-adaptive method, which is included into EC8/2 (and to Eurocode 8/1 – CEN 2004b) the N2 method (Fajfar 1999) is analysed first. As it was mentioned before, it was developed primarily for the analysis of buildings. When it is applied to bridges it can be used in the unmodified way only when the analysis is performed in the longitudinal direction. In the transverse direction, the structural system of bridges and consequently their response is considerably different from buildings. Therefore some modifications of the method are needed. They are described in Sect. 6.3.1.

Since the N2 method is simplified, it has certain limitations. They are presented in Sect. 6.3.2 and illustrated with the appropriate numerical examples. Section 6.3.3 includes a brief overview of two alternative methods: multimode non-adaptive MPA method (Chopra and Goel 2002), and multimode adaptive IRSA (Aydinoğlu 2003) method, which can be used when the N2 method is not suitable for the analysis. Others can be found e.g. in Kappos et al. (2012) or FEMA-440 (2005).

6.3.1 Specifics of the N2 Method When Applied to the Analysis of Bridges

The N2 method was initially proposed and developed for the design of buildings (Fajfar and Fischinger 1987; Fajfar 1999). Later it has been subsequently improved and generalized. It has been applied for special types of buildings like infilled frames (Dolšek and Fajfar 2005) and for 3D analysis (Fajfar et al. 2005). First applications for bridges were published in mid-90s (Fajfar et al. 1997).

The name N2 method describes its basic features. N stands for the nonlinear analysis, and 2 for the two models and two types of analysis: (1) nonlinear static analysis of the actual multi-degree-of-freedom model (MDOF model) of the structure and (2) nonlinear dynamic analysis of corresponding simplified single-degree-of-freedom model (SDOF model). The nonlinear static analysis is used to define the basic effective properties of the structure, such as e.g. effective stiffness, which are further needed to define an equivalent SDOF model, used for the nonlinear dynamic analysis.

It has been realized (i.e. Isakovic and Fischinger 2006), that in the application of the N2 method as well as all other similar procedures, which were originally developed for buildings, one should take into account special properties of the bridge structural system. Before these specifics are described, let us overview the basic steps of the method, first (see Fig. 6.5):

1. First, the multi-degree-of-freedom (MDOF) model of structure is defined.
2. The MDOF model is subjected to the lateral static (inertial) load, which is gradually increased and the displacement of the superstructure is monitored (pushover analysis is performed),
3. Based on the analysis performed in the second step, the force-displacement relationship is defined (the relationship total base shear versus displacement at the chosen position is defined; pushover curve is constructed),
4. The relationship determined in the third step is used to define an equivalent SDOF model of the structure, which is further used for the nonlinear dynamic analysis,
5. The nonlinear dynamic analysis is performed using the nonlinear response spectra that can be defined based on the standard elastic acceleration spectra.
6. The result of the nonlinear dynamic analysis is the maximum displacement of the bridge at the chosen position, corresponding to the certain seismic intensity.
7. Considering the maximum displacement, defined by the nonlinear dynamic analysis, the MDOF model is pushed again with forces defined in the 2nd step and different aspects of the bridge response is analysed

The modifications of the N2 method, which are needed when it is applied to bridges, are related to:

1. The distribution of the lateral forces along the superstructure (see 2nd step above)

1. MDOF model (nonlinear properties of columns are taken into account)

2. Static nonlinear analysis of MDOF model – pushover analysis

3. Construction of the pushover curve (base shear – displacement relationship)

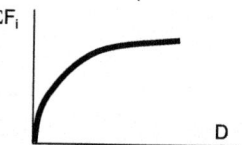

4. Construction of the capacity curve and definiton of the SDOF model

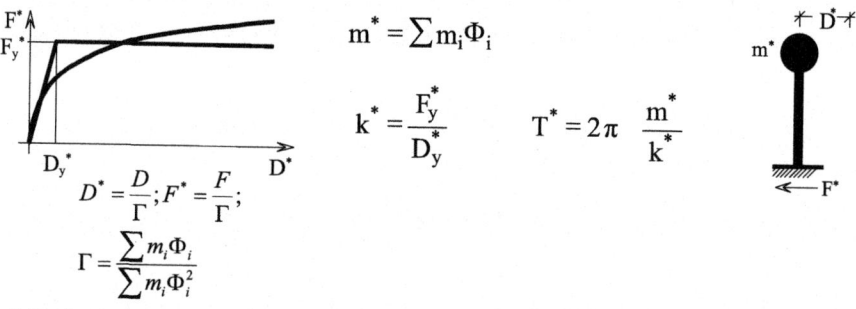

5. Nonlinear dynamic analysis of SDOF model using the nonlinear response spectra, which can be defined based on the elastic response spectra from EC8/2

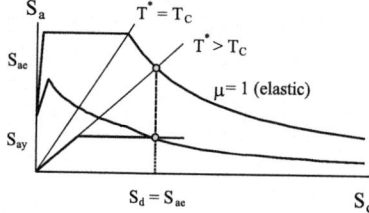

6. Transformation of the maximum displacement of the SDOF model, obtained in the 5th step, to the maximum displacement of the MDOF model using the transformation presented in the 4th step.

7. MDOF model is pushed again (nonlinear static analysis) with forces, defined in the 1st step, considering the maximum displacement obtained in the 6th step in order to analyze different aspects of the bridge seismic response.

Fig. 6.5 The scheme of the N2 method

Fig. 6.6 Distributions of the lateral load, appropriate for bridges that are pinned at the abutments

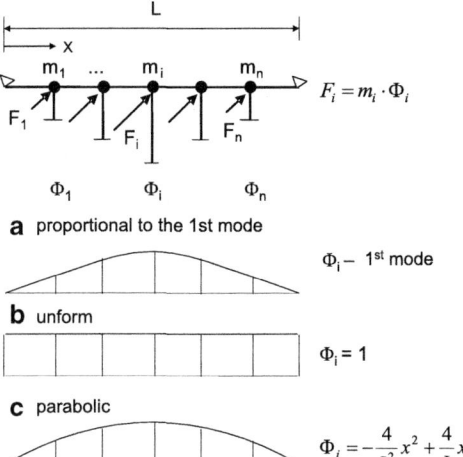

2. The choice of the reference point on the structure where the displacements are monitored in order to obtain the force-displacement relationship (see 3rd step),
3. Idealization of the force-displacement curve, and calculation of yielding force F_y^* and yielding displacement D_y^* (see 4th step).

6.3.1.1 Distribution of the Lateral Load

In the 2nd step of the N2 method (see Sect. 6.3.1) the MDOF model of the structure is subjected to the static lateral load (inertial forces). The distribution of the inertial forces (lateral load) should be assumed before the nonlinear static analysis is performed. In the Annex H (informative annex) of Standard EC8/2 two possible distributions are proposed: (a) distribution proportional to the 1st mode of the bridge in the elastic range, and (b) uniform distribution (see Figs. 6.6a, b and 6.7a, b). The first distribution can be defined based on the simple modal analysis with some of the standard programs for elastic modal analysis.

In the previous research (Isakovic and Fischinger 2006), it was found that the parabolic distribution (Fig. 6.6c) was appropriate for bridges that were pinned at the abutments. This distribution is simpler to define than that proportional to the first mode. Using the parabolic distribution, in many bridges the response can be estimated better than in the case of the uniform distribution. For more details see Isakovic et al. (2008a) and Kappos et al. (2012).

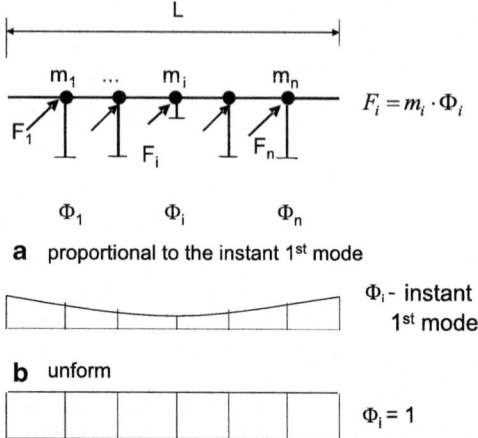

Fig. 6.7 Distributions of the lateral load, appropriate for bridges with roller supports at the abutments

6.3.1.2 The Choice of the Reference Point

One of the crucial steps in the application of the N2 method is the static nonlinear analysis of the MDOF system. Based on this analysis the force-displacement relationship is determined, which is further used to define the properties of the equivalent SDOF system.

The force-displacement relationship is determined observing changes of displacement at the certain position in the structure (reference point) due to the gradual increase of the lateral load. The top of buildings is typically selected as the reference point, since at this position the maximum displacement is typically observed in the majority of cases. In bridges this choice is not so straightforward.

In EC8/2 the centre of the mass of the deformed deck is proposed as the reference point. An alternative solution could be the top of a certain column. However, in irregular bridges both of these solutions could be inadequate.

In highly irregular bridges the influence of higher modes is typically large and variation of mode shapes is substantial (especially if the structure is torsionally sensitive). Consequently, the station of maximum displacement varies and it depends on the intensity of the load. This can quite complicate the construction of the pushover curve. The question arises, which point is the reference point. The authors of the paper believe that the pushover curve should be constructed using the maximum displacement of the superstructure regardless its position, since the maximum displacement is a measure of stiffness of the superstructure. In other words the station of the reference point is not always constant throughout the analysis.

Let's analyse the response of the viaduct V213P, presented in Fig. 6.8. Considering displacements at the top of three different columns, three very different pushover curves were obtained (curves 1–3 in Fig. 6.9a). Consequently, very different stiffness of the equivalent SDOF model was obtained, resulting in very

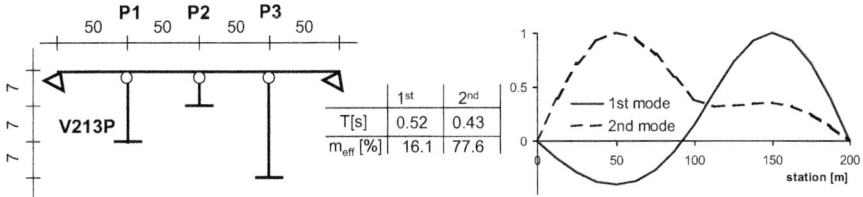

Fig. 6.8 An example of a highly irregular bridge

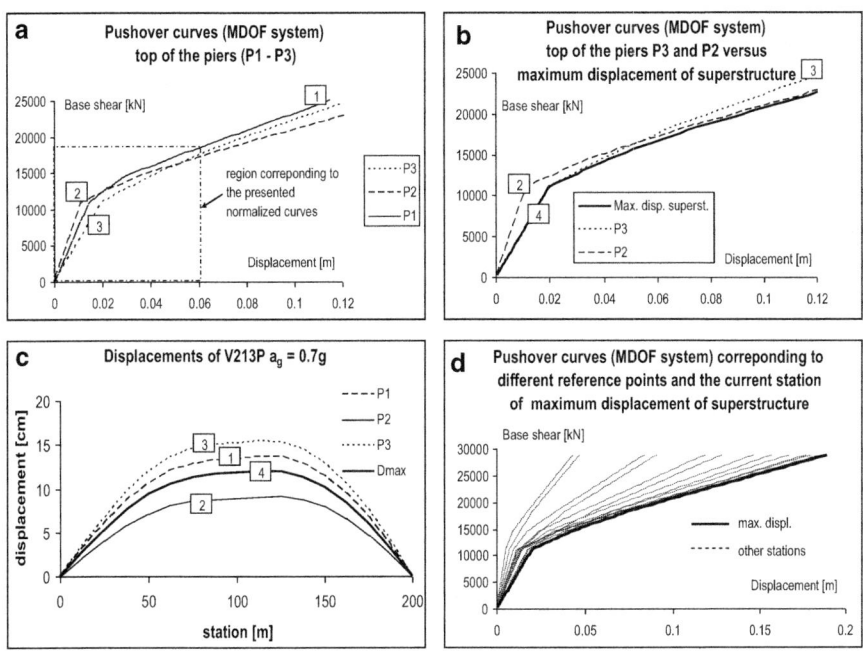

Fig. 6.9 Pushover curves, and displacement envelopes, obtained based on different reference points

different estimation of dynamic properties of equivalent SDOF system and significantly different displacements of the structure (see curves 1–3 in Fig. 6.9c). One can conclude that the pushover curve corresponding to the column with maximum displacement at the top should be evidently used. This is true so far this is the station of the maximum displacement of the superstructure, too. The station of maximum displacement of the superstructure in viaduct V213P does not coincide with the position of any column. Moreover it changes depending on the level of the load. Therefore, the corresponding pushover curve (see curve 4 in Fig. 6.9b) does not coincide with any of the pushover curves constructed based on the displacements monitored at the top of some column. Consequently, the corresponding displacements of the viaduct also differ from those, calculated using the top of the columns as the reference points (see Fig. 6.9c).

The analysed viaduct is highly irregular structure, where the mode shapes, their importance and ratios are changing depending on the seismic intensity. When the seismic load is low and the structure respond elastically the maximum displacement is above the right column. When the load is increased the position of the maximum displacement gradually moves toward the centre of the bridge. Station of the maximum displacement gradually shifts for about 40 m (20 % of the bridge length). Thus, the maximum displacement occurs at the centre of mass only at stronger seismic intensities.

The reason for such behaviour is a significant variation of shape, order and importance of modes. The authors believe that the proper pushover curve is the lowest possible one (bold line in Fig. 6.9d), corresponding to the current maximum displacement of the superstructure.

6.3.1.3 Idealization of the Pushover Curve, Target Displacement

Idealization of the base shear-displacement relationship is one of the basic steps of the N2 method, since it significantly influences the stiffness of the equivalent SDOF model and the value of the maximum displacement. When this stiffness is not adequately estimated, the actual and estimated maximum displacement can be significantly different (Isakovic and Fischinger 2006; Isakovic et al. 2008a).

Elasto-plastic idealisation is typically used. This solution is also proposed in EC8/2. However, in viaducts, which are pinned at the abutments, this idealization can be inappropriate, since an underestimated equivalent stiffness of the SDOF system, and overestimated maximum displacement (see Fig. 6.10) can be obtained. Namely, in bridges with pinned abutments where the elastic response of the superstructure is expected, the pushover curve exhibits considerable strain hardening slope, which should be properly taken into account. This is illustrated in Fig. 6.10.

The force-displacement relationship is usually idealized using the equal energy principle of idealized and actual curve. Since the energy depends on the reached maximum displacement, which is not known at the moment of the idealization, the authors' opinion is that iterations are necessary. In the majority of cases, only one iteration is needed.

In the annex H of the EC8/2 it is proposed that the maximum displacement is estimated using the results of the elastic analysis. This solution is very convenient at the first glance. However, to estimate these displacements properly, the pre-yielding effective stiffness of the whole structure corresponding to a certain level of the seismic load should be also defined. That means that (more) iterations are also needed (see Sect. 6.2).

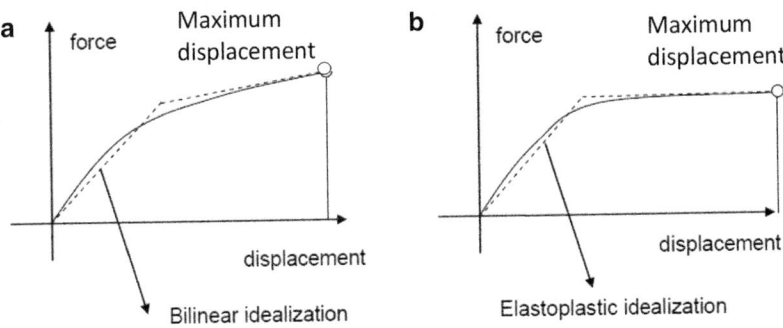

Fig. 6.10 Idealization of the pushover curve. (**a**) Bridges pinned at the abutments. (**b**) Bridges with roller supports at the abutments

6.3.2 Applicability of the N2 Method

The N2 method is a typical single mode non-adaptive pushover method. Although it is appropriate for the analysis of many bridges, it has certain limitations. Since it is single-mode method, it can take into account the predominant influence of only one vibration mode. Therefore, it is appropriate for the analysis of bridges, where the influence of the higher modes is not very important. This is the case where the effective mass of the predominant mode exceeds 80 % of the total mass.

The method is non-adaptive, which means that it cannot take into account significant variations of the predominant mode of vibration. Therefore, it is suitable for the analysis of bridges where the predominant mode does not significantly change.

The N2 method can be efficiently used for the estimation of the seismic response of the majority of the short and medium length bridges. An example of the good estimation of the bridge seismic response is presented in Fig. 6.11, where the displacements calculated by the N2 method and NRHA are compared. The response of the presented bridge is influenced by one predominant mode, which does not considerably change with the seismic intensity.

In short bridges and bridges of medium length, the accuracy of the N2 method can depend on the seismic intensity. Usually the higher intensity means better accuracy.

The example of such bridge is presented in Fig. 6.8. In the elastic range the response is influenced by two modes (Fig. 6.8). Consequently, the results of the N2 method (see dashed line in Fig. 6.12a) does not agree very well with the results of the nonlinear response-history analysis – NRHA (see solid line in Fig. 6.12a). However, when the seismic intensity is increased, the response is influenced by only one predominant mode. Consequently, the results of the N2 method agree better with the results of the nonlinear response-history analysis (see Fig. 6.12b).

Fig. 6.11 In bridges, where the response is influenced by one predominant mode, the response is estimated very well by the N2 method

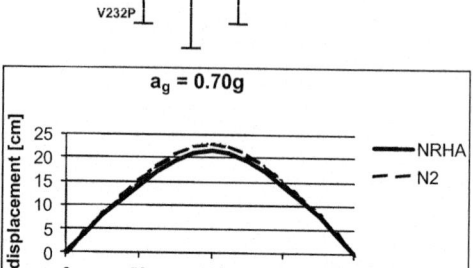

Fig. 6.12 The accuracy of the N2 method in some bridges depends on the seismic intensity

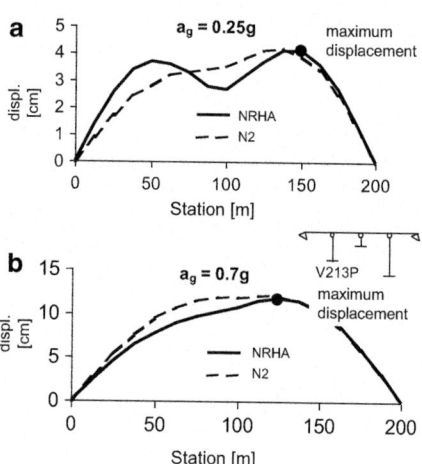

NRHA – nonlinear time history analysis, N2 – the N2 method

However, this is not a rule. There are certain types of bridges, reported in Isakovic and Fischinger (2011) where the accuracy of the method decreases with the intensity of the seismic load.

The N2 method is, in general, less accurate in the case of long bridges. It was found (Isakovic et al. 2008a) that in long bridges (e.g. the length is over 500 m), due to the large flexibility of the superstructure (due to the large length), the response is very often significantly influenced by higher modes even if they are supported by relatively flexible columns. For the analysis of such bridges multimode pushover methods can be used (see next subsection) or they can be analysed by the nonlinear response-history analysis.

Let us summarize the previous findings. The N2 method can be used in bridges where:

(a) The stiffness of the superstructure is large comparing to that of the columns. In such bridges the superstructure governs the response. This is typical for viaducts which are not too long and which are not supported by very short columns.
(b) The stiffness of the columns does not change abruptly. Namely, if a bridge is supported by columns of very different heights, each column tends to move in its natural mode. Therefore, when the superstructure is not stiff enough to control the overall response, the response is considerably influenced by higher modes.

More details about the applicability of the N2 method can be found elsewhere (Isakovic and Fischinger 2006; Isakovic et al 2008a).

6.3.3 Alternative Pushover Methods of Analysis

When the higher modes have an important role in the response of a bridge, two solutions are available: (a) the multimode pushover methods can be employed, or (b) the NRHA is performed. The choice depends again on the complexity of the bridge, experiences, available software, etc. It is worthy to note that the more refined methods demand also the more refined analysis tools. As it has been mentioned before different multimode pushover methods are available. Here, two of them: (a) non-adaptive MPA and (b) adaptive IRSA are briefly summarized.

6.3.3.1 The MPA Method

The MPA method has been developed by Chopra and Goel (2002). Later it has been modified by the authors (Goel and Chopra 2005) and other researchers, e.g. (Paraskeva et al. 2006; Paraskeva and Kappos 2009), who have been focused on the seismic response of bridges. It is simplified nonlinear pushover method, which can take into account the influence of the higher modes to the seismic response of structures.

In the MPA method the number of pushover analyses depends on the number of the important modes of vibration. Each analysis is preformed taking into account the lateral load proportional to corresponding elastic mode shape. The calculation procedure is similar to that described in Sect. 6.3.1. It is repeated taking into account each important mode, separately. Then the contributions of individual modes are combined using the SRSS or CQC combination rule.

One of the differences between the N2 method and the MPA method is related to the choice of the reference point. In the MPA, the displacements can be monitored anywhere along the superstructure, so far the mode shapes do not considerably change, because in the MPA method the shape factor is taken into account. However, when the mode shapes considerably depends on the load intensity, the

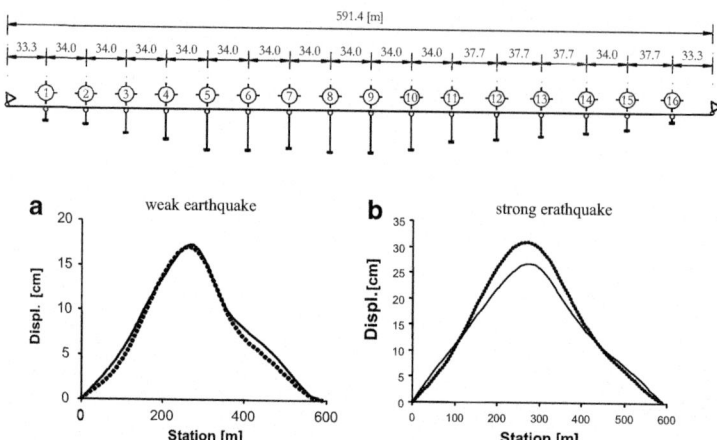

Fig. 6.13 In long bridges with common pier configuration, the accuracy of the MPA (*dotted line*) is very well (results of the NRHA are presented with the *solid line*)

appropriate choice of the monitoring point is as important as in the N2 method (Isakovic and Fischinger 2006). In such cases the ratio of displacements along the superstructure is variable and the constant shape factor used in the method cannot take into account these changes. Therefore, in such bridges it is recommended to consider the maximum displacement of the superstructure at its current (variable) position (as it is proposed for the N2 method – see the comment in Sect. 6.3.1). The results of the MPA can be considerably improved taking into account modifications proposed by Paraskeva et al. (2006) and Paraskeva and Kappos (2009).

The analysis with the MPA method is reasonable when the higher modes have considerable influence to the response of the bridge (when N2 method is less accurate), e.g. in very long bridges (e.g. when the length of the bridge is 500 m or more). In such bridges the influence of the higher modes is usually important, particularly when they are supported by short (very stiff) columns. The accuracy is good when the mode shapes do not considerably depend on the seismic intensity.

An example is presented in Fig. 6.13. The displacements of the bridge calculated by the MPA and the NRHA method are compared for two seismic intensity levels. The match between the MPA and NRHA is quite good, particularly for the weak seismic intensity, since the mode shapes are close to the initial mode shapes corresponding to the elastic range. For the strong earthquake, the results of the MPA and NRHA method still agree well, since the mode shapes do not considerably change comparing to the elastic range.

If the modes of vibrations are variable, then the MPA method is not feasible enough, like in the bridge presented in Fig. 6.14. In such cases adaptive methods can be employed, or the NRHA preformed.

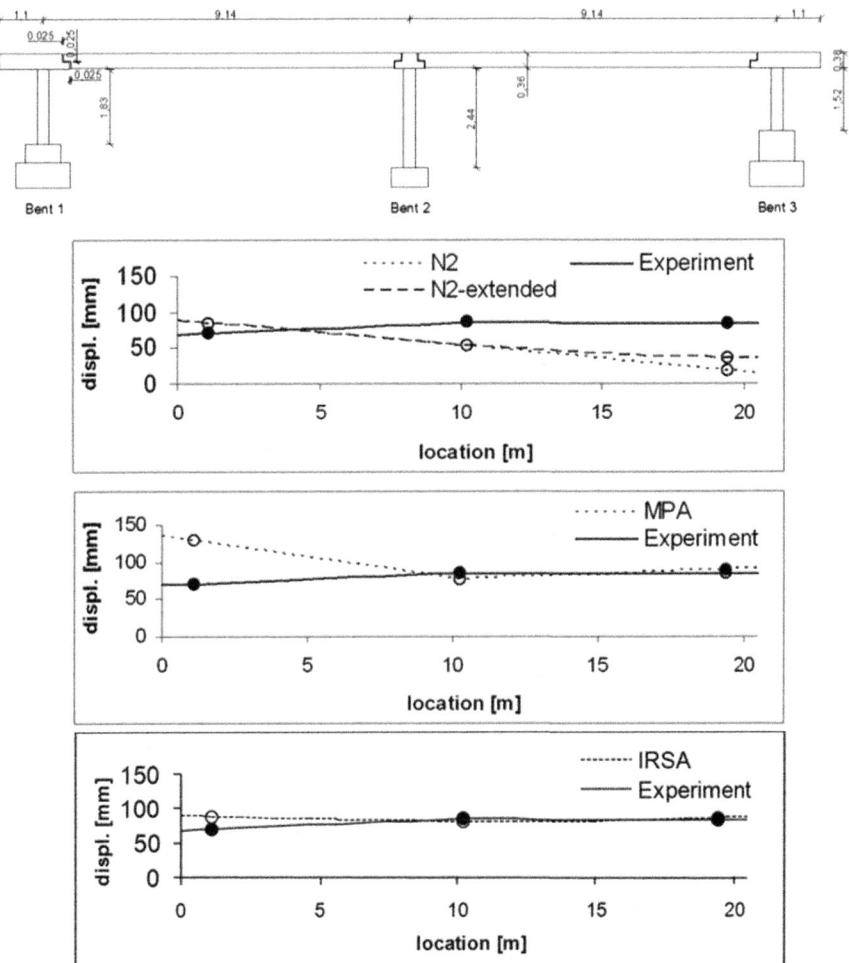

Fig. 6.14 Response of the experimentally tested bridge, where the modes of vibration changes depending on the seismic intensity

6.3.3.2 The IRSA Method

The IRSA method, proposed by Aydinoğlu (2003) is multimode adaptive pushover method. This means that it takes into account changes of the dynamic properties of the structure each time when the new plastic hinge is formed. Changes of both, modal shapes and the corresponding participation factors are considered each time the dynamic properties of the structure are changed. Contrary to the MPA method, all changes in the structure are coupled. Since it can take into account the changes of the mode shapes it can describe the response of the bridge, presented in Fig. 6.14, more accurately then both previously presented methods.

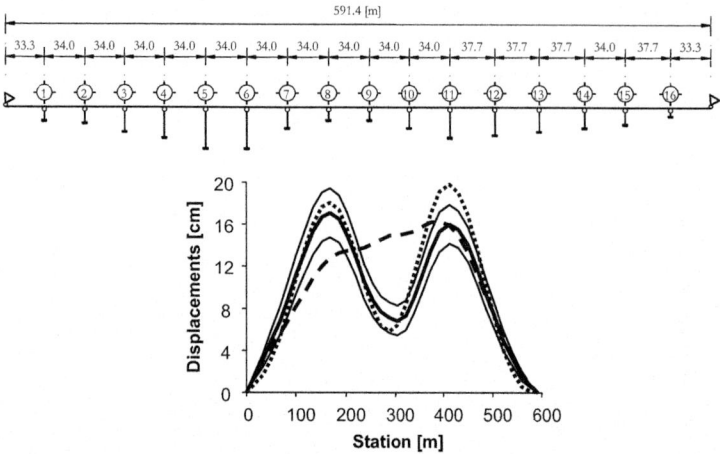

Fig. 6.15 The response of the highly irregular viaduct obtained by MPA (*dotted line*), IRSA (*dashed*) and NRHA (*solid lines*)

Since the method is more complex than the other two, the details will be skipped. They can be found in Aydinoğlu (2003), Kappos et al. (2012), and Isakovic et al. (2008a), as well as the appropriate numerical examples. It is worthy to note, that in spite of the complexity this method is not universal and cannot always replace the NRHA, particularly in the most complex bridges, similar to the one, presented in Fig. 6.15.

6.4 The Shear Strength of RC Columns

According to the EC8/2 the shear demand in RC columns is defined using the capacity design procedure. It should be less or equal to the shear capacity. In EC8/2 the shear capacity of RC columns is estimated based on the requirements of the standard EC2. According to this standard the contribution of the concrete without shear reinforcement (including the beneficial contribution of the compression stresses) should be neglected in all cases where the demand exceeds this value.

In EC8/2 the value of the shear strength, estimated in this way, is additionally reduced. In bridges, designed as limited ductile structures, it is recommended to reduce the shear strength by factor of 1.25. In ductile structures this reduction depends on the expected value of the shear demand corresponding to the elastic response and the shear demand defined based on the capacity design. The reduction factor is in a range between 1 and 1.25. When the shear resistance of the plastic hinges in ductile structures is estimated, the angle between the concrete compression strut and the main tension chord shall be assumed to be equal to 45°.

In general, the requirements of EC2 are adjusted to structural components of buildings, which have quite different dimensions of bridge columns. Consequently, different mechanisms that contribute to the shear strength, can have different importance than those in bridge columns. Due to the larger dimensions of bridge columns, the contribution of the concrete to the total shear strength can be quite important. Thus the approach, defined in EC2, can result in a quite conservative design. It is worthy to note that certain level of the conservatism is certainly needed for the shear design (since the type of the failure is brittle and the damage is difficult to repair), however the excessive conservatism can result in a very large required amount of the shear reinforcement, which is difficult to construct. Some balance between safety and feasibility is reasonable to achieve.

The classical truss analogy, where the angle between the compression strut and the tension reinforcement is assumed to be 45° seems to be reasonable, particularly for the case of the seismic (reversible) load and relatively low values of the shear span ratios of columns, where the shear response is particularly critical. This actually ensures the maximum amount of the shear reinforcement corresponding to certain truss configuration.

In addition to this requirement the contribution of the concrete without shear reinforcement and beneficial contribution of the compression stresses to the shear strength are neglected usually at quite low levels of the displacement demand. This can result in a quite conservative design, increasing the required shear reinforcement in some types of bridge columns to a quite large amount.

An example of such column is presented in Fig. 6.16. This is a hollow box column, which was experimentally tested in a scale 1:4. The basic properties of the column are presented in Fig. 6.16. More details can be found in Isakovic et al. (2008b) and Elnashai et al. (2011). The column was tested cyclically until the combined shear-flexural failure was achieved. The appearance of the specimen after the experiment is presented in Fig. 6.17b. The shear strength of the investigated column was 390 kN. In this particular case the EC8/2 requirement related to the angle between the compression strut and tension reinforcement was confirmed. It was 45°.

Taking into account the requirements of the EC2, considerably smaller value of 171 kN of the shear strength was obtained (see line 1 in Fig. 6.18). Note that all safety factors, defined in EC2, were excluded (e.g. the material safety factors for steel and concrete) since the actual shear strength was investigated. According to the requirements of the standard only the contribution of the shear reinforcement was taken into account, since the demand exceeded the sum of the contributions of the concrete without shear reinforcement and the contribution of the compression stresses. In the investigated column, however, these mechanisms contributed almost half of the total shear strength, 147 kN.

When all important mechanisms were taken into account, the estimated value of the total shear strength was increased to 318 kN. This value was still smaller than the experimentally observed strength (see line 2 in Fig. 6.18).

Since the actual and estimated strength were quite different, other procedures available in the literature and other standards were also employed. The UCBS

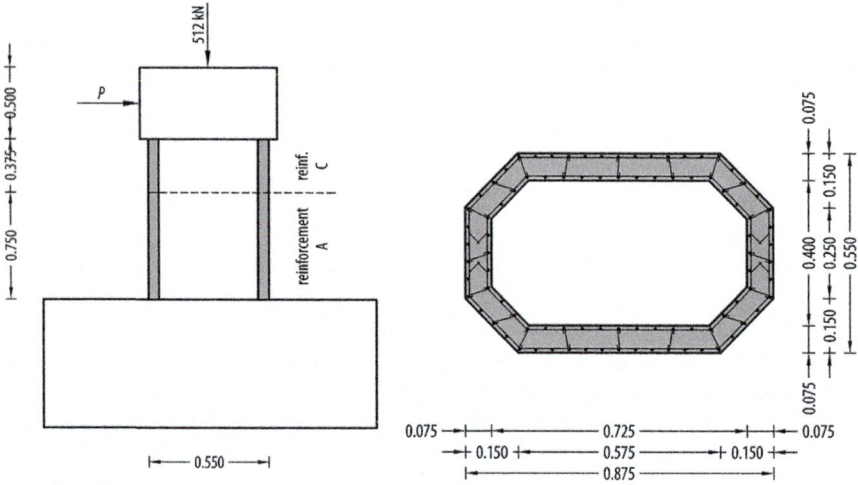

Fig. 6.16 The 1:4 scale model of the experimentally investigated column (Reinforcement type A: longitudinal reinforcement 90ϕ6 mm ($f_y = 324$ MPa), transverse reinforcement ϕ4 mm/5 cm ($f_y = 240$ MPa). Reinforcement type C: longitudinal reinforcement 90ϕ3.4 mm ($f_y = 240$ MPa), transverse reinforcement ϕ2.5 mm/5 cm ($f_y = 265$ MPa))

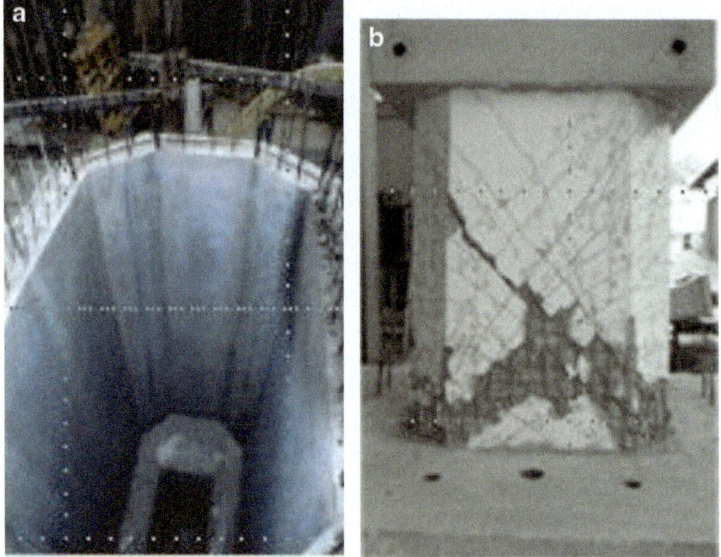

Fig. 6.17 (**a**) Casting of the tested column. (**b**) Combined shear-flexural failure of the tested column

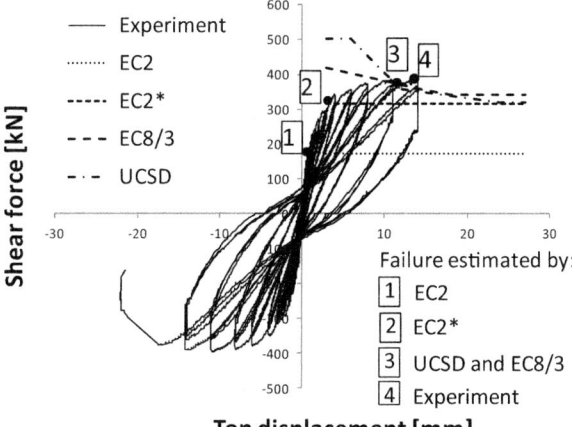

Fig. 6.18 Estimated value of the shear strength using different procedures compared to the experiment

procedure (Priestley et al. 1994) as well as the procedure included in the Eurocode 8/3 standard (CEN 2005b; Biskinis et al. 2004) was considered. Both of them predicted the shear strength of the investigated column quite well (see Fig. 6.18). Contrary to the EC2 standard, these methods define the shear strength based on the displacement ductility demand. Larger values of the shear strength correspond to smaller value of displacement demand. This approach reflects the actual response more realistically, since the reduction of the contribution of the concrete to the shear strength is gradual. In EC2 it is neglected at very small displacement demand. Thus the reduction of the shear strength is abrupt.

Consequently, the difference in the design of the column where the demand exceeds the contribution of the concrete by say 10 % and that where this contribution can be taken into account, can be unacceptably large. For example in the investigated case the difference in the amount of the shear reinforcement would be about 50 %. Therefore, it is feasible to make this transition more gradual like in the other two methods.

Further analysis of the estimated values of the shear strength, presented in Fig. 6.18, showed that EC2 approaches the other two methods in the region of large displacement demands. This is another indication that shear design in EC2 can be quite conservative.

Since the low value of shear strength was defined also for the lower displacement demand, completely misleading conclusions about the type of the failure and the corresponding displacement was obtained in the investigated case. According to the EC2 the failure of the investigated column would be pure shear corresponding to the unrealistically small displacement demand of about 3 mm (the measured displacement at the moment of the failure was about four times larger – 12 mm).

The previous discussion clearly shows that some modifications of the shear design, required in EC2, are needed. However, before any modifications are accepted, additional specialized studies, adjusted to bridge columns are needed. The alternative methods, presented in the previous paragraphs might be a suitable solution; however note that the differences between these two methods can be also quite large (at the region of the small displacement demand – see Fig. 6.18) indicating that the problem of shear is still not adequately investigated and solved. Similar conclusions can be found elsewhere in the literature (e.g. Calvi et al. 2005).

6.5 The Buckling of the Longitudinal Bars and Confinement of the Core of Cross-Sections

The lateral reinforcement has an important role in the protection of columns (bridge) against different types of brittle failure. Beside the prevention of brittle shear failure (discussed in the previous section), it should be designed to prevent also other possible types of brittle failure; to prevent buckling of the longitudinal reinforcement and also to ensure the adequate confinement of the concrete core preventing its deterioration due to the excessive lateral tensile stresses. Both functions considerably influence the ductility capacity of columns (structure). Although they are correlated, they still have to be addressed separately, since it is not always the case that these types of failure occur at the same moment.

The requirements of EC8/2, related to the confinement of the concrete core, seem to be reasonable. The minimum requirements are stringent than those included into the standard EC8/1, where the seismic design of buildings is addressed. This is, however, reasonable, since the columns have the crucial role in the seismic response of bridges, and they are typically loaded by considerable compression stresses, which reduce their ductility capacity. In general the structural system of bridges is less redundant and robust than that of buildings. Taking into account the mentioned characteristics it can be concluded that requirements related to the confinement of the concrete core are reasonable.

Several requirements of EC8/2, related to the protection of the flexural reinforcement against buckling, define the necessary amount of the lateral reinforcement, maximum distance of the lateral bars along the column as well as the maximum distance between the tie legs. These requirements prevent the two types of failure: (a) the limited maximum distance of lateral bars prevents the buckling of the longitudinal bars between two consecutive ties, and (b) the minimum amount of the lateral reinforcement prevents the buckling of the longitudinal bars between several ties.

All the requirements included into EC8/2 are known from the literature (e.g. Priestley et al. 1997). However, the one, which defines the minimum amount of transverse ties (Eq. 6.10 in the EC8/2) is misinterpreted. This requirement was defined based on the experimental investigations. An explanation can be found e.g. in Priestley et al. (1997). In the original formula the spacing of the ties in the vertical direction of column is employed. Instead of this spacing, in EC8/2 the transverse (horizontal) spacing of the tie legs in the plane of the cross-section is addressed. Thus, the use of the formula in EC8/2 should be corrected.

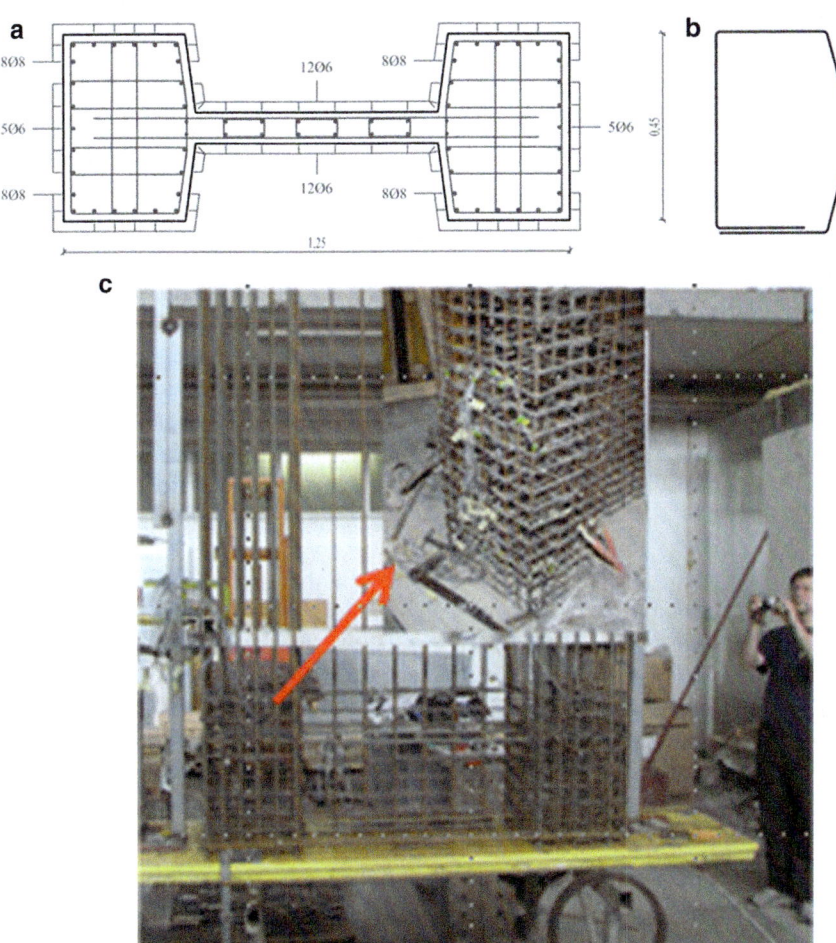

Fig. 6.19 (a) Cross-section of the tested column, (b) The shape of the outer ties, (c) The reinforcement of the specimen

Detailing of the transverse reinforcement is extremely important when the buckling of the longitudinal bars is addressed. The ties should be properly shaped with 135° hooks. Ties with 90° degree hooks usually cannot prevent buckling of the longitudinal bars, even if the proper amount of lateral reinforcement is provided. Standard EC8/2 allows cross-ties that have 90° degree hook on one side and 135° at the other side of the tie, as long as the axial force does not exceed 30 % of the characteristic compression strength of the concrete. It is the authors' opinion that 90° degree hooks should not be allowed at all, regardless of the level of the axial force.

This is illustrated on the example of the typical I shape column, presented in Fig. 6.19. This is the 1:4 scale model of the column, where the lateral reinforcement

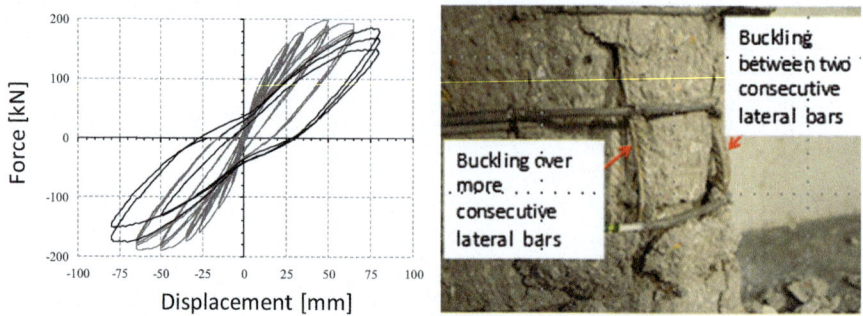

Fig. 6.20 A brittle failure was obtained due to the buckling of the longitudinal bars

fulfilled the EC8/2 requirements related to the shear strength, but the amount was insufficient considering the confinement and the buckling of the longitudinal bars. Additionally the ties were shaped according to some solutions applied in the practice, using the 90° overlapped hooks. The compression stresses due to the permanent load were relatively small (11 % of the characteristic compression strength). The column was tested cyclically until the failure occurred.

A brittle failure was obtained (see Fig. 6.20). After the spalling of the cover concrete, some of the improperly shaped ties with 90° degree hooks were opened, and could not support the longitudinal bars properly. Consequently the buckling of these bars between two consecutive ties as well as between more ties was observed. This was also the consequence of the insufficient amount of the lateral reinforcement. The failure was sudden, without any additional ductility capacity.

6.6 Conclusions and Final Remarks

During many years of use of the Eurocode 8/2 standard it was found that this standard considerably improved the seismic design of bridges, since it introduced many modern principles of the seismic engineering into design practice. This is modern standard, which is well organized, practically oriented and designer friendly.

In this paper some of the experiences, obtained when applying the standard in the practice and a critical overview of some of its requirements are presented. First the two topics related to the analysis of bridges were addressed: (a) the relationship between the pre-yielding stiffness and strength of structures as well as the application of the equal displacement rule, and (b) the nonlinear static analysis.

It was concluded that pre-yielding stiffness and strength of structures are strongly correlated. The pre-yielding stiffness is different for different levels of the selected strength. This does not negate the equal displacement rule.

The interpretation of the equal displacement rule included to the EC8/2 was compared with some different options. It has been found that certain conservatism in estimation of the seismic displacements is introduced. This conservatism has been found reasonable since the standard does not require explicit control of the displacement ductility capacity of structures. This can be particularly important in highly irregular structures, where in the nonlinear range considerable redistributions of the seismic effects can occur, and the results of the elastic analyses can be only a rough approximation of the actual response.

In general, for highly irregular structures it is strongly recommended to examine the seismic response using the nonlinear procedures. This is recognized by EC8/2 as well. It introduced the most refined nonlinear response history analysis as well as the simplified nonlinear procedures into the design practice. In the paper some issues related to the application of the single mode pushover method are discussed. The important differences between bridges and buildings related to the application of this method are analysed: (a) distribution of the lateral load, (b) the choice of the reference point and (c) the idealization of the pushover curve. Some alternatives to the procedures, defined in the standard, are proposed. The applicability of the single mode pushover methods is also briefly addressed. It was concluded that this type of methods is applicable mainly to short and medium length bridges, where the response is predominantly influenced by one invariant mode of vibration. In other cases the multimode pushover methods or the nonlinear response history analysis is recommended.

The second part of the paper is devoted to the shear and ductility capacity of RC columns. In EC8/2 the displacement ductility capacity of structures is ensured with proper structural detailing, which prevents the undesirable brittle types of failure. The brittle shear failure is prevented by a requirement that the shear strength of structural components should be at least equal to the shear demand determined based on the capacity design procedure. The shear capacity of RC columns is determined based on the requirements of the EC2. This capacity is in some cases reduced.

The procedure that is used to define the shear strength of columns can be quite conservative, since the contribution of the concrete to the shear strength is very often neglected at quite small displacement demand. In some bridge columns almost half of the total shear strength is neglected in this way. The comparison with some other procedures, available in the literature, also confirmed that the provisions of the EC2 can be quite conservative. The result can be a large required amount of transverse reinforcement, which is difficult to construct.

It has been concluded that the contribution of the concrete to the shear strength should be reduced gradually. It has been also found that the problem of the shear capacity in general is not adequately solved and that it requires further investigations. This is particularly applicable to bridge columns, since the available data are limited comparing to the structural elements in buildings.

The brittle failure due to the insufficient confinement of the concrete core is in EC8/2 prevented by proper detailing of the transverse reinforcement in columns. The required minimum amount of the transverse confinement reinforcement is larger than that in e.g. EC8/1. This was found feasible, since columns have the crucial role in the seismic response of bridges, and they are typically loaded by considerable compression stresses, which reduce their ductility capacity. It should be also noted that the bridge structures are in general less redundant than buildings.

The transverse reinforcement that protects the longitudinal reinforcement of columns against buckling is also addressed in EC8/2. The requirement related to the minimum amount of this reinforcement is, however, misinterpreted and should be corrected according to the results presented in the literature.

Acknowledgments The described experiments were performed in cooperation with the Slovenian National Building and Civil Engineering Institute (ZAG). The work was partly funded by the Company for Motorways in the Republic of Slovenia (DARS). The research was also supported by the Ministry of Higher Education, Science and Technology of the Republic of Slovenia. Some of the presented results were obtained within the work performed in the frame of Ph.D. theses by Zlatko Vidrih.

Open Access This chapter is distributed under the terms of the Creative Commons Attribution Noncommercial License, which permits any noncommercial use, distribution, and reproduction in any medium, provided the original author(s) and source are credited.

References

Aydinoğlu MN (2003) An incremental response spectrum analysis procedure based on inelastic spectral displacements for multi-mode seismic performance evaluation. Bull Earthq Eng 1 (1):3–36
Biskinis DE, Roupakias GK, Fardis MN (2004) Degradation of shear strength of reinforced concrete members with inelastic cyclic displacements. ACI Struct J 101(6):773–783
Calvi GM, Pavese A, Rasulo A, Bolognini D (2005) Experimental and numerical studies on the seismic response of R.C. Hollow Bridge Piers. Bull Earthq Eng 3:267–297
CEN (1994) Pre-standard Eurocode 8: design of structures for earthquake resistance – Part 2: bridges. Comité Européen de Normalisation, Brussels
CEN (2004a) Eurocode 2: design of concrete structures – Part 1-1: general rules and rules for buildings. EN 1992-1-1. Comité Européen de Normalisation, Brussels
CEN (2004b) Eurocode 8: design of structures for earthquake resistance – Part 1: general rules, seismic actions and rules for buildings. Comité Européen de Normalisation, Brussels
CEN (2005a) Eurocode 8: design of structures for earthquake resistance – Part 2: bridges. Comité Européen de Normalisation, Brussels
CEN (2005b) Eurocode 8: design of structures for earthquake resistance. Part 3: strengthening and repair of buildings. EN 1998-3. European Committee for Standardization, Brussels
Chopra AK, Goel RK (2002) A modal pushover analysis procedure for estimating seismic demands for buildings. Earthquake Eng Struct Dyn 31(3):561–582
Dolšek M, Fajfar P (2005) Simplified non-linear seismic analysis of infilled reinforced concrete frames. Earthquake Eng Struct Dyn 34(1):49–66
Elnashai A, Ambraseys NN, Dyke S (2011) NEEShub-JEE database. J Earthquake Eng Database. http://nees.org/resources/3166
Fajfar P (1999) Capacity spectrum method based on inelastic demand spectra. Earthquake Eng Struct Dyn 28(9):979–993

Fajfar P, Fischinger M (1987) Non-linear seismic analysis of RC buildings: implications of a case study. Eur Earthq Eng 1:31–43

Fajfar P, Gašperšič P, Drobnič D (1997) A simplified nonlinear method for seismic damage analysis of structures. Seismic design methodologies for the next generation of codes: proceedings of the international workshop, Rotterdam; Brookfield. Balkema, Bled, pp 183–194

Fajfar P, Marušić D, Peruš I (2005) Torsional effects in the pushover-based seismic analysis of buildings. J Earthq Eng 9(6):831–854

FEMA-440 (2005) Improvement of nonlinear static seismic analysis procedures. Applied Technology Council for Department of Homeland Security, Federal Emergency Management Agency, Washington, DC

Goel RK, Chopra AK (2005) Extension of modal pushover analysis to compute member forces. Earthq Spectra 21(1):125–139

Isakovic T, Fischinger M (2006) Higher modes in simplified inelastic seismic analysis of single column bent viaducts. Earthq Eng Struct Dyn 35(1):95–114. doi:10.1002/eqe.535

Isakovic T, Fischinger M (2011) Applicability of pushover methods to the seismic analyses of an RC bridge, experimentally tested on three shake tables. J Earthq Eng 15(2):303–320

Isakovic T, Popeyo Lazaro MN, Fischinger M (2008a) Applicability of pushover methods for the seismic analysis of single-column bent viaducts. Earthq Eng Struct Dyn 37(8):1185–1202. doi:10.1002/eqe.813

Isakovic T, Bevc L, Fischinger M (2008b) Modelling the cyclic flexural and shear response of the R. C. Hollow box columns of an existing viaduct. J Earthq Eng 12(7):1120–1138

Kappos AJ, Saiidi MS, Aydinoğlu MN, Isakovic T (2012) Seismic design and assessment of bridges: inelastic methods of analysis and case studies. Springer, Dordrecht/Heidelberg/New York/London

Paraskeva TS, Kappos AJ (2009) Seismic assessment of bridges with different configuration, degree of irregularity and dynamic characteristics using multimodal pushover curves. COMPDYN, ECCOMAS thematic conference on computational methods in structural dynamics and earthquake engineering, Rhodes, Greece

Paraskeva TS, Kappos AJ, Sextos AG (2006) Extension of modal pushover analysis to seismic assessment of bridges. Earthquake Eng Struct Dyn 35(3):1269–1293. doi:10.1002/eqe.582

Priestley MJN, Verma R, Xiao Y (1994) Seismic shear strength of reinforced concrete columns. ASCE J Struct Div 120(8):2310–2329

Priestley MJN, Seible F, Calvi GM (1997) Seismic design and retrofit of bridges. Wiley, New York

Priestley MJN, Calvi GM, Kowalsky MJ (2007) Displacement-based seismic design of structures. IUSS Press, Pavia

Chapter 7
From Performance- and Displacement-Based Assessment of Existing Buildings per EN1998-3 to Design of New Concrete Structures in *fib* MC2010

Michael N. Fardis

Abstract The paper traces the road to the first fully performance- and displacement-based European seismic standard, namely Part 3 of Eurocode 8 on assessment and retrofitting of existing buildings and from there to the part of the *fib* Model Code 2010 (MC2010) on performance- and displacement-based seismic design and assessment of all types of concrete structure. Performance-based seismic design is set in the broader context of performance-based engineering and European Limit State design. The major features of Part 3 of Eurocode 8 are presented, focusing on seismic demands and – mainly – on cyclic deformation capacities. Emphasis is placed on the need to use in the analysis an effective elastic stiffness which realistically represents the member secant-to-yield-point stiffness, in order to predict well the seismic deformation demands. The background of the effective stiffness and the deformation and shear capacity sides in Part 3 of Eurocode 8 is presented, with a view on developments of the State-of-Art after these aspects were finalized in Eurocode 8. The focus turns then on the seismic part of MC2010, showing the differences with Part 3 of Eurocode 8 due to recent advances in the State-of-the-Art, the difference between design of new structures and assessment of existing ones (including the need to estimate the secant-to-yield-point stiffness without knowing the reinforcement), the wider scope of MC2010 beyond buildings, etc. It is emphasised that member detailing per MC2010 is not based anymore on opaque prescriptions, but on transparent, explicit verification of inelastic deformation demands against capacities.

M.N. Fardis (✉)
Department of Civil Engineering, University of Patras, P.O. Box 1424, 26504 Patras, Greece
e-mail: fardis@upatras.gr

7.1 The European Seismic Codes Before EN-Eurocode 8

Since the early 1990s the activity of the European Earthquake Engineering community has been centred around and motivated by the drive towards a European Standard for seismic design: Eurocode 8. From early on this standard was permeated by performance-based concepts with a strong European flavour. In fact, in Europe, Performance Levels in seismic design, assessment or retrofitting have always been associated to, or identified with Limit States. The Limit State concept was introduced in the 1960s in Europe to define states of unfitness of the structure for its intended purpose (CEB 1970; Rowe 1970): Ultimate Limit States (ULS) concern safety, whilst Serviceability Limit States (SLS) the normal function and use of the structure, comfort of occupants, or damage to property; intermediate Limit States were also considered. These fundamental CEB documents and the two CEB/FIP Model Codes (CEB 1978, 1991) were the basis of Limit State design for all structural materials in the pre-Norm (CEN 1994a) and European Norm (CEN 2002) versions of the Eurocodes, and for concrete structures in particular (CEN 1991, 2004b). According to the Eurocode concerning the basis of structural design (CEN 1994a, 2002), the Limit States approach is the backbone of structural design for any type of action, including the seismic one.

Neither of the two CEB/FIP Model Codes covered seismic design. However, the CEB Model Code for seismic design of new concrete structures (CEB 1985) was meant to be a "Seismic Annex" to the CEB/FIP Model Code 1978, mainly for concrete buildings. It introduced two Limit States: (a) Structural Safety and (b) Serviceability, but design for both was for a single hazard level. The structure was to be proportioned for force resistance against elastic lateral forces derived from the 5 %-damped elastic response spectrum reduced by the "behaviour factor" q, assuming uncracked gross section stiffness. Interstorey drifts computed via the "equal displacement rule" under the same seismic action were limited to 2.5 % if only the protection of the structure is of concern, or to 1 % for Serviceability of brittle building partitions (1.5 % for non-brittle ones). Three "Ductility Levels" were included for buildings: the higher the ductility level, the larger was the q-factor and the more stringent the member detailing (albeit prescriptive). The two upper ductility levels employed "capacity design" to prevent brittle shear failure of members and soft storey plastic mechanisms in weak column-strong beam frames; the ultimate objective was global ductility.

The European Prestandard (ENV) on the seismic design of buildings of any type of material (CEN 1994b, c, d) was based on the CEB "Seismic Annex" (CEB 1985). It differed from it in that its scope covered the major structural materials, and in that distinct hazard levels were introduced for the two Limit States. The ULS against Life-threatening Collapse is checked in the same way as in the 1985 CEB seismic Model Code (except for the interstorey drift limitation) under the 475-year earthquake (10 % exceedance probability in 50 years), at least for structures of ordinary importance. For the SLS against damage and loss of use, the interstorey drift limit is 0.4 % (0.6 % for non-brittle partitions) and is checked under 50 % of the 475-year earthquake, again using uncracked gross section properties and the "equal displacement rule". The alternative options for ductility – termed now "Ductility Classes" – remained three; capacity design against shear failure of beams was limited to the higher Class.

The European Pre-standard (ENV) on repair and seismic strengthening of existing buildings (CEN 1996) did not present any conceptual advancement over its counterpart for new buildings (CEN 1994b, c, d). Except that the interstorey drift limits were not meant to be checked under the Serviceability earthquake, and that the evaluation criteria for existing buildings were limited to full conformity to the requirements of one of the three "Ductility Classes" of the ENV for new buildings (CEN 1994b, c, d) under a reduced seismic action which depends on the remaining lifetime. Retrofitting was also meant to ensure full conformity with the rules of the ENV for new buildings for one of its three "Ductility Classes".

As there was no seismic follow-up to the 1990 CEB-FIP Model Code (CEB 1991), the European Standard for seismic design of new buildings of any material (CEN 2004b) evolved from the ENV version (CEN 1994b, c, d), incorporating important developments in the State-of-the-Art which had matured in the meantime. Most of the completely new points were not specific to concrete: design with seismic isolation, capacity design of the foundation, composite (steel-concrete) buildings, design with nonlinear analysis and direct verification of deformation demands, etc. This last feature is of special importance, as it presaged the recent codification of displacement-based seismic design of new buildings in the Model Code 2010 of *fib* (2012). A very important parallel development was the European Standard for "Assessment and retrofitting of buildings" (CEN 2005a), which was the first fully and explicitly performance- and displacement-based seismic code in Europe and has formed the basis for the seismic design and assessment part of the *fib* Model Code 2010. As these two important documents will be a natural basis for the upcoming revision of the most important parts of Eurocode 8 (CEN 2004a, 2005a, b), they are the subject of the present paper, which attempts to provide some insight into their rationale, shed light onto their background and look for indications about where they may lead in the near future.

7.2 Performance-Based Earthquake Engineering

Traditionally, structural design codes have been the responsibility of Public Authorities, with public safety as the compelling consideration. Accordingly, traditional seismic design codes aim at protecting human life by preventing local or global collapse under a single level of earthquake. The no-(local-)collapse requirement normally refers to a rare seismic action, termed "design seismic action". In most present-day codes the "design seismic action" for ordinary structures has a 10 % probability to be exceeded in a conventional working life of 50 years (i.e., a mean return period of 475 years).

As early as the 1960s the international earthquake engineering community was aware of the property loss and other economic consequences due to frequent seismic events. Recognizing that it is not feasible to avoid damage under very strong earthquakes, the Structural Engineers Association of California (SEAOC) adopted in its 1968 recommendations for seismic design the requirements below:

"Structures should, in general, be able to:

1. Resist a minor level of earthquake ground motion without damage.
2. Resist a moderate level of earthquake ground motion without structural damage, but possibly experience some nonstructural damage.
3. Resist a major level of earthquake ground motion having an intensity equal to the strongest either experienced or forecast for the building site, without collapse, but possibly with some structural as well as nonstructural damage."

Major earthquakes that hit developed countries in the second half of the 1980s and the first half of the 1990s caused relatively few casualties but very large damage to property and economic loss. "Performance-based earthquake engineering" emerged, in response, in the SEAOC Vision 2000 document and developed into the single most important idea of late for seismic design or retrofitting of buildings (SEAOC 1995).

"Performance-based engineering" in general focuses on the ends; notably on the ability of the engineered facility to fulfill its intended purpose, taking into account the consequences of failure to meet it. Present-day design codes, by contrast, are process-oriented, emphasizing the means, namely prescriptive, handy, but opaque design rules, that disguise the pursuit of satisfactory performance. Such rules have developed over time into a convenient means to provide safe-sided, yet economical solutions for common combinations of structural layout, dimensions and materials. They leave limited room to judgment and creativity in conceptual design and do not lend themselves for innovation that benefits from new advances in technology or materials.

"Performance-based earthquake engineering" in particular aims to optimize the utility from the use of a facility by minimizing its expected total cost, including the short-term cost of the work and the expected value of the loss in future earthquakes (in terms of casualties, cost of repair or replacement, loss of use, etc.). In general, it should account for all possible future seismic events and their annual probability, carry out a convolution with the corresponding consequences during the working life of the facility and minimize the expected total cost. However, this is not a practical design approach. So, present-day "performance-based seismic design" just replaces the traditional single-tier design against life-threatening collapse and its prescriptive rules with transparent multi-tier seismic design or assessment, meeting more than one discrete "performance levels", each one under a different seismic event, identified through its annual probability of being exceeded (the "seismic hazard level"). Each "performance level" is identified with a physical condition of the facility and its possible consequences (likely casualties, injuries and property loss, continued functionality, cost and feasibility of repair, expected length of disruption of use, cost of relocation, etc., see Table 7.1 for the example of

Table 7.1 Seismic performance Limit States and associated seismic hazard levels for ordinary facilities and member compliance criteria: the case of *fib* MC2010 (*fib* 2012)

Limit State	Facility operation	Structural condition	Deformation limits in MC2010	Seismic action per MC2010
Operational (OP) SLS	Continued use; any nonstructural damage is repaired later	No structural damage	Mean value of yield deformation	Frequent: ~70 % probability of been exceeded in service life
Immediate Use (IU) SLS	Safe; temporary interruption of normal use	Light structural damage (localised bar yielding, concrete cracking or spalling)	Mean value of yield deformation may be exceeded by a factor of 2	Occasional: ~40 % probability of been exceeded in service life
Life Safety (LS) ULS	Only emergency or temporary use but unsafe for normal; no threat to life during the earthquake; repair feasible, possibly uneconomic	Serious structural damage, but far from collapse; sufficient capacity for gravity loads; adequate seismic strength and stiffness for life safety till repair	Safety factor, γ^*_R, of 1.35 against the lower 5 %-fractile of plastic rotation capacity	Rare: 10 % probability of been exceeded in service life
Near-Collapse (NC) ULS	Unsafe for emergency use; life safety during the earthquake mostly ensured, but not guaranteed (falling debris hazard)	Heavy structural damage, at the verge of collapse; strength barely sufficient for gravity loads, but not for aftershocks	Lower 5 %-fractile of plastic rotation capacity may be reached ($\gamma^*_R = 1$)	Very rare: 2–5 % probability of been exceeded in service life

fib Model Code 2010). The "performance objective" is then a requirement to meet a set of "performance levels" under the associated "seismic hazard level". A four-tier "performance objective" similar to the one reflected in the first three and the last column of Table 7.1 was introduced for ordinary buildings by the Vision 2000 document (SEAOC 1995); it has served ever since as the basis for "Performance-based earthquake engineering".

7.3 Displacement-Based Seismic Design or Assessment

The earthquake is a dynamic action, introducing to the structure a certain energy input and imposing certain displacement and deformation demands, but not specific forces. The forces are generated by the structure in response to the seismic displacements and depend on its resistance. It is the deformations that make a structural member lose its lateral load resistance and it is the lateral displacements (not the lateral forces) that cause a structure to collapse in an earthquake under its own weight due to second-order (P-Δ) effects. So, deformations and displacements represent a much more rational basis than forces for the seismic design, assessment or retrofitting of structures. For this reason, displacement-based seismic design has been proposed by Moehle (1992) and Priestley (1993) as a more rational alternative to the traditional forced-based design approach.

For new structures, procedures for direct dimensioning of RC members on the basis of given deformation demands were not available early on; hence in displacement-based design, dimensioning of new members has often been reduced to familiar force-based dimensioning (Priestley et al. 2007). In seismic assessment, though, which is an analysis rather than a synthesis problem, the deformation capacities of members can be easily computed for given dimensions, material properties and reinforcement. So, seismic assessment of existing structures provides better grounds than the design of new ones for deformation- and displacement-based verification. Retrofitting interventions may also be conceived as a means to reduce the seismic deformation demands on the existing members below their current deformation capacities. For these reasons, a holistic displacement- and performance-based approach was first introduced in seismic assessment, not in design, through the pioneering "NEHRP guidelines for the seismic rehabilitation of buildings" (ATC 1997), which soon became the reference for displacement-based seismic assessment and developed fairly recently into an ASCE Standard (ASCE 2007).

7.4 Performance- and Displacement-Based Seismic Assessment of Existing Buildings in Part 3 of EN-Eurocode 8

7.4.1 The Context

Part 3 of EN-Eurocode 8 (CEN 2005a, 2009) broke completely with its force-based predecessor for existing buildings (CEN 1996) and its companion for new ones (CEN 2004a) and developed in the footsteps of (ATC 1997) into a full-fledged performance- and displacement-based seismic standard for existing buildings – the first one in Europe and the only one in the suite of 58 EN-Eurocodes of the first generation which deals with existing structures.

Unlike all other EN-Eurocodes, which apply to all structures within their scope, namely to all new ones, Part 3 of EN-Eurocode 8 does not apply to all existing buildings, but only to the ones which their owner or competent Authorities decide to seismically assess and retrofit. Part 3 of EN-Eurocode 8 addresses only the structural aspects of seismic assessment and retrofitting and will apply once the requirement to assess a particular building has been established. The conditions under which seismic assessment of individual buildings – possibly leading to retrofitting – may be required are beyond its scope. The initiative for seismic assessment and retrofitting lies with the owner, unless a national or local program is undertaken for seismic risk mitigation through seismic assessment and retrofitting. The differentiation between "active" and "passive" seismic assessment and retrofitting programs should be noted in this respect. "Active" programs may require owners of certain categories of buildings to meet specific deadlines for the completion of the seismic assessment and – depending on its outcome – of the retrofitting. The categories of buildings to be targeted may depend on the seismicity and ground conditions, the importance class and occupancy and the perceived vulnerability of the building (as influenced by the type of material and construction, the number of storeys, the date of construction relative to those of older code enforcement, etc.). "Passive" programs associate seismic assessment – possibly leading to retrofitting – with other events or activities related to the use of the building and its continuity, such as a change in use that increases occupancy or importance class, remodelling above certain limits (as a percentage of the building area or of the total building value), repair of damage after an earthquake, etc. The choice of Performance levels – "Limit States" in (CEN 2004a) – to be checked, as well as the return periods of the seismic action ascribed to them, may depend on the adopted program for assessment and retrofitting, which is more stringent in "passive" programs than in "active" ones. For example, in "passive" programs triggered by remodelling, the requirements may escalate as the extent and cost of the remodelling increases.

7.4.2 Performance Objectives

Part 3 of Eurocode 8 introduces three "performance levels", called "Limit States":

- "Damage Limitation" (DL), similar to "Immediate Occupancy" in (SEAOC 1995; ATC 1997; ASCE 2007) and the first Limit State in Table 7.1.
- "Significant Damage" (SD), which corresponds to "Life Safety" in (SEAOC 1995; ATC 1997; ASCE 2007), to the third Limit State in Table 7.1 and to the (local-)collapse prevention requirement which applies to new buildings per Part 1 of EN-Eurocode 8 (CEN 2004a).
- "Near Collapse" (NC), similar to "Collapse Prevention" in (SEAOC 1995), (ATC 1997) or (ASCE 2007) and the third Limit State in Table 7.1.

In line with the policy of EN-Eurocodes to allow decision at national level regarding all safety-related issues, the "Seismic Hazard" levels for which the three "Limit States" above are to be met are Nationally-Determined-Parameters (NDPs) specified by National Authorities. National Authorities may also specify whether all three "Limit States" shall be met under the corresponding "Seismic Hazard" level, or whether verification of just one or two of them at the corresponding "Seismic Hazard" levels suffices. National Authorities may choose these levels so that the number of buildings that need retrofitting is acceptable to society and the national economy and/or retrofitting is not economically prohibitive, increasing the chances that owners will retrofit deficient property at their own initiative.

7.4.3 Compliance Criteria

A distinction is made in Part 3 of Eurocode 8 between "primary" and "secondary" structural elements, depending on their role and importance in the lateral-force-resisting system. There is no restriction on the number of "secondary" elements or their collective contribution to the total lateral resistance or stiffness. More relaxed compliance criteria apply for them. So the engineer may designate elements of the existing or the retrofitted building as "secondary", depending on the outcome of the verifications and his/her judgment on the importance of these elements. What he/she may not do is to deliberately choose the plan- or heightwise distribution of "secondary" elements to change the classification of the structural system from irregular to regular (which in turn determines the method of analysis allowed).

A distinction is also made between "ductile" and "brittle" mechanisms: for RC members and joints, flexure (with or without axial load) or shear, respectively. Verifications and compliance criteria of "ductile" mechanisms are expressed in terms of deformations; "brittle" ones are checked in terms of forces.

Local material failure (even a bar rupture) does not constitute by itself member failure under seismic loading: the member is considered to have failed if it has lost a good part of its force resistance owing to gradual accumulation of local material failures during cyclic loading. Loss of resistance takes place in flexural plastic hinges forming under seismic loading at member ends. Following proposals by Fardis (1998, 2001) and Fardis et al. (2003), compliance of RC members in flexure is checked using the chord-rotation, θ, at the two ends of the member as the relevant deformation measure (or, its plastic part, which is equivalent to the plastic hinge rotation, θ_{pl}). Recall that the chord rotation at a member end is the angle between the normal to the member section there and the chord connecting the two member ends in the deformed configuration; in the elastic regime the chord rotations at member ends A and B, θ_A and θ_B, determine alone the two bending moments M_A and M_B through the member stiffness matrix.

For the three Limit States mentioned above, Annex A of Eurocode 8-Part 3 specifies for RC members the performance requirements in Table 7.2.

- At the "Damage Limitation" (DL) Limit State, ductile mechanisms are required to remain elastic (below yielding).
- At the other extreme, the "Near Collapse" (NC) Limit State, ductile elements are allowed to reach their ultimate deformation capacity (its expected value for "secondary" elements, mean-minus-standard deviation for "primary" ones).

7 From Performance- and Displacement-Based Assessment of Existing Buildings...

Table 7.2 Compliance criteria for assessment/retrofitting of RC members in Eurocode 8-Part 3

Mechanism	Member	Damage limitation (DL)	Significant damage (SD)	Near collapse (NC)
Flexure (ductile)	Primary	$M_E^{(1)} \leq M_y^{(2)}$ or	$\theta_E^{(1)} \leq 0.75\theta_{u,m-\sigma}^{(3)}$	$\theta_E^{(1)} \leq \theta_{u,m-\sigma}^{(3)}$
	Secondary	$\theta_E^{(1)} \leq \theta_y^{(2)}$	$\theta_E^{(1)} \leq 0.75\theta_{u,m}^{(4)}$	$\theta_E^{(1)} \leq \theta_{u,m}^{(4)}$
Shear (brittle)	Primary		V_E or $V_{CD}^{(5)} \leq V_{Rd,EC2}^{(6)}$, $V_{CD} \leq V_{Rdj,EC8}^{(8)}$	V_E or $V_{CD}^{(5)} \leq V_{Rd,EC8}/1.15^{(7)}$; joints:
	Secondary		V_E or $V_{CD}^{(5)} \leq V_{Rm,EC2}^{(9)}$, $V_{CD} \leq V_{Rmj,EC8}^{(9)}$	V_E or $V_{CD}^{(5)} \leq V_{Rm,EC8}^{(9)}$; joints:

(1) M_E, θ_E: moment or chord-rotation demand from the analysis
(2) M_y, θ_y: chord-rotation at yielding per Sect. 7.4.4.2
(3) $\theta_{u,m-\sigma}$: mean-minus-standard deviation chord-rotation supply:
 • $\theta_{u,m-\sigma} = \theta_{u,m}/1.7$ for $\theta_{u,m}$ from Option 1 in Sect. 7.4.5.1, $\theta_{u,m-\sigma} = \theta_{u,m}/2$ for Option 2;
 • $\theta_{u,m-\sigma} = \theta_{u,m}/1.5$ with $\theta_{u,m}$ from Eq. (7.5a) and $\theta_{u,m-\sigma} = \theta_y + \theta^{pl}_{u,m}/1.8$ with θ_y per Sect. 7.4.4.2
(points 1–3) and $\theta^{pl}_{u,m}$ from Eq. (7.5b) (for poor detailing and/or lap-splicing, $\theta_{u,m}$, $\theta^{pl}_{u,m}$ are modified per Sect. 7.4.5.2 – points 1, 2 or 3, 4, respectively; θ_y is amended for lap splices per Sect. 7.4.4.2 points a, b)
(4) $\theta_{u,m}$: mean chord-rotation supply per (3) above, or $\theta_{u,m} = \theta_y + \theta^{pl}_{u,m}$ with θ_y, $\theta^{pl}_{u,m}$ according to (3) above
(5) V_E, V_{CD}: shear force demand from analysis per Sect. 7.4.4.3 or from capacity design per Sect. 7.4.4.4, respectively
(6) $V_{Rd,EC2}$: shear resistance before flexural yielding for monotonic loading per Eurocode 2 (CEN 2004b), using design material strengths (mean divided by partial factor of material)
(7) $V_{Rd,EC8}$: cyclic shear resistance in plastic hinge after flexural yielding per EN1998-3, from Eqs. (7.8, 7.9, 7.10a, 7.10b and 7.11), with design material strengths (mean divided by partial factor)
(8) $V_{Rdj,EC8}$: shear resistance of joints per EN1998-1 (CEN 2004a)
(9) As in (6)–(8) above, but using mean material strengths

- At the "Significant Damage" (SD) Limit State, the deformations (chord rotations at member ends) of "ductile" elements are limited to 75 % of the deformation limit above in the "Near Collapse" (NC) level.

Force demands on "brittle" mechanisms are required to remain below their force resistance at all Limit States. The value of force resistance of "primary" elements used in this verification is computed applying appropriate partial safety factors on the characteristic material strengths; the values of these factors depend also on the level of available knowledge for the existing structure. For "secondary" elements, the force resistance is computed without partial safety factors on the characteristic material strengths.

The ultimate chord rotation, θ_u, or plastic hinge rotation, θ^{pl}_u, under cyclic loading is conventionally identified with a 20 %-drop in moment resistance; in other words, increasing the imposed deformation beyond θ_u or θ^{pl}_u cannot increase the moment resistance above 80 % of its maximum ever value.

Annex A to Part 3 of Eurocode 8 (CEN 2005a, 2009) gives expressions and rules for the calculation of the mean value of the chord rotation at yielding, θ_y, or at ultimate, $\theta_{u,m}$, highlighted in Sects. 7.4.4.2 and 7.4.5, respectively. The cyclic shear resistance after flexural yielding, $V_{R,EC8}$, is also given in Annex A to Part 3 of Eurocode 8, to supplement the relevant rules in Eurocode 2 that address only the shear resistance in monotonic loading, $V_{R,EC2}$, and do not reflect the reduction of shear resistance with increasing cyclic ductility demands. Outside flexural plastic hinges the shear force resistance, V_R, is determined per Eurocode 2 (CEN 2004b), as for monotonic loading. The special rules for V_R in flexural plastic hinges under cyclic loading are highlighted in Sect. 7.4.6.

Deformation action effects, θ_E or θ^{pl}_E, are determined via nonlinear analysis for the applicable seismic action combined with the quasi-permanent gravity loads, or – under certain conditions – via linear analysis (see Sect. 7.4.4.4). Shear force action effects, V_E, are computed by nonlinear analysis for the combination of the applicable seismic action and the quasi-permanent gravity loads, or, if linear analysis is used, by capacity design calculations (see Sect. 7.4.4.4).

7.4.4 Analysis for the Determination of Seismic Action Effects

7.4.4.1 General Principles

The prime objective of a seismic analysis carried out for the purposes of displacement-based assessment or retrofitting is to estimate the inelastic seismic deformation demands, which are compared to the corresponding deformation capacities. To meet this goal, the structural analysis model should use realistic values of member elastic stiffness. This aspect is more important than the sophistication and refinement of the structural model. If anything, possible missestimations of the elastic stiffness should be on the safe-side: it is better from this point of view to underestimate the stiffness than to overestimate it.

Another important point is that, if calculated with member stiffness values representative of elastic response up to yielding, the fundamental period of a concrete structure normally comes out longer than the corner period between the acceleration- and the velocity-controlled ranges of the spectrum, T_C. Therefore, the "equal displacement" rule applies well on average, at least to a Single-Degree-of-Freedom (SDoF) approximation of concrete structures: their global inelastic displacement demand may be estimated by linear elastic analysis for 5 % damping.

Any analysis, linear or nonlinear, should be based on mean values of material properties, as inferred from the documentation of the as-built structure, combined with in-situ measurements. For new materials, added for retrofitting, the mean strength is higher than the nominal values: according to Eurocode 2, for concrete f_{cm} exceeds f_{ck} by 8 MPa; concerning steel, f_{ym} is in the order of $1.15 f_{yk}$.

Sections 7.4.4.2 and 7.4.4.4 elaborate further the points raised in the first two paragraphs, in the context of Part 3 of Eurocode 8.

7.4.4.2 Effective Elastic Stiffness for the Analysis

In force- and strength-based seismic design of new structures according to present day codes, it is safe-sided to overestimate the effective stiffness, as the computed natural periods are reduced and the resulting spectral accelerations and design forces increase. Eurocode 8 recommends in Part 1 (CEN 2004a) to use for RC members 50 % of the uncracked gross section stiffness, $(EI)_c$. On average, this still is about double the experimental secant stiffness at yielding. An overestimated effective stiffness and the ensuing reduction of natural periods underestimate the spectral displacements and seismic deformation demands and is unsafe in the context of displacement-based seismic design or assessment with direct verification of member deformation capacities against deformation demands. So, the model for the analysis should use realistic values of the effective cracked stiffness of concrete members at yielding, accounting for all sources of flexibility:

- fully cracked sections should be used for members expected to yield at the Limit State considered, without tension stiffening (which is diminished by load cycling), and
- the fixed-end-rotation of the member's end section due to slippage of longitudinal bars from their anchorage zone outside the member (in a joint or the foundation) should be taken into account, as per Fig. 7.1 and Eq. (7.2):

$$\theta_{slip} = \frac{\varphi d_{bL} \sigma_s}{8\tau_b} \qquad (7.1)$$

where:

- φ is the curvature at the end section and σ_s the stress in the tension bars there,
- d_{bL} is the tension bars' mean diameter and τ_b the mean bond stress along their straight anchorage length outside the member length.

At yielding of the end section, φ and σ_s may be taken in Eq. (7.2) equal to their yield values, φ_y and f_y; along ribbed bars τ_b (in MPa) may be taken equal to $\sqrt{f_c}$(MPa) (Biskinis and Fardis 2004, 2010a). The value of θ_{slip} at yielding at the end section is denoted by $\theta_{slip,y}$.

For members which yield at the limit state of interest, the analysis should use as effective elastic stiffness, EI_{eff}, the secant stiffness to the yield-point. According to Part 3 of Eurocode 8, in prismatic RC members (including slender walls) which may yield at one or both ends where the member frames into another component or in the foundation, the secant stiffness to yield-point of the full member between its two ends may be estimated as proposed by Fardis (1998, 2001) and Fardis et al. (2003):

Fig. 7.1 Fixed-end-rotation due to bar slippage from a joint

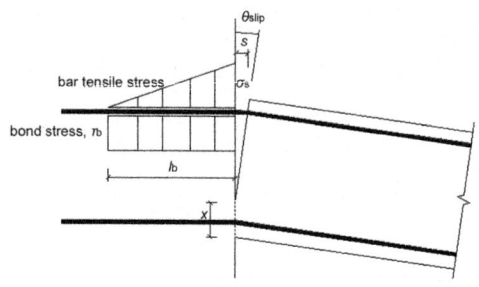

$$EI_{eff} = \frac{M_y L_s}{3\theta_y} \quad (7.2)$$

where M_y is the yield moment from section analysis with linear σ-ε laws until the tension bars yield (over one-third of the tension zone in circular columns), or a certain strain limit is exceeded by concrete (Biskinis and Fardis 2010a, 2013a, b); θ_y is the chord rotation at yielding, calculated as highlighted below; $L_s = M/V$ is the shear span at the yielding end section under seismic loading. In a beam, L_s may be taken as one-half of the clear length between columns; in a column, as one-half the clear height between beams in the plane of bending; the same for a bridge pier column fixed at the top against rotation in the plane of bending. In the strong direction of a building wall, the value of L_s in a storey is about one-half the height from the wall base in the storey to the top of the wall. In members cantilevering in the plane of bending, L_s is the member clear length. For asymmetric section and/or reinforcement, the mean value of EI_{eff} for positive and negative bending should be used. For walls or cantilevering members, the EI_{eff}-value at the base section should be used; in all other cases the average EI_{eff}-value at the two member ends applies.

According to Biskinis and Fardis (2010a, 2013a), the value of θ_y to be used in Eq. (7.2) as well as in the verification of the DL Limit State, is the sum of:

1. a flexural component, equal to $\varphi_y(L_s+z)/3$ if ribbed bars are used and 45°-cracking of the member precedes flexural yielding of its end section (see Fig. 7.2), or to $\varphi_y L_s/3$ otherwise; 45°-cracking near the member end precedes flexural yielding if the shear force at flexural yielding, M_y/L_s, exceeds the shear resistance without shear reinforcement per Eurocode 2;
2. a shear deformation, about equal to $0.0014(1 + 1.5 h/L_s)$ in beams or rectangular columns, $0.0027\max[0; 1 - L_s/(7.5D)]$ in circular piers or columns, or 0.0013 in rectangular, T-, H- or U-walls and hollow rectangular members – where h or D is the full section depth; and
3. the fixed-end-rotation due to the slippage of longitudinal bars from the anchorage past the member length, obtained as $\theta_{slip,y}$ from Eq. (7.1) for $\varphi = \varphi_y$, $\sigma_s = f_y$.

The above have been adopted in Part 3 of Eurocode 8 for the calculation of θ_y of beams, rectangular columns or walls and non-rectangular walls. Note that, in the

Fig. 7.2 Definition of chord rotation, θ, at the base of a cantilever column; effect of "tension shift" due to diagonal cracking on distribution of flexural deformations along the column

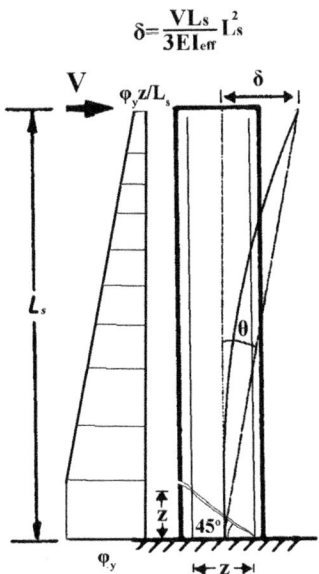

light of more recent data, better overall agreement for rectangular or non-rectangular walls and hollow rectangular members is obtained, if the constant term 0.0013 in point 2 is replaced by $0.0007[1+(4/3)h/L_s]$ (cf. (a) and (b) in Fig. 7.4).

At the end sections of T- or L-beams, slab bars parallel to the beam and within an effective slab width, b_{eff}, count as longitudinal reinforcement of the beam end section, provided they are well-anchored past it. Part 1 of Eurocode 8 (CEN 2004a) specifies an unrealistically small size of b_{eff}, intended for safe-sided design. A realistic estimate is 25 % of the beam span or the mid-distance to the adjacent parallel beam, whichever is smaller, on each side of the beam web.

Figures 7.3, 7.4 and 7.5 compare the predictions from this Section's approach to the dataset used for their calibration (Biskinis and Fardis 2010a, 2013a). Their captions give also the Coefficient of Variation (CoV) of the test-to-prediction ratio of EI_{eff}; to be compared in Fig. 7.9 with that for the empirical prediction per Eq. (7.14). Not included in this database, nor in Fig. 7.3, are columns with smooth bars (common in old buildings).

Lap splices at floor levels are common. Tests of 92 such columns with ribbed (deformed) bars and another 36 with smooth bars show certain effects of lap-splicing (Biskinis and Fardis 2010a), taken into account in Eurocode 8, Part 3:

(a) Both bars in a pair of lapped bars in compression count fully in the compression reinforcement ratio. This positive effect refers to M_y, φ_y, θ_y, as well to all properties at ultimate deformation (see Sect. 7.4.5.2);

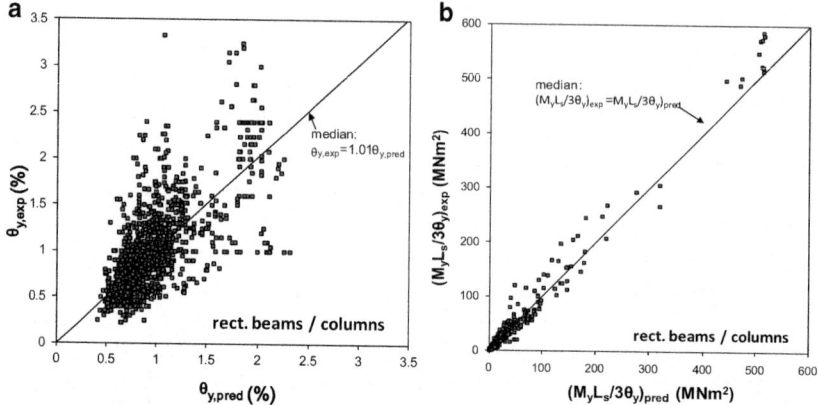

Fig. 7.3 Rectangular beams/columns: (**a**) experimental chord rotation at flexural yielding, θ_y, v predictions per (Biskinis and Fardis 2010a; CEN 2005a, 2009; *fib* 2012) in 1,674 tests; (**b**) experimental secant stiffness to yield point, EI_{eff}, v result of Eq. (7.2) in 1,637 tests – CoV 32 %

(b) In the calculation of the properties (M_y, φ_y, θ_y, etc), the yield stress, f_y, of lap-spliced ribbed tension bars with mean diameter d_{bL}, is multiplied by $l_o/l_{oy,min}$, where l_o is the lapping and $l_{oy,min}$ is given by Eq. (7.3), if l_o is less than $l_{oy,min}$:

$$l_{oy,min} = \frac{0.3 d_{bL} f_y}{\sqrt{f_c}} \quad (f_y, f_c \text{ in MPa}) \tag{7.3}$$

(c) The full yield stress may be used for hooked smooth tension bars lapped over at least $15 d_{bL}$ (there are no data for shorter lapping). If the lapped ends of the bars are straight without hooks, (b) above applies, with 50 % longer $l_{oy,min}$.

7.4.4.3 Nonlinear Analysis

Nonlinear analysis is the reference analysis method in Part 3 of Eurocode 8, applicable to all cases. Although nonlinear dynamic (response-history) analysis, with solution of the equations of motion in the time-domain, is included, the emphasis is placed on nonlinear static ("pushover") analysis.

Part 3 of Eurocode 8 requires two lateral load patterns in "pushover analysis": one produced by uniform lateral accelerations; the other from first-mode ones, which is taken as heightwise linear as in linear static (lateral force) analysis, if such analysis is applicable, or from eigenvalue analysis, if it is not. It adopts the N2 procedure (Fajfar 2000), summarised in an Informative Annex to Part 1 of Eurocode 8 (CEN 2004a). For a fundamental period in the direction of pushover analysis, T_1, longer than the corner period T_C (see Sect. 7.5.1), the target

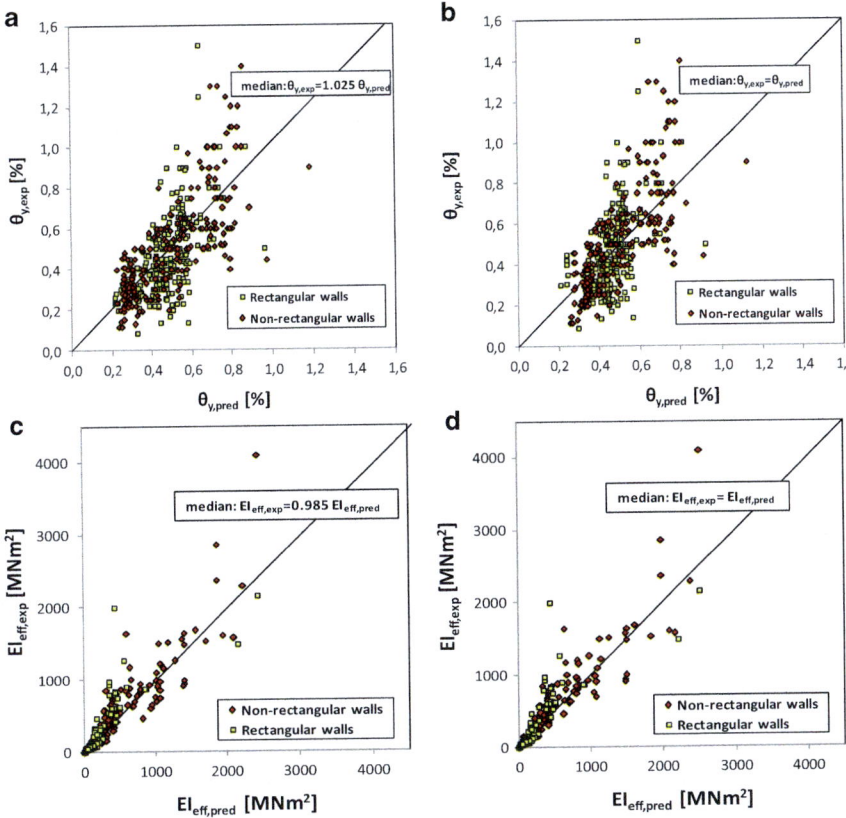

Fig. 7.4 Dataset of 520 rectangular, T-, H- or U-walls or hollow rectangular members: (**a**), (**b**) experimental v predicted chord rotation at flexural yielding, θ_y; (**c**), (**d**) experimental secant stiffness to yield point, EI_{eff}, v result of Eq. (7.2); (**a**), (**c**): prediction of θ_y per Sect. 7.4.4.2 (Biskinis and Fardis 2010a; CEN 2005a, 2009; *fib* 2012); (**b**), (**d**): prediction of θ_y per Sect. 7.4.4.2 with constant term 0.0013 replaced by $0.0007[1 + 4h/(3L_s)]$; CoV 43 % in (**c**), 41 % in (**d**)

displacement is equal to the elastic spectral one for 5 % damping; for shorter periods, the elastic displacement is multiplied by $\mu = 1 + (q-1)T_C/T_1$ (Vidic et al 1994), where the available value of the behaviour factor q may be taken equal to the ratio of the elastic base shear to the one corresponding to a plastic mechanism, i.e., the lateral force resistance of the building. As we will see in Sect. 7.5.3, apart from nonlinear dynamic analysis, this multiplication is the only departure from the "equal displacement rule" in Part 3 of Eurocode 8.

Nonlinear analysis should use the EI_{eff}-value from Eq. (7.2) as member elastic stiffness, except possibly in members confirmed to stay uncracked under the seismic action considered. Viscous damping equal to 5 % of critical is used, to model energy dissipation until member yielding.

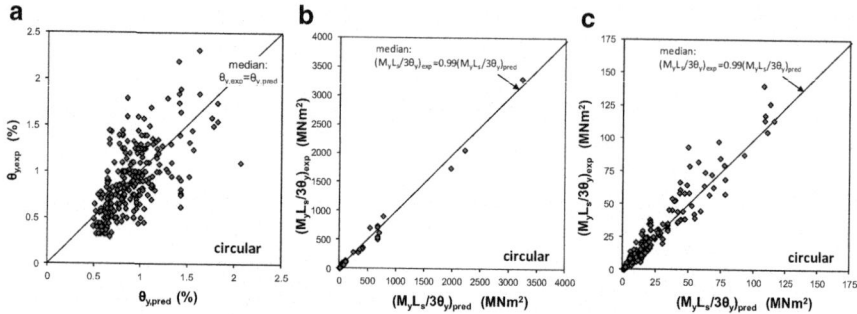

Fig. 7.5 Dataset of 291 circular columns: (**a**) experimental chord rotation at flexural yielding, θ_y, vs. predictions per (Biskinis and Fardis 2013a; *fib* 2012); (**b**) experimental secant stiffness to yield point, EI_{eff}, vs. prediction from Eq. (7.2) – CoV: 31 %; (**c**) detail of (**b**)

Linear models may be used for those structural components expected – and confirmed – to stay in the elastic domain for the seismic action of interest. Nonlinear modeling may then be limited to the rest. Nonlinear models of 1D members (including slender walls) should, as a minimum, employ a nonlinear moment-rotation relation for any flexural plastic hinge that may form at an end where the member frames into another component; if bending is mainly within a single plane, a uniaxial moment-rotation relation in that plane is sufficient.

As a minimum, nonlinear member models should use a bilinear generalised force-deformation (e.g. moment-rotation) law in primary (i.e. monotonic) loading:

- positive post-yield stiffness (due to strain-hardening) may be neglected; elastic-perfectly plastic behaviour may be assumed instead.
- significant post-yield softening due to strong strength degradation with cycling should be included via negative post-yield stiffness; however the normal reduction in resistance after ultimate strength may be neglected (after all, at the end of a design or a successful assessment-cum-retrofitting, brittle mechanisms are normally verified to remain elastic and ductile ones to have a margin against ultimate deformation – after which the drop in resistance is significant).

The requirement on hysteresis rules to be used in nonlinear response-history analysis is just to reflect realistically the post-yield energy dissipation in the range of displacement amplitudes expected.

Unlike linear elastic analysis described next, which may be relied upon, under certain conditions, to estimate seismic deformation but not internal force demands, nonlinear analysis may be used to determine all types of seismic action effects.

7.4.4.4 Linear Analysis for the Calculation of Seismic Deformations

Member seismic inelastic deformations may be determined from linear analysis with 5 % damping, provided that they are not concentrated at certain parts (e.g., at one side of the building in plan, in one or few storeys, etc.) but are spread fairly uniformly throughout the structure. This potential of linear analysis under such conditions is supported by several studies (e.g., Panagiotakos and Fardis 1999a, b, Kosmopoulos and Fardis 2007 for concrete buildings; Bardakis and Fardis 2011b, for concrete bridges with monolithic deck-pier connections). The nonlinear moment-deformation relations at member ends may be used then to determine the end moments from the inelastic flexural deformations estimated with linear analysis; shear forces are calculated from these moments by equilibrium.

A convenient way to check whether inelastic deformation demands are indeed uniformly distributed, without carrying out a nonlinear seismic analysis just for that purpose, is by looking at the spatial distribution of the ratio of the moment from linear analysis, M_E, at member end sections to the corresponding moment resistance, M_R (the M_E/M_R-ratio is an approximation to the chord-rotation ductility ratio). Part 3 of Eurocode 8 recommends a range of 2.5 between the maximum and the minimum values of M_E/M_R over all end sections in a building where plastic hinges may form (i.e., those sections where $M_E > M_R$ and plastic hinging at column or beam ends around a joint is not prevented by their higher aggregate moment resistance, $\sum M_{Rc}$, $\sum M_{Rb}$, compared to the beam or column ends, respectively).

If linear seismic analysis is allowed and adopted for the estimation of inelastic deformations, linear response-history analysis with 5 % damping – carried out simultaneously for all seismic action components of interest, or separately for each one and superposition of the results – is an option. However, as only the maximum values of these deformations are of interest, the method of choice is modal response spectrum analysis with the 5 %-damped elastic response spectrum, according to the rules set out in Part 1 of Eurocode 8 (CEN 2004a): total effective modal mass of the included modes at least 90 % of the total mass along any seismic action component considered; combination of peak modal deformations via the Complete-Quadratic-Combination (CQC) rule (Wilson et al 1981); peak values of seismic deformations due to separate application of the concurrent seismic action components combined via the Square Root of Sum of Squares (SRSS) rule, or its linear approximation in proportion 1: 0.3: 0.3 (Smebby and Der Kiureghian 1985). The values and signs of other action effects (e.g. the column deformation in the orthogonal direction), expected to take place concurrently with the peak value of the action effect obtained via the SRSS rule, may be obtained from probability-based models (Gupta and Singh 1977; Fardis 2009).

Under conditions set out in Part 1 of Eurocode 8 (CEN 2004a) and summarised below, modal response spectrum analysis may be simplified into separate linear static analyses under "equivalent" forces in the direction of each relevant seismic action component, with the structure taken as a SDoF having the period of the dominant mode, T_1, in that direction. This simplification may not be made in only one of the two horizontal directions, but may be applied to the vertical alone. For buildings with more than two storeys and period T_1 less than $2T_C$, the resultant "equivalent" force along the seismic action component of interest may be reduced by 15 % over the product of the elastic response spectral acceleration at T_1 and the total mass, to account for the smaller effective modal mass of the first mode.

"Equivalent static" analysis is allowed under the conditions set out for new buildings in Part 1 of Eurocode 8:

(a) No significant heightwise irregularity in geometry, mass and lateral stiffness or storey strength.
(b) $T_1 \leq 2$ sec, $T_1 \leq 4T_C$.

Linear analysis carried out to estimate the seismic deformation demands overestimates the internal forces. Nonlinear moment-deformation relations may be used in that case to compute the moments at member ends from the linearly estimated chord rotations; the shear forces in a component are computed then from equilibrium with the moments delivered to it at its connections to the rest of the structure. For simplification, these moments may be obtained from the moment capacities of the critical plastic hinges (multiplied by a "confidence factor" greater than 1.0, which depends on the amount and reliability of the information available or collected about the as-built structure), but not to exceed the moments from the linear analysis. Around beam-column joints, the plastic hinges are taken to form at the faces of the joint where the aggregately weaker elements frame (e.g., in the beams of a weak beam/strong column frame); the moments at the face of the non-hinging elements are estimated from moment equilibrium, as in "capacity design" of concrete beams or columns in shear per Part 1 of Eurocode 8 (CEN 2004a).

7.4.5 Cyclic Plastic (Chord) Rotation Capacity for Verification of Flexural Deformations

7.4.5.1 "Physical Model" Using Curvatures and Plastic Hinge Length

Annex A to Part 3 of Eurocode 8 includes a "physical" model for the expected (mean) value of the plastic part of the ultimate chord rotation at a member end, for use in the verification of flexural deformations at the "Significant Damage" and "Near Collapse" Limit States summarised in Table 7.1. It is a classical plastic hinge model, which assumes that, after yielding, the plastic part of the curvature is uniform within a finite "plastic hinge length", L_{pl}, from the end section:

$$\theta_{u,m}^{pl} = (\varphi_u - \varphi_y)L_{pl}\left(1 - \frac{L_{pl}}{2L_s}\right) \tag{7.4}$$

where $L_s = M/V$ is the shear span at the member end and φ_u, φ_y are the ultimate and the yield curvature, respectively, of the end section, from section analysis, using:

- for φ_y: linear σ-ε laws, until yielding of the tension or the compression chord;
- for φ_u: a bilinear σ-ε diagram for the reinforcement with or without linear strain-hardening; a parabolic-rectangular one for the concrete in compression.

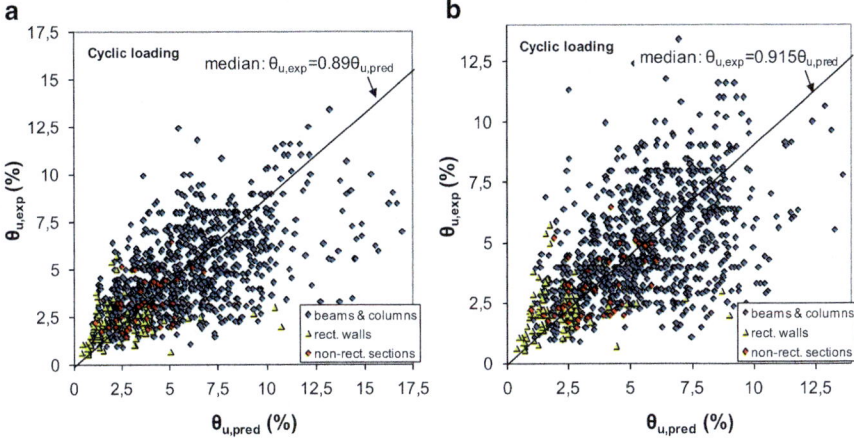

Fig. 7.6 Experimental ultimate chord rotation in cyclic flexure in 1,125 cyclic tests vs result of Eq. (7.4a), using the expressions for "plastic hinge length", L_{pl}, in (CEN 2005a) for confinement: (**a**) per (CEN 2004b; CEB 1991) – Option 1; (**b**) per (CEN 2005a) – Option 2

For the ultimate strain of reinforcing steel, ε_{su}, the 10 %-fractile limits in Annex C to Eurocode 2 are accepted in the calculation of φ_u: 2.5, 5, 6 % for steel class A, B, C. For the concrete and its confined core after spalling there are two options:

1. the Eurocode 2 model, taken from the CEB/FIP Model Code 90 (CEB 1991);
2. the strength model by Newman and Newman (1971), supplemented with a model for the ultimate strain, $\varepsilon_{cu,c}$, specifically fitted for the purposes of Part 3 of Eurocode 8 to the then available measurements of φ_u in cyclic loading (starting from a value of 0.004 for the unconfined concrete cover).

Option 1 underestimates the presently available test results by one-third in the median, whereas option 2 is almost unbiased.

Empirical expressions (different for Options 1 or 2) for the "plastic hinge length", L_{pl}, were fitted specifically for Part 3 of Eurocode 8 to the cyclic test results available then. They indirectly reflect the additional fixed-end rotation due to slippage of longitudinal bars from their anchorage in the joint or footing, including "yield penetration" in it. However, as shown in Fig. 7.6, they give large scatter (hence the large factors of 1.7 and 2 by which $\theta_{u,m} = \theta_y + \theta^{pl}_{u,m}$ is divided, in order to convert it to $\theta_{u,m-\sigma}$, see footnote (3) in Table 7.2) and marked overestimation of the presently available cyclic test results.

7.4.5.2 Empirical Rotation Capacity: Sections with Rectangular Parts

For well-detailed beams, rectangular columns or walls and members of T-, H-, U- or hollow rectangular section with continuous ribbed bars, (Biskinis and Fardis 2010b) proposed empirical expressions for the expected value of the ultimate chord rotation at a member end under cyclic loading (total $\theta_{u,m}$, or plastic part, $\theta^{pl}_{u,m} = \theta_{u,m} - \theta_y$). This option, Eqs. (7.5), is unbiased and has less scatter – hence model uncertainty – than the approach in Sect. 7.4.5.1.

$$\theta_{u,m} = a_{st}(1 - 0.42a_{w,r})\left(1 - \frac{2}{7}a_{w,nr}\right)(0.3^{\nu})\left[\frac{\max\ (0.01;\omega_2)}{\max\ (0.01;\omega_1)}f_c\right]^{0.225}$$

$$\left[\min\left(9;\frac{L_s}{h}\right)\right]^{0.35} 25^{\left(\frac{a\rho_s f_{yw}}{f_c}\right)} 1.25^{100\rho_d} \quad (7.5\text{a})$$

$$\theta^{pl}_{u,m} =$$

$$a^{pl}_{st}(1 - 0.44a_{w,r})\left(1 - \frac{a_{w,nr}}{4}\right)(0.25)^{\nu}f_c(MPa)^{0.2}\left(\frac{\max(0.01;\omega_2)}{\max(0.01;\omega_1)}\right)^{0.3}$$

$$\left(\min\left(9;\frac{L_s}{h}\right)\right)^{0.35} 25^{\left(\frac{a\rho_s f_{yw}}{f_c}\right)} 1.275^{100\rho_d} \quad (7.5\text{b})$$

In Eqs. (7.5):

- a_{st}, a^{pl}_{st} are coefficients for the type of steel, with values:
  - For ductile steel: $a_{st} = 0.0158$, $a^{pl}_{st} = 0.0143$;
  - For brittle steel: $a_{st} = 0.0098$, $a^{pl}_{st} = 0.007$.
- $a_{w,r}$ is a zero-one variable for rectangular walls:
  - $a_{w,r} = 1$ for a rectangular wall,
  - $a_{w,r} = 0$ otherwise;
- $a_{w,nr}$ is a zero-one variable for non-rectangular sections:
  - $a_{w,nr} = 1$ for a T-, H-, U- or hollow rectangular section,
  - $a_{w,nr} = 0$ for a rectangular one;
- $\nu = N/bhf_c$, with b the width of the rectangular compression zone and N the axial force ($N > 0$ for compression);
- $\omega_1 = (\rho_1 f_{y1} + \rho_v f_{yv})/f_c$ is the mechanical ratio of reinforcement in the entire tension zone (with 1 indexing the tension chord and v the web longitudinal bars);
- $\omega_2 = \rho_2 f_{y2}/f_c$ is the mechanical reinforcement ratio of the compression zone;
- $L_s/h = M/Vh$ is the shear-span-to-depth ratio at the section of maximum moment;
- ρ_d is the steel ratio of any bars in each diagonal direction of the member;

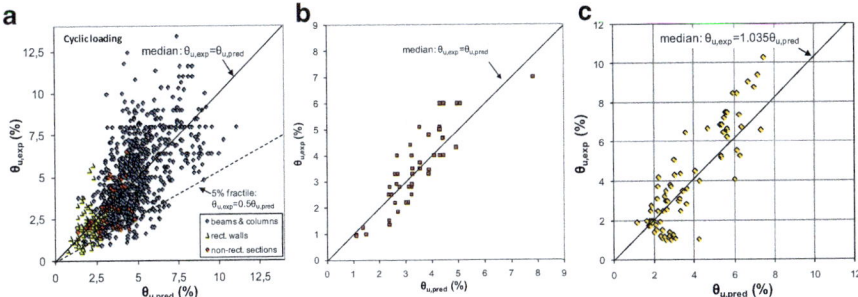

Fig. 7.7 Cyclic ultimate chord rotation of members with rectangular, T-, H-, U- or hollow rectangular section vs. empirical predictions per Sect. 7.4.5.2: (**a**) 1125 tests of well detailed members with continuous ribbed bars vs. Eq. (7.5b); (**b**) 48 tests of members with poor detailing and continuous ribbed bars vs. Eq. (7.5a) modified per point 1; (**c**) 82 tests of members with poor detailing and lap-spliced ribbed bars vs. Eq. (7.5b) modified per point 3 and Eq. (7.7)

- $\rho_s = A_{sh}/bs_h$ is the ratio of confinement steel in the compression zone parallel to the plane of bending and the shear force;
- α is the confinement effectiveness factor:

$$\alpha = \left(1 - \frac{s_h}{2b_o}\right)\left(1 - \frac{s_h}{2h_o}\right)\left(1 - \frac{\sum b_i^2/6}{b_o h_o}\right) \quad (7.6a)$$

with:

- s_h: centreline spacing of stirrups,
- b_o, h_o: confined core dimensions to the centreline of the hoop;
- b_i: centreline spacing of longitudinal bars (index: i) engaged by a stirrup corner or cross-tie along the perimeter of the section.

Part 3 of Eurocode 8 (CEN 2005a, 2009) has adopted an earlier version of Eqs. (7.5a) and (7.5b), with coefficients a_{st}, a^{pl}_{st} rounded up by about 1.3 % and a common reduction factor $a_{w,r}$ and $a_{w,nr}$ for walls, rectangular or not, equal to 0.375 in Eq. (7.5a) and 0.4 for Eq. (7.5b).

The two versions of Eq. (7.5) are equivalent, as far as bias and scatter are concerned. The comparison of experimental to predicted values in Fig. 7.7a is indicative. A further comparison with Fig. 7.6 shows that they are superior to the more fundamental approach in Sect. 7.4.5.1. They also have a wider scope and are easier to extend, in the ways suggested in (Biskinis and Fardis 2010b) and adopted in Part 3 of Eurocode 8 (CEN 2005a, 2009):

1. In members with continuous bars but poor detailing, not conforming to modern seismic design codes (e.g., with sparse, 90°-hooked ties), the confinement effect is neglected ($\alpha\rho_s = 0$ in the second term from the end) and $\theta_{u,m}$ from Eq. (7.5a), or $\theta^{pl}_{u,m}$ from Eq. (7.5b) is divided by 1.2 (see Fig. 7.7b).

2. If the bars are smooth but continuous, rule 1 above is modified to further reduce $\theta_{u,m}$ from Eq. (7.5a) by 5 % (multiplication by $0.95/1.2 \sim 0.8$) or $\theta^{pl}_{u,m}$ from Eq. (7.5b) by 10 % (multiplied by $0.90/1.2 = 0.75$). With the increase of the number of tests from 34 – on which the rule was based (Biskinis and Fardis 2010b) – to 46, no further reduction of $\theta_{u,m}$ beyond rule 1 above seems necessary, while the reduction of $\theta^{pl}_{u,m}$ from Eq. (7.5b) should be limited to 5 % (i.e., it should be multiplied by $0.95/1.2 \sim 0.8$).
3. Equation 7.5b can be extended to members with ribbed bars lap-spliced over a length l_o in the plastic hinge region (see Fig. 7.7c):
   - by applying rules (a) and (b) of Sect. 7.4.4.2 (at the end) in calculating θ_y;
   - by applying the same rule (a) to ω_2 (doubling it, if all compression bars are lapped);
   - by multiplying the outcome of Eq. (7.5b) for $\theta^{pl}_{u,m}$ by $l_o/l_{ou,min}$, if l_o is less than $l_{ou,min}$ given by:

$$l_{ou,min} = \frac{d_b f_y}{\left(1.05 + 14.5 a_{l,s} \frac{\rho_s f_{yw}}{f_c}\right) \sqrt{f_c}} \quad \left(f_y, f_{yw}, f_c \text{ in MPa}\right) \quad (7.7)$$

where:

- ρ_s is the ratio of the transverse steel parallel to the plane of bending, and

$$a_{l,s} = (1 - 0.5 s_h/b_o)(1 - 0.5 s_h/h_o) n_{restr}/n_{tot}, \quad (7.8)$$

with:

- s_h, b_o, h_o, as defined for Eq. (7.6a),
- n_{tot}: total number of lapped bars along perimeter of the section and
- n_{restr}: number of lapped bars engaged by a stirrup corner or cross-tie.

For smooth bars, with hooked ends lap-spliced over a length $l_o \geq 15 d_b$, (CEN 2009) reduces $\theta_{u,m}$ from rule 2 above by multiplying it with $0.019[10 + \min(40; l_o/d_b)]$ (which gives the reduction factor of 0.95 for continuous bars), or $\theta^{pl}_{u,m}$ from the same rule by multiplying it with $0.019[40; l_o/d_b)]$ – giving a reduction factor of 0.76 for continuous bars. The 17 tests now available – v 11 on which that rule was based (Biskinis and Fardis 2010b) – show smaller reduction of $\theta_{u,m}$ and $\theta^{pl}_{u,m}$ than the modified rule 2 above, namely to multiply them by $[60 + \min(40; l_o/d_b)]/100$.

Wrapping the plastic hinge region with Fibre Reinforced Polymers (FRP) to improve deformation capacity may be considered by including confinement by the FRP in the exponent of the second term from the end of Eqs. (7.5). If the vertical bars are lap-spliced in that region, Eqs. (7.7) and (7.8) are modified to reflect the beneficial effect of confinement by the FRP. However, in the light of newly available test results, the relevant rules in Part 3 of Eurocode 8 need improvement (see Biskinis and Fardis 2013a, b for proposals).

7.4.6 Cyclic Shear Resistance

7.4.6.1 Diagonal Tension Strength After Flexural Yielding

Shear strength decays faster than flexural strength with load cycling. So, members that first yield in flexure may, under cyclic loading, ultimately fail in shear in the plastic hinge. The shear resistance in static loading per Eurocode 2 does not apply to regions which have already yielded in flexure and have developed a certain amount of inelastic deformation in the tensile chord. After all, if loading is static and proportional, a flexural plastic hinge will not fail in shear, as its internal forces (including the shear force) do not increase much after flexural yielding.

For seismic loading shear failure of flexural plastic hinges is normally described through a shear resistance of the plastic hinge in diagonal tension, V_R, which decreases with increasing plastic rotations under cyclic loading. Part 3 of Eurocode 8 has adopted a model in (Biskinis et al 2004) giving V_R as the sum of:

- the transverse component of the diagonal strut transferring the axial load N from the compression zone of the section of maximum moment to the centre of the zero-moment section, i.e., over a distance $L_s = M/V$, as in (CEB 1991);
- a non-zero concrete contribution term, V_c; and
- the contribution of transverse reinforcement, V_w, for a 45°-truss inclination.

V_c and V_w are taken to decrease with increasing plastic rotation ductility ratio, $\mu^{pl}_\theta = \theta^{pl}/\theta_y$, where $\theta^{pl} = \theta - \theta_y$ is the plastic (chord) rotation demand and θ_y is determined according to Sect. 7.4.4.2, points 1 to 3:

$$V_R = \frac{h-x}{2L_s}\min(N; 0.55 A_c f_c) + \left(1 - 0.05\min\left(5; \mu^{pl}_\theta\right)\right)$$
$$\left[0.16\max(0.5; 100\rho_{tot})\left(1 - 0.16\min\left(5; \frac{L_s}{h}\right)\right)\sqrt{f_c} A_c + V_w\right] \quad (7.9)$$

with:

h: depth of the cross-section (equal to the diameter D for circular sections);
x: compression zone depth;
N: compressive axial force (positive, taken as zero for tension);
$L_s/h = M/Vh$: shear span ratio at the yielding member end;
f_c: concrete strength (MPa);
ρ_{tot}: total longitudinal reinforcement ratio;
A_c: cross-section area, equal to $b_w d$ for cross-sections with rectangular web of width b_w and effective depth d, or to $\pi D_o^2/4$ for circular sections (D_o: diameter of the concrete core to the centreline of the hoops);
V_w: contribution of transverse reinforcement to shear resistance:

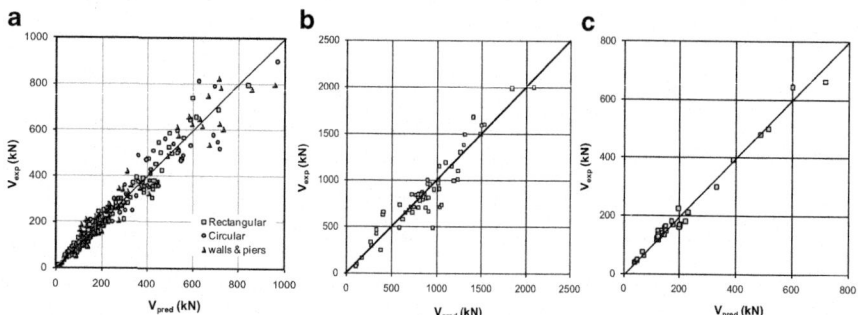

Fig. 7.8 Cyclic shear resistance v prediction: (**a**) 334 cyclic tests with diagonal tension failure after flexural yielding vs. Eq. (7.9); (**b**) 63 cyclic tests of rectangular or non-rectangular walls or hollow rectangular members with $1.0 \leq L_s/h \leq 2.5$ failing in diagonal compression vs. Eq. (7.11); (**c**) 48 cyclic tests of columns with $L_s/h \leq 2.0$ failing in diagonal compression vs. Eq. (7.12).

For cross-sections with rectangular web width b_w, having transverse reinforcement with ratio ρ_w and yield stress f_{yw}, internal lever arm z equal to $0.8h$ in rectangular walls or $d-d_1$ in columns and hollow, H-, U- or T-sections:

$$V_w = \rho_w b_w z f_{yw} \quad (7.10a)$$

For circular sections with diameter of the concrete core D_o, cross-sectional area of circular stirrups A_{sw} and centreline spacing of stirrups s_h:

$$V_w = \frac{\pi}{2} \frac{A_{sw}}{s_h} f_{yw} D_o \quad (7.10b)$$

The database of RC tests leading to failure of the type described here has considerably increased since the development of Eq. (7.9) in (Biskinis et al 2004). As depicted in Fig. 7.8a, the broader dataset agrees well with Eq. (7.9).

For assessment, the value of $\mu^{pl}_\theta = (\theta - \theta_y)/\theta_y$ at which $V_R(\mu^{pl}_\theta)$ from Eq. (7.9) becomes equal to the shear at flexural yielding, M_y/L_s, is translated into a chord rotation $\theta = (\mu^{pl}_\theta + 1)\theta_y$ for which this type of failure is expected to take place; if this value of θ is less than the expected ultimate chord rotation in flexure from Eqs. (7.5), θ_{um}, failure will most likely be in shear at $\theta = (\mu^{pl}_\theta + 1)\theta_y$, rather than by flexure at θ_{um}.

7.4.6.2 Diagonal Compression Strength of Squat Walls and Columns

Walls with $L_s/h \leq 2.5$ may fail under cyclic loading by diagonal compression at a shear force less than the predictions of Eq. (7.9) and a chord rotation much less than the value at flexure-controlled failure per Eqs. (7.5). It is now recognised that walls with $L_s/h \leq 1.0$ follow a different pattern and model (Grammatikou et al. 2014), but, as demonstrated in Fig. 7.8b, those with $1.0 < L_s/h \leq 2.5$ do confirm a model proposed by Biskinis et al. (2004) and adopted in Part 3 of Eurocode 8 for the cyclic resistance of walls with $L_s/h \leq 2.5$ against web crushing:

$$V_{R,\max} = 0.85\left(1 - 0.06\min\left(5;\ \mu_\theta^{pl}\right)\right)\left(1 + 1.8\min\left(0.15;\frac{N}{A_c f_c}\right)\right)$$
$$\cdot (1 + 0.25\max(1.75;\ 100\rho_{tot}))\left(1 - 0.2\min\left(2;\frac{L_s}{h}\right)\right)\sqrt{\min(100MPa; f_c)}b_w z \quad (7.11)$$

where all symbols have been defined above for Eq. (7.9). If $\mu^{pl}_\theta = 0$ Eq. (7.11) gives the cyclic resistance in diagonal compression before flexural yielding.

Columns with $L_s/h \leq 2.0$ under cyclic loading often fail in compression along the diagonal in elevation after flexural yielding. Part 3 of Eurocode 8 adopted for them the empirical model by Biskinis et al. (2004):

$$V_{R,\max} = \frac{4}{7}\left(1 - 0.02\min\left(5;\ \mu_\theta^{pl}\right)\right)\left(1 + 1.35\frac{N}{A_c f_c}\right)(1 + 0.45(100\rho_{tot}))$$
$$\sqrt{\min(40MPa; f_c)}b_w z \sin 2\delta \quad (7.12)$$

where:

- δ: angle of the column diagonal in elevation to the column axis: $\tan\delta = h/2L_s$;
- all other parameters have been defined above for Eq. (7.9).

Figure 7.8c shows that the current dataset, broader than the one to which Eq. (7.12) was fitted in (Biskinis et al. 2004), still confirms this model.

The procedure in the last paragraph of Sect. 7.4.6.1 can be applied to Eq. (7.11) for walls with $1.0 < L_s/h \leq 2.0$, or to Eq. (7.12) for columns with $L_s/h \leq 2.0$, to identify the failure mode most likely to occur among those in Sects. 7.4.5.2, 7.4.6.1, and 7.4.6.2.

7.5 Performance- and Displacement-Based Seismic Design of New Concrete Structures in the 2010 Model Code of *fib*

7.5.1 Introduction

Seismic design of new structures according to present day codes, including Eurocode 8 (CEN 2004a, 2005b), is force-based; members are dimensioned at the ULS against internal forces computed via elastic analysis for external ("seismic") forces derived from a "design" response spectrum, which results from dividing the ordinates of the 5 %-damped elastic response spectrum by an empirical behaviour (or force reduction) factor. Prescriptive, opaque and, by and large, arbitrary detailing rules for members are presumed to provide ductility commensurate with the behaviour factor employed in the analysis. A single level of seismic action is normally considered (the "design seismic action", chosen in general to have a 10 % probability of being exceeded in 50 years). The damage induced to non-structural elements (e.g., partitions) by a frequent ("serviceability") seismic action is sometimes checked (CEN 2004a), but this is a non-structural verification, independent of the structural material. This design approach is opaque concerning the achieved seismic performance and overall sub-optimal.

The Model Code 2010 of *fib* (2012) – in short MC2010 – is meant to serve as a guidance document to future codes for design of concrete structures (Walraven 2013). Its predecessors (CEB 1978; CEB 1991) were the basis of the European design standard for concrete structures, as pre-Norm (CEN 1991) or Norm (CEN 2004b), respectively. Those CEB Model Codes did not cover seismic design. However (CEB 1978) was supplemented by the CEB seismic Model Code (CEB 1985) for (mainly) concrete buildings, which served as the basis for the pre-Norm version (CEN 1994b, c, d) of the European seismic design standard, especially for its parts on concrete buildings. As the 1990 Model Code (CEB 1991) did not have a seismic part or follow-up, the first European standards for earthquake resistant structures (CEN 2004a, 2005a, b) developed independently.

MC2010 includes full-fledged performance-based seismic design and assessment, targeting specific and measurable performance under several levels of seismic action (Fardis 2013). Moreover, it uses deformations as the basis for verifications, and not internal forces. In these two fundamental features, as well as in many details, it follows Part 3 of Eurocode 8 (CEN 2005a). Note that this European standard concerns existing buildings, while MC2010 covers seamlessly assessment of seismic performance of existing, as well as design of new buildings and other structures (notably bridges). The introduction of performance- and displacement-based seismic design of new structures in the footsteps of a standard for seismic assessment of existing ones is a reversal of the past tradition, where procedures and codification for existing structures followed and emulated those for new.

The rest of Sect. 7.5 has the same structure as Sect. 7.4, but only points out the differences of MC2010 from Part 3 of Eurocode 8. Wherever no difference is mentioned, whatever has been said in Sect. 7.4 applies to MC2010 too.

7.5.2 Performance Objectives

MC2010 identifies four "performance levels", termed Limit States. They are listed in Table 7.1 alongside the corresponding structural condition and functionality of the facility, the compliance criteria and the appropriate "seismic hazard level" for ordinary facilities. The first two are Serviceability Limit States (SLS), the last two are Ultimate Limit States (ULS). According to MC2010, the "Performance Objective" should at least include one SLS and one ULS; the owner or competent authorities are meant to choose which ones and the level of the corresponding seismic action, depending on the use and importance of the facility.

As emphasised in Sects. 7.4.4.1 and 7.4.4.2, even though the seismic response may go well into the inelastic range, seismic deformation demands are about proportional to the intensity of the ground motion. So, the deformation limits in the second to last column of Table 7.1 show that normally just one of the two SLSs and one of the two ULSs control the design or assessment and need to be explicitly verified. For example, in a certain project the IU SLS will most likely control the design instead of the OP, if its seismic action exceeds that of the OP by more than a factor of 2.0; the NC ULS may govern over the LS one, if its seismic action exceeds that of the latter by more than a factor of 1.5.

7.5.3 Compliance Criteria

The compliance criteria in MC2010 do not distinguish "primary" from "secondary" members. The distinction between "ductile" mechanisms, checked in terms of deformations, and "brittle" ones, checked in terms of forces, is retained.

As shown in the second to last column in Table 7.1, at the two SLSs deformations are verified by comparing the chord rotation demand at each member end, θ_{Ed}, to:

1. the chord rotation at yielding of that end, θ_y, at the OP SLS; or
2. twice that value, $2\theta_y$, if the IU SLS is being verified.

So, the verification and the compliance criteria at the OP SLS are the same as for DL in Part 3 of Eurocode 8 (cf. Table 7.2)

At each ULS, deformations are checked by comparing the plastic part of chord rotation demand at a member end (equivalently the plastic hinge rotation) $\theta^{pl}_{E,d}$, to:

3. the lower 5 %-fractile of the ultimate plastic hinge rotation (or, equivalently, of the plastic part of ultimate chord rotation), $\theta^{pl}_{u,k}$, divided by a global safety factor $\gamma^*_R = 1.35$, if the Life Safety (LS) ULS is being checked; or
4. just $\theta^{pl}_{u,k}$, if Near-Collapse (NC) is being verified.

The lower 5 %-fractile of θ^{pl}_u is obtained from its mean value, $\theta^{pl}_{u,m}$, as:

$$\theta^{pl}_{u,k} = \theta^{pl}_{u,m}/\gamma_{Rd} \quad (7.13)$$

where γ_{Rd} is a model uncertainty factor, depending on the model used to determine $\theta^{pl}_{u,m}$. Sect. 7.5.5 gives its values for the models described there for $\theta^{pl}_{u,m}$.

Note that the ratio of the deformation limits against which the plastic rotation demands are checked in the NC and LS Limit States, i.e., $\gamma^*_R = 1.35$, is essentially the same as the one specified in Annex A to Part 3 of Eurocode 8 between the chord rotation demands (the ratio of the values at the intersection of the fourth and third column and the first and second row of Table 7.2 is $1/0.75 = 1.33$).

7.5.4 Analysis for the Determination of Seismic Action Effects

7.5.4.1 Effective Elastic Stiffness for the Analysis

In order to apply Eq. (7.2), the longitudinal reinforcement at member ends should be known. In new structures, it may be pre-dimensioned for the non-seismic actions and the corresponding minimum reinforcement and other detailing rules. It may be increased afterwards, if it is considered likely that it will later be necessary, in order to meet the seismic design checks. However, as EI_{eff} depends weakly on the amount of longitudinal reinforcement, MC2010 allows the use of empirical expressions, which give the ratio of EI_{eff} to the uncracked gross section stiffness, $(EI)_c$, depending on the type of member, the shear span ratio, L_s/h, the mean axial stress, N/A_c, the ratio of longitudinal bar diameter to depth, d_{bL}/h, etc. Such an expression has been fitted to experimental values of EI_{eff} in (Biskinis and Fardis 2010a, 2013a) and is presented as Eq. (7.14), with the value of a modified for walls and hollow rectangular piers in the light of more recent data:

$$\frac{EI_{eff}}{(EI)_c} = a\left(0.8 + \ln\left[\max\left(\frac{L_s}{h};\ 0.6\right)\right]\right)\left(1 + 0.048\ \min\left(50 MPa; \frac{N}{A_c}\right)\right) \quad (7.14)$$

where N/A_c is in MPa, and

- $a = 0.10$ for beams;
- $a = 0.081$ for rectangular columns;
- $a = 0.12$ for circular columns and rectangular walls;
- $a = 0.092$ for walls with T-, U-, H-section or hollow rectangular piers.

Figure 7.9 compares the predictions of Eq. (7.14) to the test results used in its fitting. The Coefficient of Variation (CoV) value of the test-to-prediction ratio of EI_{eff} given in the caption is not much larger than in Figs. 7.3, 7.4 and 7.5; this small difference conceals the lack of fit with respect to the steel ratio.

7.5.4.2 Nonlinear Analysis

The reference method in Part 3 of Eurocode 8, namely nonlinear static ("pushover") analysis, is not mentioned in MC2010. Reflecting the current design practice of tall buildings and long bridges in seismic regions worldwide, the reference method is nonlinear response-history analysis. The seismic action is specified as a suite of

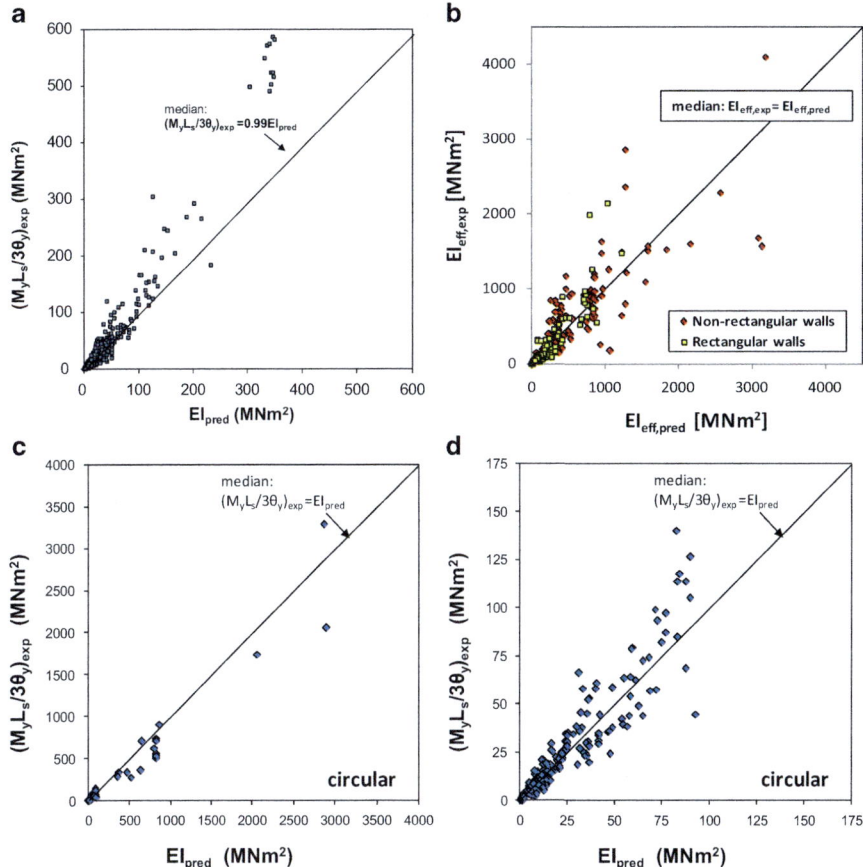

Fig. 7.9 Experimental secant stiffness to yield point, EI_{eff}, vs. empirical prediction from Eq. (7.14): (**a**) 1637 beams and rectangular columns – CoV: 36 %; (**b**) 517 walls or hollow rectangular piers – CoV 45 %; (**c**) 273 circular columns – CoV: 31 %; (**d**) detail of (**c**)

independent seismic events in terms of acceleration time-histories of the three translational ground motion components, all three applied simultaneously and together with the quasi-permanent gravity loads. The number of seismic events should be sufficient to derive robust statistics of action effects. To estimate peak response quantities, the minimum numbers are those specified in Part 1 of Eurocode: at least seven, if their results are averaged; at least three, if the most adverse peak response from the analyses is used. It is pointed out in MC2010 that more than these minimum numbers are necessary to estimate residual deformations or displacements.

The impact of significant variations in the axial force during the response (as, e.g., in exterior columns of tall frames, in the individual piers of coupled walls, or in tall bridge piers with two columns in the vertical plane of bending) on the moment-rotation behaviour should be taken into account. However, coupling between the two directions of bending of vertical elements in 3-D models may be treated in a simplified way.

The unloading-reloading ("hysteresis") rules supplementing the force-deformation law in primary loading should realistically reflect the post-yield hysteretic energy dissipation and the reduction of unloading and reloading stiffness with increasing peak deformation of a cycle ("stiffness degradation"), a characteristic of concrete components. If significant, the degradation of resistance with load cycling should be included (notably in brittle or poorly detailed components). If a significant part of the deformation is due to bond-slip (e.g., from a joint) or shear (e.g., in members with low shear-span-to-depth ratio), the hysteresis loops should be "pinched" (as an inverted-S) and the hysteretic energy dissipation reduced. The hysteresis rules are important, if we want to estimate residual deformations of members (for local damage), or of the structure as a whole (permanent tilt) after the earthquake; they affect much less the prediction of peak deformation demands during the response.

7.5.4.3 Linear Analysis for the Calculation of Seismic Deformations

MC2010 retains the relatively uniform distribution of inelastic deformations in the plastic hinges as the condition for using 5 %-damped linear analysis to estimate inelastic seismic deformations. It also keeps the ratio of the moment from linear analysis, M_E, at member end sections to the corresponding moment resistance, M_R, as the means through which this condition is checked. However, unlike Part 3 of Eurocode 8, it does not give quantitative limits for the M_E/M_R-ratio.

MC2010 promotes the application in new buildings of "capacity design" of columns, so as to be stronger in flexure than the beams and therefore to serve as a strong and stiff spine, spreading the seismic deformation demands throughout the building and preventing concentration in a (soft) storey. Application of this rule produces favourable conditions for the applicability of linear analysis to estimate inelastic seismic deformations.

The peak values of seismic deformations due to separate linear analyses for the seismic action components in X, Y, Z are always combined via the SRSS rule; the linear approximation in a 1:0.3:0.3 proportion is not mentioned. Note that the combination of modal contributions through the CQC rule and of the peak effects of the seismic action components via SRSS can be done in a single modal response spectrum analysis covering all relevant seismic action components. This renders the resulting expected value of peak seismic action effects under concurrent seismic action components along X, Y (and Z) independent of the choice of horizontal directions X and Y.

The SRSS rule is also applied to combine the peak action effects of the vertical component, Z, from "equivalent static" analysis along Z alone with the outcome of the combination of the peak action effects of horizontal components, X and Y, in a single modal response spectrum analysis for these two components.

7 From Performance- and Displacement-Based Assessment of Existing Buildings... 257

The applicability conditions of "equivalent static" linear analysis are more general than in Eurocode 8, Parts 1 and 3, covering non-building structures too:

- The dominant normal mode along the seismic action component in question should account for at least 75 % of the total mass, and
- the response spectral displacements for this mode are much larger than those of any other mode with significant effective modal mass in the same direction.

MC2010 specifies a wider portfolio of rules than Eurocode 8 for the calculation of shears or other internal forces, when linear analysis is used to estimate the seismic deformations. The scope covers the cases when equilibrium does not suffice to determine the shears or other internal forces solely from the moment capacities at plastic hinges. In such cases MC2010 estimates these forces assuming that the seismic action effects at the instant the moment capacities at plastic hinges are reached are proportional to the corresponding outcomes of linear seismic analysis. This is the approach in Parts 1 and 2 of Eurocode 8 for:

1. The independent foundation of a single vertical element, where the seismic action effects in the foundation element and the ground from linear analysis are multiplied by the minimum ratio between the two orthogonal transverse directions at the base of the vertical element of (a) the uniaxial moment resistance under the axial load due to gravity loads, to (b) the moment from linear analysis for the seismic action (with this ratio not taken greater than 1.0).
2. Multistorey walls, including the amplification of shears in slender walls (those taller than twice their horizontal length), presuming that higher modes (i.e. with a collective participating mass of about 30 % of that of the fundamental mode and with periods in the constant-spectral-acceleration range) remain elastic and increase the wall shears after yielding at the base.
3. Brittle or sensitive components of bridges forming plastic hinges in the piers, which should remain elastic after such plastic hinging (the deck, fixed bearings, abutments flexibly connected to the deck, seismic links consisting of shear keys, buffers and/or linkage bolts, etc.). The action effects from linear analysis for the seismic action component of interest are multiplied by the ratio of (a) the sum of capacity-design shears along the seismic action component to (b) the sum of seismic shear forces from linear analysis, with both sums extending over all vertical supports where plastic hinges form.

The most important extension of the approach above is to the common foundation of many vertical elements: the seismic action effects from linear analysis are multiplied by the weighted-average of the factors computed per 1 above at the base of each individual vertical element. As weight is used the moment component from linear analysis that gives the minimum ratio per 1 above between the two directions of its base section and governs plastic hinging.

7.5.5 Cyclic Plastic (Chord) Rotation Capacity

7.5.5.1 "Physical Model" Using Curvatures and Plastic Hinge Length

Equation (7.4) is modified as follows:

$$\theta_{u,m}^{pl} = (\varphi_u - \varphi_y)L_{pl}\left(1 - \frac{L_{pl}}{2L_s}\right) + \Delta\theta_{slip,u-y} \quad (7.4a)$$

to explicitly include the post-yield fixed-end-rotation due to slippage of the tension bars (with mean diameter d_bL) from their anchorage outside the member length, associated with penetration of yielding into the anchorage zone; until attainment of the ultimate curvature at the end section under cyclic loading, the fixed-end-rotation increases, per (Biskinis and Fardis 2010b), by:

$$\Delta\theta_{slip,u-y} = 5.5 d_{bL}\varphi_u \quad (7.15)$$

Moreover, MC2010 states that the calculation of φ_u should account for all possible failure modes:

(a) rupture of tension bars in the full, unspalled section;
(b) exceedance of the concrete ultimate strain ε_{cu2} at the extreme compression fibres of the unspalled section;
(c) rupture of tension bars in the confined core after spalling of the cover;
(d) exceedance of the ultimate strain $\varepsilon_{cu2,c}$ of the confined core after spalling.

Failure modes (c) or (d) govern over (b), if the moment resistance of the confined core exceeds 80 % of that of the full, unspalled, unconfined section; this percentage is associated with the conventional definition of ultimate deformation.

The calculation of φ_u under cyclic loading uses the following σ-ε parameters:

- The rupture strain of ribbed tension bars under cyclic loading taken per (Biskinis and Fardis 2010b):

$$\varepsilon_{su,cyc} = (3/8)\varepsilon_{u,k} \quad (7.16)$$

- A new expression for the ultimate strength of confined concrete:

$$f_{cc} = \left[1 + 3.5\left(\frac{\alpha\rho_s f_{yw}}{f_c}\right)^{\frac{3}{4}}\right]f_c \quad (7.17)$$

where:

- ρ_s is the ratio of transverse reinforcement in the direction of bending (or the minimum in the two transverse directions for biaxial bending) and f_{yw} its yield stress;
- α is the confinement effectiveness factor:

- in rectangular sections, according to Eq. (7.6a):
- in circular sections with circular hoops:

$$\alpha = \left(1 - \frac{s_h}{2D_o}\right)^2 \qquad (7.18)$$

(without the exponent 2 in circular sections with spiral reinforcement).

- The increased strain of confined concrete at ultimate strength over that of unconfined, ε_{c2}, per (Richart et al 1928), adopted in (CEN 2005a):

$$\varepsilon_{c2,c} = \varepsilon_{c2}\left[1 + 5\left(\frac{f_{cc}}{f_c} - 1\right)\right] \qquad (7.19)$$

- The ultimate strain of the extreme compression fibres in a concrete core confined by closed ties, according to (Biskinis and Fardis 2010b):

$$\varepsilon_{cu2,c} = 0.0035 + 0.4\frac{\alpha\rho_s f_{yw}}{f_{cc}} \qquad (7.20)$$

MC2010 notes that a term may be added to Eq. (7.20) to express a size-effect on the plastic rotation capacity clearly found in experiments; it is equal to $(10/h)^2$ – with h (in mm) denoting the depth/diameter of the full section or of the confined core, wherever Eq. (7.20) is applied (Biskinis and Fardis 2010b); its effect is minor in real-size members.

Figure 7.10 compares the predictions of the procedure above to the curvature associated with a 20 % drop in the moment resistance after its peak (conventional "ultimate" curvature) in a dataset of 205 cyclic and 269 monotonic tests of rectangular members. For monotonic loading, term 3/8 in Eq. (7.16) is replaced by $1-(\sqrt{\ln N_{b,tension}})/3$, where $N_{b,tension}$ is the number of bars in the tension zone (Biskinis and Fardis 2010b); besides, factor 0.4 in the last term of Eq. (7.20) is replaced by 0.57 and the yield penetration length at ultimate curvature in Eq. (7.15) increases to $9.5d_{bL}$. Points denoted in Fig. 7.10 as "slip" are data where the fixed-end-rotation due to slippage of tension bars from their anchorage had to be subtracted from the rotation measurements.

If φ_u and $\Delta\theta_{slip,u-y}$ are determined as above, the plastic hinge length, L_{pl}, under cyclic loading should be taken as follows:

- in beams, rectangular columns or walls and members of T-, H-, U- or hollow rectangular section (Biskinis and Fardis 2010b):

Fig. 7.10 Experimental "ultimate" curvatures in 474 rectangular members compared to those predicted per Sect. 7.5.5.1 and the "failure" criteria of Eqs. (7.16) and (7.20).

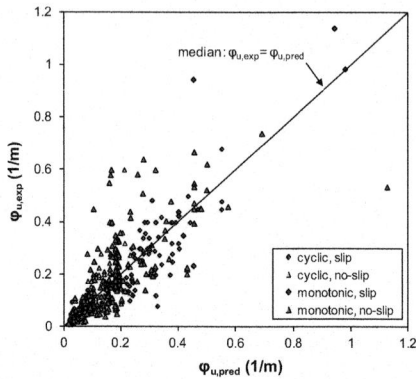

$$L_{pl} = 0.2h\left[1 + \frac{1}{3}\min\left(9; \frac{L_s}{h}\right)\right] \quad (7.21a)$$

- in circular columns with diameter D (Biskinis and Fardis 2013a):

$$L_{pl} = 0.6D\left[1 + \frac{1}{6}\min\left(9; \frac{L_s}{D}\right)\right] \quad (7.21b)$$

Predictions from the above procedure are compared in Fig. 7.11 to the ultimate chord rotation in the cyclic tests to which Eqs. (7.21) were fitted.

For a so-computed value of $\theta^{pl}_{u,m}$, the safety factor for its conversion to a lower-5 %-fractile via Eq. (7.13) is $\gamma_{Rd} = 2.0$.

7.5.5.2 Empirical Rotation Capacity for Sections of Rectangular Parts

MC2010 adopted Eq. (7.5b) and the following from (Biskinis and Fardis 2010b):

$$\theta^{pl}_{u,m} = a^{hbw}_{st}\left(1 - 0.052\max\left(1.5; \min\left(10; \frac{h}{b_w}\right)\right)\right)(0.2)^\nu$$

$$f_c(MPa)^{0.2}\left(\frac{\max(0.01; \omega_2)}{\max(0.01; \omega_1)}\min\left(9; \frac{L_s}{h}\right)\right)^{\frac{1}{3}} 25^{\left(\frac{\alpha\rho_s f_{yw}}{f_c}\right)} 1.225^{100\rho_d} \quad (7.5c)$$

In Eq. (7.5c):

- b_w is the minimum single width among all the webs which are parallel to the shear force (not the total width).
- a^{hbw}_{st} a coefficient for the type of steel with value:

  - For ductile steel: $a^{hbw}_{st} = 0.017$;
  - For brittle steel: $a^{hbw}_{st} = 0.0073$.

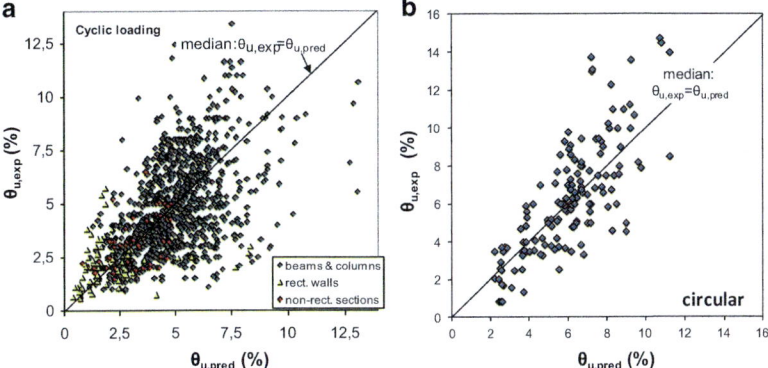

Fig. 7.11 Ultimate chord rotation per Sect. 7.5.5.1 compared to the experimental in: (**a**) 1,125 cyclic tests on members with rectangular, T-, H-, U- or hollow rectangular section; (**b**) 143 cyclic tests of circular columns

Equation (7.5c) suggests that the systematically lower plastic rotation capacity of walls with rectangular, T-, H- or U- section and of hollow rectangular piers is due to the large slenderness, h/b_w, of their webs, which makes them susceptible to lateral instability under cyclic loading. Even in columns with $h/b_w > 1.5$, the strong direction is at a disadvantage in this respect compared to the weak one.

Equation (7.5c) and (7.5b) are equivalent in lack of bias, scatter and the role they have in all modifications in Sect. 7.4.5.2 for lap-splicing, poor details and/or smooth bars; so, they may be used interchangeably for the latter purposes. They also have much less prediction uncertainty than the Sect. 7.5.5.1 procedure. Accordingly, they can be used in Eq. (7.13) with a smaller model uncertainty factor, $\gamma_{Rd} = 1.75$. The same factor applies if Eq. (7.5a) is used instead.

7.5.5.3 Comparison with the Deformation Limits in Part 3 of Eurocode 8

The only expression for $\theta^{pl}_{u,m}$ which MC2010 and Part 3 of Eurocode 8 have in common is Eq. (7.5b). The intention of Part 3 of Eurocode 8 is to base the verification of "primary" members on $\theta_{u,m-\sigma}$ (see Table 7.2), while MC2010 uses the lower 5 %-fractile, $\theta^{pl}_{u,k}$, without distinguishing between "primary" and "secondary" members (see Sect. 7.5.3). Eurocode 8 divides $\theta_{u,m}$ from Eq. (7.5b) by 1.8 to convert it to $\theta_{u,m-\sigma}$ (see second bullet point in footnote (3) of Table 7.2), whereas MC2010 divides it by $\gamma_{Rd} = 1.75$ to convert it to $\theta^{pl}_{u,k}$. The similarity of these factors, despite the different probabilities associated with $m-\sigma$ values and 5 %-fractiles, is due to the reduction in scatter and model uncertainty of Eq. (7.5b) thanks to the increase in the experimental database in the meantime. So, given the similar margins between the NC and LS ULS of MC2010 on one hand and between the SD and NC ones in Part 3 of Eurocode 8 on the other, the two codes appear to be practically equivalent in this respect. However, the safety margin provided by Part 3 of Eurocode 8 for all members designated as "secondary" is reduced by 1.8.

The above conclusion also holds, if Eq. (7.5a) is used in Part 3 of Eurocode 8 or Eq. (7.5b) in MC2010. The safety margins offered by the application of factors 1.7 or 2 on $\theta_{u,m}$ from Sect. 7.4.5.1 (cf. first bullet point in footnote (3) of Table 7.2) appear to be smaller than those provided by applying $\gamma_{Rd} = 2.0$ on the outcome of Sect. 7.5.5.1 for $\theta^{pl}_{u,m}$ in Eq. (7.14). However, because the approach in Sect. 7.4.5.1 underestimates $\theta^{pl}_{u,m}$ by about 10 % (cf. Fig. 7.6), the overall margin is not so different.

7.5.6 Cyclic Shear Resistance

Under seismic loading, the problem of shear is most acute in vertical elements; so ULS design for shear under seismic actions focuses more on columns and walls than on beams. However, the shear provisions in MC2010 refer almost exclusively to elements under high shear due to gravity loads (beams or slabs), which normally have zero axial force. The axial compression, N, of a vertical element contributes to its shear resistance the transverse component of the diagonal strut transferring N from one end section to the other (cf. first term in Eq. (7.9)), as clearly acknowledged in the CEB Model Code 1990 and in Part 3 of Eurocode 8, but only indirectly in MC2010 or EN-Eurocode 2 (through the favourable effect of any inclined compression chords or prestressing tendons). At any rate, this contribution of N to shear resistance in diagonal tension should be explicitly added to the contributions of concrete, $V_{R,c}$, and of shear reinforcement, $V_{R,s}$, as in Eq. (7.9). Indeed, the monotonic shear resistance models in MC2010 at Level of Approximation I, II and III seriously underestimate the experimental resistance in the database of cyclic tests leading to diagonal tension failure, which was used to develop Eq. (7.9) and appears in Fig. 7.9. The underestimation and the scatter are materially reduced if the first term of Eq. (7.9) is added to the value of V_R per MC2010 for diagonal tension. By contrast, the value of $V_{R,max}$ for static loading in MC2010 materially overestimates the experimental resistance of the wall- or hollow-rectangular-specimens in the database which fail by cyclic diagonal compression.

To alleviate the shortcomings of its shear provisions for static loading, MC2010 specifies for seismic loading a strut inclination of 45° if $\theta^{pl} > 2\theta_y$, or of arccot $(2.5) = 21.8°$ if $\theta^{pl} = 0$ (i.e., for elastic response), with interpolation for $2\theta_y > \theta^{pl} > 0$. Besides, for walls $V_{R,max}$ is reduced to 45 % of the value applying per MC2010 for static loading. Although not specifically stated in MC2010, this reduction should apply also to squat hollow rectangular piers (with $L_s/h < 2.5$) and squat columns (with $L_s/h < 2$). With these modifications, the cyclic shear resistance per MC2010 does not overestimate the experimental one, but in most cases underestimates it. It is necessary, in this respect, to improve the MC2010 approach to cyclic shear resistance, in order to achieve the level of agreement with tests that the models by (Biskinis et al. 2004), adopted in Part 3 of Eurocode 8, show in Fig. 7.8. Hopefully, the upcoming revision of Eurocodes 2 and 8 will make this possible.

7.6 Concluding Remarks

The seismic design part of MC2010 follows the footsteps of Part 3 of Eurocode 8, applicable only to existing buildings. MC2010 has evolved from Part 3 of Eurocode 8 to: (a) include recent developments in the State-of-the-Art, (b) face the greater challenges in the design of new structures compared to the assessment of existing ones (including the need to estimate the secant-to-yield-point stiffness without knowing the reinforcement), and (c) to include in its scope bridges and other non-building structures.

The same concept and the same or very similar rules as in MC2010 have been applied to design new concrete buildings in (Panagiotakos and Fardis 1998, 1999b, 2001), or bridges in (Bardakis and Fardis 2011a), with linear analysis for the inelastic deformation demands and Capacity Design for the shears per Sect. 7.4.4.4. This demonstrated the feasibility and cost-effectiveness of the MC2010 approach for new construction. In such cases nonlinear dynamic analysis per Sect. 7.5.4.3 may take place in the end for evaluation and possible revision of the design. As emphasised by (Fardis 2009) and (Bardakis and Fardis 2011a), this design procedure for new structures will be most fruitful and efficient if, as a first step, it aims at uniform chord-rotation ductility ratios at the IU SLS (about 2) or at the OP SLS (about 1) at all sections where plastic hinges will form at the LS or NC ULS. If this goal is attained, inelastic deformations will be fairly uniform throughout the structure and linear seismic analysis will be acceptable across the board, thus avoiding evaluation of the design via nonlinear dynamic analysis.

Acknowledgments Special thanks and appreciation are due to the Author's co-worker Dr DE Biskinis, as the expressions presented for the deformations at yielding and ultimate and the cyclic shear resistance are based on his Doctoral Thesis and subsequent joint work with the Author.

Open Access This chapter is distributed under the terms of the Creative Commons Attribution Noncommercial License, which permits any noncommercial use, distribution, and reproduction in any medium, provided the original author(s) and source are credited.

References

ASCE (2007) Seismic rehabilitation of existing buildings. ASCE/SEI Standard no. 41-06. American Society of Civil Engineers, Reston

ATC (1997) NEHRP guidelines for the seismic rehabilitation of buildings. Applied Technology Council for the Building Seismic Safety Council and the Federal Emergency Management Agency, FEMA reports 273, 274, Washington DC

Bardakis VG, Fardis MN (2011a) A displacement-based seismic design procedure for concrete bridges having deck integral with the piers. Bull Earthq Eng 9(2):537–560

Bardakis VG, Fardis MN (2011b) Nonlinear dynamic v elastic analysis for seismic deformation demands in concrete bridges having deck integral with the piers. Bull Earthq Eng 9(2):519–536

Biskinis DE, Fardis MN (2004) Cyclic strength and deformation capacity of RC members, including members retrofitted for earthquake resistance. In: Proceedings of 5th international *fib* PhD symposium in civil engineering. Balkema, Delft, pp 1125–1133

Biskinis DE, Fardis MN (2010a) Deformations at flexural yielding of members with continuous or lap-spliced bars. Struct Concr 11(3):127–138

Biskinis DE, Fardis MN (2010b) Flexure-controlled ultimate deformations of members with continuous or lap-spliced bars. Struct Concr 11(2):93–108

Biskinis DE, Fardis MN (2013a) Stiffness and cyclic deformation capacity of circular RC columns with or without lap-splices and FRP-wrapping. Bull Earthq Eng 11:1447–1466

Biskinis DE, Fardis MN (2013b) Models for FRP-wrapped rectangular RC columns with continuous or lap-spliced bars under cyclic lateral loading. Eng Struct 57:199–212

Biskinis DE, Roupakias G, Fardis MN (2004) Degradation of shear strength of RC members with inelastic cyclic displacements. ACI Struct J 101(6):773–783

CEB (1970) CEB-FIP international recommendations for the design and construction of concrete structures: 1 principles and recommendations. Bulletin no. 72. Comite Euro-international du Beton, Lausanne

CEB (1978) International system for unified standard codes of practice for structures, vol I: common unified rules for different types of construction and material; II: CEB-FIP model code for concrete structures. Bulletins no. 124/125. Comite Euro-international du Beton, Paris

CEB (1985) CEB model code for seismic design of concrete structures. Bulletin no. 165. Comite Euro-international du Beton, Lausanne

CEB (1991) CEB-FIP model Code 1990 – Final draft. Bulletins no. 203/204/205. Comite Euro-international du Beton, Lausanne

CEN (1991) European prestandard ENV 1992-1: Eurocode 2: design of concrete structures – part 1: general rules and rules for buildings. Comite Europeen de Normalisation, Brussels

CEN (1994a) European prestandard ENV 1991-1:1994: Eurocode 1: basis of design and actions on structures – basis of design. Comite Europeen de Normalisation, Brussels

CEN (1994b) European prestandard ENV 1998-1-1:1994: Eurocode 8: design provisions for earthquake resistance of structures. Part 1-1: general rules – seismic actions and general requirements for structures. Comite Europeen de Normalisation, Brussels

CEN (1994c) European prestandard ENV 1998-1-2:1994: Eurocode 8: design provisions for earthquake resistance of structures. Part 1-2: general rules – general rules for buildings. Comite Europeen de Normalisation, Brussels

CEN (1994d) European prestandard ENV 1998-1-3:1994: Eurocode 8: design provisions for earthquake resistance of structures. Part 1-3: general rules – specific rules for various materials and elements. Comite Europeen de Normalisation, Brussels

CEN (1996) European prestandard ENV 1998-1-4: 1996: Eurocode 8: design provisions for earthquake resistance of structures. Part 1-4: strengthening and repair of buildings. Comite Europeen de Normalisation, Brussels

CEN (2002) European standard EN 1990: Eurocode: basis of structural design. Comite Europeen de Normalisation, Brussels

CEN (2004a) European standard EN 1998-1:2004: Eurocode 8: design of structures for earthquake resistance. Part 1: general rules, seismic actions and rules for buildings. Comite Europeen de Normalisation, Brussels

CEN (2004b) European standard EN 1992-1-1:2004: Eurocode 2: design of concrete structures – part 1: general rules and rules for buildings. Comite Europeen de Normalisation, Brussels

CEN (2005a) European standard EN 1998-3:2005: Eurocode 8 – design of structures for earthquake resistance – part 3: assessment and retrofitting of buildings. Comite Europeen de Normalisation, Brussels

CEN (2005b) European standard EN 1998-2:2005: Eurocode 8 – design of structures for earthquake resistance – part 2: bridges. Comite Europeen de Normalisation, Brussels

CEN (2009) European standard EN 1998-3:2005: Eurocode 8 – design of structures for earthquake resistance – part 3: assessment and retrofitting of buildings. . Comite Europeen de Normalisation, Brussels

Fajfar P (2000) A nonlinear analysis method for performance-based seismic design. Earthq Spectra 16(3):573–593

Fardis MN (1998) Seismic assessment and retrofit of RC structures. Invited state-of-the-art lecture. In: Proceedings 11th European conference on earthquake engineering, Paris

Fardis MN (2001) Displacement-based seismic assessment and retrofit of reinforced concrete buildings. In: Proceedings of 20th European regional earthquake engineering seminar, European Association of Earthquake Engineering, Sion

Fardis MN (2009) Seismic design, assessment and retrofitting of concrete buildings: based on EN-Eurocode 8. Springer, Dordrecht

Fardis MN (2013) Performance- and displacement-based seismic design and assessment of concrete structures in the model code 2010. Struct Concr 14(3):215–229

Fardis MN, Panagiotakos TB, Biskinis DE, Kosmopoulos A (2003) Seismic assessment of existing RC buildings. In: Wasti ST, Ozcebe G (eds) Seismic assessment and rehabilitation of existing buildings, vol 29, NATO science series, IV. Earth and environmental sciences. Kluwer Academic, Dordrecht

fib (2012) Model code 2010 – final draft. Bulletins 65/66, Federation Internationale du Beton, Lausanne

Grammatikou S, Biskinis DE, Fardis MN (2014) Strength, deformation capacity and failure mode of RC walls in cyclic loading. 4th *fib* Congress, Mumbai, paper no 408

Gupta AK, Singh MP (1977) Design of column sections subjected to three components of earthquake. Nucl Eng Des 41:129–133

Kosmopoulos A, Fardis MN (2007) Estimation of inelastic seismic deformations in asymmetric multistory RC buildings. Earthq Eng Struct Dyn 36(9):1209–1234

Moehle JP (1992) Displacement-based design of RC structures subjected to earthquakes. Earthq Spectra 8(3):403–428

Newman K, Newman JB (1971) Failure theories and design criteria for plain concrete. In: Te'eni (ed) Structure, solid mechanics and engineering design. J Wiley-Interscience, New York

Panagiotakos TB, Fardis MN (1998) Deformation-controlled seismic design of RC structures. In: Proceedings 11th European conference on earthquake engineering, Paris

Panagiotakos TB, Fardis MN (1999a) Estimation of inelastic deformation demands in multistory RC buildings. Earthq Eng Struct Dyn 28:501–528

Panagiotakos TB, Fardis MN (1999b) Deformation-controlled earthquake resistant design of RC buildings. J Earthq Eng 3(4):495–518

Panagiotakos TB, Fardis MN (2001) A displacement-based seismic design procedure of RC buildings and comparison with EC8. Earthq Eng Struct Dyn 30:1439–1462

Priestley MJN (1993) Myths and fallacies in earthquake engineering – conflicts between design and reality. In: Proceedings of T. Paulay symposium recent developments in lateral force transfer in buildings, La Jolla

Priestley MJN, Calvi GM, Kowalsky MJ (2007) Displacement-based seismic design of structures. IUSS Press, Pavia

Richart FE, Brandtzaeg A and Brown RL (1928) A study of the failure of concrete under combined compressive stresses. Bulletin 185, Univ. of Illinois Engineering Experimental Station, Champaign

Rowe RE (1970) Current European views on structural safety. ASCE J Struct Div 96(ST3):461–467

SEAOC (1995) Performance based seismic engineering of buildings: vision 2000. Structural Engineers Association of California, Sacramento

Smebby W, Der Kiureghian A (1985) Modal combination rules for multicomponent earthquake excitation. Earthq Eng Struct Dyn 13(1):1–12

Vidic T, Fajfar P, Fischinger M (1994) Consistent inelastic design spectra: strength and displacement. Earthq Eng Struct Dyn 23:502–521

Walraven J (2013) Model code 2010: mastering challenges and encountering new ones. Struct Concr 14(1)

Wilson EL, Der Kiureghian A, Bayo EP (1981) A replacement for the SRSS method in seismic analysis. Earthq Eng Struct Dyn 9:187–194

Chapter 8
Testing Historic Masonry Elements and/or Building Models

Elizabeth Vintzileou

Abstract This paper provides an overview of the Literature on the behaviour of historic masonry elements and building models. The purpose of this paper is to identify the main parameters affecting the seismic behaviour of historic masonry buildings, as illustrated through the experimental campaigns carried out by numerous researchers. Furthermore, aspects of the seismic behaviour that are not sufficiently studied to-date are identified. Thus, selected publications are evaluated related to the behaviour of historic masonry elements in compression, in diagonal compression, in in-plane shear and simultaneous compression, out-of-plane bending, as well as publications related to the behaviour of subassemblies and building models subjected to monotonic, pseudo-dynamic or dynamic tests on earthquake simulator. The available experimental results illustrate the main weaknesses of historic masonry elements and bearing systems, namely the vulnerability to in-plane shear and to out-of-plane bending, the limited ductility, the negative effect of the flexibility of timber floors and roofs, etc. On the other hand, the beneficial effect of adequate connection between horizontal and vertical elements, as well as the connection among walls is also evident. Moreover, the variety of the construction types of masonry tested by various researchers, the scale of the models, the variety of experimental setups and loading histories do not allow, in most cases, a direct comparison of the experimental results. This is so especially as far as properties related to the deformations of masonry elements are concerned. Thus, the effort to develop sound physical models and to calibrate them is not yet satisfactorily assisted by the available experimental results. Yet, this is a prerequisite for a reliable assessment of the current state of historic structures and, by way of consequence, for the selection of adequate intervention techniques for their preservation.

E. Vintzileou (✉)
Faculty of Civil Engineering, National Technical University of Athens, Athens, Greece
e-mail: elvintz@central.ntua.gr

8.1 Introduction

Structural Engineers involved in the preservation of the built Cultural Heritage have to overcome a major contradiction (between safety requirements and internationally accepted Principles of preservation) in their mission: They have to ensure "adequate" seismic behaviour of the structures, without altering the values of the cultural heritage structures. On the other hand, even the scope of interventions (i.e. to ensure "adequate" seismic behaviour) is far from being well determined. Actually, the combination of the uncertainties related to the phenomenon of earthquakes and the still limited knowledge about the seismic behaviour of masonry structures with the inadequate education of our profession in the Mechanics of masonry structures, has led in the past, quite frequently, to an empiricism that is not for the benefit of the preservation of the built cultural heritage.

The weapons of the Structural Engineers in their work for the preservation of the built cultural heritage are: (a) The-as exhaustive as possible-documentation of the existing structure (in terms of geometry, materials, structural system and behaviour), (b) The understanding of the function of the structural system and, hence, the qualitative interpretation of its pathology and decay, (c) The numerical verification of (b) and, hence, the diagnosis and assessment of the current state. All these steps are a prerequisite for the identification of the weaknesses of the system and, hence, for the selection of adequate intervention techniques that may contribute to the improvement of the seismic behaviour of the structure.

The purpose of this paper is to present an overview of the literature that may contribute to the understanding of historic structural systems and to the interpretation of their behaviour. Due to the fact that historic structures like bridges, aqueducts, temples, churches, etc. (a) require specific studies, whereas, (b) the general principles of Mechanics are valid for special structures as well, this paper is limited to research results which regard historic buildings.

The evaluation of experimental data related to the assessment of basic properties of masonry and masonry structural elements, as well as to the seismic behaviour of entire masonry buildings (the effect of connections among the walls, of the flexibility of floors and roofs, etc.), allows also for the identification of lacunae in the knowledge of the international community and, hence, for subjects that need to be further investigated.

The international literature includes results from tests on individual structural members, on subassemblies, as well as on models of entire buildings. Results of monotonic, static cyclic or dynamic tests (on earthquake simulators) are reported. Each category of tests serves a different main purpose: Tests of individual structural members (in compression, shear, out-of-plane bending or a combination of them) provides valuable information on the respective bearing capacity and the deformation properties of the elements. Thus, design models may be adequately validated and calibrated and, hence, used in practice. On the other hand, tests on subassemblies, as well as on models of entire buildings (mainly, under dynamic actions) do provide information about aspects that characterize the overall behaviour of buildings, such as, the effect of the flexibility of floors and roofs, the effect of connection between horizontal and vertical elements, the effect of the connection between walls, the deformation capacity of the entire building, their hysteretic behaviour, etc. Although in more complex configurations of specimens, it is not possible to record the detailed behaviour of each separate structural member, the experimental results are of major significance for the identification of inherent weaknesses of the investigated structural system. Thus, the Engineer is guided in the selection of adequate intervention techniques that may lead to the improvement of the seismic behaviour of historic structural systems.

On the other hand, tests of subassemblies or of building models are frequently carried out on scaled models. Therefore, dynamic similitude laws, as well as scale effects need to be taken into account very carefully, both at the stage of planning the tests and at the stage of interpretation of the experimental results.

It should be noted that a synthesis of the available experimental data is not an easy task: The characteristics of the specimens (in terms of construction materials, geometry of specimens, etc.), of the experimental setups, as well as of the investigated parameters present a vast variety, thus making impossible the direct comparison of the experimental results. However, several valuable conclusions can be drawn, even at a qualitative level. Thus, in this paper, an exhaustive presentation of the totality of the available valuable experimental data is not attempted; only the results of a rather limited number of publications are discussed upon with the aim to identify general trends of behaviour or major lacunae in the Literature.

8.2 Masonry and Masonry Elements in Compression

8.2.1 *Compressive Strength and Deformability of Masonry*

The compressive strength is undoubtedly the more basic mechanical property of masonry, although seemingly not directly related to the seismic behaviour of buildings. Actually, one may argue that the reliable assessment of the compressive strength of masonry is not necessary, since it is known by experience that masonry structures do not fail in compression. This is normally correct, when the structure is subject to vertical loads (although there are exceptions, e.g. the collapse of the Civic Tower in Pavia, Italy-Binda 2008). When, on the contrary, the building is subject to seismic actions, compression may be significantly increased in vertical elements (due to the alternation of actions). Furthermore, in shear walls subjected to in-plane shear, a mechanism of failure of the oblique strut may be generated (Silva et al. 2014). For this specific case, the compressive strength of masonry under oblique forces should also be assessed. On the other hand, the deformations that masonry can sustain before and after the attainment of its compressive strength constitute a characteristic that is significant for the survival of buildings.

It is well known that the compressive strength of masonry depends on many parameters (Tassios 2013), namely, the mechanical properties of the constituent materials (stones, bricks, mortar), on the bonding of blocks (on the faces and within the thickness of masonry), on the volume of mortar over the volume of masonry, on the construction type of masonry, on the existence of timber reinforcement, etc. Therefore, in order to evaluate the in situ compressive strength of masonry, (a) one should perform in situ investigations to obtain information on how masonry is constructed along all three axes (length, height and thickness) and (b) physical models should be available to allow for the calculation of the compressive strength of masonry, taking into account the main influencing parameters. Alternatively, (c) experimental data (for the specific type of masonry) could be used to assess the compressive strength.

To the best of author's knowledge, a general model describing the behaviour of masonry in compression is not available. Actually, such a model should be able to describe the mechanical properties of various types of historic masonries, some of which are shown in Fig. 8.1.

It is worth noting that, even for modern masonries, Eurocode 6 (CEN-EN1996-1-1, 2005) proposes empirical formulae, valid for masonry construction conform to specific rules (limits for the thickness of masonry joints, requirements for the bond of blocks, transverse connection of leaves-in case of cavity walls, etc.). It is obvious that almost none of the constraints of EC6 are fulfilled by historic masonries. Therefore, empirical formulae, adequate for historic masonries should be applied. Actually, there are several empirical formulae in the literature, based on the evaluation of test results. However, most of them refer either to brickwork or to good quality solid stone masonry. Formula by Tassios and Chronopoulos (1986), followed by the formula proposed by Tassios (2004) allow for the estimation of the compressive strength of historic single and three-leaf masonries. The formulae were applied by Vintzileou (2011b), to predict the compressive strength of wallettes made of three-leaf stone and brick masonry with quite satisfactory results (Fig. 8.2), taking into account the scatter of the experimental values.

Marcari et al. (2010) offer an overview of measured values of compressive strengths of single and three-leaf tuff and calcareous stone masonries. The evaluation of the available experimental results shows that the compressive strength of (a) single leaf tuff stone masonry with good quality mortar varies between 3.15 and 5.40 MPa, whereas (b) single leaf tuff stone masonry with poor quality mortar have a compressive strength varying between 2.03 and 3.60 MPa. Finally, (c) for three-leaf masonry, the experimental values are quite scattered (between 1.0 and 3.70 MPa) depending on the quality of materials, as well as on whether the exterior leaves are/are not transversely connected. It has to be noted that, given the significant differences from one test series to another, the Authors do not propose empirical formulae for the estimation of the compressive strength of various types of stone masonry.

The systematic documentation of historic masonry buildings in various Italian regions (Binda and Saisi 2001) shows that single leaf rubble stone masonry is a quite general term in the sense that the total volume of mortar may vary between 11 and 37 % of the volume of masonry. Furthermore, in case of three-leaf masonry,

Fig. 8.1 Examples of types of historic masonries. (a) Double-leaf stone masonry with sporadic header stones. (b) Three-leaf stone masonry- thick interior leaf with very large voids. (c) Poor quality three-leaf rubble stone masonry. (d) Three-leaf rubble stone masonry. (e) Multi-leaf masonry with high mortar volume. (f) Mixed stone-brick masonry with large mortar volume. (g) Mixed brick-stone masonry of good quality. (h) Timber reinforced rubble stone masonry. (i) Timber reinforced adobe. (j) Timber reinforced multiple leaf masonry

usually, the ratio between the thickness of each exterior leaf and that of the infill is approximately equal to 1:0.50 (Binda et al. 1999). The survey carried out by the Politecnico of Milan, together with the evaluation of the data reported in (da Porto et al. 2003), led to the following geometrical data for three-leaf masonries: Percentage of stones/mortar/voids: 55–85 %/12–36 %/0.4–15 %. It is obvious (see also Fig. 8.1) that even if those masonries were made of exactly the same materials, their compressive strengths would result significantly different. Actually, according to the evaluation of experimental data and in-situ measurements, da Porto et al. (2003), the compressive strength of three-leaf stone masonry varies between 0.60 and 2.40 MPa.

Fig. 8.2 Comparison between experimental and predicted (Tassios 2004) values of the compressive strength of three-leaf masonry (Vintzileou 2011b)

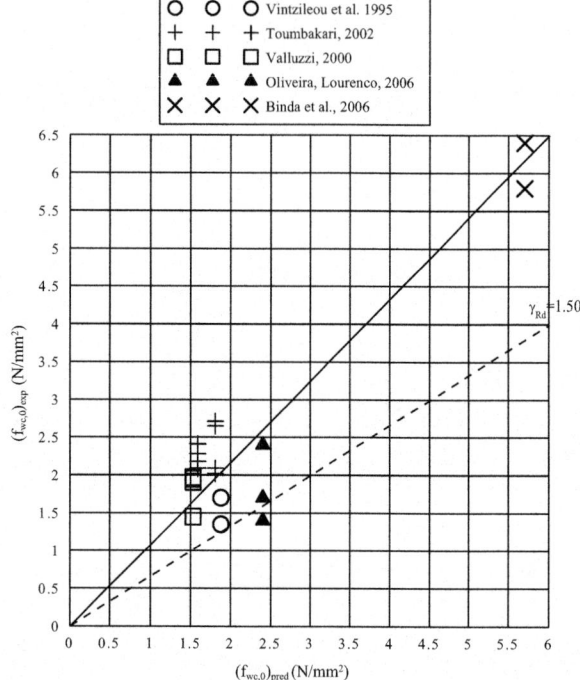

The deformation properties of historic masonries are quite scattered as well. As reported by Marcari et al. (2010), as well as by Tassios (2013), the strain corresponding to the compressive strength of single- or three-leaf masonry may vary between 0.20 and 0.80 %. Similar large scatter is observed in case of the elastic modulus of elasticity (Fig. 8.3).

It seems, therefore, that when data representative of a specific type of masonry are needed, the available experimental results are not sufficient. In such cases, an alternative to laboratory tests and to the application of empirical formulae (whenever available for the construction type under examination), would be to perform in-situ tests on masonry. However, it seems that this is a rather costly and time consuming alternative. It may be a sensible solution either in case of an important monument or in case such tests are carried out in the framework of a study concerning, for example, an entire historic centre.

As a conclusion, one may say that the evaluation of the available data show that (a) the experimental results (from in situ and in laboratory tests) are limited to few types of historic masonry, (b) there is no general physical model describing the behaviour of historic masonry in compression, not to mention that (c) it is quite uncertain to predict the elastic modulus of elasticity, as well as the deformation at failure of masonry in compression.

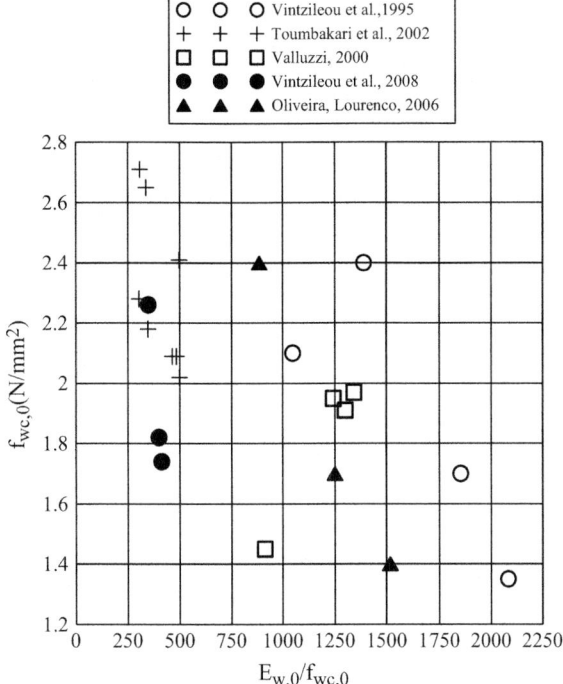

Fig. 8.3 The initial elastic modulus of elasticity reported to the compressive strength of three-leaf stone masonry as a function of the compressive strength of masonry (Vintzileou 2011b)

8.2.2 The Bearing Capacity of Masonry Elements in Compression

Provided that a vertical masonry element is (a) made of solid masonry, (b) it is axially loaded and (c) there are no significant second order effects, its bearing capacity may be calculated as the product of its cross sectional area and its compressive strength. Nevertheless, this is practically never the case:

(i) Typically, in historic structures, part of the vertical loads (weight of pavements, live loads, etc.) are eccentrically applied to masonry walls both when there is a timber floor or roof and when a curved element covers the building (Fig. 8.4). Therefore, even without the occurrence of a seismic event, masonry walls are subject to simultaneous vertical compression and out-of-plane bending.

(ii) In the most frequent types of historic masonry (double or three-leaf masonry with loose connection between leaves), there is a more or less continuous vertical joint within the thickness of masonry (Fig. 8.5). The failure of those types of masonry in compression is characterized by the occurrence of vertical cracks on the faces of masonry, as well as within their thickness (Fig. 8.6), the latter being critical (Pina-Henriques et al. 2005; Oliveira et al. 2006), Vintzileou and Miltiadou (2008).

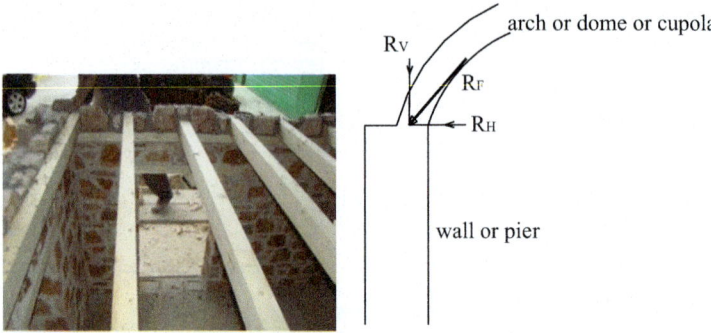

Fig. 8.4 Eccentric application of vertical loads to masonry walls

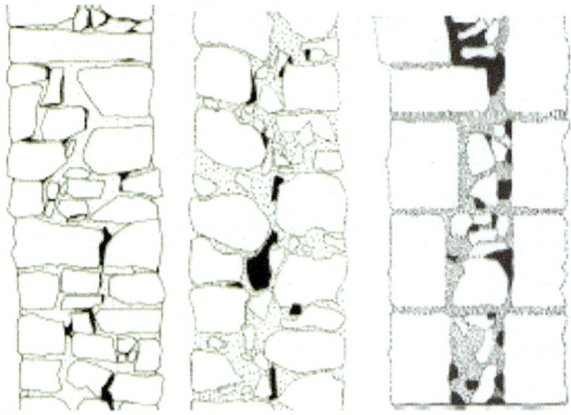

Fig. 8.5 Surveyed types of historic masonries (Binda and Saisi 2001)

Actually, although the two families of cracks open at almost the same vertical load, the transverse ones grow faster. Thus, the failure of masonry is due to simultaneous compression and out-of-plane flexure of the leaves.

It should be noted that cracks within the thickness of masonry are not visible or detectable (unless significant out-of-plane deformation of masonry has occurred). Such cracks may be due to decay of materials, as well as to previous normal and seismic actions on the structure (Fig. 8.7). Therefore, instead of a solid masonry, separated leaves may be asked to resist vertical and horizontal actions. Needless to say that due to the separation between the leaves of masonry, (a) the real slenderness of the walls may be significantly increased, (b) the bearing capacity of walls both to compression and to out-of-plane bending are significantly reduced compared to the bearing capacity of solid walls.

In conclusion, one could say that the estimation of the bearing capacity of masonry walls in compression has to be based on the real geometry, state and

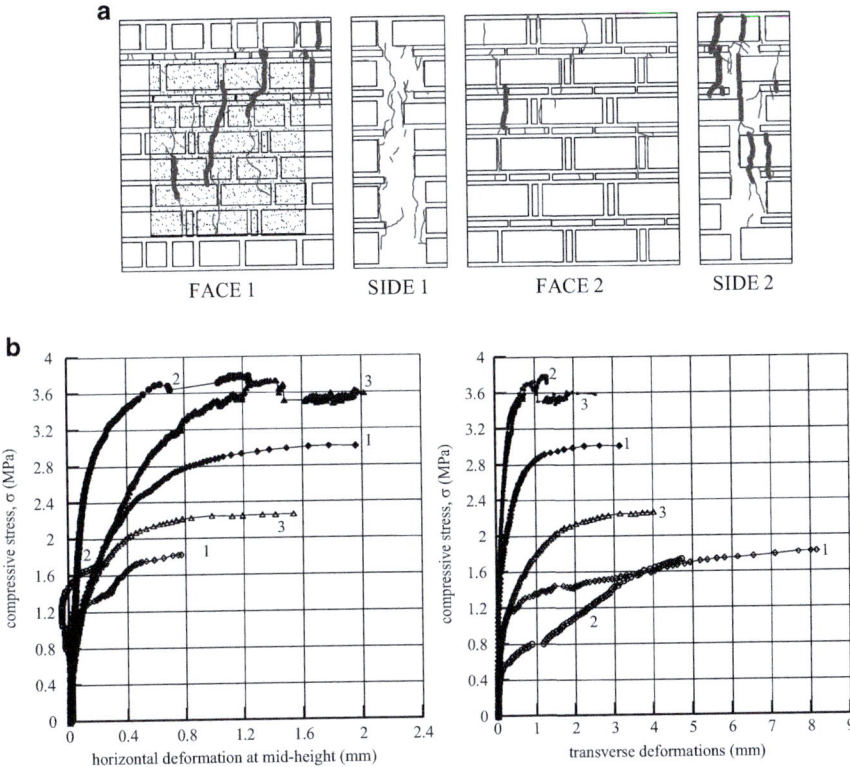

Fig. 8.6 (**a**) Typical crack pattern for three-leaf masonry in compression, (**b**) opening of vertical cracks on the faces of masonry, as well as within its thickness as a function of compressive stresses (Vintzileou and Miltiadou 2008)

Fig. 8.7 Separation of the leaves of masonry walls during tests on the shaking table (Mouzakis et al. 2012a)

Kastoria (Macedonia, Greece)

Antalya, Turkey

Lefkada island, Greece

Casa pombalina, Portugal

Fig. 8.8 Various types of timber reinforced structures in Europe (Source: https://www.google.gr/search?q=casa+pombalina+lisboa)

boundary conditions of the walls. For that purpose, the structural system has to be documented in terms of geometry, construction type of masonry and pathology.

8.2.3 The Case of Timber Reinforced Masonry

In earthquake prone areas around the globe (around the Mediterranean, in Asia, as well as in Latin America), systematic reinforcement of masonry is observed (Fig. 8.8). Although there is a vast variety of structural systems involving timber within masonry, there are clear signs testifying that those structural systems were developed with the purpose of resisting seismic actions (see e.g. Vintzileou 2011a).

Although the contribution of the timber reinforcement to the compressive strength of masonry is the least significant aspect of those structural systems, test results (Vintzileou 2008) have shown that

(a) Horizontal timber laces provide confinement to rubble stone masonry, thus, leading to a moderate enhancement (by 15–20 %) of its compressive strength. More importantly,
(b) Timber laces lead to a significant enhancement of the deformation (vertical strain) masonry can sustain without being disintegrated (Fig. 8.9).

8.3 Masonry Elements Subjected to In-Plane Shear

The behaviour of masonry elements under in-plane shear is of major significance for the seismic response of buildings, as documented by typical damage, i.e. diagonal or bi-diagonal cracks in walls and spandrels (Fig. 8.10). Thus, numerous research works were devoted to the behaviour of masonry under shear.

Fig. 8.9 Compressive stress-vertical strain for masonry wallettes: Wallette 1-plain masonry, Wallettes 2 and 3-timber laced masonry (Vintzileou 2008)

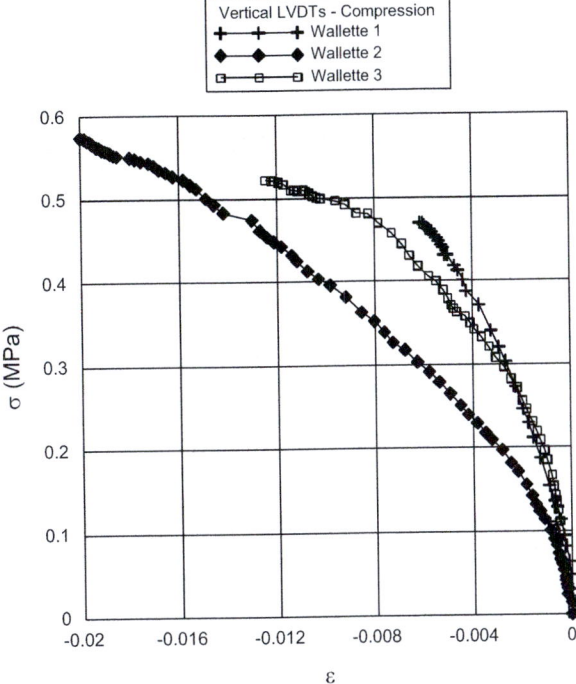

Fig. 8.10 Severe damages to shear walls and spandrels (Onna, after the 2009 earthquake of L'Aquila, Italy)

In laboratory and in situ tests were carried out on wallettes subjected to diagonal compression, with the purpose of assessing the shear strength of masonry under zero normal stress. The results show strength values depending on the mechanical properties of materials, as well as on the bond between blocks and mortar. For example, Shahzada et al. (2012) have tested solid brick masonry wallettes in diagonal compression. The shear strength under zero normal stress was very low (the Authors do not even mention its value). On the other extreme, Ali et al. (2012) have tested several wallettes made of bricks and mortars (typical for Pakistan).

The compressive strength of the mortar was varying between 3.0 and 27.0 MPa. For those, rather unusually strong mortars, they have measured shear strengths varying between 0.30 and 1.70 MPa. Brignola et al. (2006) report the results of in situ tests in several historic buildings in Tuscany. For the stone masonries tested by the authors, low shear strength values were obtained (varying between 0.04 and 0.067 MPa).

In situ diagonal compression tests on stone masonry walls by Chiostrini et al. (2000) yielded values of shear strength varying between 0.061 and 0.16 MPa. Corradi et al. (2003, 2008) have measured similar values of shear strength. Similar (low) values were measured in laboratory on wallettes made of three-leaf stone masonry to diagonal compression by Vintzileou and Tassios (1995)–0.15 MPa, as well as by Vintzileou and Miltiadou (2008)–0.10 MPa. The results obtained by Milosevic et al. (2012) on rubble stone masonry were quite scattered (between 0.024 and 0.313 MPa), irrespectively of the compressive strength of the mortar.

Limited in number test results are available for timber laced masonry (Vintzileou 2008). The presence of timber laces led to a shear strength under zero normal stress almost 5.0 times that of the plain three-leaf rubble stone masonry. More importantly, the strain at strength was by almost an order of magnitude larger.

However, the strength at zero normal stress is only one of the components of the bearing capacity of a masonry element subjected to in-plane shear, when failure is due to the occurrence of diagonal or bi-diagonal cracks. On the other hand, the in-plane behaviour of walls failing in bending or vulnerable to rocking needs to be investigated through testing under simultaneous in-plane shear and vertical load. Actually, several researchers have conducted tests on masonry walls under monotonic or cyclic shear (see i.a. Chiostrini et al. 2000; Corradi et al. 2003, 2008; Vasconcelos and Lourenco 2006; Costa et al. 2012a, b, c; Capozucca 2011; Silva et al. 2014).

Tests on individual structural members allow for the behaviour of full-scale elements to be investigated in detail (failure modes, deformations along three axes, failure load, ductility, etc.). Furthermore, the effect of various intervention techniques can be investigated. Tests on individual members provide data that are necessary for the development and the calibration of models to be applied for the assessment of the bearing capacity of existing elements, as well as for the design of the intervention techniques. It should be noted that due to the differences in materials, in geometry, in applied time-history, etc., it is impossible to provide a synthesis of the experimental results and to draw general conclusions. Finally, a large part of the tests on individual walls refer to modern brick and block masonry. Therefore, the experimental results on various construction types of historic masonry are still rather limited in number.

The available experimental data regard shear walls made of a variety of materials (mostly clay and concrete blocks, stones and mortar). Full scale walls or scaled models are tested. The walls are subject to simultaneous vertical load (either constant or varying during the lateral loading). The walls are either cantilevers or fixed at both ends. The aspect ratio (height to length) varies between 1:2 and 2:1.

Fig. 8.11 Typical shear failure of stone masonry walls under simultaneous vertical load (Vasconcelos and Lourenco 2009)

In some cases, there are also openings in the walls. The specimens are subjected either to monotonically increasing lateral load or to static cyclic lateral loading or (in a limited number of cases) to dynamic in-plane actions. The prevailing failure mode is due to the formation of diagonal or bi-diagonal cracks (Fig. 8.11), involving-in some cases-also compression failure close to the base of the wall (Silva et al. 2014). Flexural failure or mixed shear-flexural failure was observed for rather high aspect ratio values. Rocking was also observed in some cases (especially, under low vertical load, Silva et al. 2014). Typically, after the attainment of the maximum resisting shear force, significant force-response degradation is recorded (Fig. 8.12a). Deformations (vertical and horizontal) are recorded during testing. However, due to the differences among tested models, the author of this paper is unable to provide a comparison of the relevant experimental data. It should be noted that several researchers have worked on modeling of the behaviour of shear walls (see i.a. Vasconcelos and Lourenco 2006; Costa et al. 2012a; Magenes and Calvi 1997; Brencich and Lagomarsino 1998), developing either sophisticated models or simple ones, adequate for use by practitioners as well.

8.4 Masonry Elements Subjected to Out-of-Plane Bending

It is well known that in historic buildings subjected to seismic actions, the out-of-plane behaviour of (solid or with openings) walls may be critical (Fig. 8.13).

The vulnerability to out-of-plane actions is due to typical characteristics of historic masonry buildings, namely, the flexible floor and roof diaphragms (Fig. 8.13b), as well as the defective connection between floors/roof and walls (allowing for significant out-of-plane deformations of walls), the defective connection of walls at building corners (Fig. 8.13a, c), the presence of openings close to the corners of the building and, last but not least, the frequent construction type of

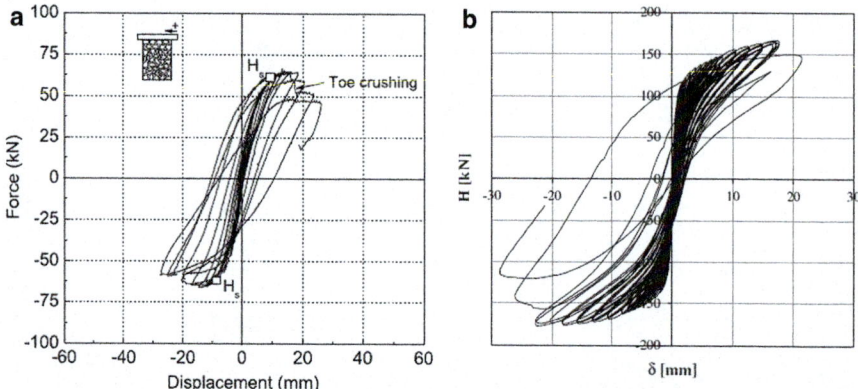

Fig. 8.12 (a) Typical hysteresis loops for stone masonry walls failed in shear (Vasconcelos and Lourenco 2009), (b) Typical hysteresis loops for rocking stone masonry wall (Silva et al. 2014)

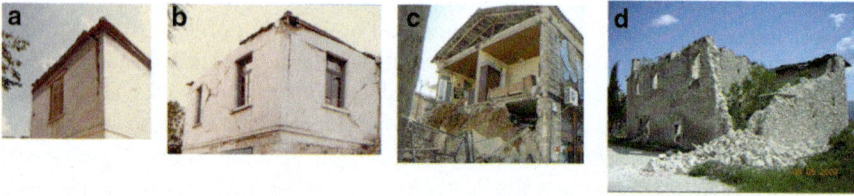

Fig. 8.13 Typical damages due to out-of-plane seismic actions. (a) Typical vertical crack due to out-of-plane bending of the solid wall. (b) Typical vertical crack at mid-length of the wall and separation of walls at the corner of the building. (c) Out-of-plane collapse of wall. (d) Collapse of the corner of a building (due to combined in-and out-of-plane action)

historic masonry (double- or three-leaf). Actually, the separation between leaves (due either to decay or to previous actions) leads to significant reduction of the out-of-plane stiffness of walls, whereas the masonry cross section is also significantly reduced (Fig. 8.13d) and Giuffrè et al. (1993).

It is obvious that the testing of individual walls out of their plane cannot describe the behaviour of walls belonging to a building. Furthermore, available test results are almost exclusively dealing with brick or concrete block masonry walls, whereas various testing procedures are applied. Some of the relevant publications are briefly presented herein: One of the earlier experimental campaigns was carried out at ABK (1981). Several construction types of masonry were tested, among them also multi-leaf brick masonry walls. The aim of this work was to assess the effectiveness of various intervention techniques, taking into account the slenderness ratio and the boundary conditions of the panels. 20 full-scale masonry panels were subjected to about 200 seismic inputs, covering the full range of USA seismicity. The walls were full height (floor to floor) and were not laterally supported along the vertical edges. The work provided data that were used both (a) to calibrate mathematical models developed by the authors for the prediction of the failure mode and (b) to draft guidelines for the design of various strengthening techniques. A finding to note is that the collapse mechanism was found to depend more on the induced peak velocities (at the top and the bottom of the panels) rather than on the relative deformation between the top and the bottom of the panels.

Griffith et al. 2004 investigated the response of unreinforced brick masonry wall panels subjected to out-of-plane loading. For this purpose, fourteen specimens, having different slenderness ratios (13.6 and 30.0), were constructed and tested. The test program included static, free-vibration, and dynamic tests (with induced harmonic, or impulse or seismic motions). However, the slenderness ratios of the walls are not typical for historic masonry. Simsir et al. (2004) carried out dynamic tests on four half-scale masonry walls made of lightweight concrete hollow blocks. The experimental set-up allowed testing walls in the free-standing boundary conditions as shown in Fig. 8.14. Two of the walls were tested in-plane, while the other two were subjected to out-of-plane seismic actions, (Fig. 8.14b). The aim of the experiment was to investigate the influence of the boundary conditions, namely the horizontal structures at top and constrains at the bottom of wall panels, simulating the real conditions of a wall panel. Differently from other similar tests, specimens did not exhibit a mid-height failure that leads to collapse, except for the cases where the panel was subjected to low axial load. Furthermore, it was proven that the flexibility of diaphragms can significantly enhance the out-of-plane displacements.

Tominaga and Nishimura (2008) have tested brick masonry walls out-of-their plane, by applying two concentrated loads at the thirds of the span. No vertical load was applied. Failure along mortar joints was observed. The maximum resistance was mobilized for very small deflection (of the order of few mm), but the residual resistance was significant, due to friction along the failed mortar joints.

Cavaleri et al. (2006) report the results of an experimental campaign on four (4) single leaf calcareous stone masonry walls (0.74 m long, 2.10 m high and 0.21 m thick). The walls were under constant compression load (equal to 0.12 the bearing capacity of walls to compression). Deformations were applied to the walls (by moving horizontally the base of the walls). The curvatures at the base region of the walls were also recorded. Failure was due to the occurrence of horizontal cracks along the mortar joints close to the base.

Meisl et al. (2006) have tested four multi-leaf plain masonry walls. The effect of the quality of construction (in terms of strength of mortar) and that of the soil conditions (one soft and one more firm substrate) were investigated. The results have shown little effect of the quality of construction on the overall behaviour of specimens. On the contrary, walls founded on soft soil exhibited more damages (for the same input) than those founded on firm soil.

Manoledaki et al. (2012) have tested piers made of three-leaf stone masonry. The piers were sitting on either a loose ($D_r = 33\ \%$) or a dense ($D_r = 92\ \%$) sand, through a rectangular RC footing (Fig. 8.15). The walls, either constrained or free at their top, were subjected to horizontal displacements at their mid-height.

The tests showed that the out-of-plane seismic performance of the masonry walls was substantially affected by soil–foundation–structure interaction (SFSI).

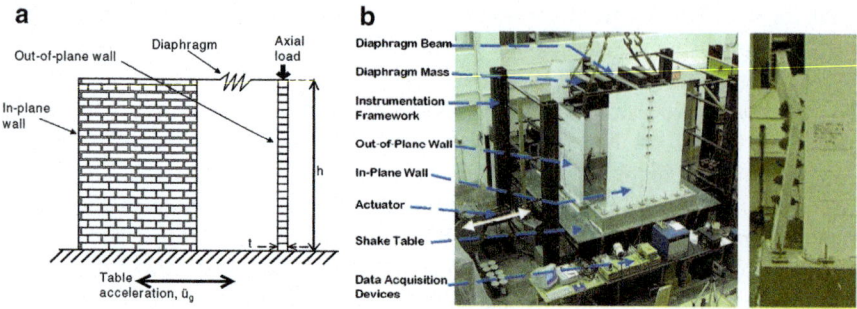

Fig. 8.14 (**a**) Test set-up and (**b**) specimen at the shaking table (Simsir et al. 2004)

Fig. 8.15 Three-leaf stone masonry piers and experimental setup (Manoledaki et al. 2012)

As indicated by the two examined cases, soil resilience had a significant influence on system response. Foundation rocking resulted in a reduction of the soil–footing contact zone in the case of dense sand, whereas, in loose sand the response was governed by sinking (Fig. 8.16). The essential influence of the boundary conditions on the out-of-plane response of the walls is also amongst the key observations made from the tests. In the cases where the elongation of the wall was partially obstructed by the top support, the induced axial load led to significant enhancement of the out-of-plane capacity. The walls generally exhibited the typical cracking pattern associated with one-way vertical out-of-plane bending. Material crushing was restricted to the weak mortar joints.

Recently, within the EU funded project NIKER, tests were carried out (Valluzzi et al. 2013) on three-leaf rubble stone masonry full scale panels (Fig. 8.17) subjected to out-of-plane excitations on a shaking table. The panels (1.30 m long, 2.60 m high and 0.50 m thick) were subjected to adequately scaled real accelerograms.

8 Testing Historic Masonry Elements and/or Building Models

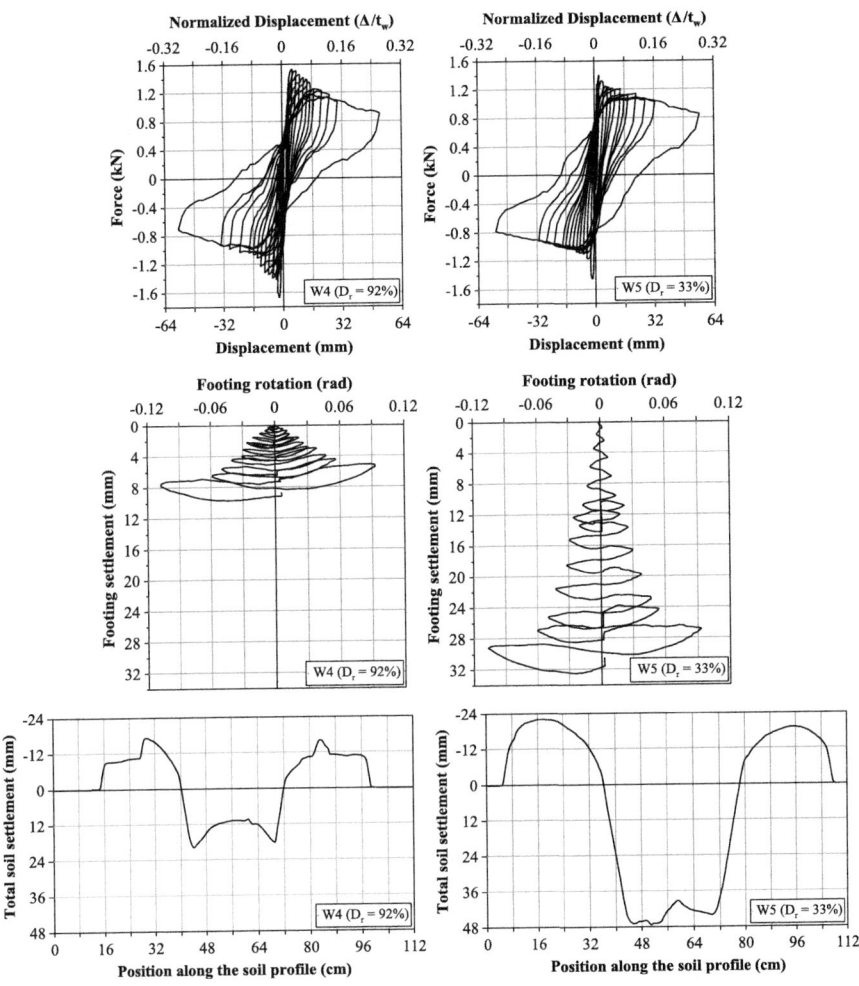

Fig. 8.16 Test results obtained by Manoledaki et al. (2012)

The panels failed under acceleration approximately equal to 0.30 g. As shown in Fig. 8.18, cracks typical for out-of-plane bending have occurred. Failure was due to the separation of the leaves of masonry and to the collapse of one of the two exterior leaves (Fig. 8.19).

The detailed data obtained during testing (accelerations, frequencies, displacements, etc.) allowed for full documentation of the behaviour of the panels. They have also served the purpose of prediction of the observed behaviour by means of modeling.

Tests on subassemblies (e.g. façade wall with portions of transverse walls) are also reported in the literature. Those tests are presented and commented upon in the following Sections.

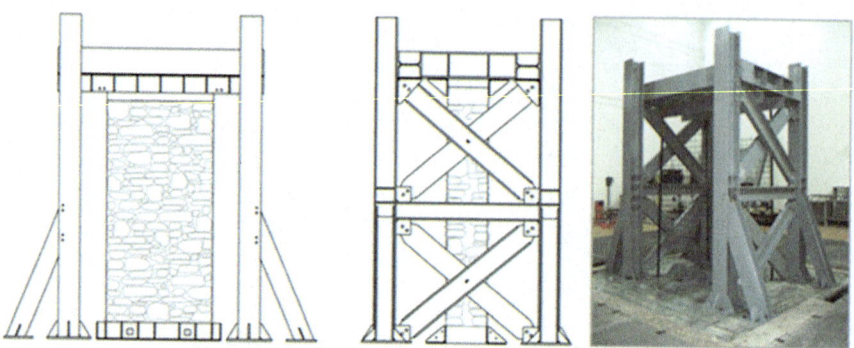

Fig. 8.17 Test specimens and experimental setup (Valluzzi et al. 2013)

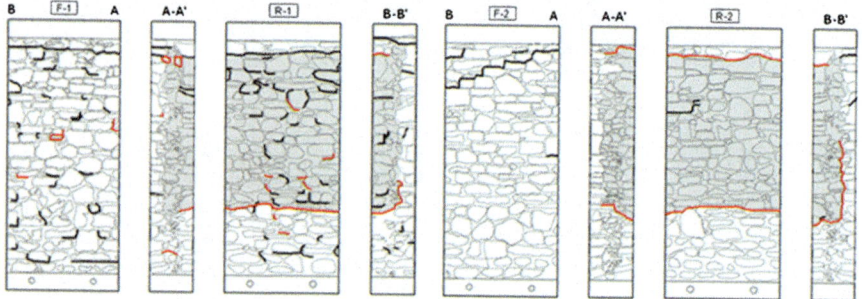

Fig. 8.18 Typical failure mode of three-leaf stone masonry walls subjected to out-of-plane seismic actions (Valluzzi et al. 2013)

8.5 Tests on Subassemblies and Building Models

8.5.1 Tests on Subassemblies

As shown in the previous Sections, tests on individual bearing elements (under monotonic or cyclic actions) provide valuable information regarding the failure mode under compression, shear or out-of-plane bending, as well as on properties like bearing capacity, deformability, hysteretic damping, stiffness, force-response degradation due to cycling, etc. Nevertheless, there are significant aspects of the seismic behaviour of masonry buildings that cannot be modelled and experimentally reproduced by testing individual bearing elements. Actually, the effect of the in-plane stiffness of floors and roofs, the effect of the connection between bearing walls, the behaviour of masonry elements subjected to simultaneous shear and out-of-plane bending, the capacity of masonry buildings to redistribute actions among bearing elements need to be identified through testing of subassemblies or models of entire buildings. Another important issue is the capacity of historic masonry buildings to undergo large post-elastic deformations, i.e. their ductility. Finally, the

Fig. 8.19 Failure of walls 1 and 2 (Valluzzi et al. 2013)

effect of several interventions applied with the purpose of improving the seismic behaviour of historic buildings, namely, the enhancement of diaphragm action of floors and roof, the improvement of the connection of walls by means of ties, etc. can only be exhaustively investigated on specimens simulating at least part of the entire building.

In the Literature, there are results obtained from quasi-static or dynamic tests on subassemblies. Some of them are related to the study of specific monuments (e.g. Pinto et al. 1999a, b, c, 2001). The valuable results of those tests are hardly offered to generalization. Therefore, they are not presented herein. There are also tests on subassemblies investigating the behaviour of arches and vaults (see i.a. Baratta and Corbi 2007; Taranu et al. 2010; Mouzakis et al. 2012b). Those experimental works are presented neither.

Al Shawa et al. (2009) have tested full scale subassemblies made of tuff masonry (Fig. 8.20), with the purpose of investigating the out-of-plane behaviour of walls connected with transverse walls. The research includes subassemblies before and after strengthening. The tested wall (3.40 m high, 0.25 m thick) was either free standing or connected to the transverse ones along a mortar joint. A third case was also considered, in which the walls were connected through bonding of stones, as well as through steel bars. The subassemblies were subjected to forced vibrations, following adequately scaled accelerograms of real earthquakes. The tests have proven the major significance of the connection between walls. Actually, in terms of maximum acceleration sustained before failure (or collapse), the free standing wall, as well as that connected to the transverse ones through a mortar joint, were able to sustain an acceleration approximately equal to 0.30 g. On the contrary, the

Fig. 8.20 Photo of a specimen after the test: case of out-of-plane loaded wall connected to the transversal walls through a mortar joint (Al Shawa et al. 2009)

proper connection between the walls, allowed for a peak ground acceleration equal to 0.60 g to be sustained.

The mechanisms of out-of-plane failure of a wall connected with transverse ones was studied by Restrepo-Vélez (2004) and Restrepo-Vélez and Magenes (2004) through testing of subassemblies made of dry stack masonry (Fig. 8.21). The models (scale 1:5) allowed for identification of the two possible failure modes, i.e. detachment of the out-of-plane loaded wall from the transverse ones and out-of-plane collapse of the wall.

The same mechanisms were detected also by Bui et al. (2010). The subassemblies they have tested were subjected to monotonically increasing uniformly distributed load on the longitudinal wall (Fig. 8.22).

A full scale shaking table test on a 3-D specimen made of three-leaf masonry was performed by Costa et al. (2012a). The subassembly-simulating a typical façade of historic buildings in the Azores-exhibited the same failure mechanisms, together with detachment of the leaves of masonry (Fig. 8.23).

Costa et al. (2012b) carried out an in situ test on a building severely damaged during the Azores earthquake in 1998 (Fig. 8.24). The building was made of double leaf stone masonry.

Cyclic tests were performed, not to collapse though due to the limitations of the equipment, as well as for safety reasons. Valuable data were collected regarding the dynamic properties of the structure, the sustained deformations, hysteretic behaviour, etc. The behaviour of the structure was tested also after the application of reinforced plaster on the walls.

A 3D subassembly was tested within the EU funded project NIKER (Vintzileou et al. 2012a, b). The subassembly (made of three-leaf stone masonry) consists in one wall with a portion of a transverse wall at its mid-length and a parallel wall of rectangular section (Fig. 8.25). A timber floor (typical for historic buildings) is

8 Testing Historic Masonry Elements and/or Building Models

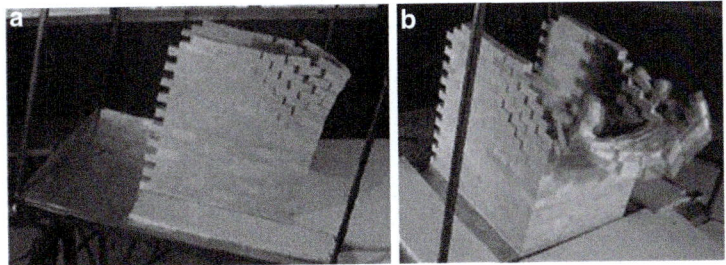

Fig. 8.21 Failure mode of out-of-plane loaded walls connected to portions of transverse walls Restrepo-Vélez (2004) and Restrepo-Vélez and Magenes (2004)

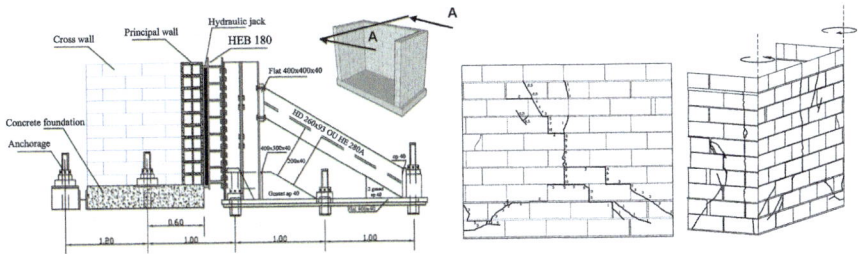

Fig. 8.22 Test setup and failure mode of walls with flanges (Bui et al. 2010)

Fig. 8.23 Failure mode of subassembly (Costa et al. 2012a)

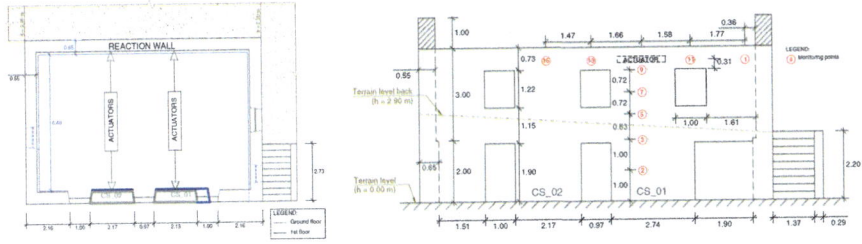

Fig. 8.24 Testing arrangement and instrumentation of a building tested on situ (Costa et al. 2012b)

Fig. 8.25 The specimen and its construction details (Vintzileou et al. 2012)

provided. For dynamic similitude purposes, additional masses are fixed on the floor before testing on the earthquake simulator. The subassembly was subjected to a series of adequately scaled accelerograms of the Irpinia, Italy 1980 earthquake out-of-the plane of the parallel walls. Figure 8.26 shows the crack pattern (at PGA ~ 0.50 g). The effect of the portion of the transverse wall (failed in shear) on the longitudinal wall to which it is connected is shown. Furthermore, the flexible wall of rectangular section was not severely damaged. It exhibited, however, extensive detachment of the masonry leaves. It should be noted that the same subassembly after strengthening (grouting of masonry, enhancement of the diaphragm action of the floor and connection thereof with the walls) exhibited a clear rocking behaviour.

The experimental works briefly presented herein have provided valuable information on several aspects of the out-of-plane behaviour of masonry walls under realistic boundary conditions. It should be noted, however, that in most of the laboratory tests there was no vertical load on the out-of-plane loaded walls. Similarly, with one exception, there was no diaphragm at floor(s) levels. The presence of vertical load plays a positive role on the out-of-plane behaviour of walls, whereas the effect of a more or less flexible diaphragm may affect significantly the seismic behaviour of the structure. Nevertheless, the obtained results are valuable and, in the opinion of the author of this paper, there is a need for systematic analytical work (with simulation of the test specimens), for the international community to take the maximum possible profit of the experimental data.

8.5.2 Tests on Building Models

Testing models of entire buildings (either under monotonic or under seismic actions) has the advantage of simulating parameters that cannot be simulated through testing of subassemblies or individual bearing elements. This is of major significance, due to some typical characteristics of existing masonry buildings that govern their seismic behaviour, namely, the presence of more or less flexible floors and roofs (that allow the vertical elements to deform independently from one another), the connection between horizontal and vertical elements, as well as the connection between longitudinal and transverse walls (its quality affecting

Fig. 8.26 Crack pattern of the subassembly (Vintzileou et al. 2012)

significantly the box action of the building and, hence, the magnitude of the imposed deformations). Furthermore, basic parameters like dynamic properties (and their modifications during the seismic event), hysteretic properties and overall ductility cannot be realistically assessed unless the entire structure is considered. It is mentioned, as an example, that it is typically assumed that unreinfroced masonry buildings are very brittle. However, inspection after seismic events shows that many structures survive (damaged, of course) in contradiction with our calculations. Last but not least, the efficiency of several intervention techniques cannot be assessed on the basis of tests on individual members. Actually, those techniques that aim at improving the overall behaviour of buildings (e.g. enhancement of the diaphragm action of floors or the arrangement of ties to improve the connection between walls) need to be assessed on the basis of large scale tests. In recognition of the above advantages of testing building models, several researchers have performed tests either on shaking tables or quasi-static tests on building models.

It should be noted, however, that the interpretation of the results obtained from testing building models is not as self evident as one could possibly think. Actually, due to several constraints related to this type of tests, very careful design of testing campaign is needed, along with systematic analytical work on both prototypes and models. In fact, shaking table tests are quite expensive (in terms of construction, instrumentation, use of the facility, etc.). Thus, within each testing campaign, the number of models that are tested is limited. By way of consequence, several parameters are usually simultaneously modeled and, hence, frequently, it is not possible to directly assess the effect of each of them. On the other hand, in order to take the maximum profit out of those tests, building models are subject to series of input motions (of increasing magnitude). Thus, the behaviour of the model subjected to a series of seismic inputs may be different than the behaviour to be exhibited by a model directly subjected to high intensity actions.

In shaking table tests, there are also limitations related to the capacity of the facility (in terms of plan dimensions, degrees of freedom, total height of the model coupled with total weight, maximum acceleration and maximum displacement that can be imposed to the model). Those limitations lead to either small scale models or to testing of rather simple in configuration buildings. In the first case, there are scale effects (to be taken into account when assessing the experimental results), the detailed discussion on which is beyond the scope of this paper. Furthermore, for dynamic similitude reasons, additional masses need to be arranged. The fact that those masses are inevitably located on floors and roof (instead of being distributed along the height of the model), as well as the fact that additional masses are transferred to part of the cross section of masonry (e.g. through the timber beams of floors resting on the interior leaf of a double- or three-leaf masonry) may affect the behaviour exhibited by the model. Last but not least, the foundation of the model cannot be realistically modeled (the models are fixed on a rigid base) and, hence, also soil structure interaction cannot be studied.

In this Section, a brief presentation of the results obtained from tests on building models in the last three decades is attempted. The overview of the experimental data is limited to tests on historic masonry. Still, an exhaustive presentation of all the available data being impossible, selected works are presented, those that allow for the identification of the effect of major features on the seismic behaviour of historic buildings. Although this paper does not cover the effect of intervention techniques to historic buildings, some selected results are included herein. Those results concern the effect of some techniques that could be termed as "systemic" interventions, in the sense that they affect the overall behaviour of historic buildings (e.g. enhancement of diaphragm action of floors and roofs, improvement of connection among the walls, etc.).

8.5.2.1 Short Presentation of Tested Models

The models that were subjected either to pseudo-dynamic or to dynamic tests on a shaking table have quite different characteristics in terms of scale (1:1 to 1:10), in dimensions, in number of storeys (1, 2 or 4), in arrangement of openings (doors and windows), in the flexibility of floors and roofs, in materials and construction type of masonry, etc. Therefore, a direct comparison of the experimental results is not possible. Nevertheless, a qualitative comparison of the data is attempted wherever possible.

Benedetti (1980) performed a series of pseudo-dynamic tests on scale 1:2, one-storey multi-leaf stone masonry model buildings (plan dimensions 1.90×2.20 m, Fig. 8.27a). Seismic excitation was simulated by static lateral loads via actuators. No roof was provided to the models. Three of the models were tested unstrengthened, whereas two models were tested after the application of a cement grout. One model was fully grouted, the other was partially grouted.

Tomaževič et al. (1990, 1991, 1993) report the results of two series of shaking table tests on reduced scale (1:4) stone masonry building models. The two-storey models (Fig. 8.27b) were 1.0×1.10 m in plan. The total height was equal to 1.50 m, whereas the thickness of walls was equal to 0.12 m. The models were subjected along one direction, parallel to the walls without openings, to an adequately scaled acceleration record (Montenegro 1979 earthquake). The purpose of the research was to investigate the effect of the rigidity of floors. Thus, Model A was provided with timber floors (simply resting on the walls without openings), Model B was provided with RC slabs, whereas in Model C, prestressed steel ties (located underneath the timber beams) were used to improve the connection between

8 Testing Historic Masonry Elements and/or Building Models

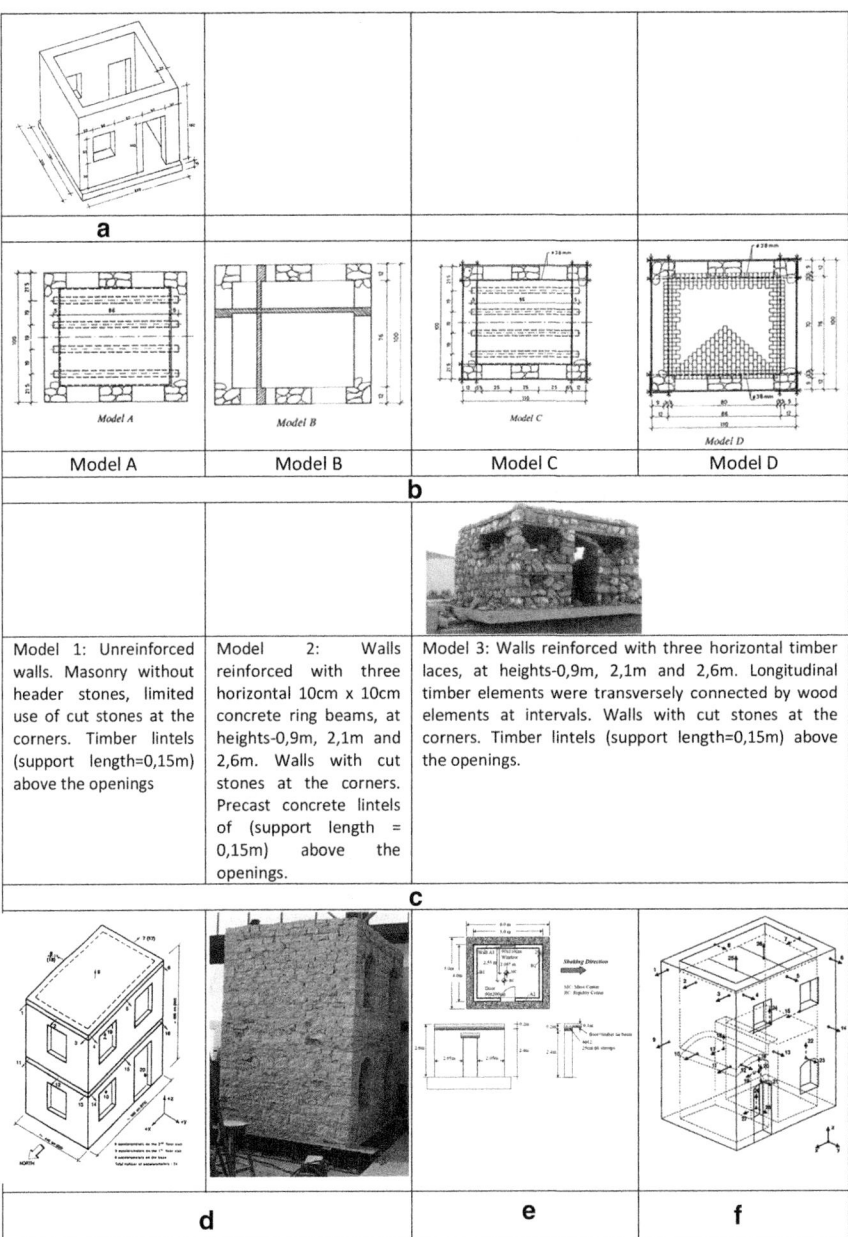

Model A	Model B	Model C	Model D

b

Model 1: Unreinforced walls. Masonry without header stones, limited use of cut stones at the corners. Timber lintels (support length=0,15m) above the openings	Model 2: Walls reinforced with three horizontal 10cm x 10cm concrete ring beams, at heights-0,9m, 2,1m and 2,6m. Walls with cut stones at the corners. Precast concrete lintels of (support length = 0,15m) above the openings.	Model 3: Walls reinforced with three horizontal timber laces, at heights-0,9m, 2,1m and 2,6m. Longitudinal timber elements were transversely connected by wood elements at intervals. Walls with cut stones at the corners. Timber lintels (support length=0,15m) above the openings.

Fig. 8.27 (continued)

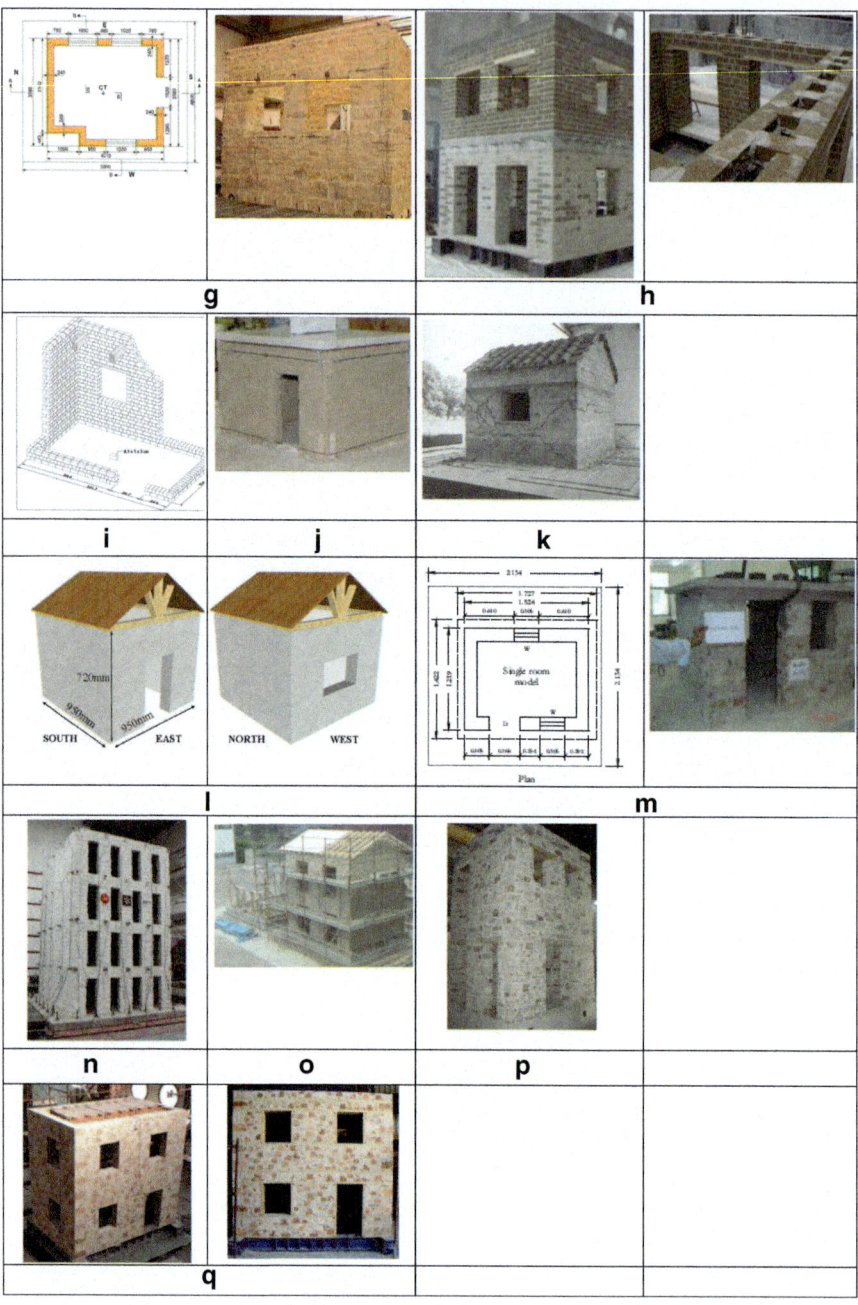

Fig. 8.27 An overview of building models subjected to dynamic testing. (**a**) Benedetti (1980). (**b**) Tomaževič et al. (1990, 1991, 1993). (**c**) Spence and Coburn (1987, 1992). (**d**) Models tested at ISMES and at LEE/Athens (Benedetti et al. 1998). (**e**) Bayülke et al. (2000). (**f**) Juhásová et al. (2002). (**g**) Juhásová et al. (2008). (**h**) Bergamo et al. (2006). (**i**) Tomaževič et al. (2009). (**j**) Ersubasi and Korkmaz (2010). (**k**) Shashi and Pankaj (2000). (**l**) Meguro et al. (2012). (**m**) Ahmad et al. (2010, 2012). (**n**) Mendes and Lourenco (2010). (**o**) Magenes et al. (2010, 2012a, b). (**p**) Mazzon et al. (2009). (**q**) Mouzakis et al. (2012a) and Adami et al. (2012)

longitudinal and transverse walls and to connect the floors to the walls. Finally, in Model D, the ground storey floor was a brick vault, whereas a timber floor was provided to the upper storey (Fig. 8.27b). Poor quality materials, typical for old buildings were used for the construction of the models.

Spence and Coburn (1987, 1992) conducted an experimental program on three full scale single storey masonry building models, simulating the structural system that is typical for Eastern Turkey. The models, subjected to uni-directional impulse tests were 4.50 × 4.50 m in plan, 2.60 m high, whereas the (rubble stone masonry) walls were 0.60 m thick. All models were provided with a typical timber roof (made of timber beams and timber planking). On top of the planking a layer of 0.20 m thick compacted soil was added. Some characteristics of the models are shown in Fig. 8.27c. The models were subjected to gradually increasing impulse load until failure.

The most extensive experimental programme reported in the literature is the one carried out at ISMES (Italy) and NTUA (Greece). Fourteen two-storey models (before and after interventions) at scale 1:2 were tested (Benedetti et al. 1998). Eight models were tested at ISMES (4 brick masonry and 4 stone masonry), and six models were tested at the Laboratory of Earthquake Engineering, Athens (3 brick masonry and 3 stone masonry) (Fig. 8.27d). The lintels were either arched or horizontal beams. All models were provided with timber floors and planking. In general, poor quality mortar was used. In the models tested at ISMES, the connection between orthogonal walls was rather defective. The models were subjected to scaled accelerograms along two orthogonal axes before and after the application of interventions. Unfortunately, the investigated parameters are so many (in terms of applied interventions) and interrelated that it is rather hard to detect the effect of each separate remedial measure.

One building model was tested on a unidirectional impulse table by Bayülke et al. (2000). The model made of pumiced bimsblock masonry was single storey. It was 4.00 × 5.00 m in plan, 2.60 m high. Masonry walls were 0.20 m thick. A peculiar characteristic was that the compressive strength of the blocks was significantly smaller than the compressive strength of the mortar. A concrete slab was constructed at the top of the model. The model employed a concrete slab (made with ready mix concrete) and timber tie beams at roof level (Fig. 8.27e).

Juhásová et al. (2002) conducted a series of shaking table tests at ISMES (Bergamo). Two-storey, scale 1:2, brick masonry models were tested before and after interventions. The peculiarity of those models is that they have quite pronounced asymmetry. Some of the characteristics of the model are shown in Fig. 8.27f. The model was initially tested as built until severely damaged. Then, it was retrofitted using lime cement fibre plaster reinforced with plastic grids.

Juhásová et al. (2008) carried out shaking table tests on a full scale single storey stone masonry building (Fig. 8.27g) at LNEC in Lisbon. The asymmetrical model (non-provided with roof) was 3.58 m wide, 4.01 m long, 3.60 m high. The thickness of walls was equal to 0.24 m. The model was subject to the adequately scaled accelerogram of the Montenegro earthquake.

Bergamo et al. (2006) carried out shaking table tests on a 2-storey tuff masonry model building with three-leaf walls (Fig. 8.27h), before and after retrofit, at the facilities of CESI in Bergamo. The model was built in reduced scale of 1:2, and was 2.85 m long, 2.60 m wide and 3.30 m high. The walls (0.30 m thick) were three-leaf (exterior leaves 0.10 m thick, filling with small pieces of tuff and mortar). The two exterior leaves were connected with header stones. Timber floors with plywood pavement were provided at both floor levels. Concrete tie beams were constructed at both floor levels. Timber lintels were provided to the openings. The model was tested both as-built and after strengthening using GFRP strips. The model was subjected to adequately scaled real accelerogram along two orthogonal axes.

Tomaževič et al. (2009) have tested five two-storey brick masonry building models on the shaking table. The models (scale 1:4), with timber floors (Fig. 8.27i) were tested before and after strengthening using CFRP laminates. Seismic isolation was also considered in some cases. The models were 1.32 m long, 0.76 m wide and 1.71 m high. The walls were 0.063 m thick. The 1979 Montenegro earthquake accelerogram was imposed along x, y and z axes.

Ersubasi and Korkmaz (2010) have tested ten small scale models (scale 1:10) on a shaking table (Fig. 8.27j). The dimensions of the single storey models were quite small (0.35 m long, 0.26 m wide and 0.30 m high). A marble plate, positioned at the top of the models was simulating a RC slab. One model was tested as-built. The other nine models were tested strengthened using various intervention techniques. Constant amplitude sinusoidal displacement was applied during tests and the frequency (and acceleration) of the motion were gradually increased.

Shashi and Pankaj (2000) have tested (on an impulse table) two full scale models of single storey stone masonry (Fig. 8.27k). The models were first tested strengthened using various techniques. Then, they were repaired and retested. The models were 2.90 m long and 2.60 m wide. The roof is described as "gable type" without any further information. The quality of materials and the construction type of masonry are not given in the publication.

Meguro et al. (2012) conducted an experimental research on two scaled (1:4) single storey models with timber roof (Fig. 8.27l). The models (0.95 m long, 0.95 m wide and 0.72 m high) having walls 0.10 m thick made of stone masonry were subjected to unidirectional motions. The models were tested both as-built and retrofitted after damage.

Ahmad et al. (2010, 2012) performed a series of tests on one single storey stone masonry model with a reinforced concrete slab, simulating typical rural buildings in Pakistan (Fig. 8.27m). The model (1.52 m long, 1.22 m wide and 1.04 m high) made of double-leaf masonry, it was scaled to 1:3 and it was subjected to a series of motions along its weak direction.

Mendes and Lourenco (2010) have tested two 4-storey models at the LNEC facility (Fig. 8.27n). The models were subjected to artificial accelerograms along two orthogonal directions. One of them was tested as-built, the other after interventions. The two models, typical for houses in Lisbon, were at 1:3 scale. The models (4.8 m long, 3.15 m wide and 4.8 m high) were made of single leaf stone masonry 0.17 m thick and timber floors (timber beams and MDF panels as pavement). The panels of the pavement were positioned leaving 1 mm joints among them, in order to reduce the in-plane stiffness of the diaphragms. The intervention techniques that were applied, aimed at increasing the diaphragm action of the floors and at improving the connection between floors and walls (to prevent out-of-plane collapse of the latter).

Magenes et al. (2010, 2012a, b) report the results of a series of shaking table tests on full scale stone masonry models carried out at the Eucentre facility, Pavia. The three models were 2-storey buildings with timber floor and roof (Fig. 8.27o). They were made of double-leaf stone masonry, 0.32 m thick. The models (provided with additional masses for dynamic similitude reasons and adequately instrumented) were subjected to series of scaled accelerograms (1979 Montenegro earthquake). One of the models was tested as built, the others after the application of intervention techniques (such as enhancement of the diaphragm action by means of a second planking, improvement of the connection of horizontal and vertical members, substitution of the floor by a reinforced concrete slab etc.).

Mazzon et al. (2009) and Mazzon (2010) report the results of shaking table tests on two storey three-leaf stone masonry building models (scale 2:3). The models (Fig. 8.27p) were provided with timber floors with double planking (for improved diaphragm action). One of the models was tested before the application of grouting to masonries, it was grouted and retested, whereas the second model was tested grouted. The purpose of those tests, with models subjected to a series of motions along two orthogonal axes, was among others, to detect the effect of grouting on the dynamic properties of buildings. Finally,

Two two-storey building models were tested at the facility of the Laboratory of Earthquake Engineering, Athens. The models, made of three-leaf rubble stone masonry (Mouzakis et al. 2012a and Adami et al. 2012), were identical in geometry, materials, construction details, etc. Their only difference was that one was made of plain masonry, whereas the other was provided with timber-laces, to simulate structural systems that are very common in earthquake prone areas around the Mediterranean. The two models (Fig. 8.27q) were subjected to a series of scaled accelerograms (Kalamata, Greece, 1986 and Irpinia, Italy, 1980) along two orthogonal axes, until they are severely damaged. Subsequently, they were strengthened (enhancement of diaphragm action and grouting of masonry) and retested to failure.

The short presentation of the Literature related to dynamic testing of building models shows the variety of the parameters investigated by various researchers and, hence, the difficulties in making comparisons and draw general conclusions. However, an attempt for such a comparison is presented herein, together with an effort to draw qualitative conclusions that may be of interest for the Reader of this paper.

8.5.2.2 The Overall Behaviour of Building Models at Their as-Built State

Although, as depicted in the previous paragraph and in Fig. 8.27, there were significant differences between the models tested by various researchers (in terms of scale, materials, construction type of masonry, number of storeys, loading history, etc.), Fig. 8.28 shows the similar results obtained by almost all experimental campaigns in terms of failure mode of the models tested on a shaking table. Actually, the models shown in the photographs and sketches of Fig. 8.28 have common characteristics, typical for historic buildings, namely, rather flexible in their plane diaphragms, a more or less good connection between perimeter walls at the corners of the building, small to medium size openings (windows and doors) and piers of rather small aspect ratio. Thus, the experimental results reproduce the damages that are usually observed to masonry buildings after seismic events, i.e.: (a) Diagonal or bi-diagonal cracks in walls subjected to in-plane shear, (b) Diagonal or bi-diagonal cracks to the masonry plates between openings of the two storeys (very vulnerable to shear, as they are usually under simultaneous horizontal tension), (c) Cracks attributed to the out-of-plane or in-plane bending of walls, i.e. almost vertical cracks close to the corners of the buildings and horizontal cracks at top and bottom of piers. In some cases, when openings are located close to the corners of a building, partial or total collapse of that region is observed. Finally, (d) in case of three-leaf masonry, separation between leaves and partial collapse of the exterior leaf of masonry was observed.

It should also be noted that in the model tested by Adami et al. (2012), in which masonry was provided with horizontal timber laces, significant improvement of the behaviour was observed. Actually, the damages occurred to the timber laced model due to a seismic motion by 30 % higher (in terms of PGA) than in the unreinforced masonry model were significantly lighter (in terms of width of cracks), whereas separation between the leaves of masonry was practically prevented. Figure 8.28m shows a splice of longitudinal timber elements, as well as the timber laces at one corner of the building: The relative movement of the timber elements at their connections proves that the timber laces were mobilized and they have prevented the opening of wide cracks in masonry. Furthermore, the presence of timber laces has reduced the out-of-plane vulnerability of walls. As shown in Fig. 8.29, the displacements of the long walls of the timber laced model were almost equal to those of the unreinforced masonry model subjected to 30 % smaller PGA.

More detailed direct evaluation of the experimental results reported in the literature would require the availability of measured data, as well as systematic analytical work. Such an assessment is obviously beyond the scope of this paper. However, the fact that most of the testing campaigns reproduce the real behaviour of historic buildings subjected to seismic actions is a clear indication of the reliability of the obtained data. Thus, it can also be assumed that testing building models on a shaking table may provide reliable results on the effect of various intervention techniques. Although the study of the effect of repair and strengthening techniques on the seismic behaviour of historic buildings is out of the scope of this paper, the author would like to comment on selected experimental results that demonstrate the effect of two intervention techniques frequently applied to historic buildings and widely accepted also by Architects involved in the preservation of the built cultural heritage, namely grouting of masonry and enhancement of the diaphragm action of floors and roofs.

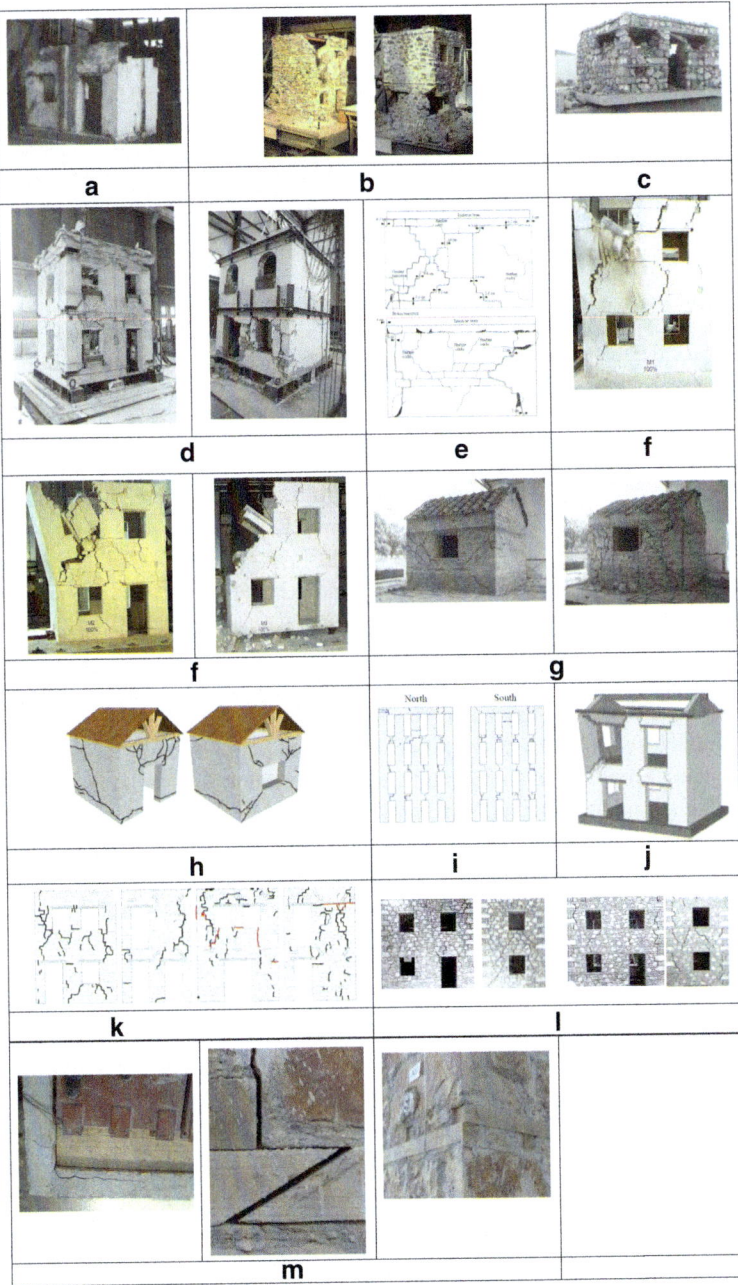

Fig. 8.28 Overview of the general behaviour of building models. (**a**) Benedetti 1980. (**b**) Tomaževič et al. 1990, 1991, 1993. (**c**) Spence and Coburn 1987, 1992. (**d**) Benedetti et al. 1998. (**e**) Bayülke et al. (2001). (**f**) Tomaževič et al. 2009. (**f**) Tomaževič et al. 2009. (**g**) Shashi and Pankaj 2000. (**h**) Meguro et al. (2012). (**i**) Mendes and Lourenco (2010). (**j**) Magenes et al. (2010, 2012a, b). (**k**) Mazzon et al. (2009). (**l**) Mouzakis et al. (2012a) and Adami et al. (2012). (**m**) Adami et al. (2012)

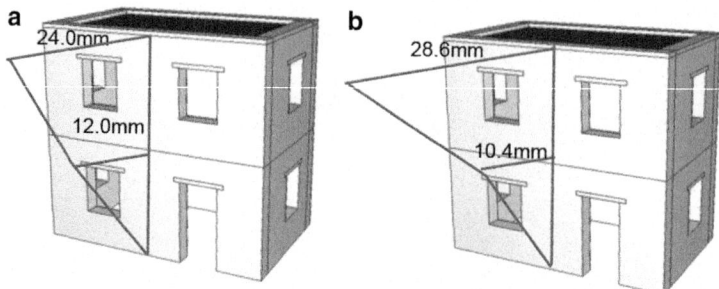

Fig. 8.29 (a) Plain masonry model: Out-of-plane displacements of the long wall for an input motion equal to 90 % Kalamata earthquake, (b) timber laced model: Out-of-plane displacements of the long wall for an input motion equal to 120 % Kalamata earthquake (Mouzakis et al. 2012a, b; Adami et al. 2012)

8.5.2.3 The Effect of Grouting and of Enhancement of the Diaphragm Action on the Behaviour of Historic Buildings

Due to the fact that masonry is a brittle material and, by way of consequence, masonry elements reach their maximum resistance at rather small imposed deformation, it is desirable to ensure to masonry buildings sufficient box-action. In such a case, the deformations to be sustained by the building are significantly reduced (for the same seismic input) and the building can sustain even strong motions without collapse. Along the same line, vulnerable construction types of masonry (double- and three-leaf masonries) that become "monolithic" through interventions, can sustain seismic actions without significant separation of their leaves and, hence, without local or more generalized collapse of the exterior leaf. Among the techniques available for enhancing the box action of masonry buildings and making the masonry behave in a more or less monolithic way, this section focuses on the enhancement of the diaphragm action of floors and roofs, as well as on the grouting of masonry.

In the past decades, the replacement of timber floors and roofs by RC (horizontal or inclined) slabs was quite frequent. However, in addition to the fact that such a replacement is rather invasive (as it alters significantly the original structural system), there is evidence of catastrophic effects of this intervention. Actually, when-stiff in their plane and quite heavy-RC slabs are simply supported by masonry (in many cases, not strengthened), they may act as a hammer during the earthquake, thus causing non-repairable damages to masonry (Fig. 8.30). Thus, the possibility to ensure sufficiently stiff diaphragms without replacing the original timber floors and roofs was experimentally investigated by several researchers.

Piazza et al. (2008), Valluzzi et al. (2010), Wilson et al. (2011), Zaopo (2011) have tested timber diaphragms either as-built or stiffened using various techniques (e.g. double board, FRP strips, diagonal steel ties, plywood panels, RC slab, etc.). They have tested single span diaphragms in their plane (monotonically or cyclically) and they have recorded both the deflection of the diaphragm and the

8 Testing Historic Masonry Elements and/or Building Models

Fig. 8.30 Catastrophic effect of RC slabs on poor quality (unstrengthened) masonry (courtesy of Prof. C.Modena)

respective in-plane load. As shown in Fig. 8.31, the use of double board may lead to an increase of the in-plane stiffness of the floor, almost by an order of magnitude. The use of plywood as pavement provides similar stiffness with a reinforced concrete slab. Similar results are shown in Fig. 8.32, where the lower curves correspond to floors typical for historic buildings. It is evident that, in all cases, significant enhancement of the in-plane bearing capacity of the floors was also recorded. On the basis of the available results, one may conclude that the addition of a second layer of boards (preferably, at an angle with respect to the original layer of boards) may render the diaphragms sufficiently stiff in their plane. This is a very promising result, since this technique is reversible and acceptable even for high value historic structures.

The effect of the enhanced in-plane stiffness of diaphragms was also tested through shaking table tests of entire building models. Actually, Tomaževič et al. (1991, 1993) have tested four building models with four different types of floors (model A: typical timber floor, model B: RC slab, model C: timber floor with prestressed steel ties, model 4: vaulted floor, see also Fig. 8.27b). As shown in Fig. 8.33, the typical timber floor is rather flexible in its plane. Actually, the mid-span displacement is almost double the displacement at the supports of the floor. On the contrary, in the other three models, the floors did perform quite satisfactorily, thus forcing the supporting walls to sustain practically equal displacements. The results were similar in the case of the building models tested by Magenes et al. (2010, 2012a, b). The authors did also draw a very significant conclusion by stating that "...the improvement on the seismic performance appears to be related more to the improvement of the floor-to-wall and roof-to-wall connections, rather than to a strong in-plane stiffening of the diaphragms".

Mouzakis et al. (2012a) in their shaking table tests have provided to the building models a second layer of boards (at an angle of 45° with respect to the original pavement, Fig. 8.34). Natural frequency measurements along x and y axes have shown a significant difference along the two axes (6.05 and 4.21 Hz respectively), due to the significantly smaller stiffness perpendicular to the long side of the model.

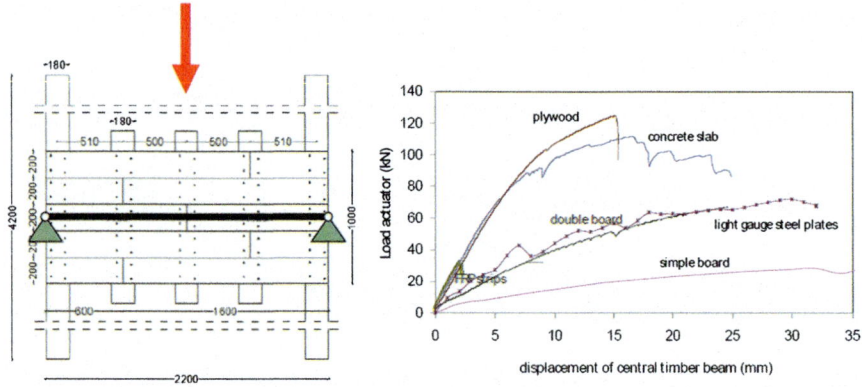

Fig. 8.31 Experimental setup and main results (Piazza et al. 2008)

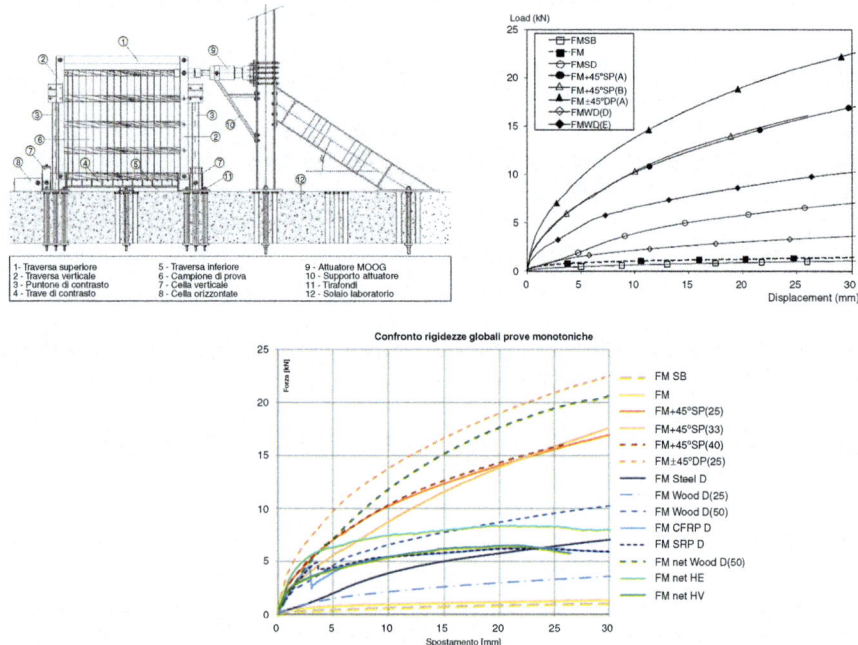

Fig. 8.32 Test setup and main results (Valluzzi et al. 2010; Zaopo 2011)

After strengthening, the two values were substantially larger (10.36 and 9.95 Hz respectively) indicating a significant overall increase of the stiffness of the model. More importantly, the two frequency values are almost equal along the two axes, indicating that the stiffness of the strengthened diaphragms was able to ensure the box action of the model.

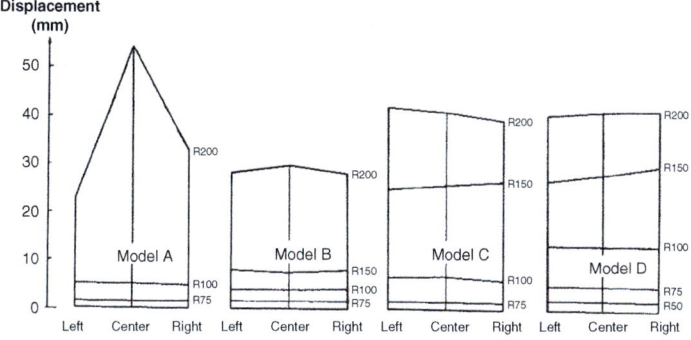

Fig. 8.33 Out-of-plane displacements of walls for various alternative floor types (Tomaževič et al. 1991, 1993)

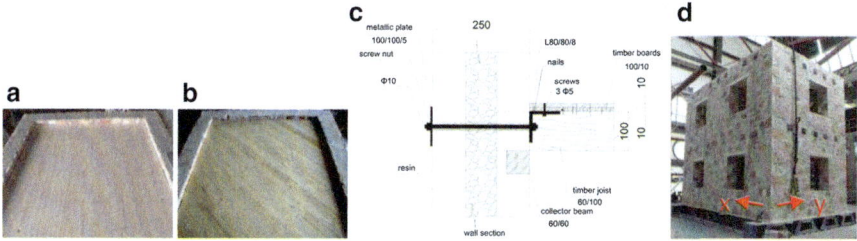

Fig. 8.34 Enhancement of the diaphragm action of floors (Mouzakis et al. 2012a): (**a**) the original pavement, (**b**) the pavement with the second layer of boards, (**c**) detail of the connection between diaphragm and walls, (**d**) the model after strengthening

The effect of grouting on the seismic behaviour of building models was investigated by several researchers, in most cases combined with other intervention techniques as well. Mazzon et al. (2009) and Mazzon (2010) have investigated the effect of grouting alone. One of the main findings of their research is that grouting provides a significant enhancement of the seismic resistance of masonry buildings without altering their dynamic properties. Grouting prevents the separation of masonry leaves and, hence, it reduces their seismic vulnerability.

Tests by Adami et al. (2012) on a timber laced masonry model before and after the application of grouting have shown that under the same input motion that led to significant damages of the unstrengthened model, the grouted model did not suffer any damage.

A final observation that, in the opinion of the author, needs to be further investigated and discussed upon is illustrated in Fig. 8.35. Shaking table tests by Mouzakis et al. (2012a) have shown that, although masonry is a brittle material and masonry elements are also brittle, masonry buildings may exhibit significant ductility, even at their as-built state. Although this result is reported with caution and it definitely needs to be confirmed by further experimental data, it may insinuate that historic masonry buildings avail of reserves-not easily detectable

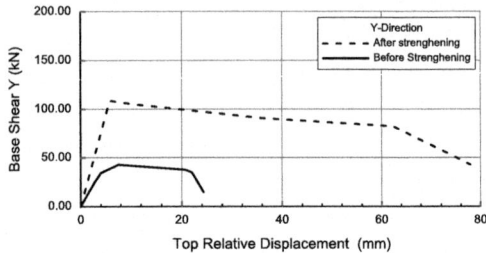

Fig. 8.35 Envelop of hysteresis loops for a plain masonry building model tested on the shaking table before and after strengthening (Mouzakis et al. 2012a)

by calculations-which ensure a significantly better behaviour than usually assumed. It should be noted that this observation seems not to contradict reality, as many historic structures survived several earthquakes, although according to our calculations they should have failed.

8.6 Concluding Remarks

This paper provides an overview of the results obtained from testing masonry elements, subassemblies and building models. Although this overview is clearly incomplete, both because it does not cover but a part of the available experimental results and because it does not offer but some general comments on the data, it allows for some qualitative conclusions to be drawn:

(a) The international Literature is rich in results of experimental campaigns related to the behaviour of masonry elements (in compression, shear or out-of-plane flexure), of subassemblies, as well as of models of entire buildings. A vast variety of combinations of building materials, construction types of masonry, geometry of specimens, experimental setups, types of loading, scale of tested specimens, etc. can be found in publications. Valuable data are available on the mechanical properties of various types of masonry, as well as on stiffness and bearing capacity of elements, on the hysteretic behaviour of elements or assemblies. However,
(b) Due to the variety of parameters investigated by various researchers, a direct comparison among seemingly comparable experimental results is in many cases not possible. In order to take the maximum profit of the available valuable data, the development of sound physical models is necessary. Furthermore, systematic analytical work is needed, in order to validate and calibrate physical models and propose design models adequate for practical use.
(c) Although numerous experimental campaigns were carried out throughout the globe, there are still several aspects of the behaviour of historic masonry structures that remain insufficiently investigated. For example, out of the frequent types of masonry found in historic structures, only a limited variety has been investigated to date. Even their behaviour under compression is not adequately documented (in terms of strength, deformability, post-peak

behaviour, etc.). Furthermore, the behaviour of historic masonry under a combination of in- and out-of-plane actions is not sufficiently investigated at the level of structural member or subassemblies. Taking into account the vulnerability of historic masonry to out-of-plane actions and the subsequent effect of that vulnerability to the in-plane behaviour of structural members, this lacuna is quite significant. Moreover,

(d) The experimental results are not presented in a form that would allow for comparisons at a large scale. Even the definition of some terms differs from publication to publication. Thus, although valuable qualitative conclusions can be drawn, the need for results liable to quantitative assessment is-in general-not satisfied. If one adds to those difficulties the inherent scatter of the experimental results, it becomes obvious that an exhaustive assessment of the Literature is a task with rather dubious outcome.

(e) Thus, the author of this paper would like to make a proposal for future work within the European Association: An international group of experts both in experimental work and in the preservation of the built cultural heritage could collect all relevant publications. The group could establish a model for the presentation of experimental data, rendering the data liable to quantitative evaluation. The model forms should be filled for each publication. Obviously, in many cases it would be necessary for the group to contact the researchers asking for more data or for data in the adequate form. The final step would be the assessment of the experimental results and the creation of a database that could be made available to the Public. Such a database could also allow for the identification of open issues and, thus, guide further research on the subject. The author of this paper is conscious of the fact that such an operation is quite ambitious. However, it is strongly believed that this is a necessary step for the rationalization of the work of Engineers and, hence, for the preservation of the wealth the built cultural heritage constitutes for Europe.

Open Access This chapter is distributed under the terms of the Creative Commons Attribution Noncommercial License, which permits any noncommercial use, distribution, and reproduction in any medium, provided the original author(s) and source are credited.

References

ABK (1981) Methodology for mitigation of seismic hazards in existing unreinforced masonry buildings: wall testing, Out-of-plane. ABK Topical Report 04. Technical report, ABK, California, USA

Adami CE, Vintzileou E, Mouzakis C, Karapitta L (2012) Three-leaf stone masonry models with timber ties: the effect of timber ties and strengthening techniques on seismic response. SAHC 2012, Wroclaw (electronic source)

Ahmad N, Ali Q, Crowley H, Pinho R (2010) Displacement-based seismic risk assessment of stone masonry buildings of Pakistan. In: 3rd Asia conference on earthquake engineering Bangkok, Thailand (electronic source)

Al Shawa O, Benedetti S, Bellisario M, de Felice G, Mauro A, Paolacci F, Ranieri N, Roselli I, Sorrentino L (2009) Prove sperimentali su tavola vibrante di pareti murarie sollecitate fuori dal piano. ReLUIS final deliverable, Annex PF-1.2-UR09_15 (in Italy)

Ali Q, Badrashi YI, Ahmad N, Alam B, Rehman S, Banori FAS (2012) Experimental investigation on the characterization of solid brick masonry for lateral shear strength evaluation. Int J Earth Sci Eng 4(4):782–791

Baratta A, Corbi O (2007) Stress analysis of masonry vaults and static efficacy of FRP repairs. Int J Solids Struct 44:8028–8056

Bayülke N, Gençer O, Başyiğit C, Terzi S (2000) Impulse table test of a "bimsblock" masonry structure. In: International conference on the seismic performance of traditional buildings, Istanbul, Turkey

Benedetti D (1980) Repairing and strengthening stone masonry buildings. In: International research conference on earthquake engineering. Skopje, pp 503–515

Benedetti D, Carydis P, Pezzoli P (1998) Shaking table tests on 24 simple masonry buildings. J Earthq Eng Struct Dyn 27(1):67–90

Bergamo G, Eusebio M, Manfredi G, Prota A (2006) Shake table tests on a tuff masonry building. In: VIII U.S. National conference on earthquake engineering, San Fransisco, California, USA

Binda L (ed) (2008) Learning from failure-Long-term behaviour of heavy masonry structures. WIT Press, Southampton, p 228

Binda L, Saisi A (2001) State of the art of research on historic structures in Italy. Department of Structural Engineering, Politecnico of Milan, Milano

Binda L, Baronio G, Mirabella Roberti G, Penazzi D (1999) Caratteristiche morfologiche e meccaniche di alcune murature di Catania. L'ingegneria sismica in Italia, 9° Convegno Nazionale Torino. Electronic source (in Italian)

Brencich A, Lagomarsino S (1998) A macroelement dynamic model for masonry walls. In: Pande GN, Middeton J, Kralj B (eds) Computer methods in structural masonry 4: fourth international symposium. E&FN SPON, Routledge

Brignola A, Podesta S, Lagomarsino S (2006) Experimental results of shear strength and stiffness of existing masonry walls. In: Roca P, Modena C, Agrawal S (eds) Structural analysis of historical constructions-lourenco. Taylor and Francis Group, London (electronic source)

Bui TT, Limam A, Bertrand D, Ferrier D, Brun M (2010). Masonry walls submitted to out-of-plane loading: experimental and numerical study. In: Proceedings of 8th international masonry conference, Dresden (electronic source)

Capozucca R (2011) Shear behaviour of historic masonry made of clay bricks. Open Constr Build Technol J 5(Suppl 1-M6):89–96

Cavaleri L, Fossetti M, La Mendola L, Papia M (2006) Roca P, Modena C, Agrawal S (eds) Structural analysis of historical constructions-lourenco. Taylor and Francis Group, London (electronic source)

CEN-EN1996-1-1 (2005) Design of masonry structures-General rules for reinforced and unreinforced masonry structures, Brussels, 123pp.

Chiostrini S, Galano L, Vignoli A (2000) On the determination of strength of ancient masonry walls via experimental tests. In: 12th WCEE, Auckland, New Zealand (electronic source)

Corradi M, Borri A, Vignoli A (2003) Experimental study on the determination of strength of masonry walls. In: Construction and building materials, vol 17. Elsevier, pp 325–337

Corradi M, Tedeschi C, Binda L, Borri A (2008) Experimental evaluation of shear and compression strength of masonry wall before and after reinforcement: deep rejointing. In: Construction and building materials, vol 22. Elsevier, pp 463–472

Costa AA, Arede A, Costa A, Guedes J, Silva B (2012a) Experimental testing, numerical modelling and seismic strengthening of traditional stone masonry: comprehensive study of a real Azorian pier. Bull Earthq Eng 10:135–159

Costa AA, Arede A, Costa A, Oliveira CS (2012b) Out-of-plane behaviour of existing stone masonry buildings: experimental evaluation. Bull Earthq Eng 10:93–111

Costa AA, Arede A, Costa AC, Penna A, Costa A (2012c) Out-of-plane shaking table test of a full scale stone masonry façade. In: 15th WCEE, Lisbon (electronic source)

Da Porto F, Valluzzi MR, Modena C (2003) Investigations for the knowledge of multi-leaf stone masonry walls. In: First international congress on construction history, Madrid, Spain, 20–24 January 2003, vol II. pp 713–722

Ersubasi F, Korkmaz H (2010) Shaking table tests on strengthening of masonry structures against earthquake hazard. J Nat Hazards Earth Syst Sci 10:1209–1220

Giuffrè A, Baggio C, Carocci C (1993) Sicurezza e conservazione dei centri storici. Laterza, Bari (In Italian)

Griffith MC, Lam NTK, Wilson JL, Doherty K (2004) Experimental investigation of unreinforced brick masonry walls in flexure. J Struct Eng 130(3):423–432

Juhásová E, Hurák M, Zembaty Z (2002) Assessment of seismic resistance of masonry structures including boundary conditions. J Soil Dyn Earthq Eng 22:1193–1197

Juhásová E, Sofronie R, Bairrão R (2008) Stone masonry in historical buildings – Ways to increase their resistance and durability. J Eng Struct 30:2194–2205

Magenes G, Calvi GM (1997) In-plane seismic response of brick masonry walls. Earthq Eng Struct Dyn 26:1091–112

Magenes G, Penna A, Galasco A (2010) A full-scale shaking table test on a two-storey stone masonry building. In: 14th ECEE, Beijing (electronic source)

Magenes G, Penna A, Rota M, Galasco A, Senaldi I (2012a) Shaking table test of a strengthened full scale stone masonry building with flexible diaphragms. In: 8th SAHC conference, Wroclaw, Poland (electronic source)

Magenes G, Penna A, Rota M, Galasco A, Senaldi I (2012b) Shaking table test of a full scale stone masonry building with stiffened floor and roof diaphragms. In: 15th WCEE (electronic source)

Manoledaki AA, Drosos V, Anastasopoulos I, Vintzileou E, Gazetas G (2012) Out-of-plane response of three-leaf stone masonry walls taking account of soil-structure interaction: an experimental study. In: Jasienko J (ed) Structural analysis of historical constructions. Wroclaw (electronic source)

Marcari G, Fabbrocino G, Lourenco PB (2010) Mechanical properties of tuff and calcarenite stone masonry panels under compression. In: Proceedings of 8th international masonry conference, Dresden, pp 1083–1092

Mazzon N (2010) Influence of grout injection on the dynamic behaviour of stone masonry buildings. Ph.D. thesis, University of Padova, Padova, Italy

Mazzon N, Valluzzi MR, Aoki T, Garbin E, De Canio G, Ranieri N, Modena C (2009) Shaking table tests on two multi-leaf stone masonry buildings. In: 11th Canadian masonry symposium (electronic source)

Meguro K, Navaratnaraj S, Sakurai K, Numada M (2012) Shaking table tests on 1:4 scaled shapeless stone masonry houses with and without retrofit by polypropylene band meshes. In: 15WCEE, Lisbon (electronic source)

Meisl CS, Elwood KJ, Mattman DW, Ventura CE (2006) Out-of-plane seismic performance of unreinforced clay brick masonry walls. In: Proceedings of 8th U.S. national conference on earthquake engineering, San Francisco, California, (electronic source)

Mendes N, Lourenco P (2010) Seismic assessment of masonry "gaioleiro" buildings in Lisbon, Portugal. J Earthq Eng 14(1):80–101

Milosevic J, Gago AS, Lopes M, Bento R (2012) Rubble stone masonry walls – evaluation of shear strength by diagonal compression tests In: 8th international conference on structural analysis of historical constructions. Wroclaw (electronic source)

Mouzakis C, Vintzileou E, Adami CE, Karapitta L (2012a) Dynamic tests on three-leaf stone masonry building model without timber ties before and after interventions. SAHC 2012, Wroclaw (electronic source)

Mouzakis C, Adami CE, Karapitta L, Vintzileou E (2012b) Seismic behaviour of a rehabilitated cross vault. SAHC 2012, Wroclaw (electronic source)

Oliveira DV, Lourenco PB, Garbin E, Valluzzi MR, Modena C (2006) Experimental investigation of the structural behaviour and strengthening of three-leaf stone masonry walls. In: Roca P, Modena C, Agrawal S (eds) Structural analysis of historical constructions-lourenco. Taylor and Francis Group, London (electronic source)

Piazza M, Baldessari C, Tomasi R (2008) The role of in-plane floor stiffness in the seismic behaviour of traditional buildings. In: 14th World conference on earthquake engineering, Beijing, China (electronic source)

Pina-Henriques J, Lourenco PB, Binda L, Anzani A (2005) Testing and modeling of multiple-leaf masonry walls under shear and compression. In: Modena C, Lourenco PB, Roca P (eds) Structural analysis of historical constructions. Taylor and Francis Group, London, pp 299–310

Pinto A, Gago A, Verzeletti G, Mollina FJ (1999a) Seismic tests on the S. Vicente de Fora model: assessment and retrofitting – part I. In: Eurodyn Conference 1999, Prague, Czech Republic (electronic source)

Pinto A, Gago A, Verzeletti G, Mollina FJ (1999b) Seismic tests on the S. Vicente de Fora model – assessment and retrofitting – part II. In: Eurodyn Conference 1999, Prague, Czech Republic (electronic source)

Pinto A, Gago A, Verzeletti G, Mollina FJ (1999c) Tests on the S. Vicente de Fora model – assessment and retrofitting. Workshop on seismic performance of built heritage in small historic centers, Assisi

Pinto A, Molina J, Pegon P, Renda V (2001) Protection of the cultural heritage at the ELSA Laboratory. In: Lourenço PB, Roca P (eds) Historical Constructions 2001 Possibilities of numerical and experimental techniques. Proceedings of the 3rd international seminar Guimarães. University of Minho, Guimarães, pp. 973–982

Restrepo-Vélez LF (2004) Seismic risk of unreinforced masonry buildings, PhD thesis, European school for advanced studies in reduction of seismic risk – ROSE School, University of Pavia, Italy

Restrepo-Vélez LF, Magenes G (2004) Experimental testing in support of a mechanics-based procedure for the seismic risk evaluation of unreinforced masonry buildings. In: Proceeding of IV international Seminar SAHC, Padua, Italy, pp 1079–1089

Shahzada K, Khan AN, Elnashai AS, Ashraf M, Javed M, Naseer A, Alam B (2012) Experimental seismic performance evaluation of unreinforced brick masonry buildings. Earthquake Spectra 28(3):1269–1290

Shashi TK, Pankaj A (2000) Seismic evaluation of earthquake resistant and retrofitting measures of stone masonry houses, 12WCEE, Auckland, New Zealand (electronic source)

Silva B, Dalla BM, da Porto F, Modena C (2014) Experimental assessment of in-plane behaviour of three-leaf stone masonry walls. Constr Build Mater 53:149–161

Simsir CC, Aschheim MA, Abrams DP (2004) Out-of-plane dynamic response of unreinforced masonry bearing walls attached to flexible diaphragms. In: 13th World conference on earthquake engineering, Vancouver, Canada (electronic source)

Spence R, Coburn A (1987) Reducing earthquake losses in rural areas. Report to the Overseas Development Administration, The Martin Centre for Architectural and Urban Studies, Cambridge, UK

Spence R, Coburn A (1992) Strengthening buildings of stone masonry to resist earthquakes. Meccanica 27:213–221

Taranu N, Oprisan G, Budescu M, Taranu G, Bejan L (2010) Improving structural response of masonry vaults strengthened with polymeric textile composite strips. In: Proceeding of the 3rd WSEAS International conference on engineering mechanics, structures, engineering geology. Greece (electronic source)

Tassios TP (2004) Rehabilitation of three-leaf masonry. In: Evoluzione nella sperimentazione per le costruzioni, Seminario Internazionale, 26 Sept – 3 Oct, Centro Internationale di Aggiornamento Sperimentale – Scientifico (CIAS), Sicily

Tassios TP (2013) Parameters affecting the compressive strength and critical strain of masonry. Centro Internazionale di aggiornamento sperimentale-Scientifico, Seminario sul tema Evoluzione nella sperimentazione per le costruzioni, Crete, pp 191–212

Tassios TP, Chronopoulos M (1986) Aseismic dimensioning of interventions on low-strength masonry buildings. In: Middle east and mediterranean regional conference on low strength masonry in seismic areas. Middle East Technical University, Ankara

Tomaževič M, Weiss P, Velechovsky T (1990) The influence of rigidity of floors on the seismic resistance of old masonry buildings: shaking table tests of model houses A and B. Institute for Testing and Research in Materials and Structures, Ljubljana

Tomaževič M, Weiss P, Velechovsky T (1991) The influence of rigidity of floors on the seismic behavior of old stone-masonry buildings. J Eur Earthq Eng 5(3):28–41

Tomaževič M, Lutman M, Velechovsky T (1993) Aseismic strengthening of old stone-masonry buildings: is the replacement of wooden floors with R.C. slabs always necessary? Eur Earthq Eng 7(2):34–46

Tomaževič M, Klemenc I, Weiss P (2009) Seismic upgrading of old masonry buildings by seismic isolation and CFRP laminates: a shaking-table study of reduced scale models. Bull Earthq Eng 7:293–321

Tominaga Y, Nishimura Y (2008) Experimental study on structural performance of historic brick masonry buildings. In: 14th WCEE, Beijing, China, (electronic source)

Valluzzi MR, Garbin E, Dalla Benetta M, Modena C (2010) In-plane strengthening of timber floors for the seismic improvement of masonry buildings. World conference on timber engineering, Riva del Guarda, Italy (electronic source)

Valluzzi MR, Mazzon N, Garbin E, Modena C (2013) Experimental characterization of out-of-plane seismic response of strengthened three-leaf stone masonry walls by shaking table tests. XV Convegno L'Ingegneria Sismica in Italia, Padova, Italy (electronic source)

Vasconcelos G, Lourenco PB (2006) Assessment of in-plane shear strength of stone masonry walls by simplified models. Roca P, Modena C, Agrawal S (eds) Structural analysis of historical constructions-lourenco. Taylor and Francis Group, London (electronic source)

Vasconcelos G, Lourenco PB (2009) In-plane experimental behavior of stone masonry walls under cyclic loading. ASCE J Struct Eng 135:1269–1277

Vintzileou E (2008) The effect of timber ties on the behaviour of historic masonry. ASCE J Struct Eng 134(6):961–972

Vintzileou E (2011a) Timber-reinforced structures in Greece: 2500 BC–1900 AD. Proc ICE Struct Build 164(3):167–180

Vintzileou E (2011b) Three-leaf masonry in compression, before and after grouting: a review of literature. Int J Archit Herit 5(4–5):513–538

Vintzileou E, Miltiadou-Fezans A (2008) Mechanical properties of three-leaf stone masonry grouted with ternary or hydraulic lime based grouts. Eng Struct 30(8):2265–2276

Vintzileou E, Tassios TP (1995) Three-leaf stone masonry strengthened by injecting cement-grouts. J Struct Eng Struct Div ASCE 121(5):848–856

Vintzileou E, Adami CE, Mouzakis Ch, Karapitta L (2012) Testing a three-leaf rubble stone masonry subassembly (T-structure) on the shaking table before and after strengthening. NTUA Research Report, Athens, Greece (in Greek)

Vintzileou E, Mouzakis H, Karapitta L, Adami CE (2012) Assessment of dynamic behaviour of three leaf stone masonry building models: Seismic enhancement by grouting and improvement of box behaviour. International conference Cultural heritage heritage protection in times of risk: challenges and opportunities. Istanbul (electronic source)

Wilson, AW, Quenneville PJH, Ingham JM (2011) Assessment of timber floor diaphragms in historic unreinforced masonry buildings. Presented at international conference on structural health assessment of timber structures, Lisbon, Portugal

Zaopo N (2011) Valutazione sperimentale dell'efficacia di interventi di miglioramento sismico di solai in legno rinforzati nel piano. Diploma dissertation, University of Padova, Italy, 212 pp. (in Italian)

Chapter 9
Earthquake Risk Reduction: From Scenario Simulators Including Systemic Interdependency to Impact Indicators

Carlos Sousa Oliveira, Mónica A. Ferreira, and F. Mota Sá

Abstract Earthquakes have a strong effect on the socio-economic well-being of countries; the consequences can lead to a complex cascade of related incidents, expanding across sectors and borders, and in a more serious context, to our basic survivability. An urban area consists on several complex and highly connected systems. A significant loss of housing, education, power outages or other component would have substantial negative impacts. How would constrains in residential areas affect the residential distribution of the region? How would a general change in accessibility due to severe damage affect the population or the economy (employment changes)?

Disasters are still predominantly seen as exogenous events, unexpected and unforeseen shocks that affect normally functioning economic systems and societies rather than as endogenous indicators, an integrated, and mutually influencing process where financial, health, economic and social risks are considered as both facets and at the same time contributing factors in an interdependent process of risk creation, accumulation, mitigation, and transference.

Seismic scenario simulators have been used as tools to estimate damages inflicted by earthquakes in a region. Up to now this powerful simulators calculate and maps the direct damages on urban environment such as the building stock and infrastructures, not including the propagation effects among these components. This paper presents a novel approach to study in a macro scale an urban region, including the systemic interdependencies among urban elements. The methodology allows the observation of urban disruptions caused by the interdependencies and measured through a Disruption index. This index permits to identify the most vulnerable elements, being essential for the risk reduction.

C.S. Oliveira (✉) • M.A. Ferreira • F. Mota Sá
Instituto Superior Técnico, Universidade de Lisboa, Lisbon, Portugal
e-mail: csoliv@civil.ist.utl.pt

9.1 Introduction

Natural disasters, namely earthquakes have clearly demonstrated that preparedness and disaster management are dynamic processes that require a holistic analysis of critical interdependencies among core infrastructures in order to mitigate the impact of extreme events and improve survivability of our society.

This paper, after a first analysis of the earthquake activity since 1900 and, in particular, in the last 20 years, in relation to the impacts caused to society, describes the main successes achieved to estimate the impact of future events and present a new indicator based on the disruption caused to the population due to not only the direct impact of shaking but also including the effect of interdependences among the various urban systems.

9.1.1 Trends of Natural Disasters

"The so called *natural* disasters, that is, those related to phenomena of Nature, have caused throughout the centuries great convolutions in the process of human development. Even though advances in science and technology have produced a great deal of knowledge on the causes of those disasters, human deaths in the world per million inhabitants are only slightly decreasing with time, but the economic losses have dramatically increased in the last decades. The rise in world population and the complexity of societal organization, among others, are factors that may explain this unfortunate fact. Inadequate non-sustainable use of the territory and present day inadequate construction practices, especially in developing countries, are clear causes of the too frequent "natural" disasters" (Oliveira et al. 2006).

The economic and livelihood losses associated with damaged and destroyed housing, infrastructure, public buildings, businesses and agriculture have been rising at a rapid rate as well as the mortality associated with geological hazards such as earthquakes and tsunamis. How is it possible that progress, which should lead to reduced losses, is actually being accompanied by rising losses?

The concentration of people and values in large conurbations as well as settlement in and industrialization of extremely exposed regions are some reasons to globally increase losses. It is estimated that by 2030 some 60 % of the world's population will live in urban areas and by 2050 this will have risen to 70 % (UN-HABITAT 2008; WDR 2010). Figure 9.1 shows the urban agglomerations with more than five million inhabitants in 2010 together with the zones of higher seismic hazard.

As known, seismic risk is a convolution of Hazard, Exposition and Vulnerability. Looking to the history of earthquakes it is very clear that the higher of one of this variables, the higher the risk.

Figure 9.2 present the evolution of number of victims (a) and economic losses (b) per decade during the twentieth century due to seismic activity. The two decades

9 Earthquake Risk Reduction: From Scenario Simulators Including Systemic... 311

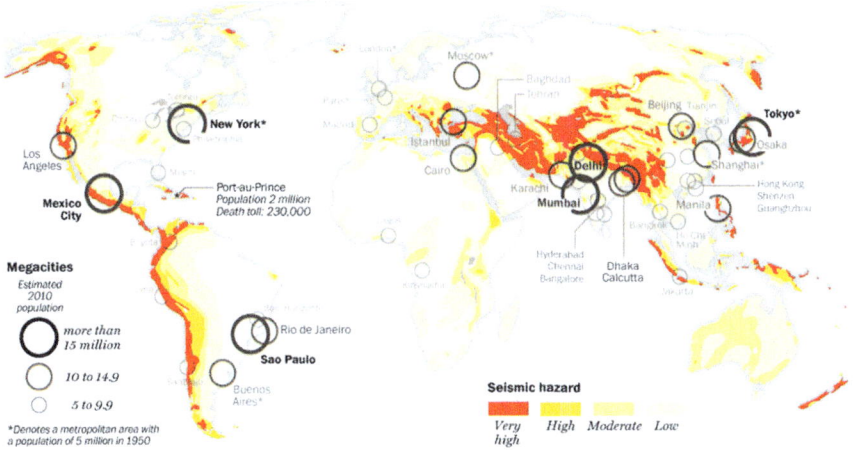

Fig. 9.1 Urban agglomerations with more than five million inhabitants in 2010 and seismic hazard regions (Karklis 2010)

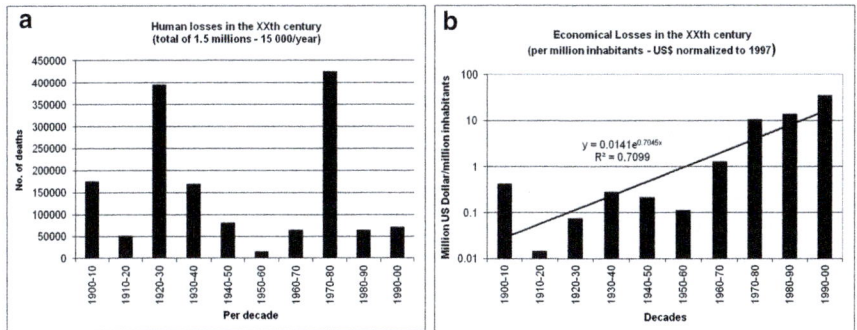

Fig. 9.2 Evolution of losses during the twentieth century: (**a**) to the population; (**b**) economic (normalized to 1997, per million inhabitants)

with more victims were the 1920–1930 with the Japan Kantō earthquake and the 1970–1980 with the Chinese Tangshan earthquake. The yearly average of events is three and the average of victims is 15,000 per year. Looking to Fig. 9.2 one observes that even though the number of victims is not a stationary process, the economical losses have increased steadily over the years in an exponential way. This increase is explained by the fact that each time a destructive earthquake strikes the larger the impact in the society, due to the larger assets involved and to the cascade effects of our modern society.

In the last 15 years a similar trend has occurred. Earthquakes and tsunamis such as Sumatra 2004, Sichuan 2008, Haiti 2010 and Tohoku 2011 are extreme events in terms of consequences as shown in Fig. 9.3. In relation to economic losses, the increase trend of Fig. 9.2(b) is similar for the first decade of the twenty-first century.

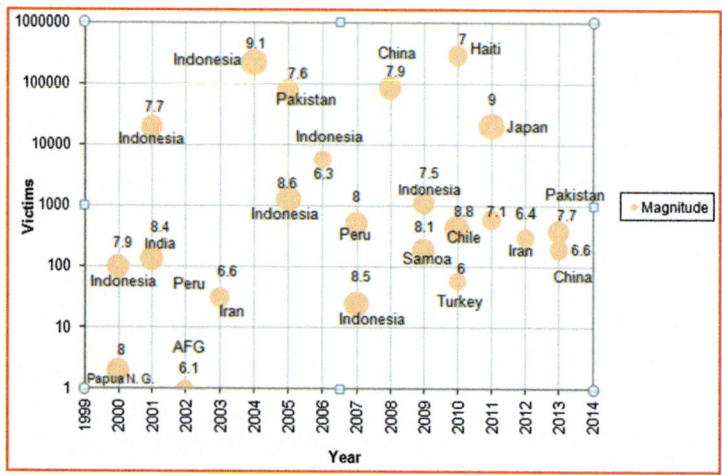

Fig. 9.3 Victims from earthquakes (and tsunamis) in the last 15 years and corresponding magnitude values

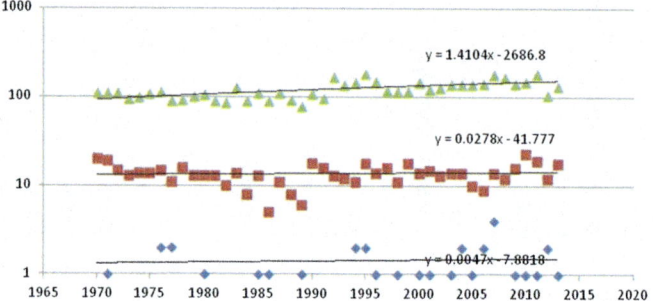

Fig. 9.4 Annual number of earthquakes in the last 43 years in the World by classes of magnitude; *green triangle* – $6 < M < 6.9$; *red squares* – $7 < M < 7.9$; *blue diamonds* – $8 < M$. (USGS 2012; EMSC 2014)

Figure 9.3 also shows that victims are not only caused by the very large magnitude events. Sometimes a M7, as the case of Haiti, can cause such a large number of victims, naturally due to the way, in many regions, society has dealt with the earthquake threat.

The pattern of earthquake impact cannot be attributed to an increase of seismic activity. In fact in Fig. 9.4 we plot the annual number of earthquakes in the world by classes of magnitude ($6 < M < 6.9$; $7 < M < 7.9$; $8 < M$), and it does not seems that seismic activity by itself has been increasing in the last 43 years. The earthquake impact is much more dependent on the increase of assets and of its vulnerability in many urban regions.

We will try to show that the impact of earthquakes is a multi-facet problem with consequences on the population and on the built environment, and demonstrate that in modern societies the effect of systemic dependences is marking with great power the traditional way to look at earthquakes and society.

The figures above show the losses in terms of victims and economic impact, but they not reflect the livelihood impact, the business disruption, the red zone areas or the number of years that this impact will last. Can we get an indicator capable of telling us the "disruption" in a society caused by an earthquake, measuring the state of disorder that was caused?

9.2 Scenario Earthquake Simulators. An Evolution

Many different software packages have been produced by different schools around the world in order to provide accurate seismic risk estimates. Table 9.1 presents a review of recent open source software packages.

These powerful seismic risk simulators can compute loss and damage estimate, risk scenarios or the associated benefit by cost of retrofitting, but they do not include the propagation and cascade effects existing in an urban area.

9.2.1 QuakeIST®

An earthquake scenario simulator is produced to assess the impact of future earthquakes on a defined area of exposure, which may be a city, region or country, or a portfolio of buildings and facilities within such a geographical area. This is an ambitious aim since the problem is very complex and there is major uncertainty related to several elements: the ground-motion prediction equations; the ground conditions (site effects); the characterisation of the building stock and infrastructure exposure; the definition of the vulnerability of the exposed elements; the modelling of propagation effects, the estimation of repair costs and human casualties.

Figure 9.5 shows the main modules that constitute an earthquake scenario simulator from hazard definition, exposure, vulnerability, to the loss assessment.

Various approaches exist regarding the damage appraisal, such as financial and economic valuation based on market values (i.e. based on historical values or replacement values) – Today the typical approach is the economic estimation of direct damage. Earthquake scenario simulators developed until now show direct physical damage in terms of victims, buildings, essential facilities and transportation systems, without including estimations of indirect losses or propagated effects (functional interdependencies). For a consistent decision analysis it is desirable to include an holistic approach.

The QuakeIST® is an integrated simulator developed by Instituto Superior Técnico (Mota de Sá, 2012, QuakeIST® software, personal communication), to

Table 9.1 Synopsis of seismic risk software packages

Software	Institution	Programming language	Applicability	Availability	Graphical user interface	Type of calculators
ELENA	NORSAR	MATLAB/C	User-defined	OS	Yes	SCN/SDA/PEB
EQRM	GA	Python	User-defined	OS	No	SCN/SDA/PEB
ELER	KOERI	MATLAB	User-defined	SA	Yes	SCN/SDA
QLARM	WAPMERR	Java	World	SC	Yes	SCN/SDA
CEDIM	CEDIM	Visual Basic	User-defined	SC	Yes	SCN/SDA/CPB
CAPRA	World Bank	Visual Basic	Central America	SC	Yes	SCN/PEB
RiskScape	GNS	Java	New Zealand	SA	Yes	SCN/SDA
LNECLoss	LNEC	Fortran	Portugal	SC	No	SCN/SDA
MAEviz	MAE Center	Java	User-defined	OS	Yes	SCN/SDA/CPB
OpenRisk	SPA Risk	Java	USA	SA	Yes	CPB/BCR
OpenQuake	GEM	Phyton	World	OS	Yes	SCN/SDA/PEB/CPB/BCR

Adapted from Silva et al. (2013)

OS open-source (code on a public repository), *SA* standard application (available under request), *SC* source code (available under request)

SCN scenario risk, *DAS* scenario damage assessment, *PEB* probabilistic event-based risk, *CPB* classical PSHA-based risk, *BCR* benefit-cost ratio

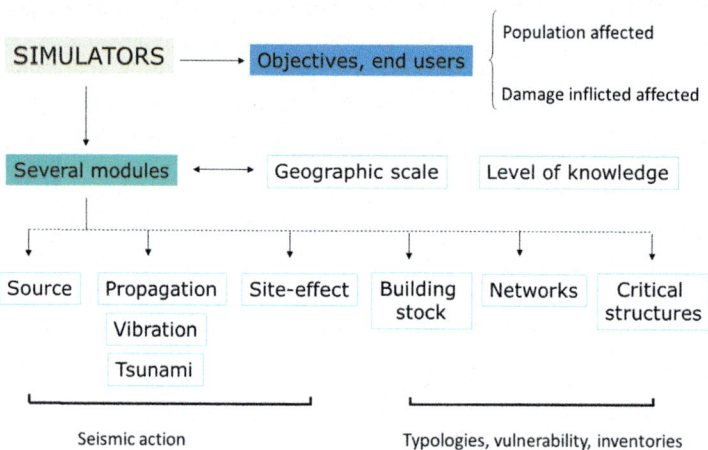

Fig. 9.5 Schematic structure of earthquake scenario simulators

provide assistance in risk assessment and disaster management to decision makers and other people concerned to take the right decisions related with the topic. This sophisticated software can model physical risk assessment and is the first earthquake risk simulator that offer an integrated cascade-effect approach and a global impact at urban or regional scale (the DI, Disruption index: Oliveira et al. 2012; Ferreira et al. 2014). The results provided by QuakeIST® are capable to identify important factors and systems which contribute to main urban disruptions, providing plans and guidance for short-, medium-, and long-term investment projects to reduce risk.

QuakeIST® software was applied in several countries (Italy, Portugal, Spain and Iceland) during the UPStrat-MAFA Project (2012), to generate and measure risk, quantify the impacts, and improve the capacity to define strategies to address adverse natural events. The locations under study were very important to calibrate and validate several parameters of the model, using real earthquakes (Lorca, Spain 2011; Faial, Azores 1998; Mount Etna, Italy 1914; and Hverageroi, Iceland 2000 and 2008).

A brief description of the key features of the QuakeIST® software is presented below:

- The simulator (QuakeIST®) can handle different ground motion scenarios provided by the user, referring the ground motion values to coordinates, using external scenarios obtained from different software's like SASHA (D'Amico and Albarello 2008), PROSCEN (Rotondi and Zonno 2010), or any historical seismic scenario.
- QuakeIST® contains several well-known attenuation relationships that the user may select or adapt to their own conveniences, in order to calculate ground shaking based on an epicentral position (coordinates) and magnitude.
- QuakeIST® requires shaking intensity, PGA, PGV or PGD as an input parameter to some objects. Conversion between PGA, PGV, PGD and different macroseismic intensity scales was implemented. Soil information can be handled through EC8 soil classes (EC-8 2004), and there are several possible options the user can choose to manage site effects (soil amplification/deamplification).
- QuakeIST® is written in C++ and interacts (but do not rely on them) with virtually all platforms of geographical information system software (GIS), such as ESRI, QuantumGIS, and others, to create maps and measure the possible impact caused by earthquakes in urban systems.
- Various models to calculate direct damages (macroseismic model -Giovinazzi and Lagomarsino 2004), the capacity spectrum (Freeman 2004), N2 (Fajfar 1999) or fragility functions) are included and users can upload their own vulnerability parameters or include new ones.
- Different types of assets can be modeled (buildings, schools, bridges, various types of networks – water, power-electricity, gas, communications-telecom, population, etc.).

- QuakeIST® contains algorithms for propagation effects and earthquake impact assessment.
- Losses maps and maps illustrating the cascade effects can be plotted for a given asset typology.
- The Disruption index can be presented for a city, a region or plotted in a geographic environment. This latest option is very important to share information to general public (people without a scientific background).

Earthquake insurance and compensation systems are important parts of strategies for dealing with seismic risks. They use sophisticated models to price earthquake risk. By using QuakeIST® with the DI calculator can assist in analyzing the damage correlation and interdependence damage propagation; DI can certainly contribute to the development of innovative earthquake insurance systems reducing some of the existing "blind spots" (http://insurance.about.com/).

9.3 New Advancements: Interdependences and Cascade Effects

9.3.1 Disruption Index

Where risk analysis looks at the impacts of catastrophic events, the analysis is generally restricted to the immediate effects and impacts rather than to identification of how economic processes generated the risk in the first place and how direct and indirect impacts then run through the economy affecting future development in diverse ways.

Damages and the magnitude of adverse impacts can be categorized as shown below:

- Direct losses: losses resulting from direct impact to buildings and infrastructures.
- Indirect losses: losses resulting from the event but not from its direct impact, for example, transport disruption, business losses that can't be made up, losses of family income, etc.

In both loss categories, there are two sub-categories:

- Tangible losses: loss of things that have a monetary (replacement) value, for example, buildings, livestock, infrastructure, etc.
- Intangible losses: loss of things that cannot be bought and sold, for example, lives and injuries, heritage, and others.

The larger the city, the greater its complexity and the potential for disruptions when facing an adverse event. For example, damage or non-functioning of infrastructure facilities also causes long-term impacts, such as disruptions to clean water

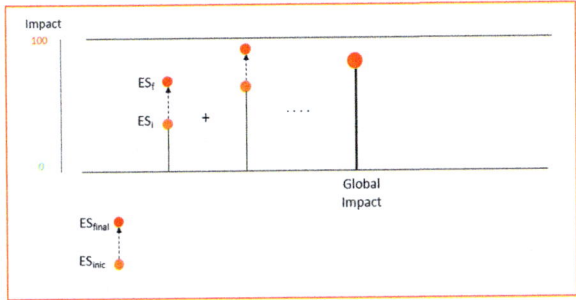

Fig. 9.6 Effect of interdependences in each sub-system (ES) and on the global impact

and electricity, deterioration of health condition owing to waterborne diseases. Loss of livelihoods, production and other prolonged economic impacts can trigger mass migration or population displacement.

The Disruption Index (DI) was constructed to quantify the state of disorder induced by the disruption of urban structure and its systemic functions. Figure 9.6 presents schematically the earthquake global impacts taking into account the various subsystems and interdependencies among them.

This general model considers a number of subsystems which deals with the allocation of activities and components and their interaction and interdependencies. Crucial to the modelling process of DI was capturing and analysing the systems dependencies and the chain of influences and effects that cross multiple systems (Ferreira 2012).

An urban area consists on complex, dynamic and highly interrelated systems. As mentioned significant loss of housing, education, power or other component would have substantial negative impacts. How would constrains in residential areas affect the residential distribution of the region? How would a general change in accessibility due to severe damage affect the population or the economy (employment changes)?

9.3.1.1 Structuring Disruption Index Model

When experimenting with urban systems, a first difficulty is to define what type of elements can be studied. A crucial part of the modelling process is to develop a general framework capable to clearly identify, capture and analyze each level of organization, the systems dependencies and the chain of influences and failures due to system/component interactions (Ferreira 2012; Ferreira et al. 2014).

In order to identify the most important effects on a society, its economy and other sectors, more than 70 "primary concerns" were found as systematically present in all texts and reports. Following some fundamental rules of decision problem structuring, these primary elements were aggregated in 14 Fundamental Criteria (Fig. 9.7) translating critical dimensions (urban functions) that cooperate in an interdependent fashion. Those dimensions encompass six fundamental human needs: "*Environment, Housing, Healthcare, Education, Employment* and *Food*",

	Urban system functional dependencies	1	2	3	4	5	6	7	8	9	10	11	12	13	14	15	16	17	18	19	20	21	22	23	24	25
		DI	Environment	Housing	Food	Healthcare	Education	Employment	Mobility	Power	Telecom	Transportation	Debris	Water	Sanitation	Security	Dangerous facilities	Electrictic facilities & components	Transportation facilities & components	Water facilities & components	Sanitation facilities & components	Telecom facilities & components	Schools	Health care facilities	Security facilities & components	Building stock
1	DI	1	1	1	1	1	1	1																		
2	Environment		-											1	1		1									
3	Housing			-					1	1				1	1											1
4	Food				-				1							1										
5	Healthcare					-			1	1	1			1	1									1		
6	Education						-		1	1	1			1	1								1			
7	Employment							-	1	1	1			1	1											1
8	Mobility											1	1													
9	Power																	1								
10	Telecom									1												1				
11	Transportation									1	1								1							
12	Debris																									1
13	Water									1										1						
14	Sanitation									1				1							1					
15	Security								1	1	1														1	

Fig. 9.7 Disruption index, the adjacency matrix A. In columns, we represent the graph elements. The square matrix contains the six criteria; the other *black rows* contain the services and components, and the right columns (*blue*) show the elements that supports all other functions (Ferreira 2012)

and are conditioned by several other main functions/systems such as mobility, electricity, water, telecoms and others, which in turn are dependent by the reliability of several buildings, equipments and critical or dangerous facilities. To give an example, from Fig. 9.7 we can say that the dependencies of Environment are Water, Sanitation and Dangerous facilities.

Water depends on the operation of the Water system equipment and of the Electricity supply, which depends in turn on the Electric system equipment, and we have a chain of dependencies and interdependencies.

Propagation and cascading effects can be calculated in a bottom – up sequence, starting with the physical damages directly suffered by the exposed assets (nodes with the lowest topological order), proceeding with the impacts that each node has in the functional performance of nodes that depends on them, until reaching the top node, DI (which is the one with higher topological order). Mathematically, the DI can be represented by its Adjacency Matrix of a Directed Graph [G], in which the element Gij equals 1 if row i depends on column j and is zero otherwise.

9.3.1.2 Impact Assessment

It is possible to associate qualitative impacts to each urban function and element (criteria), using a scale, describing as objective as possible all the plausible impacts that may presents.

Table 9.2 Criteria (*Human needs*) and respective consequence descriptors

Criteria	Descriptors
Environment	Identify materials or elements that can pose a substantial or potential hazard to human health or the environment when improperly managed: soil and water contamination, radiation, radioactive waste, oil spills, etc. It also assess the impact of service disruption of urban hygiene/public health from debris storage (building materials, personal property, and sediment from mudslides), contamination of water (unsafe drinking water and sanitation) and the high concentration of people in the same space
Housing	Evaluates whether a particular area may or may not be occupied for housing function as a result of the damage, also indicates alternative housing/shelter
Food	Evaluates if the food is accessible to the majority of the population and identifies alternatives to their supply (coping strategies)
Healthcare	Determines if the population is served by a sufficient number of health facilities
Education	Measures the discontinuity of education and the number of people without school lessons and identifies alternatives for recovery
Employment	Evaluates whether a certain area retains its activity as a result of the damage after the earthquake and identify new clusters of jobs that can be generated

Table 9.2, presents the descriptors associated with each criterion of *human needs*.

The impacts associated with a certain criterion are restricted to a range of plausible levels of impact (Roy 1985), from the more desirable level (normal or I) to a less desirable level (exceptional or IV–V). Taking into account the whole family of criteria, it is possible to define the overall response of the system, originating in the Disruption index, as the result of the interactions between the various systems (the results of sequencing actions are determined by individual actions). The values given for each criterion provide a single value to DI between I and V, a range of impacts of the earthquake in urban systems (Table 9.3).

It is worth noting that these levels have no *cardinal* meaning; these impact scales are only *ordinal* (neither interval nor ratio scales). For example, we can say that impact V is greater than impact IV and that impact III is greater than impact I but, we cannot say that impact IV is twice impact II nor that the difference between impacts IV and III is α times the difference between impact III and impact II.

Each level of DI conveys which are the disruptions and influences (physical, functional, social, economic and environmental) that a given geographic area is subjected when exposed to an adverse event. The enumeration of impact levels of each sub-system is provided in Table 9.4. Using the aforementioned DI, QuakeIST® can also compute impact and plot the respective maps. This is the first time that all the components for impact assessment are integrated and work seamlessly in just one software platform.

Table 9.3 Qualitative descriptor of disruption index, DI (impact levels are numbered in decreasing order of urban disruption/dysfunction)

Impact level	Description of the impact level
V	From serious disruption at physical and functional level to paralysis of the entire system: buildings, population, infrastructure, health, mobility, administrative and political structures, among others. Lack of conditions for the exercise of the functions and activities of daily life. High cost for recover
IV	Starts the paralysis of main buildings, housing, administrative and political systems. The region affected by the disaster presents moderate damage and a slice percentage of total collapse of buildings, as well as victims and injuries and a considerable number of homeless because their houses have been damaged, which, although not collapse, are enough to lose its function of housing. Normal daily activities are disrupted; school activities are suspended; economic activities are at a stand-still
III	Part of the population may permanently lose their property and need to permanent be relocated, which means strong disturbances of everyday life. This level is determined by significant dysfunction in terms of equipment's, critical infrastructures and losses of some assets and certain disorders involving the conduct of professional activities for some time. The most affected areas show significant problems in mobility due to the existence of debris or damage to the road network. Starts significant problems in providing food and water, which must be ensured by the Civil Protection
II	The region affected by the disaster presents few homeless (about 5 %) due to the occurrence of some damage to buildings, affecting the habitability of a given geographical area. Some people may experience problems of access to water, electricity and/or gas. Some cases require temporary relocation
I	The region affected by the disaster continues with their normal functions. No injured, killed or displaced people are registered. Some light damage may occur (non-structural damage) that can be repaired in a short time and sometimes exists a temporary service interruption. The political process begins with an awareness that the problem exists as well as some investments in strengthening policy and risk mitigation is/should be made

9.3.1.3 DI Application: Portugal

After the briefly description of Disruption index we are able to assess and calculate the earthquake impacts in a holistic approach. The results here presented highlight the potential importance of incorporating dependencies and cascading failures into such models. DI provides the basis for understanding the resource requirements, not only for recovery after events but also to identify, prior to events, the physical elements that contribute most to severe disruptions.

1755 Earthquake Scenario (M 8.7): Algarve Region in Portugal

On November 1st of 1755, three very large earthquakes, centred southwest of the Algarve region (southern Continental Portugal), devastated Algarve and Lisbon regions and was felt throughout Europe and North Africa. Hundreds of aftershocks,

Table 9.4 Table of impact

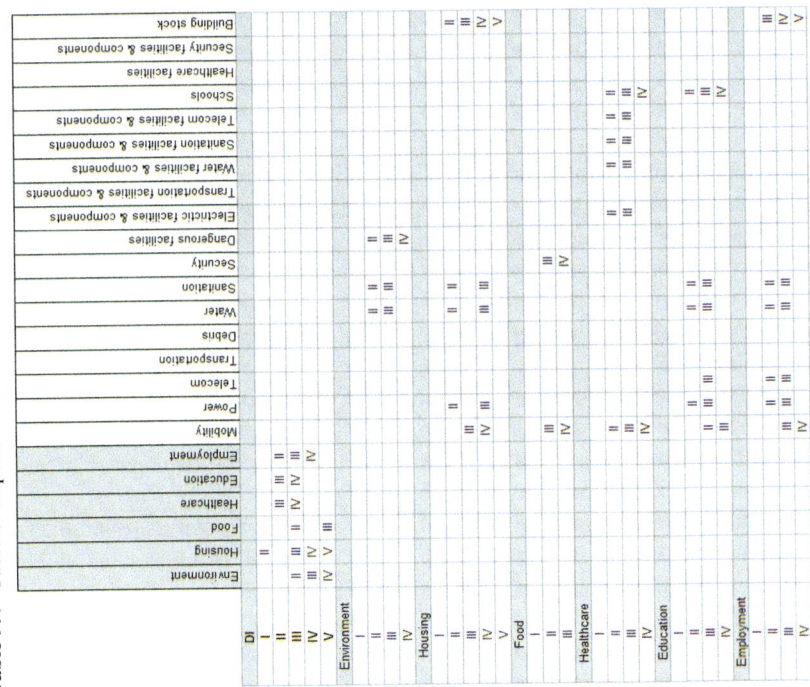

Fig. 9.8 QuakeIST® intensity distribution for 1755 earthquake scenario

some severely damaging by themselves, continued for years. A devastating fire following the earthquake destroyed a large part of Lisbon, and a very strong tsunami caused heavy destruction along the coasts of Portugal, southwest Spain, and western Morocco (Oliveira 2008).

The Algarve region was selected to demonstrate the regional impact assessment. QuakeIST® software contains detailed information on the geological surface layers, on the building inventory and on population data of the Census (INE 2002), using the statistical sub-section (Census track) as work unit. Soil influence was included through the analysis of upper soil layers classified into several categories; and vulnerability of the building stock was obtained through the analysis of different classes of construction types (55 classes in total). Finally, a pair of coordinates (longitude and latitude) was provided to define the location of each asset (ERSTA 2010).

A vulnerability index was assigned to each typology using the approach of EMS-98 scale based method to calculate expected damages in buildings. The first level of analysis of the QuakeIST® software is based on obtaining intensity (or PGA) distributions analytically (Fig. 9.8) and estimating spatial distribution of the losses (building and lifeline damages) throughout the region of interest. Second level of analysis is intended to propagate effects and earthquake impacts, using DI (Figs. 9.7 and 9.9).

The next figures illustrate how all the referred concepts should be applied and interpreted in our case study areas. Figure 9.10 shows the mean damage grade obtained for each census tract, and Fig. 9.11 illustrates the damages inflicted to bridges and the extend of their sphere of influence.

The obtained results indicate that if we gather together the debris (obtained from the building stock) and the bridges damages, we obtain an impact on Mobility, according to DI methodology (Oliveira et al. 2012; Ferreira et al. 2014) equal to II and III (Fig. 9.12). Mobility equal to III means "Local perturbation on mobility linked with landslide or major damages. Used only by recovery teams. Disruptions to commuting trips, work and nonworking trips" (Ferreira et al. 2014).

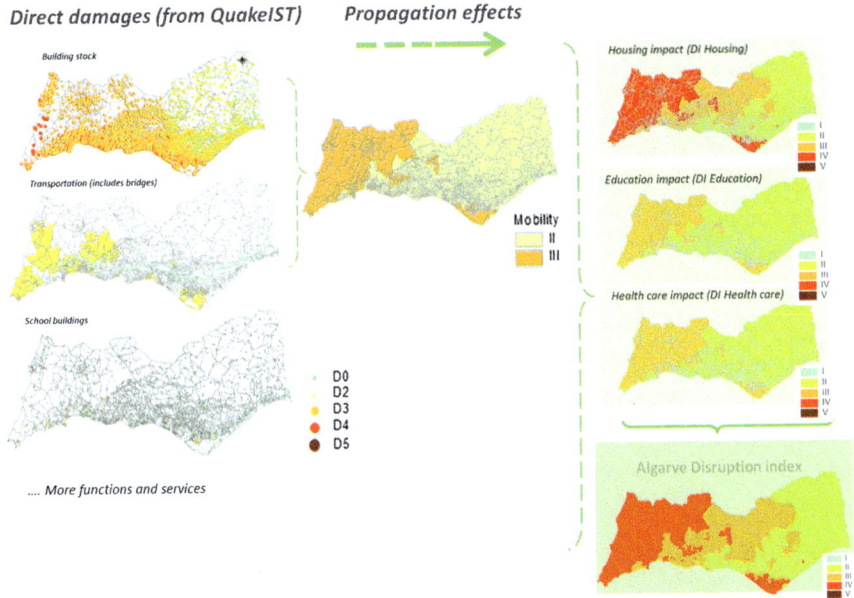

Fig. 9.9 Disruption index: earthquake impact based on the systemic analysis of the urban components

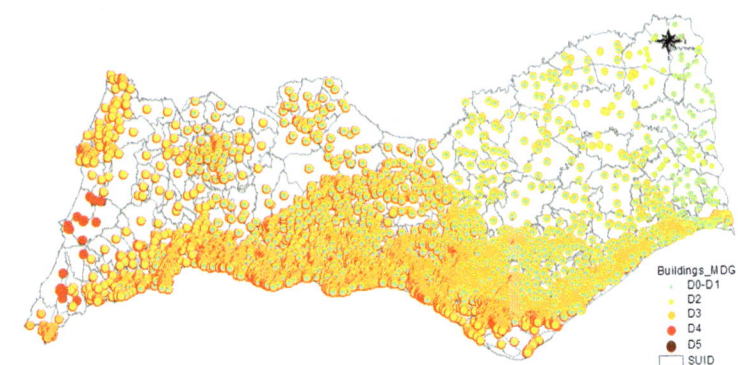

Fig. 9.10 Distribution of all damaged buildings

The expected school damages associate with the 1755 risk scenario is presented in Fig. 9.13. As shown, most of the school buildings are not affected or present at maximum "moderate damages".

Each impact level is correlated with a severity or grade of damage to either the equipment or function connected with the Education function (Fig. 9.14). By combining the conditions using the logical function OR, we are able to categorize and plot the impact level on Education system (Figs. 9.14 and 9.15).

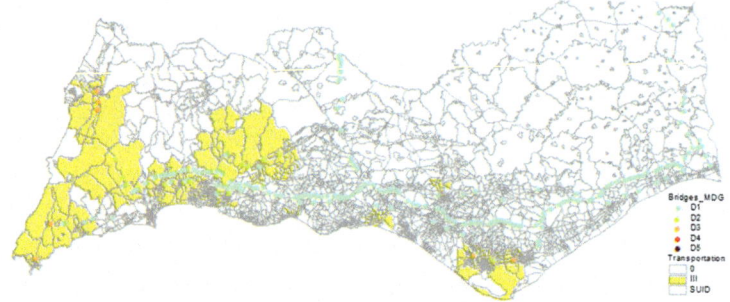

Fig. 9.11 Intensity-based distribution of all bridges damaged

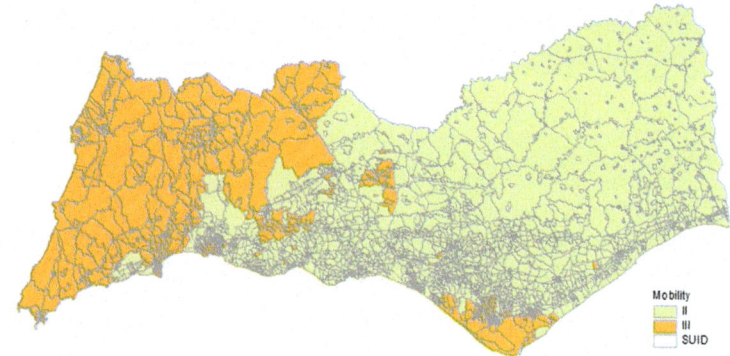

Fig. 9.12 Impact on mobility

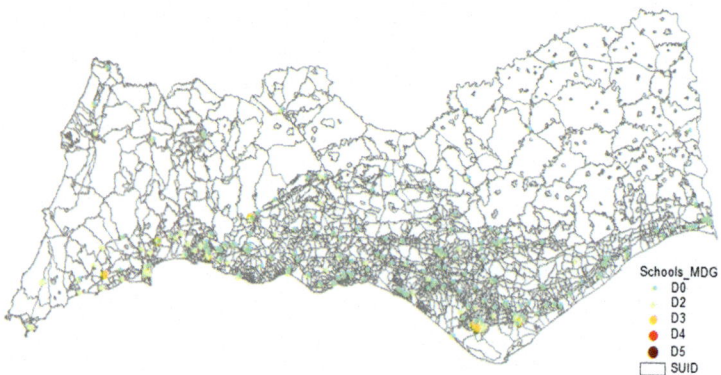

Fig. 9.13 Direct damages obtained from QuakeIST® – School buildings

The extent of damage to schools and problems on mobility, and the ensuing relocation of people (due to buildings damages), means we cannot restore the education network to its previous state. Figure 9.15 suggests that in this region

9 Earthquake Risk Reduction: From Scenario Simulators Including Systemic...

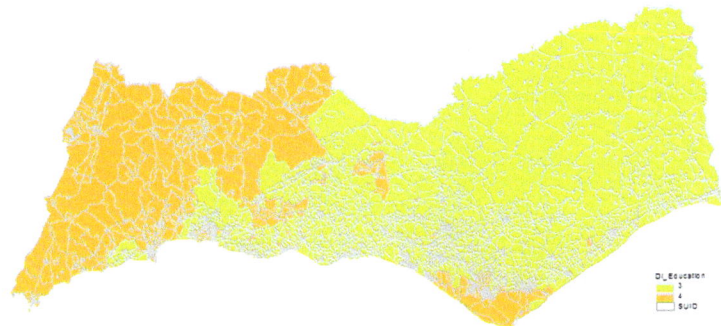

Fig. 9.14 Impact on education

Impact level	Impact descriptor *Mede a descontinuidade do ensino, o número de pessoas privadas de ensino e identifica alternativas para o retomo do mesmo*	Educational facilities		Mobility		Power supply		Telecom supply		Water supply		Sanitation supply
IV	There would be educational facilities with severe damage or collapse. Disruption of educational continuity, schools inaccessible for long periods. Students are relocated to other areas of the country. Families sometimes are not able to carry the burden of fees because of	>II	OR	>II		·		·				
III	Difficult access to education. There would be educational facilities with severe damage or collapse or restricted access due to debris. Teachers could not access, materials have been destroyed. Necessary temporary relocation or share their site with another school, until completion of rehabilitation works.	>II	OR	>I	OR	>II	OR	>II	OR	>II	OR	>II
II	Momentary disruption with resumption of classes after inspection and assessment of security conditions (weeks).	>I	OR	·	OR	>I	OR	·	OR	>I	OR	>I
I	No significant impact on function.	·		·		·		·		·		·

Fig. 9.15 Education disruption

Fig. 9.16 Direct damages obtained from QuakeIST® – Healthcare buildings

students will experience a prolonged interruption in their education and large numbers of families with school-age children will be forced to relocate either temporarily or permanently as a result of the earthquake.

In terms of physical damage to hospitals and primary health centres, Fig. 9.16 illustrates that were minor damage (D2) and one building with moderate damage (D3). However, the adverse impacts on healthcare system take a large proportion

Fig. 9.17 Impact on healthcare

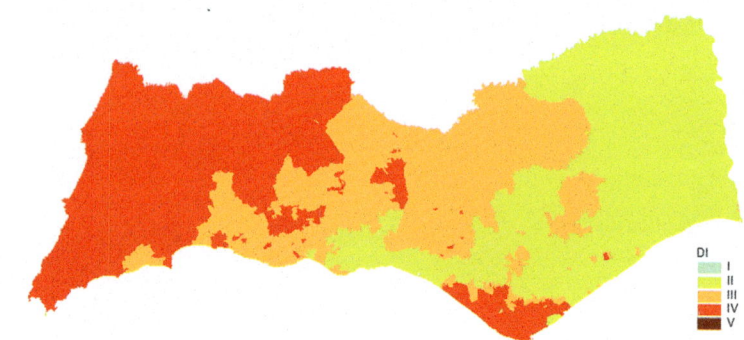

Fig. 9.18 Global disruption in Algarve region

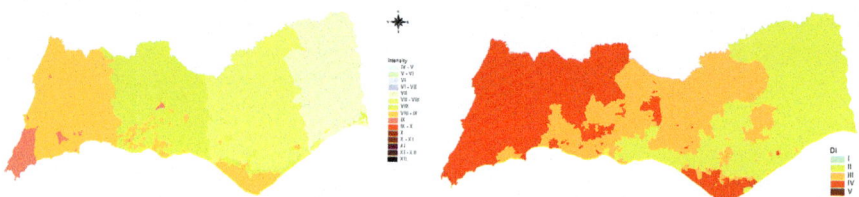

Fig. 9.19 Comparison between Intensity map (*left* – Fig. 9.8) and DI map (*right* – Fig. 9.18)

due to propagation effects in other important lifelines like power and water systems, and due to the problems on mobility.

As seen on Fig. 9.17 propagation effects severely disrupt the functioning of the health system, being unable to provide emergency services in the region. These impacts may be short- or long-term (DI equal to II or III, respectively), based on the magnitude of the damage to the community and the ability of local resources to readily address and meet the healthcare needs of the community.

It is important to notice that despite high exposure and vulnerability of building and facilities to earthquakes, the propagation effects and the number of chain

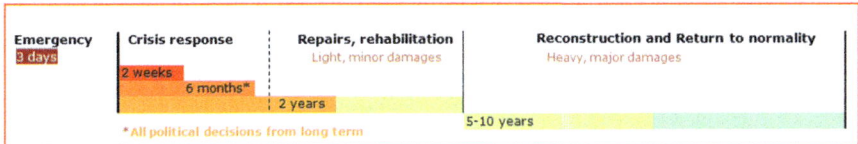

Fig. 9.20 Crises evolution in time

disruptions must be underlined in risk scenarios studies. According to Fig. 9.18, the 1755 Earthquake (not including the possible tsunami) has potential to cause disruption on infrastructure and production capacity of entire Algarve region and consequently to the national level.

Figure 9.19 compares the maps of intensity and of DI if a scenario similar to the 1755 earthquake would happen today, emphasizing the importance of including interdependencies and cascading effects in the analysis of earthquake impact.

From the above maps it is important to highlight that propagation effects due to interdependencies, largely extend the geographical scope and amplify the degree of earthquake impacts (measured by DI). As so, we can expect that zones where macroseismic intensity is low or even very low can be subjected to large disruptions. This situation happens when, for example, a pipeline feeding a region is broken in a section away from it.

The combination of this seismically active area, dense population centres, and aging or fragile infrastructure has the potential to create a massive catastrophe for urban activities (education, business, and so on) located in Algarve region. Loss of life and property damage are the first and foremost concerns for businesses, but the ripple effects of a major seismic event, including business and educational interruption as well as supply chain disruption, could take months or even years of recover.

Looking at the time component, all post-earthquake activities occur in three major phases – during response, recovery and reconstruction – as shown in Fig. 9.20. The DI concept can show the time evolution of decreasing or increasing disruptions according to decisions and reconstruction policies.

The DI concept although developed for a given deterministic seismic scenario can be extended to a set of scenarios representing the seismic activity in the region (hazard, de-aggregation, etc.).

9.4 Final Remarks

Living with earthquake risk is a devastating reality for a large and growing number of people in the world. Risk should be seen as a normal and inseparable part of economic activities and development. The construction of disaster risk reduction as an autonomous sector, concerned with protecting economic sectors and society from the impact of exogenous and extreme shocks has isolated it from the mainstream concerns of government in general, including economic growth and employment, or in the case of local governments, water and power supplies, transport and waste management. The lack of real political and economic commitment to disaster risk reduction in many countries reflects its isolation from real political and economic imperatives. This requires awareness of the impact on sectors or territories of any other given sector's policies and/or changes in strategy. As so an important issue should be highlighted; is important to identify communities at direct risk but also those indirectly affected.

The concept of disruption index presented herein can be extended to other earthquake-induced phenomena such as landslides, mudflows, tsunamis, liquefaction and fires, and to other natural or man-made hazards such as typhoons (Ferreira 2012), avalanche, floods and so on.

Fire following earthquake is a significant problem in seismic countries and many people are not aware of this hazard (urban and industrial) resulting from earthquakes and tsunamis. These fires can be classified in "earthquake-induced fires" caused directly by the earthquake, such as fires in oil and gas tanks or in urban areas, and in "tsunami-induced fires", caused by ignition of buildings by burning buildings or debris carried by the tsunami, for example, as observed on 2011 Tohoku earthquake (Yamada et al. 2012) or in the 1755 earthquake (Oliveira 2008).

In the case of insurance policies the various hazards should be taken into account, not only hazard by hazard but also considering the interdependencies among them. Both in urban tissues and in industrial areas the interdependencies are very critical.

The key messages from this work are:

– earthquakes are having a major impact on millions of people every year and therefore earthquake risk management measures need to be implemented in the short term;
– failure to enforce and implement appropriate measures could increase the impact of earthquake events and undermine the resilience of a system;
– promote a risk management approach in dealing with earthquakes, including prevention, mitigation and response;
– continuous communication to raise awareness and reinforce preparedness is necessary.

"Risk communication is successful only if it adequately informs the decision maker" (US Food and Drug Administration 2009). The DI methodology provides useful information in risk perception and risk communication as well as in developing strategies to reduce the consequences of earthquakes and benefits of a decision. This concept offers a comprehensive description of real observed scenarios and permits: (i) to identify the urban system and critical services or elements; (ii) to rank the order of priority of services or elements for continuous operations or rapid recovery; and (iii) to identify internal and external impacts of disruptions. This approach can be also extended to other natural and man-made disasters, and may be used as a tool for optimization resources of system components.

Decisions regarding earthquake risk management are complex and require wide participation and a clear vision of the alternatives from technical personnel and non-specialists. There are now many methods to assist them in making choices. The most popular focuses on evaluating costs and benefits in monetary terms using Cost Benefit Analysis (CBA). However, city managers, urban planners and risk professionals must take a broader view and consider multiple aspects – some of which cannot be quantified. This need can be addressed by the use of Multi-Criteria Analysis (MCA). These decision and risk approaches, capable to use only ordinal scales are very important in order to avoid the endless discussions about the relative weights and utility functions that are the standard procedure used nowadays to assign tangible and intangible values, which may have to be considered in the evaluation of consequences.

Finally, there are many reasons that may result in the priority of earthquake risk management being ignored in favor of more immediate demands. There are financial, practical and psychological factors that come into play here, including the common perception that earthquakes will not happen.

Acknowledgements The preparation of this paper was supported in part by FCT PhD grant SFRH/BD/71198/2010 (Francisco Mota de Sá) and was co-financed by the EU—Civil Protection Financial Instrument in the framework of the European Project "Urban disaster Prevention Strategies using MAcroseismic Fields and FAult Sources" (UPStrat-MAFA-Num.230301/2011/613486/SUB/A5), DG ECHO Unit A5. The authors acknowledge to Instituto de Engenharia de Estruturas, Território e Construção (ICIST), research unit of Instituto Superior Técnico.

Open Access This chapter is distributed under the terms of the Creative Commons Attribution Noncommercial License, which permits any noncommercial use, distribution, and reproduction in any medium, provided the original author(s) and source are credited.

References

D'Amico V, Albarello D (2008) SASHA: a computer program to assess seismic hazard from intensity data. Seismol Res Lett 79(5):663–671

EC-8 (2004) Part 1. General rules, seismic actions and rules for buildings. CEN, European Committee for Standardization

EMSC (2014) www.emsc-csem.org (consulted in Jan 2014)

ERSTA (2010) Estudo do risco sísmico e de tsunamis do Algarve, Autoridade Nacional de Protecção Civil (ANPC), Lisboa. (in portuguese)

Fajfar P (1999) Capacity spectrum method based on inelastic demand spectra. Earthq Eng Struct Dyn 28:979–993

Ferreira MA (2012) Risco sísmico em sistemas urbanos. PhD thesis, Instituto Superior Técnico, Universidade Técnica de Lisboa, 295 pp (in portuguese). http://insurance.about.com/ (consulted in Feb 2014)

Ferreira MA, Mota de Sá F, Oliveira CS (2014) Disruption Index, DI: an approach for assessing seismic risk in urban systems (theoretical aspects). Bull Earthq Eng. doi:10.1007/s10518-013-9578-5

Freeman SA (2004) Review of the development of the capacity spectrum method. ISET J Earthq Technol, Paper No. 438, 41(1):1–13

Giovinazzi S, Lagomarsino S (2004) A macroseismic method for the vulnerability assessment of buildings. In: Proceedings, 13th world conference on earthquake engineering. Vancouver, Canada, 1–6 August. Paper no. 896

INE (2002) Instituto Nacional de Estatística, Censos 2001 – XIV Recenseamentos Geral da População – XIV Recenseamento Geral da Habitação, Lisboa, Portugal

Karklis L (2010) Global seismic hazard assessment program, United Nations Population Division, Laris Karklis/The Washington Post – 23 February 2010

Oliveira CS (2008) Review of the 1755 Lisbon earthquake based on recent analyses of historical observations. In: Frechet J et al (eds) Historical seismology. Springer Science+Business Media B.V, Dordrecht

Oliveira CS, Roca A, Goula X (2006) Chapter 1: An introduction. In: Oliveira CS, Roca A, Goula X (eds) Assessing and managing earthquake risk, vol 2, Geotechnical, geological and earthquake engineering. Springer, Dordrecht, pp 1–14

Oliveira CS, Ferreira MA, Mota de Sá F (2012) The concept of a disruption index: application to the overall impact of the July 9, 1998 Faial earthquake (Azores islands). Bull Earthquake Eng 10(1):7–25. doi:10.1007/s10518-011-9333-8

Rotondi R, Zonno G (2010) Guidelines to use the software PROSCEN. Open Archives Earthprints Repository, INGV, Reports. http://hdl.handle.net/2122/6726

Roy B (1985) Méthodologie multicritère d'aide à la décision. Economica, Paris

Silva V, Crowley H, Pagani M, Monelli D, Pinho R (2013) Development of the OpenQuake engine, the Global Earthquake Model's open-source software for seismic risk assessment. Nat Hazards. doi:10.1007/s11069-013-0618-x

UN-HABITAT (2008) State of the world's cities 2008/2009: harmonious cities. Earthscan, London/Sterling

UPStrat-MAFA (2012) Urban disaster prevention strategies using MAcroseismic Fields and FAult Sources (UPStrat-MAFA – EU Project Num. 230301/2011/613486/SUB/A5), DG ECHO Unit A5

US Food and Drug Administration (2009) Strategic plan for risk communication. http://www.fda.gov/downloads/AboutFDA/ReportsManualsForms/Reports/UCM183683.pdf

USGS (2012) www.usgs.org (consulted Dec 2012)

WDR (World Development Report) (2010) Development and climate change. World Bank, Washington, DC

Yamada T, Hiroi U, Sakamoto N (2012) Aspects of fire occurrences by tsunami. In: Proceedings of the international symposium on engineering lessons learned from the 2011 Great East Japan Earthquake, 1–4 March 2012, Tokyo, Japan

Chapter 10
Physics-Based Earthquake Ground Shaking Scenarios in Large Urban Areas

Roberto Paolucci, Ilario Mazzieri, Chiara Smerzini, and Marco Stupazzini

Abstract With the ongoing progress of computing power made available not only by large supercomputer facilities but also by relatively common workstations and desktops, physics-based source-to-site 3D numerical simulations of seismic ground motion will likely become the leading and most reliable tool to construct ground shaking scenarios from future earthquakes. This paper aims at providing an overview of recent progress on this subject, by taking advantage of the experience gained during a recent research contract between Politecnico di Milano, Italy, and Munich RE, Germany, with the objective to construct ground shaking scenarios from hypothetical earthquakes in large urban areas worldwide. Within this contract, the SPEED computer code was developed, based on a spectral element formulation enhanced by the Discontinuous Galerkin approach to treat non-conforming meshes. After illustrating the SPEED code, different case studies are overviewed, while the construction of shaking scenarios in the Po river Plain, Italy, is considered in more detail. Referring, in fact, to this case study, the comparison with strong motion records allows one to derive some interesting considerations on the pros and on the present limitations of such approach.

R. Paolucci (✉)
Department of Civil and Environmental Engineering, Politecnico di Milano, P.za L. da Vinci 32, 20133 Milano, Italy
e-mail: roberto.paolucci@polimi.it

I. Mazzieri
Department of Mathematics, Politecnico di Milano, P.za L. da Vinci 32, 20133 Milano, Italy

C. Smerzini
D'Appolonia S.p.A, Via S. Nazaro 19, 16145 Genova, Italy

M. Stupazzini
MunichRE, Munich, Germany

10.1 Introduction

Tools for earthquake ground motion prediction (EGMP) are one of the key ingredients in seismic hazard analysis, both within probabilistic and deterministic frameworks, with the seminal objective to provide estimates of the expected ground motion at a site, given an earthquake of known magnitude, distance, faulting style, etc. A variety of procedures for EGMP has been proposed in the past four or five decades (Fig. 10.1), relying, on one side, on different information detail on the seismic source and propagation path, and, on the other side, providing different levels of output, either in terms of peak values of ground motion or of an entire time history. The level itself of complexity of the proposed procedures ranges from the empirical ground motion prediction equations, typically calibrated on the instrumental observations from real earthquakes, up to complex 3D numerical models, involving as a whole the system including source - propagation path – shallow soil layers. A comprehensive review of techniques for EGMP was recently published by Douglas and Aochi (2008).

In the absence of suitable and performing numerical tools for physics-based modelling of source and path effects, research has been mainly directed in the past towards statistical processing of available records to provide empirical equations for EGMP. A recent compilation by Douglas (2011) has reported about 300 such equations to estimate peak ground acceleration (PGA) since 1964, and about 200 to estimate the response spectral ordinates. More recently, the ever increasing availability of high-quality records throughout the world, coupled with the improvement of the meta-files associated to the strong motion databases, has stimulated a further development of empirical tools for EGMP, both in the United States with the NGA West2 project (Boore et al. 2013) and in Europe with the calibration of updated pan-European ground motion prediction equations (Douglas et al. 2014).

Still, in spite of such a substantial effort, empirical ground motion prediction equations suffer of intrinsic limitations, such as: (1) the available records hardly cover the range of major potential interest for engineering applications (see Fig. 10.2), with relatively few records available in the near-field of large earthquakes; (2) they refer to generic site conditions, in the best cases represented in terms of $V_{S,30}$; (3) they only provide peak values of ground motion, without the entire time history, which would be instead of major relevance in terms of input motion for engineering applications; (4) they are not suitable to be used for seismic scenario studies where the realistic representation of spatial variability of ground motion is an issue.

Physics-based numerical simulations of earthquake ground motion are often advocated as an alternative tool to cope with the previous limitations, since they provide, according to different methodologies, synthetic ground motion time histories compatible with a more or less detailed model of the seismic source process, of the propagation path, and of the local site response. Deterministic approaches rely on the rigorous numerical solution of the seismic wave propagation problem, based on detailed 3D models both of the seismic source and of the source-to-site

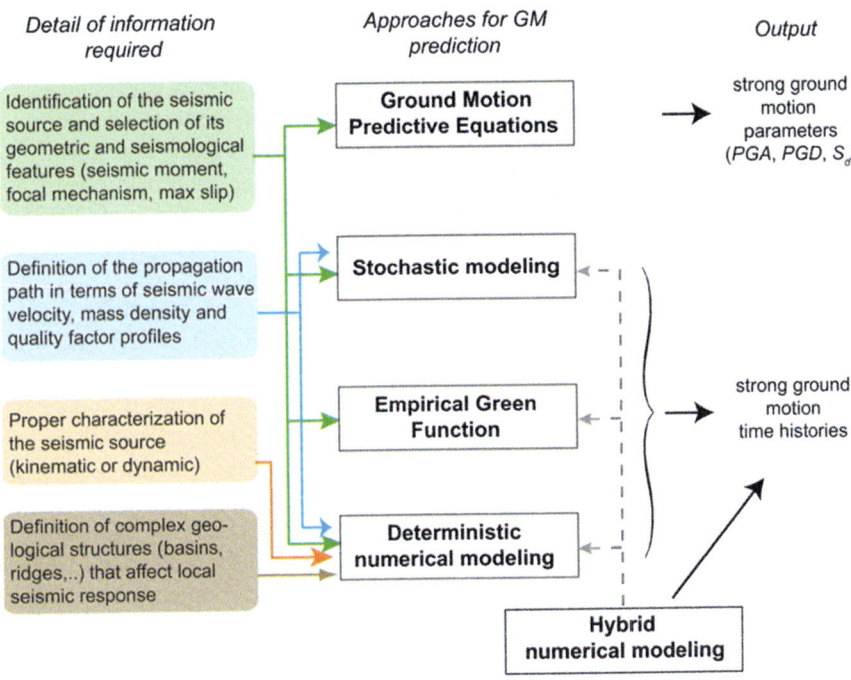

Fig. 10.1 Overview of approaches for earthquake ground motion prediction

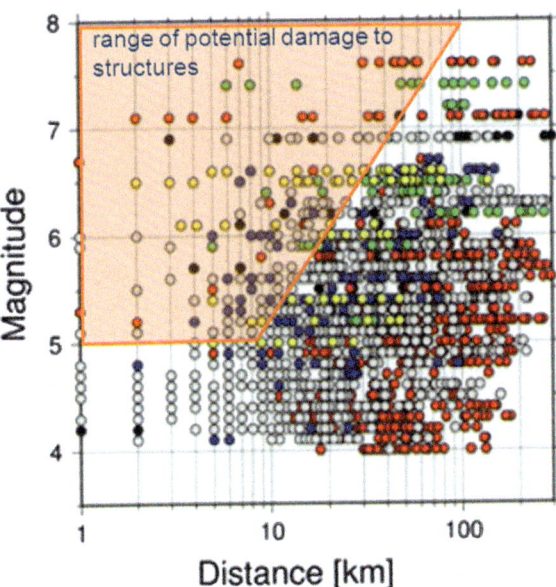

Fig. 10.2 Magnitude and distance range covered by the strong motion database for calibration of pan-European ground motion prediction equations. Records are colour coded according to the network: *red* (Turkey); *gray* (Italy); *blue* (Greece); *green* (Iran); *yellow* (Iceland); *black* (other countries) (Adapted after Bindi et al. 2014)

propagation path. However, limited by the large computational requirements on one side, and, on the other side, by the insufficient information on the local seismic source features and on the local geology, the reliability range of such numerical solutions is most often limited to 1 or 2 Hz. For this reason, the frequency range of the numerical simulations is often enlarged to produce broadband waveforms, by considering hybrid approaches where high-frequency source and path effects are either modelled by stochastic or semi-stochastic processes (Seyhan et al. 2013) or random processes are introduced within a deterministic model to provide a realistic frequency-dependent spatial incoherency of ground motion.

The *dream* behind physics-based numerical simulations of earthquake ground motion is that they may become the engine to produce, effectively and with reasonable computing efforts, plausible realizations of future earthquakes. This is for example the idea behind the *ShakeOut* experiment in California, where the physics-based simulations of a hypothetical $M_W 7.8$ earthquake on the Southern San Andreas Fault (Porter et al. 2011) were the basis to construct a comprehensive earthquake risk scenario including costs evaluations and planning of emergency response activities.

The need for such advanced tools for EGMP was made clear by the consequences of the series of earthquakes from 2010 to 2012, started with Haiti in January 2010, followed by Chile in February 2010, by the Canterbury earthquake series in New Zealand in 2010–2011, by the gigantic Tohoku earthquake in Japan in March 2011, up to the Emilia, Italy, earthquakes of May 2012. All of them illustrating, in different terms and different scales, the increasing loss potential of seismic disasters. As a matter of fact, losses in the two-digit billion dollar range have become a reality, even outside the leading industrialized countries, and nowadays a much higher fraction of these losses is insured than in the past. Before the 2010 Chile earthquake, Santiago was last time affected by the 1985 Valparaiso M8 earthquake. Whereas the total economic loss in 2010 was about 25 times higher than 1985, the insured loss increased by a factor of 100. Furthermore, comparing the 1995 Kobe and the 2011 Tohoku earthquakes, the loss statistics shows a factor 3 increase for the economic loss, and a factor 13 for the insured loss.

Therefore, these recent disasters stimulated a re-thinking of several aspects of natural disaster risk management, which has not yet produced final conclusions, but shattered what may be called a false sense of security or complacency about how to assess and manage risk, including identification, evaluation, control and financing.

In the perspective of improving tools for seismic hazard identification, Munich RE funded a research activity with Politecnico di Milano, having the main objectives, on one side, of developing a certified computer code to run effectively numerical simulations of seismic wave propagation in large-scale models within high-performance computing architectures, and, on the other side, of applying this code to produce preliminary sets of physics-based earthquake ground shaking scenarios within large urban areas. This paper provides an overview of the progress within this research activity.

10.2 Numerical Approaches for Physics-Based Earthquake Ground Shaking Scenarios

Stimulated by the ever increasing power of large parallel computer architectures, numerical codes for seismic wave propagation have considerably evolved in the last decade and are presently becoming an appealing alternative to produce reliable physics-based earthquake ground motion scenarios in the presence of realistic 3D configurations of seismic source, complex basin structures and topographic features. Two major experiments of verification of such numerical codes were conducted in the second half of the last decade, namely within the ShakeOut (Bielak et al. 2010) and the Grenoble (Chaljub et al. 2010) benchmarks, while a further experiment is in progress (E2VP) based on the Euroseistest configuration (Chaljub et al. 2013).

Relatively few numerical codes exist for this purpose, mostly belonging to the classical finite difference (e.g., Graves 1996) and finite element (e.g., Bielak et al. 2005) schemes, while spectral element methods (e.g., Faccioli et al. 1997; Komatitsch and Vilotte 1998) have emerged subsequently as an alternative powerful technique, relying on a right balance between accuracy, ease of implementation and parallel efficiency. It is not surprising that three open source codes recently made available belong to the SE family. Namely, these are SPECFEM3D,[1] EFISPEC[2] and SPEED,[3] the latter one being illustrated in the next section.

As a matter of fact, considering Table 10.1 which illustrates an overview of recent studies to produce physics-based earthquake ground shaking scenarios in large urban areas, most numerical methods included in this selection belong to the previous FD, FE or SE classes. Table 10.1 addresses as well further important issues of particular relevance:

- model sizes are very large, typically extending up to few hundreds of km size and few tens of km depth;
- the maximum frequency propagated, f_{max}, is only very seldom exceeding 1 Hz. However, even with such frequency limitation, the number of nodes of the numerical meshes exceeds as a rule 10 millions, implying a huge requirement in terms of computer time and memory requirement, so that these numerical simulations are typically carried out in parallel computer architectures;
- as we move to recent years, there is an increasing trend in terms of number of simulations per case study, clearly showing that the computing power is presently opening this world to a much wider set of applications, including parametric analyses and production of large series of scenarios.

---

[1] www.geodynamics.org/cig/software/specfem3d

[2] efispec.free.fr

[3] mox.polimi.it/it/progetti/speed

Table 10.1 Selection of recent studies to produce physics-based earthquake ground shaking scenarios in large urban areas

Reference	Study area	Method	Model Size L × W × H [km³]	f_{max} [Hz]	No. nodes (× 10^6)	No. of Sim.
Wald and Graves (1998)	Los Angeles, US	FD	232 × 112 × 44	0.5	17	3
Pitarka et al. (1998)	Kobe, Japan	FD	60 × 10 × 22	0.8	N/A	1
Stidham et al. (1999)	San Francisco bay, US	FD	200 × 100 × 36	0.5	46	4
Kim et al. (2003)	Los Angeles, US	FE	80 × 80 × 30	1	100	1
Aochi and Madariaga (2003)	Izmit, Turkey	BE-FD	N/A	1	N/A	5
Aagaard et al. (2004)[a]	Taiwan	FE	160 × 80 × 40	0.5	N/A	10
Benites and Olsen (2005)	Wellington, New Zealand	FD	30 × 10 × 20	1.5	66	2
Asano et al. (2005)	Denali fault sys, Alaska	FD	480 × 360 × 42	0.5	34	1
Ewald et al. (2006)	Lower Rhine, Germany	FD	140 × 140 × 30	1	417	4
Olsen et al. (2006, 2008)[a]	Southern California, US	FD	600 × 300 × 80	0.5	1,800	3
Furumura and Hayakawa (2007)	Kanto basin, Japan	PSF+FD	440 × 250 × 160	1	N/A	1
Day et al. (2008)	Southern California, US	FD-FE	100 × 100 × 30	0.5	N/A	60
Lee et al. (2008)[b]	Taipei, Taiwan	SE	102 × 88 × 106	1	297	10
Koketsu et al. (2009)	Kanto basin, Japan	FD	154 × 154 × 38	0.5	N/A	1
Bielak et al. (2010)	Southern California, US	FD/FE	500÷600 × 250÷300 × 50÷84	0.5	78.5 ÷ 180.6	3
Stupazzini et al. (2009)	Grenoble, France	SE	40.7 × 50 × 8	2	5.7	18
Gallovič et al. (2010)	Parkfield, US	FV/ADER-DG	100 × 60 × 28	1	N/A	6
Graves et al. (2011)	Southern California, US	FD	N/A	0.5	1,500	840,000
Villani et al. (2014)	Sulmona, Italy	SE	48 × 40 × 15	2	7.7	91
Smerzini et al. (2011)	Gubbio, Italy	SE	62 × 85 × 10	2.5	23.5	3
Smerzini and Villani (2012)	L'Aquila, Italy	SE	62 × 63 × 17.7	2.5	10.5	32
Roten et al. (2011, 2012)	Salt Lake basin, USA	FD	60 × 45 × 30	1	1,266	6

Reference	Region	Method	Dimensions (km)	Col4	Col5	Col6
Molnar et al. (2014)	Vancouver region, Canada	FD	$150 \times 180 \times 25$	0.5	N/A	8
This study	Santiago	SE	$97.4 \times 76.6 \times 19$	2	31.05	19
	Po Plain	SE	$74 \div 89 \times 51 \div 62 \times 20$	1.5	$50.85 \div 6213$	23
	Christchurch	SE	$45 \times 45 \times 20$	2	$25.22 \div 311$	3
	Wellington	SE	$80 \times 50 \times 45$	1.5	15	50

Kaser and Dumbser (2006)
[a] Dynamic rupture model
[b] Point source model
FD finite differences, *FE* finite elements, *SE* spectral elements, *BE* boundary elements, *PSF* Pseudo-spectral Fourier, *FV* finite volume, *ADER-DG* High-order Discontinuous Galerkin

10.3 SPEED: SPectral Elements in Elastodynamics with Discontinuous Galerkin

10.3.1 Development of the Numerical Code

In the framework of the joint research activity between Munich RE and Politecnico di Milano, the SPEED code (SPectral Elements in Elastodynamics with Discontinuous Galerkin) was developed, as an open-source numerical code suitable to address the general problem of elastodynamics in arbitrarily complex media (Mazzieri et al. 2013). SPEED is designed for the simulation of large-scale seismic wave propagation problems including the coupled effects of a seismic fault rupture, the propagation path through Earth's layers, localized geological irregularities such as alluvial basins and topographic irregularities. Some examples of applications with the additional presence of extended structures, such as railway viaducts, can be found in the SPEED web site.

Treating numerical problems with such a wide range of spatial dimensions is allowed by a non-conforming mesh strategy implemented through a Discontinuous Galerkin (DG) approach (Antonietti et al. 2012). More specifically, the numerical algorithm can be summarized in the following steps (Fig. 10.3): consider an elastic heterogeneous 3D medium, (i) make a partition of the computational domain based on the involved materials and/or structures to be simulated, (ii) select a suitable spectral-element discretization in each non-overlapping sub-region, and (iii) enforce the continuity of the numerical solution at the internal interfaces by treating the jumps of the displacements through a suitable DG algorithm of the interior penalty type (De Basabe et al. 2008).

SPEED allows one to use non-conforming meshes (h-adaptivity) and different polynomial approximation degrees (N-adaptivity) in the numerical model. This makes mesh design more flexible (since grid elements do not have to match across interfaces) and permits to select the best-fitted discretization parameters in each subregion, while controlling the overall accuracy of the approximation. More specifically, the numerical mesh may consist of smaller elements and low-order polynomials where wave speeds are slowest, and of larger elements and high-order polynomial where wave speeds are fastest. Moreover, since the DG approach is applied only at a subdomain level, the complexity of the numerical model and the computational cost can be kept under control, avoiding the proliferation of unknowns, a drawback that is typical of classical DG discretizations.

Taking advantage of the built-in flexibility of the underlying discretization method and of the increasing computational power of parallel computer architecture, the code provides a versatile way to handle multi-scale earthquake engineering studies in a new "from-source-to-site" philosophy. This has been addressed in the recent years only by a few studies (Krishnan et al. 2006; Taborda et al. 2012; Isbiliroglu et al. 2013), due to the related intrinsic complexities of reproducing such phenomena in a single conforming model. A sketch of potential applications of SPEED is illustrated in Fig. 10.4.

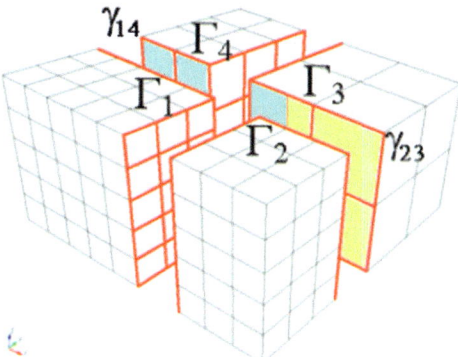

Fig. 10.3 Non-conforming Spectral Element mesh (different element sizes and spectral degrees in each sub-domain) and its partition, with jumps along the interfaces treated according to a DG approach

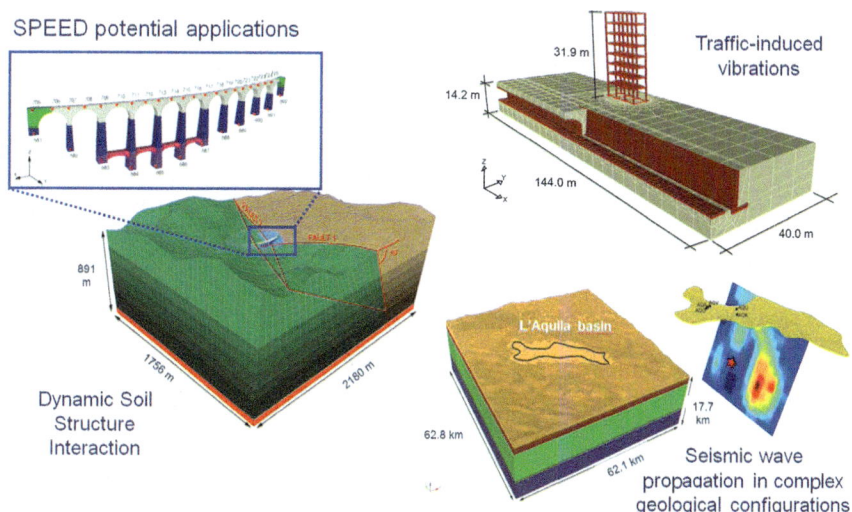

Fig. 10.4 Sketch of potential applications in elastodynamics of the SPEED code

The code is naturally designed for multi-core computers or large clusters, but it can run as well on a single processor machine. It is written in Fortran90 with full portability in mind and conforms strictly the Fortran95 standards. It takes advantage of the hybrid parallel programming based upon the Message Passing Interface (MPI) library relying on the domain decomposition paradigm and the OpenMP library for multi-threading operations on shared memory. The mesh generation may be accomplished using a third party software, e.g. CUBIT (http://cubit.sandia.gov/)

and load balancing is facilitated by graph partitioning based on the METIS library (glaros.dtc.umn.edu/) included in the package.

The code has been verified over different benchmarks, including that of Grenoble (Chaljub et al. 2010), and a further comparison with an independent solution is described in the following.

Physical discontinuities can be modeled either by the DG approach (creating physical interfaces) or by a not-honoring technique (where material properties are given node by node). The time-integration is performed either by the explicit second-order accurate leap-frog scheme or by the explicit fourth-order accurate Runge-Kutta method (Quarteroni et al. 2007).

Despite its short life-time, SPEED was awarded among the emerging applications with industrial relevance within the project PRACE-2IP (WP 9.3) and received substantial fundings for HPC resources (2012: ISCRA project MAgNITUd 500 k core hours, 2013: LISA project SISMAURB 2000 k core hours, PRACE project DN4RISC 40000 k core hours). Within the framework of PRACE-2IP, SPEED was optimized for use on the FERMI cluster at CINECA (Tier-0 machine), and optimal performances in term of efficiency, scalability and speed-up were obtained (see Dagna 2013).

10.3.2 Main Features

In its present version, SPEED allows the users to treat different seismic excitation modes, including: (i) kinematic seismic fault models (see below) (ii) plane wave load; (iii) Neumann surface load; (iv) volume force load. Dirichlet and/or Neumann boundary conditions can be set into the model; furthermore, first-order absorbing paraxial boundary conditions (Stacey 1988) have been implemented in order to prevent the propagation of spurious reflections from the external boundaries of the computational domain. The upgrade of the paraxial conditions to Perfectly Matched Layers (PML) is planned for the next release of the code.

Post-processing tools are available to produce ground shaking maps in a standard format that can be read by a variety of software, such as ArcGIS (www.esri.com), GID (gid.cimne.upc.es) and PARAVIEW (www.paraview.org).

10.3.2.1 Treatment of Kinematic Finite-Fault Models

SPEED adopts a kinematic description of the seismic source in terms of a distribution of double-couple point sources, whose mathematical representation is given by the seismic moment tensor density, i.e., $m_{ij}(\underline{x};t) = \frac{M_0(\underline{x};t)}{V}(\nu_i \cdot n_j + \nu_j \cdot n_i)$, where $M_0(\underline{x};t)$ is the time history of release at the source point $\underline{x}$ inside the elementary source volume V, n and ν denote the fault normal unit vector and the unit slip vector, respectively (Faccioli et al. 1997).

The code features a number of options for the kinematic modelling of an arbitrarily complex seismic source, by assigning realistic distributions of co-seismic slip along an extended fault plane through ad hoc pre-processing tools. These tools allow one to reproduce in a semi-automatic way realistic fault rupture models as compiled in the on-line Finite Source Rupture Models Database (Mai 2004) or computed by other methods using a specific format. Furthermore, it is also possible to define stochastically correlated random source parameters, in terms of slip pattern, rise time, rupture velocity and rupture velocity distribution along the fault plane, which may be crucial in deterministic simulations to excite high frequency components of ground motion (Smerzini and Villani 2012).

10.3.2.2 Attenuation Model

Modelling of visco-elastic media is handled by modifying the equation of motion according to the approach of Kosloff and Kosloff (1986). For this purpose, the inertial term $\rho \frac{\partial^2 u}{\partial t^2}$ of the wave equation is replaced by $\rho \frac{\partial^2 u}{\partial t^2} + 2\zeta \frac{\partial u}{\partial t} + \zeta^2 u$, where u is the generic displacement component, ρ is mass density and ζ is an attenuation parameter. It can be shown that, with this substitution, all frequency components are equally attenuated with distance, resulting in a frequency proportional quality factor $Q = Q_0 \frac{f}{f_0}$, where $Q_0 = \frac{\pi f_0}{\zeta}$ and f_0 is a reference value within the frequency range to be propagated.

This model is in agreement with numerous seismological observations supporting a frequency dependent law $Q = Q_0 f^\alpha$, with $\alpha \sim 1$ (e.g., Castro et al. 2004; Morozov 2008). Implementation of new rheological models is in progress, starting from the classical Rayleigh and Caughey damping.

10.3.2.3 Non-Linear Elastic Soil Behavior

A simple Non-Linear Elastic (NLE) soil model is implemented as a generalization to 3D load conditions of the classical modulus reduction ($G-\gamma$) and damping ($D-\gamma$) curves used within 1D linear-equivalent approaches (e.g. Kramer 1996), where G, D and γ are the shear modulus, damping ratio and 1D shear strain, respectively. Namely, to extend those curves to the 3D case, a scalar measure of shear strain amplitude is considered as follows:

$$\gamma_{max}(\underline{x},t) = \max\left[\left|\varepsilon_I(\underline{x},t) - \varepsilon_{II}(\underline{x},t)\right|, \left|\varepsilon_I(\underline{x},t) - \varepsilon_{III}(\underline{x},t)\right|, \left|\varepsilon_{II}(\underline{x},t) - \varepsilon_{III}(\underline{x},t)\right|\right] \tag{10.1}$$

where ε_I, ε_{II} and ε_{III} are the principal values of the strain tensor. Once the value of γ_{max} is calculated at the generic position $\underline{x}$ and generic instant of time t, this value is introduced in the $G-\gamma$ and $D-\gamma$ curves and the corresponding parameters are updated for the following time step. Therefore, unlike the classical linear-equivalent approach, G and D values are updated step by step, so that the initial values of the dynamic soil properties are recovered at the end of the excitation. Application of this approach can be found in Stupazzini et al. (2009) for the case of Grenoble, France.

10.3.2.4 Hybrid Approach for the Generation of Broadband Synthetics

In spite of the increasing computer resources and tools, as shown in Table 10.1, 3D numerical simulations are still restricted to the low frequency range, up to about 1–2 Hz, mainly due to computational limitations as well as insufficient resolution of geologic and seismic source models. On the other hand, earthquake engineering applications need realistic ground motion time histories in the entire frequency range of interest for the analysis of structural response and damage assessment, say between 0 and 25 Hz.

A hybrid scheme is presently the best approach to generate broadband (BB) ground motions. In this work, Low Frequency (LF) waveforms from numerical simulations are combined by means of matching filters with the High Frequency (HF) synthetics computed by other independent approaches. Namely, the method of Sabetta and Pugliese (1996) was selected because of ease to treat in the post-processing phase the huge set of synthetics of the 3D numerical simulations. On the other side, it has the disadvantage of accounting neither of detailed kinematic fault rupture models, nor of specific 1D site amplification functions. Examples of other approaches for generation of synthetics, such as EXSIM (Motazedian and Atkinson 2005), are presented by Smerzini and Villani (2012) for the case of the 2009 L'Aquila earthquake.

The procedure adopted in this work to generate BB acceleration time histories at a given site can be summarized as follows (see Fig. 10.5): (i) compute $N=20$ stochastic realizations by SP96 for each ground-motion component (EW and NS); (ii) for each stochastic realization, synchronize the LF and HF time histories in the time domain, so to have the same value for the time $t_{5\%}$ at which the normalized Arias intensity $I_a = 5\%$ is reached both by the LF and HF synthetic; (iii) for each stochastic realization, combine HF and LF waveforms in the frequency domain by applying a match filter, defined as follows:

$$BB(f) = w_{LF} \cdot A_{LF}(f) + w_{HF} \cdot A_{HF}(f) \tag{10.2}$$

where $A_{LF}(f)$ and $A_{HF}(f)$ denote the Fourier transform of the LF and HF acceleration time histories, respectively; w_{LF} and w_{HF} are the corresponding weighting cosine-shape functions and $BB(f)$ is the Fourier transform of the output BB signal.

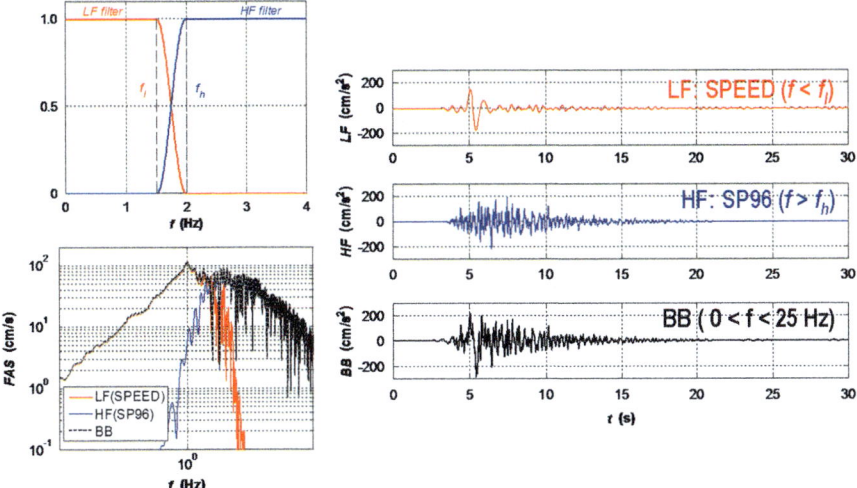

Fig. 10.5 Generation of broadband ground motions (*black*) combining the LF waveforms from SPEED (*red*) with the HF synthetic accelerograms of Sabetta and Pugliese (1996), SP96 (*blue*), through suitable matching filters

10.4 Overview of Case Studies

Hybrid deterministic-stochastic ground shaking scenarios were generated in the following areas: (i) Santiago de Chile; (ii) Po Plain, NorthEastern Italy; (iii) Christchurch and (iv) Wellington, New Zealand. Besides a relevant interest from the economic loss exposure viewpoint, all of these sites were chosen because of availability of sufficiently detailed information both of the active faults surrounding the sites and of the shallow and deep geology structures, along with a significant amount of records, notably in the Christchurch and Po Plain cases.

The last rows of Table 10.1 summarize the main features of the adopted numerical models and the associated scenarios, so that a comparison with previous case studies can be made. All numerical meshes were built by the software CUBIT (cubit.sandia.gov/) and the numerical simulations were performed on parallel computer architectures, namely, the FERMI BlueGene/Q system, at CINECA (www.hpc.cineca.it/).

10.4.1 Santiago de Chile

Different earthquake rupture scenarios along the San Ramon fault, an active thrust structure crossing the eastern outskirts of the city of Santiago, were addressed. Recent works (e.g. Armijo et al. 2010) have shown that the San Ramon fault has a key role for the seismic hazard of the city.

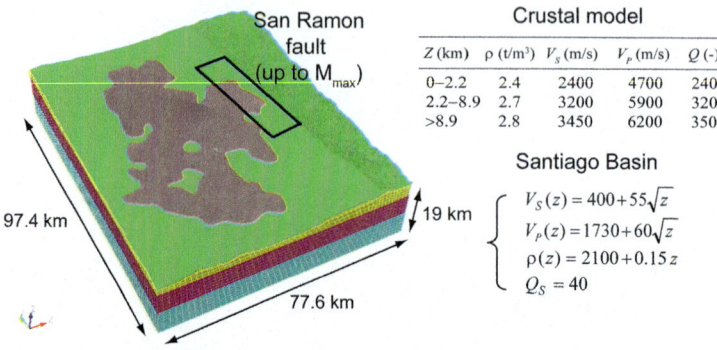

Fig. 10.6 Computational domain for the Santiago region, Chile

The numerical model (see Fig. 10.6) was built by including: (i) surface topography; (ii) 3D model of the 80-km-long and 30-km-wide Santiago valley (Pilz et al. 2010); (iii) kinematic representation of potential ruptures breaking the San Ramon fault; (iv) linear visco-elastic soil behavior. 19 scenarios were considered by varying the magnitude (range from 5.5 to 7), slip pattern (7 different distributions) and hypocenters (8 different locations).

To appreciate the potential interest of these numerical simulations, Fig. 10.7 shows at top two representative scenarios in terms of *PGV* distribution in Santiago, and, at bottom, the simulated *PGV* variation with distance compared to the one predicted according to the empirical equation of Akkar and Bommer (2007). While, for the $M_w 6$ scenario, there is an overall agreement of ground motion predicted by both approaches at stiff sites within the basin (EC8 class B), for the M_w 7 scenario the empirical equations are not fit to predict neither the very high near-fault PGV values, related to a fault slip mechanism affected by directivity, nor the high amplification levels at the edges of the basin in the vicinity of the fault (shaded areas in Fig. 10.7).

10.4.2 Po Plain, Italy

Stimulated by the major seismic sequence that struck the Emilia-Romagna region, Italy, from May to June 2012, a program for 3D numerical simulations of earthquake ground motion within the Po Plain was initiated. The model was constructed (Fig. 10.8) to include the seismogenic structures responsible of the M_W 6.1 May 20 and M_W 6.0 May 29 earthquakes (Ferrara and Mirandola faults, respectively). The irregular shape of the submerged bedrock topography in the Po Plain was also modelled, as derived by the isobaths of the basement of the Pliocene formations of the structural map of Italy (Bigi et al. 1992). Further details on the shear wave velocity model inside the Po Plain can be found in the sequel. A suite of 23 earthquake scenarios, characterized by magnitude ranging from 5.5 to 6.5, and different

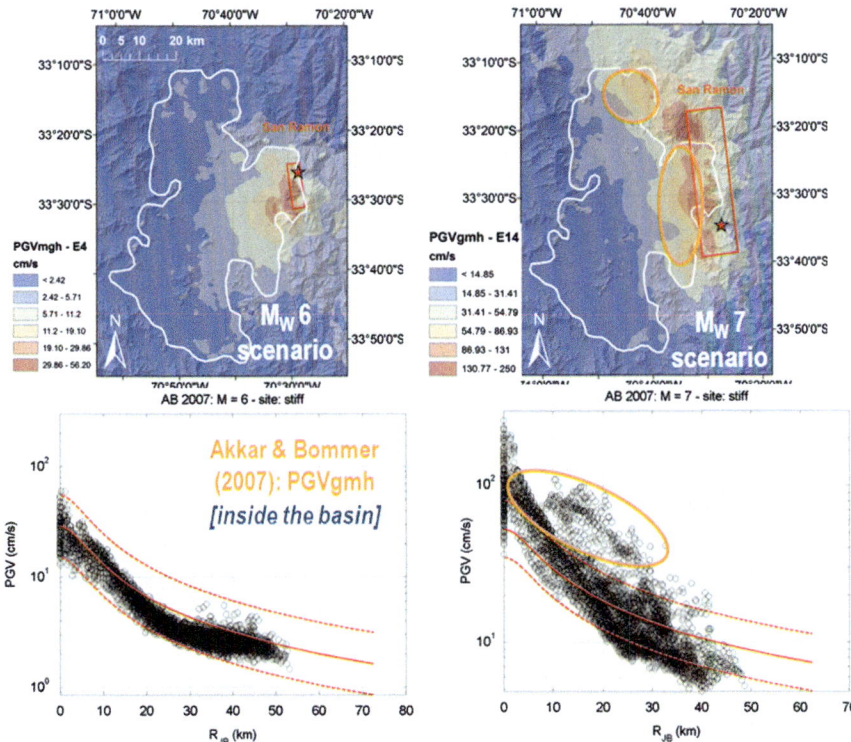

Fig. 10.7 *Top*: Horizontal PGV (geometric mean) scenario maps for 2 out of 19 scenarios considered for Santiago de Chile. *Bottom*: comparison of simulated PGV values inside the basin with the empirical prediction based on the Akkar and Bommer (2007) equations. The superimposed ellipses on the right hand side denote areas where significant deviations from the GMPEs are found

co-seismic slip distribution, focal mechanism, rupture velocity and rise time, was generated along both faults.

Results of this case study will be explored in more detail in the next section, by comparison with the observed records.

10.4.3 The Canterbury Plains, New Zealand

A 3D numerical model of the Canterbury Plains, New Zealand, was constructed which includes the city of Christchurch, part of the Canterbury Plains and of the Banks Peninsula, extending over an area of about $45 \times 45 \times 20$ km, and combines: (i) a horizontally layered deep crustal model as well as a reliable description of the alluvial-bedrock interface based on the available geological map and studies in the literature (Forsyth et al. 2008; Bradley 2012); (ii) the surface topography; (iii) a simplified velocity model of the Canterbury plain, filled with Quaternary deposits,

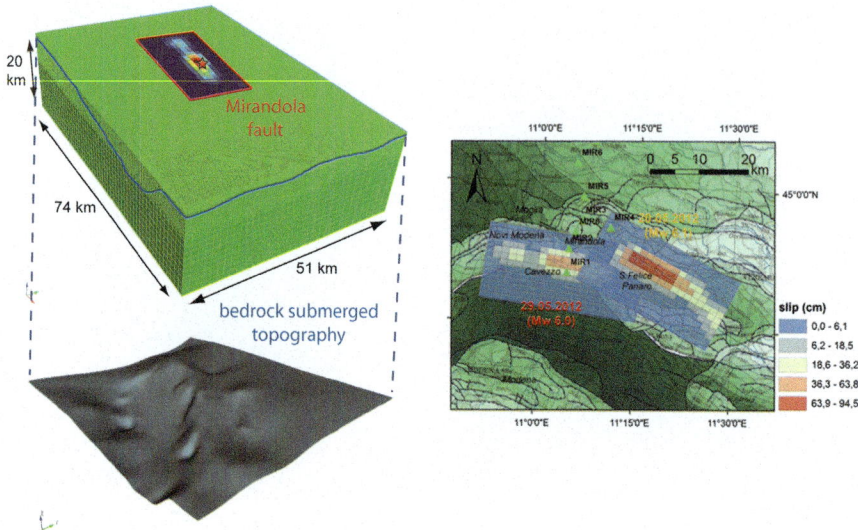

Fig. 10.8 Po Plain, Italy: 3D numerical model including the seismic faults responsible of the M_W 6.0 May 29 and M_W 6.1 May 20 May earthquakes and the submerged topography. On the *right*, the assumed slip mechanisms to model the earthquakes

constrained at shallow depths by extended MASW results within the Central Business District of Christchurch[4]; (iv) the kinematic fault models for the mainshocks of Feb 22 (M_W 6.2), June 13 (M_W 6.0) and Dec 23 2011 (M_W 6.1), proposed by Beavan et al. (2012).

For sake of brevity, we refer the reader to the results published by Guidotti et al. (2011), based on a preliminary numerical model of the basin. We limit ourselves here to show in (Fig. 10.9) the PGV maps of the EW and NS components for the Feb 22, 2011 earthquake.

10.4.4 Wellington, New Zealand

Seismic hazard in the metropolitan area of Wellington is dominated by several major active fault systems, i.e., from West to East, the Ohariu, Wellington–Hutt and Wairarapa faults, as indicated by the superimposed red lines in Fig. 10.10, top-left panel. Although all these faults were incorporated in the numerical model, the scenarios are produced only for the Wellington–Hutt fault. This is a 75-km long strike-slip fault, characterized by a return period between 420 and 780 years for a magnitude between M 7.0 and 7.8 (Benites and Olsen 2005).

---

[4] Canterbury geotechnical database, Orbit Project: canterburygeotechnicaldatabase.projectorbit.com

10 Physics-Based Earthquake Ground Shaking Scenarios in Large Urban Areas   347

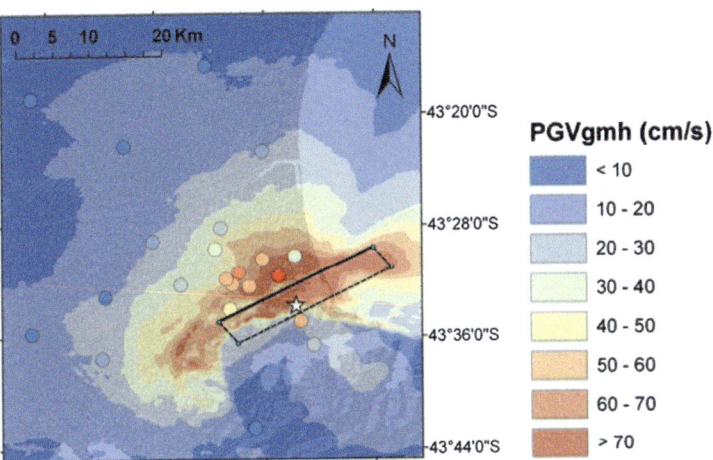

Fig. 10.9 Christchurch, New Zealand: PGV map of the geometric mean of the horizontal components for the Feb 22, 2011 M_w6.2 earthquake. Coloured dots denote the corresponding observed values from earthquake records

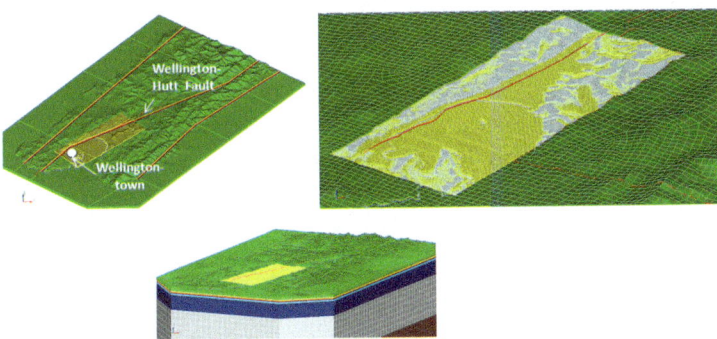

Fig. 10.10 3D numerical model of the Wellington metropolitan area (*bottom*) with indications of the main active faults (*top-left*) and the Wellington Lower-Hutt basin (*top-right*) meshed by a non-conforming strategy

Besides these faults, we incorporated in the numerical model (see Fig. 10.10, bottom panel) the most important geological features of the area, i.e., the 3D basin bedrock topography along with the 3D irregular soil layers deposited over the bedrock. This information is integrated based on the available geological and geophysical data (borehole, bathymetry, gravity, seismic), down to about 800 m depth (R. Benites, personal communication, 2013). To better describe such geological discontinuities, a non-conforming strategy was adopted to model the Wellington Valley, as depicted in Fig. 10.10, right-top panel. Note that the free-surface topography of the region is taken into account. Numerical simulations for this case study are presently in progress.

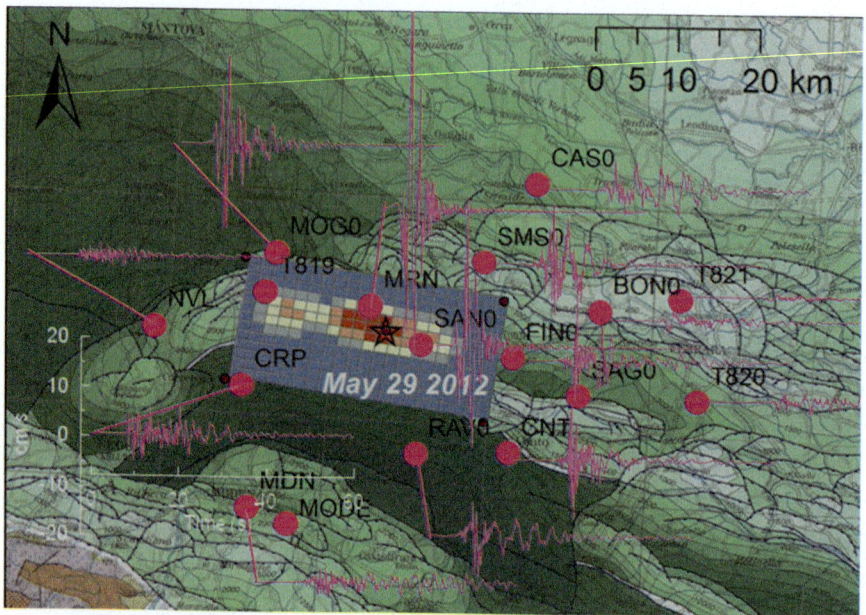

Fig. 10.11 NS components of a selected set of velocity records of May 29 Emilia earthquake. Superimposed in the slip model assumed based on Atzori et al. (2012)

10.5 Insight of a Case Study: Earthquake Ground-Shaking Scenarios in the Po Plain

Among the previous case studies, we overview in this Section the numerical simulations of the seismic response of the Po Plain, with emphasis on the sites affected by the Emilia-Romagna earthquakes of May-June 2012, for which an exceptional set of strong motion records is available, especially for the M_W 6.0 May 29 event. In addition to the simulation of this real earthquake, various fault rupture scenarios were produced, considering different hypothetical breaking mechanisms of the faults responsible of the May 20 and 29 earthquakes.

Leaving to other publications (e.g., Tizzani et al. 2013) an insight of the seismotectonic and geological environment, a critical step of this work is the validation of simulated results against strong motion records obtained during both earthquakes. While on May 20 the Mirandola (MRN) station alone was in operation, the number of near-source records from the May 29 event is much larger, mainly from temporary arrays (Fig. 10.11).

The near-source records show similar features, with large velocity pulses in the fault normal direction reaching up to about 60 cm/s, while, at larger distance from the fault, peak values rapidly decrease and records tend to be dominated by surface waves generated by the complex subsoil structure of the Po Plain (Luzi et al. 2013). Peak values of horizontal acceleration reach about 0.3 g, while on May 29 the vertical acceleration at MRN reached a remarkable 0.9 g.

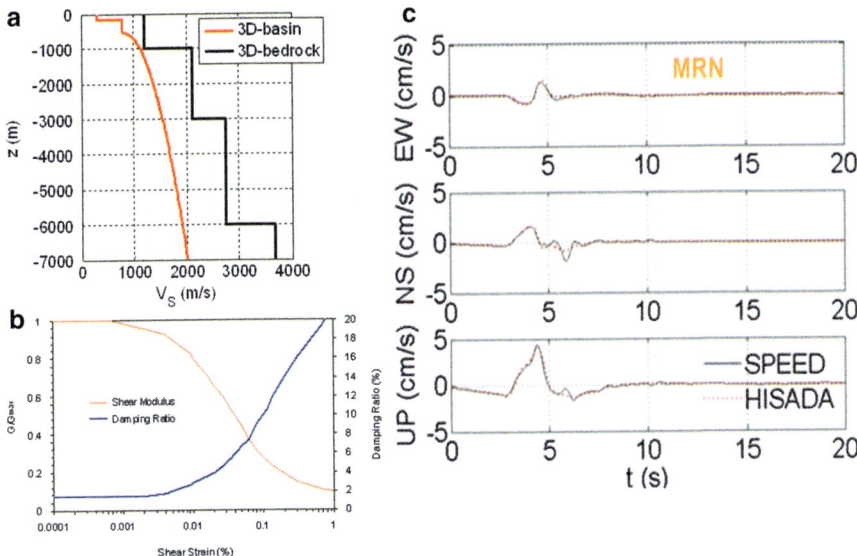

Fig. 10.12 (a) V_S profile adopted for the 3D numerical simulations (*red*: Po Plain sediments; *black*: Miocene bedrock formations). (b) G-γ and D-γ curves adopted for the first 150 m in a non-linear elastic approach. (c) Validation of SPEED numerical simulations with the Hisada (1994) code, assuming a 1D V_S soil profile, and the finite-fault of May 29 earthquake

10.5.1 3D Numerical Simulations of the 29 May 2012 Earthquake

Numerical modelling of the 29 May earthquake was addressed, being this earthquake the best constrained in terms of strong motion recordings as well as of source inversion studies. For the shear velocity model (see Fig. 10.12a), a homogenous average soil profile was defined for the Po Plain sediments, while a horizontally layered model was assumed in the rock Miocene formations. These profiles were calibrated merging the information from the available V_S profiles and published works (e.g. Margheriti et al. 2000; Martelli and Molinari 2008), along with the Down-Hole and Cross-Hole surveys (Project S2 2013). The resulting subsoil model has been found in reasonable agreement with the results recently published by Milana et al. (2013). The kinematic fault solution proposed by Atzori et al. (2012) has been adopted in the numerical simulations (see superimposed map in Fig. 10.11).

Both a linear visco-elastic and non-linear elastic soil behavior has been adopted for the numerical simulations, as discussed in the sequel. The $G-\gamma$ and $D-\gamma$ curves as derived by Fioravante and Giretti (2012) were used for the top 150 m layers.

Prior to the numerical simulations with the 3D model, we carried out a validation with the results of the Hisada (1994) code, by assuming a 1D Vs soil profile and the finite-fault of May 29 earthquake. The very good agreement of the two solutions (Fig. 10.12c) demonstrates the accuracy of SPEED.

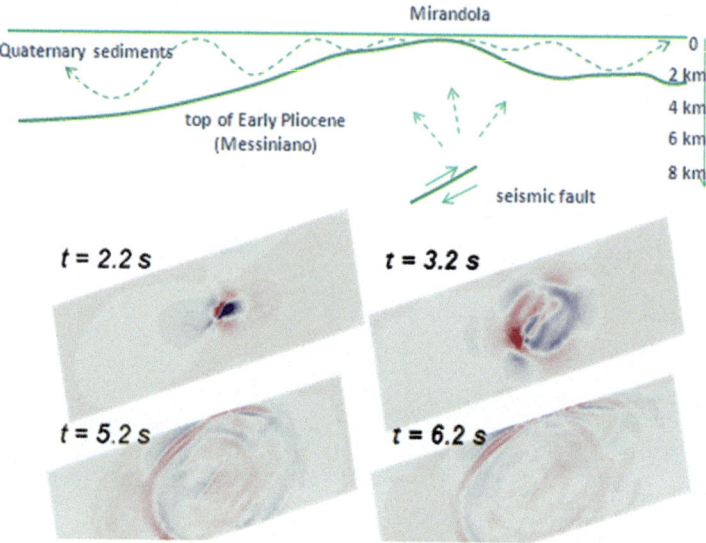

Fig. 10.13 *Top*: Simplified sketch of a NS cross-section of the Po-plain across Mirandola and the seismic source of the May 29 earthquake. *Bottom*: snapshots of the NS displacement wavefield along this cross-section

Figure 10.13 shows some snapshots of the displacement wavefield (NS component) through a NS cross-section including the seismic fault, clearly showing the key role of the submerged topography to produce prominent surface wave trains affecting seismic ground response both at short and at large distance from the epicenter.

Figure 10.14 shows the comparison between synthetics and recordings in terms of three-component displacement waveforms at 12 representative strong motion stations, distributed about uniformly around the epicenter. Both recorded and simulated waveforms were band-passed filtered between 0.1 and 1.5 Hz, the latter being the frequency limit of the numerical model. The agreement between synthetics and recordings is satisfactory, especially for stations distant from the epicenter. On the other hand, at stations in the near-field region, such as MRN and SAN0, the numerical model tends to underestimate significantly the observed horizontal ground motion amplitudes, while a good agreement is found for the vertical component. This points to one of the most critical problems to be faced when physics-based simulations of real earthquakes are compared with observations, that is, near-fault records depend on details of the source slip mechanism and rupture propagation that are hardly predicted and are often beyond the frequency range on which earthquake source inversions are provided. While, the larger the epicentral distance is, the smaller is the relevance of such details.

The most significant effects of non-linear soil behavior are found at those stations where the thickness of soft sediments reaches considerable values of a

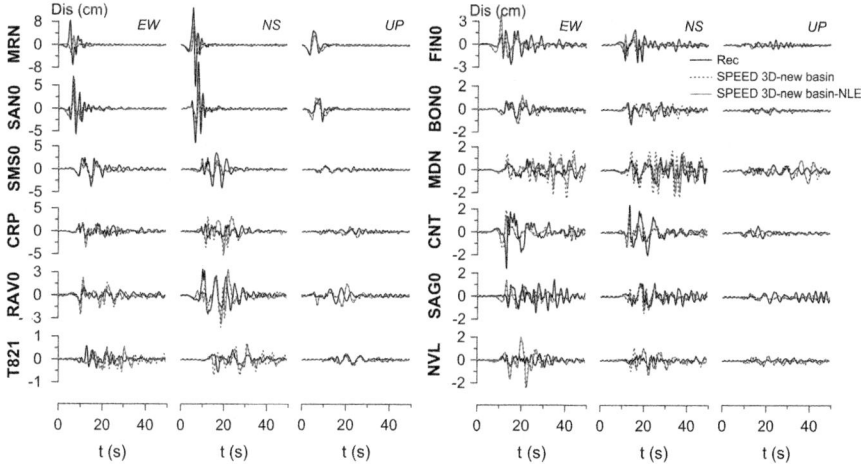

Fig. 10.14 Recorded (*black*) and simulated displacement waveforms (0.1–1.5 Hz). Results for both linear (*blue dotted*) and non-linear (*red*) visco-elastic soil behavior are shown. The location of stations is illustrated in Fig. 10.11

few thousands of kilometers (see e.g. MDN station). Given the low frequency range propagated by the model (<1.5 Hz), the overall impact of soil nonlinearity is small especially for the stations in the near-fault region.

10.5.2 Ground Shaking Scenarios in the Po Plain

Starting from the 3D models developed for the May 20 and May 29 earthquakes, different hypothetical seismic rupture scenarios were assumed, all of them breaking either the Mirandola or the Ferrara faults, with magnitude ranging from 5.5 to 6.5. Realistic slip models along the faults were obtained either by source inversion of real earthquakes with similar fault mechanisms or they were computed using a self-similar k-square model (Herrero and Bernard 1994; Gallovič and Brokešová 2007). Twelve rupture scenarios are produced along the Ferrara fault (May 20) and eleven are activated along the Mirandola fault (May 29).

An overview of the ground shaking map in terms of spatial distribution of *PGV* (geometric mean of horizontal components), for eight selected scenarios, is shown in Fig. 10.15. For each scenario, the surface projection of the seismic fault is superimposed on the *PGV* map and the corresponding kinematic source model is displayed on the right hand side. It is interesting to note that the computed seismic response is strongly affected by the combination of directivity and radiation pattern effects, with near-fault *PGV* values that appear to be only slightly dependent on magnitude, in agreement with several theoretical and experimental studies (see e.g., McGarr and Fletcher 2007).

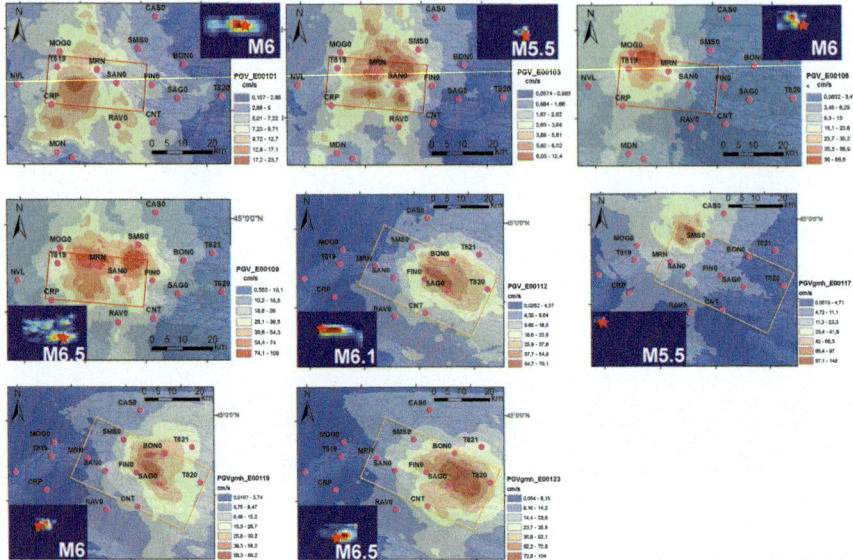

Fig. 10.15 PGV (geometric mean of horizontal components) maps for selected ground shaking scenarios in the Po Plain

Finally, we compare in Fig. 10.16 the *PGV* maps obtained through the hybrid BB approach outlined previously, by injecting high frequency components into the results of the physics-based numerical simulations, with those provided by ShakeMap tools (shakemap.rm.ingv.it) based on suitable interpolation procedures of available records. It can be noted that there is a qualitative agreement in terms of spatial distribution, but the near-fault peak values are significantly underestimated, by a factor of about 2. Namely, with the adopted kinematic fault solution and hypocenter location, it was not possible to reproduce the large recorded near-fault velocity peaks (see Fig. 10.14, stations MRN and SAN0).

10.6 A Web-Repository for Ground-Shaking Scenarios

One of the main outcomes of the cooperation of PoliMi with MunichRe was the development of a web-repository of the synthetic seismic scenarios produced in the urban areas considered, in a format suitable for risk assessment studies. The data structure of the web-repository is handled as a relational Access database, so that any standard/advanced query can be easily performed.

It is worth to remark here that the database is not constrained to SPEED results, rather it was envisioned as an open repository aiming at collecting the results of different complex scenarios, both from the simulation method and model description viewpoint.

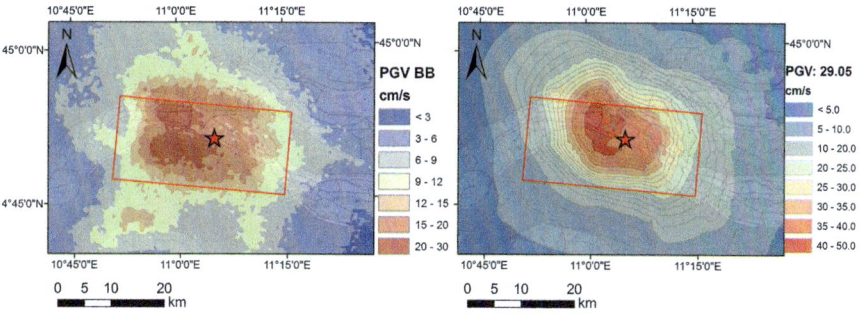

Fig. 10.16 PGV maps from BB (SPEED + SP96) numerical simulations (*left*) and from ShakeMap tool (*right*)

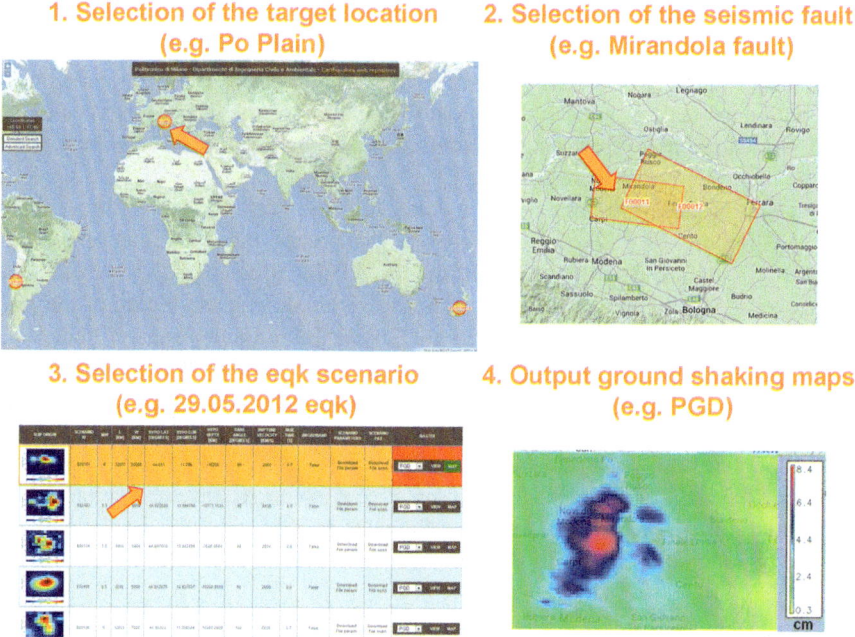

Fig. 10.17 Web-repository of earthquake ground shaking scenarios

Figure 10.17 illustrates the conceptual scheme adopted as a basis of the archive of synthetic seismic scenarios, with reference to the Po Plain case study: (1) the user first selects the target location; (2) then, the seismic fault is picked among those available for the location under study; (3) the target scenario is adopted, uniquely defined by magnitude, location and size of the broken fault, and by the additional parameters such as co-seismic slip distribution, nucleation point, rupture velocity, rise time and rake angle; (4) output ground shaking maps are downloaded. In a future version of the web site, BB time-histories at selected locations will also be downloadable.

Output maps are stored in a standard format, on a regular grid of Latitude and Longitude in terms of the following strong motion parameters (geometric mean of horizontal components): Peak Ground Displacement (*PGD*), Peak Ground Velocity (*PGV*), Peak Ground Acceleration (*PGA*), response spectral Pseudo-Acceleration (*PSA*) at 0.3, 1.0, 3.0, and 5.0 s.

10.6.1 Conclusions

In the framework of a research contract between Politecnico di Milano, Italy, and Munich Re, Germany, we generated physics-based ground shaking scenarios from hypothetical earthquakes in large urban areas worldwide. These scenarios were obtained by the open-source high-performance computer code SPEED, based on a Discontinuous Galerkin spectral element formulation of the elastodynamics equations, allowing one to treat non-conforming meshes as well as non-uniform polynomial approximation degrees.

The case studies encompass Santiago de Chile, the Po Plain, Italy, Christchurch and Wellington, New Zealand. Taking advantage of the large set of records obtained in the near-fault region of the Po Plain, affected by the earthquake sequence of May-June 2012, results from the 3D numerical modelling of the M_W 6 May 29 2012 earthquake were illustrated, under both assumptions of linear and non-linear visco-elastic materials. Comparisons with records were addressed, to highlight potential limitations of this numerical approach to obtain realistic ground shaking scenarios.

Although results for this case study were not fully satisfactory when compared to records, this simulation experiment pointed out some of the key points to be accounted for when physics-based earthquake ground motion simulations are carried out and compared with real records:

- given the complexity of the numerical model, preliminary validation tests with independent numerical codes on simplified configurations (as shown in Fig. 10.12c) are recommended;
- the accuracy of input data for finite-fault modelling is crucial, especially in the near-field region, where details on the asperity distribution along the fault, together with the relative position of the nucleation point with respect to the slip pattern, affect dramatically the ground motion computations;
- if the input seismic source and geological models are sufficiently detailed to excite seismic ground motion within a sufficiently wide frequency range, physics-based numerical simulations are capable of providing realistic ground shaking scenarios and of capturing some features of ground motion variability (such as spatial coherency, dependence on the lateral variation of soil properties, basin edge effects, surface or submerged topographic irregularities), which are not taken into account by any other tool for EGMP.

As shown in Table 10.1, much progress has been done in the last 15 years in the production of realistic physics-based earthquake ground shaking scenarios in large urban areas. Several verification benchmarks of the numerical codes against independent solutions and/or cross-validation among codes have demonstrated that a satisfactory level of reliability of results has been reached. Furthermore, the computational progress allows one presently to run numerical meshes of hundreds of millions nodes in few hours, or tens of minutes, even without having access to very powerful computer architectures.

However, in order for such numerical approaches to be accepted confidently by the engineering community as alternative and reliable tools to empirical approaches for EGMP, physics-based numerical simulations of source-to-site earthquake ground motion prediction still need to convincingly provide answers to the following questions:

- what is the level of detail required on the seismic source to excite ground motions in a large enough frequency range?
- what is the level of detail required on the local geology to produce realistic ground motion scenarios useful for seismic risk evaluations?
- how many numerical simulations are required to produce a sufficiently representative and reliable picture of the earthquake ground motion and of its spatial variability?

Answers to the previous questions will be by far more convincing if these methods will be proven to provide explanations of observed ground motions, especially in the near-source region, more satisfactory than conventional tools for EGMP.

Acknowledgments This work was partially funded by MunichRe, with the cooperation of A. Smolka and A. Allmann. The development of the SPEED code is currently in progress, under the supervision of A. Quarteroni, P. Antonietti, L. Formaggia, the cooperation of which is gratefully acknowledged. R. Guidotti performed the numerical simulations of the Christchurch earthquakes and contributed to the construction of the Wellington model. The information on the latter model was kindly provided by R. Benites, GNS. The authors are also grateful to M. Beretta for constructing the web site for the scenario repository. Numerical simulations took advantage of the Prace TIER0 DN4RISC (n. 1551) and LISA SISMAURB projects (call 1), with the support of the SuperComputing Applications and Innovations department at CINECA, Italy. Partial funding from the DPC-Reluis Project RS2 2014-18 is also gratefully acknowledged.

Open Access This chapter is distributed under the terms of the Creative Commons Attribution Noncommercial License, which permits any noncommercial use, distribution, and reproduction in any medium, provided the original author(s) and source are credited.

References

Aagaard BT, Hall JF, Heaton TH (2004) Effects of fault dip and slip rake angles on near-source ground motions: why rupture directivity was minimal in the 1999 Chi-Chi, Taiwan, earthquake. Bull Seismol Soc Am 94:155–170

Akkar S, Bommer JJ (2007) Empirical prediction equations for peak ground velocity derived from strong motion records from Europe and the Middle East. Bull Seismol Soc Am 97:511–530

Antonietti PF, Mazzieri I, Quarteroni A, Rapetti F (2012) Non-conforming high order approximations of the elastodynamics equation. Comput Meth Appl Mech Eng 209–212:212–238

Aochi H, Madariaga R (2003) The 1999 Izmit, Turkey, earthquake: nonplanar fault structure, dynamic rupture process, and strong ground motion. Bull Seismol Soc Am 93:1249–1266

Armijo R, Rauld R, Thiele R, Vargas G, Campos J, Lacassin R, Kausel E (2010) The West Andean Thrust (WAT), the San Ramon Fault and the seismic hazard for Santiago (Chile). Tectonics 29, TC2007. doi:10.1029/2008TC002427

Asano K, Iwata T, Irikura K (2005) Estimation of source rupture process and strong ground motion simulation of the 2002 Denali, Alaska, earthquake. Bull Seismol Soc Am 95:1701–1715

Atzori S, Merryman Boncori J, Pezzo G, Tolomeri C, Salvi S (2012) Secondo Rapporto sulla analisi dati SAR e modellazione della sorgente del terremoto dell'Emilia. INGV (in Italian)

Beavan J, Motagh M, Fielding EJ, Donnelly N, Collett D (2012) Fault slip models of the 2010–2011 Canterbury, New Zealand, earthquakes from geodetic data and observations of postseismic ground deformation. New Zeal J Geol Geophys 55:207–221

Benites R, Olsen KB (2005) Modeling strong ground motion in the Wellington Metropolitan Area, New Zealand. Bull Seismol Soc Am 95(6):2180–2196

Bielak J, Ghattas O, Kim EJ (2005) Parallel octree-based finite element method for large-scale earthquake ground motion simulation. Comput Model Eng Sci 10:99–112

Bielak J, Graves RW, Olsen KB, Taborda R, Ramirez-Guzman L, Day SM, Ely GP, Roten D, Jordan TH, Maechling PJ, Urbanic J, Cui Y, Juve G (2010) The ShakeOut earthquake scenario: verification of three simulation sets. Geophys J Int 180:375–404

Bigi G, Bonardi G, Catalano R, Cosentino D, Lentini F, Parotto M, Sartori R, Scandone P, Turco E (eds) (1992) Structural Model of Italy 1:500,000, CNR Progetto Finalizzato Geodinamica

Bindi D, Massa M, Luzi L, Ameri G, Pacor F, Puglia R, Augliera P (2014) Pan-European ground-motion prediction equations for the average horizontal component of PGA, PGV, and 5%-damped PSA at spectral periods up to 3.0 s using the RESORCE dataset. Bull Earthquake Eng 12:391–430

Boore DM, Stewart JP, Seyhan E, Atkinson G (2013) NGA-West2 equations for predicting response spectral accelerations for shallow crustal earthquakes. PEER Report 2013/05

Bradley BA (2012) Ground motion and seismicity aspects of the 4 September 2010 Darfield and 22 February 2011 Christchurch Earthquakes. Technical Report Prepared for the Canterbury Earthquakes Royal Commission. http://canterbury.royalcommission.govt.nz/documents-by-key/20120116.2087

Castro R, Pacor F, Bindi D, Franceschina G, Luzi L (2004) Site response of strong motion stations in the Umbria, Central Italy, Region. Bull Seismol Soc Am 94(2):576–590

Chaljub E, Moczo P, Tsuno S, Bard PY, Kristek J, Kaser M, Stupazzini M, Kristekova M (2010) Quantitative comparison of four numerical predictions of 3D ground motion in the Grenoble valley, France. Bull Seismol Soc Am 100:1427–1455

Chaljub E, Maufroy E, Moczo P, Kristek J, Priolo E, Klin P, De Martin F, Zhang Z, Hollender F, Bard PY (2013) Identifying the origin of differences between 3D numerical simulations of ground motion in sedimentatry basins: lessons from stringent canonical test models in the E2VP framework. EGU Geophysical Research Abstracts, Vienna, Austria, pp 111–118

Dagna P (2013) Enabling SPEED for near real-time earthquake simulations, PRACE Report

Day SM, Graves R, Bielak J, Dreger D, Larsen S, Olsen KB, Pitarka A, Ramirez-Guzman L (2008) Model for basin effects on long-period response spectra in Southern California. Earthq Spectra 24:257–277

De Basabe JD, Sen MK, Wheeler MF (2008) The interior penalty discontinuous Galerkin method for elastic wave propagation: grid dispersion. Geophys J Int 175(1):83–93

Douglas J (2011) Ground-motion prediction equations 1964–2010. PEER report 2011/102

Douglas J, Aochi H (2008) A survey of techniques for predicting earthquake ground motions for engineering purposes. Surv Geophys 29:187–220

Douglas J, Akkar S, Ameri G, Bard P-Y, Bindi D, Bommer JJ, Bora SS, Cotton F, Derras B, Hermkes M, Kuehn NM, Luzi L, Massa M, Pacor F, Riggelsen C, Sandıkkaya MA, Scherbaum F, Stafford PJ, Traversa P (2014) Comparisons among the five ground-motion models 1 developed using RESORCE for the prediction of response spectral accelerations due to earthquakes in Europe and the Middle East. Bull Earthquake Eng 12:341–358

Ewald M, Igel H, Hinzen KG, Scherbaum F (2006) Basin-related effects on ground motion for earthquake scenarios in the Lower Rhine Embayment. Geophys J Int 166(1):197–212

Faccioli E, Maggio F, Paolucci R, Quarteroni A (1997) 2D and 3D elastic wave propagation by a pseudo-spectral domain decomposition method. J Seismol 1:237–251

Fioravante V, Giretti D (2012). Amplificazione sismica locale e prove dinamiche in centrifuga. Parma, 11 aprile 2012 (in Italian)

Forsyth PJ, Barrell DJA, Jongens R (2008) Geology of the Christchurch Area. Institute of Geological and Nuclear Sciences. 1:250 000 geological map 16, 1 sheet + 67 pp. GNS Science, Lower Hutt, New Zealand

Furumura T, Hayakawa T (2007) Anomalous propagation of long-period ground motions recorded in Tokyo during the 23 October 2004 Mw 6.6 Niigata-ken Chuetsu, Japan, earthquake. Bull Seismol Soc Am 97(3):863–880

Gallovič F, Brokešová J (2007) Hybrid k-squared source model for strong ground motion simulations: introduction. Phys Earth Planet Int 160:34–50

Gallovič F, Kaser M, Burjanek J, Papaioannou C (2010) Three-dimensional modeling of near-fault ground motions with nonplanar rupture models and topography: case of the 2004 Parkfield earthquake. J Geophys Res 115(B3), B03308

Graves RW (1996) Simulating seismic wave propagation in 3D elastic media using staggered-grid finite differences. Bull Seismol Soc Am 86:1091–1106

Graves R, Jordan T, Callaghan S, Deelman E, Field E, Juve G, Kesselman C, Maechling P, Mehta G, Milner K, Okaya D, Small P, Vahi K (2011) CyberShake: a physics-based seismic hazard model for Southern California. Pure Appl Geophys 168(3–4):367–381

Guidotti R, Stupazzini M, Smerzini C, Paolucci R, Ramieri P (2011) Numerical Study on the Role of Basin Geometry and Kinematic Seismic Source in 3D Ground Motion Simulation of the 22 February 2011 Mw 6.2 Christchurch Earthquake. Seismol Res Lett 82(6):767–782

Hisada Y (1994) An efficient method for computing Green's function for a layered half-space with sources and receivers at close depths. Bull Seismol Soc Am 84:1456–1472

Herrero A, Bernard P (1994) A kinematic self-similar rupture process for earthquakes. Bull Seismol Soc Am 84:1216–1228.

Isbiliroglu Y, Taborda R, Bielak J (2013) Coupled soil-structure interaction effects of building clusters during earthquakes. Earthq Spectra. doi:10.1193/102412EQS315M

Kaser M, Dumbser M (2006) An arbitrary high order discontinuous Galerkin method for elastic waves on unstructured meshes I: the two-dimensional isotropic case with external source terms. Geophys J Int 166:855–877

Kim E, Bielak J, Ghattas O (2003) Large-scale Northridge earthquake simulation using octree-based muti-resolution mesh method. In: Proceedings of the 16th ASCE engineering mechanics conference. University of Washington, Seattle

Koketsu K, Miyake H, Afnimar, Tanaka Y (2009) A proposal for a standard procedure of modeling 3D velocity structures and its application to the Tokyo Metropolitan area, Japan. Tectonophysics 472:290–300

Komatitsch D, Vilotte JP (1998) The spectral-element method: an efficient tool to simulate the seismic response of 2D and 3D geological structures. Bull Seismol Soc Am 88:368–392

Kosloff R, Kosloff D (1986) Absorbing boundaries for wave propagation problems. J Comput Phys 63(2):363–376

Kramer SL (1996) Geotechnical earthquake engineering. Prentice-Hall, Upper Saddle River

Krishnan SJC, Komatitsch D, Tromp J (2006) Case studies of damage to tall steel moment-frame building in Southern California during large San Andreas earthquakes. Bull Seismol Soc Am 96(4A):1523–1537

Lee SJ, Chen HW, Liu Q, Komatitsch D, Huang BS, Tromp J (2008) Three-dimensional simulations of seismic-wave propagation in the Taipei basin with realistic topography based upon the Spectral Element method. Bull Seismol Soc Am 98(1):253–264

Luzi L, Pacor F, Ameri G, Puglia R, Burrato P, Massa M, Augliera P, Franceschina G, Lovati S, Castro R. (2013) Overview on the strong-motion data recorded during the May–June 2012 Emilia Seismic Sequence. Seismol Res Lett 84:4. doi:10.1785/0220120154

Mai M (2004) SRCMOD: online database of finite source rupture models. http://www.seismo.ethz.ch/srcmod/

Margheriti L, Azzara RM, Cocco M, Delladio A, Nardi A (2000) Analysis of borehole broadband recordings: test site in the Po Basin, northern Italy. Bull Seismol Soc Am 90(6):1454–1463

Martelli L, Molinari FC (2008) Studio geologico finalizzato alla ricerca di potenziali serbatoi geotermici nel sottosuolo del comune di mirandola. Technical Report, Regione Emilia-Romagna, Servizio geologico sismico e dei suolo (*in Italian*)

Mazzieri I, Stupazzini M, Guidotti R, Smerzini C (2013) SPEED: SPectral Elements in Elastodynamics with Discontinuous Galerkin: a non-conforming approach for 3D multi-scale problems. Int J Numer Meth Eng 95(12):991–1010

McGarr A, Fletcher JB (2007) Near-fault peak ground velocity from earthquake and laboratory data. Bull Seismol Soc Am 97:1502–1510

Milana G, Bordoni P, Cara F, Di Giulio G, Hailemikael S, Rovelli A (2013) 1D velocity structure of the Po River plain (Northern Italy) assessed by combining strong motion and ambient noise data. Bull Earthquake Eng. doi:10.1007/s10518-013-9483-y

Molnar S, Cassidy JF, Olsen KB, Dosso SE, He J (2014) Earthquake ground motion and 3D Georgia Basin amplification in Southwest British Columbia: shallow blind-thrust scenario earthquakes. Bull Seismol Soc Am 104:321–335

Morozov IB (2008) Geometrical attenuation, frequency dependence of Q, and the absorption band problem. Geophys J Int 175:239–252

Motazedian D, Atkinson GM (2005) Stochastic finite-fault modeling based on a dynamic corner frequency. Bull Seismol Soc Am 95:995–1010

Olsen KB, Day SM, Minster JB, Cui Y, Chourasia A, Faerman M, Moore R, Maechling P, Jordan T (2006) Strong shaking in Los Angeles expected from southern San Andreas earthquake. Geophys Res Lett 33(7), L07305

Olsen KB, Day SM, Minster JB, Cui Y, Chourasia A, Okaya D, Maechling P, Jordan T (2008) TeraShake2: Spontaneous rupture simulations of Mw 7.7 earthquakes on the Southern San Andreas fault. Bull Seismol Soc Am 98(3):1162–1185

Pilz M, Parolai S, Picozzi M, Wang R, Leyton F, Campos J, Zschau J (2010) Shear wave velocity model of the Santiago de Chile basin from ambient noise measurements: a comparison of proxies for seismic site conditions and amplification. Geophys J Int 182:355–367

Pitarka A, Irikura K, Iwata T, Sekiguchi H (1998) Three-dimensional simulation of the near-fault ground motion for the 1995 Hyogo-Ken Nanbu (Kobe), Japan, earthquake. Bull Seismol Soc Am 88(2):428–440

Porter K, Jones L, Cox D, Goltz J, Hudnut K, Mileti D, Perry S, Ponti D, Reichle M, Rose AZ, Scawthorn CR, Seligson HA, Shoaf KI, Treiman J, Wein A (2011) The ShakeOut Scenario: a hypothetical Mw7.8 earthquake on the southern San Andreas fault. Earthq Spectra 27:239–261

Project S2 (2013) https://sites.google.com/site/ingvdpc2012progettos2/risks-1

Quarteroni A, Sacco R, Saleri F (2007). Numerical mathematics, volume 37 of texts in applied mathematics, 2nd edn. Springer, Berlin

Roten D, Olsen KB, Pechmann JC, Cruz-Atienza VM, Magistrale H (2011) 3D simulations of M 7 earthquakes on the Wasatch Fault, Utah, Part I: Long-period (0–1 Hz) ground motion. Bull Seismol Soc Am 101:2045–2063

Roten D, Olsen KB, Pechmann JC (2012) 3D simulations of M 7 earthquakes on the Wasatch fault, Utah, Part II: Broadband (0–10 Hz) ground motions and Nonlinear soil behavior. Bull Seismol Soc Am 102:2008–2030

Sabetta F, Pugliese A (1996) Estimation of response spectra and simulation of nonstationary earthquake ground motions. Bull Seismol Soc Am 86:337–352

Seyhan E, Stewart JP, Graves RW (2013) Calibration of semi-stochastic procedure for simulating high-frequency ground motions. Earthq Spectra 29:1495–1519

Smerzini C, Villani M (2012) Broadband numerical simulations in complex near field geological configurations: the case of the MW 6.3 2009 L'Aquila earthquake. Bull Seismol Soc Am 102:2436–2451

Smerzini C, Paolucci R, Stupazzini M (2011) Comparison of 3D, 2D and 1D approaches to predict long period earthquake ground motion in the Gubbio Plain, Central Italy. Bull Earthquake Eng 9(6):2007–2029

Stacey R (1988) Improved transparent boundary formulations for the elastic-wave equation. Bull Seismol Soc Am 78:2089–2097

Stidham C, Antolik M, Dreger D, Larsen S, Romanowicz B (1999) Three-dimensional structure influences on the strong-motion wavefield of the 1989 Loma Prieta earthquake. Bull Seismol Soc Am 89(5):1184–1202

Stupazzini M, Paolucci R, Igel H (2009) Near-fault earthquake ground-motion simulation in the Grenoble valley by a high-performance Spectral Element code. Bull Seismol Soc Am 99:286–301

Taborda R, Bielak J, Restrepo D (2012) Earthquake ground-motion simulation including nonlinear soil effects under idealized conditions with application to two case studies. Seismol Res Lett 83 (6):1047–1060

Tizzani P, Castaldo R, Solaro G, Pepe S, Bonano M, Casu F, Manunta M, Manzo M, Pepe A, Samsonov S, Lanari R, Sansosti E (2013) New insights into the 2012 Emilia (Italy) seismic sequence through advanced numerical modeling of ground deformation InSAR measurements. Geophys Res Lett 40:1–7

Villani M, Faccioli E, Ordaz M, Stupazzini M (2014) High resolution seismic hazard analysis in a complex geological configuration: the case of Sulmona (Central Italy) basin. Earthq Spectra (in press). doi:10.1193/112911EQS288M

Wald DJ, Graves RW (1998) The seismic response of the Los Angeles basin, California. Bull Seismol Soc Am 88(2):337–356

Chapter 11
A Seismic Performance Classification Framework to Provide Increased Seismic Resilience

Gian Michele Calvi, T.J. Sullivan, and D.P. Welch

Abstract Several performance measures are being used in modern seismic engineering applications, suggesting that seismic performance could be classified a number of ways. This paper reviews a range of performance measures currently being adopted and then proposes a new seismic performance classification framework based on expected annual losses (EAL). The motivation for an EAL-based performance framework stems from the observation that, in addition to limiting lives lost during earthquakes, changes are needed to improve the resilience of our societies, and it is proposed that increased resilience in developed countries could be achieved by limiting monetary losses. In order to set suitable preliminary values of EAL for performance classification, values of EAL reported in the literature are reviewed. Uncertainties in current EAL estimates are discussed and then an EAL-based seismic performance classification framework is proposed. The proposal is made that the EAL should be computed on a storey-by-storey basis in recognition that EAL for different storeys of a building could vary significantly and also recognizing that a single building may have multiple owners.

A number of tools for the estimation of EAL are reviewed in this paper and the argument is made that simplified methods for the prediction of EAL are required as engineers transition to this new performance parameter. In order to illustrate the potential value of an EAL-based classification scheme, a three storey RC frame building is examined using a simplified displacement-based loss assessment procedure and performance classifications are made for three different retrofit options. The results show that even if only limited non-structural interventions are made to the case study, the EAL could be significantly reduced. It is also argued that overall,

G.M. Calvi (✉)
IUSS Pavia, Pavia, Italy
e-mail: gm.calvi@eucentre.it

T.J. Sullivan
Department of Civil Engineering and Architecture, University of Pavia, Pavia, Italy

D.P. Welch
ROSE Programme, UME School, IUSS Pavia, Pavia, Italy

such a performance classification, coupled with some form of government or insurance-driven incentive scheme, may provide an effective means of reducing the risk, and increasing the resilience, of our societies.

11.1 Introduction

Looking back at how the subject of earthquake engineering has developed, we have observed what went wrong in earthquakes, learnt from these events and subsequently developed an engineering approach (building codes, analysis tools and construction techniques) that one could argue provides our communities with an acceptable level of seismic risk. However, as communities develop, it is also apparent that the definition of what is an acceptable level of risk changes. Some 40 years ago, it would appear that the intention of seismic design and retrofit was solely to ensure that the probability of loss of life during an earthquake was acceptably low. However, following earthquakes such as the Northridge earthquake in 1994 and the more recent 2011 Christchurch earthquake, it is becoming increasingly clear that the protection of lives is not enough. Financial losses associated with repair, disruption to businesses and the time lost to clean up and reinstate services and activities, are just a number of important factors that need to be considered in a modern definition of seismic risk, and which are already entering into performance-based earthquake engineering procedures, as will be discussed shortly.

Another means of considering performance and risk is to focus on disaster resilience. Also here, as has been discussed by experts in the field (e.g. Comerio 2012), even if the number of lives lost in an earthquake are low, individuals and communities cannot return to their normal way-of-life unless they have jobs and housing, and if the community services (transport systems, schools, hospitals, banks, businesses and governments) are functioning properly. The best means of quantifying resilience is arguably still to be identified, with various resilience indicators in the literature (see Comerio 2012). However, it is clear that an engineering approach that focusses solely on the concept of life-safety will not ensure resilient communities.

With the above points in mind, this paper will review modern measures of performance and propose a new performance classification scheme that is based only on expected monetary losses. It will be argued that, whilst the important issue of life safety should not be forgotten, a monetary loss-based performance scheme could offer an effective means of reducing risk and increasing resilience, provided that it is used together with suitable government incentive schemes to motivate retrofit and improvements.

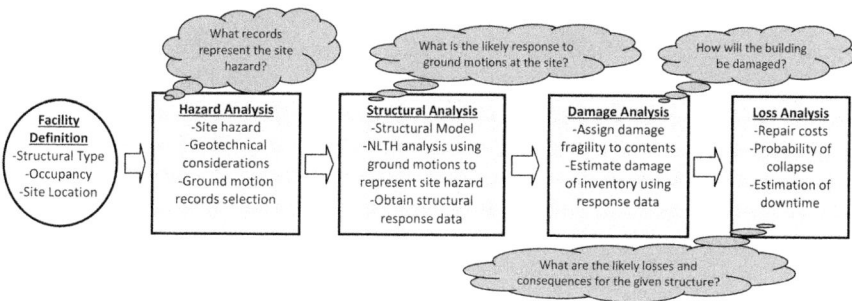

Fig. 11.1 Overview of the four stages of the PEER PBEE framework

11.2 Modern Measures of Performance

Performance measures offer engineers a means of quantifying and communicating risk. As explained in the introduction, until recently the main concern for seismic engineers was the risk of loss of life. However, since the nineties (and arguably before that time in some parts of the world where serviceability limit state checks were in place since the seventies), a need for additional performance measures has arisen, in response to the need to reduce other risks posed by earthquakes, including the high repair costs and disruption (loss of time and social upset) that earthquakes can cause. In response to this there have been a series of initiatives (SEAOC 1995; ATC 2011a) aimed at developing performance-based earthquake engineering (PBEE) approaches. The most refined PBEE procedure currently available appears to be the framework developed for the PEER PBEE methodology (Porter 2003) which offers engineers a means of quantifying performance measures of deaths, dollars and downtime (the "3 D's") by following the approach outlined in Fig. 11.1. Referring to Fig. 11.1, the PEER PBEE framework consists of defining the facility type and location followed by four analysis stages: hazard analysis, structural analysis, damage analysis and loss (decision) analysis.

The four stages allow for each aspect of the seismic assessment to be treated in a probabilistic manner where inherent uncertainties are incorporated within a given stage and carried through to subsequent stages of the assessment process. In order to better illustrate how this is performed, a mathematical relationship in the form of a triple integral is shown in (11.1). Notably, the terms in (11.1) are displayed for the calculation of consequences from damage across all seismic intensities, yet a similar form is applicable to other consequences or decision variables (DV).

$$\lambda[DV|D] = \iiint p[DV|DM]p[DM|EDP]p[EDP|IM]\lambda[IM]dIMdEDPdDM \quad (11.1)$$

The terms $\lambda[x|y]$ and $p[x|y]$ represent the mean annual occurrence rate and probability density of x given y. The design, D, represents the structure and site to be assessed, where all building details are specific to D and site hazard characteristics are addressed in order to obtain the occurrence relationship of a given intensity measure, $\lambda[IM]$. Site hazard is typically defined by a Probabilistic Seismic Hazard Analysis (PSHA) which allows for the site hazard to be related to an *IM* of interest (e.g. 1st mode spectral acceleration, $S_a(T_1)$) via proper selection of accelerograms for input into the structural analysis stage. The structural analysis stage is perhaps the most familiar to the engineering community where a model of the structure is developed in order to run nonlinear time history analyses (NLTH) to obtain likely response quantities; defined here as engineering demand parameters (*EDP*s). The output of the structural analysis stage results in probabilistic distributions of *EDP*s such as inter-storey drift and floor acceleration that are associated with a given level of seismic intensity, $p[EDP|IM]$. These *EDP*s are then used to estimate the damage of various assemblies within a building within the damage analysis stage. The relationship between structural response (*EDP*) and a given damage measure (*DM*) is represented by fragility functions (cumulative distribution of $p[DM|EDP]$) that are assigned to various components within the building (e.g. columns, partitions and ceilings). Each set of *DM*s for a given component are sufficiently separated to represent distinct methods and extent of repair; with each *DM* having an associated decision variable distribution ($p[DV|DM]$), in this case repair cost, associated with it. Remaining consistent with the formulation of (11.1), the final result of the triple integral would represent the mean annual occurrence of repair cost for the given building and site, $\lambda[DV|D]$.

The previous description of the PEER PBEE methodology represents only one metric of performance (annualized repair cost due to damage), yet the seismic performance can consider numerous sources of loss (e.g. the 3 D's) expressed in a variety of metrics. These metrics can be annualized, such as expected annual loss (EAL), to allow losses to be treated as an expense within cash flow analysis (Porter et al. 2004), based on a given intensity such as that corresponding to a design level event, or based on a given scenario possibly recreating a previous or anticipated event of known magnitude and distance (ATC 2011a). Further, loss metrics can be expressed based on input from decision makers such as the annual or 50 year probability that losses will exceed a given value, such as probable maximum loss (PML).

The PEER framework for performance assessment is attractive since it is quite clear and very flexible, noting that no restrictions are imposed on the approach used to quantify hazard, to undertake the structural analysis, relate EDPs to losses and other performance measures. To this extent, it is also apparent that the results of a performance-assessment conducted using the PEER PBEE procedure will currently lead to quite different measures of performance depending on the assumptions made in applying the procedure and the risk parameters of interest. The following sub-sections review considerations currently made when estimating life-safety, monetary losses and downtime, and identify some of the factors that will affect their quantification.

11.2.1 Life-Safety and Probability of Collapse

The inherent risk of a structure to collapse and subsequently endanger lives has been the primary concern of earthquake engineering since the earliest seismic provisions were adopted. Further, the ongoing efforts within the field of seismic design over the past four decades have made great strides in controlling the collapse risk of structures. However, only until recently have advances in computing power, experimental testing and engineering seismology allowed analysts to quantify life safety and collapse risks probabilistically. Conceptually, the estimation of the likelihood of loss-of-life is explained by three basic requirements: (i) determine the ways in which a structure can endanger life, (ii) relate critical structural conditions to the likelihood of the seismic hazard producing them and (iii) establish an estimate of the number of lives exposed to the dangerous conditions. However, numerous factors challenge the estimation of collapse probability and consequential risk of loss-of-life.

Rather intuitively, a majority of fatalities occur when at least a portion of a structure collapses (Hengjiam et al. 2003). However, although small in comparison, there are still a number of fatalities that can be attributed to the damage of non-structural elements (e.g. masonry partitions, large equipment, failed exteriors) or building contents (e.g. furniture) (Durkin and Thiel 1992; Stojanovski and Dong 1994; Hengjiam et al. 2003). Alternatively, as non-structural damage may not be a significant source of fatalities, resulting injuries may be substantial (Porter et al. 2006) which leads to another, at least viable, consideration in seismic risk assessment. Further discussion of life and injury risks associated with non-structural hazards is omitted for the sake of brevity, yet it is noted that this source of risk has received wide attention in recent years (Charleson 2007; ICC-ES 2010; FEMA 2011).

Given the complexity of the physical interactions of a building at imminent collapse, the first major challenge lies within capturing these complexities in a reliable manner within mathematical models for computer simulations of earthquake demands. For more modern (ductile) structures, current seismic provisions mandate that certain strength hierarchy be followed (e.g. SCWB ratio, flexure-controlled members) to ensure a ductile response and indirectly force a sideway or global collapse mechanism. Although numerous methods and tools have been made available for the modelling of structural members, as a result of countless experimental campaigns (Ibarra et al. 2005; Berry et al. 2004; Lignos 2013; Lignos and Krawinkler 2011; among others) the intricacy associated with even a "ductile" collapse mode require that numerous uncertainties must be accounted for. In light of state-of-the-art assessment methods such as the PEER PBEE methodology, the probability of global collapse of a structure is addressed with a collapse fragility function (typically a cumulative lognormal distribution) requiring that the median collapse intensity be estimated and the corresponding dispersion to represent uncertainty. Estimation of the median collapse intensity can be performed by various methods (ATC 2011a; FEMA 2009; Mohammadjavad et al. 2013;

Vamvatsikos and Cornell 2006). The collapse dispersion must address uncertainty involved in both demand (record-to-record) and capacity (modelling) with the former requiring a large number of time history simulations (e.g. IDA, Vamvatsikos and Cornell 2002) or reliable approximation (Perus et al. 2013). The latter source of uncertainty is typically benchmarked through parametric studies (e.g. Haselton and Deierlein 2007) and then adjusted based on the judgment of the analyst in terms of level of knowledge of the structure (e.g. details, materials, construction quality) adequacy of the structural model (ATC 2011a; FEMA 2009).

When dealing with older structures that lack strength hierarchy provisions and proper detailing, numerous additional modes of failure can be expected (e.g. joint failure, shear failure, punching shear of slab-column connections) other than a global sidesway collapse. This combined with current limitations of modelling and simulation capabilities (Liel and Deierlein 2008) requires that the collapse probability become a two staged problem. Initially the probability of a sidesway collapse is estimated using methods similar to ductile structures, and then a subsequent assessment must be made with simulations that did not produce collapse in order to estimate the probability of brittle or non-simulated modes of failure. Taking the shear failure of a column as an example, the expected deformation capacity of the column corresponding to a brittle shear failure would be estimated based on structural properties (e.g. material, axial load, detailing) and available experimental data in order obtain a fragility function similar to that used to estimate global collapse (Aslani and Miranda 2005). Further, the influence of joint deterioration could be captured in the structural model (Altoonash 2004; Pampanin et al. 2003) which would affect the expected structural deformation and subsequently influence the likelihood of a brittle collapse mode.

An additional challenge of estimating the collapse risk of a structure lies within associating a given structural demand to a proper representation of seismic hazard in order to convey collapse risk. As current assessment methods rely heavily on NLTH analysis, accelerograms must be selected to represent the expected seismic demands. Although numerous factors must be considered with record selection in general (e.g. Baker and Cornell 2006a; Iervolino et al. 2006; Kalkan and Kunnath 2006), the use of accelerograms in collapse studies becomes an even more daunting task as recorded data from very large events is just as rare as the events that produce them; with the recent improvements in seismic design producing structures that are expected to have median collapse intensities on the order of 2–3 times that expected for the 2 % in 50 year probability of exceedence intensity which typically corresponds to the maximum credible earthquake (Haselton and Deierlein 2007). As such, the proper treatment of the uncertainty associated with these rare events is critical when conducting collapse assessments. A very important characteristic of very rare ground motions is that of spectral shape; an importance that is a result of structural analysts' use of first-mode spectral acceleration as an intensity measure in collapse assessments. Briefly, spectral shape for rare ground motions (e.g. 2 % in 50 year intensity) must be properly considered because they can significantly differ from the corresponding uniform hazard (UHS) or design spectra (Baker and Cornell 2006b). The main issue relating to the prediction of collapse is that rare ground motions have a much longer return period, T_R, (e.g. 2,475 years) compared to the return period of the events that cause them (e.g. 150–500 years in the Western U.S.) requiring that this rarity be accounted for (FEMA 2009). This is typically done with an epsilon factor, ε, that relates the number of standard deviations above (or below) a median hazard spectrum for a given T_R and structural period (Baker and Cornell 2006b). Although this concept is not the most recent development, it is deemed important in the context of collapse assessment where failing to incorporate some procedure to consider epsilon (i.e. Haselton et al. 2011) has lead to collapse capacities to be underestimated by 30–80 % (FEMA 2009).

In order to estimate the number of fatalities due to the collapse of a structure, the type of failure mode must be considered with respect to how many building occupants will be exposed to dangerous or lethal conditions. This has been quantified previously as a collapsed volume ratio (CVR) expressed as a percentage of the building that completely collapses in previous efforts to estimate life safety risk; where reconnaissance data has shown it to be a good indicator of the level of fatalities within a structure (Coburn et al. 1992; Yeo and Cornell 2003). The uncertainties in estimating this parameter are even more difficult that assessing the collapse probability due to the lack of data on the subject and typically must rely on judgment. To illustrate the different considerations for estimating CVR the assumptions made by Liel and Deierlein (2008) in the assessment of reinforced concrete (RC) frame buildings are used as an example.

The data in Tables 11.1 and 11.2 illustrate how the CVR is estimated provided that a global side-sway collapse is expected. The initial CVR is estimated via NLTH analysis in terms of the number of stories involved in the collapse mechanism which can vary significantly depending on the number of stories and expected ductility of the building as shown in Table 11.1. Additionally, the likelihood of a side-sway collapse causing a complete collapse of every storey (i.e. pancake collapse) must also be estimated. An example set of values for the likelihood of a pancake collapse provided that side-sway collapse occurs is presented in Table 11.2.

Notably, the values are based on judgment, yet reflect two basic principles: i) ductile structures have a higher deformation capacity which could involve more stories in the collapse mechanism and ii) taller structures are more susceptible to secondary effects (e.g. P-delta) as shown with respect to the expected ductility and height of the building in Table 11.2 (Liel and Deierlein 2008).

When collapse is conditioned on a local brittle failure (e.g. shear) the fact that a soft-storey mechanism involving only one storey initially may lead to subsequent failure of additional stories (i.e. progressive collapse) must also be considered. The event tree shown in Fig. 11.2 shows how different modes of collapse may lead to different estimations of the collapsed volume ratio (CVR).

Once the likely percentage of the building that has collapsed in estimated, the fatality probability is calculated by estimating the number of lives expected within that area of the building. This is currently achieved by attributing a population model to the structure. Population models vary according to the use or occupancy of the building. Two examples are provided in Fig. 11.3 for a commercial office

Table 11.1 Example of variations in collapsed volume ratio for RC frame buildings (abridged from Liel and Deierlein 2008)

# of stories	Ductility of RC frame	Collapsed volume ratio[a]
4	Ductile	0.38–0.52
	Non-ductile	0.5–0.62
8	Ductile	0.15–0.28
	Non-ductile	0.27–0.43
12	Ductile	0.08–0.24
	Non-ductile	0.2–0.29

[a] Estimated from nonlinear time history analyses

Table 11.2 Assumed probability of side-sway collapse triggering pancake collapse based on height and ductility (Liel and Deierlein 2008)

# of stories	Ductility of RC frame	Probability side-sway collapse leads to pancaking P[Pancake|C]
≤4	Ductile	0.3
	Non-ductile	0.15
≥8	Ductile	0.6
	Non-ductile	0.3

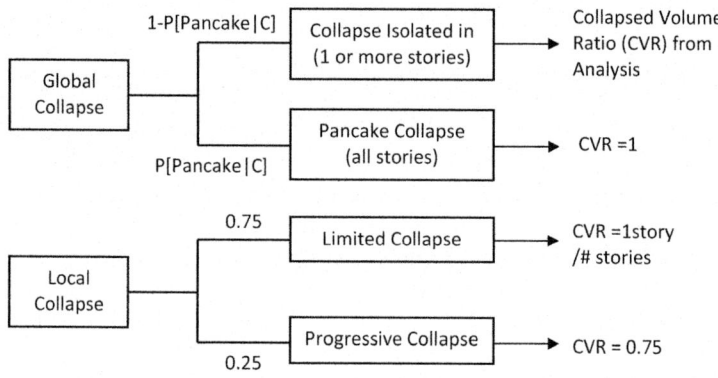

Fig. 11.2 Example of an event tree to determine the collapsed volume ratio of a structure conditioned on either a global or local collapse for the estimation of fatalities (Adapted from Liel and Deierlein 2008)

building and a healthcare facility (e.g. hospital). The figure shows that it is likely that the office building will be vacant overnight and the occupancy is drastically reduced on the weekend. Conversely, the hospital model expects a minimum of 2 people per 1,000 ft^2 (93 m^2) at all times and only a small reduction in population on the weekend. Notably the population models represent expected values and additional uncertainty may be incorporated as well as additional time frames for population variation (e.g. monthly).

Although the probability of the loss-of-life may be estimated, it may be in the decision-makers best interest to also estimate the economic impact of the expected

11 A Seismic Performance Classification Framework to Provide Increased...

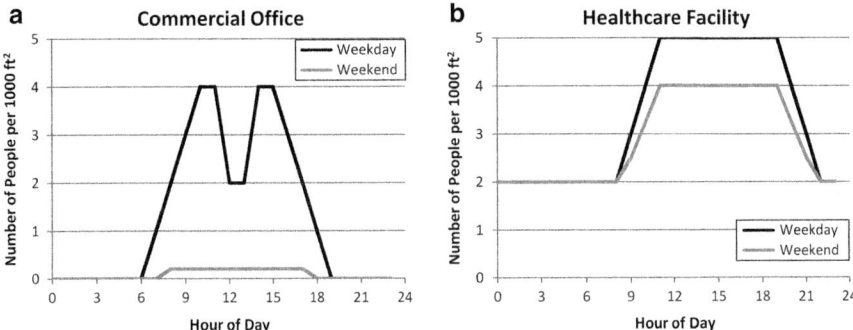

Fig. 11.3 Illustration of different population models used for life safety assessment: (**a**) commercial office, (**b**) healthcare facility (Values taken from ATC 2011b)

life safety risk of a structure or facility. Attributing a price to human life comes with both moral and economic challenges, yet this is usually necessary in order to compare the benefits of allocating monetary resources to protect public welfare; both by municipalities and decision makers within the private sector. This is typically done by estimating the value of a statistical life, VSL (FHWA 1994; Mrozek and Taylor 2002). Values can depend on the amount an industry is willing to pay to preserve life safety for a particular type of risk (Liel and Deierlein 2008) or even considering a life quality index based on a country's gross domestic product (per capita) and life expectancy (Rackwitz 2004).

11.2.2 Direct Monetary Losses

The calculation of seismic losses can have numerous sources as previously mentioned (e.g. the 3 D's). However, it is useful to make a distinction between the types of losses based on how they may affect decision making. The term direct loss is typically attributed to monetary loss from repair costs due to damage and full replacement costs in the case of a structural collapse (Mitrani-Reiser 2007; Welch et al. 2014). The remaining losses associated with other sources of loss are termed indirect losses herein. It is noted that the damage of building contents (e.g. furniture, office equipment) can also be a significant source of direct loss (Comerio et al. 2001), yet the current discussion will be limited to only the structure and its non-structural components.

The calculation of direct losses due to repair costs requires that (ideally) each damageable component within a building has a specific damage fragility and consequence function attributed to it in order to transition from structural response to damage and then repair cost in line with the progression shown in Fig. 11.1. A sample set of fragility and consequence functions are shown in Figs. 11.4 and 11.5 for a ductile interior RC beam-column joint. Figure 11.4 illustrates that as inter-

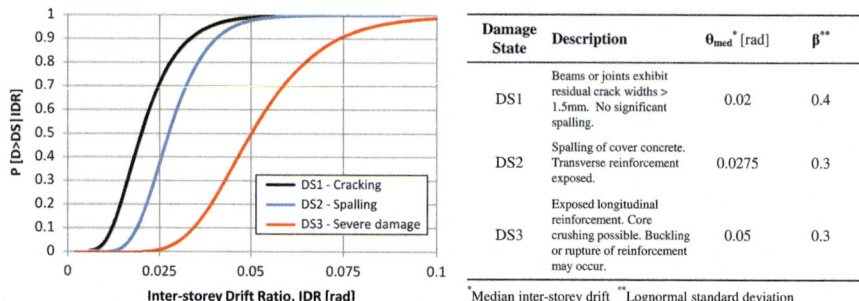

Fig. 11.4 Sample fragility function (*left*) and damage state parameters (*right*) for a modern interior RC beam-column joint (Values taken from ATC 2011b)

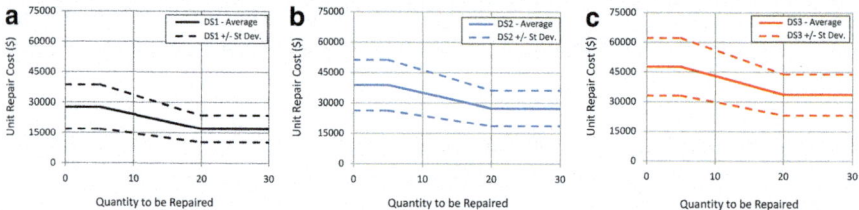

Fig. 11.5 Repair costs for various damage states of a modern interior RC beam-column joint: (**a**) significant cracking, (**b**) spalling and (**c**) severe damage (Values in 2011 USD from ATC 2011b)

storey drift ratio (IDR) increases the likelihood of each successive (more damaging) damage state also increases; where an IDR of 5.0 % will return that almost certainly the element has significant cracking and spalling and there is a 50 % probability that the element has suffered severe damage.

To estimate the repair cost associated with a given damage state, the corresponding consequence function (Fig. 11.5) is used. Notably, Fig. 11.5 displays the mean estimated repair cost (solid line) as well as the plus and minus one standard deviation bounds (dashed lines) which highlights the uncertainty associated with estimating repair costs following a seismic event. Further, the cost functions relate the unit repair cost to the total number of units to be repaired, showing a reduction in unit cost as the total increases which represents the reduction in labor required (e.g. set-up time, transport of materials) to repair numerous elements in the same building. Further, the availability of materials and human resources may fluctuate significantly, yet these types of factors will be discussed more thoroughly in the following section.

Aside from the need for additional experimental testing in order to produce more reliable and component-specific fragility and consequence functions, the next greatest challenge in estimating repair costs could be the appropriate consideration of the damageable assemblies within a building. Since structural elements are of manageable quantities within a structure the largest source of this difficulty is rooted in repairs associated with non-structural elements. Although a vast range

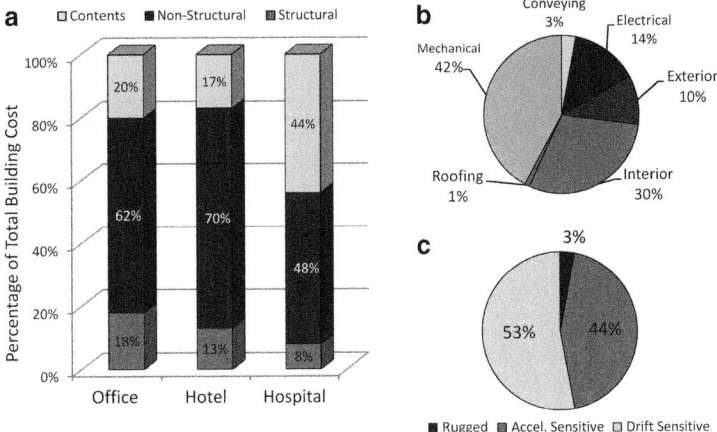

Fig. 11.6 (a) Summary of relative value of non-structural elements for three different occupancies, (b) Relative contribution of different non-structural element classes for a given building and (c) Example EDP sensitivity of non-structural elements within a building (Values from Taghavi and Miranda 2003)

of components complete a fully functional facility it is not only their quantities that make non-structural elements a critical part of estimating direct losses due to repair costs.

The importance of non-structural damage in direct loss assessment is mostly derived from the fact that non-structural elements comprise a significant portion (or majority) of the total construction costs of a building (see Fig. 11.6a) and many non-structural elements are damaged at seismic intensities much lower than structural elements. This importance is reflected in the tremendous losses associated with non-structural damage in previous seismic events (Miranda et al. 2012; Filiatrault et al. 2001; Reitherman and Sabol 1995).

In order to incorporate non-structural elements into a comprehensive loss framework, the various types of non-structural components that compose the inventory of a building (Fig. 11.6b) must be assigned engineering demand parameter (EDP) sensitivity. Typical sensitivities are (but are not limited to) inter-storey drift ratio (IDR) and peak floor acceleration (PFA). Additionally, many components within the building may not be affected by building response and are only treated as a loss in the event of collapse; these components are typically termed "rugged". An example sensitivity distribution is shown in Fig. 11.6c.

There are numerous ways in which this discretization of non-structural elements can be carried out. First, there is the component-based (or assembly-based) approach where the damageable assemblies are identified and assigned fragility and consequence functions based on available information (Mitrani-Reiser 2007; Porter et al. 2001). Additionally recent studies (Ramirez and Miranda 2009, 2012; Welch et al. 2012) have also implemented a storey-based loss model developed by Ramirez and Miranda (2009) which combines the likely structural and non-structural inventory into a set of engineering demand parameter to decision variable functions (EDP-DV). The two loss modelling aproaches differ significantly and each has its own inherent benefits and drawbacks.

The component-based model is advantageous in that it allows the actual component inventory to be represented (e.g. 12 beams/floor, 600 m^2 of ceiling/floor) whereas the storey-based model relies on relative inventories based on construction estimating documents. The storey-based approach is advantageous not only due to its simplicity (provided that EDP-DV functions have been constructed) but also eliminates the need to select the type and number of damageable assemblies. This can lead to repair costs that may or may not reflect the total damaged inventory, yet other component-based studies (Krawinkler 2005) have used "generic" fragility functions in order to consider components that do not have available fragilities based on experimental results. Further, the storey-based model avoids allocating repair cost to an element that must also be repaired in order to repair another or "double counting"; with the simplest example being the replacement of partition walls in order access structural members for repair, where considered separately the partition cost could be counted twice. However, this problem can be overcome by careful formulation of a component-based model which would indeed consider the building most accurately if formulated properly.

The allocation of direct losses based on collapse typically attribute the building replacement cost to the probability of collapse for a given intensity. However, there are a number of additional factors that may be considered when estimating direct losses due to collapse. The influence of residual displacements can significantly affect loss estimates (Ramirez and Miranda 2012) and their consideration could prove critical to accurately represent post-event conditions; based on previous reconnaissance where significant residual drifts can render a structure a complete loss without actually collapsing (Mahin and Bertero 1981; Rosenbluth and Meli 1986; Anderson and Fillipou 1995). Additonally, the direct loss based on collapse assumes a total loss in monetary terms, yet it may be difficult to properly consider expected increases in cost due to demolition before new construction can begin or even the increased cost to tear down a building that has experienced excessive residual deformation.

11.2.3 Indirect Losses and Downtime

The third and final source of seismic loss is downtime. The estimation of downtime is perhaps the most difficult to achieve of all of the 3 D's. Predominately since this metric not only involves the numerous considerations that have been discussed thus far, but because it depends on many additional external factors; not only involving a structure experiencing an earthquake, but an entire region or community.

The basic contributions to downtime following a seismic event can be broken up into two components: rational and irrational downtime as defined by Comerio (2006). Rational downtime represents the time needed to repair damage of replace a building.

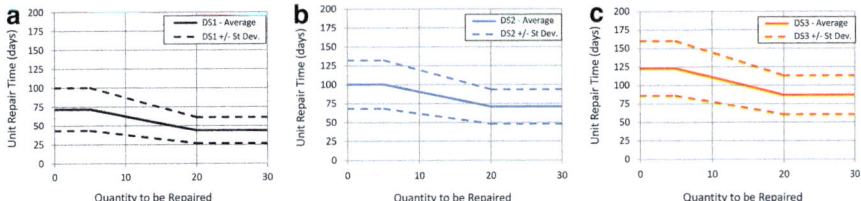

Fig. 11.7 Repair times for various damage states of a modern interior RC beam-column joint: (**a**) significant cracking, (**b**) spalling and (**c**) severe damage (Values from ATC 2011b)

Irrational downtime includes a number of factors including financing and human resources, as well as economic and regulatory uncertainty (Comerio 2006).

The concept of estimating rational downtime is quite similar to the manner in which repair costs are estimated. Using the previous example of an RC beam-column joint, as sample set of expected repair times are shown for three damage states in Fig. 11.7.

The figure shows that the estimated repair time is proportional to the level of damage for the component which is logical. However, noting that the ranges defined by the standard deviation bands (dashed lines) are giving estimates differing by a factor of two which highlights the large uncertainty involved with repair time estimation. Further, considering an entire building requiring repair, these uncertainties would be expected to exacerbate. For the repair of an entire facility, the rational component of downtime relating to mean repair time is a function of: building size (e.g. number of floors, plan area), the number of different trades that are involved (e.g. electrician, drywall installer/finisher) and, similar to the component level, the number of assemblies and the extent of damage. The downtime associated with the number of trades involved also contributes to what is termed change of trade delay where certain tradesman will not be able to access the building until others have completed their tasks. This type of delay can vary significantly depending on the repair scheme adopted (Mitrani-Reiser 2007; Beck et al. 1999). Repair schemes vary in efficiency between the lower bound of a slow-track scheme where all trades are performed in series to a fast-track scheme where (ideally) all trades are performed in parallel. A summary of the rational components of downtime is shown in Fig. 11.8.

The various contributions of irrational downtime are very difficult to estimate. Economic factors such as municipal buildings waiting for a decision on government funding or private facilities negotiating a loan for repairs could vary significantly depending on individuals and the condition of the surrounding area. Similarly, another component of the irrational downtime would be, upon acquisition of funds, the delay for the start up of construction which could involve the development of drawings and repair schemes, bidding for construction, and various levels of engineering assessments; factors that would greatly depend on the relationship of the owner with the engineers, architects and contractors (Comerio 2006). The various components of downtime are summarized in Fig. 11.8.

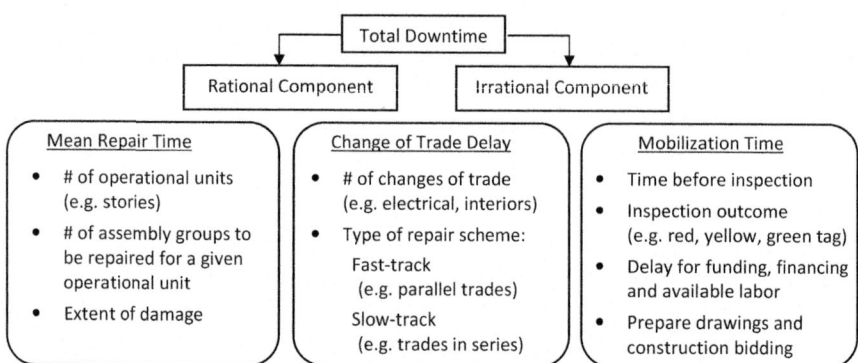

Fig. 11.8 Various aspects that can contribute to the downtime of a building following a seismic event

The outcome of initial engineering inspections has been the primary metric for the estimation of downtime in recent loss assessment studies (Mitrani-Reiser 2007). The procedure for carrying out post-earthquake inspections typically implements a "tagging" system by which buildings can be quickly identified with a commonly adopted green, yellow and red system such as the ATC-20 guidelines (ATC 2005) where:

- Green signifies that the building is "inspected" and occupancy is permitted (bearing in mind that the use of the word permitted here would suggest that the undamaged building was deemed safe),
- Yellow represents the presence of some hazard within the building and receives a "restricted use" placard typically with notes describing the risks and extent of entry and
- Red represents the case of a clear hazard to human life and returns an "unsafe" placard that prohibits any re-entry or occupation of the building.

In order to quantify downtime, Mitrani-Reiser (2007) developed a "virtual inspector" algorithm which simulates the engineering inspection process. As an example of the differences in downtime due to engineering inspection outcomes, Mitrani-Reiser (2007) assumed that the mobilization time associated with a green, yellow and red tag were 10 days, 1 month, and 6 months respectively. Notably, when considering a building that is damaged beyond repair, a downtime of 38 months was attributed. Further, although some estimations must be made in order to quantify downtime, it is mentioned that the time associated with a yellow tag can vary significantly as the purpose of the yellow tag is to provide more in depth inspections to arrive at a final decision of a red tag or possible repair requirements before the issuance of a green tag.

Despite the difficulties in its estimation, downtime following a seismic event can be orders of magnitude in importance above all other sources of seismic loss depending on the scenario. For example, some lease agreements for commercial real estate in seismic areas, such as California, include a window period (typically 270 days) in which building owners must repair damages to avoid a break of the lease agreement (Comerio 2006). Similarly, tenants of the same commercial real estate may be losing valuable clients or contracts for every week or even day they are out of operation. This would be a similar case for industrial buildings that produce a certain product or provide a service. Although building repair is different than business recovery (Chang and Falit-Baiamonte 2002), property owners and tenants will likely be forced to compete within the same pool of (possibly scarce) services and resources which could significantly affect resulting downtime. The concept of "demand surge" for human resources and materials to restore an entire city (or region) facing these types of dilemmas becomes much more apparent.

In light of the importance of downtime, as well as the other sources of seismic loss, mitigation of this risk may be a cumbersome task, yet even small reductions in seismic risk in terms of direct losses or life-safety could translate into tremendous benefits when considering the indirect loss associated with downtime.

11.3 Proposal to Use EAL for Seismic Performance Classification

This section proposes a performance-classification scheme that is based on direct expected annual monetary losses (EAL), with no consideration of life safety or indirect losses. The motivation for the classification scheme is first provided, some limitations with the EAL performance measure are discussed and then a tentative classification framework is proposed.

11.3.1 Motivation for EAL-Based Performance Classification

At first it might appear that a good performance classification scheme should be all-encompassing, considering life-safety, monetary losses and downtime, as well as the other factors considered in the definition of community resilience. However, it is argued here that best performance classification parameter really depends on the intended use of the classification scheme. In this paper it is proposed that an EAL-based performance classification can provide suitable means of motivating retrofit measures that help build community resilience and reduce losses and downtime due to earthquakes. It is argued that the issue of life-safety should be separately addressed by code-requirements; buildings should satisfy minimum requirements for what regards the probability of loss of life, but that these do not form the basis of a performance-classification scheme.

This concept of separating life safety from EAL performance could be considered somewhat analogous to the way that the performance of washing machines and refrigerators is currently quantified; the energy performance rating scheme gives us an idea of the performance of the fridge (or washing machine) in terms of running-costs (energy use) but does not provide any indication of the likelihood that the machine will break down or not. Instead, we tend to rely on brand-names and guarantees to ensure that the likelihood of breakdown is not too high. The benefit of the establishment of the energy-rating performance scheme for home appliances is that it is saving our communities (as well as the individual) money and energy (which is a sustainable initiative important for the environment). In the context of earthquake engineering, such savings are vital as they could help reduce household and business disruption and social impacts of earthquakes. Even though the 2011 Christchurch earthquakes (and other events in modern engineered societies) only caused limited loss of life, the upheaval on the community has been extensive and has taken a long time to recover from. Fortunately, in the case of Christchurch a large proportion of the damage was insured and therefore recovery is easier but it is still taking a long time and the earthquake has clearly caused widespread upset. In other parts of the world, such as Italy, the majority of homeowners and many businesses don't have earthquake insurance and therefore the government either steps in or the local community suffers hugely (or both).

In order to be effective, it is also argued that a performance classification index needs to be coupled with some sort of incentive scheme. In the case of home-appliances the benefit of energy-efficiency to homeowners is clear and immediate. In the case of low-risk building solutions the benefit of improved performance may only become apparent after an intense earthquake event, which has a low probability of occurrence and may never in fact occur during the building owner's lifetime. As such, it is considered that government incentive schemes could provide the suitable motivation to building owners and this could consist of tax-rebates, discounted bank loans or even subsidized building materials. Another possibility is to engage the insurance industry more effectively, ensuring that insurance premiums can be tailored according to the building-specific seismic risk, rather than generic fragility functions for broad building typologies. However, this will require more dialogue with insurance companies who ideally would have some input in defining final performance-classification schemes such as those defined shortly in this paper.

11.3.2 Observed Trends in Expected Annual Loss Estimates

As the implementation of advanced loss assessments is still somewhat rare in the current literature, the results of the PEER benchmark study on modern RC moment-resisting frame (MRF) buildings is the largest source of building-specific loss data currently available. The EAL for thirty 2003 International Building Code (IBC) conforming RC MRF buildings was estimated using two different loss model

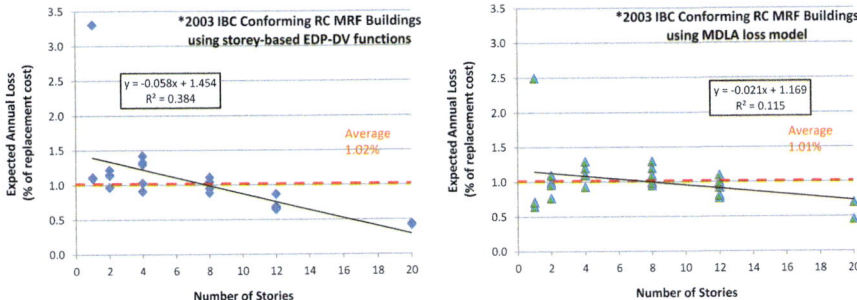

Fig. 11.9 Expected annual loss estimates for 30 different 2003 IBC conforming RC moment frame buildings conducted by Ramirez and Miranda (2009) (*left*) and Ramirez et al. (2012) (*right*)

formulations. Taking the same site hazard and structural analysis as input, the buildings were assessed using a storey-based loss model by Ramirez and Miranda (2009) and a component-based MDLA (Matlab Damage and Loss Analysis) toolbox (Mitrani-Reiser 2007; Beck et al. 2002) reported within Ramirez et al. (2012). The buildings range from one to twenty stories and consider either space-frame or perimeter-frame lateral load systems. Buildings also consider a variety of foundation modelling assumptions (e.g. pinned, fixed, grade beams modelled). The EAL results are shown for the two different loss models in Fig. 11.9. The figure shows that code conforming RC MRF designs have EAL values between 0.5 % and 1.5 % of replacement cost which is a plausible initial benchmark for standard buildings designed to modern seismic codes. Notably, the one story building with higher EAL was treated as an outlier.

The figure also shows a general trend of decreasing EAL with story height. This is quite easily explained by the concentration of damage in only a few stories of taller, more expensive, buildings. Conversely, shorter buildings will have a larger percentage of its stories damaged which can result in larger losses in terms of the percentage of replacement cost. This relationship with height may need to be considered before making further assumptions of generalized EAL values for code conforming buildings. However, the range of 0.5–1.5 % is supported by the previous results for variations of modern 4-storey RC MRF frames reported by Haselton et al. (2008) who found EAL in the range of 0.55–1.07 % of replacement cost.

As part of a continuing effort, Liel and Deierlein (2008) essentially extended the previous benchmark study to include non-ductile structures. The study examines eight different non-ductile 1967 IBC conforming RC MRF designs and compares them with the equivalent 2003 IBC designs that were discussed in the previous section. The buildings consist of perimeter and space frame designs ranging from two to twelve stories. The EAL results are shown in Table 11.3 in comparison with the corresponding 2003 IBC conforming design from other PEER studies.

The table shows that the EAL values range from 1.6 % to 5.2 % with an average of 2.5 % of replacement cost for non-ductile RC frame buildings. These values

Table 11.3 Comparison of expected annual loss for ductile 2003 and non-ductile 1967 RC moment-resisting frame buildings (Liel and Deierlein 2008)

# stories	Framing	1967 RC frames Expected annual loss EAL [% repl.]	2003 RC frames Expected annual loss EAL [% repl.]
2	Perimeter	3.2 %	1.0 %
	Space	5.2 %	1.0 %
4	Perimeter	2.3 %	1.2 %
	Space	2.3 %	1.1 %
8	Perimeter	2.1 %	1.0 %
	Space	1.8 %	1.3 %
12	Perimeter	1.6 %	0.8 %
	Space	1.6 %	1.1 %
Min		1.6 %	0.8 %
Average		**2.5 %**	**1.1 %**
Max		5.2 %	1.3 %

suggest that a possible "non-ductile" range of EAL could be 1.5–3.0 %. However, the resulting values show an even stronger dependence on height which suggests that EAL classification ranges should distinguish between low-rise (say 1–4 stories), mid-rise (5–12) and high rise (>12 stories) in order to consider this difference, yet furture research is needed to confirm these trends.

The study by Krawinkler (2005) on the Van Nuys hotel building, which is a 7-storey RC perimeter frame building located in California, is an additional case study involving non-ductile structures. The structure was constructed in 1966 in the San Fernando Valley and can be confidently labeled as a "non-ductile" structure based on the witnessed performance in the 1971 San Fernando and 1994 Northridge events; the latter of which causing brittle shear failures of columns and beam column joints (Trifunac and Hao 2001). As Krawinkler (2005) estimated an EAL of 2.2 % of the replacement cost ($198,000 of $9 M replacement in 2002 USD), the generalization of non-ductile buildings having an expected annual loss on the order of 1.5–3 % is supported. However, additional work with this case study building has shown different results and this will be discussed along with other concerning points about generalizing EAL values to classify seismic risk categories.

11.3.3 Uncertainties with Expected Annual Loss Estimates

A number of inherent difficulties in implementing expected annual loss (EAL) as a seismic risk classification metric are addressed in this section. It is shown that even while using a normalized loss value (e.g. percentage of replacement cost) there are still various aspects of the loss estimation procedure that must, ideally, also be "normalized" before EAL could be expected to give reliable results for various structural typologies.

General trends, thus far, have shown expected annual loss (EAL) to be on the order of 0.5–1.5 % of replacement cost for 2003 IBC conforming MRF buildings (Haselton et al. 2008; Liel and Deierlein 2008; Ramirez and Miranda 2009) and non-ductile RC MRF buildings exhibiting EAL values on the order of 1.5–3.0 % of the replacement cost (Liel and Deierlein 2008). However, the manner in which the replacement cost of these structures has been calculated has been somewhat controlled (typically with the current version of the RS Means estimating manual at the time the study was conducted). Liel and Deierlein (2008) point out that the replacement cost estimates using RS Means (Balboni 2007) are expected to be at least 25 % lower than the actual cost of construction and that total project costs can be underestimated by as much as $200/ft^2 (2006 USD). Further, Liel and Deierlein (2008) state that these discrepancies from actual repair costs can still produce unbiased loss estimates provided that *both* replacement cost (e.g. entire structure) and repair costs (e.g. non-structural damage) are calculated using the same estimating reference (e.g. RS Means). The implications that deviation from this caveat can have on obtaining consistent EAL estimates to classify the seismic risk of a structure are illustrated with a previous case study performed on base isolated buildings.

The work of Sayani (2009) implemented the PEER PBEE methodology on two variations of a three storey steel moment frame building located in Southern California: (i) a typical special moment-resisting frame (SMRF) and (ii) an isolated ordinary moment-resisting frame building (IMRF). The buildings are designed to modern U.S. seismic code provisions, assume typical office occupancy and consider similar non-structural typologies and fragilities as studies that have been previously discussed (e.g. Mitrani-Reiser 2007; Beck et al. 2002). Assuming similar site hazard (e.g. Los Angeles area), the reported values of EAL were 0.134 % and 0.194 % of replacement cost for the IMRF and SMRF respectively; assuming the "total building and site" estimate for replacement cost (refer Sayani 2009).

Initially, the EAL estimate of 0.134 % for the isolated building suggests a continuation of the general trend of a traditional modern building giving results on the order of 0.5–1.5 % of replacement with the drastic reduction stemming from the intuitive "protection" that base isolation can provide. However, the traditional steel building (SMRF) gave EAL results (0.194 %) less than half of the lower bound (0.55 %) value reported from PEER studies which implies that the manner in which the replacement cost was calculated is inconsistent with previous studies conducted in the PEER benchmark study. Opposite of the suggestion to use the same costing reference for both replacement and repair costs set by Liel and Deierlein (2008), the work of Sayani (2009) used a professional cost estimator for the replacement and construction costs while repair costs were adjusted based on reported values within RS Means (Balboni 2007). Notably, the possible underestimation of up to $200/ft^2 when using RS Means for replacement cost was not a terrible estimate in this case, where only by adding $200/ft^2 to the 2 and 4 storey buildings (more than doubling the cost) examined in Liel and Deierlein (2008) are the replacement costs in agreement with the 3-storey estimates made by Sayani (2009), at least in terms of storey height and gross area. This raises much concern for the results of advanced

Table 11.4 Expected annual loss estimates for the Van Nuys hotel from two different studies

Building	Study	Replacement cost [$M]	EAL [$]	EAL [% replacement]
Van Nuys hotel	Krawinkler (2005)[a]	9.0	198,000	2.20 %
	Porter et al. (2004)[b]	7.0	53,600	0.77 %

[a]2002 USD
[b]2001 USD

loss estimates as neither study estimated the replacement cost improperly as no clear guidelines for performing this step are currently in available guidelines (ATC 2011a). Further, it could be argued that the replacement estimate by Sayani (2009) was performed at a very high level of competence, yet due to the repair costs not being treated to the same level the resulting estimates are not held to the same criteria as other studies and therefore can not be compared.

In addition to problems associated with the manner in which replacement cost is estimated, the numerous decisions that must be made in order to estimate EAL will be shown to drastically affect results. Although only the selection of damageable assemblies and variation in fragility selection will be the focus, it must also be noted that selection of initial (onset of damage) intensity, consideration of downtime or fatalities, and numerous economic factors (post-event demand surge for repairs, additional costs of tear down due to residual displacements) could also drastically affect EAL.

The Van Nuys Hotel study that was discussed when describing trends with non-ductile structures is recalled. Interestingly, there are two loss estimates for this building, the aforementioned study by Krawinkler (2005) and another conducted by Porter et al. (2004). The two estimates of EAL for the Van Nuys hotel are displayed in Table 11.4 showing the estimate of Porter et al. (2004) to be approximately one third (0.77 % vs. 2.2 %) of that reported by Krawinkler (2005).

Now how could such a discrepancy exist? Certainly the large difference is not rooted in the difference in replacement cost as the higher replacement cost (1 year of inflation is negligible) from Krawinkler (2005) would give a *reduction* in EAL by the same principles discussed in the previous section concerning the base isolated steel building. The large difference is most likely attributed to the number of damageable assemblies considered in the study and the manner in which their repair costs are distributed. Reportedly, the damageable assemblies (with subsequent fragilities and consequence functions) in Porter et al. (2004) consist of *select* structural and non-structural typologies from the collection of fragility and repair cost information within Beck et al. (2002). Conversely, the fragilities for the Krawinkler (2005) study consider a, comparatively, exhaustive list of non-structural components as identified by Taghavi and Miranda (2003) as well as numerous structural elements with distinct seismic fragility and consequences. Possibly the largest distinction is that the Krawinkler (2005) study adopts fragilities for numerous non-structural typologies and includes generic drift- and acceleration-sensitive fragilities in order to consider repair implications of numerous assemblies within the building in lieu of specific experimental data.

As a final point, loss estimates conducted within Welch et al. (2012) recreated previous assessments of a four-storey RC frame building using both the component-based model developed by Mitrani-Reiser (2007) and the storey-based model by Ramirez and Miranda (2009). Even with varying modelling assumptions and discrepancies within the many steps of the PEER PBEE framework, the resulting losses tended toward the parent study which highlights the reliability in the methodology. However, since the difference in the values between the two models varied by 30 % on average, the manner in which the loss model is developed should also be regulated in order to classify seismic risk. Finally, given the that the topic is relatively new, it is expected that rigorous loss assessments would be best for internal comparisons and cost benefit analysis, where regulations in order to reduce the interpretation required by the analyst may be defeating the purpose of having such a versatile loss framework.

11.3.4 Tentative Classification Framework

The previous sections have highlighted some important uncertainties in the definition of EAL as a performance parameter. In particular, (and leaving the performance issue of life-safety aside as a matter that could be addressed through code-requirements) the following two points were made:

- EAL is currently very uncertain and the values obtained are greatly affected by the loss models adopted and the value placed on replacement.
- The total EAL for a building, expressed as a fraction of the building replacement cost, will tend to decrease as the building height increases.

For what regards the first point, this would appear to be an issue with the current state of the art and could be dealt with by more research and some consensus on a standard procedure for estimating EAL. This uncertainty need not, however, prevent the creation of an EAL-based performance classification framework (which could actually help motivate the additional research that is required into EAL) and one should recognize that the engineering community already accepts large uncertainties and variations in performance checks. For example, the Eurocode 8 (CEN 2005) currently allows the use of four different types of structural analysis (equivalent-lateral force, modal-response spectrum analysis, pushover analysis, and non-linear dynamic analyses) in order to check specific engineering performance criteria and all four methods will generally provide different response estimates. Therefore, the current uncertainties inherent in EAL need not be seen as a large deterrent for the creation of an EAL-based performance classification scheme.

The second point raised above, which notes that EAL tends to decrease with building height, should also be given some attention. As the building height increases the total EAL may well tend to decrease because deformations and damage tend to be concentrated on specific floors, which make up a smaller fraction

Table 11.5 Proposed EAL-based seismic performance rating scheme

Class	EAL (storey-specific)
A+	≤0.25 %
A	0.25–0.75 %
B	0.75–1.5 %
C	1.5–3.0 %
D	≥3.0 %

of the total building as the number of storeys increases. Nevertheless, it would appear inappropriate to tell the owner of the storey in which high losses are expected that the EAL for the whole building was very low, when in fact it is the EAL of their apartment that is of most interest and relevance to them. A logical solution to this is to define EAL not on a building level, but on a storey-by-storey basis, so that different storeys of a building might be given different performance classifications. To this extent, the proposal is not that the performance of one storey can be considered completely independent of another and clearly, if there is a soft-storey collapse at the ground floor of a building then all floors have a high loss as the building will have to be replaced. However, it is proposed that the whole building be assessed and performance ratings then assigned to different levels, recognizing that repairable damage from low to moderate intensity earthquake shaking may tend to concentrate in specific levels. Then, a given owner at a certain level of the building might recognize that by using well-detailed non-structural elements they could significantly reduce the EAL for their storey.

With the above points in mind, and considering the EAL results from the literature presented in Sect. 3.2, Table 11.5 proposes a tentative EAL-based seismic performance rating scheme. It is proposed that the EAL limits in Table 11.5 refer to storey-specific values of EAL (i.e. the expected annual loss of the storey divided by the replacement value of the storey) which is a slightly different definition of EAL than is traditionally used, but would assist in addressing bullet-point 2 made above. The next section of the paper will present some simplified tools for the estimation of the EAL which will be followed by a case-study example.

11.4 Tools for Simplified Performance Classification

For most practicing engineers the challenge of computing the EAL for a building is currently likely to appear a somewhat daunting and impractical task. As computing power improves, software develops and loss assessment concepts and procedures become more widely established, it is likely that this situation will change. However, in the interim (and to permit such change to happen), it is apparent that there is a need for simplified tools that will allow engineers to estimate losses in a relatively simplified manner, without departing too greatly from current engineering procedures. This section reviews a recent proposal by Sullivan and Calvi (2011) and Welch et al. (2014) for simplified loss assessment, which combines the

Direct-displacement based assessment (Priestley et al. 2007) and SAC-FEMA (Cornell et al. 2002) methodologies together with an evaluation of losses at specific limit states.

11.4.1 Displacement-Based Seismic Assessment

Within a text proposing Direct displacement-based design, Priestley et al. (2007) also set out a procedure for the displacement-based seismic assessment (DBA) of structures. The procedure offers an estimate of the probability of exceeding a certain limit state, which could be the collapse prevention limit state, serviceability limit state or some other intermediate limit state. The first task in the Direct DBA procedure is to establish a force-displacement response curve, such as that shown in Fig. 11.10a, for an equivalent SDOF representation of the building. Priestley et al. (2007) explain that this can be done using hand-calculations in which the relative strengths of members are first compared in order to identify the expected lateral mechanism, which is then used together with (mechanism-dependent) approximations for the displaced shape and limit-state deformation capacity (which may be linked to resistance of brittle mechanisms). Alternatively to hand-calculations, one could undertake non-linear static analyses to obtain the force-displacement response curve.

With the force-displacement curve known, the effective stiffness, effective mass and ductility demand at the assessment limit are computed for the equivalent SDOF system. Equation 11.2 is then used to compute the system's effective period:

$$T_e = 2\pi \sqrt{\frac{m_e}{K_e}} \qquad (11.2)$$

where m_e is the effective mass given, as a function of the assessed displaced shape Δ_i, by:

$$m_e = \frac{\left(\sum m_i \Delta_i\right)^2}{\sum m_i \Delta_i^2} \qquad (11.3)$$

The use of the effective period and mass stems from the substitute-structure concept of Shibata and Sozen (1976) and Gulkan and Sozen (1974) and permits the use of linear elastic spectrum analysis to gauge the impact of seismic demands, with the effect of non-linear response accounted for through the use of effective-period inelastic spectrum scaling factors. Traditionally, such spectral scaling factors are set in Direct displacement-based design as a function of an equivalent viscous damping value, which is in turn a function of the ductility demand and hysteretic properties of the building. Recent research (Pennucci et al. 2011) has indicated that there are advantages in computing the spectral scaling factor (referred to as the displacement

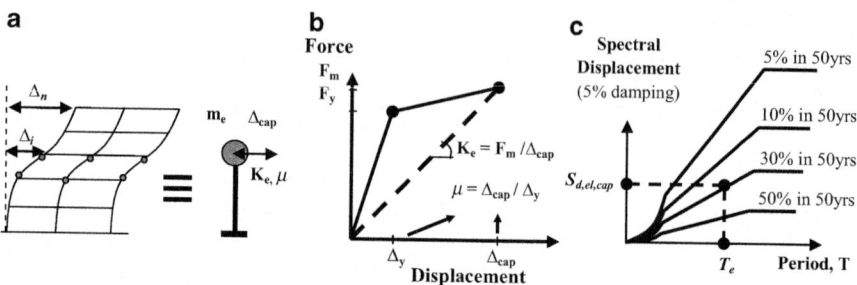

Fig. 11.10 Overview of displacement-based assessment approach (after Priestley et al. 2007). (a) Equivalent SDOF representation of structure at critical limit state. (b) Force-Displacement (pushover) curve for equivalent SDOF system. (c) Identification of seismic intensity expected to create limit state damage

reduction factor in Pennucci et al. 2011) directly as a function of the ductility demand, skipping the computation of the equivalent viscous damping. This lead to the proposal that the inelastic displacement demand, Δ_{in}, can be related to an elastic spectral displacement demand, $S_{d,el}$, using an empirical ductility-dependent expression. The resulting expression obtained for RC wall structures and bridge piers using equations proposed in Priestley et al. (2007) is:

$$\eta = \frac{\Delta_{in}}{S_{d,el}} \approx \sqrt{\frac{1}{1 + 6.34\left(\frac{\mu-1}{\mu\pi}\right)}} \qquad (11.4)$$

Note that this expression can be related back to an equivalent viscous damping value from expressions in the literature, such as that proposed in Eurocode 8 (CEN 2005) (adapted here to give ξ as a function of η):

$$\xi_{eq} = \frac{10}{\eta^2} - 5 \qquad (11.5)$$

Proceeding with the displacement-based assessment, once the effective period and system ductility demand, μ, at the limit state have been identified, an empirical spectral displacement scaling factor is computed (11.6) and divided into limit state displacement capacity to provide an equivalent elastic spectral displacement capacity, $S_{d,el,cap}$, as shown:

$$S_{d,el,cap} = \frac{\Delta_{cap}}{\eta} \qquad (11.6)$$

With knowledge of elastic spectral displacement demands at a site, for various hazard levels, the earthquake intensity required to push the structure to its limit state can then be identified using the effective period (T_e) and spectral displacement capacity ($S_{d,el,cap}$) as shown in Fig. 11.10c. Note that this relatively simple approach could also be done using a capacity-spectrum method or other non-linear static procedures.

The benefit of this type of assessment over a traditional assessment approach in which code-specified intensity levels are checked via a pass-fail type approach is that a better appreciation of the real risk can be obtained. Priestley et al. (2007) go as far as suggesting that the probability associated with the hazard level shown in Fig. 11.10c provides an indication of the probability that the assessed limit state will be exceeded. However, such a proposal does neglect the effect of dispersion in both demand and capacity which is should be accounted for in probabilistic assessment methods.

In order to extend the DBA procedure to provide a probabilistic assessment of the likelihood of exceeding a certain limit state, some consideration must be made of uncertainties in the assessment process, and more generally, for dispersion in the demand and capacity estimates. To permit a simplified probabilistic displacement-based assessment, Sullivan and Calvi (2011) and Welch et al. (2014) have recommended adaption of the SAC-FEMA approach (Cornell et al. 2002) simplified as per the suggestions of Fajfar and Dolsek (2010). According to the SAC-FEMA approach, the probability, $P_{LS,x}$, of exceeding a certain limit state can be found for an x-confidence level according to:

$$P_{LS,x} = \tilde{H}\left(S_{a,\tilde{C}}\right) C_H C_f C_x \qquad (11.7)$$

Where C_x, C_H and C_f are coefficients account for C values are coefficients accounting for the desired confidence level, differences between mean and median hazard levels, and dispersion in the demand and capacity, respectively. $\tilde{H}(S_{a,C})$ is the median value of the hazard function at the seismic intensity $S_{a,C}$, expected to cause a specific limit state to develop. Simplifying the approach according to the suggestions of Fajfar and Dolsek (2010) both the coefficients C_H and C_x are set to one, and a 50 % confidence level estimate using the mean hazard of the probability of exceedence is obtained as:

$$P_{LS,x} = \overline{H}\left(S_{a,\overline{C}}\right) C_f \qquad (11.8)$$

As shown in Fig. 11.10c, the DBA procedure as proposed by Priestley et al. (2007) provides the mean value of the hazard function, $\overline{H}\left(S_{a,\overline{C}}\right)$, expected to cause a selected limit state to develop. Subsequently, the adjustment required to arrive at a simplified estimate of the probability of exceeding a certain limit state only needs computation of the dispersion factor, C_f. According to Cornell et al. (2002), the C_f factor can be calculated, assuming log-normal distributions of demand and capacity, as:

$$C_f = \exp\left[\frac{k^2}{2b^2}\left(\beta_{DR}^2 + \beta_{CR}^2\right)\right] \qquad (11.9)$$

where the constant k is set as a function of local hazard data using a power expression to relate hazard with probability of exceedence, the constant b relates engineering demand parameters to the intensity measure and could be approximated as 1.0 (as per equal-displacement rule even if in reality more accurate values could be obtained considering different structural typologies and hysteretic systems), and β_{CR} and β_{DR} are dispersion measures for randomness in capacity (modelling) and demand (record-to-record) respectively. Indicatively, one could expect a value of $(\beta_{DR}^2 + \beta_{CR}^2) = 0.2025$ as suggested by Fajfar and Dolsek (2010), who also report that reliable data on modelling dispersion is not yet available. More refined/reliable information on dispersion appears to emerging within the recent ATC-58 document (ATC 2011a) based on recent parametric studies as described in Sect. 11.2.1.

As discussed in the *fib* Bulletin 68 (*fib* 2012), the accuracy of the SAC-FEMA approach is limited but it is very simple and therefore is considered to provide engineers with a useful approach in the transition to more rigorous probabilistic methods. The approach will be used later in Sect. 11.5 as part of an example case-study to illustrate possible application of the performance-classification scheme.

One aspect of the DBA procedure not clarified above is that in addition to checking displacement demands, one should also take care to assess demands on acceleration-sensitive non-structural elements and secondary-structural elements, particularly when assessing the serviceability limit state. In work by Welch et al. (2014) acceleration demands up the height of a building were estimated using empirical expressions from ATC-58 (ATC 2011a) but existing empirical procedures are known to possess a number of limitations. Progress towards improved estimation of floor acceleration spectra has been made by Sullivan et al. (2013), Calvi and Sullivan (2014), who provide expressions for the estimation of floor acceleration spectrum demands as a function of the non-linear response of the underlying structure and the period and damping of the supported non-structural element. However, it is still an area of the DBA procedure that requires further development.

11.4.2 Approximation of the Expected-Annual Loss

The DBA procedure described in the previous section provides an estimate of the probability of exceeding a given limit state. This approach should appear within the grasp of most practicing engineers who have become used to exercise of assessing different limit states. However, the proposal in this paper is for the performance of a building to be classified according to the expected annual monetary loss (EAL). As such, the next step in the assessment process is to convert the probability of exceeding different limit states into values of EAL. In order to do this, Welch et al. (2014) have shown that by estimating losses associated with four key limit states, and assuming that losses vary linearly with intensity between the key limit states, simple integration can be used to arrive at an estimate of EAL. This process is illustrated in Fig. 11.11 and will be explained in more detail subsequently.

Fig. 11.11 Overview of the simplified EAL estimation using displacement-based assessment as proposed by Welch et al. (2014)

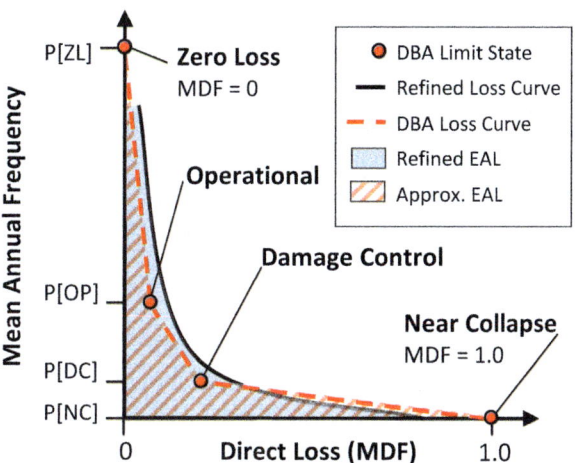

Referring to Fig. 11.11, it is shown that the smooth curve, representing a series of intensity-based assessments using refined methods (e.g. PEER PBEE), has a distinct transition region between intensities of large annual frequency (lower expected losses) and rarer events with smaller annual frequency (higher expected losses). The main concept behind the simplified method using DBA is that a refined loss curve is reasonably approximated using only four key limit states; two bounding limit states to represent the onset of damage (zero loss) and the point of total loss (near collapse), as well as two intermediate limit states (operational and damage control) that represent the transition region in the loss curve.

As discussed previously, a single DBA assessment is capable of estimating the probability of exceeding a limit state defined by a peak displacement demand (e.g. peak IDR). Therefore only limit state definition is required in order to obtain the vertical ordinates (mean annual frequency) shown in Fig. 11.11, yet the loss values associated with each of the four limit states are conditioned on a few simplifying assumptions. The zero loss limit state is assigned a mean damage factor (MDF, % of replacement cost) of zero; a similar assumption to assigning an initial intensity to begin analysis within the PEER PBEE approach. The near collapse limit state is assumed to represent the total loss threshold and is attributed a MDF of 1.0. This leaves only direct loss estimates to be calculated at the intermediate operational and damage control limit states.

In order to estimate losses at intermediate limit states, the work within Welch et al. (2014) adopted the engineering demand parameter to decision variable functions (EDP-DV) formulated by Ramirez and Miranda (2009). These functions are constructed for frame buildings based on number of stories, ductility capacity, structural system (space or perimeter frame) and occupancy (e.g. office). As part of a storey-based loss framework, EDP-DV functions directly relate the EDP's of peak inter-storey drift ratio (IDR) and peak floor acceleration (PFA) to the expected direct losses associated with structural and non-structural damage. The functions assume three performance groups considering structural (drift-sensitive),

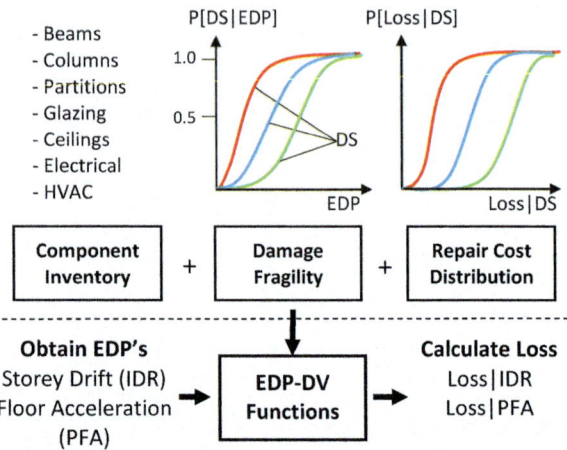

Fig. 11.12 Summary of the development of EDP-DV functions (Ramirez and Miranda 2009) used to estimate repair costs at intermediate damage states using the four-point EAL model

non-structural drift-sensitive and non-structural acceleration-sensitive components. The functions consider the variation in expected assembly inventory between ground floor, typical floors, and roof level. Notably, the EDP-DV functions consider many interactions between components in order to avoid attributing the same repair cost twice to a component that may need repair in order to access additional elements for repair. A summary of how EDP-DV functions are developed and implemented is shown in Fig. 11.12.

With the assumptions in place, the last important aspect of the simplified EAL calculation using DBA is the definition of limit states. Ideally, the zero loss limit state should represent the onset of damage of the most fragile non-structural components (e.g. partitions, infills) and this should transition to an operational limit state that would produce only light non-structural damage. Further, the damage control limit state should represent only minor structural damage and the near collapse limit state, appropriately, should consider the expected displacement demand at imminent collapse. Notably, the work within Welch et al. (2014) developed limit state criteria similar to that described in Vision 2000 (SEAOC 1995), yet a few modifications were made. Most importantly the near collapse limit state considered both the imminent collapse displacement as well as an approximation of the peak displacement corresponding to a target residual drift in order to include the possibility of a total loss due to residuals.

11.5 An Example Application

11.5.1 Assessment, Retrofit Options, Estimate of EAL

In order to illustrate how a performance classification scheme could be used in practice, the three storey office building shown in Fig. 11.13 is examined.

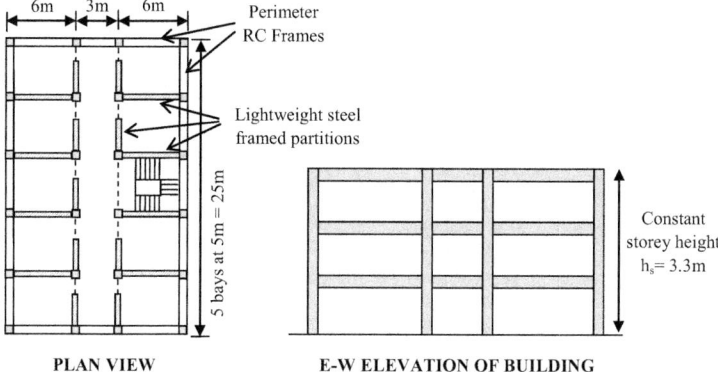

Fig. 11.13 Illustration of the case study frame building

This hypothetical case study building, assumed to be located in the city of L'Aquila, possesses features typical of construction practice in the 1980s with a ductile RC frame structure, an exterior glass façade, lightweight steel framed interior partitions and suspended ceilings. This example will consider how a performance classification scheme could be coupled with a government-funded incentive scheme to encourage retrofit and subsequently reduce likely monetary losses and disruption caused by earthquakes.

A non-linear static (pushover) seismic assessment of the building reveals that the building forms a ductile beam-sway mechanism and develops the bi-linearized force-displacement response shown in Fig. 11.14, with a (cracked) fundamental period of vibration of 1.15 s (similar responses are expected for both the E-W and N-S directions). The base shear resistance at yield of 2250kN is approximately 20 % of the full seismic weight of the building. The pushover curve is annotated to show the corresponding storey drift demands for different potentially critical response points.

As shown in Fig. 11.14, the lightweight steel framed partitions considered for this example structure are assessed as possessing a drift-capacity of 0.3 % before repairs are required (noting that 0.3 % drift capacity has been observed through experimental testing by Davies et al. (2011). The drift limit corresponds to an equivalent SDOF system displacement limit of 0.0231 m at period of 1.15 s (i.e. the cracked elastic period). The other non-structural elements in the case-study building are assessed as being less critical, with the glazing have a serviceability drift capacity of greater than 1.0 % and the ceilings expected to sustain the peak acceleration demands without damage. The frame has a yield drift of 1.0 %, which is quite typical of RC frame structures and a total drift capacity of 5.0 %

In the following paragraphs the EAL expected for the building under three different retrofit approaches will be reported:

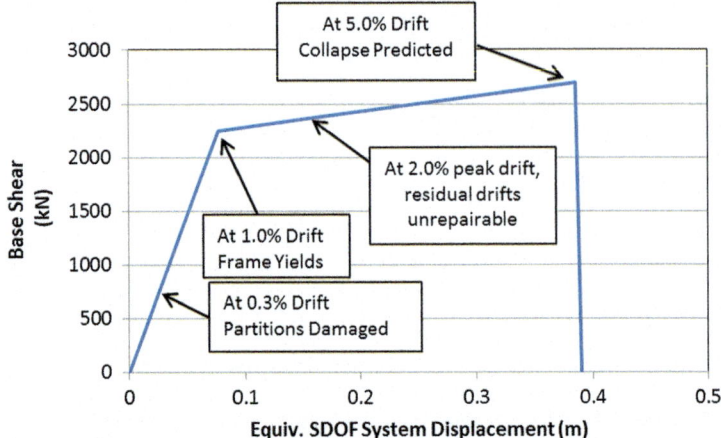

Fig. 11.14 Force-displacement response curve for the building, showing important response points

- OPTION 1: no retrofit such that the structure remains as it is;
- OPTION 2: replacement of the lightweight steel partitions with well detailed partitions that increase the drift required to exceed zero-loss limit state from 0.3 to 0.7 %;
- OPTION 3: replacement of the partitions (as per OPTION 2) and addition of viscous dampers to reduce the seismic demands at all limit states.

The retrofit options listed above will allow this study to highlight how the improvement of non-structural elements (OPTION 2) could lead to significant reductions in EAL that could represent a more feasible option for building owners to consider than the costlier OPTION 3 that would improve the performance at all limit states. Clearly other retrofit options could also be considered and the options listed above should not necessarily be considered the most effective retrofit solutions. Another retrofit possibility could have been to add a RC wall or other structural elements that increase the stiffness and strength of the system. This would have the benefit of reducing the displacement demands but would have the negative effect of increasing acceleration demands, which in the present scenario are considered to be below limit state values for the ceilings. Note therefore that in all cases the structure remains as it is, coherently with a satisfactory predicted drift capacity of 5 % at collapse.

Proceeding with the displacement-based assessment approach described in Sect. 11.4, Table 11.6 summarizes the characteristics (effective period, displacement capacity and equivalent viscous damping) for the three different retrofit scenarios at both the zero-loss and replacement limit states. Note that the replacement limit state was defined as being the point at which the peak storey drift reached 2.0 %, making the relatively conservative assumption that residual drifts would become unrepairable at this level (exceeding a residual drift limit of 0.5 %). It can

Table 11.6 Summary of key characteristics obtained from displacement-based assessment

		Retrofit OPTION 1	Retrofit OPTION 2	Retrofit OPTION 3
Displacement capacity (m)	Zero loss limit state	0.023	0.054	0.054
	Replacement limit state	0.154	0.154	0.154
Effective period (s)	Zero loss limit state	1.15	1.15	1.15
	Replacement limit state	1.59	1.59	1.59
Equivalent viscous damping	Zero loss limit state	5 %	5 %	20 %
	Replacement limit state	15 %	15 %	36 %

be seen that the effective period for the zero-loss limit state for all three retrofit options is 1.15 s (the fundamental period of the building), whereas the effective period for the replacement limit state is 1.59 s (obtained using the effective stiffness of the building at a peak drift of 2 %).

Spectral displacement demands at each value of effective period and for each value of equivalent viscous damping were then obtained from seismic hazard data for L'Aquila (NTC 2008). Subsequently, the hazard level expected to cause the limit state displacement values indicated in Table 11.6 were identified, as per the procedure described in Sect. 11.4.1. To account for dispersion, Eq. (11.8) was applied, with the constant k set to the local hazard data for the site (around the displacement response point of interest), the constant b set equal to 1.0 (which is approximate but should not affect dispersion estimates too greatly), and with estimated values of dispersion in demand and capacity equal to 0.35 respectively (as used for RC frames by Fajfar and Dolsek 2010). Table 11.7 presents values from the simplified SAC-FEMA approach used to identify the probability of exceeding different limit states. The limit states include the zero-loss limit state which (as the name suggests) corresponds to a mean damage factor (MDF) of 0.0, and the replacement which corresponds to an MDF of 1.0 (i.e. the full replacement cost). In order to be able to apply the four-point loss model described in Sect. 11.4.2, the probability of exceeding another two intermediate limit states corresponding to mean damage factors (MDFs) of 0.2 and 0.5 were also computed, making simplifying assumptions about EDP-loss values for the purpose of this example.

At this stage of the assessment one can already begin to get a feel for the impact of the different retrofit measures on the likely losses. Figure 11.15 compares the probability of exceedence of each value of MDF reported in Table 11.7 for the three different retrofit options. The increased deformation capacity offered by the new partitions in retrofit OPTION 2 leads to a considerable reduction in the probability of exceedence of the zero-loss limit state and the overall losses, which can be gauged from the area under the curves. This reduction occurs even if retrofit OPTION 1 and OPTION 2 have the same probability of exceeding the replacement limit state. By adding viscous dampers in retrofit OPTION 3, it can be seen that probability of exceeding all limit states are reduced, but considering the areas under the curves, the difference in losses between retrofit OPTION 2 to OPTION 3 do not appear as significant as those between retrofit OPTION 1 to OPTION 2.

Table 11.7 Use of SAC-FEMA procedure to identify probabilities of exceedence for application of the four-point EAL estimation

		Retrofit OPTION 1	Retrofit OPTION 2	Retrofit OPTION 3
Assessed hazard level for limit state	Zero loss limit state	93.5 % prob. exc. In 50 years	42.8 % prob. exc. in 50 years	17.1 % prob. exc. in 50 years
	Replacement limit state	5.4 % prob. exc. in 50 years	5.4 % prob. exc. in 50 years	2.3 % prob. exc. in 50 years
Local site hazard coefficient, k	Zero loss limit state	1.85	1.85	2.15
	Replacement limit state	2.10	2.10	2.20
C_f	Zero loss limit state	1.41	1.41	1.60
	Replacement limit state	1.56	1.56	1.63
Probability of exceeding L.S.	Zero loss limit state	0.0753	0.0157	0.0060
	Replacement limit state	0.0017	0.0017	0.0008

Fig. 11.15 Curves illustrating the probability of exceeding various loss levels for the three retrofit strategies

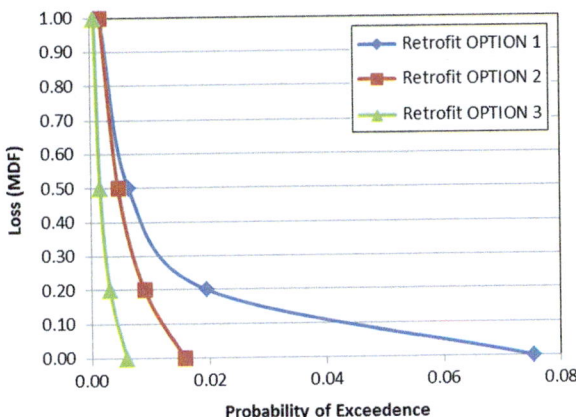

Table 11.8 Cost considerations for different retrofit strategies

	Description	EAL	Perf. CLASS	Total retrofit cost	Tax rebate (incentive)	Net retrofit cost[a]
Retrofit OPTION 1	Do nothing	1.54 %	C	0	0	0
Retrofit OPTION 2	Replace partitions	0.63 %	A	€50,000	€15,000 (30 % rebate)	€35,000
Retrofit OPTION 3	Replace partitions and add structural dampers	0.22 %	A+	€200,000	€100,000 (50 % rebate)	€100,000

[a]Figures should be adjusted to allow for inflation

The next step in the assessment is to compute the EAL for each retrofit strategy and this is done here using the approximate 4-point approach described in Sect. 11.4.2. Figure 11.16 presents the results obtained, together with the performance classification that would be assigned to the building according to the proposal made in Sect. 11.3.4. It can be seen that the existing building would be a class C building, bordering on class B (and if required, more refined loss estimates could be undertaken to confirm the final class). If the non-structural partitions are replaced, as per retrofit strategy 2, the building would become class A. If, in addition to this, viscous dampers are provided then it can be seen that a seismic performance Class A+ can be achieved.

In order to highlight the possible implications of these retrofit options, Table 11.8 presents possible costs of the different retrofit scenarios, considering also a possible tax incentive scheme that a government might provide (clearly there is no fundament on the values provided, assumed for the sake of discussion only).

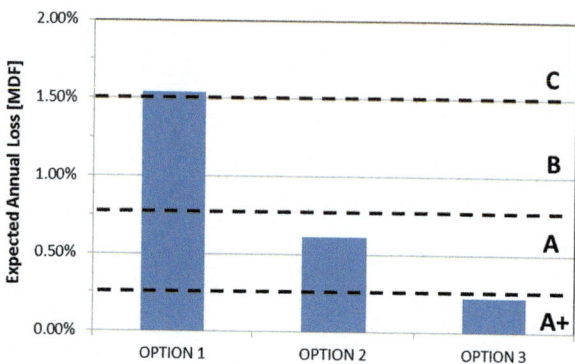

Fig. 11.16 Expected annual losses estimated for the three retrofit strategies and seismic performance classification

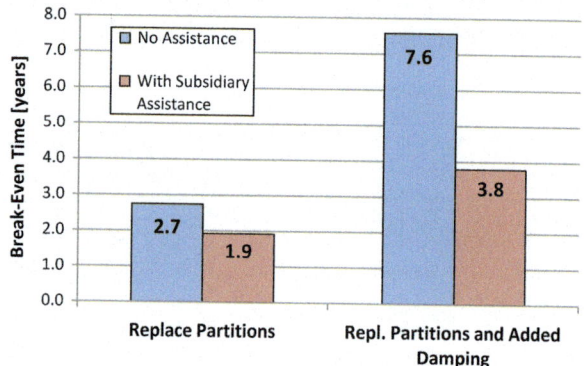

Fig. 11.17 Break-even times for the considered retrofit options showing the potential of subsidiary assistance

11.5.2 Breakeven Times

In order to further illustrate the potential benefits of the retrofit options, as well as the influence of subsidiary measures, the EAL values are presented in terms of break-even times. The break-even time, $t_{Break-Even}$, represents, probabilistically, the time necessary for the upfront cost of the retrofitting intervention to be balanced by the expected annual reduction in seismic losses as shown in (11.10):

$$t_{Break-Even} = \frac{Value}{Value/time} = \frac{Cost_{Retrofit}}{EAL_{Existing} - EAL_{Retrofit}} \qquad (11.10)$$

where the total cost of the intervention, $Cost_{Retrofit}$, could include a reduction to due subsidiary measures depending on the situation. The results of break-even times are shown in Fig. 11.17 for the example case study. Notably, the replacement cost of the structure is taken as €2,000,000 for the sake of simplicity. Actual values could vary significantly yet this cost corresponds to the more comprehensive retrofit

Table 11.9 Values for the calculation of break-even times for the considered retrofit options

Retrofit OPTION	Reduction in EAL[a]	Retrofit cost	Tax rebate (incentive)	Break-even time, $t_{Break-Even}$ (year)	
				No assistance	With subsidiary assistance
Replace partitions	€18,200	€50,000	€15,000	2.7	1.9
Replace partitions and add structural dampers	€26,400	€200,000	€100,000	7.6	3.8

[a]Replacement cost taken as €2,000,000

(e.g. added damping) to be 10 % of the replacement cost. The values used for calculation of $t_{Break-Even}$ are shown in Table 11.9.

Reflecting on the numbers shown, one can see that a significant capital outlay is required to increase the performance class to A+. Even though the government incentive for this option is assumed to be greater than for retrofit option 2, it might be deemed too expensive by the building owner to pursue. Retrofit option 2 still results in a significant retrofit cost, but is likely to be more acceptable to the building owner, particularly considering that the replacement of partitions might be undertaken as part of a refurbishment scheme. Another instance in which option 2 might be considered more attractive is the situation in which the building is owned by several different parties, as is the case for the majority of residential buildings in Italy. In such occasions it may be very difficult to obtain agreement from all building owners to proceed with retrofit option 3, owing to the costs. On the other hand, retrofit option 2 could actually be implemented only on specific floors of a building (or part of it), by owners interested in improving the seismic performance rating of their apartment. Clearly, the same cannot be said for retrofit option 3 (addition of structural dampers) which should be implemented for the entire building system.

As a closing comment to this example, note that by motivating people to make some form of retrofit, even if only to non-structural elements as for option 2, the negative impacts of earthquakes should be reduced, with reduced disruption, monetary losses and downtime in the event of an earthquake. This is considered to provide good justification for the development and implementation of a seismic performance rating system, ideally coupled with some form of incentive scheme, in the years ahead.

11.6 Conclusions

This paper has reviewed a range of performance measures that are being adopted in modern seismic engineering applications and has then proposed a seismic performance classification framework based on expected annual losses. The motivation for an EAL-based performance framework stems from the observation that, in addition to limiting lives lost during earthquakes, changes are needed to improve the resilience of our societies, and it is proposed that increased resilience could be achieved by limiting monetary losses. Typical values of EAL reported in the literature have been reviewed, uncertainties in such EAL estimates have been discussed and then a EAL-based seismic performance classification framework has been proposed. The proposal has been made that the EAL should be computed on a storey-by-storey basis in recognition that EAL for different storeys of a building could vary significantly and also recognizing that a single building may have multiple owners.

A number of tools for the estimation of EAL exist in the literature and both the PEER PBEE framework and a simplified displacement-based loss assessment (DBLA) procedure have been reviewed in this paper. It has also been argued that there is a need for simplified methods for the prediction of EAL as engineers make a transition into this new performance parameter. In order to illustrate the potential value of an EAL-based classification scheme, a three storey RC frame building is assessed using the simplified DBLA procedure and performance classifications are made for three different retrofit solutions. The results show that even if only limited non-structural interventions are made to the case study building, the EAL could be significantly reduced. As the less-expensive non-structural retrofit could be more within the grasp of building owners, it is argued that overall, such a performance classification, coupled with some form of government or insurance-driven incentive scheme, may provide an effective means of motivating (even if limited) retrofit, thereby reducing the risk and increasing the resilience of our societies.

Acknowledgements The authors would like to acknowledge that this work is a product of Research Line 7 of the 2014 RELUIS Project.

Open Access This chapter is distributed under the terms of the Creative Commons Attribution Noncommercial License, which permits any noncommercial use, distribution, and reproduction in any medium, provided the original author(s) and source are credited.

References

Anderson JC, Fillipou FC (1995) Dynamic response analysis of the 17-story canoga building, SAC technical report 95-04, pp 12-1–12-53

Altoonash A (2004) Simulation and damage models for performance assessment of reinforced beam-column joints. PhD dissertation, Stanford University

Aslani H, Miranda E (2005) Probability-based seismic response analysis. Eng Struct 27(8):1151–1163

ATC (2005) Procedures for postearthquake safety evaluation of buildings. ATC-20. Applied Technology Council, Redwood City, Published 1989, Revised 2005

ATC (2011a) Guidelines for seismic performance assessment of buildings: vol 1 – methodology. ATC 58-1. Applied Technology Council, Redwood City

ATC (2011b) Guidelines for seismic performance assessment of buildings: vol 2 – implementation guide. ATC 58-1. Applied Technology Council, Redwood City

Baker JW, Cornell CA (2006a) Which spectral acceleration are you using? Earthq Spectra 22(2):293–312

Baker JW, Cornell CA (2006b) Spectral shape, epsilon and record selection. Earthq Eng Struct Dyn 34(10):1193–1217

Balboni B (ed) (2007) Means square foot costs. R.S. Means Co., Inc., Kingston

Beck JL, Kiremidjian A, Wilkie S et al (1999) Decision support tools for earthquake recovery of business, final report for CUREe-Kajima phase III project, Consortium of Universities for Research in Earthquake Engineering, Richmond

Beck JL, Porter KA, Shaikhutdinov R, Au SK, Moroi T, Tsukada Y, Masuda M (2002) Impact of seismic risk on lifetime property values. Final report for CUREE-Kajima phase IV project, Consortium of Universities for Research in Earthquake Engineering, Richmond

Berry M, Parrish M, Eberhard M (2004) PEER structural performance database user's manual. Pacific Earthquake Engineering Research Center, University of California, Berkeley. http://nisee.berkeley.edu.spd

Calvi PM, Sullivan TJ (2014) Improved estimation of floor spectra in RC wall buildings. In: Proceedings of the 10th US national conference on earthquake engineering, Anchorage, paper 1113

CEN (2005) Eurocode 8: design of structures for earthquake resistance. EN 1998-1, Brussels

Chang SE, Falit-Baiamonte A (2002) Disaster vulnerability of businesses in the 2001 Nisqually earthquake. Environ Hazards 4(2/3):59–71

Charleson A (2007) Architectural design for earthquakes – a guide to the design of non-structural elements. New Zealand Society for Earthquake Engineering, Wellington

Coburn A, Spence R, Pomonos A (1992) Factors determining human casualty levels in earthquakes: mortality prediction in building collapse. 10th world conference on earthquake engineering, Madrid, Spain

Comerio MC, Stallmeyer JC, Holmes W, Morris P, Lau S (2001) Nonstructural loss estimation: The UC Berkeley case study, PEER report 2002/01, Pacific Earthquake Engineering Research Center, Dec 2001

Comerio MC (2006) Estimating downtime in loss modeling. Earthq Spectra 22(2):349–365

Comerio MC (2012) Resilience, recovery and community renewal, Keynote paper. In: 15th world conference of earthquake engineering, Lisbon, Portugal

Cornell CA, Jalayer F, Hamburger RO, Foutch DA (2002) Probabilistic basis for 2000 SAC federal emergency management agency steel moment frame guidelines. J Struct Eng 128(4):526–533

Davies RD, Retamales R, Mosqueda G, Filiatrault A (2011) Experimental seismic evaluation, model parameterization, and effects of cold-formed steel-framed gypsum partition walls on the seismic performance of an essential facility, MCEER-NEES nonstructural, technical report MCEER-11-0005, Buffalo, New York

Durkin ME, Thiel CC (1992) Improving measures to reduce earthquake casualties. Earthq Spectra 8(1):95–113

Fajfar P, Dolsek M (2010) A practice-oriented approach for probabilistic seismic assessment, In: Fardis MN (ed) Advances in performance-based earthquake engineering. Springer, Dordrecht, Netherlands pp 225–333

Federal Highway Administration (FHWA) (1994) Technical advisory: motor vehicle accident costs. Technical advisory #7570.2. U.S. Department of Transportation, Washington, DC

FEMA (2009) Quantification of building seismic performance factors. FEMA P695. Prepared by the Applied Technology Council for the Federal Emergency Management Agency, Washington, DC

FEMA (2011) Reducing the risks of nonstructural earthquake damage – a practical guide. FEMA E-74. Prepared by the ATC for FEMA, Washington, DC, 2011

fib (2012) Probabilistic performance-based seismic design, fib Bulletin 68. International Federation of Structural Concrete, Lausanne, Switzerland

Filiatrault A, Uang CM, Folz B, Christopoulos C, Gatto K (2001) Reconnaissance report of the 28 Feb 2001 Nisqually (Seattle-Olympia) Earthquake, report no. SSRP-2001/02. University of California, San Diego

Gulkan P, Sozen M (1974) Inelastic response of reinforced concrete structures to earthquake motions. ACI J 71(12):604–610

Haselton CB, Deierlein GG (2007) Assessing seismic collapse safety of modern reinforced concrete moment frame buildings, Technical report no. 156, John A. Blume Earthquake Engineering Research Center, Stanford University

Haselton CB, Goulet CA, Mitrani-Reiser J et al (2008) An assessment to benchmark the seismic performance of a code-conforming reinforced concrete moment frame building, Report no. PEER 2007/12, Pacific Earthquake Engineering Research Center, Aug 2008

Haselton CB, Baker JW, Liel AB, Deierlein GG (2011) Accounting for expected spectral shape (epsilon) in collapse performance assessment. ASCE J Struct Eng, Special Issue on Earthquake Ground-Motion Selection and Modification for Nonlinear Dynamic Analysis, 137(3):332–344

Hengjiam L, Kohiyama M, Horie K, Maki N, Hayashi H, Tanaka S (2003) Building damage and casualties an earthquake. Nat Hazards 29:387–403

Ibarra LF, Medina RA, Krawinkler H (2005) Hysteretic models that incorporate strength and stiffness deterioration. Earthqu Eng Struct Dyn 34:1489–1511

Iervolino I, Manfredi G, Cosenza E (2006) Ground-motion duration effects on nonlinear seismic response. Earthq Eng Struct Dyn 35:21–38

International Code Council Evaluation Services (ICC-ES) (2010) Acceptance criteria for certification by shake-table testing of nonstructural components, ICC-ES AC 156, Subsidiary of the International Code Council

Kalkan E, Kunnath SK (2006) Effects of fling step and forward directivity effects on seismic response of buildings. Earthq Spectra 22(2):367–390

Krawinkler H (2005) Van Nuys hotel building testbed report: exercising seismic performance assessment. Report PEER 2005/11. Pacific Earthquake Engineering Research Center, Richmond

Liel AB, Deierlein GG (2008) Assessing the collapse risk of California's existing reinforced concrete frame structures: metrics for seismic safety decisions. Technical report no. 166, John A. Blume Earthquake Engineering Research Center, Stanford University

Lignos D (2013) Web-based interactive tools for performance-based earthquake engineering, http://dimitrios-lignos.research.mcgill.ca/databases/index.php. Accessed 26 Jan 2014

Lignos D, Krawinkler H (2011) Deterioration modeling of steel components in support of collapse prediction of steel moment frames under earthquake loading. J Struct Eng 137(11):1291–1302

Mahin SA, Bertero VV (1981) An evaluation of inelastic seismic design spectra. J Struct Eng 107(ST9):1777–1795

Miranda E, Mosqueda G, Retamales R, Pekcan G (2012) Performance of nonstructural components during the 27 February 2010 Chile earthquake. Earthq Spectra 28(S1):S453–S471

Mitrani-Reiser J (2007) An ounce of prevention: probabilistic loss estimation for performance-based earthquake engineering. PhD dissertation, CalTech, Passadena

Mohammadjavad H, Filiatrault A, Aref A (2013) Simplified sideway collapse analysis of frame buildings. Earthq Eng Struct Dyn. doi:10.1002/eqe.2353

Mrozek JR, Taylor LO (2002) What determines the value of life? A meta-analysis. J Policy Anal Manage 21(2):253–270

NTC (2008) Norme techniche per le costruzioni, Decreto Ministrale (in Italian)

Pampanin S, Magenes G, Carr A (2003) Modelling of shear hinge mechanism in poorly detailed RC beam-column joints. In: Proceedings of the FIB 2003 symposium, Athens, Greece

Pennucci D, Sullivan TJ, Calvi GM (2011) Displacement reduction factors for the design of medium and long-period structures. J Earthq Eng 15(S1):1–29

Perus I, Klinc R, Dolenc M, Dolsek M (2013) A web-based methodology for the prediction of approximate IDA curves. Earthq Eng Struct Dyn 42(1):43–60

Porter KA (2003) An overview of PEER's performance-based earthquake engineering methodology. In: Proceedings of 9th international conference on applications of probability and statistics in engineering, San Francisco, CA

Porter KA, Kiremidjian AS, LeGrue JS (2001) Assembly-based vulnerability of buildings and its use in performance evaluation. Earthq Spectra 12(2):291–312

Porter KA, Beck JL, Shaikhutdinov R (2004) Simplified estimation of economic seismic risk for buildings. Earthq Spectra 20(4):1239–1263

Porter KA, Shoaf K, Seligson H (2006) Value of injuries in the northridge earthquake, technical note. Earthq Spectra 22(2):553–563

Priestley MJN, Calvi GM, Kowalsky MJ (2007) Displacement-based seismic design of structures. IUSS Press, Pavia

Rackwitz R (2004) Optimal and acceptable technical facilities involving risk. Risk Anal 24(3):675–695

Ramirez CM, Miranda E (2009) Building specific loss estimation methods & tools for simplified performance-based earthquake engineering. Technical report no. 171, John A. Blume Earthquake Engineering Center, Stanford University

Ramirez CM, Miranda E (2012) Significance of residual drifts in building earthquake loss estimation. Earthq Eng Struct 41(11):1477–1493

Ramirez CM, Liel AB, Mitrani-Reiser J, Haselton CB, Spear AD, Steiner J, Deierlein GG, Miranda E (2012) Expected earthquake damage and repair costs in reinforced concrete buildings. Earthq Eng Struct Dyn 41:1455–1475

Reitherman R, Sabol T (1995) Nonstructural damage. In: Hall J (ed) Northridge earthquake of 17 Jan 1994 Reconnaissance report. Earthq Spectra 11(C):453–514

Rosenbluth E, Meli R (1986) The 1985 Mexico earthquake: causes and effects in Mexico City. Concr Int 8(5):23–34

Sayani PJ (2009) Relative performance comparison and loss estimation of seismically isolated and fixed-based buildings using PBEE approach. All Graduate Theses and Dissertations, Paper 482, http://digitalcommons.usu.edu/etd/482

SEAOC (1995) Vision 2000, a framework for performance-based engineering. Structural Engineers Association of California, Sacramento

Shibata A, Sozen M (1976) Substitute structure method for seismic design in reinforced concrete. J Struct Div, ASCE 102(ST1):1–18

Stojanovski P, Dong W (1994) Simulation model for earthquake casualty estimation. In: 5th national conference on earthquake engineering, Chicago, IL

Sullivan TJ, Calvi GM (2011) Considerations for the seismic assessment of buildings using the direct displacement-based assessment approach. In: Proceedings of the 2011 ANIDIS conference, Bari, Italy

Sullivan TJ, Calvi PM, Nascimbene R (2013) Towards improved floor spectra estimates for seismic design. Earthq Struct 4(1):109–132

Taghavi S, Miranda E (2003) Response assessment of nonstructural building elements. Report PEER 2003/05. Pacific Earthquake Engineering Research Center, Richmond

Trifunac MD, Hao TY (2001) 7-storey reinforced concrete building in Van Nuys, California: photographs of the damage from the 1994 northridge earthquake. Report CE01-05, University of Southern California Department of Civil Engineering, Los Angeles, CA

Vamvatsikos D, Cornell CA (2002) Incremental dynamic analysis. Earthq Eng Struct Dyn 31(3):491–514

Vamvatsikos D, Cornell CA (2006) Direct estimation of seismic demand and capacity of oscillators with multilinear static pushovers through incremental dynamic analysis. Earthq Eng Struct Dyn 35(9):1097–1117

Welch DP, Sullivan TJ, Calvi GM (2012) Developing direct displacement-based design and assessment procedures for performance-based earthquake engineering. ROSE report 2012/03. IUSS Press, Pavia

Welch DP, Sullivan TJ, Calvi GM (2014) Developing direct displacement-based procedures for simplified loss assessment in performance-based earthquake engineering. J Earthq Eng 18(2):290–322

Yeo G, Cornell C (2003) Building-specific seismic fatality estimation methodology. In: 9th international conference on applications of statistics and probability in civil engineering, San Francisco, CA

Chapter 12
Towards Displacement-Based Seismic Design of Modern Unreinforced Masonry Structures

Katrin Beyer, S. Petry, M. Tondelli, and A. Paparo

Abstract Unreinforced masonry (URM) structures are known to be rather vulnerable to seismic loading. Modern URM buildings with reinforced concrete (RC) slabs might, however, have an acceptable seismic performance for regions of low to moderate seismicity. In particular in countries of moderate seismicity it is often difficult to demonstrate the seismic safety of modern URM buildings by means of force-based design methods. Displacement-based design methods are known to lead to more realistic and less conservative results, opening up hence new opportunities for the use of structural masonry. An effective implementation of displacement-based design approaches requires reliable estimates of the structure's force and displacement capacity. This paper contributes to this endeavour by taking a fresh look at the drift capacity of URM walls with hollow clay bricks and mortar joints of normal thickness. It discusses in particular the influence of the size of the test unit and the applied loading history and loading velocity on the drift capacities of URM walls.

12.1 Introduction

Although unreinforced masonry (URM) construction features excellent properties with regard to sustainability, durability, indoor climate and fire resistance, in most regions of moderate seismicity the total amount of structural masonry in new residential buildings has decreased over the last three decades (Magenes 2006). One reason for this decrease relates to the conservatism of force-based methods which often lead to the situation that URM buildings do not satisfy the seismic

---

K. Beyer (✉) • S. Petry • M. Tondelli • A. Paparo
Earthquake Engineering and Structural Dynamics Laboratory (EESD), School of Architecture, Civil and Environmental Engineering (ENAC), École Polytechnique Fédérale de Lausanne (EPFL), EPFL ENAC IIC EESD, GC B2 504, Station 18, CH – 1015 Lausanne, Switzerland
e-mail: katrin.beyer@epfl.ch

design check in regions of moderate seismicity. As a result alternative structural systems such as reinforced concrete (RC) walls and gravity frames are used instead. Furthermore, for RC structures already several well developed displacement-based design methods are in place, which yield more realistic and less conservative results than force-based design methods. In order to regain the URM construction's competitiveness with regard to seismic design, displacement-based design methods for URM buildings are necessary. A number of displacement-based design methods for URM structures have recently been proposed. These include applications of the capacity spectrum methods (Fajfar 1999) using inelastic (e.g. Graziotti 2013) or overdamped (e.g. Norda and Butenweg 2011) response spectra or the direct displacement-based design method (Priestley et al. 2007). A summary of these methods can be found in Graziotti (2013).

Displacement-based design methods require the force-displacement response of the structure up to failure as input. With the development of macro-elements representing the nonlinear response of URM walls (Braga and Liberatore 1990; Chen et al. 2008; Belmouden and Lestuzzi 2009; Penna et al. 2013) and their implementation in software packages (Lagomarsino et al. 2013), nonlinear static and dynamic analyses of entire URM buildings have become feasible not only in research but also in engineering practice. Macro-element models are based on pre-defined failure mechanisms and force-displacement relationships of structural components. Next to models for strength and stiffness, the drift capacities of URM walls at horizontal and axial load failure are important input parameters for such models. For RC structures the structural engineer can control the failure mechanism by providing appropriate longitudinal, vertical and confinement reinforcement ratios and layouts. In contrast, most parameters controlling the failure mechanism of URM walls, such as the geometry of the walls, the axial load carried by the walls and the boundary conditions provided by the slabs are defined by architectural considerations or other non-structural requirements (e.g. the thickness of RC slabs depends often on requirements for sound insulation and heating installation). For this reason the ability to predict the nonlinear response of URM buildings forming all kinds of failure mechanisms is a key element towards displacement-based design of URM structures.

Mechanical models for the stiffness and strength of URM walls have been proposed and successfully validated (e.g. Magenes and Calvi 1997). For the deformation capacity of URM walls, comprehensive mechanical models are, however, still lacking. Furthermore, the prediction of the deformation capacity by means of numerical tools remains a challenge although the numerical analysis of URM structures has seen significant advances (for a review see Lourenco 2008; Milani 2012). Numerical models that have been developed for the analysis of URM structures include limit analysis tools (e.g. Milani et al. 2006a, b) which aim at the prediction of failure load and failure mechanism; the simplified micro-models where joints are modelled as interface elements (e.g. Lourenço and Rots 1997; Gambarotta and Lagomarsino 1997a; Snozzi and Molinari 2013); and finite element approaches where masonry is modelled as continuum (e.g. Gambarotta and Lagomarsino 1997b; Zucchini and Lourenço 2002; Facconi et al. 2013). While most of these analysis techniques provide very good approximations of the failure load and often also the failure mechanism,

Table 12.1 Parameters considered in codes when estimating the drift capacity of URM walls

	Failure mode (shear vs. flexure)	Slenderness ratio H_0/L_w or H/L_w	Axial stress ratio	Moment profile	Shape of cross section (rectangular vs. flanged)
EC8-Part 3 (CEN 2005)	x	x			
German National Annex to EC8-Part 1 (DIN 2011)	x	x	x		
Italian code (NTC 2008; MIT 2008)	x				
New Zealand Standard for seismic assessment (NZSEE 2006, 2011)	x	x			x
FEMA 306 (ATC 1998)	x	x			
FEMA 273 (ATC 1997)	x	x			
SIA D0237 (SIA 2011)			x	x	

the deformation capacity associated with horizontal load failure (20 % drop in strength) or axial load failure (loss of axial load bearing capacity) is often difficult to predict. Both performance points lie in the post-peak branch where localisation issues render the numerical analyses particularly difficult. The displacement capacity of URM structures is therefore typically determined by drift limits established on the basis of experimental results.

The principal elements in modern URM buildings are URM walls, RC slabs and sometimes spandrel elements consisting of a masonry spandrel and a strip of the RC slab ("composite spandrels"). While the stiffness and strength of RC slabs and composite spandrels are important in order to predict the force-displacement response of the building, their deformation capacity is typically sufficiently large to be non-critical (see experimental results in Beyer and Dazio 2012). Research needs with regard to horizontal elements in URM buildings relate therefore mainly to the effective width of the slab and the stiffness and strength of composite spandrels. First attempts to address these issues are reported in Da Parè (2011), Benaboud (2013) and Marino (2013). The displacement capacity of modern URM buildings is therefore expected to be limited by the URM walls of the building rather than the horizontal elements (Salmanpour et al. 2013). Of all URM walls the first storey walls are expected to be most critical since shear demands are largest for the first storey.

A comprehensive overview on drift capacities in codes is given in Petry and Beyer (2014a). Table 12.1 summarises the different factors considered in these drift capacity models. With the exception of the Swiss guidelines for the seismic assessment of masonry structures (SIA 2011), all drift capacity models are rather similar: The main parameter is the failure mode; typical drift capacities at the "Significant Damage" (SD) limit state are 0.4 % for shear failure and 0.8 % for flexural failure. The origin of these two values is unknown to the authors but it is assumed that they were derived from results of quasi-static cyclic tests. Quasi-static cyclic tests are of course only an approximation of the loading an URM wall is subjected to during a real earthquake. However, most structural engineering laboratories do not have the capacity of conducting dynamic tests but many are equipped for quasi-static cyclic tests. As a result, the number of quasi-static tests on URM walls that has been carried out until today clearly outnumbers dynamic tests on URM walls or entire URM buildings. Hence, empirical drift capacity models will have to rely on quasi-static cyclic test results. Using experimental results from isolated URM walls under quasi-static cyclic loading as the basis for empirical drift capacity models raises a number of questions; in particular whether the drift capacity of URM walls is influenced by:

- the size of the test unit?
- the loading history applied to the wall?
- the loading velocity?

This paper attempts to shed some light on these aspects. The paper is limited to the behaviour of URM walls with hollow clay bricks and cement mortar for joints of normal thickness (walls with thin bed joints are not considered).

12.2 Tests on URM Walls: Influence of Wall Height on Drift Capacity

Many tests on URM walls have been conducted on specimens with heights between 1.2 and 1.8 m, which corresponds roughly to one half to three quarters of typical storey heights H_s. Apart from restrictions imposed by the test setup, the observation that walls with reduced free height often fail first might have influenced this choice (Fig. 12.1a). In modern URM buildings, however, the window units often reach over the entire storey height and therefore the effective height H of the walls is equal to the storey height (Fig. 12.1b). In older construction, inner walls correspond also to storey-high walls. Given the range of effective wall heights in real buildings, the question whether the size and therefore height of the test unit influences the drift capacity of URM walls is therefore pertinent.

12.2.1 Database on URM Wall Tests

Figure 12.2 shows the distribution of test unit heights from a recently published database on URM wall tests (Petry and Beyer 2014a). A large part of this database stems from the study by Frumento et al. (2009). The database includes walls constructed with full-size clay brick units and cement mortar for joints of normal thickness. The smallest test unit in the database had a height of $H = 1.17$ m (≈ 0.5 H_s) and the largest test unit had a height of $H = 3.00$ m. The database covers therefore well the effective height of walls in real buildings, but it is biased towards the walls with reduced effective heights: out of the 64 tests, 41 tests were conducted on walls with heights smaller than three quarters of a storey height ($H \leq 2.4$ m).

12 Seismic Design of Modern Unreinforced Masonry Structures

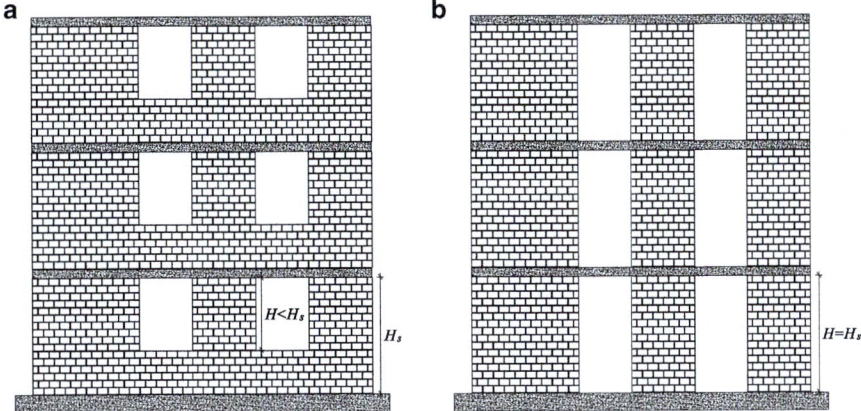

Fig. 12.1 Effective height H of walls in facades with and without masonry spandrels

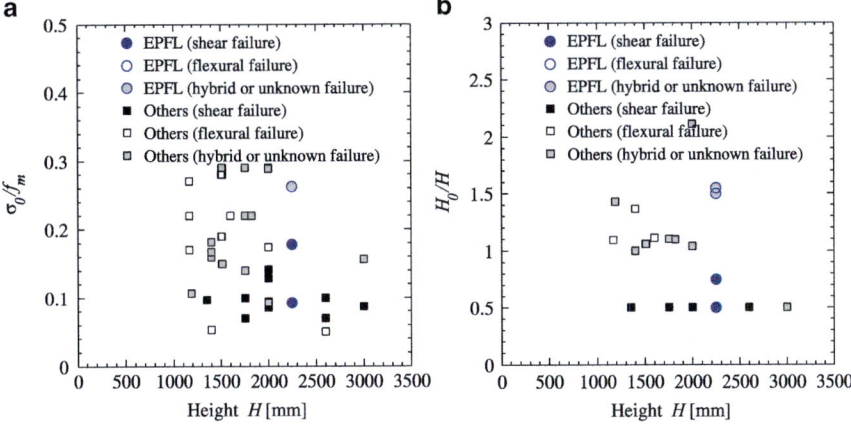

Fig. 12.2 Database on URM walls (Petry and Beyer 2014a): distribution of height H, axial stress ratio σ_0/f_m and shear span ratio H_0/H

Most of the walls were tested as cantilever walls ($H_0/H \approx 1.0$) or with fixed-fixed boundary conditions ($H_0/H \approx 0.5$, Fig. 12.2b). Apart from the EPFL-campaign, where the shear span ratio H_0/H was the key parameter investigated, three further tests featured shear span ratios other than 0.5 or 1.0. In these tests the axial force was applied eccentrically or the height of the top beam was considerable when compared to the rather small test unit. The walls with larger heights than the EPFL-walls were all subjected to fixed-fixed boundary conditions. Seventy percent of the 41 walls smaller than 1.8 m ($\approx 0.75\ H_s$) were tested as cantilevers. The database is therefore dominated by walls that have a height smaller than H_s and were tested as cantilevers. Deriving drift capacity estimates for codes by averaging the drift capacities of all test units in the database that are displaying, for example, a particular failure mode, will inevitably lead to drift capacity estimates representative for this subset. However, it is questionable if short cantilever walls are representative for walls in modern URM structures.

Before closing this section, some reflections on the testing of walls with $H < H_s$: For walls extending only over part of the storey height (Fig. 12.1a), the boundary conditions at the bottom of the wall require particular attention. In the test stand, the test unit is typically placed between concrete or steel beams in order to fix the test unit to the strong floor and apply the horizontal and vertical loads at the top. In a modern building the URM wall is framed by RC slabs. The boundary conditions in laboratories seem therefore representative if full storey high walls are tested. Walls in facades with masonry spandrels (Fig. 12.1a) would be framed by a RC slab at the top and URM masonry at the bottom. The boundary condition at the bottom should therefore be given some consideration since the steel or RC foundation in the laboratory might not be representative. In particular, the confining effect on the bottom mortar joint provided by the steel or RC foundation might be stronger than that of the masonry supporting the wall in real buildings. As a result, the lateral expansion of the brick due to the different Poisson ratios of mortar and brick might be smaller. If the failure mode includes crushing of the URM wall's toe, the confinement provided by the foundation might therefore potentially lead to an increase of the URM wall's drift capacity. To avoid this effect one could consider testing the specimen with an additional brick layer at the base that is fixed to the foundation by a high performance glue.

12.2.2 A New Empirical Drift Capacity Model for URM Walls

Figure 12.3a shows the experimentally determined drift capacity δ_u as a function of the wall height. The drift capacity is the drift capacity associated with a 20 % drop in strength. The figure shows a clear decreasing trend of drift capacity with increasing height. This holds also if the drift capacity is normalised with the shear span ratio H_0/H accounting for the fact that the drift capacity reduces with reducing shear span ratio (SIA 2011). These plots suggest that the drift capacity of URM walls is influenced by a size effect, as it has first been proposed by Lourenço (1997). Accounting for the effect that the drift capacity of walls reduces with increasing axial load ratio, the following drift capacity equation was recently proposed by Petry and Beyer (2014a):

$$\delta_{CT} = 1.3\% \cdot \left(1 - 2.2\frac{\sigma_0}{f_u}\right) \cdot \frac{H_0}{H} \cdot \left(\frac{2400mm}{H}\right)^{0.5} \quad (12.1)$$

The equation aims at predicting a mean drift capacity as obtained for quasi-static cyclic tests where the test unit is subjected to a constant axial load ratio and a

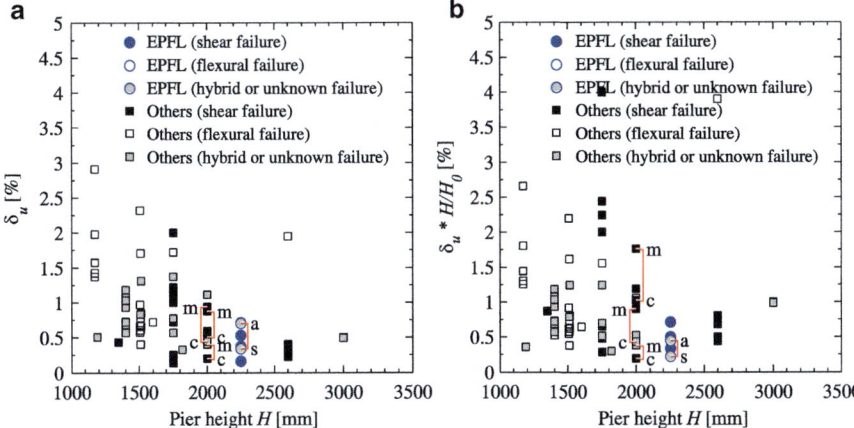

Fig. 12.3 Drift capacity δ_u (**a**) and drift capacity normalised with the shear span ratio (**b**) as a function of the wall height

constant shear span ratio throughout the test. To predict the drift capacities under real earthquake loading, where axial load ratio and shear span ratio might vary and where the wall is subjected to larger strain rates, the drift capacity equation needs to be modified by two correction factors accounting for loading history (ψ_{LH}) and strain rate (ψ_{SR}) effects respectively. The drift capacity equation at "Near Collapse" limit state therefore becomes:

$$\delta_{NC} = \delta_{CT} \cdot \psi_{LH} \cdot \psi_{SR} \qquad (12.2)$$

Section 12.3 investigates effects of the loading history on the drift capacity. In Sect. 12.4 results from static and dynamic tests are compared and conclusions regarding the importance of strain rate effects on the drift capacity are drawn.

12.3 The Effect of the Loading History on the Drift Capacity

Since reliable analytical models for predicting the drift capacity of URM walls are currently not available, the drift capacity is typically determined by quasi-static cyclic tests. The main variables that are used in these tests are:

- The axial load ratio,
- The rotational or moment restraint at the top of the wall, and
- The loading history.

In most tests reported in the literature, the axial load ratio was maintained constant throughout the test, the test unit was subjected to either cantilever or fixed-fixed boundary conditions, and a loading history with two or three cycles per amplitude level was applied. The total number of cycles until failure was often not a key parameter when defining the loading history. However, for systems susceptible to cumulative damage demands, the number of cycles can influence the force and/or displacement capacity obtained from the quasi-static cyclic test. In current testing practice, in order to capture the evolution of damage limit states, a relatively large number of cycles is often applied. The questions that arise from such testing practice are:

- Does the loading history have an influence on the key parameters of interest, i.e., the effective stiffness, maximum force capacity and drift capacity?
- If it does, is the number of applied cycles representative of the expected cumulative seismic damage demand in the region of interest?
- Are the boundary conditions representative for the critical walls in a structure?

To investigate these questions, first the results from pairs of test units are discussed where one had been subjected to monotonic and one to cyclic loading (Sect. 12.3.1), then loading protocols for cyclic tests on URM walls are reviewed (Sect. 12.3.2) and typical axial force and shear force histories of first storey URM walls are investigated (Sect. 12.3.3).

12.3.1 Monotonic vs. Cyclic Tests

When reviewing the test results on URM walls (Sect. 12.2.1), three pairs of tests on URM walls were identified where one wall had been subjected to monotonic loading and the other to cyclic loading. The first two pairs stem from the experimental campaign by Ganz and Thürlimann (1984), the third from Magenes and Calvi (1992). Ganz and Thürlimann applied always 10 cycles per amplitude level, which from today's point of view is certainly not representative since it exceeds considerably the number of cycles imposed by an earthquake. The total number of cycles applied until failure was 58 for W6 and 61 for W7. Magenes and Calvi applied a loading history which corresponds in many respects already to today's standard for URM wall testing. Until failure, the cyclic loading history comprised ~6 cycles.

Table 12.2 summarises the three main properties of the envelope curves in Fig. 12.4, i.e. the effective stiffness, the maximum force and the drift capacity. The effective stiffness is the secant stiffness at 0.75 F_{max} and the drift capacity the drift at which the force had dropped to 0.8 F_{max}. For the cyclic tests, the effective stiffness K_C and the strength $F_{max,C}$ are taken as average values obtained for the positive and negative loading direction. The drift capacity $\delta_{u,C}$, on the contrary, is defined as the minimum of the two values (see Frumento et al. 2009). From the three parameters, the maximum force is the one which is the least affected by the loading scheme. The largest influence of the loading history is observed for the drift capacity, which is in average twice as large for monotonic tests than for quasi-static cyclic tests. Somewhat surprising is the consistently larger stiffness for cyclic tests

Table 12.2 Monotonic vs. cyclic loading: comparison of effective stiffness, maximum force and drift capacity

	Cyclic test			Monotonic test			Cyclic/monotonic		
	K_C [kN/m]	$F_{max,C}$ [kN]	$\delta_{u,C}$ [%]	K_M [kN/m]	$F_{max,M}$ [kN]	$\delta_{u,M}$ [%]	K_C/K_M [−]	$F_{max,C}/F_{max,M}$ [−]	$\delta_{u,C}/\delta_{u,M}$ [−]
W1 & W6	178	256	0.45	127	266	0.94	1.40	0.96	0.48
W2 & W7	218	496	0.20	163	479	0.40	1.34	1.03	0.50
MI1m & MI1	98	263	0.28	66	258	0.78	1.49	1.02	0.35
Mean ratio							1.41	1.00	0.44

than for monotonic tests. The authors do not have an explanation for this observation. It must be assumed that it is linked to the alternating loading direction since mortar strengths and age of the test units at the day of testing were very similar for all test units by Ganz and Thürlimann; Magenes and Calvi did not report mortar strengths for the individual walls.

Despite the admittedly very limited data set, this comparison of monotonic vs. cyclic test results suggests that the loading history is not important if one is only interested in the force capacity of the URM wall. It becomes, however, significant if the displacement capacity and possibly also the effective stiffness are of interest. When results of quasi-static cyclic tests of URM walls are used to derive drift capacity limits for displacement-based design, attention should therefore be paid to the loading history that was applied in the test.

12.3.2 Loading Protocols for Cyclic Tests

For systems susceptible to strength and stiffness degradation, the strength and deformation obtained from quasi-static cyclic tests will depend on the imposed loading history. Hence, the obtained capacities are directly related to imposed demands. For this reason, loading protocols for quasi-static cyclic tests on URM walls should be given some consideration.

Tomazevic and co-workers (1996, 2000) addressed loading history effects on the response of reinforced masonry walls displaying a flexural failure mode but until today no systematic investigation on the influence of different cyclic loading protocols on the drift capacity of URM walls was carried out. The effect of the number of cycles on the performance of URM walls can therefore only be inferred indirectly via the comparison of the envelopes of first cycles with envelopes of second or third cycles. The walls of the EPFL test series were subjected to two cycles per drift level. Figure 12.5 shows for three of these walls the envelopes of the first and second cycles. The three walls developed different failure mechanisms: PUP2 a

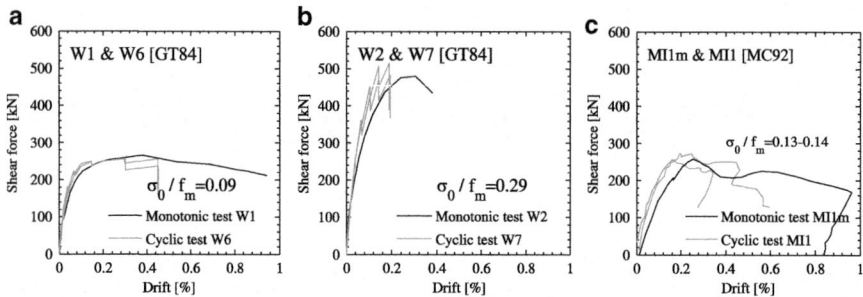

Fig. 12.4 Monotonic vs. cyclic loading: comparison of force-displacement envelopes

diagonal shear failure, PUP3 a rocking failure and PUP4 a hybrid failure mode. Up to a certain point close to the peak force, the difference between first and second cycle envelopes is negligible. As soon as the difference becomes significant, failure is imminent. The envelopes of the first and second cycle start to diverge once the first limit state inducing irreversible damage has been reached, i.e. failure of the compression zone (Limit State (LS) F3, Petry and Beyer 2014c) or concentration of shear deformations in a single diagonal crack (LS S3). Hence, before these limit states, the behaviour of the URM walls is rather insensitive to the loading history while the remaining drift capacity after one of these limit states have been reached appears strongly dependent on the loading history. Since at these limit states the maximum force capacity has already been reached, the force capacity does not seem sensitive to the loading history, while the loading history is expected to influence the drift capacity.

Since the quantitative effect of the loading history on the drift capacity is unclear, a loading history should be applied, that represents the seismic demand of the geographical region of interest as closely as possible. Existing standardized loading protocols were derived for regions of high seismicity (e.g. ATC-24 1992; FEMA-461 2007); one even specifically for masonry structures (Porter 1987). Krawinkler (2009), however, points out that the latter imposes even for high seismic regions far too many cycles. Most research projects on URM structures address construction practice in low to moderate seismic countries and hence loading protocols should be applied that impose fewer cycles until failure. Figures 12.6a and b show examples of loading protocols that represent the cumulative damage demands imposed in regions of high and regions of low to moderate seismicity for a hazard level with a 2 % probability of exceedance in 50 years (Mergos and Beyer 2014). The loading protocols were derived from nonlinear time history analysis results of a large set of single degree of freedom systems that reflect the fundamental properties of typical structural systems. To avoid excessive conservatism for particular types of structures, a set of protocols was developed that account for the different cumulative demands as a function of the structural type, fundamental period, number of cycles per amplitude level and the seismicity.

12 Seismic Design of Modern Unreinforced Masonry Structures   411

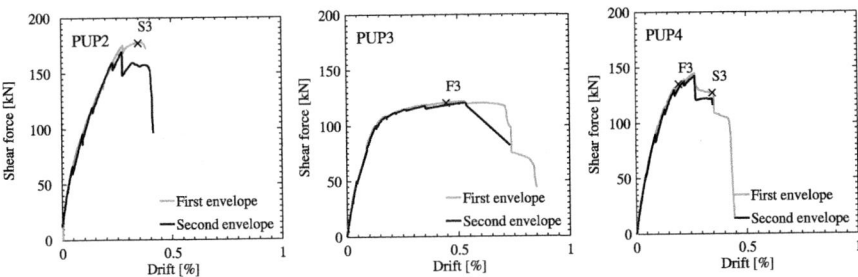

Fig. 12.5 Comparison of first and second envelopes of quasi-static cyclic tests (Petry and Beyer 2014b)

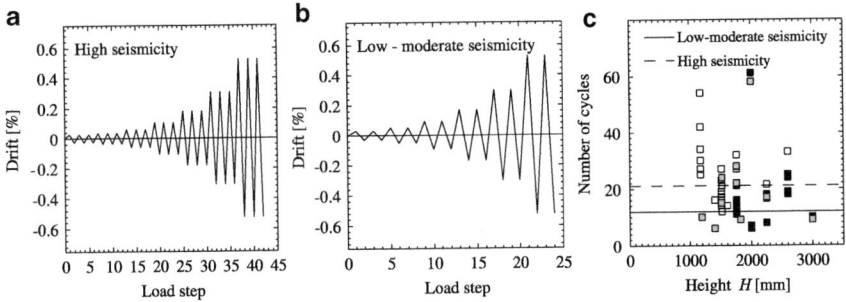

Fig. 12.6 Influence of the seismicity on a representative load protocol: drift controlled load histories for quasi-static cyclic testing of URM walls expected to fail in shear for which the cumulative cyclic demand is representative for countries of high seismicity (**a**) and low – moderate seismicity (**b**). Number of cycles applied in quasi-static cyclic tests of URM walls (**c**)

The loading protocols shown in Fig. 12.6a, b reflect the cumulative damage demands on shear dominated URM structures with a fundamental period $T = 0.2$ s. In this example, a loading history with three cycles per drift amplitude was chosen for the high seismicity case and with two cycles per drift amplitude for the low-moderate seismicity case. The expected drift capacity at the "Near Collapse" limit state was estimated according to EC8-Part 3 (CEN 2005). One sees that the number of cycles imposed on URM structures in high seismicity regions is approximately twice as large as in regions of low-moderate seismicity (21 cycles vs. 12 cycles). Figure 12.6c shows the number of cycles applied to the test units of the database (Sect. 12.2.1). For most tests the number could only be roughly estimated from hysteresis plots. However, the figure clearly shows that in many tests the number of applied cycles exceeds what would be representative for the demand on URM buildings in countries of low-moderate seismicity.

12.3.3 Inner Walls vs. Outer Walls

In an URM building with strong RC slabs, most of the damage concentrates typically in the first storey of the building (Paulay and Priestley 1992; Paparo and Beyer 2014). Quasi-static cyclic tests should therefore represent the boundary conditions of the first storey walls. Among these, the demands on inner and outer walls differ significantly with regard to axial forces and shear spans (Petry and Beyer 2014a).

Figure 12.7 shows the axial force and base shear of an outer and an inner wall for a 4-storey example building. The structure was analysed using the macro element software Tremuri (Lagomarsino et al. 2013). The input ground motion was an artificial record from the study by Priestley and Amaris (2002). In the analysis the walls were not assigned an ultimate drift capacity. The analysis results show that the inner wall is subjected to an axial force which is relatively constant throughout the duration of the earthquake and its shear force – drift hysteresis is fairly symmetric about the origin. The seismic behaviour of such a central wall seems therefore well represented in quasi-static cyclic tests where a constant axial force and a constant shear span ratio are applied to the wall.

The picture is different if an outer wall is considered: In the left wall, the axial force increases when the structure is pushed towards the left while it decreases when the structure moves towards the right. As a result, the maximum base shear is larger for the negative loading direction than for the positive loading direction. Due to the decrease in displacement capacity with increasing axial force (Lang 2002), outer walls fail therefore typically in the loading direction where the axial force increases in the wall (see, for example, Beyer et al. 2014). The question arises how the drift capacity of walls subjected to such asymmetric boundary conditions for the two loading directions compares to the drift capacity of walls subjected to the same boundary conditions in the two loading directions.

The behaviour of outer walls is less well represented by standard test configurations for URM walls and the question arises how well their displacement capacity can be estimated from standard tests. A preliminary attempt to investigate this topic has been carried out within the EPFL-series on URM walls. The sixth test PUP6 represented boundary conditions of an outer wall: It approached for the positive loading direction those of PUP5 and for the negative loading direction those of PUP4. Figure 12.8 shows the applied axial load and shear span ratio as function of the applied horizontal load and drift.

In the negative loading direction PUP6 was subjected to larger axial forces than in the positive loading direction. Hence, the wall was expected to fail for loading in the negative direction where the boundary conditions of PUP4 were approached. For horizontal and vertical load failure, the drift capacities of PUP6 were 2.0 and 1.6 times larger than those of PUP4 (Figure 12.8), i.e., the results suggest that the displacement capacity of an asymmetrically loaded wall is about twice as large as the displacement capacity of a symmetrically loaded wall. The weakest zone in symmetrically loaded walls failing in diagonal shear is the zone where the two

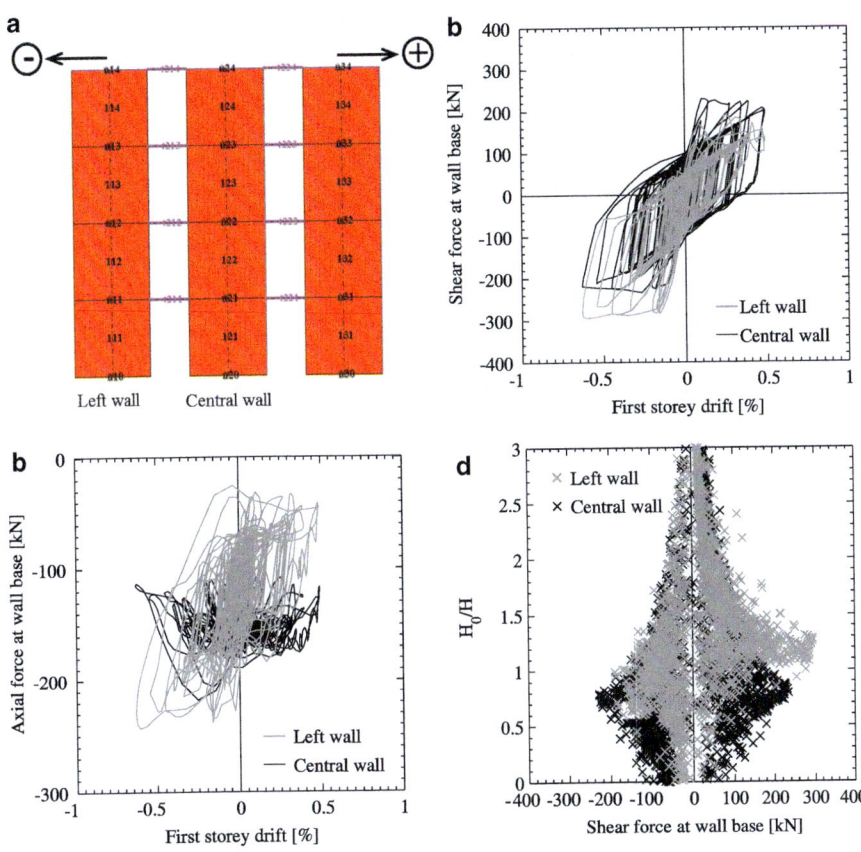

Fig. 12.7 Demand on inner and outer first storey walls in terms of axial force (**a**), shear force (**b**) and shear span ratio H_0/H (**c**)

diagonal cracks intersect and failure of this zone tends to trigger horizontal load and axial load failure. In asymmetrically loaded walls that develop a flexural mechanism in one and a shear mechanism in the other direction, such zone does not exist. If the wall develops a shear mechanism for both directions, the shear crack is typically much smaller for one direction than for the other. As a result, this heavily disaggregated zone at the intersection of two diagonal shear cracks, which often controls the drift capacity of symmetrically loaded walls, does not exist for asymmetrically loaded walls. Note that for PUP6 the definition of the critical loading direction was less clear than for walls in most real buildings. While the axial load was larger for the negative loading direction, the shear span ratio was smaller for the positive loading direction. As a result, the maximum shear forces were rather similar for the two loading direction and the onset of failure occurred in fact for the positive loading direction (Petry and Beyer 2014a).

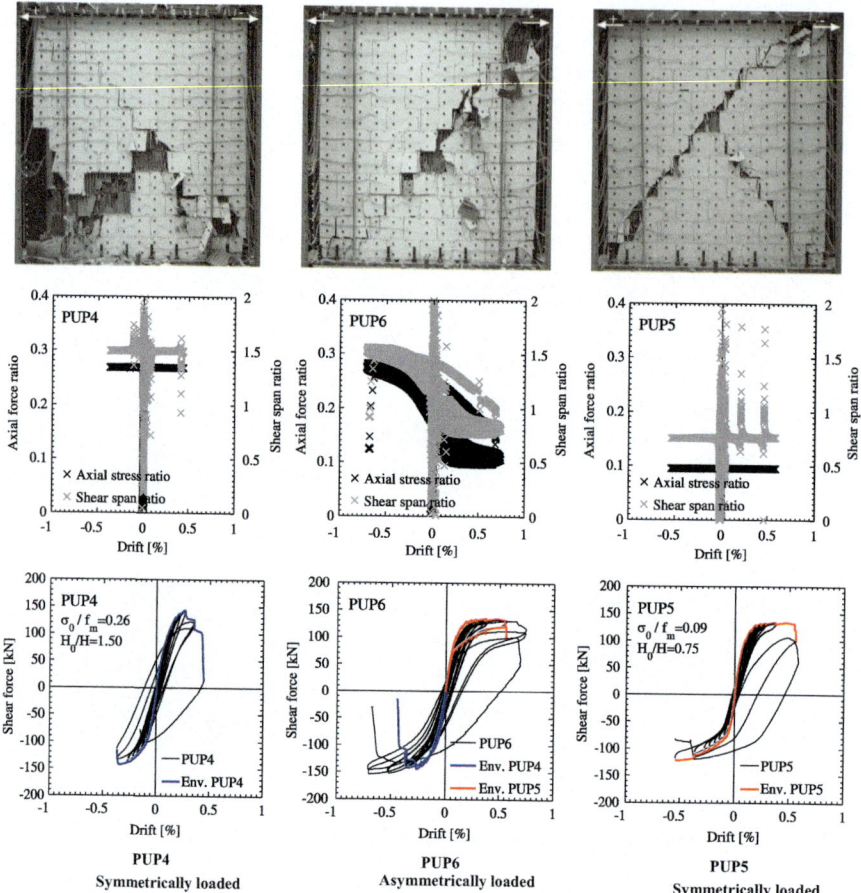

Fig. 12.8 Results of quasi-static cyclic tests on symmetrically (PUP4, PUP5) and asymmetrically (PUP6) loaded walls: failure mode, axial load ratio, shear span ratio and hystereses (Petry and Beyer 2014a)

12.3.4 Conclusions on Loading History

For systems susceptible to cumulative damage demands, the system's force and/or displacement capacities are a function of the imposed cumulative demand, i.e., the number of cycles until failure is attained. When determining these parameters from quasi-static cyclic tests, the applied horizontal loading history needs to be given due attention. Until today no systematic study of the effect of the loading history on the force and displacement capacity of URM walls has been carried out and therefore a final expression for the correction accounting for load history effects cannot be proposed. The following paragraphs summarise the preliminary trends identified in Sects. 12.3.1, 12.3.2, and 12.3.3.

Cyclic tests resulted in only half the drift capacity as monotonic tests but the loading regime had no effect on the force capacity. The reduction in drift capacity under cyclic loading is linked to the cumulative damage induced by the cycles that are applied after the limit states F3 (failure of the compressed zone) or S3 (concentration of shear deformations in a single diagonal crack) have been attained. A study comparing the behaviour of walls subjected to different cyclic loading histories could not be found in the literature. To get a first idea of the impact of the number of cycles on the force-displacement response, envelopes of first and second cycles were compared. As for the monotonic and cyclic tests, the difference between these envelopes became only significant once the limit state F3 or S3 have been passed. Since these limit states are attained after the strength plateau has been reached, only the drift capacity but not the force capacity is expected to be affected by the loading history. When determining quantities relevant for force-based design from quasi-static cyclic tests, the loading history is therefore of lesser importance. However, when drift capacity estimates are sought, due attention should be given to the number of cycles applied until failure of the wall.

Since at present the exact effect of the loading history on the drift capacity of URM walls is unknown, a loading history should be applied which reflects the expected cyclic demand on the wall during a "Near Collapse" scenario. Attention should be paid to

- The number of cycles imposed until failure,
- The boundary conditions, i.e. axial load ratio and shear span ratio,
- Whether the boundary conditions are the same for the positive and negative loading direction.

The number of cycles a system is subjected to depends on its properties (fundamental period, hysteretic behavior) and the seismicity of the case study region. URM structures are mainly constructed in low-moderate seismicity regions and therefore fewer cycles than for high seismicity regions should be applied.

Quasi-static cyclic tests applying a constant axial force to the specimen that is tested as cantilever or with fixed-fixed boundary conditions will remain the standard test since the boundary conditions are well defined and within the capabilities of many structural engineering laboratories around the world. Boundary conditions of URM walls in real buildings are, however, more diverse. This applies in particular to the shear span ratio, which can vary approximately between 0.5 and ~2.0 H for URM buildings with RC slabs and the symmetry of the boundary condition for the positive and negative loading direction. While symmetric cycles with constant shear span and axial load ratio approximate the demand on inner walls typically well, this does not hold for outer walls. For the latter the axial load and shear span ratios fluctuate with the loading direction. A first investigation into the effect of such asymmetric loading histories showed that the drift capacity of asymmetrically loaded walls might be twice as large as that of symmetrically loaded walls, i.e., similar to the drift capacities obtained from monotonic load tests. For such walls a correction factor of $\psi_{LH} = 2$ is therefore proposed (Eq. (12.1), Sect. 12.2.2).

12.4 Quasi-static vs. Dynamic Tests

It is likely that quasi-static cyclic tests will remain the standard tests for determining drift capacities of URM walls. However, the actual purpose is to find drift capacity estimates for walls subjected to earthquake loading. During an earthquake, URM walls are subjected to strain rates that are approximately 1,000 times higher than during quasi-static tests. To link static to dynamic drift capacities, strain rate effects on the drift capacity need to be quantified and expressed by means of the correction factor ψ_{SR} (Sect. 12.2, Eq. (12.2)).

Williams and Scrivener (1974) and Tomazevic et al. (1996) investigated strain rate effects on reinforced masonry. Both reported similar drift capacities for static and dynamic tests. Abrams (1996) compared the behaviour of unreinforced masonry structures under static and dynamic loading and concluded that the loading history affected the cracking pattern. However, he acknowledges that the structures were tested at different scales with different construction materials and different restraints provided to the flexible diaphragms, which made it difficult to compare them one-to-one. Elgawady et al. (2004) compared the results of URM walls with and without GFRP wrapping under static and dynamic loading. However, the shake table tests were stopped prematurely and hence no conclusions regarding the drift capacities at horizontal and vertical load failure under dynamic loads were possible. A numerical study by Snozzi and Molinari (2013) showed that the strength of URM walls is larger when subjected to higher strain rates due to a more diffuse cracking pattern. However, this study did not yield any information regarding the effect of the strain rate on the deformation capacity since the bricks were modelled as elastic.

This section addresses the effect of strain rates on drift capacities by comparing the maximum drifts attained in quasi-static cyclic tests on walls to the maximum drifts attained in a shake table test of a 4-storey building (Beyer et al. 2014; Tondelli et al. 2014). Both walls and building were constructed at half scale using the same, special fabricated half-scale bricks (Petry and Beyer 2014d). The walls had similar but not identical dimensions. The walls tested under quasi-static cyclic loading were 1.00 m long and 1.11 m high and had a rectangular cross section. The walls of the building tested on the shake table were 1.55 m long and 1.40 m high and had small flanges at the wall ends in order to increase the out-of-plane stability of the walls.

The comparison between shake table test results of an entire building and quasi-static cyclic tests will always be approximate since the exact boundary conditions and loading history of the walls in the building are unknown. In addition, the geometries of the two sets of walls differ slightly. However, in the absence of tests where only the loading velocity but none of the other parameters was varied, the comparison of results from a shake table test and quasi-static cyclic tests might allow to shed some new light on the effect of strain rates on the deformation capacity of URM walls. The following sections investigate the demand on the walls in the building on the basis of nonlinear analysis (Sect. 12.4.1), analyse the drift capacities obtained from quasi-static cyclic tests (Sect. 12.4.2) and compare drifts attained in the shake table test to the drift capacities from quasi-static cyclic tests (Sect. 12.4.3). Section 12.4.4 gives recommendations for the choice of the correction factor ψ_{SR} that accounts for strain rate effects (Eq. (12.2)).

12.4.1 Shake Table Test

The building tested on the shake table was a 4-storey structure with URM walls and RC walls. The building was subjected to uni-directional shaking and tested at the TREES laboratory in Pavia (Italy). The shaking induced in-plane loading in the facade shown in Fig. 12.9. Detailed information on the shake table test will shortly be published (Beyer et al. 2014) and the data collected during the test shared (Tondelli et al. 2014).

It is assumed that the drift capacity of the URM walls is a function of the axial load ratio, the shear span ratio and the height of the wall (see Eq. (12.1)). Hence, to estimate the drift capacities of the first storey walls, the axial load ratios and shear span ratios need to be estimated. Since the internal forces cannot be measured during a shake table test, they need to be estimated from numerical analyses. Since the building was symmetric about its longitudinal axis, a pushover analysis of a 2D simplified micro model of the facade using the software package Atena (Cervenka et al. 2010) was carried out. Details on this analysis and a comparison of experimental and numerical results are given in Beyer et al. (2014). Figures 12.10 and 12.11 show the internal force distribution in the façade when the wall is pushed towards the left (increase in axial forces in the left URM wall) and the right (increase in axial force in the RC wall). Assuming that all axial forces are carried by the in-plane loaded walls, the first storey walls are subjected to axial forces of 137 kN (left wall) and 104 kN (central wall), which correspond to axial load ratios of 0.16 and 0.12, respectively.

During the last two runs (Run 8 and 9) the building reached for both directions of loading the inelastic range and therefore it can be assumed that the internal force distribution at peak displacements was similar for both runs. The plots in Figs. 12.10 and 12.11 show the internal forces at $\delta_{avg} = -0.26$ % and 0.32 % respectively. These average drifts over the height of the building correspond to the peak drifts attained in negative and positive direction during Run 8. Table 12.3 summarises the axial stress ratios and shear span ratios of the two first storey walls for the positive and negative loading direction. The following section compares for these two URM walls the drift capacities obtained from static and dynamic tests.

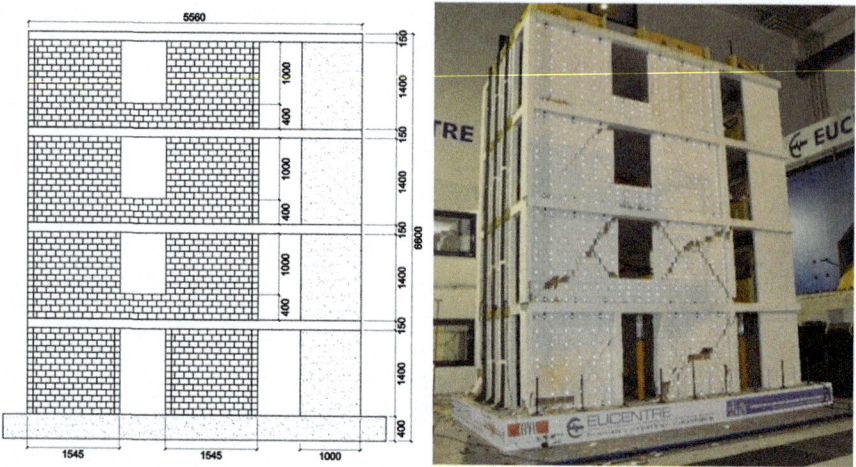

Fig. 12.9 Shake table test unit at half scale

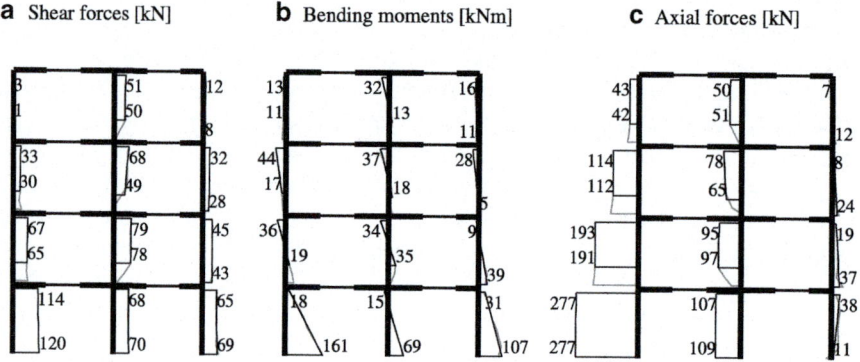

Fig. 12.10 Numerical model of shake table test unit: internal forces for negative loading direction at $\delta_{avg} = -0.26\,\%$

12.4.2 Drift Capacities Estimated from Quasi-static Cyclic Tests

To prepare the shake table test and to decide in particular on the model brick to be used, five out of the six quasi-static cyclic tests on full-scale walls were replicated at half-scale (Petry and Beyer 2014d). Figure 12.12 shows the test setups for the two test series. The half-scale walls reflected the behaviour of the full-scale walls very well in terms of stiffness, strength, drift at maximum horizontal force, drift capacity at horizontal load failure and the failure mode. Only with regard to the drift capacity at axial load failure led the half-scale walls to values which were significantly larger

12 Seismic Design of Modern Unreinforced Masonry Structures

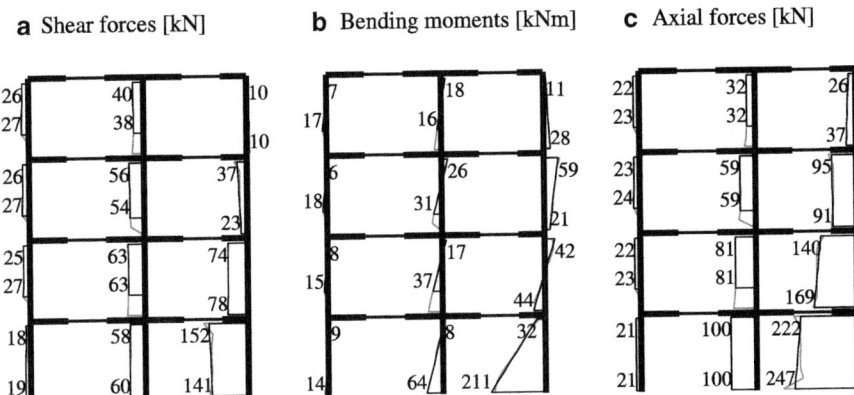

Fig. 12.11 Numerical model of shake table test unit: internal forces for positive loading direction at $\delta_{avg} = 0.32\%$

Table 12.3 Demand on first storey walls based on results of numerical model and estimated drift capacities based on quasi-static cyclic tests

First storey	Negative loading direction		Negative loading direction but drift values from test unit with max. axial stress ratio	Positive loading direction	
	Outer wall	Inner wall	Outer wall	Outer wall	Inner wall
N [kN]	277	109	277	21	100
σ [MPa]	1.94	0.76	1.94	0.15	0.70
σ/f_m	0.34	0.14	0.34	0.03	0.12
$H_0 = M/V$ [m]	1.34	0.99	1.34	0.74	1.07
H_0/H	0.96	0.70	0.96	0.53	0.76
R	0.85	0.85	0.85	0.85	0.85
ψ_{SR}	2.0	1.0	2.0	1.0	1.0
δ_{peak} [%]	0.06	0.24	0.36	0.28	0.29
δ_u [%]	−0.26	0.40	0.59	0.51	0.49
δ_{max} [%]	−0.26	0.58	0.76	0.73	0.71

than those of the full-scale walls. Figure 12.13 shows the comparison of the drift capacities at the three performance limit states including linear trend lines.

The drift capacities of the first storey walls of the building tested on the shake table are estimated from these linear trend lines in Figure 12.13 for the axial stress and shear span ratios obtained from the pushover analyses (Table 12.3). In addition, the following two effects are considered:

- The walls in the building are somewhat larger than the walls tested under quasi-static cyclic loads. To account for the size effect discussed in Sect. 12.2, the drift

Fig. 12.12 Test setups for quasi-static cyclic tests on full-scale (**a**) and half-scale (**b**) walls

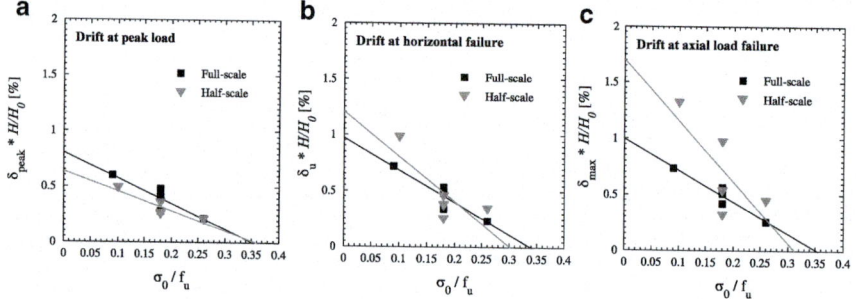

Fig. 12.13 Drift values of full- and half-scale walls at peak load (**a**), horizontal load failure (**b**), and axial load failure (**c**)

capacities obtained from the quasi-static cyclic tests are reduced by the following factor (see Eq. (12.1)):

$$R = \left(\frac{1110mm}{1400mm}\right)^{0.5} = 0.89 \qquad (12.3)$$

- For the outer wall, when loaded in the negative direction, the correction factor accounting for the load history was assumed as $\psi_{LH} = 2$ (see Eq. (12.2) and Sect. 12.3.3). For all other walls/loading directions $\psi_{LH} = 1$ was assumed.

The correction factor for strain rate effects ψ_{SR} was set to unity. Table 12.3 reports the resulting drift capacities at peak load (δ_{peak}), horizontal load failure (δ_u) and axial load failure (δ_{max}). One problem becomes immediately apparent: The axial load ratio of the outer wall for loading in the negative direction is outside the range of axial load ratios covered in the quasi-static cyclic tests. With the linear trend model the drift capacities at this axial load ratio are negligible or even negative. This is of course not meaningful. Furthermore, it is probable that the axial load ratio in the outer wall was overestimated by the 2D model which neither included out-of-plane walls nor the flanges of the in-plane loaded walls. For these reasons the drift capacities obtained for the wall that had been subjected to the largest axial stress ratio ($\sigma/f_m = 0.27$) will be used to derive the drift capacity of the outer wall for the negative loading direction. As outlined above, the reduction factor $R = 0.89$ accounting for the size effect and the correction factor $\psi_{LH} = 2$ for the load history effect will be considered. The resulting drift capacity of an outer wall is therefore computed as follows:

$$\delta = \delta_{PUM4} \cdot \frac{(H_0/H)}{(H_0/H)_{PUM4}} \cdot R \cdot \psi_{LH} \tag{12.3}$$

The drifts of PUM4 at peak load, horizontal load failure and axial load failure were 0.31 %, 0.52 % and 0.67 % respectively; the shear span ratio was 1.5. The drift capacities resulting for the outer wall and the negative loading direction are summarized in the central column of Table 12.3.

12.4.3 Comparison of Drift Histories from Shake Table Test with Drift Capacities from Quasi-static Cyclic Tests

The shake table test unit was subjected to nine runs; only the last two induced significant damage. In the following, the drift histories measured at the centre line of the outer and inner URM walls of the first storey are compared to the drift limits derived in the previous section from quasi-static cyclic tests (Table 12.3). For details of the computation of the drift histories from the optical measurements see Beyer et al. (2014).

After Run 8 the damage in the URM panels started concentrating in one diagonal crack. From quasi-static cyclic tests on URM walls it is known that this indicates that the post peak branch has been reached and failure is rather imminent (Petry and Beyer 2014d). The drift histories of the outer and inner wall exceeded just the drift limits corresponding to the peak force (Fig. 12.14). Hence, for this limit state, the drift limits derived from the quasi-static cyclic tests seem to correspond well with the observed behaviour of the shake table test unit.

In Run 9 all walls of the first and second storey lost their axial load bearing capacity. Also this observation agrees with the findings when comparing drift histories from the shake table test with the drift limits obtained from quasi-static cyclic tests (Fig. 12.15): Both walls exceeded the drift limit for axial load failure for the negative loading direction. The inner wall touched the same limit also for the positive loading direction. Figure 12.16 shows the damage of the URM walls after this final run.

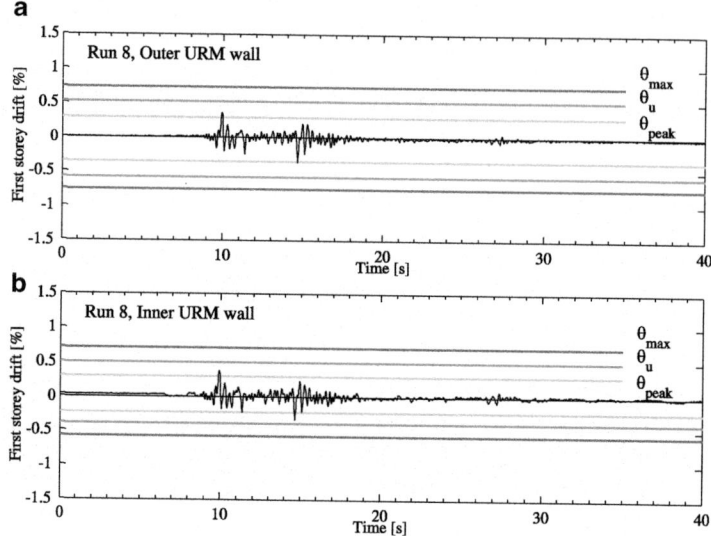

Fig. 12.14 Shake table test, Run 8: comparison of drift histories of first storey walls with drift limits derived from quasi-static cyclic tests

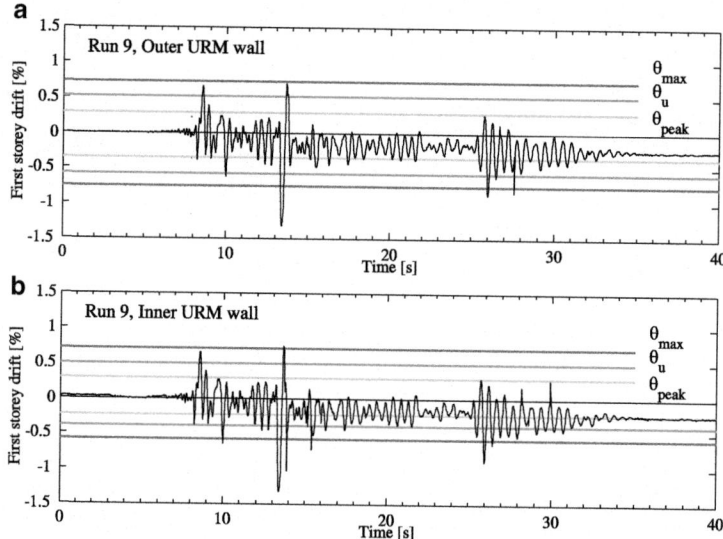

Fig. 12.15 Shake table test, Run 9: comparison of drift histories of first storey walls with drift limits derived from quasi-static cyclic tests

Fig. 12.16 URM walls of shake table test unit after Run 9 (Beyer et al. 2014)

12.4.4 Conclusions on Comparison of Drift Capacities from Static and Dynamic Tests

The comparison of drift limits derived from quasi-static cyclic tests with drift histories recorded for a shake table test showed that the former estimate the limit states of the dynamic test rather well. This suggests that the difference in strain rates between quasi-static and dynamic tests does not influence significantly the drift capacities of URM walls associated with different limit states. Hence, the correction factor ψ_{SR} accounting for strain rate effects should be set to unity. In this study, two limit states could be addressed: The limit state at peak force, which corresponds typically with the onset of localization of the damage in a single crack or row of bricks, and the limit state where the walls lost their axial load bearing capacity.

While the comparison has certain advantages over existing studies comparing the results of static and dynamic tests on URM walls (e.g. same construction material for static and dynamic tests, testing until axial load failure), it still suffers from a number of disadvantages: The size and cross section of the walls subjected to static and dynamic loads was not identical, neither were the boundary conditions the walls were subjected to. In order to investigate strain rate effects systematically, tests at different strain rates using the same test setup should be carried out.

12.5 Summary and Outlook

To promote the application of displacement-based approaches for the design of modern URM buildings, reliable estimates of deformation capacities of key structural elements inherent in these buildings are of paramount importance. URM walls, RC slabs and often spandrel elements consisting of a URM spandrel and a strip of the RC slab (referred to as "composite spandrels") are the key elements determining seismic vulnerability of URM buildings. Although stiffness and strength of slab and composite spandrels significantly influence the seismic response of these buildings, their deformation capacity is likely to be sufficiently large to be non-critical. The displacement capacity of the building is therefore likely to be controlled by the drift capacity of the first storey URM walls where the shear demand is largest, which is typically determined from empirical equations derived from quasi-static cyclic tests on URM walls.

Recognizing that the accurate estimation of the deformation capacity of URM building elements will lead to displacement-based design approaches that allow to safely and cost effectively design URM buildings, this paper has addressed the influence of the (i) size of the test unit, (ii) loading history, (iii) loading velocity, and (iv) the boundary conditions on the drift capacities of URM walls in the context of results obtained from quasi-static cyclic tests, thereby predicting their true drift limits under earthquake loads. Based on results from 64 monotonic and cyclic tests on URM walls with clay bricks, cement mortar and joints of normal thickness, it was found that the drift capacity decreases with increasing wall height. A drift capacity equation that was recently proposed accounts for this effect as well as the influence of the axial stress ratio and the shear span ratio on the drift capacity of URM walls (Petry and Beyer 2014a, b, c).

Comprehensive studies on the effect of the loading history on URM walls are limited. However, existing studies show that monotonic tests led to drift capacities that were approximately twice as much as those obtained from cyclic tests. Furthermore, load histories commonly applied to quasi-static cyclic tests were too excessive and place much higher demands than those expected for low to moderate seismic earthquakes. This discrepancy in load histories is unlikely to influence the strength of URM walls, but affects their drift capacity. A comparison of response envelopes of first and second cycles of quasi-static cyclic tests of URM walls showed that cumulative damage causes negligible effect on stiffness and strength until a limit state responsible for irreversible wall damage is reached, i.e., onset of crushing of the compression zone or concentration of shear deformations in a single diagonal crack. At the onset of crushing or the concentration of damage in one diagonal crack, the peak force has been attained but not horizontal or axial load failure. Apart from the number of cycles, a potential asymmetry of the boundary conditions (e.g., axial load ratio and shear span ratio) for loading in positive and negative direction can influence the drift capacity. Such conditions are representative for outer walls where slabs and spandrels frame into the wall from only one side of the wall, which can lead to large variations in the axial load of the URM wall under seismic excitation. Since the drift capacity decreases with increasing axial load ratio, the critical loading direction is the one for which the axial load increases. The available experimental data suggest that the drift capacity of a wall subjected to an axial stress ratio three times as high in one direction than in the other is approximately twice that of a wall subjected to large axial forces in both loading directions.

Ideally the effect of the loading velocity should be investigated using the same test setup and loading histories but applying the latter at different speeds as Tomazevic and his co-workers have done it for reinforced masonry developing flexural failure modes (Tomazevic et al. 1996; Tomazevic 2000). Unfortunately for

URM walls such results are not yet available. To get a first idea, the drift limits obtained from quasi-static cyclic tests were compared to drifts recorded for a 4-storey building tested on a shake table and in general a good agreement was found. To apply the drift limits obtained from quasi-static cyclic tests to the walls in the shake table test unit, a couple of assumptions regarding their axial forces and shear spans were required, which were derived from nonlinear static analysis. Furthermore, the cross section and dimensions of the walls in the shake table test unit and the walls tested quasi-statically were not the same. All walls were, however, constructed using the same type of half-scale bricks and cement mortar.

The results presented in this paper therefore confirm that empirical drift capacity models derived from results of quasi-static cyclic tests can be applied to predict the performance of URM buildings under real earthquake loading—although attention should be paid to the effect of asymmetric boundary conditions of outer piers (captured by the correction factor for the loading history). Future research should, however, also target the development of mechanical drift capacity models as such models will foster the understanding of the behaviour of URM walls, allow to extrapolate with confidence to new configurations of parameters and potentially also allow to develop masonry types with improved performance. For flexural behaviour modes such models have recently been proposed (Priestley et al. 2007; Benedetti and Steli 2006; Petry and Beyer 2014e) but models that address walls developing shear and hybrid modes are currently lacking.

Acknowledgements The shake table test referenced in Sect. 12.4.1 of this paper received funding from the European Community's Seventh Framework Programme [FP7/2007-2013] for access to TREES laboratory of EUCENTRE under the grant agreement n^0 227887. Additional financial support was received from the Office Fédéral de l'Environnement (OFEV) in Switzerland. The reduced scale bricks were fabricated and donated by Morandi Frères SA, Switzerland. The authors appreciate and gratefully acknowledge all contributions.

Open Access This chapter is distributed under the terms of the Creative Commons Attribution Noncommercial License, which permits any noncommercial use, distribution, and reproduction in any medium, provided the original author(s) and source are credited.

References

Abrams D (1996) Effects of scale and loading rate with tests on concrete and masonry structures. Earthq Spectra 12(1):13–28
ATC (1997) FEMA-273: NEHRP Guidelines for the seismic rehabilitation of buildings. Basic Procedures Manual. Applied Technology Council (ATC), Washington, DC
ATC (1998) FEMA-306: evaluation of earthquake damaged concrete and masonry wall buildings. Basic Procedures Manual. Applied Technology Council (ATC), Washington, DC
ATC-24 (1992) Guidelines for cyclic seismic testing of components of steel structures for buildings. Applied Technology Council, California
Belmouden Y, Lestuzzi P (2009) An equivalent frame model for seismic analysis of masonry and reinforced concrete buildings. Construct Build Mater 23(1):40–53

Benaboud H (2013) Effective slab width in URM buildings with RC slabs. M.Sc. Dissertation, EPFL, 2013

Benedetti A, Steli E (2006) Analytical models for shear-displacement curves of unreinforced and FRP reinforced masonry walls. Construct Build Mater 22:175–785

Beyer K, Dazio A (2012) Quasi-static monotonic and cyclic tests on composite spandrels. Earthq Spectra 28(3):885–906

Beyer K, Tondelli M, Petry S, Peloso S (2014) Dynamic testing of a 4-storey building with reinforced concrete and unreinforced masonry walls. Bull Earthquake Eng (submitted)

Braga F, Liberatore D (1990) A finite element for the analysis of the response of masonry buildings. Proceedings of the 5th North American Masonry conference, University of Illinois at Urbana-Champaign

CEN (2005) Eurocode 8: design of structures for earthquake resistance – Part 3: General rules, seismic actions and rules for buildings. Design Code EN 1998-3, European Committee for Standardisation (CEN), Brussels, Belgium

Cervenka V, Jendele L, Cervenka J (2010) Atena – computer program for nonlinear finite element analysis of reinforced concrete structures, Theory and User Manual, Prague, Czech Republic

Chen SY, Moon FL, Yi T (2008) A macroelement for the nonlinear analysis of in-plane unreinforced masonry walls. Eng Struct 30:2242–2252

Da Parè M (2011) The role of spandrel beams on the seismic response of masonry buildings. M.Sc. dissertation, MEEES-ROSE School, Pavia, Italy

DIN (2011) National annex – nationally determined parameters – Eurocode 8: design of structures for earthquake resistance – Part 1: General rules, Seismic actions and rules for buildings. National Annex of Germany, DIN EN 1998-1/NA: 2011–01, Berlin, Germany

Elgawady E, Lestuzzi P, Badoux M (2004) Dynamic versus static cyclic tests of masonry walls before and after retrofitting with GFRP. Proceedings of the 13th world conference on earthquake engineering, Vancouver, Canada

Facconi L, Plizzari G, Vecchio F (2014) Disturbed Stress Field Model for Unreinforced Masonry. *J. Struct. Eng.*, 140(4), 04013085

Fajfar P (1999) Capacity spectrum method based on inelastic demand spectra. Earthquake Eng Struct Dyn 28:979–993

FEMA-461 (2007) Interim protocols for determining seismic performance characteristics of structural and non-structural components through laboratory testing. Federal Emergency Management Agency, Washington, DC

Frumento S, Magenes G, Morandi P, Calvi GM (2009) Interpretation of experimental shear tests on clay brick masonry walls and evaluation of q-factors for seismic design. Technical report, IUSS PRESS, Pavia, Italy

Gambarotta L, Lagomarsino S (1997a) Damage models for the seismic response of brick masonry shear walls. Part I: The mortar joint model and its applications. Earthquake Eng Struct Dyn 26:423–439

Gambarotta L, Lagomarsino S (1997b) Damage models for the seismic response of brick masonry shear walls. Part II: The continuum model and its applications. Earthquake Eng Struct Dyn 26:441–462

Ganz HR, Thürlimann B (1984) Versuche an Mauerwerksscheiben unter Normalkraft und Querkraft. Test Report 7502-4, ETH Zürich, Switzerland, (in German)

Graziotti F (2013) Contributions towards a displacement-based seismic assessment of masonry structures. PhD thesis, Istituto Universitario di Studi Superiori di Pavia, Pavia, Italy

Krawinkler H (2009) Loading histories for cyclic tests in support of performance assessment of structural components. 3rd international conference on advances in experimental structural engineering, San Francisco

Lagomarsino S, Penna A, Galasco A, Cattari S (2013) TREMURI program: an equivalent frame model for the nonlinear seismic analysis of masonry buildings. Eng Struct 56:1787–1799

Lang K (2002) Seismic vulnerability of existing buildings. PhD thesis, ETH Zurich, Zurich, Switzerland

Lourenço PB (1997) Two aspects related to the analysis of masonry structures: size effect and parameter sensitivity. Technical report TU-DELFT No 03.21.1.31.25, Faculty of Engineering, TU Delft, Delft, The Netherlands

Lourenco PB (2008) Structural masonry analysis: recent developments and prospects. Keynote at the 14th international brick and block masonry conference, Sydney, Australia

Lourenço PB, Rots JG (1997) Multisurface interface model for the analysis of masonry structures. J Eng Mech ASCE 123(7):660–668

Magenes G (2006) Masonry building design in seismic areas: Recent experiences. Keynote at the 1st European conference on earthquake engineering and seismology, Geneva, Switzerland

Magenes G, Calvi GM (1992) Cyclic behaviour of brick masonry walls. Proceedings of the 10th world conference on earthquake engineering, Madrid, Spain

Magenes G, Calvi GM (1997) In-plane seismic response of brick masonry walls. Earthquake Eng Struct Dyn 26:1091–1112

Marino S (2013) Force-deformation characteristics for composite spandrels. M.Sc. Dissertation, EPFL/University of Bologna, Italy

Mergos P, Beyer K (2014) Loading protocols for European regions of low to moderate seismicity. Bull Earthquake Eng online

Milani G (2012) Preface Special Issue: New trends in the numerical analysis of masonry structures. Open Civil Eng J Suppl 1-M1:119–120

Milani G, Lourenço PB, Tralli A (2006a) Homogenised limit analysis of masonry walls, Part I: Failure surfaces. Comput Struct 84(3):166–180

Milani G, Lourenço PB, Tralli A (2006b) Homogenised limit analysis of masonry walls, Part II: Structural examples. Comput Struct 84(3):181–195

MIT (2008) Ministry of Infrastructures and Transportation, Circ. C.S.Ll.Pp. No. 617 of 2/2/2009: Istruzioni per l'applicazione delle nuove norme tecniche per le costruzioni di cui al Decreto Ministeriale 14 Gennaio 2008, G.U.S.O. n.27 of 26/2/2009, No. 47, 2008 (in Italian)

Norda H, Butenweg C (2011) Möglichkeiten und Grenzen statisch nichtlinearer Verfahren nach DIN EN 1998-1. Der Bauingenieur 86:S13–S21

NTC (2008) Decreto Ministeriale 14/1/2008: Norme tecniche per le costruzioni. Ministry of Infrastructures and Transportations. G.U.S.O. n.30 on 4/2/2008, 2008 (in Italian)

NZSEE (2006) Assessment and improvement of the structural performance of buildings in earthquakes. New Zealand Society of Earthquake Engineering, University of Auckland, Auckland, New Zealand

NZSEE (2011) Assessment and improvement of unreinforced masonry buildings for earthquake resistance. New Zealand Society of Earthquake Engineering, supplement to "Assessment and improvement of the structural performance of buildings in earthquakes", University of Auckland, Auckland, New Zealand

Paparo A, Beyer K (2014) Quasi-static tests of two mixed reinforced concrete – unreinforced masonry wall structures. Eng Struct online

Paulay T, Priestley MJN (1992) Seismic design of reinforced concrete and masonry buildings. Wiley, New York

Penna A, Lagomarsino S, Galasco A (2013) A nonlinear macroelement model for the seismic analysis of masonry buildings. Earthquake Eng Struct Dyn. doi:10.1002/eqe.2335

Petry S, Beyer K (2014b) Cyclic test data of six unreinforced masonry walls with different boundary conditions. Earthq Spectra (accepted)

Petry S, Beyer K (2014d) Scaling unreinforced masonry for reduced-scale testing. Bull Earthquake Eng online

Petry S, Beyer K (2014a) Influence of boundary conditions and size effect on the drift capacity of URM walls. Eng Struct 65:76–88

Petry S, Beyer K (2014c) Limit states of URM piers subjected to seismic in-plane loading. Bull Earthquake Eng (submitted to VEESD-special issue)

Petry S, Beyer K (2014e) Review and improvement of simple mechanical models for predicting the force-displacement response of URM piers subjected to in-plane loading. Proceedings of the 2nd European conferences on earthquake engineering and seismology, Istanbul, Turkey

Porter ML (1987) Sequential phased displacement (SPD) procedure for TCCMAR testing. Proceedings of the 3rd meeting of the joint technical coordinating committee on masonry research, US-Japan Coordinated Research Program

Priestley MJN, Amaris AD (2002) Dynamic amplification of seismic moments and shear forces in cantilever walls. Research report ROSE 2002/01, Roseschool, Pavia, Italy

Priestley MJN, Calvi GM, Kowalsky MJ (2007) Displacement-based seismic design of structures. IUSS Press, Pavia, Italy

Salmanpour AH, Mojsilovic N, Schwartz J (2013) Deformation capacity of unreinforced masonry walls subjected to in-plane loading: a state-of-the-art review. Int J Adv Struct Eng 5(1):1–12

SIA (2011) SIA D0237: Evaluation de la sécurité parasismique des bâtiments en maçonnerie. Swiss Society of Engineers and Architects SIA, Zürich, Switzerland (in French)

Snozzi L, Molinari JF (2013) A cohesive element model for mixed mode loading with frictional contact capability. Int J Numer Meth Eng 93:510–526

Tomazevic M (2000) Some aspects of experimental testing of seismic behaviour of masonry walls and models of masonry buildings. ISET J Earthquake Tech 37(4):101–117

Tomazevic M, Lutman M, Petkovic L (1996) Seismic behavior of masonry walls: experimental simulation. J Struct Eng 122:1040–1047

Tondelli M, Petry S, Beyer K, Peloso S (2014) Data set of a shake table test on a four storey structure with reinforced concrete and unreinforced masonry walls. Bull Earthquake Eng (submitted)

Williams D, Scrivener JC (1974) Response of reinforced masonry shear walls to static and dynamic cyclic loading. In: Proceedings of the 5th world conference on earthquake engineering, Rome, Italy

Zucchini A, Lourenço PB (2002) A micro-mechanical model for the homogenisation of masonry. Int J Solid Struct 39:3233–3255

Chapter 13
Pushover Analysis for Plan Irregular Building Structures

Mario De Stefano and Valentina Mariani

Abstract Nonlinear static procedures (NSPs), also known as "pushover methods", represent the most used tool in the professional practice for assessment of seismic performance of building structures. Most of the methods subscribed by major seismic codes for seismic analysis of new or existing buildings have been originally defined for simple regular structures.

Nevertheless, perfect regularity is an idealization that very rarely occurs and, in principle, the concept of irregularity itself is a *fuzzy* one. Most codes attempt to give a definition to the concept of "regularity", considering issues related to the distribution of mass, stiffness and strength in the building, both in plan and in elevation. Real buildings rarely comply with these regularity requirements, resulting in a barely reliable application of the basic NSPs. Code specifications concerning irregular structures are in need of improvement and they do not provide for clear and specific guidelines for the seismic analysis of such structures. Therefore the problem of the seismic evaluation of irregular structures is still an open one and basic issues need to be further explored.

The present paper aims at providing a wide outlook on the problem of the seismic assessment of plan irregular building structures. Firstly, a brief review of the elastic and inelastic methods for the assessment of the torsional effects induced by in-plan irregularity is presented, mainly aimed at the definition of the variables governing the problem. Then, the basic features of the most important NSPs are discussed, followed by the description of the recent improvements developed for irregular structures. Since there is not yet a fully satisfactory solution, pros and cons of the various approaches are outlined, highlighting the most promising methods and the issues that are yet to be investigated. Finally, recommendations for code improvement are suggested.

M. De Stefano (✉) • V. Mariani
Department of Architecture, University of Florence, Piazza Brunelleschi 6, 50121 Firenze, Italy
e-mail: mario.destefano@unifi.it; valentina.mariani@unifi.it

13.1 Introduction

Structural irregularities are one of the major causes of damage amplification under seismic action. Past earthquakes, indeed, have shown that buildings with irregular configuration or asymmetrical distribution of structural properties are subjected to an increase in seismic demand, causing greater damages. The sources of irregularity in a building configuration can be multiple and of different kinds and are usually classified in two major categories: irregularities in plan and in elevation. The first type is related to in plan asymmetrical mass, stiffness and/or strength distributions, causing a substantial increase of the torsional effects when the structure is subjected to lateral forces. The second one involves variation of geometrical and/or structural properties along the height of the building, generally leading to an increase of the seismic demand in specific storeys. Both these types of irregularity often entail the development of brittle collapse mechanisms due to a local increase of the seismic demand in specific elements that are not always provided with sufficient strength and ductility.

Most seismic codes, including EC8-1 (2004), provide empirical criteria for the classification of buildings into regular and irregular categories with reference to: mass and lateral stiffness variations in plan and in elevation (and related eccentricities), shape of the plan configuration, presence of set-backs, in-plan stiffness of the floors (rigid diaphragm condition), continuity of the structural system from the foundations to the top of the building. This list is not comprehensive of all the possible causes of irregularity and there is no definition for the degree of irregularity of the overall three-dimensional system. Code definitions fail to capture some irregularities, especially those resulting from the combination of both plan and vertical irregularities. Moreover, system irregularity does not solely depend on geometrical and structural properties of the building, but can also be induced by the features of the earthquake excitation and increased by the progressive damage of the structure.

Considering this scenario, there is an urgent need to define and measure structural irregularity with a more rational approach, to deeply understand its effect on the seismic behavior and consequently upgrade seismic codes with specific and effective prescriptions for irregular buildings.

Among the two aforementioned types of structural irregularity, in-plan irregularity appears to have the most adverse effects on the applicability of the classical nonlinear static procedures (NSPs), precisely because such methods have been developed for the seismic assessment of structures whose behavior is primarily translational. This is the reason why, in recent years, the extension of NSPs to plan irregular building structures has been widely investigated by specialists in this field.

13.2 Brief Review of the Assessment Methods of Induced Torsional Effects in Plan Irregular Structures

The dawn of the studies concerning the torsional effects featuring irregular buildings dates back to the 30s of the last century (Ayre 1938), due to an increasing awareness of the complexity of the response of non-symmetric buildings to seismic actions, that is not purely translational, but involves torsional deformations that in most cases adversely affect their seismic behavior.

In the early studies (Housner and Outinen 1958; Bustamante and Rosenblueth 1960; Kan and Chopra 1977; Reinhorn et al. 1977) the problem has been faced in the elastic range, referring to simplified one-storey or multistorey models. Some research is still under development in this field (for state-of-the-art reports, refer to Anagnostopoulos et al. 2013 and to previous reviews by Rutenberg 1992, 2002; Rutenberg and De Stefano 1997; De Stefano and Pintucchi 2008), even if the assumptions made for formulating such models involve many simplifications.

Nevertheless, these studies mainly succeeded in underlining the parametric nature of the problem. The main identified parameters that play a crucial role in the definition of the torsional behavior of irregular structures are the uncoupled natural periods, the stiffness eccentricity and the stiffness radius of gyration (non-dimensionalized with respect to the mass radius of gyration). These parameters, for a one-storey building, are defined as follows, with reference to the x direction (Fig. 13.1). Similar equations apply to the y direction.

- Uncoupled natural period $T_x = 2\pi \sqrt{\frac{m}{K}}$

  where m and K are the total mass and stiffness in x direction respectively;

- Stiffness eccentricity $e_{sx} = \frac{1}{L} \frac{\sum_{i=1}^{N} k_{yi} x_i}{K}$

  i.e. the distance (along x direction) between the stiffness centre C_S and the mass centre C_M;

- Torsional stiffness $I_{p,k} = \sum_{i=1}^{N} \left[k_{yi}(x_{i,C_s})^2 + k_{xi}(y_{i,C_s})^2 \right]$

  i.e. the polar moment of inertia of system stiffness computed with respect to the axes parallel to the z direction and passing through C_S;

- Stiffness radius of gyration $d_s = \frac{1}{\rho L} \sqrt{\frac{I_{p,k}}{K}}$

  non-dimensionalized with respect to the mass radius of gyration ρ.

Lately, the problem has been widely faced even in the inelastic range, introducing parameters related to resistance, i.e. strength eccentricity and strength radius of gyration. These parameters, for a one-storey building, are defined as follows, with reference to the x direction (Fig. 13.1). Similar equations apply to the y direction.

- Strength eccentricity $e_{rx} = \frac{1}{L} \frac{\sum_{i=1}^{N} F_{yi} x_i}{F}$

  i.e. the distance (along x direction) between the strength centre C_R and the mass centre C_M;

Fig. 13.1 Simplified scheme of a one-storey building, for the identification of the key parameters characterizing the torsional behavior of plan irregular structures

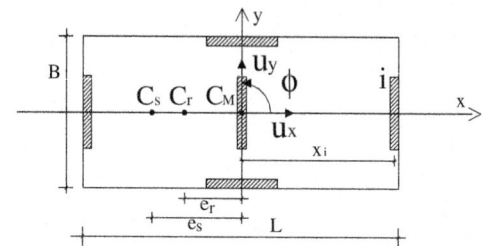

- Torsional strength $I_{p,f} = \sum_{i=1}^{N} \left[F_{yi}(x_{i,C_r})^2 + F_{xi}(y_{i,C_r})^2 \right]$

  as defined by De Stefano and Pintucchi (2010), i.e. the polar moment of inertia of system strength computed with respect to the axes parallel to the z direction and passing through C_R;

- Strength radius of gyration $d_r = \frac{1}{\rho L}\sqrt{\frac{I_{p,f}}{F}}$

  non-dimensionalized with respect to the mass radius of gyration ρ.

The studies in the inelastic range have been conducted by analyzing both the one-storey and the multistorey models. In the former case, methods considering uni-directional eccentricity, strength and ground motion were developed, subsequently improved considering these parameters in both principal directions. Concerning the multistorey models, some simplified shear-type models have been developed as well as detailed plastic hinge type models (see reviews by Rutenberg 1992, 2002; Rutenberg and De Stefano 1997; De Stefano and Pintucchi 2008; Anagnostopoulos et al. 2013).

Shifting from elastic to inelastic range, the parametric dependence of the problem become more complex and less analytically determined. One key-aspect is for example the assumption of a proportional relationship between stiffness and strength, that can be considered valid for pre-normative existing structures not designed for torsional effects, but not for more recent buildings designed according to modern seismic codes. Other issues are related to the evaluation of the effect of level of ductility of the structure, assumption of different nonlinear constitutive laws etc.

This large amount of studies has not yet led to general conclusions. Indeed, since many parameters affect the problem, different combinations of assumptions have often led to conflicting conclusions. Moreover, both one-storey and multi-storey models still suffer from several shortcomings related to many simplifying assumptions, that often make very difficult the generalization of obtained results.

13.3 Fundamentals of Classical Nonlinear Static Procedures

The formulation of the nonlinear static analysis, often defined as "pushover analysis", dates back to the 70s of the last century. Although it has only recently been included in seismic code provisions, the procedure itself has been already largely applied in the past, in research and design applications. With the coming of the performance-based (PB) design philosophy, pushover analysis turned to be the most used approach for the seismic assessment and design of structures, and became the starting point of all the so-called nonlinear static procedures (NSPs). PB design focuses on the actual performance of the structure under earthquake conditions, defining multiple performance objectives related to multiple seismic action levels. The modern PB design/assessment methods generally refer to displacements and deformations as performance targets.

The best way to evaluate the seismic performance of a structure is the nonlinear dynamic analysis (NLDA) that represents the most rigorous and accurate approach, as it directly provides the behavior of the structure under a series of seismic records. Nevertheless it should be kept in mind that the response is sensitive to the input ground motion, therefore several analyses are required with increased complexity, computational costs and time consumption. This is the reason why NLDA is still far from an extensive application in common practice.

Given the aforementioned limitations in the use of NLDA, in the last decades the NSPs have been brought to the forefront of seismic design/assessment of structures. Basically, the methods are based on the evaluation of three key quantities: seismic capacity, seismic demand and performance. In all the NSPs, the seismic capacity is evaluated through pushover analysis, that consists of "pushing" the structure with an increasing lateral load pattern, in combination with gravity loads, until the attainment of the structure collapse. As the load increases, the structure shifts from elastic to inelastic field and the overall behavior can be expressed in terms of global significant quantities, e.g. base shear and displacement of a control point (generally the top of the structure). The plot of the top displacement versus the total base shear is currently known as "capacity curve".

The seismic demand is a representation of the expected earthquake action through acceleration and displacement spectra. Generally in the NSPs the seismic demand is expressed in terms of "target displacement", that represents the maximum inelastic displacement that the structure should be able to undergo.

Finally, the performance, very clearly defined in ATC-40 (1996), "is dependent on the manner that the capacity is able to handle the demand. In other words, the structure must have the capacity to resist the demand of the earthquake such that the performance of the structure is compatible with the objectives of the design". This definition represents the core meaning of PB design/assessment methods.

The various NSPs mainly differ in the evaluation of the seismic demand, that represents a key aspect, because of the need to account for the inelastic response of the structure. Several approaches are available. The most well-known NSPs, suggested also by the most important worldwide seismic codes, are briefly described in the following.

13.3.1 Capacity Spectrum Method

The Capacity Spectrum Method (CSM) has been firstly proposed by Freeman et al. (1975) and Freeman (1998, 2004) as a rapid seismic evaluation procedure and then developed into a seismic design/assessment method adopted by the California Seismic Safety Commission through the ATC-40 (1996) guidelines, lately improved considering innovative features suggested in the FEMA-440 (2005) report. The CSM is a graphical procedure that compares the capacity of the structure, in terms of capacity (pushover) curve of an equivalent Single-Degree-Of-Freedom (SDOF) system, with the seismic demand, in the form of a response spectrum. Both capacity and demand are expressed in the Acceleration-Displacement Response Spectrum (ADRS) format.

The pushover curve of the MDOF system is converted in the equivalent pushover curve of a SDOF system and then bi-linearized according to the equal energy or equal displacement rules. Finally it is expressed in terms of spectral acceleration S_a and spectral displacement S_d obtaining the capacity spectrum. The seismic demand is represented by several spectra with different values of equivalent viscous damping ratio ξ. The graphical verification consists in checking if the capacity spectrum can extend through the envelope of the demand spectrum. If yes, the building is able to undergo the seismic demand action. Otherwise, if the capacity spectrum has no intersection with the demand spectrum, the structure does not resist the design earthquake. The intersection between capacity and demand spectra represents a performance point in terms of maximum acceleration and displacement for the SDOF system.

Once defined a certain performance point on the capacity curve, in order to quantify the deficiency (or the exceedance) of the capacity with respect to demand, the elastic spectrum has to be iteratively scaled until it intersects the capacity curve in correspondence of the assumed capacity (performance) point. The scaling procedure is done through spectral reduction factors related to equivalent viscous damping values, that account for the inherent viscous damping of the structure (generally assumed as 5 %) and hysteretic damping. Therefore the seismic capacity evaluation is done through damped elastic spectra.

The main limitation of the CSM is that the inelastic response of the structure is represented with over-damped elastic spectra, characterized by modified values of damping. This issue will be lately overcome with the development of the N2 method by Fajfar and Fischinger (1988), which considers the use of constant-ductility inelastic spectra, rather than over-damped elastic spectra.

13.3.2 N2 Method

The N2 method, firstly proposed by Fajfar and Fischinger (1988) and then developed in Fajfar and Gašperšić (1996), Fajfar (1999, 2000), is the NSP adopted by the Eurocode 8 (EC8-1 2004) and represents a modified version of the CSM. In the N2 method indeed the evaluation of the seismic demand is based on the use of inelastic spectra, instead of highly damped elastic spectra, as done through the CSM.

Therefore this method maintains the clarity of a visual graphical representation of the capacity-demand comparison, in combination with a more consistent approach related to the use of inelastic demand spectra as an alternative to highly damped elastic spectra, that indeed, have no physical basis. The inelastic spectra are derived reducing the elastic spectrum by a reduction factor R_μ, directly related to the hysteretic dissipative capacity of the structure, expressed by the ductility factor μ, i.e. the ratio between the maximum displacement and the yield displacement of the SDOF bilinear capacity curve.

The target displacement is determined referring to the equal displacement rule for medium and long period range, while for short period range, the target displacement is larger than the one associated to the corresponding equivalent elastic system (Fig. 13.3). More in details, the method assumes that in the medium/long period range ($T^* \geq T_C$) the equal displacement rule applies, i.e. the displacement of the inelastic system S_d is equal to the displacement of the associated elastic system S_{de} characterized by the same period T^* (Fig. 13.2a). This means that in the above mentioned range of periods $R\mu = \mu$. Therefore the seismic demand in terms of inelastic displacement, can be obtained by intersecting the radial line corresponding to the period of the SDOF system with the elastic demand spectrum.

On the other hand, in the case of short-period structures ($T^* < T_C$) the inelastic displacement is larger than the elastic one and the equal displacement rule does not apply anymore (Fig. 13.2b). Consequently $R_\mu < \mu$ and it can be determined as the ratio between the elastic acceleration demand S_{ae} and the inelastic acceleration capacity S_{ay}. The inelastic displacement demand is, in this case, equal to $S_d = \mu \cdot D^*_y$, being D^*_y the yielding displacement of the SDOF system. The ductility factor can be derived from the reduction factor by the relation:

$$\mu = (R_\mu - 1)\frac{T_C}{T^*} + 1$$

In both cases ($T^* \geq T_C$ and $T^* < T_C$) the inelastic acceleration demand S_a is equal to the elastic one S_{ae} and it can be determined at the intersection of the radial line corresponding to the period of the SDOF system with the elastic demand spectrum.

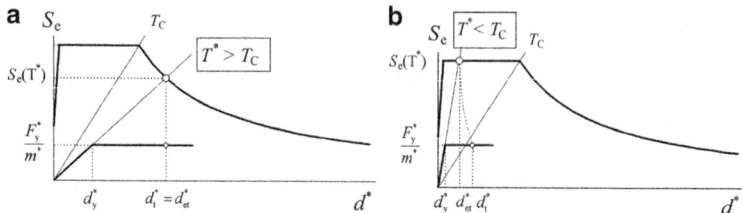

Fig. 13.2 Evaluation of the inelastic displacement demand S_d for (**a**) short-period structures ($T^* < TC$) and (**b**) medium/long period structures ($T^* \geq TC$), according to EC8-1 Annex C "Determination of the target displacement of the equivalent SDOF system"

13.3.3 Displacement Coefficient Method

The Displacement Coefficient Method (DCM), adopted by FEMA 356 (2000), is a simplified procedure for the estimation of the seismic demand, that applies a series of corrective coefficients to the elastic spectral displacement demand so as to obtain a target displacement, i.e. the maximum inelastic displacement demand. The following relation applies for the determination of the target displacement δ_t:

$$\delta_t = C_0 C_1 C_2 C_3 S_a \frac{T_e^2}{4\pi^2} g$$

The four modification coefficients (C_0, C_1, C_2, C_3) have been evaluated through a statistical approach based on time history analyses of SDOF models of different types. They account for: the difference between the roof displacement of a MDOF building and the displacement of the equivalent SDOF system, i.e. the amplification of displacement with respect to the spectral one; observed difference in peak displacement response amplitude for nonlinear response as compared with linear response, as observed for buildings with relatively short initial vibration periods (validity limits of the equal-displacement approximation); the effect of hysteresis type on the maximum displacement response; second order effects.

13.4 Extension of NSPs to Plan-Irregular Buildings

The current trends in research concerning the improvement of the NSPs are primarily focused on two main issues: (i) the effects of stiffness degradation and changes in dynamic properties related to progressive damage with the need for an update of inertial forces to be applied as a function of the level of inelasticity; (ii) the contribution of higher modes of vibration, intended to account for the effects of vertical and in-plan irregularity.

Within the first issue a lot of research contributions have been produced in recent years, introducing the adaptive pushover methods (APM). The first procedures, initially applied to concrete frames, have been developed by Reinhorn (1997) and Bracci et al. (1997) that used inelastic storey forces of the previously equilibrated load step to update the lateral load pattern. Afterwards Gupta and Kunnath (2000) proposed a constantly updated load pattern depending on the results of an eigenvalue analysis performed each step, assuming the tangent or secant stiffness related to the deformations of the previous load step.

Concerning the second issue, a large research effort has been devoted to the improvement of the pushover methods so as to consider the contribution of higher modes. This aspect is strictly related to structural irregularity, because irregular structures are generally characterized by significant participating mass ratio of higher modes. The basic NSPs indeed, relate the dynamic behavior of the structure, assumed as a multi-degree of freedom (MDOF) system to an equivalent one-degree of freedom (SDOF) system, which considers the contribution of the main translational mode only.

From the dynamic point of view, a plan irregular building is that for which one or more rotational modes have a significant participating mass ratio. Therefore the dynamic behavior of the structure cannot be defined referring to one translational mode only. The basic NSP approach is not reliable for plan irregular buildings, for which the first translational mode is not representative of a more complex dynamic behavior, that involves both translational and rotational components.

Among the many proposed methods developed in this research field, two main approaches can be recognized: the first one aims to take into account the contribution of more eigenmodes. One of the first attempts has been done by Paret et al. (1996) and it is known as multi-modal pushover (MMP) procedure. Structure's capacity for each mode is then compared with earthquake demand using CSM. Chopra and Goel (2002) developed a similar approach known as modal pushover analysis (MPA), in which several independent pushover analyses are carried out, considering different load patterns associated to different modal shapes. Specifically, in the case of plane irregular structures, the method involves the application of both lateral forces and torque at each level of the building. The results are finally combined by the square-root-of-sum-of-squares (SRSS) rule or the complete-quadratic-combination (CQC) rules. Afterwards, Chopra et al. (2004) proposed the modified modal pushover analysis (MMPA) in which the inelastic response associated to the first mode is combined with the elastic contribution of higher modes. Extensions of this approach with the adaptive load formulation have also been proposed in Shakeri et al. (2012) and Tabatabaei and Saffari (2011).

These methods involve the running of several analyses, one for each modal shape considered and the results are then combined with SRSS or CQC. Moreover the use of quadratic combination rules to sum up the effects of the different modes, like in the linear range, is not strictly correct. Therefore Elnashai (2001) proposed an adaptive pushover procedure able to include, in a single analysis run rather than combining results from more analyses, all features mentioned above. The method uses the combination rules to update the force distribution each step, rather than combining the effects. However this approach has the disadvantage that the use of quadratic combination rules of modal contributions for the definition of load pattern at each step leads inevitably to positive increments, and hence to a monotonic increase in the load vector.

The inability to reproduce sign change in the applied load patterns has been overcame by the definition of adaptive procedures where the load patterns are based on displacements. This approach, namely displacement-based adaptive pushover (DAP), has been firstly proposed by Antoniou and Pinho (2004) and is based on prescribed adaptive displacement patterns from which the lateral loads are derived. In this way it is possible to capture changes in the sign of lateral loads, even if the displacement increment remains always positive. This approach has been also adopted within the Adaptive Capacity Spectrum Method (ACSM) by Casarotti and Pinho (2007).

On the other hand, the second approach is still based on the first modal shape, but with the awareness that a single target displacement is no longer sufficient to describe the overall dynamic behavior of irregular buildings, because torsional effects entail amplifications and reductions of the displacement demand at the two opposite ends of the storey. In this framework, Tso and Moghadam (1997) and Moghadam and Tso (2000a, b) defined a procedure for monosymmetric structures subjected to one component excitation. The method consists in the evaluation of target displacements in the different resisting elements through elastic response spectrum analysis; consequently the load patterns are determined and several 2D pushover analyses are performed for the different resisting elements. The method has been applied for the evaluation of the seismic progressive collapse of 3-storey RC moment resisting buildings with different levels of plan eccentricity (Karimiyan et al. 2013).

With a similar approach, an extended version of the N2 method has been proposed by Fajfar et al. (2005a, b) for the application to plan irregular building structures. In the extended N2 method, linear elastic analysis is used to define the torsional amplification of lateral displacements to account for the torsional response, on the assumption that the elastic envelope is conservative with respect to the inelastic one.

Another method has been proposed by Bosco et al. (2012), on the bases of previous studies by Ghersi and Rossi (2000), Calderoni et al. (2002) and Ghersi et al. (2007), who introduced the use of "corrective eccentricities" related to the elastic and inelastic parameters that define the torsional behavior of the building. These eccentricities are then used to define the application points of the load vectors, on either sides of the CM so as to obtain an envelope of plan distribution of maximum displacements.

In the following sections, the basic features of the methods addressing to the two main approaches for the seismic assessment of plan irregular building structures will be described, outlining the advantages and drawbacks of each single approach and trying to identify the most promising methods and the issues that are yet to be more deeply investigated.

13.4.1 Modal Pushover Analysis

One of the main approaches in the developing of NSPs for the analysis of irregular building structures involves the evaluation of the contribution of more eigenmodes in the analysis. Within this approach, the major contribution has been given by Chopra and Goel (2004) who extended the previously defined MPA to unsymmetric-plan buildings. The fundamentals of the method remained the same of the original version of MPA (Chopra and Goel 2002), based on structural dynamics theory, in which the seismic demand due to individual terms in the modal expansion of the effective earthquake forces is determined by a pushover analysis using the inertia force distribution associated to each single mode. The total seismic demand of the inelastic system is then determined combining the modal demands associated to multiple modes with the SRSS or the CQC rules. Actually this superposition of effects is valid in the linear range, therefore the use of these combination rules represents the first approximation of the method. The second one is the neglecting of coupling among modal coordinates associated with the modes of the corresponding linear system arising from yielding of the system. The original method has been then improved in Goel and Chopra (2004) with three major enhancements: inclusion of P-Δ effects due to gravity loads for all modes (initially it was included only for the first mode); computation of plastic rotations of elements from the total storey drift and not through combination rules; idealization of the pushover curve of nth mode at the peak roof displacement obtained from inelastic SDOF system for the selected ground motion, leading to a reduction of the dependence on the ground motion.

The application of the method to unsymmetrical-plane building structures involves no particular changes in the general approach, except that two lateral forces and a torque are applied at each floor level. The CQC rule is suggested in this case, more suitable for unsymmetric-plan buildings, which may have closely-spaced frequencies of vibration.

Further developments are provided by Reyes and Chopra (2011a, b) who extended the method to 3D eccentric buildings subjected to two components motion and defined the practical modal pushover analysis (PMPA), introducing another simplification: the seismic demands are estimated directly from the elastic design spectrum without performing any NLDA of the modal SDOF systems for each ground motion, thus avoiding the complications of selecting and scaling ground motions.

All the improved versions of the MPA appear to perform rather well, the adopted approximations does not overly affect the results, with respect to those obtained by NLDA, with the exception of cases in which the analyzed structure has close modal periods and strong coupling of the lateral and torsional motions. In this case the individual modal responses attain their peaks almost simultaneously and consequently the CQC modal combination rule become not valid anymore, especially for lightly damped systems. Significant discrepancies with NLDA are also found as the structure experiences high levels of inelasticity with significant degradation in lateral capacity.

13.4.2 Extended N2 Method

In recent years, an important step toward the inclusion of torsional effects into pushover analysis has been done by Fajfar et al. (2005a, b) with the definition of an extended version of the N2 method, based on a combination of results of a pushover analysis performed on a 3D model of the structure, that controls the target displacement distribution at the center of mass along the height of the building, with a dynamic modal analysis which controls lateral displacement distribution due to torsional effects. Therefore, modal analysis is used to estimate the displacement amplification due to torsional behavior, that cannot be captured with the standard NSPs.

The displacements obtained by pushover analysis are amplified through a corrective factor, given by the ratio of the normalized displacement obtained by modal analysis and that coming from pushover analysis. The normalized displacement is the displacement in a specific point of the horizontal plane divided by the displacement in the center of mass. Only amplifications of target displacement are considered, whilst reductions in lateral displacements, typical at the stiff edge of the structure, are neglected, with the assumption of a "no-reduction rule". In this way, it is assumed that the elastic envelope of lateral displacements is conservative with respect to the inelastic one and therefore dynamic modal analysis provides an upper bound of the torsional amplification. Such assumption is supported by findings from several studies demonstrating that displacement amplifications decrease at the flexible side as the structure experiences larger inelasticity, i.e. torsional effects decrease in the inelastic range. This behavior has been observed both for torsionally flexible structures (Fig. 13.3a), i.e. structures characterized by a ratio between the uncoupled torsional frequency and the uncoupled lateral frequency lower than 1, and torsionally stiff structures (Fig. 13.3b), i.e. structures for which the same ratio is larger than 1. On the other hand, the behavior at the stiff side resulted less predictable, influenced by several modes of vibration and by the ground motion in the transverse direction. For torsionally flexible structures, displacement amplification can be found also at the stiff side, although decreasing with plastic deformation. In extreme cases the behavior becomes similar to that of torsionally stiff structures (de-amplification at the stiff side). Typical qualitative behavior of torsionally stiff and flexible structures is represented in Fig. 13.4 which shows the variations of lateral displacement demands at both flexible and stiff sides, with respect to a torsionally balanced structure.

The extended N2 method appears to be a very promising approach aimed at the application of pushover analysis to irregular building structures, because it combines conceptual clarity with simplicity of application. Nevertheless, the basic assumption of the conservativeness of the elastic envelope of lateral displacements with respect to the inelastic one surely needs to be further investigated. De Stefano and Pintucchi (2010) performed a wide parametric analysis on one-storey models and found that the method lose its conservativeness for very torsionally stiff structures, such as shear-walled buildings, for which the inelastic response almost

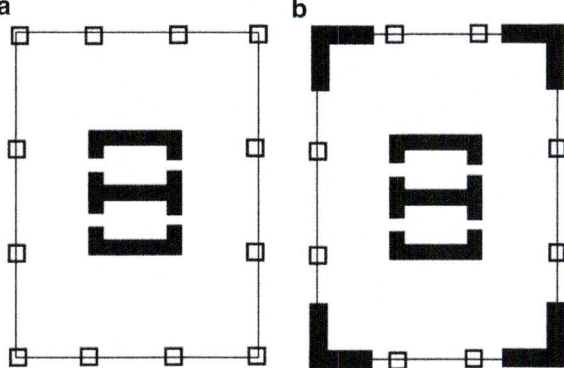

Fig. 13.3 (a) Torsionally flexible structure; (b) torsionally stiff structure (FEMA 274 (1997))

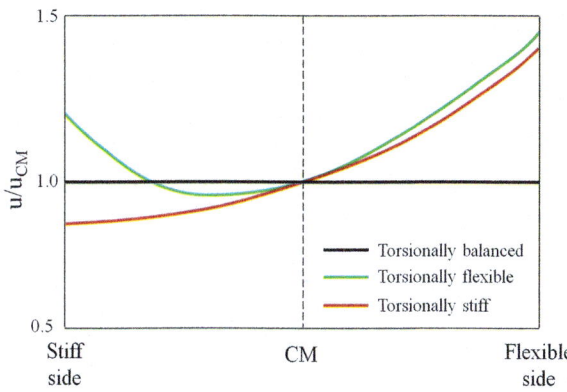

Fig. 13.4 Qualitative comparison of plan-wise shapes of lateral displacement envelopes

always exceeds the elastic one. This is mainly due to the development of a strength eccentricity related to the failure of one or more components of the structural system, leading to a significant increase of the inelastic torsional effects.

Some preliminary parametric boundaries to applicability of the extended N2 method have been defined in terms of stiffness an strength radii of gyration and of behaviour factor q. The procedure resulted effective for values of d_s and d_r lower than 1.3, characterizing most building framed structures, and q values higher than 2.

Other authors tested the procedure on sample multi-storey buildings structures. Bhatt and Bento (2012) applied the extended N2 procedure, together with the CSM, the MPA and the ACSM to two case studies of real existing plan irregular structures. They found that the extended N2 method was the most suitable method, among all the evaluated NSPs, because it was the only one to present always conservative results with respect to average time-history analysis results, both at the flexible edge (S1 in Fig. 13.5) and stiff edge (S23 in Fig. 13.5). Bosco et al. (2013) made a comparative evaluation of the N2, the extended N2 and the corrective eccentricities methods on a set of asymmetric single-storey systems and a set of 12 multi-storey buildings. Authors defined the extended N2 method as the easier to apply and the one always giving conservative results, though sometimes overly conservative.

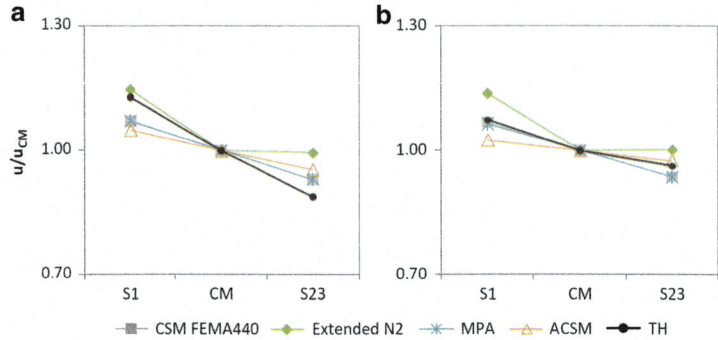

Fig. 13.5 Normalized top displacements obtained with several NSPs and time-history analysis (TH): (**a**) seismic intensity of 0.2 g; (**b**) seismic intensity of 0.4 g (Bhatt and Bento 2012)

Recently the extended N2 method has been improved to take into account higher modes effects both in plan and elevation (Kreslin and Fajfar 2011, 2012) and has been applied, as an alternative to incremental dynamic analysis, to determine the relationship between seismic demand and seismic intensity for different values of the seismic intensity measure. In this case the method has been called Incremental N2 method (Dolsek and Fajfar 2004, 2007).

13.4.3 Specifications of Major Seismic Codes on Applicability of NSPs to Irregular Buildings

Despite the large efforts of researchers aimed at a better understanding of the seismic behavior of irregular building structures and at the enhancement of the current NSPs, most regulatory bodies appear to have not yet translated the achieved research developments into seismic codes.

Even the criteria for the definition of plan and vertical irregularity are still not exhaustive, as underlined by a statement in EC8-1 (2004), where it is asserted that "it shall be verified that the assumed regularity of the building structure is not impaired by other characteristics, not included in these criteria". However, American codes provides for a more accurate and analytical definition of torsional irregularity, based on results of numerical analysis and not only on geometrical and qualitative evaluations on the structural features of the building. ASCE7-10 (2010), indeed, defines that a torsional irregularity exists when the ratio of the maximum storey drift at one end of the structure δ_{max} ($\delta_{max} = max\{\delta_A, \delta_B\}$) and the average of the two storey drifts at the two ends A and B of the structure δ_{avg} is larger than 1.2 (Fig. 13.6).

For the purpose of this paper, in the following only specifications related to applicability of NSPs to irregular buildings are summarized, based on the current major seismic American and European codes. The basic American reference codes

Fig. 13.6 Definition of torsional irregularity according to ASCE7-10 (2010)

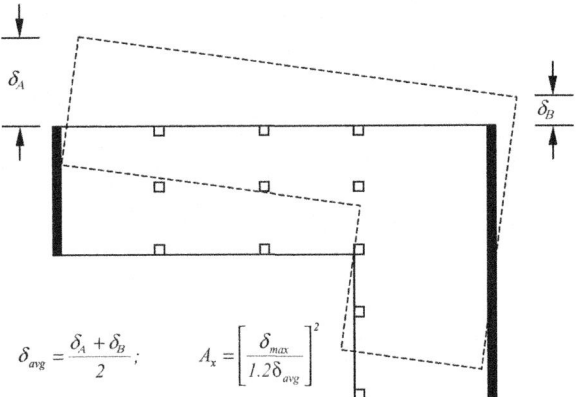

are International Building Code (IBC 2012), ASCE 7-10 (2010) for general building structures and International Existing Building Code (IEBC 2012), ASCE 31-03 (2003) (Seismic Evaluation of Existing Buildings), ASCE 41-06 (2006) (Seismic Rehabilitation of Existing Buildings), recently joined and implemented in the ASCE 41-13 (2013) (Evaluation and Retrofit of Existing Buildings). Basically ASCE 41-13 (2013) retains the three-tired approach found in ASCE 31-03 (2003), while relying on the technical provisions in ASCE 41-06 (2006) as the basis for all the analytical procedures.

Concerning European codes, the Eurocode 8 part 1 (EC8-1 2004), containing general rules for seismic design of buildings and Eurocode 8 part 3 (EC8-3 2005), concerning seismic assessment and retrofit of buildings, are considered.

The IBC mostly recalls ASCE7-10 (2010) for earthquake design. ASCE 7-10 (2010) does not require any form of nonlinear analysis for traditional buildings that do not incorporate seismic isolation or passive energy systems. The permitted analytical procedures are: equivalent lateral force analysis, modal response spectrum analysis and seismic response history procedures. Therefore it does not contain specific prescriptions on the use of NSPs. The only limitation on the choice of the analysis type with reference to torsional irregularity, is that equivalent lateral force analysis is not allowed for torsionally irregular structures.

American seismic codes for existing buildings (IEBC, ASCE 31-03 (2003), ASCE 41-06 (2006) and ASCE 41-13 (2013)) also define limitations at the use of linear analyses based on the existence of structural irregularities and to excessive values of DCR (Demand-Capacity Ratio) evaluated through linear static or dynamic analysis. If one or more structural components are characterized by DCR higher than 2 and any kind of structural irregularity exists (in-plane and out-of-plane discontinuities, weak storey, torsional strength/stiffness irregularity), then linear procedures are not applicable and shall not be used. More restrictive criteria are also defined for the application of linear static analysis. According to IEBC, NSPs are the fundamental tool to perform a Tier 3 analysis, i.e. the most advanced phase of an existing building evaluation process ASCE 31-03 (2003), necessary when the previously performed two phases (Tier 1 and 2) have evidenced potential deficiencies of the building.

NSPs are considered acceptable in most cases, but should be used in conjunction with a linear dynamic procedure, if higher modes effects are significant. This condition should be verified performing two modal response spectrum analyses: one considering sufficient modes to produce 90 % mass participation, another considering only the first mode participation. If the shear in any storey resulting from the first analysis exceeds 130 % of the corresponding storey shear considering only the first mode response, higher modes effects have to be considered significant.

The combined use of pushover analysis and response spectrum analysis appears as a precursor to the basic idea in the development of the extended N2 method. Moreover FEMA 273 (1997) prescribes that the effects of torsion shall not be used to reduce force and deformation demands on components and elements, someway recalling the no-reduction rule of the extended N2 method. Notwithstanding the conceptual connections with the N2 method, most of the current American seismic codes and guidelines (IBC 2012, ASCE 41-13 (2013), ATC 40 (1996) and FEMA 440 (2005)) refer to the CSM and to the DCM as analysis procedures for the evaluation of seismic capacity of building structures.

Even in EC8-3 (2005) the prescription to take into account higher modes effects is defined for buildings with a fundamental period higher than *2 s* or *$4T_c$*. In this case the code requires to perform a NLDA or "special versions" of NSPs. Nevertheless the code does not provide any suggestion concerning which kind of upgraded NSPs should be used and refer to national codes for more specific provisions.

EC8-1 (2004) provides for the application of the N2 method, although it recognizes the absence of a full suitability for irregular building structures. Nevertheless, no restriction to the use of this method for irregular structures is defined. EC8-1 (2004) declares that the conventional procedure "may significantly underestimate deformations at the stiff/strong side of a torsionally flexible structure, ... For such structures, displacements at the stiff/strong side shall be increased, compared to those in the corresponding torsionally balanced structure". To do that, EC8-1 (2004) implicitly refers to the extended N2 method, as it prescribes to evaluate the amplification factor to be applied to the displacements of the stiff/strong side through an elastic modal analysis of the spatial model. Nevertheless, no specific prescriptions are provided to account for displacement amplifications at the flexible side, observed for both torsionally stiff and flexible structures. Therefore, the extended N2 method is only partially adopted, highlighting how EC8 provisions for the application of pushover analysis to irregular building structures are still lacking and not satisfactory.

Another weak point of the code is that it allows the use of two separate planar models even for plan irregular structures that comply with some other prescriptions: well-distribution and sufficient rigidity of cladding and partitions, building height lower than 10 m, diaphragm behavior of the floors, centres of lateral stiffness and mass approximately on a vertical line and adequate torsional stiffness. The assumption of this simplification has been questioned by Athanatopoulou and Avrimidis

(2008) that demonstrated how the use of two planar models for the nonlinear static analysis of a sample plan irregular building complying the aforementioned conditions, led to unconservative results with respect to NLDA.

Even ASCE 41-13 (2013) in some cases allows for the use of NSPs on two-dimensional models, when the building has rigid diaphragms and the displacement multiplier η, i.e. the ratio of the maximum displacement at any point on the floor diaphragm to the average displacement, does not exceeds 1.5. When NSP is applied to two-dimensional models, the target displacement shall be amplified by the maximum value of η calculated for the building.

13.5 Conclusions

Classical NSPs for the evaluation of seismic vulnerability of buildings have been originally defined for symmetric, regular structures and it is demonstrated that torsional behavior calls into question their validity for the seismic evaluation of torsionally sensitive structures. Therefore there is the urgent need for an update of such methods aimed at a reliable application to irregular building structures. Two major approaches have been identified concerning the improvement of NSPs: the first one is based on the inclusion of the contribution of higher modes in the analysis and has led, among others, to the development of MMP and MPA procedures; the second one focuses on the need to account for amplification of displacement demand, through corrective factors to be applied to the target displacement. Under this perspective, the most promising developed procedure is the extended N2 method, that combines in a synergic way the results coming from pushover analysis and response spectrum modal analysis. The procedure appears to be the most effective in the evaluation of displacement amplification due to torsional effects while maintaining simplicity and clarity for practical applications. The main assumption is that the elastic displacement pattern is conservative with respect to the inelastic one and this aspect need further investigations, because it cannot be the case for very torsionally stiff structures and for low ductility values.

Despite the large efforts of researchers aimed at the improvement of the classical NSPs for a reliable application to irregular building structures, most regulatory bodies appear to have not yet transposed the achieved developments into major seismic codes. Both European and American codes are still in need of improvement regarding specific prescriptions concerning the seismic analysis of irregular structures. There is the awareness of a partial unsuitability of classical NSPs, some improved solutions have been proposed by researchers, but a comprehensive and always suitable set of rules to extend NSPs to plan irregular buildings has not yet been established.

Open Access This chapter is distributed under the terms of the Creative Commons Attribution Noncommercial License, which permits any noncommercial use, distribution, and reproduction in any medium, provided the original author(s) and source are credited.

References

American Society of Civil Engineering (ASCE 31-03) (2003) Seismic evaluation of existing buildings. American Society of Civil Engineers, Reston, VA

American Society of Civil Engineering (ASCE 41-06) (2006) Seismic rehabilitation of existing buildings. American Society of Civil Engineers, Reston, VA

American Society of Civil Engineering (ASCE 7-10) (2010) Minimum design loads for buildings and other structures. American Society of Civil Engineers, Reston, VA

Anagnostopoulos SA, Kyrkos MT, Stathopoulos KG (2013) Earthquake induced torsion in buildings: critical review and state of the art. Plenary keynote lecture in 2013 World Congress on Advances in structural engineering and mechanics (2013 ASEM Congress), 2013, Techno-Press Journals, International Convention Center-Jeju, Jeju Island, Korea (in Press)

Antoniou S, Pinho R (2004) Development and verification of a displacement-based adaptive pushover procedure. J Earthq Eng 16(2):367–391

Applied Technology Council (ATC) (1996) Seismic evaluation and retrofit of concrete buildings. Report no. ATC-40 vols 1 and 2, Redwood City, CA

Athanatopoulou AM, Avrimidis IE (2008) Evaluation of EC8 provisions concerning the analysis of spatial building structures using two planar models. Proceedings of the 5th European workshop on the seismic behaviour of irregular and complex structure, 5EWICS, 16–17 Sept 2008, Catania, Italy, pp 285–297

Ayre RS (1938) Interconnection of translational and torsional vibration in structures. Bull Seism Soc Am 28:89–130

Bhatt C, Bento R (2012) Comparison of nonlinear static methods for the seismic assessment of plan irregular frame buildings with non seismic details. J Earthq Eng 16:15–39

Bosco M, Ghersi A, Marino EM (2012) Corrective eccentricities for assessment by nonlinear static method of 3D structures subjected to bidirectional ground motion. Earthq Eng Struct Dyn 41 (13):1751–1773

Bosco M, Ghersi A, Marino EM, Rossi PP (2013) Comparison of nonlinear static methods for the assessment of asymmetric buildings. Bull Earthquake Eng (in press)

Bracci JM, Kunnath SK, Reinhorn AM (1997) Seismic performance and retrofit evaluation of reinforced concrete structures. ASCE J Struct Eng 123(1):3–10

Bustamante JI, Rosenblueth E (1960) Building code provisions on torsional oscillation. Proceedings 2nd world conference earthquake engineering, Japan

Calderoni B, D'Aveni A, Ghersi A, Rinaldi Z (2002) Static vs. modal analysis of asymmetric buildings: effectiveness of dynamic eccentricity formulations. Earthq Spectra 18(2):219–231

Casarotti C, Pinho R (2007) An adaptive capacity spectrum method for assessment of bridges subjected to earthquake action. Bull Earthq Eng 5(3):377–390

CEN. Eurocode 8 (EC8-1) (2004) Design of structures for earthquake resistance. Part 1: General rules, seismic actions and rules for buildings. EN 1998-1:2004 Comité Européen de Normalisation, Brussels, Belgium

CEN. Eurocode 8 (EC8-3) (2005) Design of structures for earthquake resistance. Part 3: Assessment and retrofitting of buildings. EN 1998-3:2005 Comité Européen de Normalisation, Brussels, Belgium

Chopra AK, Goel RK (2002) A modal pushover analysis procedure for estimating seismic demands for buildings. Earthq Eng Eng Vib 31(3):561–582

Chopra AK, Goel RK (2004) A modal pushover analysis procedure for estimating seismic demands for unsymmetric-plan buildings. Earthq Eng Eng Vib 33:903–927

Chopra AK, Goel RK, Chintanapakdee C (2004) Evaluation of a modified MPA procedure assuming higher modes as elastic to estimate seismic demands. Earthq Spectra 20:757–778

De Stefano M, Pintucchi B (2008) A review of research on seismic behaviour of irregular building structures since 2002. Bull Earthq Eng 6:285–308

De Stefano M, Pintucchi B (2010) Predicting torsion-induced lateral displacements for pushover analysis: influence of torsional system characteristics. Earthq Eng Struct Dyn 39(12):1369–1394

Dolsek M, Fajfar P (2004) IN2—a simple alternative for IDA. Proceedings of the 13th world conference on earthquake engineering, Vancouver, Canada, 2004; Paper 3353

Dolsek M, Fajfar P (2007) Simplified probabilistic seismic performance assessment of plan-asymmetric buildings. Earthq Eng Struct Dyn 36(13):2021–2041

Elnashai AS (2001) Advanced inelastic static (pushover) analysis for earthquake applications. Struct Eng Mech 12(1):51–69

Fajfar P (1999) Capacity spectrum method based on inelastic demand spectra. Earthq Eng Struct Dyn 28(9):979–993

Fajfar P (2000) A nonlinear analysis method for performance based seismic design. Earthq Spectra 16(3):573–592

Fajfar P, Fischinger M (1988) N2 – a method for non-linear seismic damage analysis of RC buildings. Proceedings of the 9th world conference on earthquake engineering, Kyoto, 1988, Maruzen, Towkyo, vol V, pp 111–116

Fajfar P, Gašperšič P (1996) The N2 method for the seismic damage analysis of RC buildings. Earthq Eng Struct Dyn 25:23–67

Fajfar P, Marušić D, Perus I (2005a) Torsional effects in the pushover-based seismic analysis of buildings. J Earthq Eng 9(6):831–854

Fajfar P, Marušić D, Perus I (2005b) The extension of N2 method to asymmetric buildings. Proceedings of the 4th European workshop on the seismic behaviour of irregular and complex structures, Thessaloniki

Federal Emergency Management Agency (FEMA 273) (1997) Guidelines to the seismic rehabilitation of existing buildings, Washington, DC

Federal Emergency Management Agency (FEMA 274) (1997) Commentary on the guidelines for the seismic rehabilitation of buildings, Washington, DC

Federal Emergency Management Agency (FEMA 356) (2000) Prestandard and commentary for the seismic rehabilitation of buildings, Washington, DC

Federal Emergency Management Agency (FEMA 440) (2005) Improvements of nonlinear static seismic analysis procedures, Washington, DC

Freeman SA (1998) Development and use of capacity spectrum method. Proceedings of the sixth U.S. national conference on earthquake engineering, Seattle, Paper no. 269

Freeman SA (2004) Review of the development of the capacity spectrum method. J Earthq Tech 41(1):1–13

Freeman SA, Nicoletti JP, Tyrell JV (1975) Evaluation of existing buildings for seismic risk – a case study of Puget Sound Naval Shipyard, Bremerton, Washington. Proceedings of U.S. National conference on earthquake engineering, Berkley, CApp 113–122

Ghersi A, Rossi PP (2000) Formulation of design eccentricity to reduce ductility demand in asymmetric buildings. Eng Struct 22(7):857–871

Ghersi A, Marino EM, Rossi PP (2007) Static versus modal analysis: influence on inelastic response of multi-storey asymmetric buildings. Bull Earthq Eng 5(4):511–532

Goel RK, Chopra AK (2004) Evaluation of modal and FEMA pushover analyses: SAC buildings. Earthq Spectra 20:225–254

Gupta B, Kunnath SK (2000) Adaptive spectral-based pushover procedure for seismic evaluation of structures. Earthq Spectra 16(2):367–391

Housner GW, Outinen H (1958) The effect of torsional oscillations on earthquake stresses. Bull Seism Soc Am 48:221–229

International Building Code – IBC (2012) International Code Council, Washington DC

International Existing Building Code – IEBC (2012) International Code Council, Washington DC

Kan CL, Chopra AK (1977) Effects of torsional coupling on earthquake forces in buildings. J Struct Div 103(4):805–819

Karimiyan S, Moghdam AS, Vetr MG (2013) Seismic progressive collapse assessment of 3-story RC moment resisting buildings with different levels of eccentricity in plan. Earthq Struct 5:277–296

Kreslin M, Fajfar P (2011) The extended N2 method taking into account higher mode effects in elevation. Earthq Eng Struct Dyn 40(14):1571–1589

Kreslin M, Fajfar P (2012) The extended N2 method considering higher mode effects in both plan and elevation. Bull Earthq Eng 10(2):695–715

Moghadam AS, Tso WK (2000a) Pushover analysis for asymmetric and set-back multi-story buildings. Proceedings of the 12th world conference on earthquake engineering, Auckland, New Zeeland.

Moghdam AS, Tso WK (2000b) 3-D push-over analysis for damage assessment of buildings. J Seism Earthq Eng 2(3):23–31

Paret TF, Sasaki KK, Elibeck DK, Freeman SA (1996) Approximate inelastic procedures to identify failure mechanism from higher modes effects. Proceedings of the 11th world conference on earthquake engineering, Acapulco

Reinhorn AM (1997) Inelastic analysis techniques in seismic evaluations. In: Krawinkler H, Fajfar P (eds) Seismic design methodologies for the next generation of codes. Balkema, Rotterdam, pp 277–287

Reinhorn A, Rutenberg A, Gluck J (1977) Dynamic torsional coupling in asymmetric buildings structures. Build Environ 12:251–260

Reyes JC, Chopra AK (2011a) Three-dimensional modal pushover analysis for buildings subjected to two components of ground motion, including its evaluation for tall buildings. Earthq Eng Eng Vib 40:789–806

Reyes JC, Chopra AK (2011b) Evaluation of three-dimensional modal pushover analysis for unsymmetric-plan buildings subjected to two components of ground motion. Earthq Eng Eng Vib 40:1475–1494

Rutenberg A (1992) Nonlinear response of asymmetric building structures and seismic codes: a state of the art. Eur Earthquake Eng VI(2):3–19

Rutenberg A (2002) EAEE Task Group (TG) 8: behavior of irregular and complex structures – asymmetric structures – progress since 1998. Proceedings of the 12th European conference earthquake engineering, London

Rutenberg A, De Stefano M (1997) On the seismic performance of yielding asymmetric multistorey buildings: a review and a case study. In: Fajfar and Krawinkler (eds) Seismic design methodologies for the next generation of codes. Bakelma, pp 299–310, ISBN 90.54.10.928.9

Shakeri K, Tarbali K, Mohhebi M (2012) An adaptive modal pushover procedure for asymmetric-plan buildings. Eng Struct 36:160–172

Tabatabaei R, Saffari H (2011) Evaluation of the torsional response of multistory buildings using equivalent static eccentricity. J Struct Eng 137(8):862–868

Tso WK, Moghadam AS (1997) Seismic response of asymmetrical buildings using push-over analysis. Proceedings of workshop on seismic design methodologies for the next generation of codes, Bled, Slovenia

Chapter 14
Recent Development and Application of Seismic Isolation and Energy Dissipation and Conditions for Their Correct Use

Alessandro Martelli, Paolo Clemente, Alessandro De Stefano, Massimo Forni, and Antonello Salvatori

Abstract More than 23,000 structures, located in over 30 countries, have been so far protected by passive anti-seismic (AS) systems, mainly by the seismic isolation (SI) and energy dissipation (ED) ones. The use of such systems is going on increasing everywhere, although its extent is strongly influenced by earthquake lessons and the features of the design rules used. As to the latter, SI is considered as an additional safety measure (with consequent significant additional construction costs) in some countries (Japan, USA, etc.), while, in others (including Italy), the codes allow to partly take into account the reduction of the seismic forces acting on the superstructure that is induced by SI. Applications of the AS systems have been made to both new and existing civil and industrial structures of all kinds. The latter include some high risk (HR) plants (nuclear reactors and chemical installations). The applications in a civil context already include not only strategic and

---

A. Martelli (✉)
ENEA, Via Venezia 34, San Lazzaro di Savena (Bologna) 40068, Italy
e-mail: marteisso1@gmail.com

P. Clemente
Laboratory on Prevention and Mitigation of Natural Risks, Casaccia Research Centre, ENEA, Via Anguillarese 301, 00123, Santa Maria di Galeria (Roma), Italy
e-mail: paolo.clemente@enea.it

A. De Stefano
Dipartimento di Ingegneria Strutturale, Edile e Geotecnica, Polytechnic of Torino, GLIS and ASSISi, Corso Duca degli Abruzzi 24, Torino 10129, Italy
e-mail: alessandro.destefano@polito.it

M. Forni
Technical Unit on Seismic Engineering, Bologna Research Centre, ENEA, Via Martiri di Monte Sole 4, Bologna 40129, Italy
e-mail: massimo.forni@enea.it

A. Salvatori
Dipartimento di Ingegneria Civile, Edile – Architettura e Ambientale, GLIS and ASSISi, University of L'Aquila, Via Campo di Pile, Zona Industriale di Pile, L'Aquila 67100, Italy
e-mail: antonello.salvatori@alice.it

public structures, but also residential buildings and even many small private houses. In Italy, the use of the AS systems has become more and more popular especially after the 2009 *Abruzzo* earthquake (nowadays more than 400 Italian buildings are seismically isolated). Based on the information provided by the authors at the *ASSISi* 13th World Conference, held in Sendai (Japan) in September 2013, and on more recent data, the paper summarizes the state-of-the-art of the development and application of the AS systems and devices at worldwide level, by devoting particular attention to SI of buildings in Italy, in the context of recent seismic events. Moreover, it outlines the benefits of the aforesaid systems for ensuring the indispensable absolute integrity of strategic and public structures, as, primarily, schools, hospitals and HR plants, but also (for an adequate protection of cultural heritage) museums. Finally, based on Italian experience, it provides some remarks on costs of SI, stresses the conditions for the correct use of this technique and mentions some recent initiatives of the Italian Parliament to ensure such a correct use and to widely extend such an use to the HR chemical plants too (for which only very few applications already exist in Italy).

14.1 Introduction

On September 24–26, 2013, the 13th event of the *Anti-Seismic Systems International Society (ASSISi)*, namely the 13th World Conference on Seismic Isolation, Energy Dissipation and Active Vibrations Control of Structures & JSSI 20th Anniversary International Symposium, took place in Sendai (Japan). This conference (JSSI 2013) was organized jointly with the Japan Society of Seismic Isolation (JSSI) and with the collaboration of the Italian association GLIS ("GLIS – Isolamento ed altre Strategie di Progettazione Antisismica", namely "GLIS – Isolation and Other Anti-Seismic Design Strategies"), which are both *ASSISi* corporate members. The first author of this paper was a member of the Scientific Committee of the conference (as GLIS President and *ASSISi* Founding President and present Vice-President), as well as key-note and invited lecturer for Italy (Martelli et al. 2013b, c). The text of this paper is partly based on the aforesaid contributions provided at Sendai by all its authors, but it also includes some further and more updated information.

According to the data made available the Sendai conference and to subsequent information received by the first author of this paper (Martelli 2013b), more than 23,000 structures in the world have been protected by passive anti-seismic (AS) techniques, such as seismic isolation (SI) or energy dissipation (ED) systems, shape memory alloy devices (SMADs), or shock transmitter units (STUs). They are located in more than 30 countries (see Fig. 14.1 and Table 14.1) and concern both new constructions and retrofits of existing structures of all kinds: bridges and viaducts, civil and industrial buildings, cultural heritage and industrial components and installations, including some High Risk (HR) nuclear and chemical plants and components. Buildings are made of all types of materials: reinforced

Fig. 14.1 Numbers of seismically isolated buildings in the most active countries (data of September 2013)

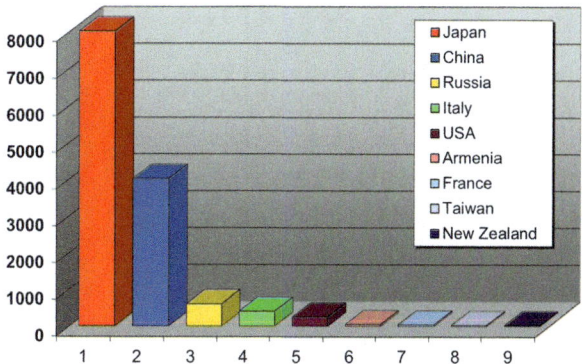

concrete (r.c.), steel and even wood (Eisenberg et al. 2011; JSSI 2013; Martelli and Forni 2010, 2011a, b; Martelli et al. 2008, 2011; Martelli et al. 2012a, b, c, 2013a, b, c, Mazzolani and Herrera 2012).

As shown by Fig. 14.1 and Table, 14.1, Japan is the leading country for the overall number of applications of the AS systems; it is followed by the Peoples' Republic (P.R.) of China, the USA, the Russian Federation and Italy (Fig. 14.2).

The use of the AS systems and devices in a civil context already includes not only the strategic structures (civil defence centres, hospitals) and the public ones (schools, churches, museums, commercial centres, hotels, airports), but also residential buildings and even many small and light private houses. Everywhere, the number of such applications is increasing, although it is strongly influenced by earthquake lessons and the availability and features of the design rules used.

As stressed by Martelli et al. (2013a, b), most SI systems rely on the use of rubber bearings (RBs), such as the High Damping natural Rubber Bearings (HDRBs), Neoprene Bearings (NBs), Lead Rubber Bearings (LRBs), or (especially in Japan) Low Damping Rubber Bearings (LDRBs) in parallel with dampers; in buildings, some plane surfaces steel-Teflon (PTFE) Sliding Devices (SDs) are frequently added to the RBs to support their light parts without unnecessarily stiffening the SI system (which would make it less effective) and (if they are significantly asymmetric in the horizontal plane) to minimize the torsion effects (the effects of the vertical asymmetries are drastically reduced by the quasi "rigid body motion" of the seismically isolated superstructure).

Another type of isolators, which has been used in Italy after the 2009 Abruzzo earthquake, is the so called Curved Surface Slider (CSS), which derived from the US Friction Pendulum (FPS) and the subsequent German Seismic Isolation Pendulum (SIP).

Finally, rolling isolators (in particular Ball Bearings, BBs, and Sphere Bearings) are also applied: as mentioned by Martelli et al. (2013b), they are very effective and find numerous applications (more than 200) to protect buildings in Japan, but not in Italy, because there they have been judged too expensive (however, they have already been used, even in Italy, to protect precious masterpieces and costly equipment, including operating-rooms in hospitals).

Table 14.1 Application of anti-seismic systems (data provided at the ASSISi Sendai Conference of September 24–26, 2013 and, for Turkey, later – see Martelli 2013b) [n.a. = not available; (?) = to be checked]

Country	Seismic isolation				Energy dissipation and other AS systems			
	Large buildings	Houses	Bridges & viaducts	Industrial structures	Large buildings	Houses	Bridges & viaducts	Industrial structures
Japan	3,000	5,000	>1,000 (?)	Some	1,000	>5,000	n.a.	n.a.
P.R. China	4,000	0	400	50	500		200	50
Russian Federation	600		>100	0 (?)	8	0	>100	0 (?)
USA	≈250		Hundreds	Some	Hundreds		n.a.	n.a.
Italy	>400 (?)		Tens (?)	3	Tens (?)		>300 (?)	0
Taiwan	n.a. (>29)	n.a.	n.a. (>20)	n.a.	n.a. (>85)	n.a.	n.a.	n.a.
Armenia	43	2	≈10	0	3	0	0	0
New Zealand	15 (?)	0 (?)	3 (?)	2 (?)	2 (?)	0 (?)	1 (?)	2 (?)
Turkey	18	0	11	2	2	0	0	0
Other countries	Tens (?)		n.a.	Tens	Tens (?)		n.a.	n.a.
Totals	**>14,000**		**>>1,500**	**≈80**	**>>6,500**		**>>600**	**>50**

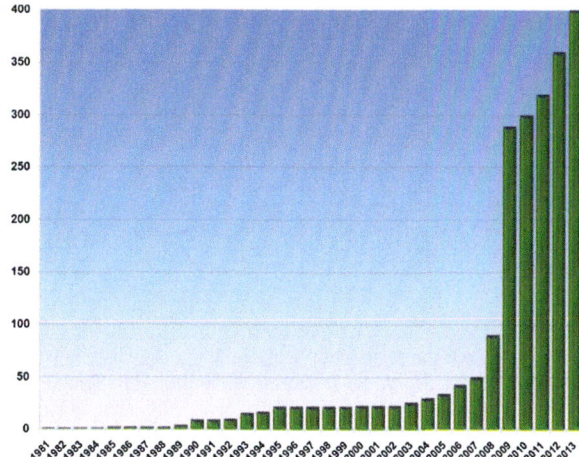

Fig. 14.2 Overall number of Italian buildings protected by seismic isolation during years

It shall be stressed that, to the knowledge of the authors, all structures protected by RBs that were located in areas hit by even severe earthquakes (including those struck by the 2011 *Tohoku* event in Japan, e.g. Figs. 14.3 and 14.4) exhibited an excellent behaviour, in spite of the fact that the violence of such earthquakes was frequently underestimated (Eisenberg et al. 2011; JSSI 2013; Martelli 2013b, 2014; Martelli and Forni 2010, 2011a, b; Martelli et al. 2011, 2012a, b, c, 2013a, b, c, 2014; Mazzolani and Herrera 2012; Zhou et al. 2013).

14.2 Application of the Anti-seismic (AS) Systems

14.2.1 Application in Japan

In Japan the first application of base SI dates back to 1983 (Eisenberg et al. 2011). Thanks to the availability of an adequate specific code since 2000, the free adoption of SI since 2001 (Martelli and Forni 2010) and the excellent behaviour of isolated buildings in violent earthquakes (Martelli et al. 2013a), this country is more and more consolidating its worldwide leadership on the use of the AS systems and devices. As shown by Fig. 14.1 and Table 14.1, the Japanese buildings or houses protected by SI, which were over 6,600 in 2011 (Martelli et al. 2013a), are now approximately 8,000, while the large buildings provided with other type control systems are now approximately 1,000, against the 900 that had been reported by Eisenberg et al. (2011) in 2011 (in that year the latter systems were active or semi-active in 70 cases). Moreover, there are now more than 5,000 houses where ED systems have been installed (they were 2,000 in 2009, as reminded by Martelli et al. 2011).

More precisely, Japan is continuing the extensive adoption of the AS systems initiated after the excellent behaviour of two isolated buildings near Kobe during the 1995 *Hyogo-ken Nanbu* earthquake, of magnitude M = 7.3. This behaviour was

Fig. 14.3 *Left*: the seismically isolated 10-storey Hachinohe City Hall, near Sendai, isolated by means of LRBs. *Right*: the 18-storey MT Building in Sendai, isolated by means of RBs and SDs. Both buildings withstood the 2011 Tohoku earthquake undamaged (Eisenberg et al. 2011)

Fig. 14.4 The 4-storey National Western Art (Le Corbusier) Museum in Tokyo, retrofitted by inserting HDRBs in a sub-foundation in 1999. During the 2011 Tohoku earthquake the SI system reduced the PGA values in the two horizontal directions from 0.19 to 0.27 g at the base to 0.08 and 0.10 g at the top (Eisenberg et al. 2011)

confirmed by all Japanese buildings protected by SI during all severe events which followed that of 1995, namely those of *Tokachi Offshore* (M = 8.0, 2003), *Niigata Chuetsu* (M = 6.8, 2004), *Fukuoka West Offshore* (M = 7.0, 2005), *Niigata Chuetsu Offshore* (M = 6.8, 2007), *Iwate-Miyagi Inland* (M = 7.2, 2008) and *Tohoku* (M = 9.0, 2011) (Eisenberg et al. 2011; Martelli et al. 2011, 2012a; Mazzolani and Herrera 2012).

Several buildings which withstood violent earthquakes prior to that of *Tohuku* were listed by JSSI (Kani 2008). With regard to the *Tohoku* earthquake, it is noted that the related seismic hazard was considerably underestimated, as for several previous violent events all over the world (Martelli et al. 2011, 2013a). In spite of this, most of the 118 isolated buildings located in the *Tohoku* area or erected in other Japanese sites behaved well, at least without considering the effects of the subsequent tsunami (see Figs. 14.3 and 14.4, as well as Martelli et al. 2013a). A similar behaviour was shown, for the isolated bridges and viaducts, by most of those protected by RBs (LRBs and HDRBs), although a certain number of them was later

Fig. 14.5 The 91 m tall 20-storey building, with steel structure, of the Suzukakedai Campus of Tokyo Institute of Technology, protected by 16 RBs, 58 steel or oil dampers and by mega X-shape braces, visible on the façade (Martelli et al. 2012a)

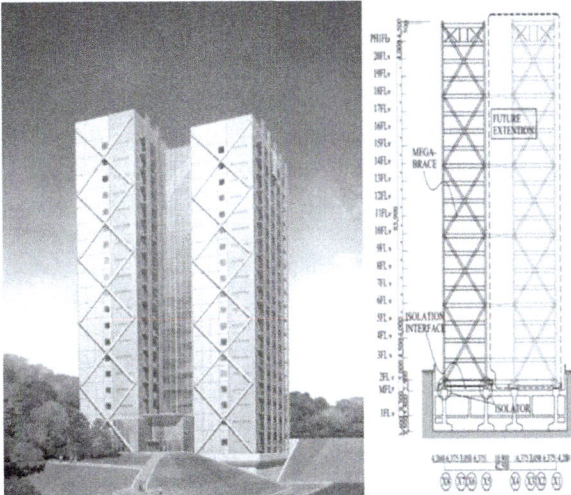

Fig. 14.6 Lateral view of a complex of twenty one 6- to 14-storey buildings, all erected on an unique "artificial ground" isolated at Sagamihara (Tokyo area) with 48 LRBs, 103 SDs and 83 BBs (Martelli and Forni 2010). This was the first Japanese application of "artificial grounds"

destroyed by the tsunami (due to deck rotation toward the upstream side, resulted from the uplifting force – see Martelli et al. 2013a).

Japanese, on the one hand, have confirmed the trend, initiated years ago, to isolate even high-rise buildings (Fig. 14.5) and sets of buildings supported by a common isolated r.c. structure (the so called *artificial ground*, a solution which enables large savings of construction costs – see Fig. 14.6) and, on the other hand, are more and more increasing the number of even very small private houses protected by SI (Martelli and Forni 2010). Based on data provided by JSSI, the isolated high-rise buildings are now rather numerous (they included 250 condominiums in the middle of 2011, as mentioned by Eisenberg et al. 2011). Furthermore, the isolated houses are already about 5,000; the latter were about 3,000 at the end of 2009 (Martelli and Forni 2010) and about 4,000 in the middle of 2011 (Eisenberg et al. 2011). More generally

(Martelli et al. 2013a), in the middle of 2011 46 % (1,100) of the Japanese isolated large buildings (e.g. excluding houses) were condominiums, 20 % offices, 12 % hospitals and 2 % schools and most of these large buildings were new constructions (the retrofits of those existing were 90 some months ago).

The Japanese structures provided with ED systems include several high-rise buildings; these and the similarly protected private houses make use of various kinds of dampers (Martelli and Forni 2010): for instance, the applications of Buckling-Restrained Braces (BRBs) were already more than 250 in 2003. Moreover, approximately 40 Japanese buildings were seismically controlled by Tuned Mass Dampers (TMDs), of active or hybrid types, in June 2007, and so-called *Active Damping Bridges* (ADBs) were installed between pairs of adjacent high-rise buildings to reduce the seismic response of both of them, based on their different vibrational behaviours (Martelli and Forni 2010).

The use of the AS systems and devices also recently increased in Japan for the protection of cultural heritage and for that of bridges and viaducts (Martelli and Forni 2010; Martelli et al. 2011). For the latter it had began rather later than for buildings; it is being largely based on the use of HDRBs and LRBs and considerably extended especially after the *Hygo-ken Nanbu* earthquake, which struck Kobe in 1995 (by becoming obligatory for overpasses in this town).

Finally, as to the industrial installations, besides detailed studies for the SI (even with three-dimensional – 3D – systems) of various kinds of nuclear reactors, the *Nuclear Fuel Related Facility* was erected on 32 LDRBs and LRBs (Martelli and Forni 2010). Application of SI to large industrial factories also began in 2006: the first, concerning the fabrication of semi-conductors, was a 5-storey steel and SRC (mixed steel-concrete system) structure, with a height of 24.23 m and a total floor area of about 27,000 m^2, which was built on LRBs, Viscous Dampers (VDs) and oil dampers; at least 2 further similar factories were also already in use at the end of 2009 (Martelli and Forni 2010; Martelli et al. 2012a).

14.2.2 Application in the P.R. China

In the P.R. China very ancient monasteries, temples and bridges, protected by means of rough sliding SI systems, are still standing, although they had to face numerous earthquakes, including very violent events, up to M = 8.2; however, the application of modern SI systems began only in 1991 (Dolce et al. 2006). In any case, initially the SI systems, then the ED ones too have rapidly got a footing since that year, so that the isolated buildings were already 490 in June 2005, by leading the P.R. China to the third place at worldwide level for the number of applications, only slightly after the Russian Federation. Many of these applications were to dwelling buildings and no less than 270 to the masonry ones (Dolce et al. 2006).

At the end of 2006 the number of the Chinese isolated buildings had increased to more than 550 and included even rather tall constructions. Moreover: SI had already been applied to 5 further large span structures and 20 road and railway bridges or viaducts; 30 buildings were already protected by ED devices; 5 buildings and 6 bridges were already been provided with hybrid or semi-active seismic vibration control systems. SI had also already been used, for the first time in the P.R. China, to protect *Liquefied Natural Gas* (LNG) tanks (see Fig. 14.32 and Martelli and Forni 2010).

In 2007 the P.R. China passed the Russian Federation (Martelli and Forni 2010): in fact, the Chinese isolated buildings were 610 in May 2007 (against the approximately 600 declared at that time by the Russian experts – see Martelli and Forni 2010) and those protected by ED systems 45. The first included the so-called *Isolation House Building on Subway Hub*, completed near the centre of Beijing in 2006, which consists of 20 7- to 9-storey buildings, all separately isolated above an unique huge 2-storey isolated structure that contains all services and infrastructures, including railways and subways (Dolce et al. 2006). The objective of this application had been to optimize the use of a wide and valuable central area, which was previously occupied only by railway junctions and the subway, by also minimizing the consequent vibrations and noise: SI enabled saving 25 % of construction costs, which made it possible to use the available budget for funding an average 3 storey rising of the 50 buildings. In the same years, Chinese application of 3D SI systems to civil buildings and of isolators or SMADs to cultural heritage had also begun (Martelli and Forni 2010).

In October 2008, the number of isolated Chinese buildings was about 650. In November 2009 a further significant extension of the applications of the AS systems was reported in the P.R. China; in particular, the number of the newly erected isolated buildings per year doubled there after the violent *Wenchuan* earthquake of May 12, 2008 (moment magnitude $M_W = 7.9$), by increasing from 50 to 100 per year (Martelli and Forni 2010).

This more rapid increase of the number of building applications of SI was due to both the excellent behaviour of two r.c. isolated buildings and even a 6-storey masonry one during the aforesaid earthquake (although its violence had been largely underestimated, by a factor 10 for the Peak Ground Acceleration – PGA) and the fact that the Chinese code (which still required the submission of the designs of the isolated buildings to the approval of a special commission) permitted to reduce the seismic loads acting on the superstructure and foundations of such buildings (Martelli and Forni 2010).

In November 2009 SI systems had been installed in the P.R. China in 32 bridges and 690 buildings, while 83 buildings had been protected by ED devices such as Elastic–plastic Dampers (EPDs), VDs or Visco-Elastic Dampers (VEDs), 16 by TMDs or other type dampers and 5 by semi-active or hybrid systems (Martelli and Forni 2010). The latter had also been installed in 8 bridges. SI has been applied in the P.R. China not only at the building base or at the top of the lowest floor, but also on more elevated floors (for risings or for erecting highly vertically asymmetric constructions), or at the building top (to sustain, in the case of retrofit, one or more new floors acting as a TMD), or also on structures that join adjacent buildings having different vibrational behaviours. More recent applications also included sets of buildings on *artificial ground*, base and roof SI of stadiums and the protection of valuable objects (e.g. electronic equipment and art objects) by means of SI tables (Martelli and Forni 2010).

Fig. 14.7 Complex of dwelling buildings in South-Western P.R. China, formed by 82 4- to 16-storey seismically isolated buildings (overall floor area = 210,000 m^2) (Eisenberg et al. 2011)

Fig. 14.8 Complex of dwelling buildings in South-Western P.R. China, formed by 94 4- to 6-storey seismically isolated buildings (overall floor area = 280,000 m^2) (Eisenberg et al. 2011)

More recently, the number of applications of the AS systems really strongly increased in the P.R. China. In fact, it was reported by Eisenberg et al. (2011) that about 2,500 buildings and 80 bridges and viaducts had already been protected there by SI in September 2011, in addition to about 400 buildings and 50 bridges provided with dampers, 36 buildings or towers equipped with passive TMDs and 8 bridges and 5 towers with active or hybrid systems (note, for instance, the extensive building applications of SI shown by Figs. 14.7 and 14.8). Moreover, in the last 2 years, there was a further very significant increase of the aforesaid numbers: as shown by Table 14.1, SI has already been used in the P.R. China to protect 4,000 buildings, 400 bridges and viaducts and even 50 industrial structures (e.g. Fig. 14.32), while ED devices have already been installed in 500 buildings, 200 bridges and viaducts and further 50 industrial structures (Zhou et al. 2013; Martelli 2013b).

It is worthwhile stressing that the effectiveness of SI for the protection of buildings and, in particular, of schools and hospitals was recently confirmed in the P.R. China during the *Lushan* earthquake ($M_W = 7.0$) of April 20, 2013 (Zhou et al. 2013). This earthquake occurred in an area that had already been affected by the violent *Wenchuan* event in 2008 (the distance between the two epicentres was 150 km) and was characterized by PGA values that reached 0.4–0.6 g, compared to the design value of 0.3 g (note, for this earthquake too, similar to previous events mentioned by Martelli et al. 2013a, the inadequacy of the probabilistic approach used

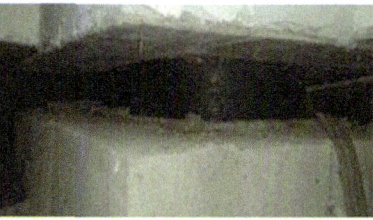

Fig. 14.9 View of the Lushan county hospital (China), before the earthquake of April 20, 2013 (Zhou et al. 2013)

to define this design value, and the fact that the new event occurred only 5 years after that of *Wenchuan*, while the latter was preceded by an earthquake of magnitude comparable nearly 80 years earlier).

The *Lushan* event caused 196 deaths (besides 21 missing people) and the wounding of 250,000 people. About 40,000 buildings (i.e. about 75 % of those in the area affected by the earthquake) collapsed or were damaged. The heavily damaged buildings included numerous strategic and public buildings (including schools and hospitals), even constructed or reconstructed after the *Wenchuan* event. However, where it was used, SI showed, once again, an excellent effectiveness.

Particularly interesting were, in Lushan, two cases of r.c. structures:

- that of two primary schools, the first conventionally founded, the other base isolated, both provided with a seismic monitoring system;
- that the county hospital (7 floors above ground and one basement), consisting of two buildings with conventional foundations and one with base SI.

About the two schools, while for that conventionally founded the PGA value of 0.2 g was amplified, at the roof, to 0.72 g, for that with SI the aforesaid value was reduced to 0.12 g. Thus, the effectiveness of SI can be quantified in a reduction factor of the roof maximum acceleration equal to 6.

As to the county hospital (Fig. 14.9), the two buildings with conventional foundations suffered damage to both partitions, roof and equipment contained, which made them unusable after the earthquake (Fig. 14.10); on the contrary, the seismically isolated block was the only hospital building of the county to be remain fully undamaged and operational (Fig. 14.11): this allowed to heal thousands injured people, which was impossible in other hospitals in Lushan.

Fig. 14.10 Damage suffered by the two conventionally founded buildings of the hospital of Fig. 14.9 (Zhou et al. 2013)

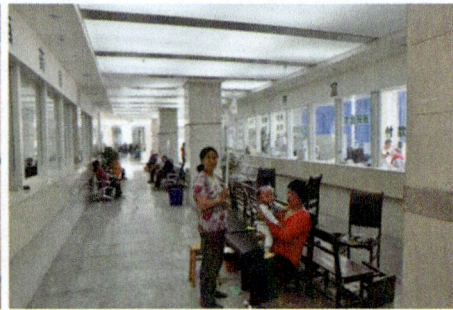

Fig. 14.11 Full integrity and operability of the isolated building of the hospital of Fig. 14.9 (Zhou et al. 2013)

14.2.3 Application in the Russian Federation

The Russian Federation is third for the number of seismically isolated structures. As shown by Table 14.1, there are about 600 applications of SI to buildings and more than 100 to bridges and viaducts (according to the information provided by Eisenberg et al. 2011, the overall number of isolated structures was about 550 in September 2011). In addition, there is already a significant number of structures (especially bridges) protected by ED systems (Table 14.1). The use of modern SI systems, formed by RBs, frequently in conjunction with SDs and/or dampers (similar to those adopted in the other countries), is going on replacing that of the previous so called *low cost* isolators (reversed mushroom-shaped r.c. elements), which had been installed since the years 1970s. After the retrofits of some important historical buildings (Dolce et al. 2006; Martelli et al. 2008; Sannino et al. 2009), recent Russian application includes even high-rise buildings, in particular in Sochi, the site of the 2014 Winter Olympic Games. For some of these, Italian HDRBs have also been used, as shown by Figs. 14.12 and 14.13 (Eisenberg et al. 2011; Martelli et al. 2011).

Fig. 14.12 Design of the base-isolated Hayat (Sea Plaza) Hotel, a 28-storey r.c. building (with 2 underground floors), 93.6 m tall and with a total floor area of 40,000 m², erected in Sochi (Russia) (Eisenberg et al. 2011)

Fig. 14.13 Some of the 193 HDRBs, manufactured in Italy, which protect the building of Fig. 14.12 (picture taken on September 23, 2011, during the 2011 Sochi Conference) (Martelli et al. 2012a)

14.2.4 Application in the USA

As shown by Table 14.1, the USA remain at the third place, after Japan and the P.R. China, for the overall number of applications of the AS systems and devices (Eisenberg et al. 2011; JSSI 2013). In this country, however, such applications go on being satisfactorily progressing only for bridges and viaducts and for buildings protected by ED systems. They concern both new constructions and retrofits. More precisely, at the end of 2009, HDRBs, LRBs and, more recently, ED devices and STUs had already been installed in about 1,000 U.S. bridges and viaducts, located in all U.S. states, while over 1,000 buildings had been provided with dampers of various kinds (Martelli and Forni 2010): VDs and friction dampers (FDs) already protected approximately 40 and, respectively, 12 buildings in 2001 and BRBs 39 further buildings in 2003 (Dolce et al. 2006).

On the contrary, as far as SI of buildings is concerned, the number of new applications remains still limited (recently 3 or 4 per year), in spite of the excellent behaviour of some important U.S. isolated buildings during the 1994 *Northridge* earthquake (Dolce et al. 2006) and the long experience of application of this technique to such structures (since 1985). This is a consequence of the very penalizing design code in force in the USA for the isolated buildings: these were not more than 200 in September 2011 and are now approximately 250, although the related applications are mostly very important and half of them are retrofits, even of monumental buildings (Martelli and Forni 2010).

SI of US buildings has been performed using HDRBs, LRBs (in some cases in conjunction with LDRBs, SDs, VDs and other ED devices) and, later, the FPS too. With regards to the design earthquake levels adopted in California, Martelli and Forni (2010) stressed that they correspond very large magnitudes (e.g. $M = 8.3$ for the new *911 Emergency Communications Centre* erected in San Francisco in the years 1990s and $M = 8.0$ for the *San Francisco City Hall* retrofitted with 530 LRBs in 2000): this imposes the use of SI (as the only possibility) for these applications, in spite of its large implementation cost in the USA.

14.2.5 Application in Italy

Fifth (after Japan, the P.R. China, the Russian Federation and the USA) and first in Western Europe for the overall number of applications of the passive AS devices remains Italy (Figs. 14.1 and 14.2 and Table 14.1). There, the use of the AS systems began in 1975 for bridges and viaducts and in 1981 for buildings (Dolce et al. 2006; Eisenberg et al. 2011; Martelli et al. 2008, 2012c). It is worthwhile stressing that the design of the first two Italian suspended buildings protected by AS systems, located in Naples, had been completed before the 1980 *Campano-Lucano* earthquake, when the Naples area was not yet considered as seismic: after such an event, that area was classified in "seismic category 3" (i.e. with moderate seismic hazard): thus, in order to avoid large modifications of the buildings designs, NBs were added on the roof of such buildings, together with other passive AS devices inside them (Dolce et al. 2006; Martelli et al. 2012a, b).

In spite of the aforesaid pioneering role of Italy in the development and application of the passive AS systems, in the years 1990s their use remained rather limited several years long, due to the lack of design rules to the end of 1998, then due to their inadequacy and a very complicated and time-consuming approval

Fig. 14.14 Francesco Jovine school of San Giuliano di Puglia, after its collapse during the 2002 Molise & Puglia earthquake (M = 5.9)

Fig. 14.15 *Left*: the isolated complex including the new Francesco Jovine primary school and "Le Tre Torri" Poly-Functional Centre in San Giuliano di Puglia (Campobasso), in 2008. *Right*: view of some the isolators supporting their common base slab

Fig. 14.16 One of the three tanks of the Company Polimeri Europa of the Italian ENI Group located in Priolo, which were seismically retrofitted using U.S. FPS devices in the years 2005–2008 and one of the isolators during and after its installation. To the knowledge of the authors, this is the only application of SI to chemical plants and components so far existing in Italy (prior to the 2009 Abruzzo earthquake, it was also the only application of CSS devices in Italy)

process to 2003 (Dolce et al. 2006). However, significant application of the passive AS systems (especially of SI) restarted in Italy about 10 years ago, initially as a consequence of the collapse of the *Francesco Jovine* primary school in San

Fig. 14.17 Main building of the Civil Defence Centre of Foligno (Perugia), former seismic zone 1, isolated by 10 HDRBs and certified as safe by A. Martelli in 2011

Fig. 14.18 The new block B of the Romita High School for scientific studies in Campobasso (former seismic zone 2), isolated by 12 HDRBs and 10 SDs, which was reconstructed with SI, after its demolition in 2010 (blocks A and B had been found unsafe by ENEA, CESI and the University of Basilicata in 2003 – see Fig. 14.29). The safety of the new building was certified by A. Martelli in 2013

Giuliano di Puglia (Campobasso) during the 2002 *Molise & Puglia* earthquake (Fig. 14.14) and the subsequent enforcement of a new national seismic code (in May 2003), which freed and simplified the adoption of the AS systems in Italy (see Figs. 14.15, 14.16, 14.17, 14.18, and 14.19, Martelli et al. 2008, 2013b; Martelli and Forni 2010).

The use of SI became particularly rapid especially after the *Abruzzo* earthquake of April 6, 2009, as a consequence of the large damage caused by this event to the conventionally founded structures and cultural heritage (Martelli and Forni 2010, 2011b). Thus, in 2009, Italy overtook the USA for the number of seismically

Fig. 14.19 Eight-storey isolated building which is nearing completion in Messina on 22 LDRs and 2 SDs. Its structural safety will be certified by A. Martelli in 2014

Fig. 14.20 *Left*: one of the 184 pre-fabricated houses (wood, or r.c., or steel structures) erected in L'Aquila to host up to 17,000 residents who remained homeless after the 2009 Abruzzo earthquake. *Right*: detail of one of the 40 CSS devices, manufactured in Italy, which have been installed at the top of columns (made of steel or r.c.) to isolate the supporting slabs of such houses

isolated buildings (although not for their importance): those in use were about 70 before the aforesaid earthquake, with further 20÷30 under construction or design, while they are now more than 400 and several further applications to new-built and retrofitted structures of these kinds are in progress (Martelli and Forni 2011b; Martelli et al. 2012a, 2013b).

The recent applications of SI include 184 wood, r.c. or steel pre-fabricated houses erected in L'Aquila, each on a large isolated r.c. slab (Fig. 14.20), to provisionally host up to 17,000 homeless residents (at least in the first years). These were seismically isolated, for the first time in Italy, using CSS devices manufactured in the country (Fig. 14.20). However, the use of the traditional HDRBs or LRBs, in conjunction with some SDs, is also going on, in both L'Aquila

Fig. 14.21 *Left*: the new Headquarters of ANAS (National Agency for Roads Construction) in L'Aquila, erected on 60 HDRBs after the 2009 Abruzzo earthquake (which had severely damaged the previous headquarters building), completed at the beginning of 2011. *Right*: view of some of its isolators

Fig. 14.22 *Left*: the dwelling building complex (3 buildings) of Via Borgo dei Tigli 6-8-10 in L'Aquila (Pianola area), which had been just completed before the 2009 Abruzzo earthquake. *Right*: damage caused to the building by this event

Fig. 14.23 Seismic retrofit of the building complex of Fig. 14.22, performed by means of 42 HDRBs and 62 SDs and connection of the originally separated three buildings. The structural safety will be certified by A. Martelli in 2014

and other Italian sites, for several new constructions and retrofits (see, for instance, Figs. 14.21, 14.22, 14.23, and 14.24, Martelli et al. 2011, 2013b; Martelli and Forni 2011a, b). In particular, the new *Francesco Jovine* school (Fig. 14.15), protected by

Fig. 14.24 *Left*: seismically isolated dwelling building complex under construction in Ozzano Emilia (Bologna), irregular in shape (61 m × 28 m; 2 buildings with 2 staircases, connected to the second floor; 4 isolated above ground isolated floors, with the isolators installed in the non-isolated basement). *Right*: some of the isolators of the building (61 HDRBs and 56 SDs)

a SI system designed with the cooperation of ENEA (Italian National Agency for New Technologies, Energy and Sustainable Economic Development) and formed by 61 HDRBs and 13 SDs, which was the first Italian isolated school (certified as safe by the first author of this paper in September 2009), has been followed by several further projects of this kind: the seismic protection of schools by means of SI, besides that of hospitals and other strategic structures, is now a "priority 1" objective of GLIS (Martelli and Forni 2010; Martelli and Forni 2011a, b).

Moreover, the use of the AS systems is going on for bridges and viaducts (those with such systems were already at least 250 in 2009), as well as for cultural heritage (Martelli 2009; Martelli and Forni 2010, 2011a, b, Martelli et al. 2008, 2011). For the latter, the application of new retrofit techniques using SI, applicable to monumental buildings (Martelli 2009; Clemente et al. 2011), has also been planned for both reconstructing L'Aquila and for enhancing the seismic protection of some ancient constructions in Sulmona, an historic town close to L'Aquila which was not damaged by the 2009 event, but is also very earthquake-prone. This method (Fig. 14.25) consists in the lateral insertion of large diameter tubes below the building, inside which the isolators will be inserted. The applications planned in Sulmona will be made in the framework of a collaboration agreement signed between ENEA and the local municipality, which will entrust ENEA with the check of the retrofit designs and supervision of the subsequent construction works (Martelli et al. 2011).

14.2.6 Application in Other Countries

The countries which follow Italy for the overall number of applications of the AS systems are South Korea, Taiwan, Armenia, New Zealand, France, Turkey, Mexico, Canada, Chile and others (Martelli and Forni 2010, 2011b): many applications in

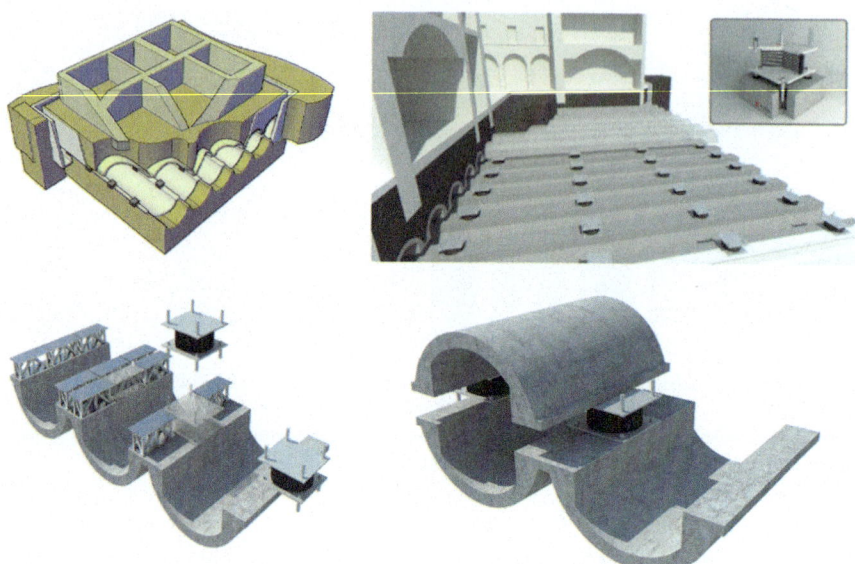

Fig. 14.25 System patented by ENEA and Polytechnic of Torino for the retrofit with SI of monumental buildings, performed by laterally inserting tubes (which will contain the isolators) below the foundations (Clemente et al. 2011; Salvatori 2013)

these countries make use of Italian AS devices (e.g. in Turkey, Greece, Portugal, Spain, Iran) and some (in Romania, Cyprus) have also been designed by Italians (Martelli et al. 2012a).

Armenia, with 45 completed isolated buildings and others under construction (see Table 14.1 and Martelli 2013b), remains second, at worldwide level, for the number of applications of such devices per number of residents, in spite of the fact that it is a still developing countries. In New Zealand, one of the motherlands of AS devices (in particular of those based on the use of lead, like LRBs and LDs) and third in the world for the number of applications of such devices per number of residents, the isolated structures had an excellent behaviour in both the 2010 *Canterbury* earthquake, of $M = 7.1$, and the 2011 *Christchurch* event, of $M = 6.3$ (Martelli and Forni 2011b; Mazzolani and Herrera 2012; Martelli et al. 2011, 2012a). Similarly, the isolated structures in Santiago had an excellent behaviour in Chile too, during the magnitude 8.8 *Maule* earthquake of February 27, 2010 (Eisenberg et al. 2011; Martelli and Forni 2011b; Martelli et al. 2011; Mazzolani and Herrera 2012).

Fig. 14.26 *Left*: the "Student House" in L'Aquila, collapsed during the Abruzzo earthquake of April 6, 2009 (M = 6.2), causing the death of eight students. *Right*: secondary school in Dujiangyan (P.R. China), collapsed during the Wenchuan earthquake of May 12, 2008 (M ≈ 8.0), causing the death of 900 students

14.3 Benefits of the as Systems for the Protection of Schools, Hospitals, Cultural Heritage and HR Plants

Schools and hospitals are the strategic and public buildings that should guarantee the highest level of safety. In fact, schools host the most valuable asset of a community, that is its future, while hospitals shall remain fully operational after all catastrophic events (Martelli et al. 2013c, d). This means that, for any accidental event that may hit them, the total integrity of schools and hospitals must be ensured, namely not only that of the structural elements, but also that of the non-structural ones (partitions, claddings, ceilings, plants, contained objects and equipment and, especially, the occupants, namely, for schools, students, teachers and school staff). To achieve this goal for existing buildings, the so-called "seismic improvement" is insufficient: it is indispensable that such buildings are put in the same safety conditions as those obtainable for the new constructions. The aforesaid remarks obviously apply to the other kind strategic and public structures too; among these, the seismic protection of museums is particularly important in Italy, because this country hosts a large part of the cultural heritage existing in the word.

However, even recent experience shows that too many schools, too many hospitals, too many museums and too many other kind strategic and public structures, both in Italy and in other countries, are very unsafe, especially (but not only) in the case of earthquakes (see, for instance, Figs. 14.14 and 14.26 for schools and Fig. 14.10 for hospitals).

In addition, as stressed by Martelli (2012, 2013c), it should be obvious to all that the High Risk (HR) plants too should be adequately protected from natural disasters, primarily from earthquakes and other accidental events that can be triggered by them (in particular by tsunamis, caused by violent earthquakes with epicentres in the sea or even, if these are close to the coast, on the ground). The HR plants include

Fig. 14.27 *Left*: rupture of a storage tank in the Yarimca Refinery (Turkey), owned by Tupras, during the Izmit earthquake of August 17, 1999 ($M_W = 7.4$), which caused 17,000 victims. At the centre: initial fire of one the two tanks (the first containing crude oil, the second naphtha) of the Tomakomai City petrochemical installation (Japan), caused by the two main shocks of the Off Tokachi earthquake of September 26 and 28, 2003 ($M = 8.0$ and $M = 7.1$), with epicenters at 220 km from the plants. *Right*: propagation of the fire in the Tomakomai City petrochemical installation, during the aforesaid quake, with the consequent damage of 45 tanks (30 severely, 29 with leakage) of the 105 present

not only the nuclear ones, but also several types of chemical installations and components: in particular the Liquefied Natural Gas (LNG) tanks, which are large in size (with volumes up to 150,000 m^3 or more), and also the smaller spherical or cylindrical storage tanks that are present, for example, in petrochemical plants, the danger of which is an increasing function of their (often large) number in each installation (Martelli 2012). A substantial amount of these tanks already suffered serious damage in several countries, during a significant number of seismic events (see Fig. 14.27 and Martelli 2012).

With regard to earthquake protection of both schools and HR plants and components (as well as of other types of structures), SI, ED and the other types AS technologies have been demonstrated to be extremely efficient (Martelli et al. 2013a, b). However, their wide use (which obviously frequently entails the acceptance of some additional construction costs) requires a correct perception of the seismic risk, which does not yet exist in countries like Italy. For this reason, besides continuing to promote the development and application of the AS systems (Martelli et al. 2013a, b), the Italian association GLIS and ENEA are devoting great efforts to raise public awareness and to stimulate institutions to start in Italy, at last, an adequate seismic prevention policy (Clemente and Martelli 2013; Martelli 2013a; Martelli et al. 2013b, c, d).

As mentioned by Martelli (2012, 2013c), this campaign was undertaken for the protection of the HR chemical plants and components several years ago, due to the presence of several installations of this kind in Italian areas that are characterized by high seismic hazard (Fig. 14.28) and was soon extended to the civil constructions, well before the *Emilia* earthquake of May 20, 2012 (Martelli 2012, 2013c; Martelli et al. 2013d). It has been brought to the attention of the Italian institutions for a long time (especially of that of the 8th Commission on Environment, Territory and Public Works of the Italian Chamber of Deputies – see Martelli et al. 2013b). This first occurred for the issues related to HR chemical plants and components

Fig. 14.28 The Italian petrochemical installations of Milazzo and Priolo (Sicily)

(Martelli 2012, 2013c), then, since the end of May 2012, for the civil structures too, thanks to the audits of the authors of this paper, as well as those of other GLIS members, which took place in the aforesaid commission in the framework of the "Survey on the State of Seismic Safety in Italy" of the Italian Chamber of Deputies in 2012 (Camera dei Deputati 2012a, b; Martelli 2013c; Martelli et al. 2013d; Martelli et al. 2013b).

In particular, the authors of this paper, on behalf of GLIS, have organized (or actively participated in) some conferences devoted to the promotion of schools safety (Martelli 2013a).

14.3.1 Safety of the Italian Schools

During the mentioned audits at the 8th Commission of the Italian Chamber of Deputies, held in the framework of the «Survey on the State of Seismic Safety in Italy», the very poor safety level of Italian schools was confirmed. In particular, it was stressed that 49 % of school buildings in Italy has no certificate of use and occupancy (Martelli et al. 2013d). It was estimated that 27,920 Italian school buildings are located in highly seismic areas: 4,856 in Sicily, 4,608 in Campania, 3,130 (100 % of the total) in Calabria, 2,864 in Tuscany and 2,521 in Lazio. In addition, 6,122 schools are located in high landslide hazard areas: 994 in Campania, 815 in Emilia-Romagna and 629 in Lombardy (Martelli et al. 2013d).

Especially during the audits of the ENEA representatives (who were the first two authors of this paper) and of the President of the Italian Major Risks Commission, it was also pointed out that more than 70 % of the Italian buildings are unable to withstand the earthquakes to which they may be subjected and that such a huge number of highly seismically vulnerable buildings includes several schools, often hosted by ancient or simply old constructions, for which seismic retrofit is impossible or overly expensive (Martelli et al. 2013d). In the above cases it is imperative to move the schools to other buildings, or existing (if they can ensure the necessary safety level or may be adequately seismically retrofitted), or ad hoc reconstructed with the best available technologies, by devoting the ancient buildings that cannot

Fig. 14.29 *Left*: the Romita High School in Campobasso (Italy), in 2003 (when it hosted 1,300 students), before the demolition of its two most unsafe A and B blocks and reconstruction of block A with SI (see Fig. 14.18). At the centre and *right*: the new school of Marzabotto (Bologna, Italy, former seismic zone 3), which was seismically isolated (with the collaboration of ENEA) by means of 28 HDRBs and 14 SDs, with 500 mm diameters; it is the first seismically isolated school in Northern Italy, which was certified as safe by A. Martelli in September 2010

Fig. 14.30 *Left*: one of the 2 isolated blocks of the new primary and secondary school of Gallicano (Lucca, Italy), former seismic zone 2, which was erected with the collaboration of GLIS and was opened to activity in September 2009. *Centre-left*: some of the 46 HDRBs installed in their underground floors. *Centre-right*: the new isolated kindergarten and primary school of Mulazzo (Massa Carrara, Italy), former seismic zone 2, certified as safe by A. Martelli and opened to activity in September 2012. *Right*: some of the 29 LRBs, which, together with 15 SDs, form its SI system

be adequately seismically retrofitted to other activities and by demolishing and rebuilding those which are just old (e.g. Figs. 14.18 and 14.29).

It is also essential to really complete the evaluations of seismic vulnerability of the Italian public buildings (including schools) within an extremely short time (according to the Decree of the President of the Council of Ministers Nr. 3274 of 2003, these evaluations should have been completed by the institutions in charge in a short time!).

With regard to the seismic protection of schools, as mentioned, the use of the AS systems (which have been developed and already significantly applied, even in Italy) ensures – in the case of SI – the absolute integrity of buildings and minimizes the panic effects, or – for example, in the case of ED systems – allows to approach to this objective (Martelli et al. 2013a, b, c). As mentioned in Sect. 14.2.5, such systems have already been used to protect a significant number of Italian schools, both of new construction and existing. In particular, after the first application of SI in the reconstruction of *Francesco Jovine* primary school in San Giuliano di Puglia, which ended and was certified as safe by A. Martelli in 2008 (Fig. 14.15), following the collapse of the previous one during the 2002 *Molise & Puglia* earthquake (Fig. 14.14), the Italian schools protected with this technique, both of new construction and seismically retrofitted, are already at least 30, even in moderate seismic hazard areas (see, for example, Figs. 14.29 and 14.30 and Martelli et al. 2013a), and others are in progress.

SI is the technology that, nowadays, should be used for both the erection of all new schools in seismic areas and (where possible) seismic retrofit of the existing ones. For the latter, the applicability of SI obviously requires the presence or possibility of realization of structural gaps of sufficient width to enable the free transverse motion of the isolated superstructure (the related displacement can reach 40–50 cm in Italy and 1 m or even more in more seismic countries, as Japan and California).

Moreover, earthquakes are not the only accidental events to be hazardous for the safety of school occupants: others are, for example, fires and collapses due to static problems of school buildings and their parts. As a consequence of the tragic events of November 22, 2008, when Vito Scafidi, a young student, lost his life in the municipality of Rivoli (Torino), due to the collapse of the ceiling of his classroom of the Darwin High School, the Italian Civil Defence Department estimated the need for at least 13 billion Euros to put Italian schools in safe conditions. For the first time the amount of resources needed to address the long standing problem of unsafe schools conditions was assessed in Italy; it is a problem that, although "difficult to sustain" in the short term, should have represented a horizon within which to set the subsequent fiscal policies.

To raise public awareness and to stimulate the institutions to start in Italy, at last, adequate prevention policies for ensuring schools safety, GLIS organized, in collaboration with other partners, a meeting on "Safe schools: right and duty of a civil society", which was held in Asti on February 16, 2013 (Martelli et al. 2013d). The objective of this meeting, held on purpose just before the 2013 general elections of the new Italian Parliament, was to stress, to its future new members, the extreme urgency to secure the existing school buildings (the history of which is often very long and poorly documented) and the need to erect the new ones by adopting the most effective available technologies (in particular, with regard to earthquake protection, by extending the use of SI as much as possible – see Figs. 14.29 and 14.30). In fact, although the Italian Parliament has turned with these goals for some years, starting with a commitment act proposed by the first author of this paper and approved by the 8th Commission of the Chamber of Deputies and the government in 2009, this act was not followed by any concrete measure (Martelli et al. 2013d).

Due to the mentioned extremely large percentage of Italian schools that are unsafe, a problem is obviously how to find the necessary funds, especially in the present very critical national economic situation. To face this problem, in April 2012, a bill had been submitted in the Italian Senate to give the opportunity to citizens to allocate 8 per thousand of their tax return to put school buildings in safe conditions (Martelli et al. 2013d). When it became clear, due to the anticipated end of the legislation, that this bill had no chance of being approved, it was turned into an amendment to the Stability Law, which, however, found a considerable resistance to its acceptance in the competent 5th Budget Commission of the Senate (Martelli et al. 2013d). Of no use was a resolution prepared with the collaboration of the first author of this paper and proposed in the 5th Commission on Budget, Treasury and Planning of the Chamber of Deputies at the beginning of December 2012, in support of the aforesaid proposals, although it had been signed by as many as 17 MPs from all parties (Martelli et al. 2013d). To help overcome these oppositions, GLIS and other partners wrote to both the representative of the government and the President of the 5th Commission of the Senate (Martelli et al. 2013d). Unfortunately, even these letters had no effect (no answer was received).

However, GLIS and its partners did not give up and have continued to stimulate the institutions, starting from the Asti meeting, where they formulated a request and a commitment: the request was addressed to the candidates of all parties in the general elections, who were invited to sign a statement of commitment to undertake the actions needed to ensure the safety of Italian schools (no electoral programs incorporated such actions within their priorities). The commitment, of the organizers, was to regularly check the implementation of such actions and to make the results of these verifications known, through Internet and by organizing special meetings: as mentioned by Martelli (2013a) and Martelli et al. (2013c, d), the first of such meetings were later held in Lanciano, Teramo, on April 19, 2013 and in Bologna on June 13 (the latter, which was organized with the collaboration of Rotary and Lions Clubs and other partners, was entitled "Safe schools: right and duty of a civil society", as in Asti).

During the aforesaid meetings it was stressed how, in spite of the shortage of economic resources, the active commitment of people of good will can result in a significant success. However, if the procedures are crippling, if the surveillance is uncoordinated and punitive rather than collaborative, if the main political and administrative choices on school safety were impromptu and irrational, even a great voluntary commitment of motivated and attentive people can produce only fragile results. In addition, it was complained that not always what appears from official documents and the certification of professionals and technicians corresponds to reality. Sometimes, modification and renovation works carried out on existing buildings are conducted in a superficial and irresponsible way. What is more serious is that there are people who are aware of poorly made and dangerous works, but do not denounce this.

With regard to the problem of economic resources, it was agreed that the proposed use of the part of the 8 per thousand of the tax revenues allocated to the state can be only an emergency and temporary solution: a comprehensive plan of budget allocation, dedicated in a structured and consistent way, is needed. Problems like school safety cannot be solved thanks to emergency procedures, which did so much harm to Italy in recent times: the rules must always apply, even if they have to be simplified, and the principle of responsibility shall apply.

As to the issue of information and the danger of generating alarm through it, participants in the Asti meeting claimed the right of the public opinion to be made aware of all, in time to pick and choose, rather than somebody still risks to die unexpectedly under a pile of rubble. The role of a full and transparent information, according to the Japanese experience, was reminded, together with the use that the population can make of this information through appropriate aggregations to increase safety of schools. For example, a case reported was that of a school on which a sign was posted which declared that the building had not been erected according to anti-seismic norms: in this way, people can choose, deciding whether to drop out of school, or pressure the administration to force it to put the building in safe conditions. The filtered information under the pretext of the danger of creating alarmism allows opacity and arbitrary decisions, especially in a corrupt country, and is a betrayal of democracy, which must not only be a proxy.

Risk evaluation and selection of priorities was a further issue that was discussed in all aforesaid meetings. There are not only the schools, not only earthquakes: there are also the hydro-geological events, floods and many other disasters. In case of limited overall economic resources, like those existing in Italy, a choice of priorities is a must. However, this choice has many reasons, including the sensitivity of the population to give consent to the choice itself. The frequent accidents and the fact that schools host our children are reasons to arouse such a sensitivity. Furthermore, as to earthquakes, the recent developments in the seismological field (such as the so-called "earthquake prediction experiments", or, more precisely, "intermediate-term middle-range earthquake predictions"), if duly considered, can be very useful for defining the intervention priorities (Martelli et al. 2013a). Finally, we remind that repair of the damage caused by an earthquake costs from three to five times the funds needed for preventive measures aimed at ensuring the safety of structures (Martelli et al. 2013d).

Luckily, it seems that now, at last, the issue of protecting the Italian schools, by devoting sufficient funds to this purpose, has been understood by some qualified representatives of the Italian institutions and political parties. We hope that adequate actions will be really urgently undertaken.

14.3.2 Safety of Italian Hospitals and Cultural Heritage

Contrary to schools, only a limited number of Italian hospitals has already been protected by SI. For schools, it was necessary to wait for the collapse of that mentioned above in San Giuliano di Puglia in 2002, before deciding to use the aforesaid technique. For a wide application of SI to the Italian hospitals too, shall we need to wait that a next earthquake destroys one of them and causes further victims? Have we learned nothing from the damages suffered by the hospital in Mormanno (Martelli 2014), knocked out by the modest *Pollino* event, on October 26, 2012, of moment magnitude $M_W = 5.2$? If the earthquake had been (or will be) more violent (as is very possible in that area), what would have happened (or what will happen)?

Moreover, as far as the protection of cultural heritage is concerned, we must unfortunately note that no Italian museums, even of new construction, have been so far protected by SI. There are only a few masterpieces that have been seismically isolated so far (Bronzes of Riace, etc. – see, for instance, Martelli 2009). However, this will be insufficient to protect them from the collapse of the museum (or parts of it, e.g. the roof), if it is unable to withstand the possible earthquakes. Have we learned nothing from the collapse of several statues in the museum of L'Aquila during the 2009 *Abruzzo* earthquake?

Finally, is the collapse of so many valuable monumental buildings during quakes acceptable, as occurred during the aforesaid *Abruzzo* event? As mentioned in Sect. 14.2.5, new retrofit techniques using SI in a sub-foundation (namely applicable to monumental buildings, because they are compatible with the conservation requirements) have been developed in Italy (Clemente et al. 2011) and attempts to use them in Abruzzo, for the reconstruction (Salvatori 2013) or as prevention measures, are in progress (Fig. 14.25). However, let's hope that will be really adopted.

14.3.3 Safety of High Risk (HR) Chemical Plants and General Remarks on Seismic Prevention

It is historically proven that a large part of the Italian territory is characterized by high or least significant seismic hazard (up to magnitude values of at least $M = 7.0$–7.5); in addition, some areas are also exposed to possible non-negligible tsunamis in case of earthquakes with epicentres in the sea (even in shallow water zones) or near it, on the coast. Nevertheless, in Italy, there are now more than 1,000 HR industrial installations subjected to the requirements of the so-called *"Seveso II"* decree, namely in which there are potentially dangerous substances in quantities that exceed certain thresholds. Many of these installations are also subjected to the so-called *"Integrated Environmental Authorization"* (AIA). Some of them are located in areas of high seismic hazard, such as, for instance, in Sicily, in those of Milazzo and Priolo-Gargallo (Fig. 14.28). It is worthwhile remembering that, in 1693, the plain of Catania, which includes the Priolo-Gargallo site, was hit by one of the most devastating earthquakes occurred in Italy, probably more violent than that of *Messina & Reggio Calabria* of 1908 ($M = 7.2$), and that (as later in 1908) such an earthquake generated a violent tsunami. It shall also be remembered that Milazzo is located in the Messina Province and that Mount Marsili, a huge submerged volcano (the biggest in Europe, 70 km long, 30 km wide and 3,000 m high), rises in front of it, in the Tyrrhenian Sea, with a crater at 450 m from the water surface: according to some geologists, this volcano might explode at any time, with the possible collapse of a large part of its flanks, by causing a violent tsunami.

In the Priolo-Gargallo and Milazzo sites, should the HR plants that are present there be inadequately protected from earthquakes, an event of $M \approx 7.0$ (which is quite possible) would trigger serious accidents, perhaps even worse than those occurred in Turkey due to the *Kocaeli* earthquake of August 17, 1999 (Fig. 14.27), with serious consequences for the population and the environment, besides the economic ones. Moreover, if the earthquake were followed by a significant tsunami, the proximity of such plants to the coast, in the absence of barriers with adequate strength and height (which is the present situation), would make these consequences even more dramatic.

As reported by some scientific publications for several years and, more recently (after the 2011 *Tohoku* earthquake and tsunami), even by the press, in Italy, in spite of the availability of maps covering both the seismic hazard and that related to tsunamis for several years, there are still no organic nor adequate legal rules for chemical installations, even for the HR ones (contrary to what happens for civil constructions on the one hand and for nuclear installations on the other), regarding their seismic design, the measures to be taken to protect them (when necessary) from tsunamis and those to make the existing plants resistant to both earthquakes and tsunamis. About the inadequacy of the rules currently in use in Italy for the HR chemical plants, we note that such plants are now designed taking, as seismic loads, those defined by the national codes, which are based essentially on the characteristics of civil buildings, namely on a probabilistic approach (*Probabilistic Seismic Hazard Assessment* or PSHA). According to some well-known seismologists (in Italy, in particular, the team led by the GLIS and *ASSISi* honorary member Prof. Giuliano Panza of the University of Trieste and the International Center of Theoretical Physics), these seismic loads can, therefore, be particularly inappropriate for constructions that are certainly much more complex (in terms of structures, systems and components) than the civil ones. The reasons for this are both that the seismic risk of the HR installations is significantly larger than that concerning the civil constructions and that the PSHA approach showed severe limits on the occasion of the most violent earthquakes recently occurred in the world (Martelli et al. 2013a, b, c): therefore, according to the aforesaid experts, the use of the PSHA approach should be combined with that of a deterministic one (e.g. the so called *Neo-Deterministic Seismic Hazard Assessment* or NDSHA), which, differently from the PSHA, is based on the physics of the phenomena involved and is proving to be more and more reliable and able to quickly adapt to the advancements of the seismological research (Martelli 2013c; Martelli et al. 2013a, b, c).

Moreover, the information available about the level of protection from earthquakes and tsunamis that characterizes the existing HR plants in Italy is still far from exhaustive, and indeed, as it has been reported by some publications for some time, there is a clear evidence of the high vulnerability (at least at the time of these publications) of HR plants and tanks that are located in areas characterized by high seismic and tsunami hazards in Italy (Martelli 2012, 2013c; Clemente and Martelli 2013).

As the first author of the paper denounced in 2012 (Martelli 2012), the warnings that he had already launched in 2011 and his suggestions remained fully ignored; similar subsequent warnings of ENEA, in particular at some important events devoted to the lessons of Tohoku earthquake, were also unsuccessful (Martelli 2013c). Anyway, this had occurred even earlier, although GLIS and ENEA had tried to bring the problem of seismic safety of the HR chemical plants and components to the attention of the institutions, media and public opinion (Martelli 2012, 2013c).

The above inattention ceased only because of the concerns on the high seismic risk of the Sicilian HR chemical plants that were stressed at the conference on "Lessons of the Tohoku earthquake", held at the ENEA headquarters in Rome on July 1st, 2011 and, especially, as a consequence of a parliamentary question, based on a proposal of GLIS, that was submitted by the president of the 8th Commission of the Chamber of Deputies on September 8, 2011, with the above-mentioned contents (Alessandri 2011; Martelli 2012, 2013c). Thanks to this question and to a meeting held in Milazzo on December 2, 2011, at last, the subject began to attract a considerable interest of media and public opinion (Martelli 2012, 2013c). Unfortunately, however, not that of the national and regional institutions, even after the serious concerns expressed and communicated to Major Risks Commission by the seismologists of University of Trieste and other well-known experts of the Russian Academy of Sciences in early January 2012, based on the results of their "earthquake prediction experiments", about the possible occurrence, in the intermediate term, of a violent earthquake in Southern Italy, in particular (according to the Russian experts) in an area including Southern Calabria and Eastern Sicily (however, in a large area, certainly not in a precise location – see Martelli 2012, 2013c; Martelli et al. 2013b, c). Therefore, on January 31, 2012, the aforesaid parliamentary question was transformed into a resolution (Alessandri 2012; Martelli 2012, 2013c; Martelli et al. 2013b, c). In addition, the issue of seismic protection of HR chemical plants was part of those examined in the "Survey on the State of Seismic Safety in Italy" held at the Italian Chamber of Deputies in 2012 (Benamati 2012; Martelli 2013c).

Despite further scientific events and information on the seismic safety of the HR chemical plants held, in particular, in Sicily (at Augusta and Messina, in February and March 2012, respectively – see Martelli 2013c) and the proliferation of newspaper articles and radio and TV reportages (especially after the beginning of the seismic events in Emilia in May 2012 and the disclosure of the fact that, at the beginning of March 2012, the seismologists of the University of Trieste had expressed their concerns for Northern Italy too – see Martelli 2012, 2013c), the Italian institutions have continued to remain idle. In the meantime, very little has been done in terms of prevention, at least to limit the severe consequences that a violent earthquake (whether or not followed by a tsunami) could have if it hits the Sicilian areas of Milazzo or Priolo; instead, many sterile and damaging controversies were made.

The aforesaid controversies, born following statements of the first author of this paper especially after the 2012 *Emilia* earthquake, misrepresented the positions he had expressed about the "earthquake prediction experiments" and about the concerns communicated by the aforementioned well known seismologists for Southern Italy too (Martelli 2013c). This unnecessarily and detrimentally exacerbated the climate, by also leading to panic situations and diverting the attention from the main goal: to urgently undertake a serious prevention policy, as regards both civil and industrial constructions, especially (but not only) in Southern Italy. The information available to date tells us that earthquakes can be predicted with high statistical significance, but with great spatial and temporal uncertainties and with the possibility of false alarms (Panza et al. 2011; Martelli et al. 2013a). The information should, therefore, be used in an appropriate manner, to take urgent preventive actions that will be essential if a strong earthquake will actually occur, but that will be useful in case of false alarm too, i.e. if the strong earthquake will not occur (such actions concern, for example, the verification and, if needed, putting in safe conditions of facilities characterized by a particular risk, the preparation of the civil protection system, information to the public opinion, etc.).

As a matter of fact, the ENEA experts limited themselves to apply their knowledge in the field of earthquake engineering in order to assess the risk concerning the HR chemical plants, which are characterized by a very high exposure and, as mentioned, a seismic vulnerability that is very often at least unknown. Obviously, to express this judgment, they could not neglect the concerns and the results of the cited "earthquake prediction experiments". In disclosing the results of such evaluations, covering Southern Italy, the only goal of ENEA was, as always, to encourage the institutions to establish the necessary measures, within their competences, and to give, as far as possible, the necessary information to the public opinion.

In the information and stimulus work that ENEA had carried out, very unheeded, for a long time (not for a few days, as can be easily checked), unfortunately, often happened that the statements of its experts were distorted, unfortunately not only by media: this, however, was a risk that had to be be taken, because silence would have been even worse. On the other hand, what ENEA and the authors of this paper actually said and are saying is certified, for example, by several TV and radio reportages transmitted since April 10, 2012 (i.e. well before the 2012 *Emilia* earthquake – see Clemente and Martelli 2013).

The fear that earthquake has aroused in the public opinion in Southern Italy (first when there was one in progress in the north of the country, then when, on October 26, 2012, the Pollino area was hit in Calabria, namely really in the south, by a $M = 5.0$ event, and more recently due to some events which are going on hitting both Southern and Central Italy) takes origin from the fact that, only now, having proof the problem, many are realizing how unsafe their homes, the schools where their children study and the places they go to can be, and that only now many are waking up to reality, becoming aware of the serious deficiencies that plague Italy in the field of the prevention of seismic risk. Creating panic shall be certainly avoided, but we must not also pass over the problem and we shall aim at transforming this fear into "claim for prevention": this is not an impossible goal, although, of course, the path is full of traps (Martelli 2013c).

The safety of HR chemical plants was discussed in detail at the conference on «Seismic safety of chemical high risk plants», jointly organized by ENEA and GLIS, which was held with great success in Rome on February 7, 2013 (Clemente and Martelli 2013; Martelli 2013c). This conference brought together representatives of all the institutions involved in the topics discussed. Despite the difficult political moment, it was considered appropriate that the event should take place before the Italian general elections of 2013, because of the importance of the issues

Fig. 14.31 *Left*: one of the two LNG tanks of Egegaz in Aliaga (Turkey), which were seismically isolated using 112 LRBs and 241 LDRBs. At the *centre*: view of the isolators during construction. *Right*: an installed LRB

Fig. 14.32 *Left*: one of the two seismically isolated LNG tanks in the Guandong Province (Southern China) during construction (each of them was isolated by means of 360 HDRBs). At the *centre*: some of the isolators after their installation. *Right*: a HDRB during acceptance tests

dealt with in it and also to stimulate the subsequent (namely, the present) government to address these issues with the necessary urgency.

At the 2013 Rome conference, it was agreed on the need and urgency to:

- adequately address the problem of seismic safety and in the face of a possible tsunami of Italian HR chemical plants and components, especially of the petrochemical installations of Priolo-Gargallo and Milazzo;
- proceed with urgency to the development of specific regulations for earthquake-resistant and anti-tsunami designs of such plants and components and, where necessary, for the retrofit of the existing ones;
- build, at least in areas of high seismic hazard, especially the LNG tanks, but also other types of HR chemical installations and components of new construction, by making an extensive use of AS systems, in particular (where possible) of SI;
- accurately assess, in the areas of significant seismic and/or tsunami hazard, the vulnerability of these HR plants and components that are already existing;
- retrofit such plants and components using, to protect them from earthquakes (where useful and possible), SI or ED systems (which, unfortunately, are very little applied in Italy in this field, contrary to other countries, as shown by Clemente and Martelli 2013; Martelli et al. 2013a, and Figs. 14.31 and 14.32);

- proceed to the identification of the situations of highest seismic or tsunami risk, for the existing HR chemical installations, so as to make the civil defence system capable of adequately addressing possible accidents caused by the collapse or damage of the plants or components present in such installations;
- start in Italy too, at last, a correct program of participated information of the population and rise the perception of seismic risk in it.

We are confident that the loud and at last unanimous message of the Italian scientific and technological community that came out of the Rome conference, has really allowed to overcome the harmful and useless controversies of 2012 (Clemente and Martelli 2013; Martelli 2013c) and is now able to stimulate the present and future Italian governments to tackle the aforesaid serious problems with the indispensable urgency, with the contribution of all the competences necessary to that purpose.

14.4 Costs of Seismic Isolation

As mentioned, SI is a technology of great interest not only for public or strategic buildings, but also for the residential ones. Indeed, in addition to confer a level of seismic safety much higher than that obtainable with conventional foundations and to allow to avoid the costs (of repair, demolition, reconstruction, relocation, etc.) that, after a significant earthquake, should be faced for the structures with conventional foundations, the use of SI entails, for new buildings, a very limited additional construction cost in Italy. In countries as Italy, were the seismic code allows to somewhat decrease the seismic loads acting on the structures, if they are protected by SI (Sect. 14.5), the aforesaid cost decreases with increasing seismic hazard of the area where the building is located, number of its floors and extent of its structural asymmetries.

Typically (as occurred, for example, for the 5 isolated building of the new San Samuele residential district in Cerignola, Foggia, which was certified as safe by the first author of this paper – see Martelli and Forni 2010), this additional construction cost vanishes in Italy for residential buildings of 5 floors, even if they are very regular, which are located in areas of medium seismicity (i.e. former Italian seismic zone 2). As a second example, for the new school of Marzabotto (opened to activity in 2010, again with safety certification of the first authors of this paper, see Fig. 14.29), even though it arose in an area considered to be of low seismicity (former Italian seismic zone 3) and despite its limited height and non-use of the underground floor where the isolators have been installed (which is considered as a "technical space"), the additional cost due to SI was of only 96,000 Euros, out of a total construction building cost of about five million Euros (Basu et al. 2014).

Moreover, for interventions on the existing buildings, the use of SI could even cause a saving, as demonstrated, for example, by the case of a dwelling building of Fabriano, Ancona, damaged by the 1997–1998 Marche and Umbria earthquake, which was also certified as safe by the first author of this paper in 2006 (Martelli and Forni 2011a): in fact, it is not necessary to "undress" the structure, so as to be able to stiffen beams and nodes, or to insert shear walls (which is often quite complicated).

Finally, if the intervention is carried out as a preventive measure (that is, before the building is damaged by an earthquake), it is often possible to keep the building in use (except, of course, for the storey at which the isolators have to be inserted and, to this end, pillars and/or load-bearing walls have to be cut and, if necessary, strengthened): this advantage is of particular importance for the retrofit of hospitals (Martelli 2014).

14.5 Remarks on the Correct Use of Anti-seismic Systems and Devices

The large effects of earthquake lessons and seismic design code features on the extent of the use of the AS systems in the various countries shall be stressed (Martelli 2010; Martelli and Forni 2010, 2011a, b). In the codes of countries like Japan, the USA and Chile SI is considered as a safety measure additional to the conventional design; consequently, the use of SI obviously always introduces additional construction costs. Nevertheless, this technique is being widely adopted by the Japanese, due to their high level of perception of the seismic risk and because violent earthquakes are very frequent in their country (Martelli et al. 2011, 2012a).

The aforesaid level of perception is much lower elsewhere: this is the reason why, to limit or even sometimes balance the additional construction costs entailed by the use of SI (and, thus, to promote a significant application of this technique), the seismic codes of other countries (Italy, P.R. China, Armenia, etc.) allow for some lowering of the seismic forces acting on the superstructure when SI is used (Clemente and Buffarini 2010). Thus, in these countries, a real safety will be ensured to the isolated structures if and only if great care is paid to:

- the selection of the SI devices (taking into account the amplitude of vertical motion and low frequency vibrations), their qualification, production quality, installation, protection, maintenance and verification that their design features remain unchanged during the entire structure life;
- some further construction details (structural gaps, their protections, interface elements – e.g. gas and other safety-related pipes, cables, stairs, lifts –, etc.).

Otherwise, the isolators, instead of largely enhancing the seismic protection, will make the structure less earthquake resistant with respect to a conventionally founded one and, thus, will expose both human life and the entire SI technology to great risks.

Last but not least, a common key requirement for the optimal performance of all kinds of AS systems and devices (but especially of the isolators) is the realistic and reliable definition of seismic input (Martelli 2010; Martelli and Forni 2010;

Martelli et al. 2011; Panza et al. 2011), which cannot rely only upon the oversimplified routine probabilistic methods, mainly when dealing with displacements definition (on which the design of isolated structures is based): thus, the ongoing rapid extension of the use of the AS systems and devices requires a considerable improvement of the PSHA approach, which is now in use in several countries (including Italy). Taking into account the mentioned unreliable results shown by PSHA for several violent earthquakes in the last decade (Martelli and Forni 2011b; Martelli and Forni 2011; Panza et al. 2011), such a change is very urgent now and can be achieved by complementing PSHA through the development and application of deterministic models, e.g. NDSHA. This particularly applies to the P.R. China, Italy, New Zealand and Japan, to ensure a safe reconstruction after the earthquakes of *Wenchuan* (2008), *Abruzzo* (2009), *Canterbury* and *Christchurch* (2010 and 2011) and *Tohoku* (2011), because SI is widely used in the concerned areas.

All said items were discussed in Italy by the 8th Commission on Environment, Territory and Public Works of the Italian Chamber of Deputies in 2010 and 2011, based on two proposals drafted by the first author of this paper, with the collaboration of other experts (Martelli et al. 2011, 2012a). Following these discussion and audits of various experts (including those of the first, second, third and fifth authors of this paper), one of such resolutions (that most detailed), was approved, with minor changes, by the aforesaid commission and by the Italian Government on June 8, 2011 (Martelli et al. 2011). The final document (Benamati et al. 2011), after some introductory remarks, contains recommendations for modifications of some parts of the existing Italian and European seismic codes that concern the AS systems and devices and the structures provided with them (experimental qualification of the new device types to be carried out on prototypes by subjecting them to at least bi-directional excitations, control of all construction phases to be performed by an expert in the field, recommendations and requirements to be provided by him in his final certificate concerning the structural safety of the structure, etc.), as well as the need for using the NDSHA together with the PSHA for the definition of seismic input (in particular for that of the design displacement, which is a key parameter for the design of isolated structures). These recommendations have been reported in detail by Martelli et al. (2011). Based on them and on the results of the "Survey on the State of Seismic Safety in Italy" (Benamati 2012), promoted with the collaboration of the first author of this paper and held at the 8th Commission on Environment, Territory and Public Works of the Italian Chamber of Deputies in 2012 (with, as mentioned, audits of the authors of this paper and other GLIS and *ASSISi* members), a new law concerning modifications of the Italian seismic code was proposed during the last legislation and was recently proposed again (Martelli et al. 2013c).

14.6 Conclusive Remarks

SI and the other AS systems have already been widely used in over 30 countries and their application is increasing more and more, for both new constructions and retrofits, for all kinds of structures and their materials. The features of the design rules used, as well as earthquake lessons, have plaid a key role for the success of the aforesaid technologies. Japan is the leading country for the number of applications of both SI and ED systems. For the overall number of applications, it is now followed by the P.R. China, the Russian Federation, the USA and Italy. Italy (where the contributions provided by ENEA and GLIS have been of fundamental importance) is the leading country at European level, with regard to both SI and ED of buildings, bridges and viaducts. In addition, it is a worldwide leading country for the use of AS systems and devices to protect cultural heritage (Martelli 2009; Martelli and Forni 2010). Its applications are being significantly extended after the 2009 *Abruzzo* earthquake. Italian passive AS devices have been installed in several other countries too.

SI is now worldwide recognized as particularly beneficial for the protection of strategic constructions like civil defence centres and hospitals (by ensuring their full integrity and operability after the earthquake) and for schools and other highly populated public buildings (also because the large values of the isolated superstructure vibration periods minimize panic). Some codes (e.g. those adopted in Italy, P.R. China, Armenia, etc.) allow for taking advantage of the reduction of seismic forces operated by SI: their use makes SI attractive for the dwelling buildings too, because the additional construction costs due to the use of this technique (if any) are frequently rather limited.

In order to really strongly enhance the seismic protection of our communities, an extensive but correct application of the AS systems is necessary (Martelli and Forni 2011b; Martelli et al. 2011, 2012a, 2013b, c). With regard to Italy, a wide-ranging use of such systems (where possible of SI) will certainly greatly contribute to enhance the seismic safety of structures, since there over 70 % of those existing are not able to withstand the earthquakes which may hit them and since this number includes many schools, other strategic or public buildings and important HR chemical plants. This is particularly necessary and urgent for schools, which (together with hospitals) are the buildings that should guarantee the highest safety level, and for the HR chemical plants, which are characterized by a very high exposure. To contribute to promote risk prevention policies, in particular for the seismic one and for schools, the *National Coordination of Voluntary Associations for Seismic and Environmental Prevention* (*Coordinamento Nazionale Associazioni di Volontariato per la Prevenzione Sismica e Ambientale* – Co. Prev.) was founded at the Bologna meeting of June 13, 2013, which was organized by the first author of this paper (who is member of the Co.Prev. Technical Committee – see Martelli et al. 2013c, and Martelli 2014).

To achieve the objective of widely extend the correct use of the AS systems, regulatory and legislative measures, such as those that were proposed in Italy by the

8th Commission on the Environment, Territory and Public Works of the Chamber of Deputies for the isolated structures in general and for protecting the high risk chemical plants in particular, may considerably contribute, especially in the countries (like Italy) where the perception of seismic risk is not yet sufficient. In fact, the use of AS systems (in particular that of SI) will hopefully strongly increase not only for the protection of civil structures, but also for that of cultural heritage and high risk plants (Martelli and Forni 2010; Martelli et al. 2013b, c). For the application of the AS systems to monumental buildings, the problem is the compatibility with the conservation requirements (Martelli 2009). For that to the high risk plants, SI has a great potential not only for nuclear structures, but also for chemical components like LNG tanks, for which, to date, only a limited number of applications exists or has been planned (in South Korea, P.R. China, Turkey, France, Greece, Mexico, Chile and Peru): in fact, detailed studies have shown that SI is indispensable for such components in highly seismic areas (Dolce et al. 2006; Martelli et al. 2011, 2013b, c; JSSI 2013).

Thus, as recommended by a parliamentary question of the President of the 8th Commission of the Italian Chamber of Deputies in September 2011 (Alessandri 2011), which was fully reported by Martelli and Forni (2011a), and by a resolution which was proposed by the same member of the Italian Parliament at the end of January 2012 (Alessandri 2012), we hope that the use of SI will soon increase for high risk chemical installations in Italy. This applies, especially, to Sicily, in sites like those of Milazzo (not far from the area destroyed by the 1908 *Messina & Reggio Calabria* earthquake and tsunami, besides being in front of the submerged Marsili Volcano) and Priolo (located in the Catania Plane, which was razed by the 1,693 event): in both sites hundreds of quite seismically vulnerable cylindrical and spherical tanks already exist (only 3 retrofitted using SI, to date, see Fig. 14.16) and, in the latter, the construction of a large re-gasification terminal with LNG tanks had been planned.

Generally speaking, however, it shall be kept in mind that the use of SI in countries as Italy, where the designers are allowed by the code to decrease the seismic forces acting on the superstructure when adopting this technology, requires:

- first of all, a reliable definition of the seismic input, namely by means of intensive use of NDSHA, as well, in addition to PSHA;
- then a very careful selection, design, manufacturing, installation, protection and maintenance of the SI devices during the entire life of the isolated structure;
- finally, particular attention to be also paid to some further construction aspects (in particular, to the design, realization, protection and maintenance of the structural gaps and the safety-related pipelines – e.g. the gas ones –, again during the entire life of the isolated structure).

Otherwise, the seismic safety of these structures would be lower than that of the conventionally founded ones.

In any case, the technologies to make buildings safe during earthquakes, in Italy and elsewhere, exist and it is foolish not to use them extensively. Certainly the goal is (at least theoretically) easier for new construction, while the difficulties to be overcome in order to make the existing buildings safe are frequently huge, from an economic standpoint. However, this does not justify the continuing inertia of the institutions, in Italy and in other countries.

In Italy it will take several decades to solve the problem of the high seismic risk of the existing buildings, but to do this we must start immediately, acting by priority and using the best available technologies as described above. If we want public opinion to acquire a correct perception of risks (in particular of the seismic one), the institutions shall set an example, by promoting, at last, proper prevention policies (Martelli et al. 2013b).

Open Access This chapter is distributed under the terms of the Creative Commons Attribution Noncommercial License, which permits any noncommercial use, distribution, and reproduction in any medium, provided the original author(s) and source are credited.

References

Alessandri A (2011) Parliamentary question requiring written answers no. 4–1360 (concerning the seismic protection of high risk chemical plants). In: Atti Parlamentari – Camera dei Deputati (Parliamentary Acts – Italian Chamber of Deputies), Rome, Italy, pp 24010–24013 (in Italian)

Alessandri A (2012) Resolution in the VIII Commission of the Chamber of Deputies no. 7–00764 (concerning the seismic protection of high risk chemical plants). In: Atti Parlamentari – Camera dei Deputati (Parliamentary Acts – Italian Chamber of Deputies), Rome, Italy, pp 27324–27327 (in Italian)

Basu B, Bursi OS, Casciati F, Casciati S, Del Grosso A, Domaneschi M, Faravelli L, Holnicki J, Irschik H, Krommer M, Lepidi M, Martelli A, Ozturk B, Pozo F, Pujol G, Rakicevic Z, Rodellar J (2014) An EACS joint perspective. Recent studies in civil structural control across Europe. Struct Control Health Monit. doi:10.1002/stc

Benamati G (proponent and rapporteur) (2012) Indagine conoscitiva sullo stato della sicurezza sismica in Italia – Programma. In: Resoconti delle Giunte e Commissioni – Resoconto dell'VIII Commissione Permanente (Ambiente, Territorio e Lavori Pubblici), Italian Chamber of Deputies, Rome, Italy, 12 Apr, pp 64–67 (in Italian)

Benamati G, Ginoble T, Alessandri A (2011) 7–00414 Benamati: In materia di isolamento sismico delle costruzioni civili e industriali (7–00414 Benamati: on seismic isolation of civil and industrial structures). Resolution approved by the VIII Commission no. 8/00124, 16th Legislature, Announcement Meeting of June 8, 2011. In: Bollettino della Camera dei Deputati (Bulletin of the Italian Chamber of Deputies), Rome, Italy 491(5):388–393 (in Italian)

Camera dei Deputati (ed) (2012a) Indagine conoscitiva sullo stato della sicurezza sismica in Italia – Audizione di rappresentanti dell'ENEA, del professor Giuliano Panza e del professor Antonello Salvatori. In: Resoconti Stenografici delle Indagini Conoscitive – Commissione VIII, Seduta di Mercoledì 30 Maggio 2012, Italian Chamber of Deputies, Rome, Italy, pp 3–22 (in Italian)

Camera dei Deputati (ed) (2012b) Indagine conoscitiva sullo stato della sicurezza sismica in Italia – Audizione di rappresentanti dell'ENEA. In: Resoconti Stenografici delle Indagini Conoscitive – Commissione VIII, Seduta di Giovedì 13 Settembre 2012, Italian Chamber of Deputies, Rome, Italy, pp 2–27 (in Italian)

Clemente P and Buffarini G (2010) Base isolation: design and optimization criteria. In: Seismic Isolation And Protection Systems (SIAPS), Mathematical Sciences Publishers (MSP), Berkeley, CA, USA 1(1):17–40. doi:10.2140/siaps.2010.1.17

Clemente P, Martelli A (eds) (2013) Atti della Giornata di Studio Sicurezza Sismica degli Impianti Chimici a Rischio di Incidente Rilevante, ENEA Report, ISBN 978-88-8286-285-5, Rome, Italy (in Italian)

Clemente P, De Stefano A, Renna S (2011) Application of seismic isolation in the retrofit of historical buildings. In: 12th World conference on seismic isolation, energy dissipation and active vibration control of structures (12WCSI), proceedings on CD, ASSISi, Sochi/Russian Federation/Moskow, 20–23 Sept 2011

Dolce M, Martelli A, Panza G (2006) Moderni Metodi di Protezione dagli Effetti dei Terremoti. In: Martelli A (ed) Special edition for the Italian Civil Defence Department, Milan: 21mo Secolo (in Italian)

Eisenberg J et al. (eds) (2011) 12th World conference on seismic isolation, energy dissipation and active vibration control of structures – conference proceedings and abstract book (CD), ASSISi, Sochi/Russian Federation/Moskow, 20–23 Sept 2011

JSSI (ed) (2013) Seismic isolation, energy dissipation and active vibration control of structures, proceedings on electronic key of the ASSISi 13th World conference (13WCSI) & JSSI 20th Anniversary International Symposium, JSSI, Sendai, Japan/Tokyo, 24–26 Sept 2013

Kani N (2008) Current state of seismic-isolation design. In: Proceedings of the 14th World conference on earthquake engineering, Beijing, 12–17 Oct 2008

Martelli A (2009) Development and application of innovative anti-seismic systems for the seismic protection of cultural heritage. Key-note lecture. In: Mazzolani FM (ed) Protection of historical buildings; Proceeding of the PROHITECH 2009 international conference, CRC Press – Taylor and Francis Group 1, Rome, Italy/Leiden, pp 43–52, 22–24 June 2009

Martelli A (2010) On the need for a reliable seismic input assessment for optimized design and retrofit of seismically isolated civil and industrial structures, equipment and cultural heritage. Pure Appl Geophys. doi:10.1007/s00024-010-0120-2

Martelli A (2012) Impianti chimici RIR italiani: le incognite terremoto e maremoto. In: Il Giornale dell'Ingegnere, Focus. Milan, Italy 7, pp 8–11 (in Italian)

Martelli A (2013a) Prosegue l'impegno del GLIS per la prevenzione sismica – Anzitutto si proteggano finalmente le scuole italiane. In: 21mo Secolo – Scienza e Tecnologia. Milan, Italy 2/2013, pp 17–23 (in Italian)

Martelli A (2013b) L'ASSISi 13th World Conference di Sendai (Giappone) e le più recenti e prossime manifestazioni del GLIS – Continua a crescere l'applicazione dei sistemi antisismici. In: 21mo Secolo – Scienza e Tecnologia, Milan, Italy 3/2013, pp 19–26

Martelli A (2013c) Impianti chimici a rischio terremoto: proteggerli si può, basta volerlo. In: 2087, Fabbriche a Rischio e Terremoti, Rome, Italy 3, pp 4–10 (in Italian)

Martelli A (2014) Scuole ed ospedali: edifici che "dovrebbero" restare totalmente integri e pienamente operativi dopo un terremoto. Come garantirlo, In: La Proprietà Edilizia 2 (in press, in Italian).

Martelli A, Forni M (2010) Seismic isolation and other anti-seismic systems: recent applications in Italy and worldwide. In: Seismic Isolation And Protection Systems (SIAPS), Mathematical Sciences Publishers (MSP), Berkeley, CA, USA 1(1):75–123. doi:10.2140/siaps.2010.1.75

Martelli A, Forni M (2011a) Seismic retrofit of existing buildings by means of seismic isolation: some remarks on the Italian experience and the new projects. Invited paper. In: Minisymposium on Innovative vs Conventional Retrofitting of Existing Buildings, 3rd International Conference on Computational Methods in Structural Dynamics and Earthquake Engineering (COMPDYN 2011), Corfu, 26–28 May 2011

Martelli A, Forni M (2011b) Recent worldwide application of seismic isolation and energy dissipation and conditions for their correct use. In: Structural engineering world congress (SEW5), proceedings on CD, Abstract Volume, p 115

Martelli A, Sannino U, Parducci A, Braga F (eds) (2008) Moderni Sistemi e Tecnologie Antisismici. Una Guida per il Progettista. In: Irsuti R (ed) Milan, 21mo Secolo

Martelli A, Clemente P, Forni M, Panza G, Salvatori A (2011) Recent development and application of seismic isolation and energy dissipation systems, in particular in Italy, conditions for their correct use and recommendations for code improvement. In: 12th World conference on seismic isolation, energy dissipation and active vibration control of structures, proceedings on CD, Russian Federation, Sochi, Abstract Volume, pp 9–11, 20–23 Sept 2011

Martelli A, Clemente P, Forni M (2012a) Recent worldwide application of seismic isolation and energy dissipation to steel and other materials structures and conditions for their correct use. In: STESSA 2012 – behaviour of steel structures in seismic areas, Santiago, Chile, pp 3–14, 9–11 Jan 2012

Martelli A, Forni M, Clemente P (2012b) Development and application of anti-seismic systems in Italy and worldwide and conditions for their correct use. In: EACS 2012 – 5th European conference on structural control, Proceeding on CD, Genova, Italy,Abstract Volume, p 160, 18–20 June

Martelli A, Forni M, Clemente P (2012c) Recent worldwide application of seismic isolation and energy dissipation and conditions for their correct use. In: Proceedings on electronic key of the 15th World conference on earthquake engineering (15WCEE), Lisbon, Conference programme, p 52, 24–28 Sept 2012

Martelli A, Forni M, Panza G (2013a) Features, recent application and conditions for the correct use of seismic isolation systems. In: Syngellakis S (ed) Seismic control systems: design and performance assessment. Wit Press, Southampton, pp 1–16

Martelli A, Clemente P, De Stefano A, Forni M, Salvatori A (2013b) Development and application of seismic isolation, energy dissipation and other vibration control techniques in Italy for the protection of civil structures, cultural heritage and industrial plants. Key-note lecture. In: Seismic isolation, energy dissipation and active vibration control of structures; Proceeding on electronic key of the ASSISi 13th World conference (13WCSI) & JSSI 20th Anniversary international symposium, JSSI, Sendai, Japan/Tokyo, 24–26 Sept 2013

Martelli A, Clemente P, De Stefano A (2013c) On the benefits of a wide use of anti-seismic systems for the seismic protection of schools and high risk chemical plants. Invited lecture. In: Seismic isolation, energy dissipation and active vibration control of structures; Proceeding on electronic key of the ASSISi 13th World conference (13WCSI) & JSSI 20th Anniversary international symposium, JSSI, Sendai, Japan/Tokyo, 24–26 Sept 2013

Martelli A, De Stefano A, Vizzaccaro A (2013d) Scuole sicure: diritto e dovere della società civile. In: Villaggio Globale, http://vglobale.it, Naples, Italy 61 (in Italian)

Mazzolani FM, Herrera R (eds) (2012) STESSA 2012 – behaviour of steel structures in seismic areas; proceeding of the STESSA 2012 conference, CRC Press – Taylor and Francis Group, Santiago, Chile/Leiden, 9–11 Jan 2012

Panza G, Irikura K, Kouteva M, Peresan A, Wang Z, Saragoni R (eds) (2011) Advanced seismic hazard assessment. In: Pageoph Topical Volume, ISBN 978-3-0348-0039-6 & ISBN: 978-3-0348-0091-4

Salvatori A (2013) Il sistema CAM nel recupero dei centri storici. Esempi applicativi. In: Come ricostruire in sicurezza sismica i centri storici, In: GLIS/ANCE/ENEA conference, L'Aquila, Italy, pdf versions of the ppt presentations, www.assisi-antiseismicsystems.org (in Italian), 13 Dec 2013

Sannino U, Sandi H, Martelli A, Vlad I (eds) (2009) Modern systems for mitigation of seismic action – proceedings of the symposium held at Bucharest, Romania, on 31 Oct 2008, AGIR Publishing House, Bucharest, ISBN 978-973-720-223-9

Zhou FL, Tan P, Heisa WLH, Xian XL (2013) Lu Shan earthquake M7.0 on 2013.4.20 and recent development on seismic isolation, energy dissipation & structural control in China. Key-note lecture. In: Seismic isolation, energy dissipation and active vibration control of structures; Proceeding on electronic key of the ASSISi 13th world conference (13WCSI) & JSSI 20th anniversary international symposium, JSSI, Sendai, Japan/Tokyo, 24–26 Sept 2013

Chapter 15
Conservation Principles and Performance Based Strengthening of Heritage Buildings in Post-event Reconstruction

Dina D'Ayala

Abstract Recommendations for repairing and strengthening historic buildings after an earthquake and before the next in modern times go back to the contribution to the ICOMOS General Assembly of 1987 by Sir Bernard Fielden "Between two Earthquakes" (Fielden 1987). In that circumstance two important points were made: the first is that failure and damage should be used to understand performance and behaviour, so as to avoid measures that do not work. The second is that the engineer work should be integrated into the architecture historical methodology. Almost 30 years later this contribution investigate to which extent these two recommendations have been fulfilled, whether there is a common understanding between the conservation and the seismic engineering community and whether lessons from past failures are informing new strengthening strategies.

15.1 Introduction

The global seismic response of historic masonry buildings is highly influenced by the integrity of the connections among vertical and horizontal structural elements, to ensure the so-called box behaviour. Such behaviour, providing the transfer of inertial and dynamic actions from elements working in flexure out-of-plane to elements working in in-plane shear, leads to a global response best suited to the strength capacity of the constitutive materials, and hence enhanced performance and lower damage levels. While, many properly designed buildings of the past demonstrated such behavior when exposed to seismic action and successfully survived ground shaking (D'Ayala 2011; Tavares et al. 2014), too often, due to

D. D'Ayala (✉)
Department of Civil Environmental and Geomatic Engineering, University College London, London, UK
e-mail: d.dayala@ucl.ac.uk

inherent defects, alterations or decay, such resilient features are not present or are not effective and lack of connections among orthogonal walls and walls and floors structures are clearly apparent. In churches with a Latin cross plan shape, delivering the box action, might result particularly difficult, due to the change in stiffness between the nave and the central crossing area and often the presence of trusting arches and domes over the central crossing pillars. The engineering community has historically remedied to such problems by developing strengthening devices, to be applied either as repair to damaged buildings or, often enough, as a retrofit and upgrading programme to improve the seismic performance of the existing building stock before the next damaging event. Such attitude towards strengthening and retrofitting is not confined to modern earthquake engineering, as retrofit programmes were promulgated around the turn of the twentieth century for instance in Turkey and Italy after major earthquakes in Istanbul (D'Ayala and Yeomans 2004) and Messina (Barucci 1990). However from recurring observation of damage in earthquakes worldwide in the past three decades, and more recently from the Pisco, Peru' 2007, L'Aquila, Italy, 2009, Maule, Chile 2010, Christchurch, New Zealand 2011, and even from the very recent 2013 Philippines event, the lack of a systematic critical approach to strengthening of historic buildings to prevent damage and casualties while preserving architectural value, clearly stands out. In general the use of materials and elements with strength and stiffness greater than the original materials is still prevalent and recommended in several guidelines. Design provisions for strengthening usually rely on capacity design approach, assuming that the retrofitted building should withstand an action proportional or equal the one decreed for new buildings of the same structural typology.

Alternatives to increase in strength and stiffness are the concepts of base isolation and introduction of damping devices aimed at modifying the response of the structure, aiming at shifting its fundamental frequencies from the frequency content of the ground shaking and increasing its damping capacity. Examples of these solutions exist in history. In modern times they have been unfrequently used from the 1980s onward, in very high profile cases, but guidelines and recommendations for application to more ordinary cases do not currently exists.

After introducing the context of structural conservation and its principles, the paper will review typical damage observed in the events listed above outlining the shortcoming of conventional strengthening approaches, strengthening interventions currently advocated by guidelines and implemented in post-earthquake retrofit programmes and proposals for alternative strategies.

15.2 Structural Conservation Principles

Seismic retrofitting intervention in heritage structures, while satisfying seismic code performance requirements, should also comply with recognized conservation principles, enshrined in international documents such as the Venice Charter of 1964 (Venice Charter 1964) and, more specifically, in the ICOMOS/ISCARSAH

Recommendations for the Analysis and Restoration of Structures of Architectural Heritage, (ICOMOS/ISCARSAH 2003), and the Annex on Heritage Structures of ISO/FDIS 13822, (ISO/TC96/SC2 2010). These criteria however do not have the same legal enforcement framework of a seismic standard and hence should be seen as guidelines useful to strike a balance between the improvement of the seismic behavior and the retention of the existing fabric and architectural and cultural value. The ISCARSAH Principles besides reconfirming the more generic conservation principles of conserve as found, minimal intervention, compatibility, and reversibility of repair, introduce concepts specific to the structural and seismic performance of buildings and have direct consequences on seismic strengthening. These are the concepts of:

- *Structural authenticity*, which should be preserved as much as the architectural authenticity, ensuring that the original mechanical and resisting principles governing the structural response are not altered and original structural elements are not made redundant.
- *Structural reliability*, relates to the necessity of striking the correct balance between the public safety requirements and the preservation requirements. Conventionally it is accepted that buildings of high cultural significance may be intervened upon so as to ensure damage limitation as a performance target, in events where for ordinary buildings, life-safety is the performance requirement. However in many occasions the attainment of such target may cause a significant loss of artistic or cultural value, maybe greater than the ones bestowed by the earthquake damage, in probabilistic terms. Hence the extent of seismic upgrading should be verified by a cost-benefit analysis including the intangible value losses. According to ISO/FDIS 13822, the solution finally adopted should consist of "an intervention that balances the safety requirements with the protection of character-defining elements, ensuring the least harm to heritage values". This is also defined as "optimal or minimal intervention".
- *Strengthening compatibility, durability, reversibility, monitorability.* These criteria influence more directly the technical choices and details of the interventions and impact upon: the suitability of "new" materials and structural elements in terms of their physical and mechanical performance when compared with original materials and structural elements; their performance in time; the possibility of removing partially or totally the intervention if monitoring proves that it is not suitable. Compatibility should be such that the new materials and elements not only do no harm to the original ones, but also they act as sacrificial elements in presence of external actions, i.e. they should act as fuses of the structural system. At the same time the new elements should be durable as to extend the expected life of the original structures as intended, but should also be non-intrusive, non-obtrusive and reversible. The concept of reversibility, or more realistically removability, is a very interesting one, as it acknowledges limitation in current practice and the possibility of finding better solutions in future. Removability is strictly correlated with the idea of monitorability, i.e. the possibility of observing and recording the performance of both the original

structure and the intervention, to ascertain its effectiveness or alert of any possible undesirable side-effect.

These criteria, although having being actively debated and applied in the international structural conservation community for at least the last 25 years, to my knowledge, they were eventually given recognized status, in 2003 with the approval of the ISCARSAH principle by the ICOMOS general assembly in Zimbabwe. It is hence worthy, a decade later, to verify on one hand how they have been incorporated into national and European seismic codes and on the other whether they had any impact on current seismic strengthening practice. A useful point to start this investigation is to review the performance in recent past earthquake of buildings strengthened with conventional force enhancing systems.

15.3 Damage of Heritage Buildings Strengthened with Conventional Capacity Enhancing Systems

In the last two decades increasing attention has been paid worldwide to the performance of historic and heritage buildings during major seismic events and specific surveys included in reconnaissance missions and reports. It is recognized that such buildings represent on one hand valuable cultural and economic assets to their country and to humanity at large, on the other they are in some cases responsible for non-negligible death tolls and casualties, hence appropriate mitigation measures need to be considered (see Blue Shield statements, after natural disaster, such as http://www.usicomos.org/international-icomos-news/blue-shield-statement-haiti-earthquake).

Well known examples of the lethality of heritage buildings are the collapse of the vaults of San Francis of Assisi basilica in the 1997 Umbria Marche earthquake (Spence and D'Ayala 1999), the collapse of several timber and mud vaulted roofs caused by the 2007 Pisco earthquake in Peru'(Cancino 2010), collapses of several adobe churches in the Colchagua Valley during the 2010, Maule Chile event in (D'Ayala and Benzoni 2012), partial collapses of several churches in the 2009 L'Aquila, Italy and the dramatic collapse of the Bell tower and spire of Christchurch Cathedral, New Zealand, in 2011. (Dizhur et al. 2011.) Following a two year long legal battle, what remains of the cathedral is now listed for demolition. A similar approach to damaged heritage was witnessed in Peru' following the 2007, Pisco earthquake and in Chile following the 2010, Maule earthquake. Indeed in Chile a generalized call for demolition of the architectural heritage damaged in the earthquake seemed to be the immediate reaction common to the people living in the small traditional communities as much as to the Governmental Authority of the Santiago Metropolitan Area. This approach is in contrast with the ICOMOS charters (Venice 1964; Cracow 2000) and with the attitude exhibited, for instance, by the communities of Bam (Fallahi 2008; Ghafory-Ashtiany and Hosseini 2008) or L'Aquila (D'Ayala and Paganoni 2011; Rossetto et al. 2014), which have seen their

Fig. 15.1 (a) Church in Lalol, Colchagua, Chile. Collapse of the lateral adobe wall, strengthened by shotcrete. (b) Church in Curepto, Maule, Chile. Collapse of the lateral adobe wall, strengthened by shotcrete

historic centers evacuated while waiting for funds and strategies to repair and rebuild. Montez and Giesen (2010), observe that the lack of provisions in Chile for the retrofit of historic buildings creates two options: to leave the building untouched or to adapt the structure to the present code, introducing reinforced concrete or steel elements. In the visited sites in the Valle de Colchagua, where historic structures experienced damages during previous earthquakes, recurring typologies of repairs and strengthening were observed. These consisted in prevalence of shotcreting of longitudinal adobe walls, although this was not always implemented in conjunction with wire mesh and through thickness ties. The shotcreting often accelerated deterioration of the original adobe wall. In general shotcreting has not been sufficient to prevent cracking and partial or total collapse of the adobe walls as evidenced by the collapses in the church in Lalol and in the church in Curepto (Fig. 15.1a, b). Current research on geo-synthetic mesh is aimed at providing a more effective alternative than wire-mesh for confinement of adobe walls (Torrealva et al. 2008), however interventions using geo-synthetic mesh on heritage buildings have yet to be reported in literature.

The general lateral stability of churches is a main issue, due to substantial difference in lateral flexibility of internal timber colonnades and external longitudinal adobe walls. This behavior is also common to churches of similar typology in Perù that were affected by the 2007, Pisco earthquake. Blondet et al. (2008) summarized the following recurring damage observed in single naves churches:

- Horizontal cracks on the lateral walls at about 1/3 of their total height. These cracks can even break through the earthen pilasters, causing the walls to collapse.
- Diagonal cracks on some of the lateral walls.
- Detachment of the choir and the altar's wall (parallel to the façade) from the church's lateral walls and cylindrical vault ceiling.
- Appearance of vertical cracks and fissures on the church towers and detachment of the towers from the rest of the church.
- Humidity related damage.

Fig. 15.2 (a) Ica cathedral, (Peru') collapse of the timber barrel vaults. (b) Collapse of the brickwork façade of the Church San Francisco of Curico', revealing the timber structure supporting the roof. Maule, Chile

The first two points highlight the out-of-plane rocking and in plane shear, respectively, of the lateral walls. All other observations describe failures of connections among macro-elements and resulting partial or total collapse. In churches with lateral aisles created by pillar-and-arch timber frames, the author observed failures due to excessive displacement of the internal pillars and collapse of the supported vaulted roof (Fig. 15.2a).

The lateral stability could be enhanced by bracing roof structures and by providing better transverse connections between the columns and walls. On the visited sites it was noted that many of the columns did not have foundations or plinths, but were simply sitting on the ground. Possible improvements in behavior could be achieved by the addition of a foundation system and the connection of the longitudinal and transverse roof structure to both the columns and the adobe walls. Use of timber wall-plates anchored to the walls by means of timber pegs should help redistribute the load of the roof structures, avoiding concentration of stresses and hence unfavorable localization of vertical cracks. Loss of the façade by overturning was not usually an issue, neither in Peru' or Chile, except for one surveyed case in Curico' (Fig. 15.2b). This show of resilience can be attributed to the relatively low horizontal and vertical slenderness ratio of the main facade the presence of two flanking bell towers, and in general the absence of very steep gables.

An extensive review of damages to churches following the earthquake in L'Aquila was conducted by Lagomarsino (2012) with the aim of correlating some constructive and strengthening features with corresponding collapse mechanisms.

Fig. 15.3 (a) Cathedral of L'Aquila, Italy; (b) The basilica of Collemaggio in L'Aquila, Italy

This study highlights the generally positive performance of wooden ties and ring beams found in heritage buildings which survived or where repaired in the aftermath of the 1709 devastating earthquake, and identifies the façade overturning and the gable overturning among the most common observed mechanisms, triggered by a general lack of connections of these macroelements to the longitudinal walls and the roof structure, but rarely resulting per-se in collapse. Indeed many detached facades were visible in the earthquake aftermath. The most recurring observation made in L'Aquila by several researchers (D'Ayala and Paganoni 2014; Augenti and Parisi 2010), refers to the pervasive substitution of historic timber roof trusses with twentieth century concrete trusses and slabs. Many of the observed collapses are directly connected with this change in stiffness and mass of the roof and are usually affecting the area of the transept and main crossing of the church. The most notorious examples are the Collemaggio basilica and the Cathedral of St. Massimo and Giorgio (Figure 15.3a and 15.3b). In both cases ring beams had been added at the top of the walls and the arches over the central crossing. However several other churches in L'Aquila had similar interventions, such as the church of St. Marco or the church of Santa Maria Paganica (Fig. 15.4), and although the roof was made with slightly less heavier solutions, the outcome was still the loss of the cover of the central crossing and of the nave. The church of Santa Giusta (Fig. 15.5), where the ring beam had been made by reinforced masonry rather than concrete performed marginally better with localized damaged but without major collapse.

An extensive survey of damaged churches was also conducted in the aftermath of Christchurch earthquake swarm of 2010–2011, by the Masonry Recovery Project (Dizhur et al. 2011). While the majority of the churches surveyed in L'Aquila were first built in the mediaeval period with poorly cut masonry stones and relatively poor lime mortar, then altered in the eighteenth century with baroque additions, the religious heritage in Christchurch mostly dates from the nineteenth Century and

Fig. 15.4 Collapse of the roof and vaults of the church of Santa Maria Paganica, L'Aquila

Fig. 15.5 Partial collapse and evidence of a reinforced masonry ring beam in the Church of Santa Giusta in L'Aquila

beginning of the twentieth. However, as just less than 50 % of the churches surveyed were built in either stone or clay brick masonry a comparison between the observed damage and any strengthening that was implemented at the time of the earthquake for the two sites might be of some interest. Statistics of damage reveal that for both typologies, brick and stone masonry, approximately 80 % of the buildings surveyed were either structurally damaged or presented partial collapses. This corresponds to either yellow or red tagging and according to New Zealand rule, implies demolition, if the structure is deemed unsafe. The two recurring mechanisms observed were partial overturning of the main façade and in-plane failure of the longitudinal walls. Although various strengthening techniques are mentioned by Dizhur et al. (2011) including shotcreting, steel strong-backs and steel moment frames, besides the use of adhesive anchors, it is not stated whether and to which extent any of these systems were used in heritage buildings. Following the New Zealand 1991 Building Act (New Zealand Parliament 1991), all unreinforced masonry buildings deemed earthquake prone (EPB) with an ultimate lateral capacity less that 1/3 of new built demand at the same site, should have been retrofitted to raise their ultimate capacity to 50 % of new built demand. NZSEE advocated this to be raised to 67 % of the ultimate demand for new design and this change was included in the 2004 Building Act (New Zealand Parliament 2004). Heritage buildings listed as EPB should be either strengthened or demolished, within a timeframe varying from 5 to 25 years, depending on enforcement provisions of the single territorial authority in relation to the perceived risk (McClean 2009).

According to Turner et al. (2012), a large proportion of retrofitted masonry buildings surveyed in the Commercial Business District of Christchurch, post February 2011 event, only had restrained gables and wall anchorage to floors and roofs, with a few cases of roof diaphragm improvements, while a minority also had installed additional vertical elements to the original lateral force resisting system. These would include concrete and steel moment frames, reinforced concrete and masonry walls, steel diagonal braces, and strongbacks. Horizontal retrofit elements included addition of plywood sheeting to roofs and floors as well as horizontal steel trusses to improve diaphragm action. In many cases was noted that irregularly spaced, insufficiently sized and too far apart anchorage proved ineffective in avoiding the separation of walls from floor structures or external wythes from internal ones, whilst regular layouts prevented out-of-plane failures. Weak mortar was also a cause of premature bond failure in the mortar joints, preventing stress transfer from the anchor to the masonry fabric (Wilkinson et al. 2013). In several cases buildings retrofitted with additional steel or concrete frame did not performed well with partial or total collapse of the masonry walls (Wilkinson et al. 2013; Turner et al. 2012). From a conservation point of view, this type of intervention is considered totally against the principle of authenticity and reliability stated in Sect. 15.2, but also against several of the strengthening criteria. In the aftermath of the Christchurch earthquake, the issue of how heritage buildings should be dealt with was brought to front by a Governmental public consultation exercise closed in March 2013, the Building Seismic Performance Consultation document, Proposals to improve the New Zealand earthquake-prone building system (Ministry of Business, Innovation and Employment (MBIE) 2012). Five questions were specifically aimed at heritage buildings, including, what factors should be considered when balancing heritage values with safety concerns, what are the deterrents for heritage building owners to proceed to strengthen their buildings, what are the cost and benefits of setting a consistent set of rules across the country for heritage building strengthening, what guidance will be needed by owners and communities to strengthen heritage buildings. SESOC (NZ Structural Engineering Society 2013) provided a very comprehensive answer to these questions in terms of expected performance target, specifically noting that "Heritage buildings in private ownership are potentially under threat due to the high cost of compliance." (SESOC 2013) On one hand if the standard is only concerned with life safety compliance "may result in buildings that are unlikely to be practically repairable after the event." (SESOC 2013) On the other hand, a higher level of protection can only come at extra cost to the owner, as the current system does not make provision for public subsidies. "SESOC supports the Historic Places Trust recommendation for the development of a National Risk Map for New Zealand's heritage. This may form the basis of a prioritization of heritage buildings requiring additional protection; and could also inform an approach to public funding (or partfunding)" (SESOC 2013). Answers to the first of the five questions are particularly relevant to this paper. SESOC viewpoint is that buildings should not be assessed in terms of percentage of capacity of new build demand, but instead specific vulnerabilities should be identified and amended. The major drawback of the current assessment approach is seen as the lack of an assessment of actual ductility reserves. A major issue felt is whether there is consistency on the application of ICOMOS principles, for instance in relation of clearly visible, external to the original fabric retrofit elements, which are less costly to implement and more likely to be effective. Finally it is not clear whether the driver for decision making should be the public safety concern or the preservation of the heritage value.

15.4 Strengthening Strategies Included in Standards and Guidelines

It was seen in the previous section that the re-instatement of continuity of load paths and the delivery of a robust global behavior are paramount for the seismic upgrade of historic buildings. A wide range of techniques and products are described in the scientific literature and used in current practice to ensure the enhancement of damaged or underperforming connections. However, as observed in the introductory section, in respect to engineered structures, heritage buildings require far more attention, especially when dealing with issues such as the compatibility between the chemical and mechanical properties of the strengthening system and the parent material. Many strengthening techniques, after an initial success and a strong commercial promotion, have proved to be unable to perform at the required level and showed unexpected drawbacks when undergoing dynamic loading outside the controlled conditions of the laboratory environment (see for instance the extensive programme of onsite testing of adhesive anchors connections conducted within a joint project of University of Auckland and University of Minnesota, Dizhur et al. 2011). On the other hand, strengthening systems can provide highly flexible applications and meet the expected requirements in terms of performance; indeed, some of these systems draw on traditional reinforcement techniques, with the addition of innovative materials and a deeper insight in the laws governing the dynamics of structures. In the following we briefly review the provisions included in the standards and Codes of practice of the countries considered, before looking at some implementation on heritage buildings observed in L'Aquila.

15.4.1 Peruvian Code

"Strengthening of structures" is ruled in the National Building Code, E.030, Section 8, in its 2014 version proposed for public approval (Comité Técnico Permanente Norma E.030 Diseño Sismorresistente NTE E.030 2014).[1] The provision are easily summarised: structures damaged by earthquakes should be evaluated and repaired so that the possible structural defects that cause the failure can be amended and they can recover their resisting capacity toward a new seismic event, according to the Earthquake –Resistant Design Philosophy of the Code. Structures affected by an event, should be evaluated by a civil engineer, to determine whether reinforcement, repair or demolition is required. This study must consider the geotechnical characteristics of the site. The repairing process should be able to give the structure an adequate combination of stiffness, resistance and ductility and should guarantee its good behaviour for future events.

The repairing or reinforcement project will include the details, procedures and constructive systems to be followed. No further details are provided in this version of the code and the document itself does not include unreinforced masonry or adobe structures. Current work undertaken by the author's research group in collaboration with Getty and PUCP aims at providing guidelines for assessment and strengthening of four common types of Peruvian heritage buildings (Ferreira et al 2014).

15.4.2 European and Italian Codes

Eurocode 8, Section 6.1 Retrofit Design Procedure for existing building (EN 1998–3:2005), states that the design process of strengthening elements should cover:

1. Selection of techniques and/or materials, as well as of type and layout of intervention;
2. Preliminary sizing of additional structural parts;
3. Preliminary calculation of stiffness of strengthened elements;
4. Analysis of strengthened structure by linear or non-linear analysis. The typology of analysis is chosen depending on the level of knowledge regarding the geometry detailing and materials of the structure;
5. Safety verifications for existing, modified and new structural elements carried out by checking that the demand at three different limit states – Damage Limitation, Significant Damage and Near Collapse – is lower than the structural capacity.

The safety checks should be carried out using mean values of mechanical properties of existing materials obtained from either in-situ tests or other available documentation, taking into account the confidence factors (CFs) specified in

---
[1] Consulted in Spanish version.

Eurocode 8 Section 3.5 (EN 1998–3:2005). Conversely, for new materials, nominal properties shall be used without modification by confidence factor. The code also states that in case the structural system, comprising both existing and new structural elements, can be made to fulfill the requirements of EN1998-1:2004, the checks may be carried out in accordance with the provisions therein (EN 1998–1, 2004). This last sentence indicates that for systems such as reinforced concrete ring beams or corner confinement, reference can be made to the specifications for RC members in the relevant sections of EC8 and other Eurocodes for new design. However, this leaves open the problem of quantifying the interaction between original and additional structural elements and the assessment of the global seismic performance of the strengthened structure will still be affected by a large number of uncertainties.

Other strengthening systems hardly feature in codes. This could be due to the fact that the sizing of the element itself, for instance a steel cross-tie with end plate, is fairly straightforward and established in the current technical know-how; furthermore, formulas can be drawn from those of other structural members, e.g. axial capacity of steel element. Still, designers are left to their own devices when assessing the interaction between old and new, the hierarchy of failure mechanism that the connection should comply with, the value of bond or slip that should correspond to a specific performance target.

In other cases the lack of standardization is caused by the recent development of techniques as well as the high level of expertise and financial resources required for their implementation. Innovative technologies haven't been extensively applied and validated in real-life situations yet and the retrofit of a complex, precious building by means of unconventional systems is a difficult task that goes beyond the standard conservation practice. In fact, looking at the current scientific literature, it is clear that many projects of restoration and upgrade of monumental buildings are carried out by organizations within the framework of specific research projects, or by large enterprises that specialize in the production and design of advanced strengthening devices. On the other hand, it could be argued that it is this lack of appropriate standards and procedures which leads to incorrect application of novel strengthening systems and lack of awareness of innovative more suitable techniques.

In some occasions, following major destructive events, ready to the market technology finds a sudden growth in popularity and implementation which pre-date the standardisation phase.

It is worth noting however, that some systems, in spite of their relatively recent development, have already been included in specific technical guidance documents, as in the case of Fibre Reinforced Polymers, whose use in retrofit of substandard structures is addressed in the CNR-DT 200 R1/13, Italian National Research Council (CNR), (CNR-DT 200 R1/13, 2013).[2] This recently re-issued Italian

---

[2] This version of the Guidance document is in Italian. A previous version CNR-DT 200 /2004 is translated in English.

15 Conservation Principles and Performance Based Strengthening of Heritage... 501

guidelines (Italian National Research Council [CNR]) for use of FRP for the "Design, installation and control of strengthening intervention with Fibre Reinforced composites", provides advice for use of such techniques to either strengthen or reconstruct some elements, or to connect the various structural elements to improve the behaviour of the whole structure. The document covers all structural materials, including masonry. The objectives that any strengthening intervention on a masonry structure should have are listed as follows:

- The masonry structural substratum should be adequately consolidated to withstand the design actions or replaced
- Orthogonal walls should be appropriately connected
- Inadequate connections between the walls and the horizontal and roof structures should be improved
- Thrust from roofs, arches and vaults should be adequately contained
- Floors should be sufficiently stiff in their plane to redistribute the horizontal action while at the same time act as constraint for out-of-plane motion of walls.

It is not openly stated whether strengthening with FRP is suitable to meet these performance criteria or whether these are prerequisites to the use of FRP in masonry structures, however some disclaimers are included:

- Interventions with FRP cannot as a rule improve or amend situations characterised by strong irregularities in terms of strength and stiffness, even though, if applied to a reduced number of elements, they can provide a more even distribution of strength
- Interventions with FRP aimed at improving local ductility such as columns or pillars confinements are always appropriate, although
- Local intervention with FRP should not reduce the overall ductility of the structure.

Besides this very specific document, the most updated relevant legislation for interventions on heritage buildings is represented by the guidance document "Linee Guida per la valutazione e riduzione del rischio sismico del patrimonio culturale – allineamento alle nuove Norme tecniche per le costruzioni", become ministerial decree as Circolare 26/2010 (Circolare 26/2010) (see also NTC, 2008). This document incorporates all aspects of the ISCARSAH guidelines mentioned in Sect. 15.2, while at the same time conforming to the performance based approach of the latest version of the technical standards for implementation of the Eurocode at national level. The specific recommendations of the Linee Guida are further described in the next section.

15.4.3 New Zealand Provisions

The New Zealand provisions for strengthening and retrofit are summarised in the NZSEE document "Assessment and Improvement of the Structural Performance of

Buildings in Earthquakes" (NZSEE 2006, revised version 2012). The document focuses on the assessment of all type of structures including unreinforced masonry buildings, but it does not distinguish for buildings of historical or cultural value. Moreover Derakhshan et al. (2009) have proven that some of the criteria used in NZSEE 2006 are over-conservative when considering the out-of-plane response of masonry walls and have proposed an alternative displacement based procedure. The strengthening strategies are confined to Sect. 13.6 and subdivided by the strengthening effect in in-plane strengthening, face load strengthening, combined face load/in-plane strengthening, diaphragm strengthening and chimneys towers and appendages.

Shotcrete is recommended for in-plane as well as out-of plane performance enhancement, as well as FRP wrapping. To prevent out-of plane failure anchoring to floors and walls is recommended, as well as buttressing and addition of columns, while the in–plane performance can be enhanced by introduction of concrete frames and v-braced frame. There is no value judgement or guidance for which intervention is most suitable to specific conditions or to which extent any of the suggested interventions contributes typically to the lateral capacity demands enhancement. Moreover no advice is given of how to choose among different strengthening options from each set that together would deliver the best integrated and overall performance. A commentary provides for each technique further details that should ensure good quality implementation and effectiveness.

15.5 Evidence from the Field: Strengthening in L'Aquila

In conjunction with a return mission to L'Aquila organized by EEFIT in November 2012, (Rossetto et al. 2014), the author had the opportunity to inspect a small number of building sites where conservation and repair projects were underway. These visits provide some insight on how retrofitting strengthening projects are implemented. The masonry fabric typologies most frequently observed in the district of L'Aquila for heritage buildings are rubble stone, roughly squared stone blocks mixed with bricks, sometimes in regular courses, brick masonry, and dressed stone blocks. Walls in a few cases appear to be massive, but most commonly are formed by the so called "muratura a sacco", namely two wythes of dressed stones poorly connected, sometimes with a rubble infill. Mortar is mainly lime mortar. Large squared stone blocks are used for quoins. A typical intervention that was observed to be extensively used at the few sites which were undergoing restoration at the time of the EEFIT mission and that could be visited is fluid mortar injection grouting of all bearing walls (Fig. 15.6). The aim of such an intervention is to improve the coherence and cohesion of existing walls by injecting them with fluid grout through a series of drilled holes regularly spaced on a 500 mm grid and proceeding from the bottom to the top, after having sealed and repointed the mortar joints. Although for material compatibility only lime-based grouts should be used, often epoxy additives or cement are included in the mix for faster setting. While

Fig. 15.6 Wall prepared for grout injection

such additives might improve the short term strength and cohesion of the masonry, they can create serious long term problems in terms of decay of the original materials due to different hygro-thermal behaviour and salt content release. One of the major issue is that such interventions are not directly monitorable. One way of verifying their effectiveness is to conduct flat jack tests of the masonry wall, before and after strengthening, although this is partially destructive.

Strengthening of floor to improve diaphragm action is recommended by the Linee Guida (Circolare n. 26/2010). This can be achieved by either nailing superimposed sets of floorboards at right angles or by adding a lightweight reinforced lime-based concrete screed above the existing set of floorboards. The reinforcement should be anchored in the perimeter masonry walls. Extensive tests campaign have been carried out at several institution in Italy in past years to devise the best technical details and performance improvement that can be obtained with such interventions (Riggio et al. 2012). The joists and beams forming the floor structures should also be anchored to the walls by means of ties. A similar approach should be followed also for roof structures (Giuriani and Marini 2008). This type of intervention was traditionally extensively applied in the past and it can be observed that in cases where the ties have been well maintained and are regularly distributed on the wall, the damage is usually no greater than airline cracks.

A common structural element of many buildings in L'Aquila is the brick vault. Brick vaults are present in lower floors of residential buildings as a load bearing structure with a typically shallow cross-shaped arch profile, as a non-loadbearing false ceiling in upper floors (built in folio) and in most religious buildings as support to the roof structure. Post-earthquake surveys have revealed partial collapse and extensive damage of these structures. The Linee Guida (Circolare n. 26/2010) recommend either the use of traditional steel ties or specifically built spandrels at the extrados (Ferrario et al. 2009) while strengthening intervention with extradossal reinforcement made of FRP strips (see Fig. 15.7) are tolerated with numerous provisos. While a body of research exists on the strength gain benefit of such

Fig. 15.7 Reinforcement of a cross vault with strips of FRP laid at the extrados

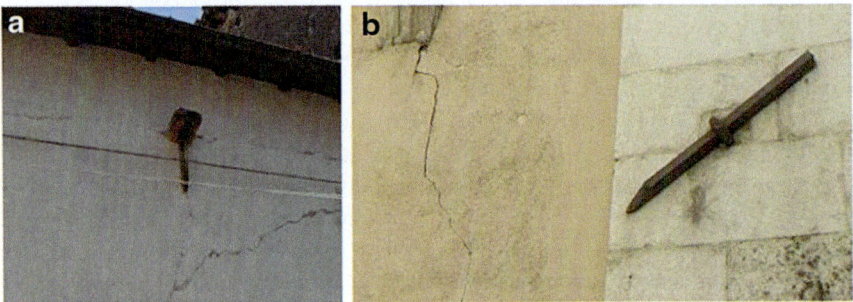

Fig. 15.8 Two examples of traditional reinforcement: (**a**) timber tie, (**b**) wrought iron cross tie inserted in a quoin

interventions, most of the experimental research conducted to date focus on static concentrated loading conditions, or support movement, rather than dynamic performance (Modena et al. 2009). Durability and breathability are the major concerns.

The Linee Guida (Circolare n. 26/2010) recommends the use of ties and anchors to connect vaults and timber floors to walls, and walls to walls. A thorough review of traditional and modern solutions, their effectiveness, shortcomings and possible improvement by use of dissipative devices is included in D'Ayala and Paganoni (2014) and some surveyed examples are illustrated in Fig. 15.8. In the few sites undergoing repair or strengthening at the time of the return mission, there was no evidence of such strengthening devices being implemented.

Fig. 15.9 Extensive use of reinforced coring with grouted injection with epoxy resins on the end wall of a 5 storey residential palace in the historic centre of L'Aquila

In one of the few on-going projects seen during the return mission, it was noticed that transversal reinforcement was applied to masonry walls by use of FRP bars, drilled through the thickness and then anchored by opening the threads as a star (Fig. 15.9). In the Guidelines issued in 2010 (Circolare n. 26/2010) it is stated that "the use of reinforced cores should be limited to cases where there is no other alternative due to the extreme alterations and disturbance produced vis a vis [its] doubtful effectiveness, especially in the presence of walls with several wythes not well connected. In any case the durability of the strengthening element, whether of stainless steel, composite plastic materials or other material, should be ensured and the grouts used should be compatible with the original materials". Moreover it is advised that this type of intervention only has at best a local effect (Circolare n. 26/2010).

15.6 Dissipating Energy as an Alternative to Strengthening

The drawbacks of strength-based systems were clearly brought to the fore by the seismic events reviewed in Sect. 15.3. Low compatibility in terms of mass and stiffness of concrete ring beams, often inadequately connected to the existing masonry, concurred to cause tragic collapses, as in the case of the Collemaggio basilica in L'Aquila (Gattulli et al. 2013). Numerous are the failures observed when traditional timber roof and floor structures are substituted with concrete ring beams and slabs in an attempt to deliver diaphragm action. The sudden change in stiffness and the difference in shear capacity of the two systems is simply too substantial to be accommodated by the interface. Shotcreting has also proven inadequate when coupled to both adobe and stone masonry due to poor bond to the parent material that can be achieved and maintained as the masonry decays for lack of proper aeration. The New Zealand approach of inserting new lateral resisting system, such as steel or concrete frame, while not always effective, is certainly, if not extremely sensitively designed, in breach of most of the ICOMOS/ISCARSAH acceptance criteria.

On the other hand, cross-ties, which have been and still are commonly applied in rehabilitation practice not just in Europe (Tomaževič 1999), but also in Latin America (D'Ayala and Benzoni 2012) and new Zealand (University of Auckland 2011), are able to restore the box-like behaviour, without a substantial increase of mass, if they are regularly distributed and properly sized. Indeed, traditional cross ties can provide connection at the joints of perpendicular sets of walls, where poor quality, previous damage, or general wear and tear facilitate crack onset and otherwise out-of-plane failure. Nonetheless, localised damage at the head of the anchorage similar to punching shear is a possible drawback, which might become a major problem when damage limitation and protection of valuable finishes should be pursued or might eventually lead to the wall overturning failure (Wilkinson et al. 2013).

The concept of reducing demand by dissipating energy in a controlled way is not novel, nor recent. With specific references to applications of the concept to masonry structures and heritage buildings in particular, Benedetti (2004, 2007) developed a series of energy absorbing devices drawing on the observation that the more energy is absorbed through damage by non-critical elements of the structure the less likely is that global failure occurs. Key feature of the devices were activation for small relative displacement (1 mm) and long displacement range (up to 10 mm), i.e. low level of damage, ability to accommodate both in plane and out of plane movements, low magnitude of forces at the interface with the parent material (0.3–0.5 KN). The devices were set in series with traditional steel ties connecting parallel walls. Martelli (Martelli 2008) also highlights a relatively conspicuous number of high-profile cultural buildings in Italy that have been strengthened, either post or prior a damaging event, using one or more energy based devices such as shock transmitter units (STUs) and shape memory alloy devices (SMADs) in the period 1997–2008 by using technologies developed within European Frameworks Programmes. It is stated that STUs were inserted as a dynamic constraint between a new stiffening truss and the original walls at a height of 8 m along the longitudinal walls of San Francis Upper Basilica in Assisi. The displacement range in the STUs is ± 20 mm with maximum forces of 220–300 KN. Among these early interventions listed by Martelli (2008) stands out the Santa Maria di Collemaggio Cathedral at L'Aquila, which was retrofitted by installing Elasto-Plastic Dampers (EPDs), within a system of diagonal cable braces in the bottom plane of the roof trusses. The aim of the intervention was to limit transmission of large forces from the nave walls to the façade and the transept due to the truss structure inserted at the roof level to ensure coupling in the vibration of the longitudinal walls. The appropriateness of this intervention, among the few being tested by a real event, was reassessed after the collapse of the central crossing (Gattulli et al. 2013). A rocking-damper system, called DIS-CAM (DISsipative Active Confinement of Masonry) was developed and installed within the framework of the project of restoration of the drum of the dome of S. Nicolò church in Catania, although the collapse in this case was due to long term decay (Di Croce et al. 2010).

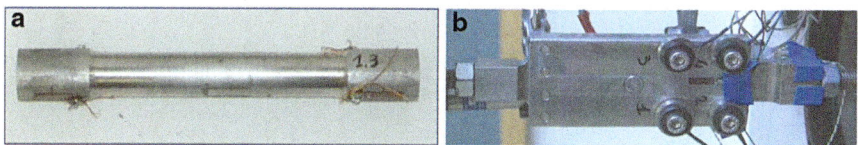

Fig. 15.10 Dissipative devices prototypes: (**a**) hysteresis based; (**b**) friction based

Drawing on the principles of performance based design, allowing and controlling modest drift and limiting damage by providing sacrificial elements able to dissipate energy, Paganoni and D'Ayala (2010) in collaboration with Cintec International developed two prototypes of dissipative anchor devices to address the problem of out-of-plane mechanisms of facades and lateral walls (Fig. 15.10).

The devices are conceived to be inserted at the connection between perpendicular walls, as part of longitudinal steel anchors grouted within the thickness of the walls. This type of installation ensures a low impact on the aesthetic of the building as it doesn't affect the finishing. The anchors can also be installed between floors elements and walls.

While the anchors improve the box-like behaviour of the building, contributing to an increase of stiffness that improves the structural response to small excitation, the devices allow small relative displacements between orthogonal sets of walls; for higher horizontal loads, they dissipate part of the energy input into the structure so that problems of localised damage can be avoided. Therefore, the design focuses on the achievement of control of displacements and reduction of accelerations and stress concentration.

Of the two developed prototypes, one is based on yielding, the other on friction. The former relies on a stainless steel element with a lower capacity in respect to the anchor, this lower capacity depending on a reduction of cross sectional area and the use of a different steel strength class. The friction prototype consists instead of a set of metallic plates able to slide past each other once a pre-set threshold of force is overcome, this been governed by controlled pressure.

The two dissipative devices, covered by patents, have been extensively validated by cyclic pseudo-static and dynamic tests on the isolated devices (Paganoni and D'Ayala 2010), and by cyclic pull-out tests on specimens modelling the T joint between two perpendicular walls connected by a passing anchor (D'Ayala and Paganoni 2014). The devices' performance has then been calibrated by using real time history obtained by obtaining from a finite element nonlinear analysis the relative motion at the crack of two orthogonal walls of a two storey house subjected to a real accelerogram from the L'Aquila earthquake. The response of the two devices is shown in Fig. 15.11.

What is relevant to the above discussion is the possibility to determine a rigorous design and dimensioning procedure, based on experimental results and on the principle of performance based seismic response. The strengthening apparatus can be seen as a relatively simple system made of a number of components in series. The objective is to determine the performance criteria of the dissipative

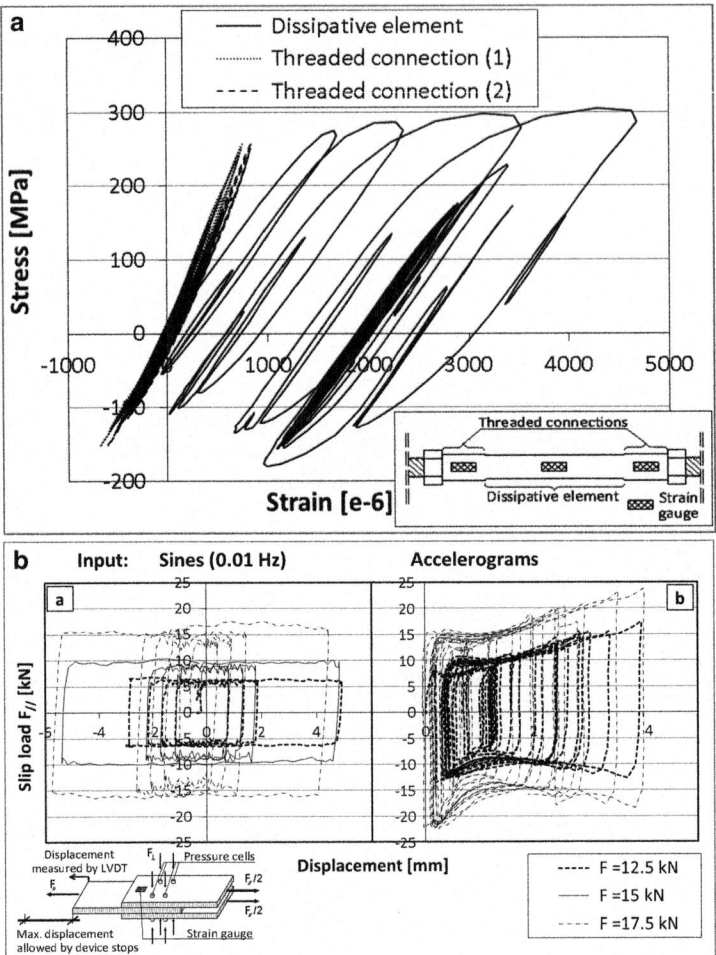

Fig. 15.11 Devices' response to accelerogramme excitation (**a**) hysteretic device and (**b**) friction device

device so that damage to the parent material can be controlled. The procedure is briefly summarised herein.

For the demand to the strengthening system, depending on the importance of the heritage building and its protection status a choice can be made to use a suite of non-linear time-history analyses of the building to determine envelop of displacement demand requirements, or to use reference drift limits from seismic code. Then use output of above analysis or modal analysis with spectrum superposition, or other simplified procedure as advised by seismic code, to determine acceleration amplification at selected heights of structure to determine the axial force on each of a set of anchors so as to determine the number of anchors required at any given storey of the structure, by using the initial assumption that failure of bond between anchor grout and parent material is prevented:

$$F_{bond,b/p} = f_{b,b/p} \cdot \pi \cdot D \cdot L \geq \gamma_D M_i \cdot a_j = \gamma_D \rho_m l_i h_i t_i a_j \quad (15.1)$$

Where $f_{b,b/p}$ is the bond strength of grout to parent material including safety coefficient; D and L are the diameter and length of grouted anchorage; γ_D, design safety coefficient; M_i: mass of portion of structure that bears on the ith anchor; ρ_m, l_i, h_i, t_i, density and dimensions of the portion of structure restrained by the ith anchor; and a_j is the horizontal acceleration at storey j of the structure, calculated on the basis of the performance target defined in BS EN 1998–3:2004 depending on the performance criteria and hazard return period defined for the structure with:

- F_{DNC}: near collapse (2 % exceedance in 50 years);
- F_{DSD}: significant damage (10 % exceedance in 50 years);
- F_{DDL}: damage limitation (20 % exceedance in 50 years).

Once the anchor is preliminary sized, the capacity of the dissipative devices can be determined by using two different approaches depending on the device. In case of grouted metallic ties with hysteretic device:

- Step 1: Determine the minimum among:

  – Yielding strength of tie,
  – Adhesion strength tie/grout
  – Adhesion strength grout/masonry
  – Punching through strength of surrounding masonry

  Hence, yielding point of hysteretic device < Minimum
  If punching through of surrounding masonry is critical, it will be necessary to improve the masonry locally with grouting, for instance.
- Step 2: Determine the ductility requirements which will lead to maximum elongation of the device, while preventing buckling.

  These two conditions will determine the yielding point of the device as well as its geometric dimension and cross section shape.
- Step 3: Verify that performance is not compromised by instability of cycles and hardening limits

In the case of grouted metallic ties with friction device

- Step 1: Determine the minimum among

  – Yielding strength of tie,
  – Adhesion strength tie/grout
  – Adhesion strength grout/masonry
  – Punching through strength of surrounding masonry

  Hence, tightening of device < Minimum

- Step 2: Determine maximum sliding requirements and energy dissipation which will determine the size of the plate and the value of friction.
- Step 3: Control stick–slip, stability of cycles, apparent hardening

The above approach requires a series of laboratory tests to determine all material characteristics and certify performance requirements of the devices before installation, and a series of onsite tests to determine quality and characteristics of the parent material and quality and strength of the bond, which can be ascertained by, for instance, static pull-out-test, aimed at ensuring also the quality of the installation.

The dissipative devices are designed to be activated at the threshold of damage limitation of the structural response, while all other components are designed to withstand the forces associated with near collapse. If the damage limitation threshold is not a requirement for the building, then the devices can be designed to greater strength capacity. In the case of the friction device it will just be a matter of determining the different activation level of the slider for different performance requirements.

But the dimensioning of the devices should not be based on the force but on the amount of energy to be dissipated and hence on the associated deformation/sliding past the force threshold. While the two values of triggering force and demand displacement are independent for the friction device this is not the case for the hysteretic device, which needs also to be dimensioned to control buckling. Hence the design will need to undergo a series of iterations to optimise the elongation of the device and its axial buckling limit. As seen in previous applications typical relative displacement is of the order of 10–20 mm leading to interstorey drifts of the order of 0.3 %, corresponding to the damage limitation threshold for historic building according to the Circolare n. 26/2010. Finally, devices need to be designed so that they can offer additional capacity at NC limit state. In particular, referring to the experimental results reported by D'Ayala and Paganoni (2014), it is important that:

- Yielding devices reach the threshold of the 5 % elongation, so that they can offer extra capacity both in terms of displacement and load capacity;
- Frictional devices reach the end of their run. This ensures that the device will offer additional load capacity, this being quantified by a safety factory of 10 (D'Ayala and Paganoni 2014).

15.7 Conclusions

A considerable amount of research has and is being conducted to improve the way in which the issue of strengthening historic buildings is approached by the engineering community. This research has led to novel assessment procedures which were not covered in details here, novel strengthening techniques which best meet the requirements of the conservation principles and attempt at maintaining both the original structure and the historic fabric, without substantial disruption. Indeed the tragic events of the last 4 years have triggered generally very good and responsible response on the part of the engineering community, clearly more sensitive to the cultural heritage agenda than not in the past.

Public cultural differences exist and cannot be ignored when devising policies. In some countries demolition is still considered in many respects a more viable option than repair and retrofit. However recent initiatives such as the ICOMOS New Zealand Charter 2010 (ICOMOS 2010) or the new regulations for earthen buildings of historic significance, which the Ministerio de Vivienda y Urbanismo of Chile is drafting in the document NTM002 (Ministerio de Vivienda y Urbanismo, 2010), currently in the pre-standard stage, show a change in perspective of the public as well as the engineering community towards historic buildings and perhaps a different acceptance of risk.

From a technical point of view however, much training and education of professional engineers is needed to ensure that the shift in design emphasise from force requirements to displacement and energy requirements is fully understood. As seen from evidence in the field far too often strengthening of historic buildings is still pursued in terms of increasing strength and stiffness, while some assessment criteria are far too conservative. A similar training is also needed among contractors.

Hurdles of other nature, related to the economics of developing and installing dissipative devices, can be overcome, as shown by the prototype devices described in the previous section which can be manufactured in small sizes and at costs which is affordable in the retrofit of residential historic buildings, as well as more prestigious landmark. However robust testing and design protocols need to be develop to gain confidence among practitioners.

Open Access This chapter is distributed under the terms of the Creative Commons Attribution Noncommercial License, which permits any noncommercial use, distribution, and reproduction in any medium, provided the original author(s) and source are credited.

References

Augenti N, Parisi F (2010) Learning from Construction Failures due to the 2009 L'Aquila, Italy, Earthquake. J Perform Constr Facil 24(6):536–555

Barucci C (1990) La casa antisismica prototipi e brevetti. Gangemi editore, Roma

Benedetti D (2004) Increasing available ductility in masonry buildings via energy absorbers. Shaking table tests. Eur Earthq Eng 3:1–29

Benedetti D (2007) La risposta fuori piano di sistemi murari: progetto e valutazione sperimentale di assorbitori di energia. Ing Sismica 24(2):7–35 (in Italian)

Blondet M, Vargas J, Tarque N (2008) Observed behaviour of earthen structures uring The Pisco (Peru) earthquake of August 15, 2007, The 14th World conference on earthquake engineering, Beijing, 12–17 Oct 2008

Cancino C (2010) Damage assessment of historic earthen sites after the 2007 earthquake in Peru. Adv Mater Res 133–134:665–670

Circolare n. 26/2010 (2010) Linee Guida per la valutazione e riduzione del rischio sismico del patrimonio culturale – allineamento alle nuove Norme tecniche per le costruzioni, Ministero per i Beni e le Attività Culturali, Prot. 10953 del 2 dicembre 2010

CNR – Italian National Research Council (2012) Technical Guidance Document 200 R1/2013, "Istruzioni per la Progettazione, l'Esecuzione ed il Controllo di Interventi di Consolidamento Statico mediante l'utilizzo di Compositi Fibrorinforzati", p 131

Comité Técnico Permanente Norma E.030 Diseño Sismorresistente (2014) Proyecto De Norma E.030 Diseño Sismorresistente, SENSICO 2014

Cracow C (2000) Signed by representatives of conservation institutions. In: Proceedings of international conference on conservation, Krakow 2000, Krakow, pp 191–193

D'Ayala D (2011) The role of connections in the seismic resilience of historic masonry structures, Invited lecture in L'ingenegneria Sismica In Italia, ANIDIS Bari 2011, ISBN 978-88-7522-040-2

D'Ayala D, Benzoni G (2012) Historic and traditional structures during the 2010 Chile earthquake: observations, codes, and conservation strategies. Earthq Spectra 28(S1):425–451, 10.1193/1.4000030

D'Ayala DF, Paganoni S (2011) Assessment and analysis of damage in L'Aquila historic city centre after 6th April 2009. Bull Earthq Eng 9(1):81–104. doi:10.1007/s10518-010-9224-4

D'Ayala DF, Paganoni S (2014) Testing and design protocol of dissipative devices for out-of-plane damage. Proc ICE-Struct Build 167:26–40

D'Ayala D, Yeomans D (2004) Assessing the seismic vulnerability of late Ottoman buildings in Istanbul. In: Modena C, Lourenço PB, Roca P (eds) Structural analysis of historical construction, Padova. 2004-01-01

Derakhshan H, Ingham JM, Griffith MC (2009) Out-of-Plane assessment of an unreinforced masonry wall: comparison with NZSEE recommendations. In: Proceedings 2009 NZSEE conference, Christchurch, 3–5 Apr 2009

Di Croce M, Ponzo FC, Dolce M (2010) Design of the Seismic Upgrading of the Tambour of the S. Nicols Church in Catania with the DIS-CAM System. VII International seminar on structural analysis of historical constructions – SAHC10, Shanghai, China

Dizhur D, Ingham J, Moon L, Griffith M, Schultz A, Senaldi I, Magenes G, Dickie J, Lissel S, Centeno J, Ventura C, Leite J, Lourenco P (2011) Performance of masonry buildings and churches in the 22 February 2011 Christchurch earthquake. Bull N Z Soc Earthq Eng 44(4): 279–296

EN 1998–1 (2004) Eurocode 8- design of structure for earthquake resistance. Part 1: General rules, seismic actions and rules for buildings

EN 1998–3 (2005) Eurocode 8- design of structure for earthquake resistance. Part 3: Assessment and retrofitting of buildings

Faculty of Engineering University of Auckland (2011) Commentary to [NZSEE 2006] assessment and improvement of unreinforced masonry buildings for earthquake resistance accesed at http://masonryretrofit.org.nz/SARM/2.pdf in June 2014

Fallahi A (2008) Bam earthquake reconstruction assessment: an interdisciplinary analytical study on the risk preparedness of Bam and its cultural landscape: a world heritage property in danger. Struct Surv 26(5):387–399

Feilden BM (1987) Between two earthquakes: the management of cultural property in seismic zones. In: Old cultures in new worlds. 8th ICOMOS General assembly and international symposium. Programme report – Compte rendu. US/ICOMOS, Washington, DC, pp 582–589

Ferrario L, Marini A, Riva P, Giuriani E (2009) Traditional and innovative techniques for the seismic strengthening of barrel vaulted structures subjected to rocking of the abutments. In: Goodno B (ed) ATC and SEI conference on improving the seismic performance of existing buildings and other structures, pp 1329–1340

Ferreira CF, Quinn N, D'Ayala D (2014) A logic-tree approach for the seismic diagnosis of historic buildings: application to adobe buildings in Peru'. In: Proceedings of Second European conference on earthquake engineering and seismology, Istanbul, 24–29 Aug 2014

Gattulli V, Antonacci E, Vestroni F (2013) Field observations and failure analysis of the Basilica S. Maria di Collemaggio after the 2009 L'Aquila earthquake. Eng Fail Anal 34(2013):715–734

Ghafory-Ashtiany M, Hosseini M (2008) Post-Bam earthquake: recovery and reconstruction. Nat Hazards 2(44):229–241. doi:10.1007/s11069-007-9108-3

Giuriani E, Marini A (2008) Wooden roof box structure for the anti-seismic strengthening of historic buildings. Int J Archit Herit 2(3):226–246

ICOMOS-ISCARSAH Committee (2003) Recommendations for the analysis, conservation and structural restoration of architectural heritage, http://iscarsah.icomos.org/content/principles/ISCARSAH_Principles_English.pdf

ICOMOS New Zealand (2010) ICOMOS New Zealand Charter, Te Pumanawa Aotearoa Hei Tiaki I Nga Taonga Whenua Heke Iho o Nehe, http://www.icomos.org.nz/nzcharters.htm

ISO 13822 (2010) Bases for design of structures – assessment of existing structures. International Organization for Standardization – ISO, Switzerland

Lagomarsino S (2012) Damage assessment of churches after L'Aquila earthquake (2009). Bull Earthq Eng 10:73–92. doi:10.1007/s10518-011-9307-x

Martelli A (2008) Recent progress of application of modern anti-seismic systems in Europe – Part 2: energy dissipation systems, shape memory alloy devices and shock transmitters. In: The 14th world conference on earthquake engineering, s.n., Beijing, China

MBIE (2012) Building seismic performance proposals to improve the New Zealand earthquake-prone building system, consultation document – Dec 2012. http://www.dbh.govt.nz/UserFiles/File/Archive/consulting/2012/building-seismic-performance-consultation-document.pdf. Accessed June 2014

McClean R (2009) Toward improved national and local action on earthquake-prone heritage buildings, Historic Heritage Research Paper No.1, New Zealand Historic Places Trust Pouhere Taonga

Ministerio de Vivienda y Urbanismo (2010) Anteproyecto de Norma NTM 002, Proyecto de Intervencion Estructural de la Construcciones Patrimoniales de Tierra

Modena C, Casarin F, DaPorto F, Garbin E, Mazzon N, Munari M, Panizza M, Valluzzi M (2009) Structural interventions on historical masonry buildings: review of eurocode 8 provisions in the light of the Italian Experience. In: Cosenza E (ed) Eurocode 8 perspectives from the Italian Standpoint workshop, 225–236, © 2009 Doppiavoce, Napoli

Montes CA, Giesen CG (2010) Recommendations for the conservation of earthen architectural heritage in Chile. Adv Mater Res 133–134(2010):1125–1130

New Zealand Parliament (1991) Building act 1991, Department of Building and Housing – Te Tari Kaupapa Whare, Ministry of Economic Development, New Zealand Government, Wellington, New Zealand

New Zealand Parliament (2004) Building Act 2004, Department of Building and Housing – Te Tari Kaupapa Whare, Ministry of Economic Development, New Zealand Government, Wellington, New Zealand, 24 Aug 2004

NTC (2008) Norme Tecniche per le Costruzioni. DM 14 gennaio 2008, Gazzetta Ufficiale, n. 29 del 4 febbraio 2008, Supplemento Ordinario no. 30, Istituto Poligrafico e Zecca dello Stato, Roma (www.cslp.it)

NZSEE (2006) Assessment and improvement of the structural performance of buildings in earthquakes. New Zealand Society of Earthquake Engineering

Paganoni S, D'Ayala D (2010) Dissipative device for the protection of heritage structures: a comparison of dynamic tests and FE models. In: 14th European conference of earthquake engineering, 2010-08-30-2010-09-03, Ohrid, Republic of Macedonia

Riggio M, Tomasi R, Piazza M (2012) Refurbishment of a traditional timber floor with a reversible technique: the importance of the investigation campaign for the design and the control of the intervention. Int J Archit Herit 8(1):74–93

Rossetto T, D Ayala D, Gori F, Persio R, Han J, Novelli V, Wilkinson SM, Alexander D, Hill M, Stephens S et al (2014) The value of multiple earthquake missions: the EEFIT L'Aquila Earthquake experience. Bull Earthq Eng 12:277–305

SESOC (2013) Building seismic performance. Submission to the Ministry of Business, Innovation and Employment 8 March 2013. Accessed 29 Jan 2014 at: http://sesoc.org.nz/documents/SESOC-Building-Seismic-Performance-Consultation-Document-final.pdf

Spence R, D'Ayala D (1999) Damage assessment and analysis of the 1997 Umbria-Marche earthquakes. Struct Eng Int J Int Assoc Bridge Struct Eng (IABSE) 9(3):229–233

Tavares A, D Ayala D, Costa A, Varum H (2014) Construction systems. In: Structural rehabilitation of old buildings. Springer, Berlin/Heidelberg, pp 1–35

Tomaževič M (1999) Earthquake-resistant design of masonry buildings. Imperial College Press, London

Torrealva D, Cerrón C, Espinoza Y (2008) Shear and out of plane bending strength of adobe walls externally reinforced with polypropylene grids, The 14th World conference on earthquake engineering, Beijing, 12–17 Oct 2008

Turner F, Elwood K, Griffith M, Ingham J, Marshall J (2012) Performance of retrofitted unreinforced masonry buildings during the Christchurch earthquake sequence, Structures Congress 2012, ASCE 2012

Venice Charter (1964) International charter for the conservation and restoration of monuments and sites. In: Proceeding, 2nd international congress of architects and technicians of historic monuments, Venice (accessed at http://www.international.icomos.org/charters/venice_e.pdf in June 2014)

Wilkinson S, Grant D, Williams E, Paganoni S, Fraser S, Boon D, Mason A, Free M (2013) Observation and implications of damage from the magnitude Mw 6.3 Christchurch, New Zealand earthquake of 22 February 2011. Bull Earthq Eng 13(1):1–35

Chapter 16
Earthquake Risk Assessment: Present Shortcomings and Future Directions

Helen Crowley

Abstract This paper looks at the current practices in regional and portfolio seismic risk assessment, discusses some of their shortcomings and presents proposals for improving the state-of-the-practice in the future. Both scenario-based and probabilistic risk assessment are addressed, and modelling practices in the hazard, fragility/vulnerability and exposure components are presented and critiqued. The subsequent recommendations for improvements to the practice and necessary future research are mainly focused on treatment and propagation of uncertainties.

16.1 Introduction

In the 1st European Conference on Earthquake Engineering and Seismology in Geneva in 2006, a keynote paper was presented by Norman Abrahamson on "Seismic hazard assessment: problems with current practice and future developments" (Abrahamson 2006). Abrahamson reviewed areas within the practice of probabilistic seismic hazard assessment (PSHA) that needed improvement and made recommendations on the direction that future research in PSHA should take. In this paper I take inspiration from Abrahamson, but will focus on the practice and development of probabilistic seismic risk assessment (PSRA), i.e. the estimation of the probability of damage and loss, for distributed buildings.

The main components of a PSRA for buildings comprise the hazard model (to get the probability of levels of ground shaking), the exposure model (location and characteristics of buildings) and physical vulnerability models (that provide the probability of loss, conditional on the level of ground shaking). An exposure model

---

H. Crowley (✉)
European Centre for Training and Research in Earthquake Engineering (EUCENTRE), Via Ferrata 1, Pavia, Italy
e-mail: helen.crowley@eucentre.it

provides information of the distribution of assets (e.g. buildings) within the region and might include the location, structural/non-structural characteristics, built area, replacement cost (new), contents value, business interruption cost, number of occupants (day/night). The buildings are grouped in terms of building classes as a function of their similar structural/non-structural characteristics, and a physical vulnerability function is developed for each building class. Vulnerability functions for structures provide the probability of loss or loss ratio (the loss as a percentage of the value, e.g. the repair cost divided by replacement cost), conditional on a level of input ground motion (Fig. 16.1), and can be derived from empirical, analytical or expert opinion based methods, or a combination of these methods (hybrid) (see e.g. Calvi et al. 2006; Rossetto et al. 2014). In empirical and expert-opinion based vulnerability modelling it is common to separate the damage distribution that is conditional on the ground motion (i.e. fragility function), from the loss distribution that is conditional on the damage (i.e. damage-loss model). In analytical vulnerability modelling, fragility functions are developed considering both the nonlinear response (in terms of parameters such as inter-storey drift) that is conditional on the input ground motion, and the damage state that is conditional on the nonlinear response. Aspects related to the application of each of the components of a PSRA are discussed in more detail herein, starting with the hazard model in the following section.

16.2 Ground-Motion Modelling

16.2.1 Scenario-Based Hazard/Risk Assessment

Abrahamson (2006) summarised both deterministic and probabilistic approaches to hazard assessment, and outlined many of the misunderstandings related to these two approaches. Abrahamson's focus was on hazard input for design and assessment, whereas herein we are interested in the hazard input for risk assessment of distributed assets. Nevertheless, the key message that Abrahamson put forward – that both deterministic and probabilistic approaches result in probabilistic statements about the ground motion – is also of relevance for risk assessment.

In fact, the use of the term "deterministic" in current hazard and risk assessment practice is misleading as it implies that there is no uncertainty involved in the process. On the contrary, it is just the event characteristics (magnitude, location, style of faulting etc.) that are commonly modelled as deterministic, whereas the ground motion as well as the damage and loss estimation all involve uncertainties. Furthermore, it is not necessarily the case that the event characteristics are deterministic (for example, the location may have an uncertainty associated with it), and it would be possible to model both aleatory and epistemic uncertainties related to the event as part of the assessment. For this reason, it is perhaps better to use the term "scenario-based" risk assessment, rather than deterministic risk assessment.

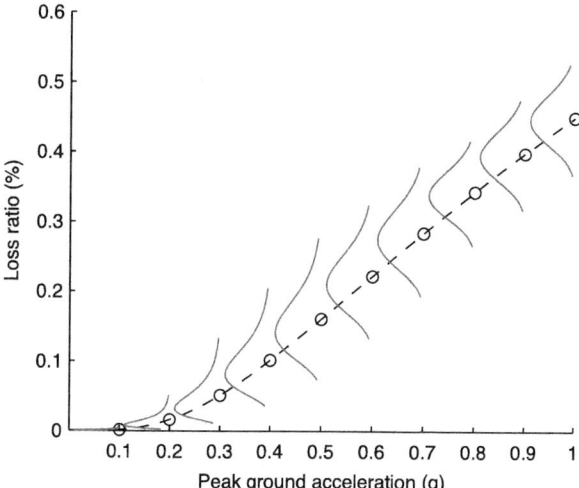

Fig. 16.1 Example of a physical vulnerability function, where the intensity measure type on the x axis is Peak Ground Acceleration (*PGA*) and the mean and distribution of loss ratio is shown at discrete levels of PGA

In a site-specific design project the current practice in "deterministic" hazard assessment is to select a certain number of standard deviations (i.e. epsilon) above or below the median ground motion for the design seismic actions (Abrahamson 2006), but in a scenario risk assessment of distributed assets (e.g. buildings, people, infrastructure), which can be useful for emergency planning as well as risk communication and awareness, the epsilon should not be modelled as fixed across the region of interest. Figure 16.2 shows the natural aleatory variability in ground motions with distance that can be observed from two different earthquakes, together with the median attenuation from both events (thick black line) and the median attenuation from each event (thin black lines). Each event has an inter-event residual ($\delta_{e,1}$ or $\delta_{e,2}$) which is given by the difference between the median curve for both events and the median curve for the specific event; this variability arises due to differences in the source mechanics of the events, such as the stress drop. Within a given event, each site, j, where ground motions have been observed, has a different intra-event residual, ($\delta_{a,1j}$ or $\delta_{a,2j}$) which arises due to the varying path characteristics from the source to the site. Many researchers (e.g. Wang and Takada 2005; Goda and Hong 2008; Jayaram and Baker 2009; Esposito and Iervolino 2011) have shown that the intra-event residuals at two different sites for a given event are correlated, as a function of their separation distance – the greater the distance, the lower the correlation between the residuals. Hence, when modelling distributed ground motions for a future potential scenario earthquake, a sample of the inter-event residual/epsilon for the event should be made and then this should be combined (through SRSS) with the intra-event residual/epsilon at each site, which should be obtained by employing a model of spatial correlation of the intra-event residuals (see e.g. Crowley et al. 2008 for a summary of this process). Figure 16.3 shows examples of ground-motion distributions, or fields based on different assumptions: median ground motion everywhere, uncorrelated ground-motion residuals, and spatially correlated ground-motion residuals.

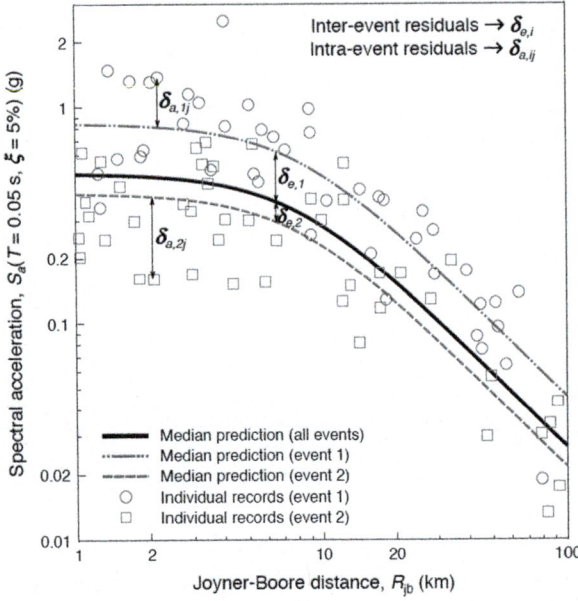

Fig. 16.2 Spatial variability from two different earthquake events (Bommer and Stafford 2008)

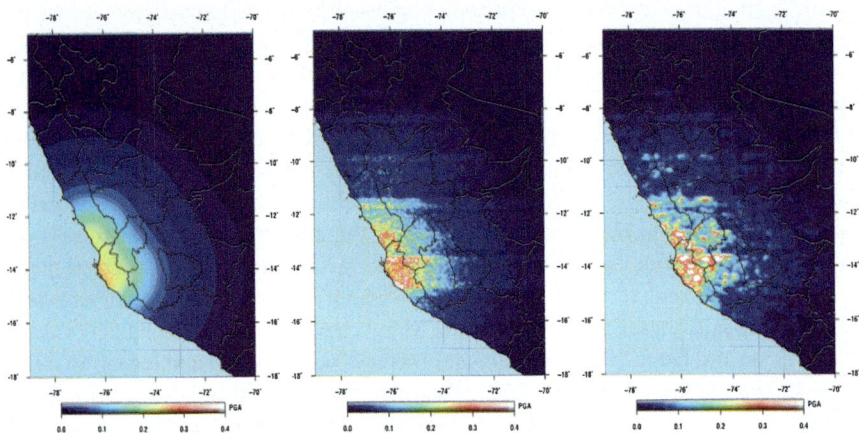

Fig. 16.3 Example of simulated ground-motion fields (PGA in g), based on median ground motion (*left*), one realization of uncorrelated ground-motion residuals (*centre*) and one realization of spatially correlated ground-motion residuals (*right*) (From Silva et al. 2014a)

For the estimation of the loss to all assets in the exposure model, the damage/loss assessment should be based on a simulation of all possible ground-motion fields that could occur, and thus the event should be repeated many times, sampling across the full inter-event variability, and then the total mean damage/loss and total standard deviation of damage/loss across all simulations can be estimated.

Nevertheless, in practice, scenario-based risk assessments are frequently based on the ground motions with a fixed epsilon (often taken as 0 or +1) applied at all sites. Such an approach assumes the unrealistic scenario of full spatial correlation of the ground-motion residuals. When epsilon is taken as +1 everywhere, the assumption being made is that the shaking at all locations has just 16 % probability of ever being exceeded, and the joint probability of occurrence of this level of ground motion at all sites will be extremely low. The resulting damage/loss thus also has an extremely low probability of occurrence, and its usefulness for communicating risk or preparing for emergency situations is questionable.

Even when the damage/loss is required at just a single location, the use of the median or even the mean ground motion should be avoided as the resulting damage/loss will often (though not always) be an underestimation of the damage/loss that would be expected, on average, should the event be repeated many times. An underestimation of damage/loss is expected when the ground motion is concentrated over the range that leads to loss ratios that are less than 50 % (from the vulnerability function), though the opposite may occur if the ground motions are concentrated in the upper 50 %. Figure 16.4 shows an example of the mean loss based on the median ground motion (A) and the mean loss and standard deviation of loss based on the ground motion with aleatory variability (B).

In order to estimate the mean damage/loss at a *single site*, an alternative procedure can be employed which does not require the added complication of separating the inter- and intra-event ground-motion variability and simulation of the ground motions, as described previously. Instead, at the chosen location, one should combine the probability of occurrence of each intensity measure level IML (by integrating the probability density function of ground motion based on the total aleatory variability) with the mean loss ratio from the vulnerability function at each IML, and sum across all IMLs. Due to the lognormal function of ground-motion variability and the nonlinear vulnerability function, the mean loss at the mean ground motion will not be the same as the mean loss considering the full range of potential ground motions at the site; in the example given in Fig. 16.5, the former is 0.098 and the latter (as shown in the workings of Table 16.1) is 0.105. Although the difference is not pronounced in this example, it can be larger and will depend on the specific ground-motion distribution and vulnerability function.

In this example the numerical integration of the ground-motion variability with the mean loss ratio has been used, but since the vulnerability function could also have an analytical form, an analytical integration is also possible, which would be based on the following formula:

$$\overline{LR} = \int_0^\infty LR|IML \times f_{IML}(IML|\mu_{IML}, \sigma_{IML}) dIML$$

where $LR|IML$ stands for the conditional loss ratio for a given an intensity measure level (IML), and $f_{IML}(IML|\mu_{IML}, \sigma_{IML})$ stands for the conditional probability density function of ground motion given a mean intensity measure level (μ_{IML}) and associated standard deviation (σ_{IML}).

Fig. 16.4 Mean loss based on the median ground motion (**a**) and the mean loss and standard deviation of loss based on the full aleatory variability of ground motion (**b**) (Silva 2013)

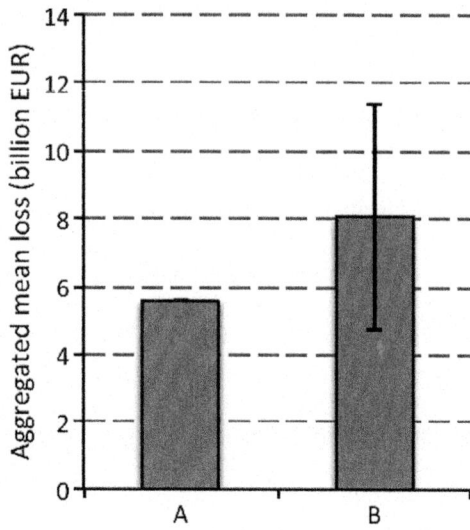

Fig. 16.5 Illustrative figure of the variability in ground motion (in this case PGA) at a given site and how this probability distribution should be integrated at intervals to get the probability of occurrence, and combined with the mean loss ratios from the vulnerability function

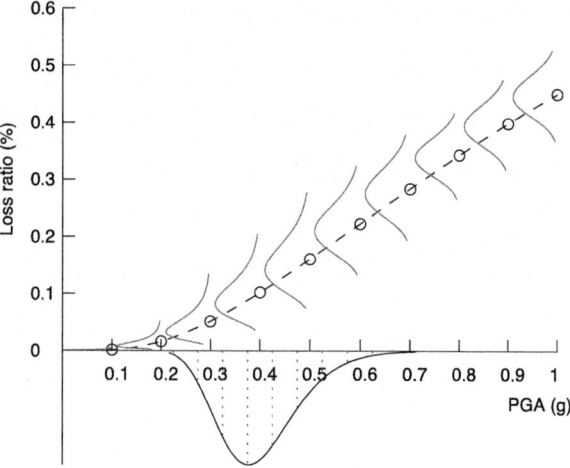

For what concerns the estimation of the standard deviation of the loss, it is also possible to do that by combining the probability density function of the loss ratio and ground-motion shaking through the employment of the total probability theorem (more details are given in Crowley et al. 2010).

16.2.2 Probabilistic Hazard/Risk Assessment

In a fully probabilistic risk assessment, where all possible and relevant deterministic earthquake scenarios are considered together with all possible ground motion

Table 16.1 Estimation of the mean loss ratio based on example shown in Fig. 16.5[a]

IML	Prob. occur PO|IML	Mean loss ratio MLR	PO|IML × MLR
0.20	0.004	0.016	0.000
0.25	0.041	0.032	0.001
0.30	0.135	0.052	0.007
0.35	0.230	0.075	0.017
0.40	0.240	0.102	0.024
0.45	0.176	0.131	0.023
0.50	0.099	0.161	0.016
0.55	0.046	0.192	0.009
0.60	0.019	0.223	0.004
0.65	0.007	0.254	0.002
0.70	0.002	0.284	0.001
			$\sum = 0.105$

[a]It is noted that the numerical integration depicted in Fig. 16.5 and in the calculations in Table 16.1 is purely demonstrative and in practice a much smaller integration interval should be employed

probability levels, there are two commonly applied approaches in practice: one based on the outputs of a PSHA (i.e. using the rate or probability of exceedance of a set of IMLs) and the other based on the simulated ground-motion fields from scenario events (which can either represent the full set of potential ruptures, or can be a reduced set of scenarios, each with an associated probability of occurrence). The use of one method over the other depends on the application, and whether there is a need to robustly model the standard deviation of damage/loss across the full set of assets, or not. If the main output of interest is the annual expected/average value of damage/loss, if the risk at a single site is required, or if a comparative analysis of the risk at different sites is required, then the outputs of classical PSHA (i.e. Cornell 1968; McGuire 1976) can be employed.

In this approach, a PSHA is carried out for the region leading to hazard maps for a given intensity measure type (e.g. spectral acceleration at 1 s) for a number of return periods. The use of PSHA hazard maps is appropriate for site-specific risk assessment and maps which present the comparative risk at different sites, but a frequent error that is made in practice is to use a single hazard map and to report that the damage/loss at each site has the same return period/probability of exceedance as the hazard map upon which it was derived. The problem with such an approach is that it ignores the uncertainty in the vulnerability assessment (e.g. from the fragility functions and the damage-loss conversion). As shown previously in Fig. 16.1, the probability of exceeding a specific loss value is conditional on a number of different intensity measure levels; from the hazard curve one can obtain the probability of occurrence of those intensity measure levels, and by multiplying the two we obtain a number of unconditional probabilities of exceeding the loss value, which are then summed to get the total probability of exceeding the loss value. We then plot the loss value against its respective probability of exceedance to produce a so-called loss exceedance curve (Fig. 16.6).

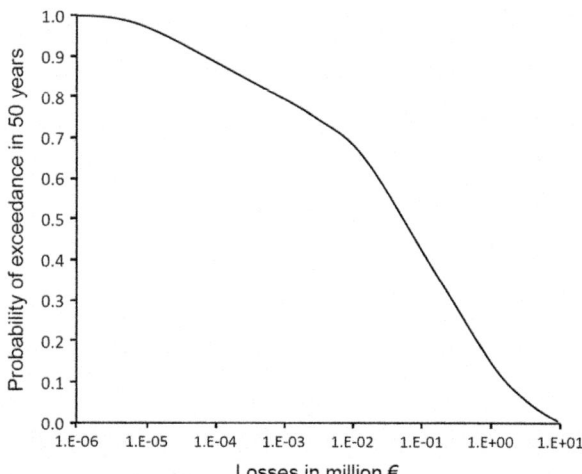

Fig. 16.6 Loss exceedance curve

An event-based approach to probabilistic risk assessment is required when the mean and standard deviation of the total, aggregated, loss to a spatially distributed portfolio of assets is to be estimated. By modelling each event separately we are able to model the spatial correlation of ground motions, as discussed previously. The way in which the ground-motion aleatory variability is spatially modelled affects the standard deviation of the loss; neglecting site-to-site ground-motion correlation leads to systematically underestimation of large, rare losses and overestimation of smaller but frequent ones (see e.g. Crowley and Bommer 2006; Park et al. 2007; Weatherill et al. 2013). Monte Carlo simulation is generally employed to simulate the seismicity of the next one hundred thousand years or so (see e.g. Pagani et al. 2014), and for each event a spatially correlated field of ground motion is simulated, and the resulting damage/loss is estimated by combing this with the exposure and vulnerability models (see e.g. Crowley and Bommer 2006; Silva et al. 2013a).

However, when different intensity measure types are used in the model (e.g. for the vulnerability functions of different assets) then they need to be cross-correlated (also known as spectrally correlated). Baker and Cornell (2006) looked at the cross-correlation between the residuals of spectral accelerations (i.e. the difference between the spectral acceleration from a record at a given period and the spectral acceleration predicted for that record using a ground-motion prediction equation) at different periods using a number of records and found that they were neither uncorrelated (Fig. 16.7a) nor fully correlated (Fig. 16.7b), but featured a correlation that varied as a function of the inter-period difference. Application of the model leads to simulated spectra like those shown in Fig. 16.7c, which are seen to be highly realistic when compared with real spectra with similar characteristics (Fig. 16.7d). It should be noted that it is not just the intra-event variability of different intensity measures that is cross-correlated but also the inter-event variability (see e.g. Goda and Atkinson 2009).

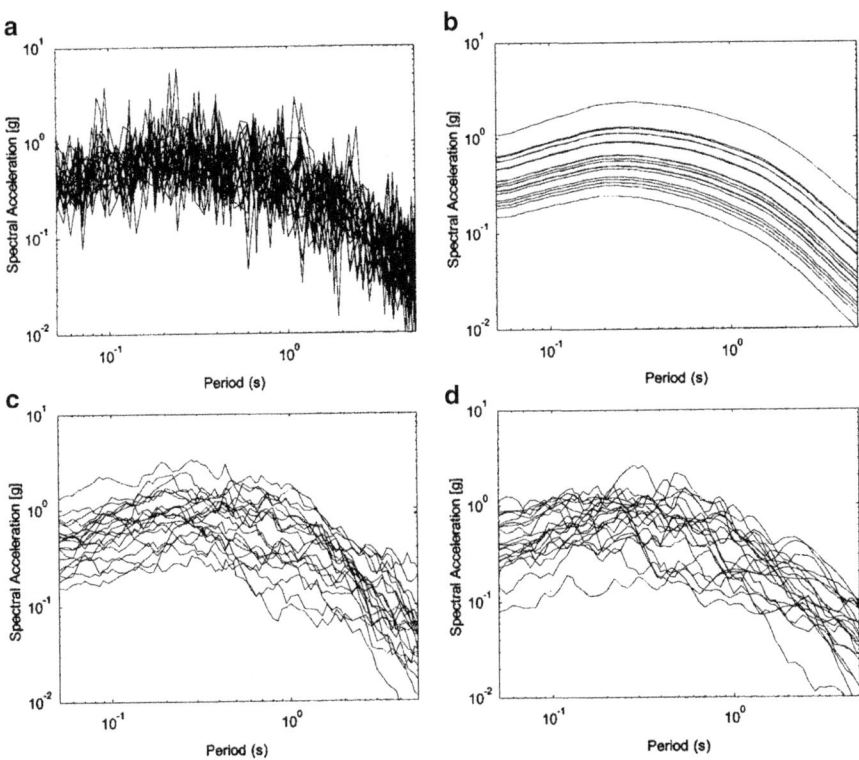

Fig. 16.7 Comparison of simulated spectra with no inter-period correlation (**a**), full inter-period correlation (**b**), modelled inter-period correlation (**c**) with real spectra (**d**), taken from Baker and Cornell (2006)

When simulating spatial distributions of ground motion for loss assessment, if cross-correlation, for example between the spectral acceleration at 0.3 s (used for the vulnerability function of a low rise building type) and that at 1.0 s (used for a mid rise building type), is not modelled, and each ground-motion field is simulated independently, the impact of the spatial correlation is eroded when the combined damage/loss to both building types is estimated. Weatherill et al. (2013) show that the impact of spatial correlation on the total loss to a heterogeneous portfolio is minimal when cross-correlation is not modelled (Fig. 16.8) but that when both spatial correlation and cross-correlation are accounted for, the impact on the losses at low probabilities of exceedance can be significant. However, it is noted that the portfolio selected by Weatherill et al. (2013) was highly heterogeneous and included building types with a very wide range of periods of vibration; should the portfolio be more clustered around a smaller range of periods of vibration then the impact of the inclusion or not of spatial correlation (without cross correlation) will have a significant effect on the resulting losses, as has been shown in other studies (e.g. Crowley et al. 2008).

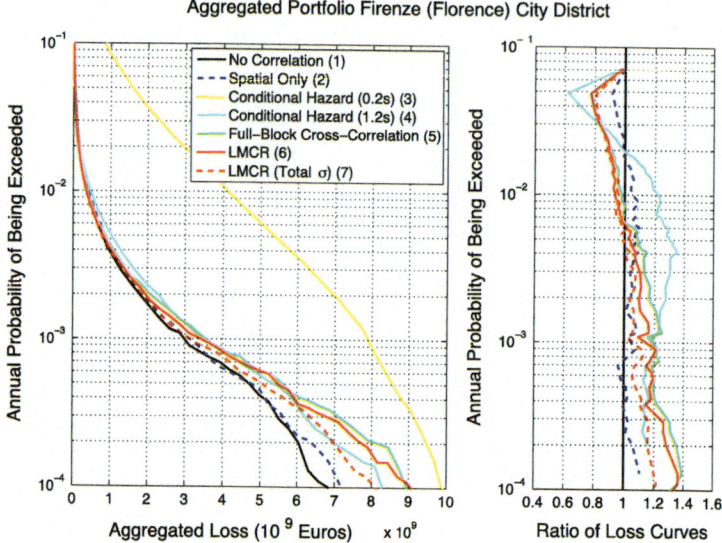

Fig. 16.8 Comparison of spatial correlation (*blue curve*) and spatial cross-correlated losses (*green and red curves*) on the total loss to a heterogeneous portfolio of losses (Weatherill et al. 2013)

16.3 Fragility and Vulnerability Modelling

16.3.1 Issues Related to Commonly Used Intensity Measure Types

The use of macroseismic intensity continues to be a popular choice for fragility and vulnerability modelling, especially when the latter is based on observed damage and loss data. One of the main reasons for this lies behind the volume of macroseismic intensity data that is available following an event, which allows us to constrain the level of shaking, and thus reduce the uncertainty in an empirical vulnerability model. It is furthermore frequently argued that the use of macroseismic intensity leads to more reliable damage/loss estimates as it is possible to carry out an internal consistency check. However, there are still a number of shortcomings in using macroseismic intensity in risk assessment. The previous section discussed the developments on the modelling of spatially correlated ground motion for the loss assessment of distributed portfolios; although state-of-the-art Intensity Prediction Equations are still being developed (e.g. Allen et al. 2012) there are currently few, if any, models of spatial correlation of the residuals of macroseismic intensity. Furthermore, when good data on the site conditions within a given area is available, the impact of site amplification on macroseismic intensity is still generally modelled in an empirical manner without explicit modelling of the uncertainties.

The use of instrumental intensity measures in vulnerability modelling is required when analytical modelling of the response of structures is employed. In this case the explicit nonlinear behaviour of structures of a given class under accelerograms with differing characteristics is evaluated. However, many analytical vulnerability models developed today do not propagate all the uncertainties from the variability in the capacity of the structures of a given class (due to varying geometrical, material and design detailing properties), to the variability in the response from records with the same intensity measure level (i.e. record to record variability), to the variability in the limit state thresholds to damage (e.g. in the values of inter-storey drift that would lead to collapse), to the uncertainties in the conversion of damage to loss (e.g. uncertainty in the cost of repairing buildings that are extensively damaged). Although these uncertainties might not necessarily be robustly and explicitly modelled at every stage of the vulnerability function derivation, an attempt should be made to include them, even just through engineering judgement. This is an area that vulnerability modellers will need to focus on further in the future.

One of the most diffused methodologies for scenario-based risk assessment includes the use of the capacity spectrum method (see e.g. Freeman et al. 1975), as proposed in ATC 40 (ATC 1996) and implemented in the HAZUS software (FEMA 2003). In this methodology the median nonlinear response of the buildings of a given class is estimated by combing the capacity curve with a response spectrum, and then fragility functions based on this nonlinear response parameter provide the damage distribution (see Fig. 16.9).

In the original HAZUS method the spectral ordinates at 0.3 and 1.0 s are estimated, and then the full response spectrum is obtained by applying a code spectral shape. With the use of a fixed spectral shape, the specific spectral characteristics of the event under consideration are not accounted for, and given that a code spectral shape attempts to reproduce a uniform hazard spectrum, enveloping both low magnitude nearby events as well as high magnitude distant events (see Fig. 16.10), the response spectrum used may be unrealistic. An improvement on this practice is to use a scenario spectrum from a ground-motion prediction equation, appropriate for the region and scenario. However, this modelling decision is not without its drawbacks as a fixed epsilon (defined in Sect. 16.2), generally taken as zero, is frequently applied in practice and thus cross-correlation is ignored. Instead, and as mentioned previously, a large number of cross-correlated scenario spectra should be simulated and used in the scenario risk analyses, after which the mean and standard deviation of damage/loss can be estimated. An alternative approach to using ground-motion prediction models for simulating realistic ground motions (with spatially cross correlated intensity measures) would be to use physics-based methods for modelling the fault rupture and wave propagation (and associated uncertainties), leading to a number of synthetic records at the sites in question (see e.g. Atkinson 2012).

When the capacity spectrum method (or any other nonlinear static procedure, NSP) is used in PSHA-based risk assessment, as has been done in many applications (e.g. in the LESSLOSS project as described in Spence 2007; in the RISK-UE

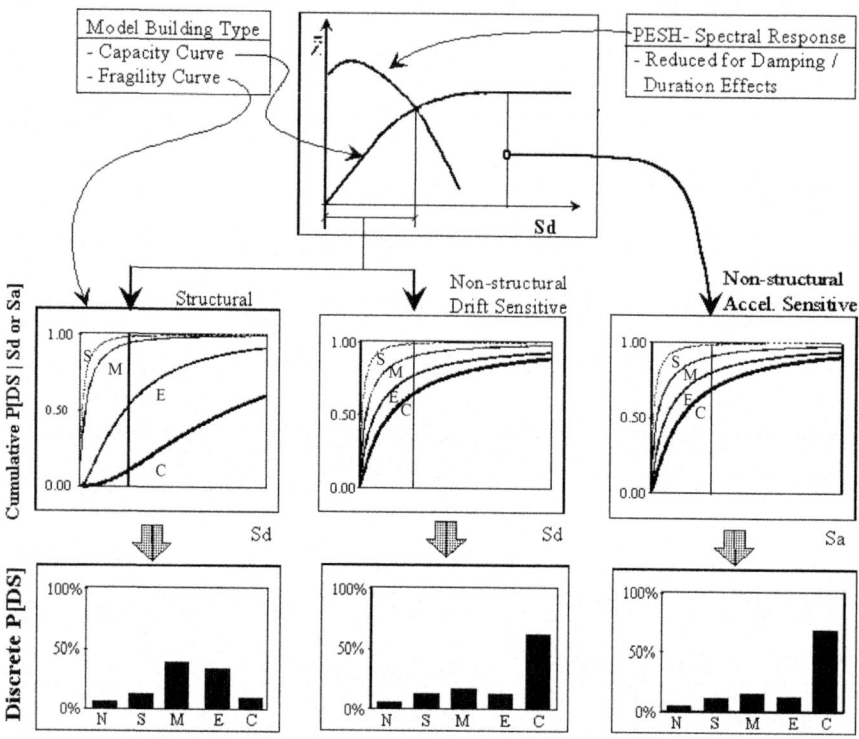

Fig. 16.9 Application of the capacity spectrum method in HAZUS (FEMA 2003)

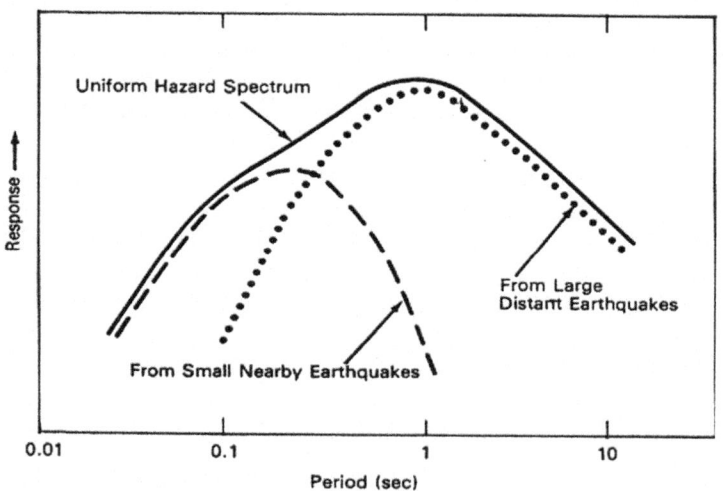

Fig. 16.10 Schematic sketch of a uniform hazard spectrum at a given return period in which the contributions to hazard at the shorter and longer periods come from different sources (Reiter 1990)

project, as described in Mouroux and Le Brun 2006) and software (see e.g. Crowley et al. 2010), the uniform hazard spectrum (UHS) at a number of different return periods needs to be employed. The problems with this approach are that, again, the spectral shape is unrealistic and all spectral ordinates are assumed to be fully correlated. A vector-based PSHA analysis (e.g. Bazzurro and Cornell 2002), where the joint probability of exceedance of spectral acceleration at multiple periods is estimated, would need to be employed to address these issues. However, applying such a method to the full response spectrum might not be feasible and it would most probably be simpler to revert to a Monte Carlo event-based approach (as mentioned earlier in Sect. 16.2).

There are other issues with the use of NSPs in risk assessment which include bias and uncertainty in the nonlinear response (due to the assumptions on the elongation of the period of vibration and the equivalent viscous damping in the structural system, which often do not have an associated uncertainty) and underestimation of the record-to-record variability (see e.g. Pinho et al. 2013; Silva et al. 2013b). Hence, the use of vulnerability functions based on nonlinear dynamic analysis and derived in terms of elastic scalar intensity measures would both simplify the hazard modelling required in the risk assessment (at least for homogeneous portfolios, as discussed in Sect. 16.2), and avoids issues of response bias and underestimation of uncertainties. The main price that is paid with the use of dynamic analysis is the computational demand, which is much higher when many structures and records are considered. Should there thus be a desire to improve the computational efficiency, NSPs could instead be used (provided the increased uncertainties and bias are both accounted for), but it is nevertheless recommended that they are used to develop scalar intensity measure-based vulnerability functions, to simplify the hazard modelling requirements (see e.g. Silva et al. 2014b).

The elastic scalar intensity measure that is most commonly applied is the spectral acceleration at the fundamental period of the structure. However, as discussed previously, different structures in the portfolio will have different periods of vibration and thus with the use of such an intensity measure type there will be a need to model vector quantities of ground motion. In order to avoid this, one option could be to use a fixed period of vibration (e.g. 0.5 s) for all buildings in the portfolio. This avoids the need to model spectral correlation, but has the drawback that the chosen period may not be the most efficient for all the building types in the exposure model. The primary advantage of an efficient intensity measure is that it should require fewer numerical analyses to achieve a desired level of confidence in the nonlinear response (Mackie and Stojadinovic 2005). Hence, it is to be expected that the use of an inefficient intensity measure type would increase the uncertainty in the vulnerability functions. A comparison of the loss exceedance curves that are produced for a heterogeneous portfolio with vulnerability models based on efficient (structure-dependent) intensity measures and cross-correlation of the ground motion should be made against the curves obtained with vulnerability functions based on a fixed intensity measure type and no cross-correlation, to assess whether the increased simplicity of the analysis is penalised by an increased uncertainty in the final loss.

16.3.2 Correlation of Vulnerability Uncertainty

When vulnerability functions for a class of structures are used in a regional risk assessment, the uncertainty needs to be sampled from the loss distribution (see Fig. 16.1). The question which then arises is whether all the buildings of a given typology within the region will respond better or worse than average, and thus whether there is a correlation in this uncertainty. For example, after the Northridge earthquake in 1994 a previously unknown design deficiency in the connections of steel structures was observed, which led to a correlation in the response of the buildings of this class, and in Turkey after the 1999 Kocaeli earthquake, there was a case where all but one mid-rise concrete frame buildings in the same complex collapsed. Currently, however, it is generally not possible to do more than estimate the losses both with and without vulnerability uncertainty correlation; more research is needed to better constrain this correlation. In the meantime, a useful practice is to run the risk model both with and without full correlation to get the bounds of the expected losses.

16.3.3 Epistemic Uncertainty

Finally, a practice that has increased recently includes the use of logic trees to model the epistemic uncertainties in vulnerability modelling (e.g. Molina et al. 2010). However, this practice is not widespread and more research is needed in order to bring this practice to the level of maturity found with the use of logic trees within PSHA studies. For example, the recent European hazard modelling project SHARE (www.share-eu.org) used a state-of-the-art methodology for developing the ground-motion logic tree that combined expert judgement with the use of strong ground-motion data for the selection, ranking and weighting (Delavaud et al. 2012). Although the data available for testing vulnerability models is sparse, initiatives such as the GEM Global Earthquake Consequences Database[1] (that is collecting damage and loss data for a number of building typologies around the world) will help improve the potential for data-driven guidance for vulnerability model selection.

---

[1] http://www.globalquakemodel.org/what/physical-integrated-risk/consequences-database/

16.4 Exposure Modelling

There are two main types of exposure models: building-by-building and aggregated. In the latter case the buildings with the same structural/non-structural characteristics (taxonomy[2]) are aggregated within the boundaries of a given area, which is often a zip code, administrative area or grid cell, and relocated to a single location (either because the locations of the individual buildings are unknown, or to increase computational efficiency of the model). This is the most common type of exposure model (e.g. Crowley et al. 2010; Campos Costa et al. 2009; Erdik et al. 2003), but is also the one that raises the most risk modelling difficulties.

As discussed in Bazzurro and Park (2007), when all of the buildings are relocated and aggregated, the same intensity measure level is input to the vulnerability model which means that a full correlation of ground motion is assumed for these buildings. In reality, however, these buildings would be distributed across the zip code/grid cell and would thus be subject to spatially variable ground motion. Furthermore, all of these buildings will have the same sample of uncertainty in the vulnerability model applied to them, further correlating the loss of these building types. If we know the number of buildings that have been aggregated we can avoid the latter correlation by sampling a number of vulnerability residuals equal to the number of buildings at the given location, and estimate the loss for each building separately, after which the statistics for the building typology can be estimated.

There are at least two options to deal with the induced ground-motion correlation due to aggregation of the buildings: random disaggregation of the buildings within the aggregation area, or modification of the ground-motion aleatory variability (see e.g. Stafford 2012). The former approach is straightforward but increases significantly the computational demands of the analysis, especially when there are millions of assets in the model. The latter approach, described in Stafford (2012), reduces the variance of the ground motion when it is taken to represent the average of a given area, rather than the ground motion of a single point (which is the case for distributed assets), following the recommendations of Vanmarcke (1983). More investigation is needed to compare these methods and to study the difference in losses and computational performance of both these two approaches together with the case that simply ignores this induced correlation, thus adding to the studies and conclusions of Bazzurro and Park (2007). The availability of more building-by-building exposure models (so-called "ground truth" models), such as those that can be produced with the tools developed by the Global Earthquake Model,[3] will allow the impact of various exposure aggregation assumptions to be further investigated.

In practice exposure models do not generally feature uncertainties, even though they are usually developed with poor data and a large number of assumptions, and are arguably the most uncertain component of the risk model. For large regions these models are often a combination of population and building census data (where

---

[2] http://www.globalquakemodel.org/what/physical-integrated-risk/building-taxonomy/

[3] http://www.globalquakemodel.org/what/physical-integrated-risk/inventory-capture-tools/

the latter might actually refer to the dwellings rather than the buildings and which often do not include the necessary structural information of the buildings), statistics on the average characteristics of dwellings/buildings in the region, expert judgement on replacement costs per square metre and so on. The assignment of uncertainty to exposure models, as well as of any correlations in the uncertainty, is certainly an area that would benefit from increased research attention.

16.5 Conclusions

This paper has looked at many commonly applied modelling assumptions in the seismic risk assessment of portfolios of distributed buildings. One of the main points that should be clear is that as the developments in ground-motion modelling continue to progress, in particular those related to the correlation of aleatory variability, these have an impact on the way in which exposure and vulnerability models are treated in risk modelling. Furthermore, the correlated uncertainties in the vulnerability and exposure models require more attention in future regional risk modelling research.

A number of research questions that require further investigation have been raised herein:

- Is the penalty for simplifying the intensity measures in vulnerability models too high in terms of the associated uncertainties in the losses?
- How can we define the correlation of vulnerability uncertainty within a given building class?
- Can we apply lessons learned from data-driven ground-motion prediction equation logic tree modelling to vulnerability models?
- How should we deal with the induced ground-motion correlation of aggregated buildings in exposure models, and what is the impact of ignoring it?
- How can we attempt to model the uncertainties in exposure models?

Hence, although the practice of seismic risk assessment is well established, there are still a number of areas that require further research and exploration by the present and next generations of risk modellers.

Acknowledgements The author is grateful to Vitor Silva and Rui Pinho for reviewing earlier drafts of this manuscript and whose comments have greatly improved the paper.

Open Access This chapter is distributed under the terms of the Creative Commons Attribution Noncommercial License, which permits any noncommercial use, distribution, and reproduction in any medium, provided the original author(s) and source are credited.

References

Abrahamson N (2006) Seismic hazard assessment: problems with current practice and future developments. In: Proceedings of 1st European conference on earthquake engineering and seismology, paper number: Keynote address K2. Geneva, Switzerland

Allen TI, Wald DJ, Worden B (2012) Intensity attenuation for active crustal regions. J Seismol 16(3):409–433

ATC-40 (1996) Seismic evaluation and retrofit of concrete buildings, vols 1 and 2. report no. ATC-40, Applied Technology Council, Redwood City, CA, USA, p 180

Atkinson GM (2012) Integrating advances in ground-motion and seismic hazard analysis. In: Proceedings of the 15th World conference on earthquake engineering, Keynote/Invited lecture, Lisbon

Baker JW, Cornell CA (2006) Correlation of response spectral values for multicomponent ground motions. Bull Seismol Soc Am 96(1):215–227

Bazzurro P, Cornell CA (2002) Vector-valued probabilistic seismic hazard assessment. In: Proceedings of 7th US National conference on earthquake engineering, Boston, MA

Bazzurro P, Park J (2007) The effects of portfolio manipulation on earthquake portfolio loss estimates. In: Proceedings of the 10th International conference of application of statistics and probability in civil engineering (ICASP10), Tokyo, Japan

Bommer JJ, Stafford PJ (2008) Seismic hazard and earthquake actions. In: Elghazouli AY Seismic design of buildings to eurocode 8, Taylor & Francis, London, UK

Calvi GM, Pinho R, Magenes G, Bommer JJ, Restrepo-Velez LF, Crowley H (2006) Development of seismic vulnerability assessment methodologies over the past 30 years. ISET J Earthq Technol, paper no. 472, 43(3):75–104

Campos Costa A, Sousa ML, Carvalho A, Coelho E (2009) Evaluation of seismic risk and mitigation strategies for the existing building stock: application of LNECLoss to the metropolitan area of Lisbon. Bull Earthq Eng 8:119–134

Cornell CA (1968) Engineering seismic risk analysis. Bull Seismol Soc Am 58:1583–1606

Crowley H, Bommer JJ (2006) Modelling seismic hazard in earthquake loss models with spatially distributed exposure. Bull Earthq Eng 4(3):249–273

Crowley H, Bommer JJ, Stafford P (2008) Recent developments in the treatment of ground-motion variability in earthquake loss models. J Earthq Eng, Special Issue, 12(S2):71–80

Crowley H, Cerisara A, Jaiswal K, Keller N, Luco N, Pagani M, Porter K, Silva V, Wald D, Wyss B (2010) GEM1 seismic risk report: part 2, GEM Technical Report 2010-5, GEM Foundation, Pavia, Italy

Delavaud E, Cotton F, Akkar S, Scherbaum F, Danciu L, Beauval C, Drouet S, Douglas J, Basili R, Sandikkaya MA, Segou M, Faccioli E, Theodoulidis N (2012) Toward a ground-motion logic tree for probabilistic seismic hazard assessment in Europe. J Seismol 16(3):451–473

Erdik M, Aydinoglu N, Fahjan Y, Sesetyan K, Demircioglu M, Siyahi B, Durukal E, Ozbey C, Biro Y, Akman H, Yuzugullu O (2003) Earthquake risk assessment for Istanbul metropolitan area. Earthq Eng Eng Vib 2(1):1–23

Esposito S, Iervolino I (2011) PGA and PGV spatial correlation models based on European multievent datasets. Bull Seismol Soc Am 101(5):2532–2541

FEMA (2003) HAZUS-MH technical manual. Federal Emergency Management Agency, Washington, DC

Freeman S, Nicoletti J, Tyrell J (1975) Evaluation of existing buildings for seismic risk – a case study of Puget sound naval shipyard, Bremerton, Washington. In: Proceedings of 1st U.S. National conference on earthquake engineering, Berkley, USA

Goda K, Atkinson GM (2009) Probabilistic characterisation of spatial correlated response spectra for earthquakes in Japan. Bull Seismol Soc Am 99(5):3003–3020

Goda K, Hong HP (2008) Spatial correlation of peak ground motions and response spectra. Bull Seismol Soc Am 98(1):354–365

Jayaram N, Baker JW (2009) Correlation model of spatially distributed ground motion intensities. Earthq Eng Struct Dyn 38:1687–1708

Mackie K, Stojadinovic B (2005) Fragility Basis for California Highway Overpass Bridge Seismic Decision Making, PEER Report 2005/12. Pacific Earthquake Engineering Research Center, University of California, Berkeley, CA

McGuire RK (1976) FORTRAN computer program for seismic risk analysis, US Geological Survey Open-File Report 76–67, Denver, US

Molina S, Lang DH, Lindholm CD (2010) SELENA: an open-source tool for seismic risk and loss assessment using a logic tree computation procedure. Comput Geosci 36:257–269

Mouroux P, Brun BTL (2006) Presentation of RISK-UE project. Bull Earthq Eng 4:323–339

Pagani M, Monelli D, Weatherill G, Danciu L, Crowley H, Silva V, Henshaw P, Bulter L, Matteo N, Panzeri L, Simionato M, Vigano D (2014) OpenQuake-engine: an open hazard (and Risk) software for the global earthquake model. Seismol Res Lett 85:692–702, in press

Park J, Bazzurro P, Baker JW (2007) Modelling spatial correlation of ground motion Intensity Measures for regional seismic hazard and portfolio loss estimation. In: Kanada J (ed) Applications of statistics and probability in Civil Engineering. Takada and Furuta, Taylor & Francis Group, London

Pinho R, Marques M, Monteiro R, Casarotti C, Delgado R (2013) Evaluation of Nonlinear Static Procedures in the assessment of building frames. Earthq Spectra 29(4):1459–1476

Reiter L (1990) Earthquake hazard analysis. Columbia University Press, New York

Rossetto T, D'Ayala D, Ioannou I, Meslem A (2014) Evaluation of existing fragility curves. In: Pitilakis KD, Crowley H, Kaynia AM (eds) SYNER-G: typology definition and fragility functions for physical elements at seismic risk. Springer, New York

Silva V (2013) Development of open-source tools for seismic risk assessment: application to Portugal, PhD Thesis, University of Aveiro, Portugal

Silva V, Crowley H, Pagani M, Monelli D, Pinho R (2013a) Development of the OpenQuake engine, the Global Earthquake Model's open-source software for seismic risk assessment. Nat Hazards. doi:10.1007/s11069-013-0618-x

Silva V, Crowley H, Pinho R, Varum H (2013b) Extending Displacement-Based Earthquake Loss Assessment (DBELA) for the computation of fragility curves. Eng Struct 56:343–356

Silva V, Crowley H, Yepes C, Pinho R (2014a) Presentation of the OpenQuake-engine, an open source software for seismic hazard and risk assessment. Proceedings of the 10th US National conference on earthquake engineering, Anchorage, Alaska

Silva V, Crowley H, Varum H, Pinho R, Sousa R (2014b) Evaluation of analytical methodologies to derive vulnerability functions. Earthq Eng Struct Dyn 43(2):181–204

Spence R (ed) (2007) Earthquake disaster scenario prediction and loss modelling for urban areas. IUSS Press, Pavia. ISBN 978-88-6198-011-2

Stafford PJ (2012) Evaluation of structural performance in the immediate aftermath of an earthquake: a case study of the 2011 Christchurch earthquake. Int J Forensic Eng 1(1):58–77

Vanmarcke E (1983) Random fields, analysis and synthesis. The MIT Press, Cambridge, MA

Wang M, Takada T (2005) Macrospatial correlation model of seismic ground motions. Earthq Spectra 21(4):1137–1156

Weatherill G, Silva V, Crowley H, Bazzurro P (2013) Exploring strategies for portfolio analysis in probabilistic seismic loss estimation. In: Proceedings of Vienna Congress on recent advances in earthquake engineering and structural dynamics (VEESD 2013), paper no. 303, Vienna, Austria

Chapter 17
The Role of Pile Diameter on Earthquake-Induced Bending

George Mylonakis, Raffaele Di Laora, and Alessandro Mandolini

Abstract Pile foundations in seismic areas should be designed against two simultaneous actions arising from kinematic and inertial soil-structure interaction, which develop as a result of soil deformations in the vicinity of the pile and inertial loads imposed at the pile head. Due to the distinct nature of these phenomena, variable resistance patterns develop along the pile, which are affected in a different manner and extent by structural, seismological and geotechnical characteristics. A theoretical study is presented in this article, which aims at exploring the importance of pile diameter in resisting these actions. It is demonstrated that (a) for large diameter piles in soft soils, kinematic interaction dominates over inertial interaction; (b) a minimum and a maximum admissible diameter can be defined, beyond which a pile under a restraining cap will inevitably yield at the head i.e., even when highest material quality and/or amount of reinforcement are employed; (c) an optimal diameter can be defined that maximizes safety against bending failure. The role of diameter in seismically-induced bending is investigated for both steel and concrete piles in homogenous soils as well as soils with stiffness increasing proportionally with depth. A number of closed-form solutions are presented, by means of which a number of design issues are discussed.

G. Mylonakis (✉)
Department of Civil Engineering, University of Bristol, Bristol, UK

Department of Civil Engineering, University of Patras, Patras, Greece

Department of Civil Engineering, University of California, Los Angeles, USA
e-mail: g.mylonakis@bristol.ac.uk

R. Di Laora • A. Mandolini
Department of Civil Engineering, Second University of Napoli, Aversa, Italy

17.1 Introduction

An increasing number of research contributions dealing with the behavior of piles under earthquake action has become available in recent times. The topic started attracting interest by researchers when theoretical studies (accompanied by a limited number of post-earthquake investigations) revealed the development of large bending moments at the head of piles restrained by rigid caps, even in absence of large soil movements such as those induced by slope instability or liquefaction. Nevertheless, the interpretation of available evidence – and thus its implementation in design – has proven to be difficult due to the lack of simple analysis methods to assess the specific type of pile bending. The simultaneous presence of kinematic and inertial interaction phenomena (Fig. 17.1), whose effects are difficult to separate, adds to the complexity of interpreting such data.

On the other hand, evaluation of kinematic moments is mandatory under certain conditions according to most modern seismic Codes. For example, Eurocode 8 prescribes that: "*bending moments developing due to kinematic interaction shall be computed only when all of the following conditions occur simultaneously: (1) the ground profile is of type D, S_1 or S_2, and contains consecutive layers of sharply differing stiffness; (2) the zone is of moderate or high seismicity, i.e. the product $a_g S$ exceeds 0.10 g; (3) the supported structure is of class III or IV*".

The first to propose a simple method for assessing the kinematic component of pile bending appear to be Margason (1975) and Margason and Holloway (1977). These articles can be credited as the first to recognize the importance of pile diameter (to be denoted in the ensuing by d) and recommend using small diameters to "conform to soil movements", though without providing rational analysis methods. While several subsequent studies investigated the problem (e.g., Kaynia and Kausel 1991; Kavvadas and Gazetas 1993; Pender 1993; Mylonakis 2001; Nikolaou et al. 2001; Castelli and Maugeri 2009; de Sanctis et al. 2010; Sica et al. 2011; Di Laora et al. 2012; Anoyatis et al. 2013; Kampitsis et al. 2013), only a handful of research efforts focused on the effect of pile diameter – mostly for bending in the vicinity of deep interfaces separating soil layers of different stiffness (Mylonakis 2001; Saitoh 2005).

Recently, Di Laora et al. (2013) explored the role of pile diameter in resisting seismic loads at the pile head under a restraining cap, with reference to steel piles in homogeneous soil. Identified key issues include a d^4 dependence of kinematic bending moment at the pile head, as opposed to a mere d^3 dependence of moment capacity. The first dependence results from pile and soil curvatures being approximately equal at the pile head, while the second stems from fundamental strength-of-materials theory. The discrepancy in the exponents suggests that moment demand on the pile increases faster with diameter than moment capacity, thus making yielding at the head unavoidable beyond a certain size (assuming pile is always a flexural element). The value of the maximum diameter was found to depend mainly on peak ground acceleration, soil stiffness and factor of safety against gravity loading. Interestingly, this behavior is not encountered in the

17 The Role of Pile Diameter on Earthquake-Induced Bending

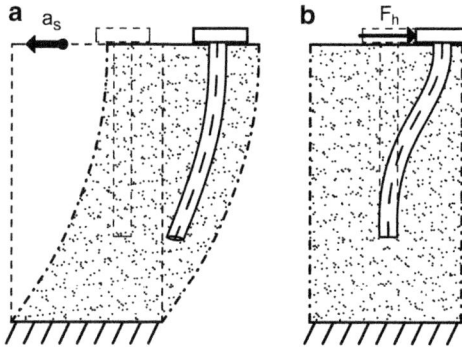

Fig. 17.1 Kinematic and inertial loading of pile foundations. (**a**) Kinematic loading (**b**) inertial loading

vicinity of deep interfaces – which is the topic most investigated in the literature (Mylonakis 2001; Maiorano et al. 2009; Dezi et al. 2010), since in those regions capacity and demand increase with the same power of pile diameter (d^3). Di Laora et al. (2013) also established that combining kinematic and inertial moment at the pile head leads to a limited range of admissible diameters, with the upper bound governed by kinematic action, and the lower one by inertial action.

Proceeding along these lines, the work at hand has the following main objectives: (i) to investigate the relative importance of kinematic and inertial components of seismic demand, and provide a number of closed-form expressions for kinematic/inertial moment demand on piles (ii) to inspect the role of pile diameter on the seismic performance of both steel and concrete piles for the soil types shown in Fig. 17.2; (iii) to provide a number of closed-form solutions for the limit diameters defining the admissible ranges; (iv) to assess the practical significance of the phenomenon through pertinent numerical studies encompassing a wide range of parameters; (v) to define an optimal diameter which maximizes safety against bending failure.

The study employs the following main assumptions: (a) foundation is designed to remain elastic during earthquake ground shaking (i.e., the force modification coefficients are set equal to one); (b) pile is long and can be idealized as a flexural beam that behaves linearly up to the point of yielding; (c) soil restraining action can be modeled using a bed of linear or equivalent-linear Winkler springs, uniformly distributed along the pile axis; (d) pile axial bearing capacity is controlled by both shaft and tip action; (e) perfect contact (i.e., no gap and slippage) exists between pile and soil; (f) group effects on bending at the pile head are minor and can be ignored from a first-order analysis viewpoint. In addition, for the sake of simplicity partial safety factors are not explicitly incorporated in the analysis; a global safety factor is employed instead. It is worth mentioning that the approach in (a) has been questioned in recent years. Under-designing foundations, however, although conceptually promising, is by no means an established design approach and will not be further discussed in this work. Also, the Winkler assumption in (c) is not essential for the subsequent analysis (a wealth of results from numerical continuum solutions do exist as well), yet it is adopted here since it yields sufficiently accurate predictions for the cases examined and allows simple closed-form expressions to be obtained.

17.2 Kinematic Versus Inertial Moment Demand

17.2.1 Kinematic Bending at Pile Head

In recent articles, de Sanctis et al. (2010) and Di Laora et al. (2013) showed that a long fixed-head pile in homogeneous soil experiences a curvature at the top, $(1/R)_s$, which is approximately equal to soil curvature at the same elevation and, thereby, can be computed as:

$$M_{kin} = E_p I_p (1/R)_p = E_p I_p (1/R)_s = E_p I_p \frac{a_s \rho_s}{G_s} \quad (17.1)$$

where $(1/R)_p$, E_p and I_p are curvature, Young's modulus and cross-sectional moment of inertia of the pile (for a circular cross section, $I_p = \pi \, d^4/64$), $(1/R)_p$ and a_s are the soil curvature and horizontal acceleration at soil surface respectively, and $G_s = E_s/2(1+\nu_s)$ is the soil shear modulus, ν_s being the corresponding Poisson ratio. For layered soil and shallow interfaces located within a few pile diameters from the surface, (17.2) provides only a conservative estimate of kinematic bending at the pile head.

Using rigorous elastodynamic Finite Element analyses, Di Laora and Mandolini (2011) derived a fitting formula for kinematic bending in soils with stiffness varying proportionally with depth:

$$M_{kin} = 1.36 a_s \rho_s \left(\frac{E_p}{\overline{E}_s} I\right)^{\frac{4}{3}} (1 + \nu_s) \quad (17.2)$$

where $\overline{E}_s$ is the gradient of soil Young's modulus with respect to depth (Fig. 17.2). Evidently, kinematic moment at the pile head increases with pile bending stiffness and surface acceleration, and decreases with soil stiffness.

17.2.2 Inertial Bending at Pile head

Inertial forces transmitted to piles from an oscillating superstructure, are inherently associated with structural mass. To relate this mass to the geotechnical parameters involved in the problem at hand, it is convenient to assume that the weight carried by each individual pile is a fraction of the pile bearing capacity against axial load, W_P. Considering a long floating cylindrical pile in fine-grained soil and neglecting the contribution of base resistance, W_p can be expressed in terms of geometry, soil properties and a global safety factor (Viggiani et al. 2011) as

$$W_p = \frac{1}{SF} \pi \alpha L d S_u \quad (17.3)$$

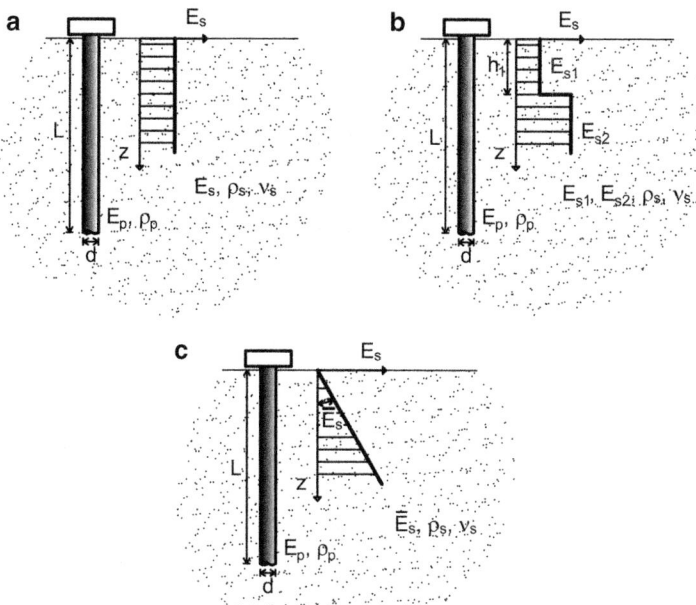

Fig. 17.2 Soil profiles considered in this study. (**a**) Homogeneous profile (**b**) two-layer profile (**c**) inhomogeneous profile

where d and L are the diameter and the length of the pile, S_u is the undrained shear strength of the soil material, α the pile-soil adhesion coefficient (typically ranging from 0.3 to 1 depending on the value of undrained shear strength S_u).

Assuming that the lateral load imposed at the pile head is proportional to the corresponding axial gravitational load W_p, it is straightforward to show from Winkler theory that the maximum seismic moment developing under a rigid cap for soils having constant stiffness near the surface is:

$$M_{in} = \frac{1}{4}\left(\frac{\pi q_I}{\delta}\right)^{\frac{1}{4}}\left(\frac{a_s}{g}\right)\left(\frac{E_p}{E_s}\right)^{\frac{1}{4}} S_a \; W_p d \qquad (17.4)$$

δ being the Winkler stiffness parameter (which varies between approximately 1–2 for inertial loading – Roesset 1980; Dobry et al. 1982, Syngros 2004), $q_I = 1 - (1 - 2t/d)^4$ a dimensionless geometric factor accounting for wall thickness t of a hollow pile, S_a a dimensionless spectral amplification parameter, and g the acceleration of gravity.

The inertial moment at the pile head for soils with stiffness varying proportionally with depth may be calculated according to the formula provided by Reese and

Matlock (1956), based on Winkler considerations, which can be expressed using the notation adopted in this paper as

$$M_{in} = 0.93 \frac{S_a W_p a_s}{g} \left(\frac{q_I E_p I}{\delta \overline{E}_s} \right)^{\frac{1}{5}} \qquad (17.5)$$

17.2.3 Kinematic Versus Inertial Bending Moments

In light of the above solutions, it is straightforward to derive the ratio of kinematic to inertial bending moments under the same seismic conditions. For a homogeneous soil profile, it is possible to calculate the ratio between the two seismic demands by dividing (17.4) and (17.5). For instance, considering a solid concrete pile ($q_I = 1$) and undrained conditions ($\nu_s = 0.5$), one obtains:

$$\frac{M_{kin}}{M_{in}} = 0.2 \left(\frac{E_p}{E_s} \right)^{\frac{3}{4}} \left(\frac{E_s}{S_u} \right) \frac{\rho_s g SF}{E_s S_a \alpha L} d^2 \qquad (17.6)$$

The above expression reveals that: (1) the relative magnitudes of kinematic and inertial bending is independent of ground acceleration. Thus, the conditions concerning importance of kinematic loads based on seismicity by the Eurocode seems to be unjustified; (2) Soil stiffness plays a major role on the relative size of the two seismic demands, with the contribution of the kinematic component increasing with decreasing soil stiffness; (3) Kinematic over inertial bending moment ratio increases with the square of pile diameter.

Equation 17.6 is depicted in Fig. 17.3 as function of soil Young's modulus for different values of spectral amplification S_a and different pile lengths and diameters. Evidently, kinematic over inertial bending moment ratio decreases with increasing soil stiffness and with decreasing pile diameter, and decreases with increasing pile length. This must be attributed to the fact that while kinematic bending of flexible piles is independent of pile length, inertial action is proportional to pile length under constant safety factor for gravitational action.

Similar trends are observed for piles in soils with stiffness proportional to depth. Equations 17.2 and 17.5 can be divided to provide the corresponding kinematic over inertial moment ratio:

$$\frac{M_{kin}}{M_{in}} = 0.24 \left(\frac{E_p}{\overline{E}_s} \right)^{\frac{3}{5}} \left(\frac{E_s}{S_u} \right) \frac{\rho_s g SF}{\overline{E}_s S_a \alpha L^2} d^{\frac{7}{5}} \qquad (17.7)$$

Compared to the homogeneous case, pile diameter exerts a weaker influence ($d^{1.4}$ over d^2 for the previous case), whereas pile length plays a more important role (L^{-2} over L^{-1} dependence).

Equation 17.7 is illustrated in Fig. 17.4 as function of soil Young's modulus gradient for different values of spectral amplification, pile diameter and pile length.

17 The Role of Pile Diameter on Earthquake-Induced Bending 539

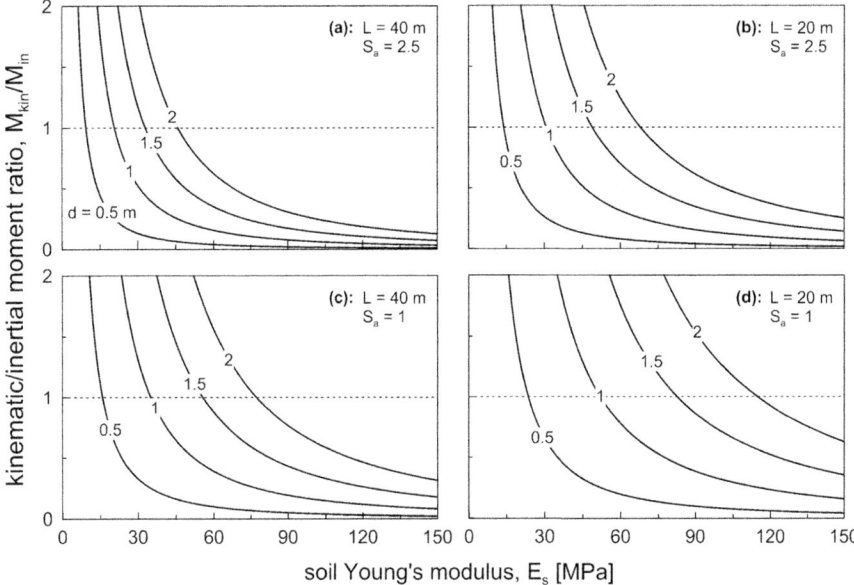

Fig. 17.3 Kinematic/inertial moment ratio for a solid concrete pile in homogeneous soil, as function of soil stiffness, for different values of spectral amplification, pile diameter and pile length

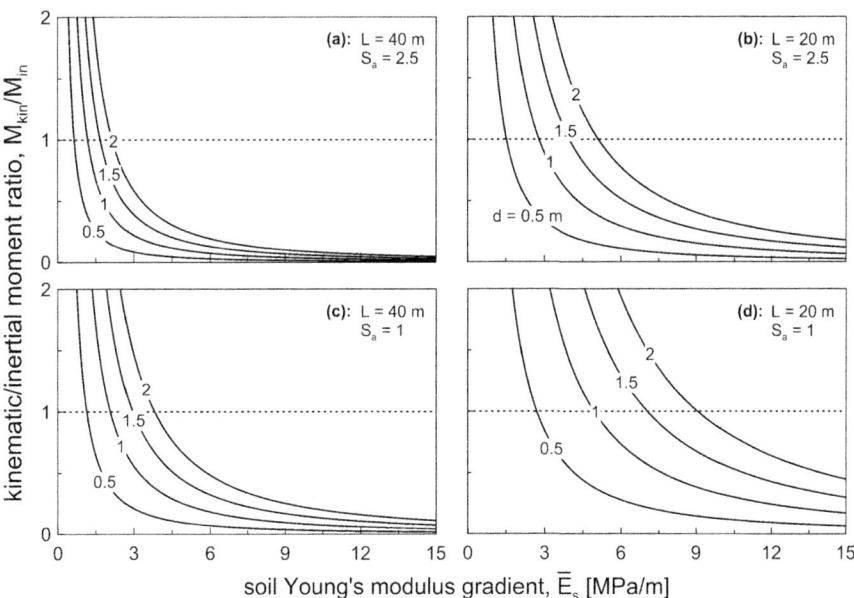

Fig. 17.4 Kinematic/inertial moment ratio for a solid concrete pile in inhomogeneous soil, as function of soil stiffness gradient, for different values of spectral amplification, pile diameter and pile length

17.3 Pile Size Limitations Under Seismic Loads

The seismic performance of piles under combined kinematic and inertial loading can be investigated by comparing the overall bending demand with the corresponding yield moment of the cross section.

With reference to a cylindrical hollow steel pile, the cross-sectional yield moment in the context of strength-of-materials theory can be computed from the well-known formula

$$M_y = E_p I_p \varepsilon_y \frac{2}{d} \left(1 - \frac{W_p}{f_y A}\right) \quad (17.8)$$

ε_y and f_y being the uniaxial yield strain and the corresponding yield stress of the steel material, and A the cross-sectional area. Note that for simplicity, no partial factors accounting for material strength have been included in the calculation.

Note that for a pile in layered soil, another critical location for the assessment of seismic demand is interface separating two consecutive layers of sharply differing stiffness. Considering deep interfaces located below the active pile length, kinematic bending may be evaluated from the approximate formula of Di Laora et al. (2012):

$$M_{kin}^{int} = E_p I_p \frac{2}{d} (\varepsilon_p/\gamma_1)\gamma_1 \simeq E_p I_p \frac{1.86}{d} \gamma_1 \left[\left(\frac{E_p}{E_{s1}}\right)^{-\frac{1}{2}} \left(\left(\frac{E_{s2}}{E_{s1}}\right)^{\frac{1}{4}} - 1\right)^{\frac{1}{2}}\right] \quad (17.9)$$

where γ_1 is the free-field soil shear strain at interface level in the first layer, ε_p/γ_1 the strain transmissibility parameter between pile and soil (Mylonakis 2001).

Clearly bending in such locations is essentially proportional to d^3. As section capacity increases with the same power of diameter, interface bending does not govern the selection of pile diameter.

17.3.1 Steel Piles in Homogeneous Soils

For friction piles in soft soil, axial stresses at the pile top are typically well below the structural capacity (i.e., the term $W_p/f_y A$ is small) so that section capacity is practically proportional to d^3. As kinematic demand is proportional to the fourth power of pile diameter (d^4), it follows that kinematic action prevails over section capacity with increasing pile size. This suggests that there exists a *maximum diameter* beyond which the pile is not able to withstand the kinematically imposed bending moments in an elastic manner. On the other hand, inertial action increases in proportion to d^2 and, therefore, withstanding this type of bending requires a

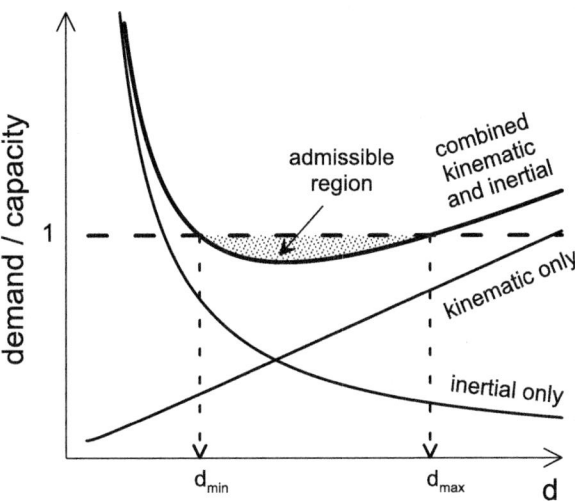

Fig. 17.5 Kinematic and inertial bending moments over corresponding capacity as function of pile diameter

minimum diameter – the opposite to the previous behaviour (Fig. 17.5). Both cases are investigated below.

17.3.1.1 Kinematic Loading

Setting the kinematic demand moment in (17.1) equal to the yield moment in (17.8) and considering the axial load W_p given by (17.3), one obtains the following dimensionless equation for pile size (Di Laora et al. 2013)

$$\frac{1}{2\varepsilon_y}\frac{a_s L}{V_s^2}\left(\frac{d}{L}\right)^2 - \left(\frac{d}{L}\right) + \frac{4\alpha}{q_A\,SF}\frac{S_u}{f_y} = 0 \qquad (17.10)$$

where $q_A = 1 - (1 - 2\,t/d)^2$ is a dimensionless geometric factor accounting for wall thickness, t, of a hollow pile.

Equation 17.10 admits the pair of solutions

$$d_{kin} = 2\varepsilon_y \frac{V_s^2}{a_s}\left[\frac{1}{2} \pm \sqrt{\frac{1}{4} - \frac{2\alpha}{\varepsilon_y q_A\,SF}\left(\frac{V_s^2}{a_s L}\right)^{-1}\left(\frac{S_u}{f_y}\right)}\right] \qquad (17.11)$$

the largest of which, corresponding to the (+) sign, defines the critical (maximum) pile diameter to withstand kinematic action.

If shear wave velocity under the square root is expressed in terms of soil Young's modulus E_s and mass density ρ_s [$v_s = 0.5 \rightarrow E_s = 2(1+v_s)\rho_s V_s^2 = 3\rho_s V_s^2$], the above solution takes the form:

$$d_{kin} = 2\varepsilon_y \frac{V_s^2}{a_s}\left[\frac{1}{2} + \sqrt{\frac{1}{4} - \frac{6\rho_s \alpha a_s L}{\varepsilon_y q_A SF}\left(\frac{E_s}{S_u}f_y\right) - 1}\right] \quad (17.12)$$

which has the advantage that the term in brackets does not depend on absolute soil stiffness and strength, but only on their ratio, E_s/S_u.

In the ideal case of a pile carrying zero axial load (which implies infinite safety against bearing capacity failure due to gravity; $SF \rightarrow \infty$), the term in brackets in (17.11) and (17.12) tends to unity and the solution reduces to the simple expression:

$$d_{kin} = 2\varepsilon_y \frac{V_s^2}{a_s} \quad (17.13)$$

which can be obtained directly from (17.1) and (17.8).

17.3.1.2 Inertial Loading

Setting the right sides of (17.4) and (17.8) equal and employing (17.3), the following solution is obtained:

$$d_{in} = \frac{8\alpha}{SF}L\left[\frac{S_a}{\varepsilon_y}\left(\frac{\pi}{\delta}\right)^{\frac{1}{4}}\left(\frac{a_s}{g}\right)\left(q_I\frac{E_p}{E_s}\right)^{-\frac{3}{4}}\left(\frac{S_u}{E_s}\right) + \frac{1}{2q_A}\left(\frac{S_u}{f_y}\right)\right] \quad (17.14)$$

Equation 17.14 defines a critical (minimum) pile diameter to withstand inertial action. In the limit case of zero ground acceleration ($a_s = 0$), (17.14) degenerates to

$$d_{in} = \frac{4\alpha L}{SF\, q_A}\left(\frac{S_u}{f_y}\right) \quad (17.15)$$

corresponding to the minimum diameter required to resist the gravitational load W_p. The same result can be obtained by setting $a_s = 0$ in (17.10).

17.3.1.3 Combined Kinematic and Inertial Loading

For the more realistic case of simultaneous kinematic and inertial loading, (17.1) and (17.4) can be combined for the overall flexural earthquake demand at the pile head through the simplified superposition formula

$$M_{tot} = M_{kin} + e_{ki}M_{in} \qquad (17.16)$$

where subscript *tot* stands for "total" and e_{ki} is a correlation coefficient accounting for the lack of simultaneity in the occurrence of maximum kinematic and inertial actions. For simplicity and as a first approximation, $e_{ki} = 1$ is assumed here.

Setting the total earthquake moment equal to the yield moment in (17.8), one obtains the second-order algebraic equation for pile size

$$\frac{1}{2}\frac{a_s L}{V_s^2}\left(\frac{d}{L}\right)^2 - \varepsilon_y \left(\frac{d}{L}\right) + \frac{4\alpha}{q_A SF}\left(\frac{S_u}{E_p}\right)\left[1 + 2\frac{q_A}{q_I}\left(\frac{\pi q_I}{\delta}\right)^{\frac{1}{4}}\left(\frac{a_s}{g}\right)\left(\frac{E_p}{E_s}\right)^{\frac{1}{4}} S_a\right] = 0 \qquad (17.17)$$

Equation 17.17 can be solved analytically for the pair of pile diameters

$$d_{1,2} = \frac{\varepsilon_y V_s^2}{a_s}\left\{1 \mp \sqrt{1 - \frac{24\alpha \rho_s a_s L}{q_A f_y \varepsilon_y SF}\left(\frac{S_u}{E_s}\right)\left[1 + 2\frac{q_A}{q_I}\left(\frac{\pi q_I}{\delta}\right)^{\frac{1}{4}}\left(\frac{a_s}{g}\right)\left(\frac{E_p}{E_s}\right)^{\frac{1}{4}} S_a\right]}\right\} \qquad (17.18)$$

which correspond to a minimum value, d_1, obtained for the negative sign, and a maximum value, d_2, obtained for the positive sign, respectively. Values between these two extremes define the range of admissible pile diameters for the conditions at hand. It will be demonstrated that d_1 is always larger than d_{in} in (17.14), and d_2 is always smaller than d_{kin} in (17.12) that is, the admissible range of pile diameters is narrower over the hypothetical case of kinematic and inertial loads acting independently.

17.3.1.4 Results

A schematic representation of the foregoing developments is depicted in Fig. 17.6, in terms of pile diameter versus soil stiffness. Diameters lying inside the hatched zone defined by (17.18) are admissible, whereas diameters lying outside the zone are not. Evidently, upper and lower bounds are sensitive to soil stiffness, E_s leading to a wider range of admissible diameters as soil becomes progressively stiffer. Naturally, the curves for purely kinematic and purely inertial action (shown by continuous curves) in (17.12) and (17.14) bound the admissible range from above

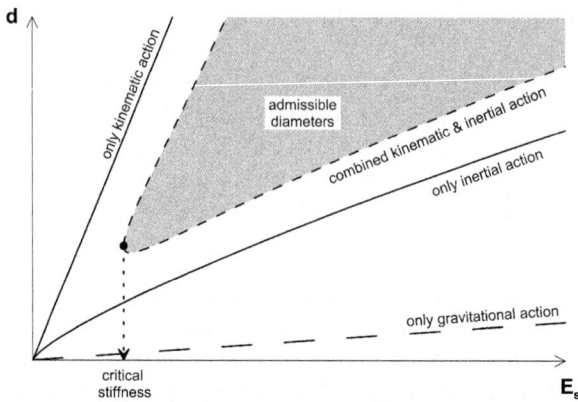

Fig. 17.6 Range of admissible diameters for different types of loading

and below, respectively, suggesting that kinematic and inertial moments interact detrimentally for pile safety. Whereas this effect becomes aggravated by the simplifying assumption of simultaneous maxima in kinematic and inertial response (e_{ki} in Eq. (17.16) equal to 1), the same pattern would be obtained for any linear combination of individual moments involving positive multipliers e_{ki}.

It is worth mentioning that there is always a minimum soil stiffness for which the admissible range collapses to a single point corresponding to a unique admissible diameter (i.e., $d_1 = d_2$). This diameter can be obtained by eliminating the term in square root in (17.18), to get

$$d_1 = d_2 = \frac{\varepsilon_y V_s^2}{a_s} \qquad (17.19)$$

which, remarkably, is equal to exactly one half of the value obtained for kinematic action alone under zero axial load (17.13). It is noteworthy that this diameter is independent of pile Young's modulus and wall thickness. Evidently, for stiffness values smaller than critical, no real-valued pile diameters can be predicted from (17.18), which suggests that it is impossible for the pile head to stay elastic under the imposed surface acceleration a_s.

With reference to a hollow steel pile, numerical results for the range of admissible diameters predicted by (17.18) is plotted in Fig. 17.7, as function of soil stiffness E_s, for different values of surface seismic acceleration (a_s/g) and pile length L. The detrimental effect resulting from the particular load combination becomes gradually more pronounced with increasing pile length and seismic acceleration, as higher inertial loads are induced at the pile head. Note that for piles in very soft soil such as peat, having E_s less than 10 MPa, maximum pile diameter may be less than 1 m, thereby severely restricting design options.

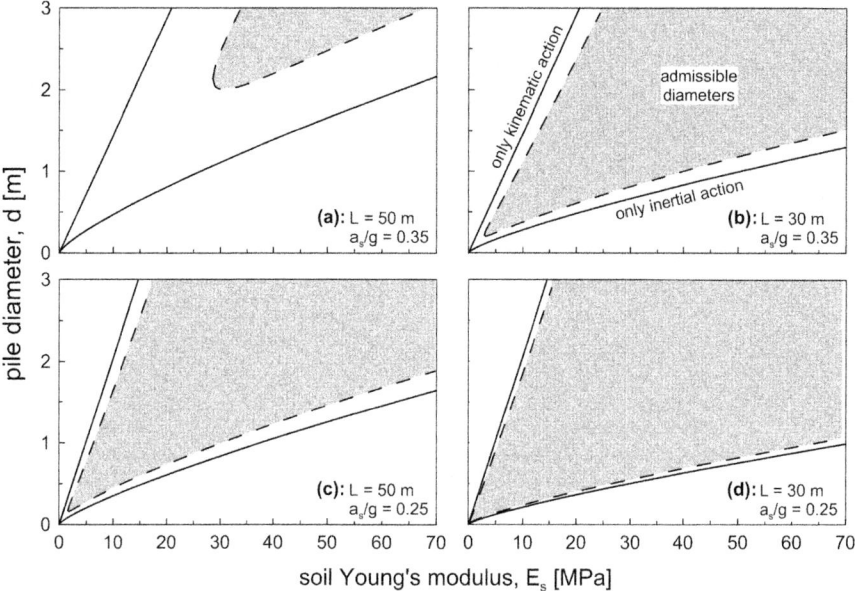

Fig. 17.7 Admissible pile diameters against soil Young's modulus ($E_s/S_u = 500$, $f_y = 275$ MPa, $E_p = 210$ GPa, $\nu_s = 0.5$, $\rho_s = 1.7$ Mg/m^3, $S_a = 2.5$, FS = 3, t/d = 0.015, $\alpha = 0.7$, $\delta = 1.2$). *Continuous lines* represent pure kinematic and inertial actions whereas *dashed lines* refer to combined action

17.3.2 Steel Piles in Inhomogeneous Soil

Kinematic and inertial demands for inhomogeneous soils in (17.2) and (17.5) may be expressed for undrained conditions, through trivial algebraic manipulation, as:

$$M_{kin} = 0.185 a_s \rho_s \left(\frac{q_I E_p}{\overline{E}_s}\right)^{\frac{4}{3}} d^{\frac{16}{5}} \quad (17.20)$$

$$M_{in} = 1.6 \frac{S_a L \alpha S_u}{SF}\left(\frac{a_s}{g}\right)\left(\frac{q_I E_p}{\delta \overline{E}_s}\right)^{\frac{1}{3}} d^{\frac{9}{5}} \quad (17.21)$$

Equation 17.20 reveals that the effect of pile diameter on peak kinematic bending moment is weaker than in homogeneous soil, as the corresponding exponent is 3.2 (=16/5) instead of 4, due to I_p in (17.1). This can be explained considering that an increase in pile diameter corresponds to an increase in pile active length which, in turn, forces a larger portion of progressively stiffer soil to control pile curvature at the head.

While the exponent of 3.2 still exceeds the corresponding exponent in capacity (3 – see 17.8), this is unlikely to create a significant design constraint.

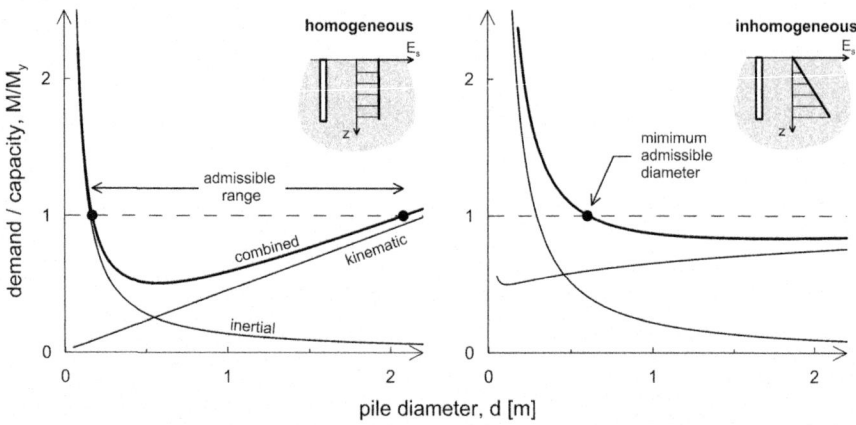

Fig. 17.8 Kinematic, inertial and combined moment vs. capacity for a homogeneous and an inhomogeneous soil profile. In both graphs, $a_s/g = 0.35$, $E_s/S_u = 500$, $f_y = 275$ MPa, $E_p = 210$ GPa, $v_s = 0.5$, $\rho_s = 1.7$ Mg/m³, $S_a = 2.5$, FS = 3, t/d = 0.015, $\alpha = 0.5$, L = 15 m, $\overline{E}_s = 2$ MPa/m, Es = $\overline{E}_s$ L/2 = 15 MPa

In a similar fashion, (17.21) reveals that the effect of pile diameter on peak inertial moment is weaker than in homogeneous soils with the power dependence on d being 1.8 (=9/5) instead of 2 in (17.4), and thereby size limitation in terms of minimum diameter will be more critical than in homogeneous soil.

Equating seismic moment demand from (17.20) and (17.21) with section capacity in (17.8), the following dimensionless algebraic equation for pile size is obtained:

$$0.185 \left(\frac{q_I E_p}{\overline{E}_s L}\right)^{\frac{4}{5}} \left(\frac{d}{L}\right)^{\frac{16}{5}} - \frac{\pi}{64} \left(\frac{q_I E_p \varepsilon_y}{a_s \rho_s L}\right) \left(\frac{d}{L}\right)^3 + \frac{\pi}{16} \frac{q_I \alpha S_u}{q_A SF a_s \rho_s L} \left(\frac{d}{L}\right)^2 +$$

$$+ 1.6 \frac{S_a \alpha S_u}{SF \gamma L} \left(\frac{q_I E_p}{\delta \overline{E}_s L}\right)^{\frac{1}{5}} \left(\frac{d}{L}\right)^{\frac{9}{5}} = 0 \qquad (17.22)$$

Due to the intrinsically non-integer nature of the exponents, no exact closed-form solutions for pile diameter can be derived from (17.22). However, a Newton-Raphson approximate scheme may be easily employed to obtain the roots (not shown here) in an iterative manner.

Comparison between size limitations in homogeneous and inhomogeneous soil is provided in Fig. 17.8, where the ranges of admissible diameters are compared for the two cases. As can be noticed, beyond a certain diameter the ratio of demand over capacity for the inhomogeneous case (solid line) becomes nearly constant.

17 The Role of Pile Diameter on Earthquake-Induced Bending 547

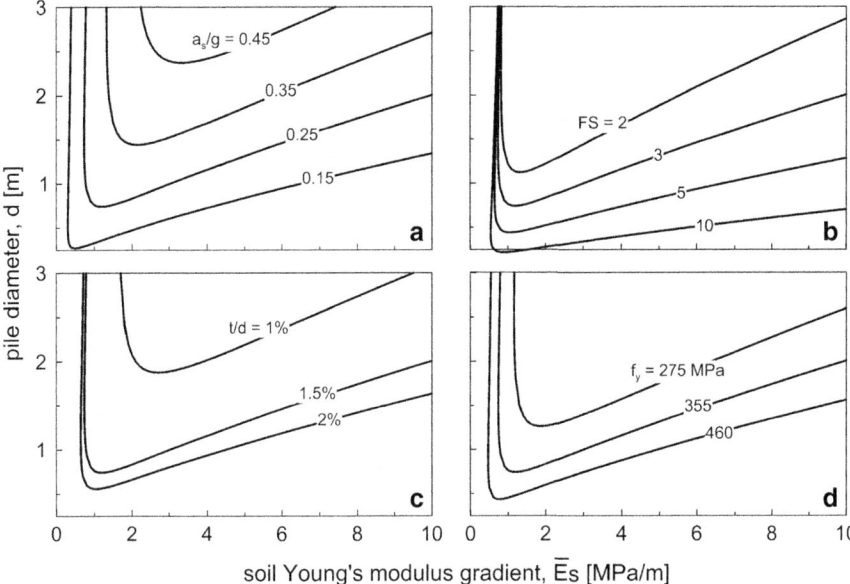

Fig. 17.9 Admissible pile diameters for a tubular steel pile in soil with stiffness proportional to depth. In all graphs, except specifically otherwise indicated, $a_s/g = 0.25$, $E_s/S_u = 500$, $f_y = 355$ MPa, $E_p = 210$ GPa, $\nu_s = 0.5$, $\rho_s = 1.7$ Mg/m^3, $S_a = 2.5$, $FS = 3$, $t/d = 0.015$, $\alpha = 0.5$, $L = 30$ m

This, however, does not indicate an overall weaker influence of kinematic interaction on size limits, as minimum diameter is strongly affected by kinematic demand. In addition, the graphs indicate that, contrary to common perception, kinematic demand is higher than inertial demand starting from relatively small pile diameters.

To further explore the role of pile size, Fig. 17.9 depicts the bounds of the admissible diameter regions for different values of problem parameters. As anticipated, no controlling maximum diameter exists, so that the upper bound consists of a nearly vertical line in $\overline{E}_s - d$ plane. Pile size limitation thus reduces to establishing a minimum diameter, which increases with increasing soil resistance due to the larger mass carried by the pile under the assumption of a constant SF.

Figure 17.9a explores the role of design acceleration on pile size. Understandably, the admissible region shrinks with increasing (a_s/g), as the latter affects both inertial and kinematic loading, and moves towards larger diameters. It is noted that for cases of moderate to strong seismicity (i.e., $a_s/g = 0.25$–0.35) and common values of design spectral amplification ($S_a = 2.5$), piles in soft clay should possess very high diameters (of the order of 2 m) to resist seismic loads without yielding at the head. This result alone might explain the considerable number of failures at the pile head observed in post-earthquake investigations around the world.

When a preliminary design carried out by axial bearing capacity considerations does not satisfy seismic structural requirements, a solution is to decrease the weight carried by the individual piles by increasing the safety factor SF. The influence of

SF on seismic performance is illustrated in Fig. 17.9b, where the minimum diameter decreases with increasing *SF*. Nevertheless, it should be kept in mind that increasing the safety factor against axial bearing capacity leads to an increase in foundation cost over the original design. Studying this aspect involves additional factors which lie beyond the scope of this work.

In Fig. 17.9c, d the role of section capacity on admissible diameters is examined. Figure 17.9c indicates that lowering the wall thickness may impose a significant restriction on the size of the admissible region, whereas the choice of material strength (Fig. 17.9d) seems to be less important.

17.3.3 Concrete Piles

The behavior of concrete piles is fundamentally different from that of steel piles, as: (1) the moment of inertia of the pile cross section is typically higher; (2) the material has negligible tensile strength, thereby moment capacity relies on steel reinforcement. The impact of these differences on the phenomena at hand is examined below.

In the same spirit as before, critical diameters may be assessed by equating capacity (Cosenza et al. 2011), and demand obtained by summing up the contributions of kinematic and inertial interaction, as shown in the foregoing.

As an example, numerical results for concrete piles in soil with stiffness varying linearly with depth are depicted in Fig. 17.10. This case leads to the narrowest regions of admissible diameters compared to those examined earlier. As in the case of hollow steel piles, maximum diameter in soils with stiffness varying proportionally with depth is not particularly important, as the curves tend to be vertical at the left side of the graphs. On the other hand, kinematic interaction has a profound role in increasing the minimum admissible diameter. Like in the other cases, concrete and steel strengths are of minor importance (Fig. 17.10c, d). On the contrary, seismicity and geometrical parameters (Fig. 17.10a, b) have a considerable effect in controlling the minimum admissible diameter.

A comparison among the four combinations of sections and soil profiles examined here is provided in Fig. 17.11, where admissible regions are plotted for steel and concrete piles, embedded in homogeneous and linear soil profiles. It is noted that curves corresponding to linearly-varying soil stiffness are somehow rotated with respect to the homogeneous case, due to the different importance of pile diameter in kinematic bending. As already mentioned, maximum diameter is of concern only for homogeneous and very soft inhomogeneous soil, while in all other cases a minimum diameter is of the main concern which may reach large values due to the detrimental interplay of kinematic and inertial components.

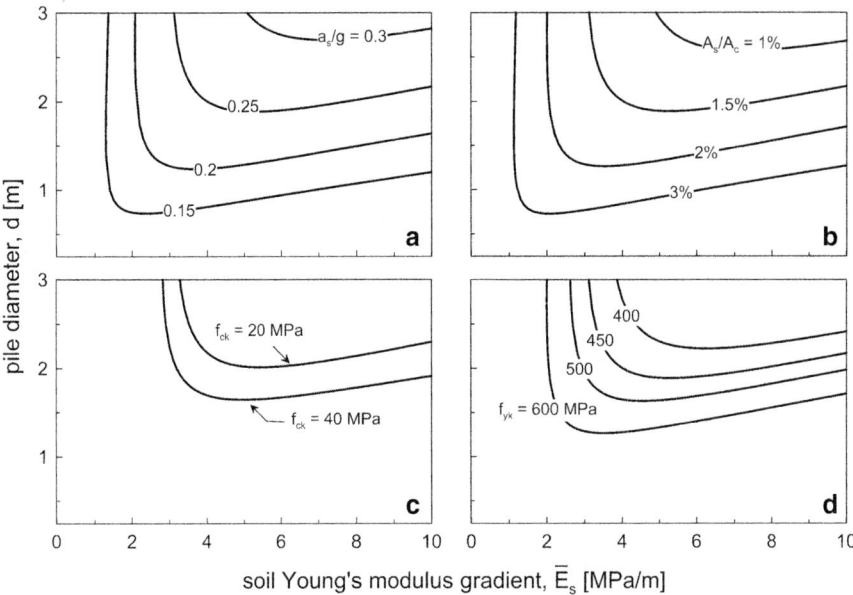

Fig. 17.10 Admissible pile diameters for a solid concrete pile in soil with stiffness proportional to depth. In all graphs, except specifically otherwise indicated, $a_s/g = 0.25$, $E_s/S_u = 500$, $E_p = 30$ GPa, $\nu_s = 0.5$, $\rho_s = 1.7$ Mg/m^3, $S_a = 1.5$, FS $= 3$, $A_s/A_c = 0.015$, $f_{ck} = 25$ MPa, $f_{yk} = 450$ MPa, c $= 5$ cm, $\alpha = 0.5$, L $= 30$ m

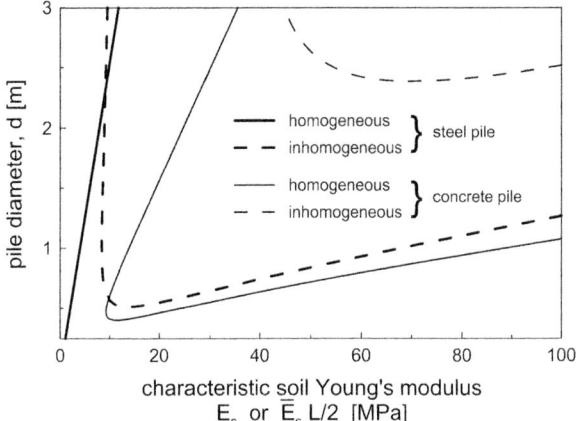

Fig. 17.11 Admissible diameters for steel and concrete piles in homogeneous and inhomogeneous soil. For all cases, $a_s/g = 0.25$, $E_s/S_u = 500$, f_y (steel) $= 355$ MPa, f_{yk} (concrete reinforcement) $= 450$ MPa, $f_{ck} = 25$ MPa, $E_p = 30$ GPa or 210 GPa (for concrete and steel, respectively), $\nu_s = 0.5$, $\rho_s = 1.7$ Mg/m^3, $S_a = 2.5$, FS $= 3$, $t/d = A_s/A_c = 0.015$, $\alpha = 0.5$, L $= 25$ m

17.4 Optimal Pile Diameter

It has already been demonstrated that for a given set of seismicity, geotechnical and structural properties, a pile possesses a limited range of admissible diameters. This means that outside this range, a pile will yield (bending safety factor $FS_b = M_{rd}/(M_{kin} + M_{in}) < 1$), whereas inside the range it will stay elastic ($FS_b > 1$). Naturally, the limits of the range correspond to $FS = 1$. It can be deduced that there exists a particular diameter, falling within the admissible range, for which bending safety factor is maximum, and thereby it represents an optimum choice from a safety viewpoint.

To derive analytical expressions for the specific diameter for a steel pile in homogeneous soil, we recall that the expressions of moment capacity, kinematic moment and inertial moment can be cast in the simple form:

$$M_y = A_1 \cdot d^3 - A_2 \cdot d^2$$
$$M_{kin} = A_3 \cdot d^4 \quad (17.23\text{a, b, c})$$
$$M_{in} = A_4 \cdot d^2$$

A_1 to A_4 being parameters that can be readily indentified from the foregoing solutions.

Neglecting the contribution of axial load on section capacity (i.e., setting $A_2 = 0$), the reciprocal of bending safety factor assumes the form:

$$\frac{1}{FS_b} = \frac{M_{kin} + M_{in}}{M_y} = \frac{1}{A_1}\left(A_3 \cdot d + A_4 \cdot \frac{1}{d}\right) \quad (17.24)$$

Differentiating this expression with respect to diameter, one obtains:

$$\frac{d\left(\frac{1}{FS_b}\right)}{dd} = \frac{1}{A_1}\left(A_3 - A_4 \cdot \frac{1}{d^2}\right) \quad (17.25)$$

The optimal diameter d_{opt} is thereby equal to:

$$d_{opt} = \sqrt{\frac{A_4}{A_3}} \quad (17.26)$$

In terms of physical parameters, we obtain the following expression:

$$d_{opt} = \sqrt{\frac{16}{3}\left(\frac{\pi q_1 E_p}{\delta E_s}\right)^{\frac{1}{4}} \frac{S_a}{SF}\left(\frac{S_u}{E_s}\right)\left(\frac{\alpha L}{E_p \rho g}\right)} \quad (17.27)$$

As evident from (17.26) and (17.27) optimal diameter, remarkably, does not depend on seismicity and section capacity.

17 The Role of Pile Diameter on Earthquake-Induced Bending

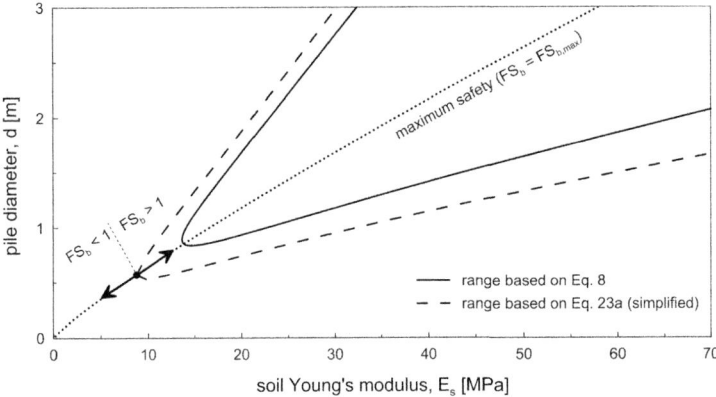

Fig. 17.12 Optimal pile diameter and admissible regions for a hollow steel pile in homogeneous soil. ($a_s/g = 0.4$, $E_s/S_u = 500$, $f_y = 275$ MPa, $E_p = 210$ GPa, $\nu_s = 0.5$, $\rho_s = 1.7$ Mg/m^3, $S_a = 2.5$, FS = 3, t/d = 0.015, $\alpha = 0.7$, L = 30 m)

Furthermore, from (17.23b, c) one obtains:

$$\left.\frac{M_{kin}}{M_{in}}\right|_{d=d_{opt}} = 1 \qquad (17.28)$$

which means that a steel pile sized at $d = d_{opt}$ balances the kinematic and inertial components of total moment demand.

Figure 17.12 provides a graphical representation of the optimal diameter, in the context of the regions of admissible diameters described earliear, obtained both in an approximate and an exact manner through (17.8) and (17.23a). Evidently, the optimal diameter curve intersects the approximate admissible region at point ($E_{s,crit}$, d_{crit}). For stiffer soils, optimal diameter naturally falls within the admissible region and bending safety factor is larger than one. For stiffness smaller than critical, optimal diameter still exists, in the sense that it defines a maximum safety factor below 1. On the other hand, critical diameter possesses the following properties: (a) it leads to a unit safety factor and (b) it balances kinematic and inertial moments. Moreover, the optimal diameter passes close to the critical point predicted from the exact analysis, so that above observations hold regardless of the method employed to evaluate the admissible regions. While, actual design choices for d will naturally involve additional considerations, it is expected that they will lie in the region between the maximum safety curve and the minimum admissible diameter.

The optimal diameter may also be derived analytically for a steel pile in inhomogeneous soil. To this end, kinematic and inertial demand may be expressed as:

$$M_{kin} = 0.185 a_s \rho_s \left(\frac{q_1 E_p}{\overline{E}_s}\right)^{\frac{4}{3}} d^{\frac{16}{5}} = B_1 d^{\frac{16}{5}} \quad (17.29a, b)$$

$$M_{in} = 1.6 \frac{S_a L \alpha S_u}{SF} \left(\frac{a_s}{g}\right) \left(\frac{q_1 E_p}{\delta \overline{E}_s}\right)^{\frac{1}{3}} d^{\frac{9}{5}} = B_2 d^{\frac{9}{5}}$$

In the same vein, the reciprocal of FS_b is:

$$\frac{1}{FS_b} = \frac{M_{kin} + M_{in}}{M_y} = \frac{1}{A_1}\left(B_1 \cdot d^{\frac{1}{5}} + B_2 \cdot d^{-\frac{6}{5}}\right) \quad (17.30)$$

The optimal diameter d_{opt} is obtained by differentiating the above expression with respect to d, to get

$$d_{opt} = \left(6\frac{B_2}{B_1}\right)^{\frac{5}{7}} \quad (17.31)$$

Contrary to the previous case, optimal diameter for the particular conditions does not balance kinematic and inertial demands. The corresponding 'equal demand' diameter is obtained from (17.29a, b) as:

$$d_{bal} = \left(\frac{B_2}{B_1}\right)^{\frac{5}{7}} = 0.278 \cdot d_{opt} \quad (17.32)$$

Figure 17.13 depicts optimal and equal seismic demand diameters for inhomogeneous soil together with rigorous admissible regions corresponding to different material strengths. As anticipated, these diameters are insensitive to seismicity and material properties, so that the curves in the figure pertain to all regions.

17.5 Discussion

It has been shown that, contrary to perceptions reflected in seismic Codes, kinematic bending at the pile head may not be negligible compared to the overall seismic demand, in soft soils and large pile diameters regardless of seismic intensity. In certain cases, kinematic interaction may even be higher than the inertial counterpart.

In addition, the simultaneous action of kinematic and inertial components of pile bending leads to a limited range of admissible pile diameters to resist seismic

17 The Role of Pile Diameter on Earthquake-Induced Bending

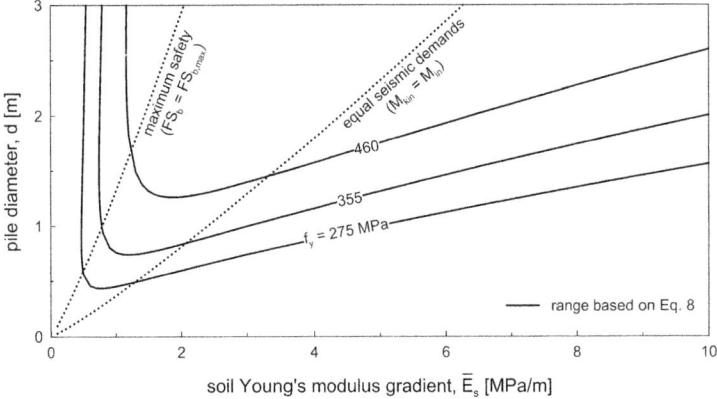

Fig. 17.13 Optimal pile diameter and admissible regions for a hollow steel pile and an inhomogeneous soil profile. ($a_s/g = 0.25$, $E_s/S_u = 500$, $E_p = 210$ GPa, $\nu_s = 0.5$, $\rho_s = 1.7$ Mg/m^3, $S_a = 1$, FS = 3, t/d = 0.015, $\alpha = 0.7$, L = 30 m)

action. For homogeneous soil, kinematic interaction requires a minimum admissible diameter whereas inertial interaction leads to a corresponding maximum. As these actions interact detrimentally with each other, the range is reduced over the ideal case of kinematic and inertial loads acting independently.

In very soft deposits, if soil stiffness close to the surface (i.e., within a depth of few pile diameters) may be assumed to be nearly constant, kinematic interaction has a dominant influence, thus leading to small maximum admissible diameter. In these cases, inertial interaction leads to smaller pile bending compared to kinematic interaction, yet may have an important effect in reducing the maximum admissible diameter obtained solely for kinematic loading. In the context of the assumptions adopted in this work, pile length has a profound effect in reducing the admissible pile diameter and increasing critical soil stiffness below which no pile diameter is admissible, so that modifications in design coed provisions might be needed.

For stiffer soils and especially for conditions involving linearly-varying stiffness with depth, the limitation in pile size essentially reduces to establishing a minimum diameter. In several cases, safety factors commonly used in classical geotechnical design for axial bearing capacity do not guarantee safety for seismic action. To overcome the problem, a solution could be to increase the number of piles, thus to make the safety factor against gravitational action larger. An alternative is to increase the capacity of the pile cross section by increasing wall thickness or reinforcement. On the contrary, increasing material strength will not substantially improve performance. In other words, for a given design acceleration, the geotechnical and geometrical properties appear to be more important than the structural properties in controlling pile safety. It is worth stressing that these remedial actions may increase foundation cost.

It was also demonstrated that among all admissible diameters for a specific set of problem parameters, there always exists an optimal value that maximizes safety against bending failure. This diameter could be of guidance in designing piles in seismically prone areas. On the other hand, the diameter that minimizes foundation cost requires taking into account additional parameters (including cost of materials and construction methods) that lie beyond the scope of this work.

17.6 Conclusions

Kinematic and inertial interaction between a pile and the surrounding soil are of different nature and, thereby, are affected by pile size in a different manner. In layered soil, bending at deep interfaces is not affected by pile size, at least from a first-order analysis viewpoint, as seismic demand and section capacity increase with the same power of diameter. On the other hand, with reference to a pile head under a restraining cap, it was shown that kinematic bending dominates over inertial bending for large-diameter piles in soft soil regardless of variation of soil stiffness with depth and, therefore, even in conditions for which Codes do not require assessment of kinematic action.

In addition, (a) kinematic interaction provides a maximum diameter beyond which the pile yields, and (b) inertial interaction provides a corresponding minimum diameter. The simultaneous presence of these actions leads to a range of admissible diameters which is narrower than that obtained for kinematic and inertial actions considered independently.

The following general conclusions were drawn from this study:

1. Concrete piles possess a narrower range of admissible diameters to withstand seismic action over hollow steel piles. This can be attributed to the higher bending stiffness of the concrete pile cross-section (which attracts higher kinematic moments), as well as the inability of the concrete material to withstand tension.
2. For soft soils of constant stiffness with depth, kinematic interaction dominates seismic demand. As a result, admissible pile sizes are essentially overbounded by a critical diameter which in some cases may be quite small (of the order of 1 m) and, hence, it may affect design. Under these circumstances, adding more piles or increasing pile length would not improve safety, as these actions will not affect kinematic demand.
3. In stiffer/stronger soils, inertial interaction is dominant due to the heavier loads carried by the pile. This yields a minimum admissible pile diameter which, in areas of moderate to high seismicity, may be quite large (of the order of 1 m or so).
4. Stiffness varying proportionally with depth essentially enforces only a lower bound in pile diameter; this may be rather large (above 2 m) especially for high stiffness gradients. Note that the absence of an upper limit is not due to weak

kinematic demand (which can be quite large), but due to a lack of dependence of kinematic moment on pile diameter.
5. The range of admissible diameters decreases with increasing ground acceleration, spectral amplification, soil strength and pile length, whereas it increases with increasing soil stiffness, pile safety factor and amount of reinforcement (or wall thickness for hollow piles). On the other hand, pile material strength plays a minor role in controlling pile size.
6. There always exists a critical soil stiffness or a critical stiffness gradient below which no pile diameter is admissible for a given ground acceleration. Below the particular threshold, a fixed-head flexible pile cannot remain elastic regardless of diameter or material strength.
7. There always exists an optimal diameter that maximizes safety against bending failure. Analytical expressions for steel piles, embedded in both in homogeneous and inhomogeneous soils, have been presented, which reveal that optimal diameter is independent of seismicity and section material properties.

It has to be stressed that the work at hand deals exclusively with the role of pile diameter in the seismic behaviour of piles themselves. The important complementary topic of the role of pile size in reducing seismic forces in the superstructure through kinematic filtering of the seismic waves is addressed elsewhere (Di Laora and de Sanctis 2013).

Despite the simplified nature of some of the assumptions adopted in this work, issues of practical importance related to pile design in seismic areas were quantitatively addressed. Nevertheless, some of the conclusions may require revision in presence of strong nonlinearities such as those associated with high-amplitude earthquake shaking, soil liquefaction and pile buckling. Additional research is required to address issues of this kind.

Acknowledgements Support from ReLUIS research program funded by DPC (Civil Protection Department) of the Italian Government and coordinated by the AGI (Italian Geotechnical Association), is acknowledged.

Open Access This chapter is distributed under the terms of the Creative Commons Attribution Noncommercial License, which permits any noncommercial use, distribution, and reproduction in any medium, provided the original author(s) and source are credited.

References

Anoyatis G, Di Laora R, Mandolini A, Mylonakis G (2013) Kinematic response of single piles for different boundary conditions: analytical solutions and normalization schemes. Soil Dyn Earthq Eng 44:183–195

Castelli F, Maugeri M (2009) Simplified approach for the seismic response of a pile foundation. J Geotech Geoenviron Eng 135(10):1440–1451

Cosenza E, Galasso C, Maddaloni G (2011) A simplified method for flexural capacity assessment of circular RC cross-sections. Eng Struct 33:942–946 de Sanctis L, Maiorano RMS, Aversa S (2010) A method for assessing bending moments at the pile head. Earthq Eng Struct Dyn 39:375–397

Dezi F, Carbonari S, Leoni G (2010) Kinematic bending moments in pile foundations. Soil Dyn Earthq Eng 30(3):119–132

Di Laora R, de Sanctis L (2013) Piles-induced filtering effect on the foundation input motion. Soil Dyn Earthq Eng 46:52–63

Di Laora R, Mandolini A (2011) Some remarks about Eurocode and Italian code about piled foundations in seismic area. ERTC-12 workshop on evaluation of EC8, Athens

Di Laora R, Mandolini A, Mylonakis G (2012) Insight on kinematic bending of flexible piles in layered soil. Soil Dyn Earthq Eng 43:309–322

Di Laora R, Mylonakis G, Mandolini A (2013) Pile-head kinematic bending in layered soil. Earthq Eng Struct Dyn 42:319–337

Dobry R, Vicente E, O'Rourke M, Roesset M (1982) Horizontal stiffness and damping of single piles. J Geotech Geoenviron Eng 108(3):439–759

Kavvadas M, Gazetas G (1993) Kinematic seismic response and bending of free-head piles in layered soil. Géotechnique 43(2):207–222

Kampitsis AE, Sapountzakis EJ, Giannakos SK, Gerolymos NA (2013) Seismic soil-pile-structure interaction – a new beam approach. Soil Dyn Earthqu Eng 55:211–224

Kaynia A, Kausel E (1991) Dynamics of piles and pile groups in layered soil media. Soil Dyn Earthq Eng 10(8):385–401

Maiorano RMS, de Sanctis L, Aversa S, Mandolini A (2009) Kinematic response analysis of piled foundations under seismic excitations. Can Geotech J 46(5):571–584

Margason E (1975) Pile bending during earthquakes. Lecture, 6 March 1975, ASCE UC/Berkeley seminar on design construction and performance of deep foundations

Margason E, Holloway DM (1977) Pile bending during earthquakes. In: Sarita Prakashan (ed) Proceedings of 6th world conference on earthquake engineering, Meerut, vol II, 1977, pp 1690–1696

Mizuno H (1987) Pile damage during earthquakes in Japan (1923–1983). In: Nogami T (ed) Dynamic response of pile foundations, vol 11, Geotechnical special publication. ASCE, New York, pp 53–78

Mylonakis G (2001) Simplified model for seismic pile bending at soil layer interfaces. Soils Found 41(3):47–58

Nikolaou A, Mylonakis G, Gazetas G, Tazoh T (2001) Kinematic pile bending during earthquakes analysis and field measurements. Géotechnique 51(5):425–440

Pender M (1993) Seismic pile foundation design analysis. Bull N Z Natl Soc Earthq Eng 26(1):49–160

Reese LC, Matlock MM (1956) Non dimensional solutions for laterally loaded piles with soil modulus assumed proportional to depth. VIII Texas conference SMFE, Spec. Publ. 29, Univ. of Texas, Austin

Roesset JM (1980) The use of simple models in soil-structure interaction. ASCE specialty conference, Knoxville, Civil Engineering and Nuclear Power, vol 2

Saitoh M (2005) Fixed-head pile bending by kinematic interaction and criteria for its minimization at optimal pile radius. J Geotech Geoenviron Eng 131(10):1243–1251

Sica S, Mylonakis G, Simonelli AL (2011) Transient kinematic pile bending in two-layer soil. Soil Dyn Earthq Eng 31(7):891–905

Syngros K (2004). Seismic response of piles and pile-supported bridge piers evaluated through case histories. PhD thesis, City University of New York

Viggiani C, Mandolini A, Russo G (2011) Piles and pile foundations. Spon Press/Taylor and Francis, London/New York

Chapter 18
Predictive Models for Earthquake Response of Clay and Sensitive Clay Slopes

Amir M. Kaynia and Gökhan Saygili

Abstract Earthquake-induced permanent displacement and shear strain are suitable indicators in assessing the seismic stability of slopes. In this paper, predictive models for the permanent displacement and shear strain as functions of the characteristics of the slope (e.g. factor of safety) and the ground motion (e.g. peak ground acceleration) are proposed. The predicted models are based on numerical simulations of seismic response of infinite slopes with realistic soil profiles and geometry parameters. Predictive models are developed for clay and sensitive clay slopes. A strain-softening soil model is used for sensitive clays. A comparison of the permanent displacement and strain predictions for clay and sensitive clays reveals that the displacement and shear strains are larger for sensitive clays for the same slope geometry and similar earthquake loading conditions. A comparison of the displacement predictive model with other predictive models published recently reveals that the displacement predictions of the proposed model fall into the low estimate bound for soft slopes and into the high estimate bound for stronger slopes. Permanent displacements from a limited number of 2D FE analyses and from predictive models compare well; however, the predictive model for shear strain tends to overly estimate the shear strains. This is a typical effect of 2D geometry, which represents a conservative situation. As the size of the slope increases, this effect is diminished, and the 2D results tend more to the 1D results as captured by the predictive models developed in this paper.

A.M. Kaynia (✉)
Computational Geomechanics, Norwegian Geotechnical Institute, Oslo, Norway

Department of Structural Engineering, Norwegian University of Science and Technology, Trondheim, Norway
e-mail: amir.m.kaynia@ngi.no

G. Saygili
Department of Civil Engineering, University of Texas at Tyler, Tyler, TX, USA

18.1 Introduction

Stability evaluation of slopes under earthquake loading is an important issue in geotechnical earthquake engineering. While slopes with low static safety margin could fail due to moderate and large earthquakes, most slopes experience only permanent displacements without failure. The displacements could be from a few millimeters to as large as a few meters depending on the slope conditions and the earthquake excitation. The seismic response of slopes is assessed using approaches that utilize limit equilibrium methods or the Finite Element Method (FEM). The limit equilibrium approach considers the shear stresses along a failure surface and computes a factor of safety (FS) based on the available shear strength and the shear stresses required for equilibrium. Failure is expected when the shear stress exceeds the shear strength. The minimum factor of safety for a slope is estimated by trial and error for a large number of assumed slip surfaces. Typically, the factor of safety is assumed to be constant along the slip surface and the same factor of safety is applied to each of the shear strength parameters (i.e., cohesion intercept and internal friction angle). A pseudostatic slope stability analysis is a limit equilibrium analysis that models earthquake shaking as a destabilizing horizontal static force. This approach significantly simplifies the problem, but it is not an accurate representation of earthquake shaking. A pseudostatic analysis does not provide any information about consequences when the pseudostatic factor of safety is less than unity. Even if the pseudostatic factor of safety is less than 1.0, the slope may have limited deformation and acceptable performance because the shear strength is exceeded only during short time intervals by the earthquake loading.

If, on the other hand, one uses the FEM to evaluate the stability of a slope, one does not need to make prior assumptions regarding the location of the critical slip surface. A dynamic FEM captures the entire nonlinear stress-strain-strength properties of the soil, and computes the deformation patterns throughout the slope under the earthquake excitation. However, robust nonlinear stress-strain-strength models of the soil are required to produce reliable numerical results.

A simple model used in slope response analysis is the Sliding Block model that was originally proposed by Newmark (1965). This model acknowledges that the horizontal force induced by earthquake shaking is variable and earthquake shaking could impart a destabilizing force sufficient to reduce temporarily the factor of safety of a slope below 1.0. This type of analysis attempts to quantify the sliding displacement of a sliding mass during these instances of instability. The original Newmark procedure models the sliding mass as a rigid block and utilizes two parameters: the yield acceleration and the acceleration-time history of the rigid foundation beneath the sliding mass. A sliding episode begins when the acceleration exceeds the yield acceleration and continues until the velocity of the sliding block and foundation again coincide. The relative velocity between the rigid block and its foundation is integrated to calculate the relative sliding displacement for each sliding episode, and the sum of the displacements in these episodes represents the cumulative sliding displacement. The original rigid sliding block procedure is applicable to thin, veneer slope failures. This failure mode is common in natural slopes, while deeper sliding surfaces are common in engineered earth structures. The magnitude of sliding displacement is strongly affected by the characteristics of the earthquake ground motion (i.e., intensity, frequency content, duration). Many researchers have proposed models that predict rigid block sliding displacement as a function of ground motion parameters. Permanent sliding displacements are generally used to evaluate the seismic stability of earth slopes such that different displacement levels represent different levels of landslide hazard (e.g. very low landslide hazard when D < 5 cm).

Biscontin et al. (2004) described three scenarios for earthquake-induced slides; (i) slope failure occurs during earthquake, (ii) post-earthquake slope failure occurs due to pore pressure redistribution, and (iii) post-earthquake failure occurs due to creep effects. The last scenario requires that significant cyclic shear strains take place during the earthquake shaking. Nadim and Kalsnes (1997) presented laboratory test results on Norwegian marine clays that revealed that if the earthquake-induced cyclic shear strains are large, slopes can undergo further creep displacements after the earthquake and experience a significant reduction of static shear strength. It was observed that creep strains and reduction of static shear strength become significant when the earthquake-induced cyclic shear strains exceed 1–2 %. Andersen (2009) showed that a slope subjected to large cyclic loading could experience delayed failure due to undrained creep. By using lab test data, he demonstrated that the permanent shear strain is a key parameter that governs this form of failure in slopes. The data and procedure by Andersen (2009) was used by Johansson et al. (2013) in the evaluation of the effect of blast vibrations on the stability of quick clay slopes.

This paper proposes predictive models for the permanent displacement and shear strain as functions of the characteristics of the slope (e.g. factor of safety) and the ground motion (e.g. peak ground acceleration). The database used for this purpose was obtained from numerical simulations of 1D slopes with different soil and geometry parameters under different levels of earthquake shaking. The predictive models were developed by using realistic parameters for clay and sensitive clay (sometimes referred to as quick clay). A strain-softening soil model was used for sensitive clays. The results are compared with the sliding-block-based predictive models available in the literature and with a limited number of 2D FEM results.

18.2 Review of Existing Predictive Models

Earthquake-induced displacement is the parameter most often used in assessing the seismic stability of slopes. Various researchers have proposed equations based on the sliding block model that predict the slope displacement as functions of ground motion parameters and slope characteristics. Bray et al. (1998) developed a prediction model for solid-waste landfills using wave propagation results in equivalent 1 – D slide masses. The model is a function of the amplitude of shaking in the sliding mass, yield acceleration, and significant duration of shaking. More recent researches have used larger ground motion datasets to develop displacement predictive models and have developed better estimates of the variability in the predictions. Watson-Lamprey and Abrahamson (2006) developed a model using a large dataset consisting of 6,158 recordings scaled with seven different scale factors and computed for three values of yield acceleration. Their displacement model is a function of various parameters including PGA, spectral acceleration at a period of 1 s ($S_{a,T=1s}$), root mean square acceleration (A_{RMS}), yield acceleration, and the duration for which the acceleration-time history is greater than the yield acceleration ($Durk_y$).

Jibson (2007) developed predictive models for rigid block displacements using 2,270 strong motion recordings from 30 earthquakes. A total of 875 values of calculated displacement, evenly distributed between four values of yield acceleration, were used. The models have been developed as functions of (i) k_y/PGA (called the critical acceleration ratio), (ii) k_y/PGA and earthquake magnitude (M), (iii) yield acceleration and Arias Intensity, and (iv) k_y/PGA and Arias Intensity. Bray and Travasarou (2007) presented a predictive relationship for earthquake-induced displacements of rigid and deformable slopes. Displacements were calculated using the equivalent-linear, fully-coupled, stick-slip sliding model of Rathje and Bray (1999, 2000). A set of 688 earthquake records (2 orthogonal components per record) obtained from 41 earthquakes were used to compute displacements for ten values of k_y and eight site geometries (i.e., fundamental site periods, T_s). the displacements for the two components of orthogonal motion were averaged and values less than 1 cm were set equal to zero because they were assumed to be of no engineering significance. The model input parameters include yield acceleration, the initial fundamental period of the sliding mass (T_s), the magnitude of the earthquake (M), and the spectral acceleration at a period equal to 1.5Ts, called $S_{a,T=1.5Ts}$

Rathje and Saygili (2009) and Saygili and Rathje (2008) presented empirical predictive models for rigid block sliding displacements. These models were developed using displacements calculated from over 2,000 acceleration time histories. The considered various single ground motion parameters and vectors of ground motion parameters to predict the sliding displacement. The scalar model presented by Rathje and Saygili (2009) predicts sliding displacement based on the parameters PGA, M, and k_y, and the vector model presented by Saygili and Rathje (2008) predicts sliding displacement based on PGA, PGV, M, and k_y. Table 18.1 summarizes the parameters used in the above predictive models.

Table 18.1 Displacement predictive models and their parameters

Model	Parameters
Bray et al. (1998)	D_{5-95} = significant duration of shaking in seconds
	k_y = yield acceleration
	k_{max} = peak demand (acceleration) coefficient
Watson-Lamprey and Abrahamson (2006)	PGA = peak ground acceleration
	$S_{a,T=1s}$ = spectral acceleration at a period of 1 s (SaT = 1 s)
	A_{RMS} = root mean square acceleration
	k_y = yield acceleration
	$Dur k_y$ = duration for which the acceleration-time history is greater than the yield acceleration
Jibson (2007)	Model 1: k_y/PGA = critical acceleration ratio
	Model 2: k_y/PGA, M
	Model 3: k_y and I_a (Arias intensity)
	Model 4: k_y/PGA, I_a
Bray and Travasarou (2007)	k_y = yield acceleration
	T_s = initial fundamental period of the sliding mass
	M = earthquake magnitude
	$S_{a,T=1.5Ts}$ = Spectral acceleration at a period equal to 1.5Ts
Rathje and Saygili (2009)	Scalar Model: PGA, k_y, M
Saygili and Rathje (2008)	Vector Model: PGA, PGV, k_y, M

18.3 Description of Simulations

18.3.1 Computational Model

The predictive models proposed in this paper are based on a database of numerically-computed responses of slopes due to earthquake loading. To this end, infinite slopes with realistic soil profiles were considered. The computer code *QUIVER_slope* (Kaynia 2011) was used for simulating one-dimesional seismic response of the slopes. The code is based on a simple nonlinear model consisting of a visco-elastic linear loading/unloading response together with strain softening and a kinematic hardening yield function post peak strength. The model is implemented in a one-dimensional slope consisting of soil layers with infinite lateral extensions under vertically propagating shear waves. The strain softening turns out to have a considerable impact on the nonlinear response of the soil once the soil reaches the peak shear strength. The advantage of *QUIVER* over other 1D codes is the inclusion of strain softening in the nonlinear soil model.

The earthquake input is defined in the form of an acceleration-time history on the half-space outcrop at the base of the model. The computational model is based on FEM using a unit soil column. Each layer is replaced by a nonlinear spring and viscous dashpot. The masses are lumped at the layer interfaces. Each layer is characterized by the following parameters:

Fig. 18.1 Parameters of strain-softening soil model

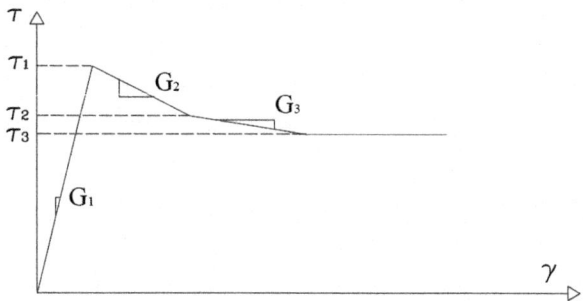

- Thickness, h
- Total unit weight, γ
- Viscous damping ratio, D
- Peak shear strength, τ_1, residual shear strength, $\tau_r = \tau_3$, and intermediate shear stress point on the strain softening branch, τ_2 (Fig. 18.1)
- Shear modulus of the loading/unloading response, G_1, together with the shear moduli of the strain softening branches, G_2 and G_3 (Fig. 18.1); alternatively, the shear strains corresponding to the three shear stresses in Fig. 18.1.

Damping in the loading/unloading cycles is simulated by the Rayleigh damping (e.g. Chopra, 2001) which is defined as $C = \alpha M + \beta K$ where M and K are the mass and stiffness matrices.

A model with N soil layers over a half space contains $N+1$ degrees of freedom corresponding to the displacements at the soil interfaces. The differential equation of motion of this model is given by:

$$M\ddot{U} + C\dot{U} + KU = -M\{I\}\ddot{u}_g(t) \qquad (18.1)$$

where M, K and C are the mass, stiffness and damping matrices of the system, U is the vector of displacements at layer interfaces relative to the base, and $\ddot{u}_g$ (t) is the earthquake acceleration on the half-space outcrop. The symbol $\{I\}$ denotes a vector of $N+1$ unit values. The equations of motion were solved by the constant acceleration method which is an implicit and unconditionally stable integration algorithm (e.g. Chopra 2001).

18.3.2 Model Parameters

The analyses included two different clay types, sensitive and ordinary clays. As shown in Fig. 18.1, a strain-softening soil model was used for the sensitive clay. The normalized small-strain shear modulus (G_{max}/S_u^{DSS}) for clay was established as a function of plasticity index (I_p) using (18.2) based on the lab test data presented

18 Predictive Models for Earthquake Response of Clay and Sensitive Clay Slopes

Table 18.2 Model parameters for sensitive clays and clays

Parameter	Sensitive clay	Clay
G_{max}/S_u^{DSS}	900	900–1700
τ_2/τ_1 at 5 % shear strain	0.9	1.0
τ_3/τ_1 at 50 % shear strain	0.5	1.0

by Andersen (2004). The soil parameters used in the analyses are summarized in Table 18.2.

$$\frac{G_{max}}{s_u^{DSS}} = 325 + \frac{55}{\left(\frac{I_p}{100}\right)^2} \qquad (18.2)$$

It is admitted that the results of this study (especially those for the sensitive clay) are dependent on the selected soil parameters. Nevertheless, it is believed that these results provide a step in the right direction in the development of more reliable predictive equations.

A normally consolidated soil profile with a normalized direct simple shear strength value $s_u^{DSS}/\sigma'_v = 0.21$ (with σ'_v being the effective vertical stress) was used for the analyses. To account for the increased strength under dynamic loading, a strain rate factor of 1.4 was applied to the static shear strength (Lunne and Andersen 2007). To capture the variation in the slope angle and soil profiles, the analyses were conducted for slope angles of 3°, 6°, 9°, and 12° and for soil profile depths of 40 m, 70 m, and 100 m. The numerical analyses were carried out for five earthquake strong motions records using PGA levels ranging from 0.05 g to 0.40 g (next section). Totally, 315 *QUIVER* analyses were performed for sensitive clay slopes and 515 analyses were conducted for clay slopes.

18.4 Selection and Scaling of Acceleration Time Histories

The acceleration response spectrum used in Norway for rock (ground type A according to Eurocode 8 terminology) was used as the target spectrum. The spectrum is shown in Fig. 18.2 for PGA = 0.05 g. The spectrum follows the standard parameterized form in Eurocode 8. Pacific Earthquake Engineering Research (PEER) Ground Motion Database Web Application (PGMD) was used for the selection of the best matching earthquake strong motion records. PGMD allows the user to select recordings for which the geometric mean of the two horizontal components provides a good match to the target spectrum. The quantitative measure of the 'good match' of the motion with respect to the target spectrum is evaluated by Mean Squared Error (MSE) of the difference between the spectral accelerations of the record and the target spectrum. Scale factors are applied to reduce the MSE over the period range of interest. The scaling factor is applied to the

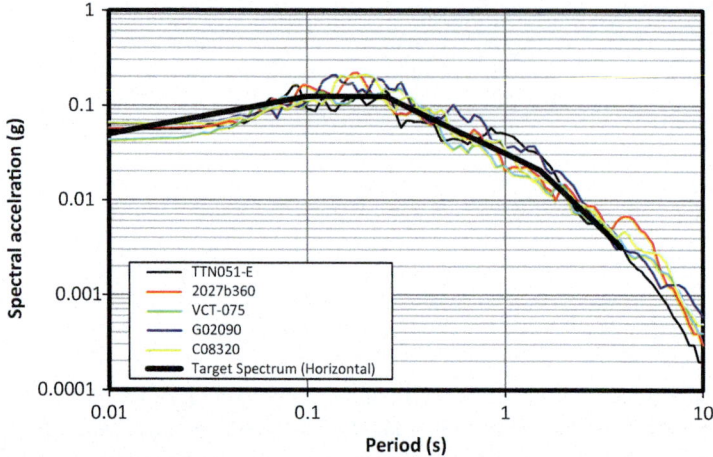

Fig. 18.2 Target acceleration spectrum corresponding to PGA = 0.05 g and response spectra of scaled acceleration time histories

Table 18.3 Main parameters of selected recorded motions

Earthquake	Designation	Magnitude	Depth (km)	Epicentral distance (km)	PGA (g)	Scale factor
Imperial valley 1979	VCT-075	6.53	10.0	43.90	0.14	0.437
Baja border 2002	2027b360	5.31	7.0	55.40	0.06	0.953
Morgan hill 1984	G02090	6.19	8.5	38.10	0.19	0.366
Parkfield 1966	C08320	6.19	10.0	34.01	0.26	0.242
Chi-Chi 1999	TTN051-E	6.20	18.0	49.99	0.07	0.766

geometric mean of two horizontal components so the same scale factor is applied over the two components for the same strong motion data.

Five horizontal components of recorded earthquake strong motions from the PEER Center strong motion database (PEER 2011) were selected as seed motions, and they were scaled to the horizontal target spectrum. Table 18.3 summarizes the relevant parameters of the selected seed motions. The scaling factors used for these motions are also presented in Table 18.3. The response spectra of the scaled time histories and the target spectrum are plotted in Fig. 18.2.

18.5 Development of Predictive Models

The two parameters, PGA and yield acceleration, have commonly been used in the earlier predictive models based on the sliding-block concept. These parameters give measures of the driving force and resistance, respectively. While PGA on bedrock has a clear role in sliding block models, it loses its significance in realistic soil profiles. A more representative parameter for the driving force is the peak acceleration on the ground surface that relates closely to the destabilizing force on the slope mass. The yield acceleration is closely related to the factor of safety, FS, and hence was replaced by this parameter in the present study. The advantage of using FS in the predictive equations is that one could readily extend the equations derived from the 1D analyses to more general 2D and 3D geometries. A limited number of 2D seismic slope analyses are used in this paper to test the validity of this idea. In applying the presented predictive equations, the value of FS should be computed by using the peak shear strength applicable to earthquake loading, for example after it is increased to account for the rate effect.

The existing predictive models give only estimates of the slope displacements. The underlying assumption is that if the computed displacement is larger than a threshold value (typically in the range 5-15 cm), the slope is considered to fail. As pointed out earlier, permanent shear strain is a more robust indicator of slope stability as compared to sliding displacement. Laboratory test data could then be used to establish the threshold shear strain for initiation of soil failure. While in clay the threshold can be as large as 15 %, for sensitive and quick clay the value is much smaller due to the possibility of undrained creep failure (e.g. Andersen 2009).

18.5.1 Permanent Slope Displacement in Sensitive Clay

Figure 18.3a, b show the computed permanent displacements as a function of the computed peak acceleration on the ground surface with different labels for slope angles and for earthquake strong motion records, respectively. Figure 18.4a, b show the histograms of the computed displacements and the peak acceleration on the ground surface from 315 seismic response analyses for sensitive clay slopes.

Equation 18.3 shows the functional form of the predictive model. In this equation, a_{max} is the peak acceleration on the ground surface in g, and D is the permanent displacement in cm. The standard deviation (σ_{lnD}) for the best fit predictive model is 1.15. Figure 18.5 shows the prediction of the model for different slope angles.

$$\ln D = 5.89 + 2.65 \ \ln(a_{max}) - 0.51 \ FS - 0.4 \ (FS - 3.11) \\ (\ln(a_{max}) + 1.4) \tag{18.3}$$

18.5.2 Permanent Slope Displacement in Clay

Figure 18.6a, b show the computed permanent displacements as function of the computed peak acceleration on the ground surface with different labels for slope angles and for earthquake strong motion records, respectively. Figures 18.6a and

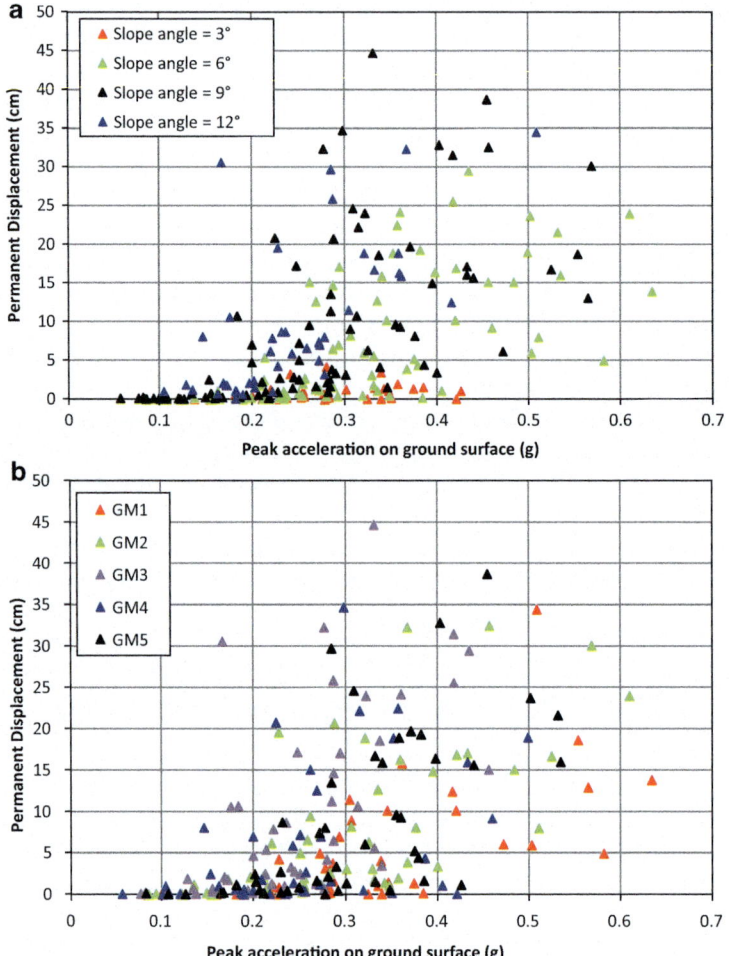

Fig. 18.3 Permanent displacement versus peak acceleration on ground surface for sensitive clay with labels (**a**) for slopes angles, and (**b**) for selected acceleration-time histories (GM stands for Ground Motion)

18.7b show the histograms of the computed permanent displacements and the peak acceleration on ground surface from 515 seismic response analyses for clay slopes.

Equation 18.4 shows the functional form of the predictive model. The standard deviation ($\sigma_{\ln D}$) for the best fit predictive model is 0.97. Figure 18.8 displays the prediction of the model for different slope angles.

$$\ln D = 5.65 + 2.57 \ \ln(a_{max}) - 0.50 \ FS - 0.3 \ (FS - 2.96) \\ (\ln(a_{max}) + 1.3) \quad (18.4)$$

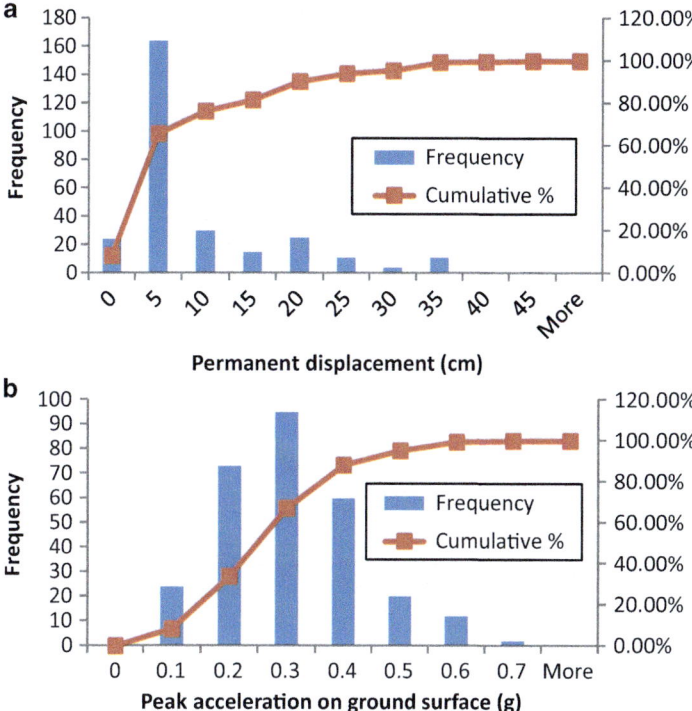

Fig. 18.4 Histograms of (**a**) permanent displacement, and (**b**) peak acceleration on ground surface in sensitive clay

18.5.3 Permanent Shear Strain in Sensitive Clay

Figure 18.9a, b display the computed permanent shear strains as function of the computed peak acceleration on the ground surface with different labels for slope angles and for earthquake strong motion records, respectively. Figure 18.10a, b present the histograms of the permanent strains and the peak acceleration on ground surface for 315 seismic slope response analyses for sensitive clay.

Equation 18.5 expresses the functional form of the predictive model. The standard deviation ($\sigma_{\ln S}$) for the best fit predictive model is 1.19. In this equation, S is the permanent shear strain in percent, and a_{max} is the peak acceleration (in g) on the ground surface. Figure 18.11 shows the prediction of the model for different slope angles.

$$\ln S = 5.75 - 0.52\ FS + 2.77\ \ln(a_{max}) + 0.076\ FS\ \ln(a_{max}) \qquad (18.5)$$

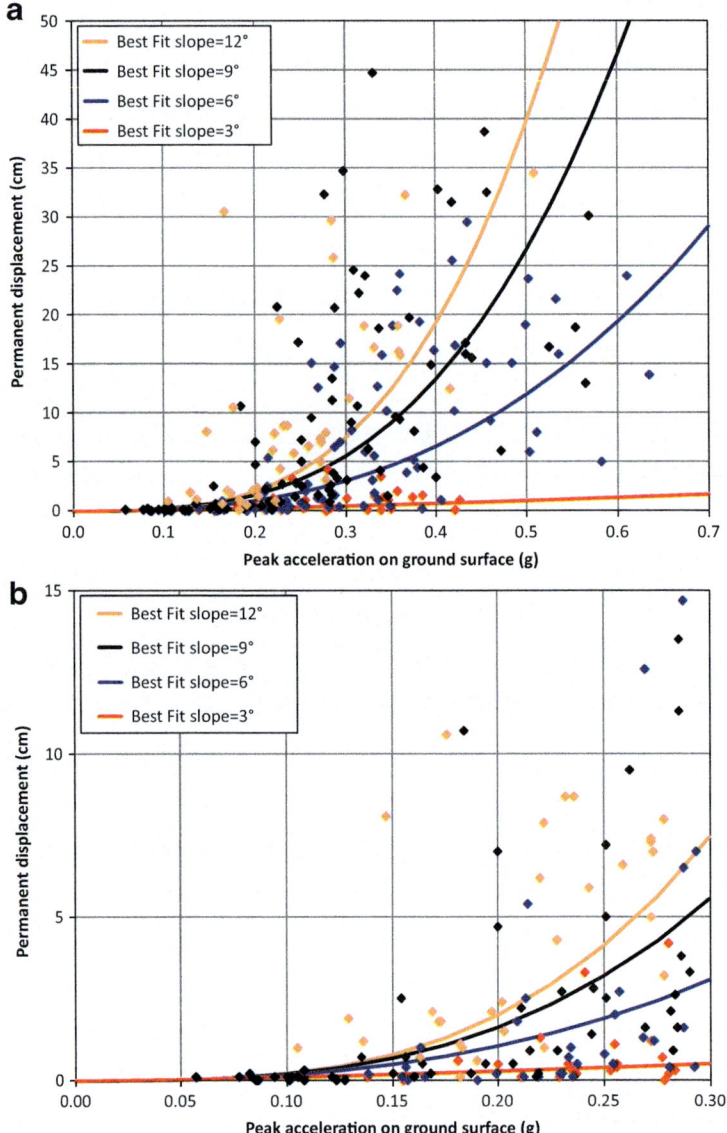

Fig. 18.5 Displacement predictions of model for sensitive clay slopes for (**a**) all displacement values, and (**b**) zoomed-in region for D < 15 cm

18.5.4 Permanent Shear Strain in Clay

Figure 18.12a, b present the computed permanent shear strains as function of the computed peak acceleration on the ground surface with different labels for slope

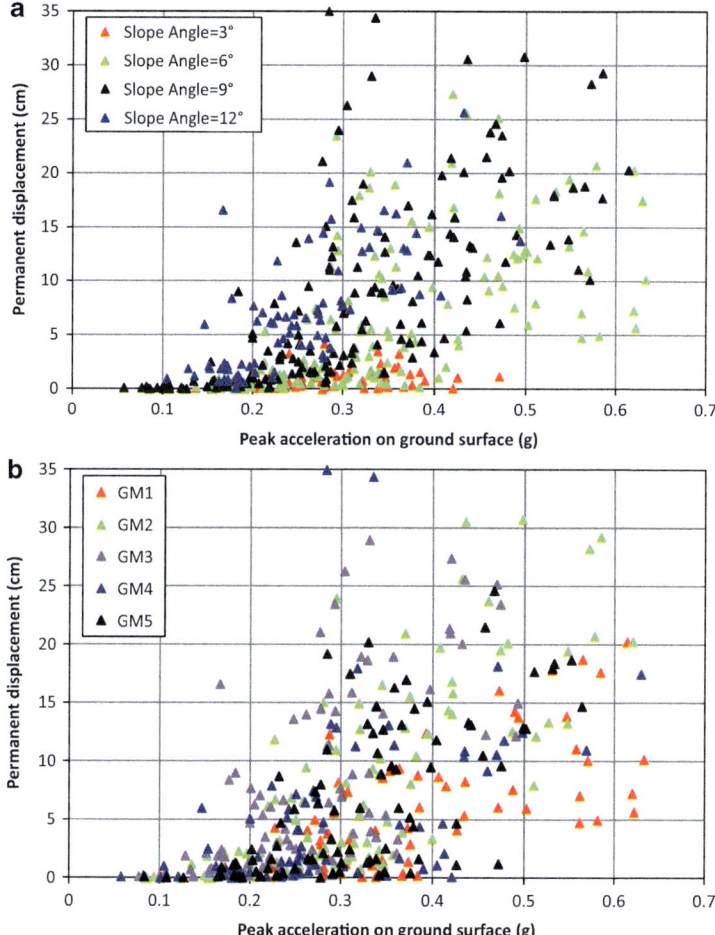

Fig. 18.6 Permanent displacement versus peak acceleration on ground surface in clay slopes with labels (**a**) for slope angles, and (**b**) for selected acceleration time histories (GM stands for Ground Motion)

angles and for earthquake strong motion records. Figure 18.13a, b show the histograms of the permanent strain and the peak acceleration on ground surface out of 515 seismic slope response analyses for clay slopes.

Equation 18.6 gives the functional form of the predictive model. The standard deviation (σ_{lnS}) for the best fit predictive model is 0.92. Figure 18.14 shows the prediction of the model for different slope angles.

$$\ln S = 4.15 - 0.30\ FS + 2.06\ \ln(a_{max}) + 0.16\ FS\ \ln(a_{max}) \qquad (18.6)$$

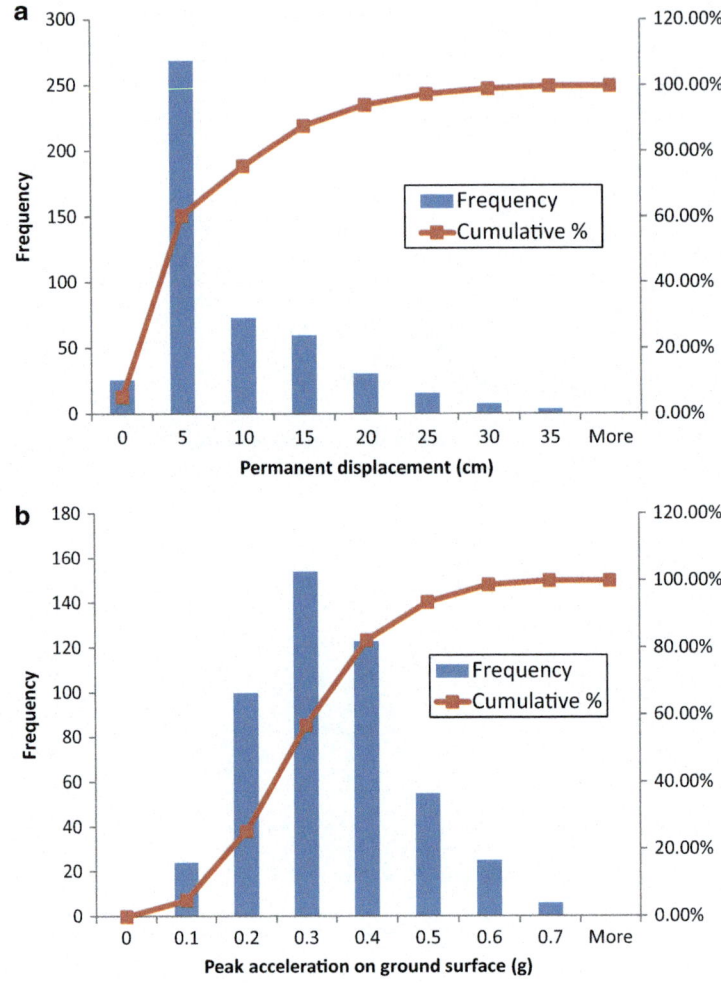

Fig. 18.7 Histograms of (**a**) permanent displacement, and (**b**) peak acceleration on ground surface in clay slopes

18.5.5 Comparisons of Displacement and Strain Predictions for Clay and Sensitive Clay

Figure 18.15 presents a comparison of the displacement predictions for clay and sensitive clay. Figure 18.16 shows a comparison of the permanent shear strain predictions for ordinary and sensitive clays. As expected, for the same slope geometry and similar earthquake loading, the displacements and shear strains are larger for the sensitive clay.

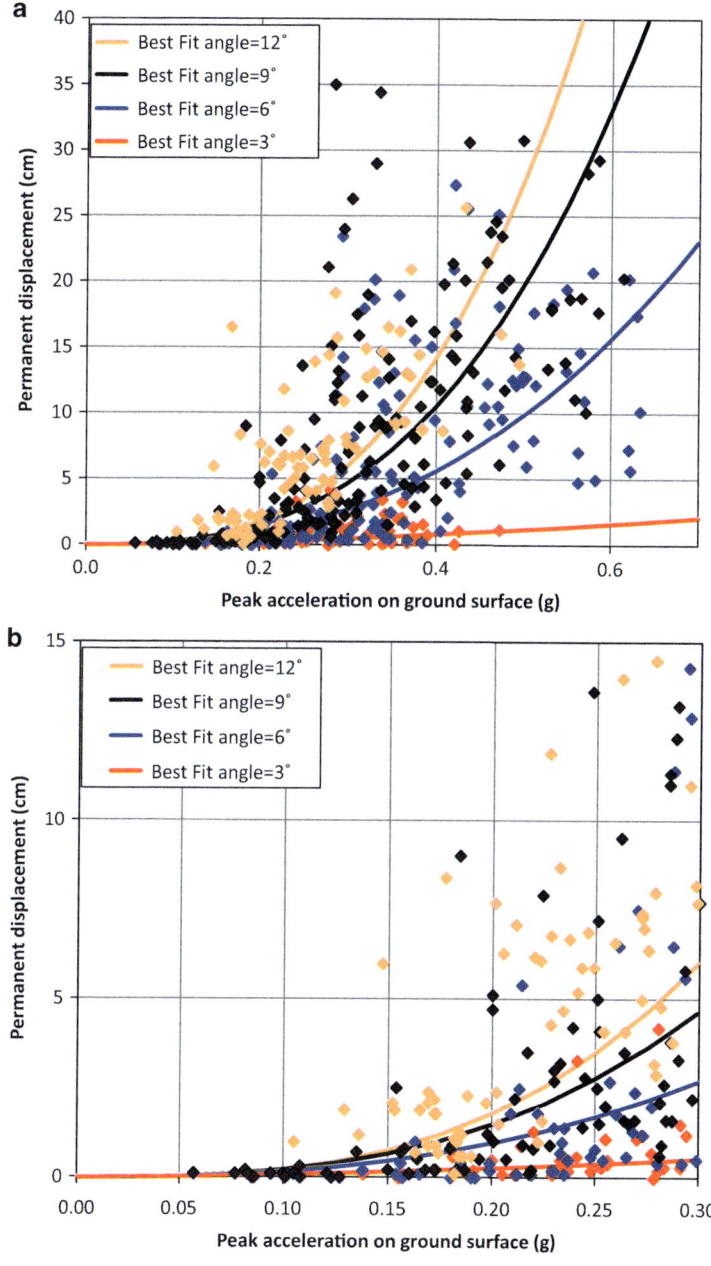

Fig. 18.8 Displacement predictions of model for clay slopes for (**a**) all displacement levels, and (**b**) zoomed-in region for D < 15 cm

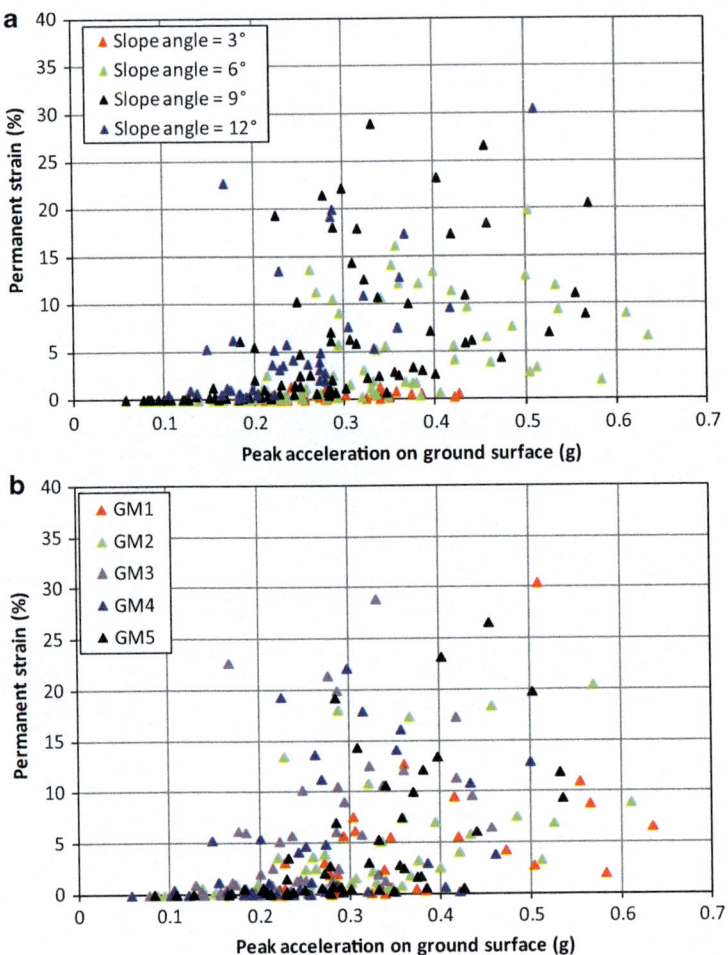

Fig. 18.9 Permanent shear strain versus peak acceleration on ground surface for sensitive clay, with labels (**a**) for slopes angles, and (**b**) for selected acceleration time histories

18.6 Comparison with Other Predictive Models for Displacement

Figure 18.17 presents a comparison of several predictive models (namely, Watson-Lamprey and Abrahamson 2006; Bray and Travasarou 2007, the Jibson 2007 k_y/PGA model, the Rathje and Saygili 2009 scalar (PGA, M) model and the Saygili and Rathje 2008 vector (PGA, PGV) model) for a deterministic earthquake scenario of $M_w = 7.5$ and $R = 5$ km for a shallow, rigid sliding mass, and rock site conditions ($V_{s30} > 760$ m/s). The predicted ground motion parameters for each scenario are listed in the figure. The values of PGA and $Sa_{T=1s}$ are from Boore and Atkinson

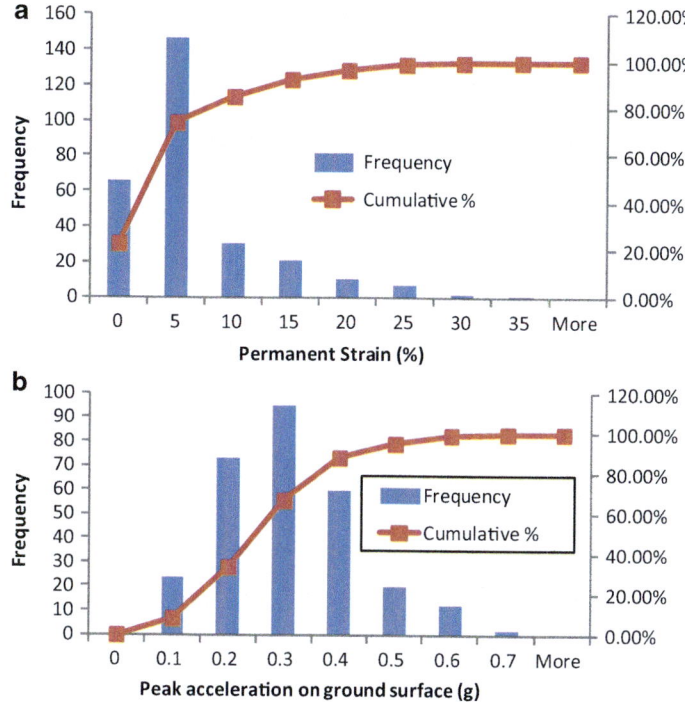

Fig. 18.10 Histograms of (**a**) permanent shear strain, and (**b**) peak acceleration on ground surface for sensitive clay

(2008), I_a is from Travasarou et al. (2003), T_m is from Rathje et al. (2004), and D_{5-95} is from Abrahamson and Silva (1997). Even though these models were developed using large datasets and rigorous regression techniques, there is more than a magnitude difference in the final predictions. The Bray et al. (1998) model predicts the largest displacement, the Watson-Lamprey and Abrahamson (2006) model predicts the smallest, and the other models fall in between. As shown in the figure, the displacement predictions of the proposed model fall into the low estimate bound for less stable slopes (e.g. $k_y = 0.05$–0.10 g) and into the high estimate bound for more stable slopes ($k_y = 0.20$–0.25 g). The proposed model uses the maximum acceleration on the ground surface whereas the other models use PGA in the equations. It should be noted that Jibson (2007), Bray and Travasarou (2007), Rathje and Saygili (2009) and the proposed model each use only one ground motion parameter (*PGA*), while Saygili and Rathje (2008) and Bray et. al. (1998) use two ground motion parameters, and the Watson-Lamprey and Abrahamson (2006) model uses four parameters (*PGA*, A_{RMS}, $Sa_{T=1s}$, and Dur_{k_y}).

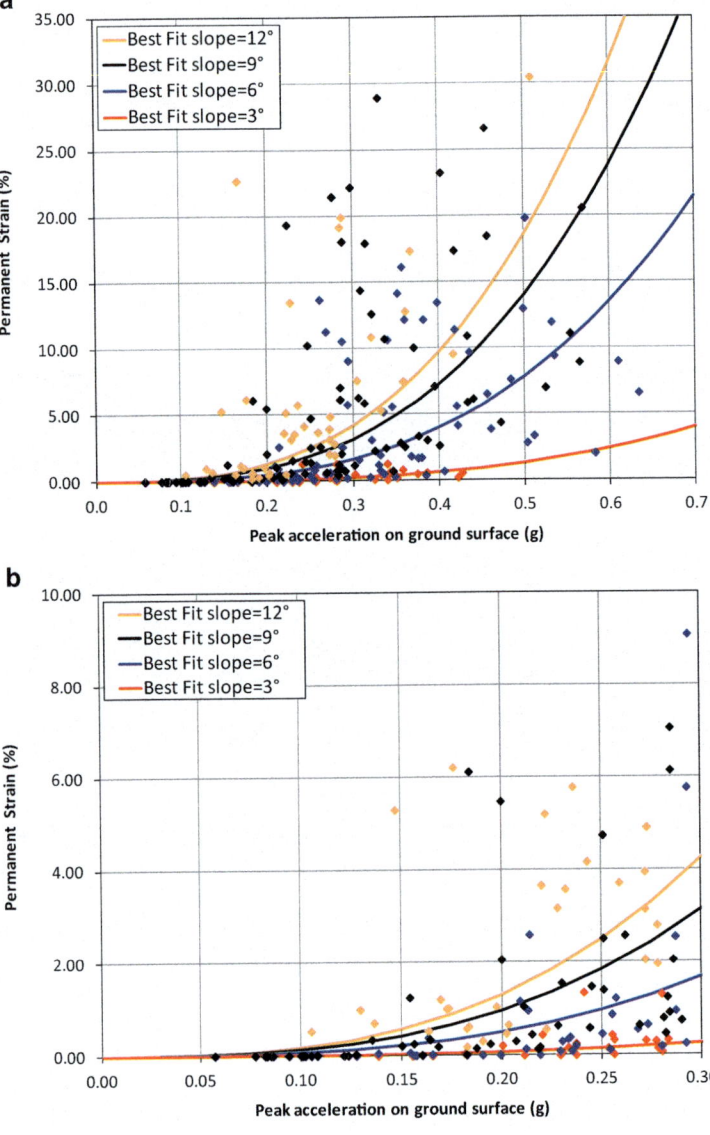

Fig. 18.11 Shear strain predictions of model for sensitive clay for (**a**) all strain levels, and (**b**) zoomed-in region for S < 10 %

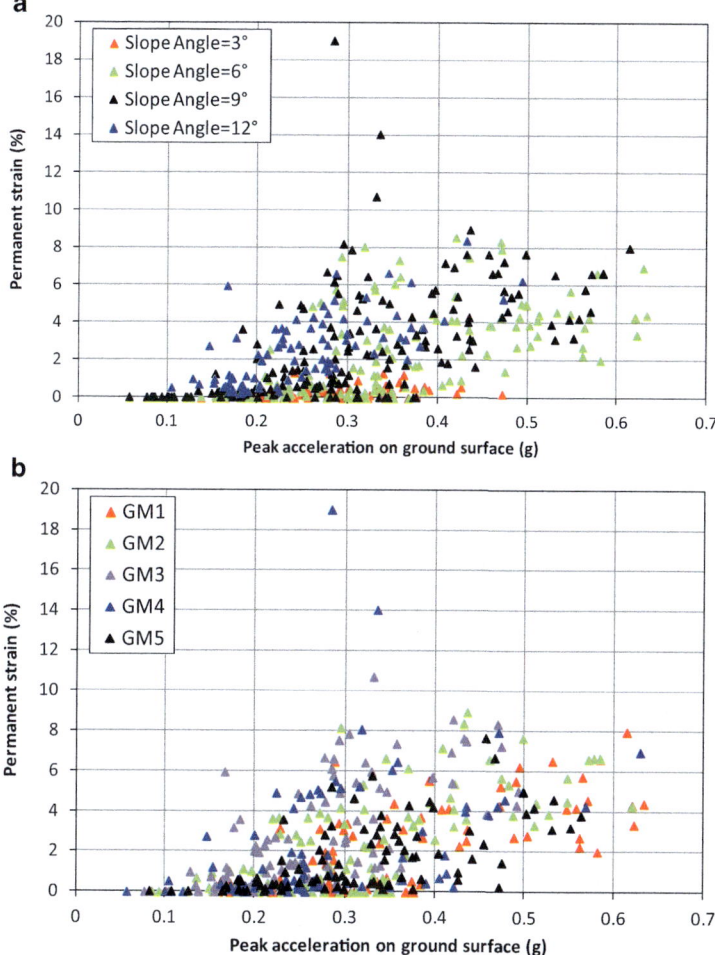

Fig. 18.12 Permanent shear strain versus peak acceleration on ground surface with labels for clay slopes (**a**) for slope angles, and (**b**) for selected acceleration time histories

18.7 Comparison of Displacement Predictions with 2D FEM Results

The predictive models were developed from a database of numerically computed response parameters using 1D earthquake analyses. The factor of safety, FS, was used in the predictive equations with the intention that these equations could be applied to more general soil types and slope geometries. A natural step along this line is to test the performance of the developed models in a two-dimensional geometry. To this end, a number of simple 2D slope models with normally-

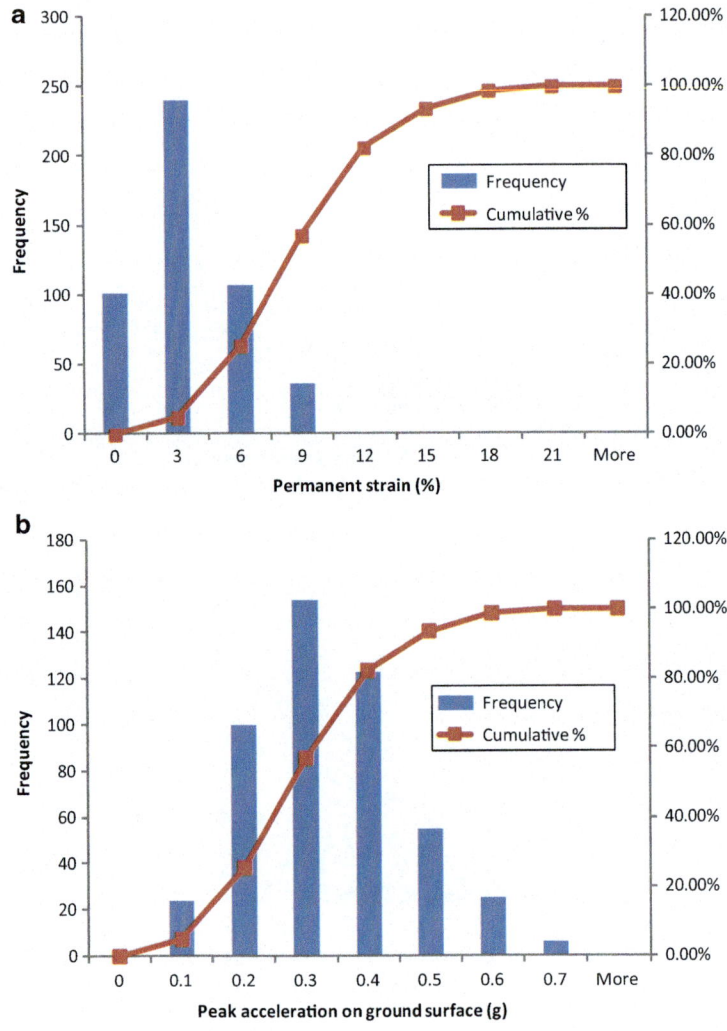

Fig. 18.13 Histograms of (**a**) permanent shear strain and (**b**) peak acceleration on ground surface in clay slopes

consolidated clay were constructed and were excited by earthquake at their bases. The permanent displacements and permanent shear strains in these slopes were computed at the end of the shaking and were compared with the predictions from the developed equations. The analyses were carried out with the FE software Plaxis.

Figure 18.18 displays part of the slope model used in the analyses together with its FE mesh. The model is 75 m deep on the downslope side and 110 m deep on the upslope side. The slope was placed in two series of analyses such that their factors

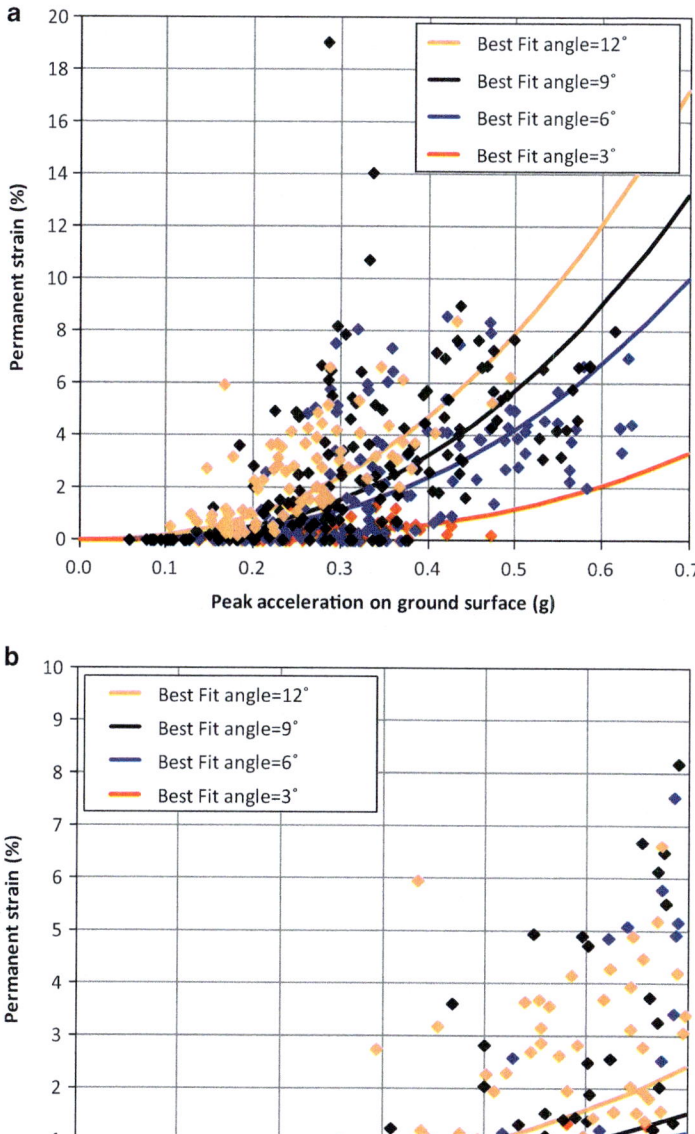

Fig. 18.14 Permanent shear strain predictions of model for clay slopes for (**a**) all strains levels, and (**b**) zoomed-in region for S < 10 %

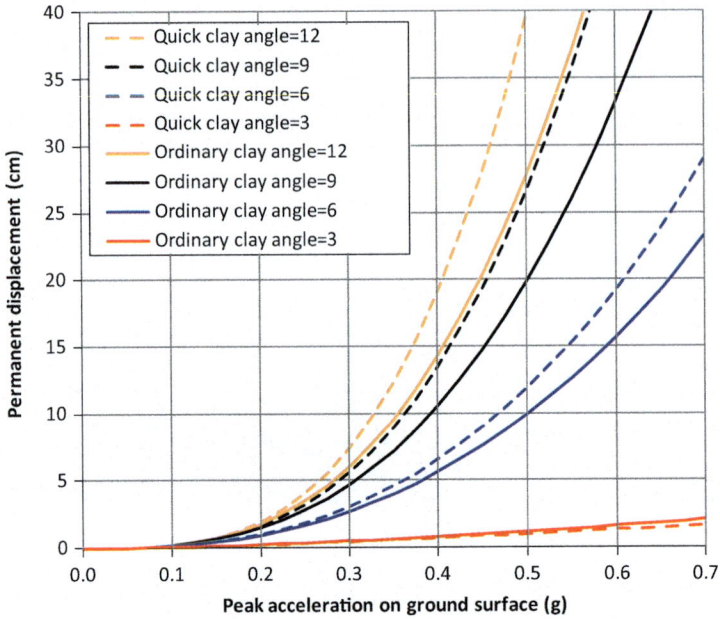

Fig. 18.15 Displacement predictions for clay and sensitive clay

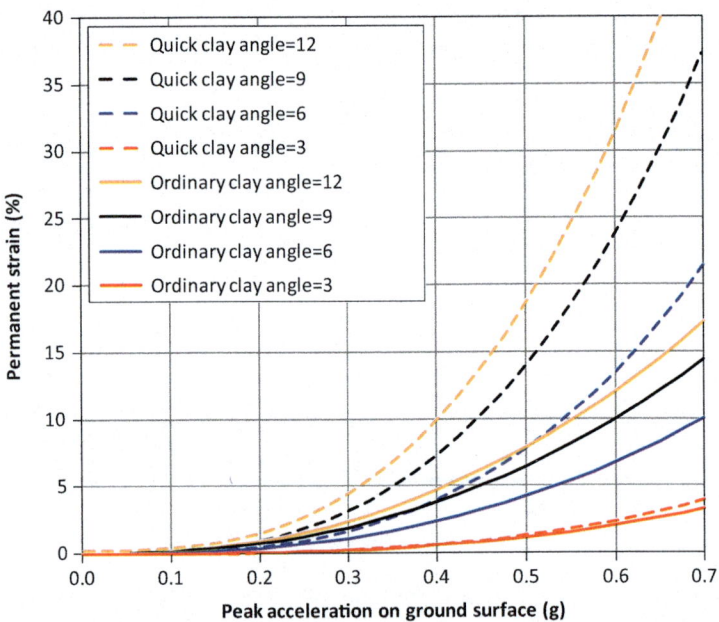

Fig. 18.16 Shear strain predictions for clay and sensitive clay

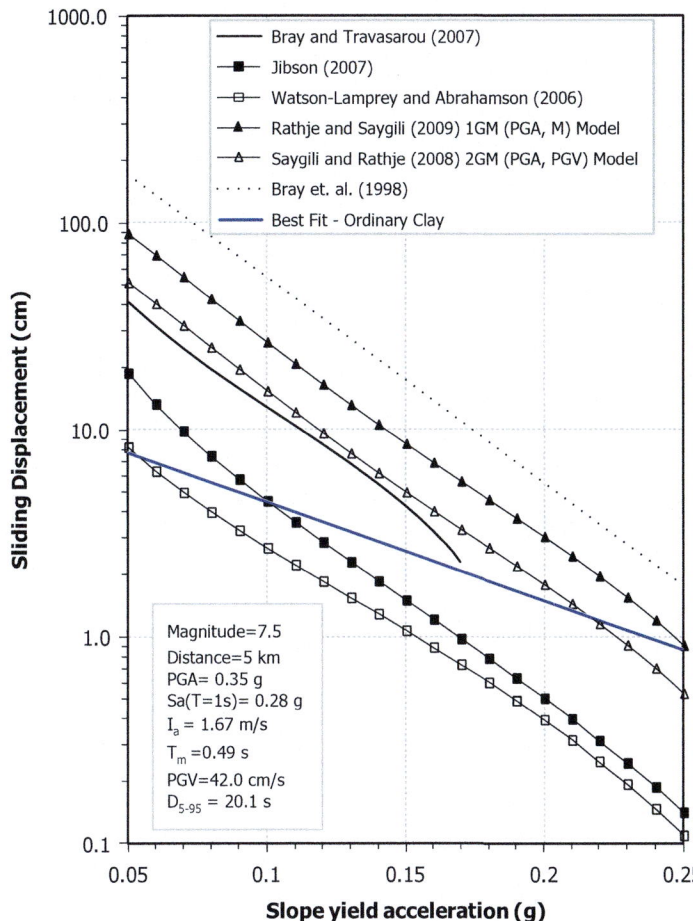

Fig. 18.17 Comparisons of predictive models for sliding displacement for a deterministic scenario of $M_w = 7.5$ and $R = 5$ km

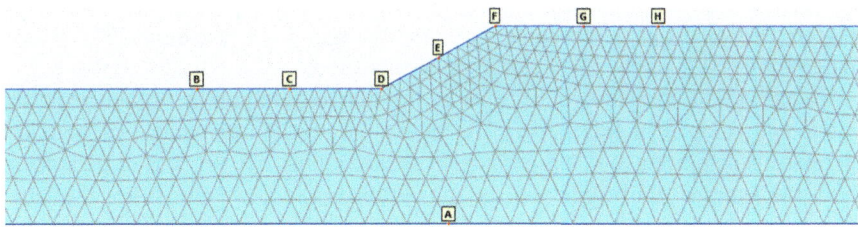

Fig. 18.18 Two-dimensional FE model, mesh detail and monitoring points on ground surface

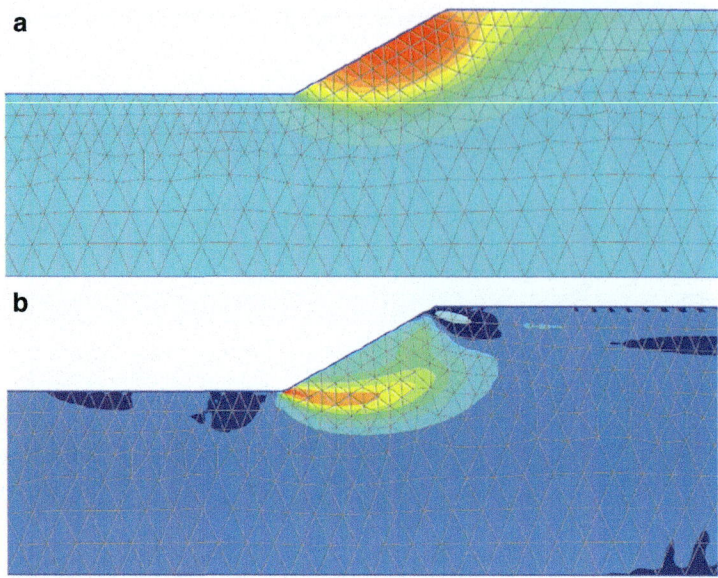

Fig. 18.19 Results of 2D FE analyses for slope with FS = 1.2 and PGA = 0.4 g: (**a**) permanent horizontal displacements with maximum value about 1.3 m, (**b**) permanent shear strains with maximum value 10 %

of safety, SF, were 1.2 and 1.5. Because the peak accelerations and permanent displacements vary on the ground surface, seven monitoring points (points B to H, as shown in Fig. 18.18), were placed on the ground surface. The slopes were excited by acceleration time histories with PGA varying from 0.05 g to 0.40 g on the bedrock (base of the model, point A in Fig. 18.18). The values of the peak accelerations and permanent displacements at the monitoring points were determined from the FE analyses and were averaged. For the permanent shear strain, the maximum value was determined from each analyses.

Figure 18.19a, b present typical results of the FE analyses for the case FS = 1.2 due to an earthquake with PGA = 0.4 g. Figure 18.19a displays the contours of permanent slope displacements. The displacement values range from 0.0 to 1.3 m. Figure 18.19b displays the contours of the permanent shear strains. The values range from 0.0 to about 10 % at the toe of the slope.

Figure 18.20 compares the results of the 2D FE analyses with the predictive models developed in this paper. The figures show the comparison of both the permanent displacements and permanent shear strains. For the former parameter, both the average 2D results and the maximum values are plotted. For the latter parameter, the maximum permanent strains from the 2D model are plotted together double the strains. The reason for this is that the shear strain is more sensitive to the FE mesh size, and there is a tendency that the maximum strain increases, as the mesh is refined. The results in both cases show fairly good agreement with those from the predictive models.

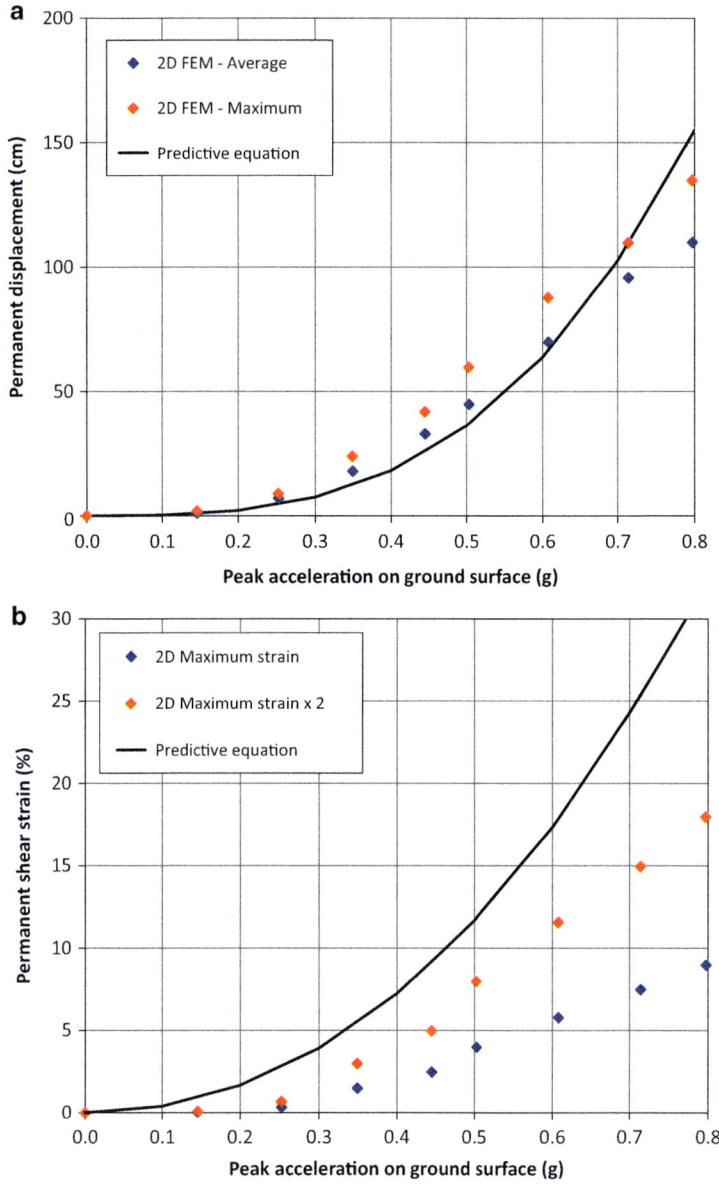

Fig. 18.20 Results from 2D FEM for FS = 1.2 versus best fit predictions, (**a**) permanent displacements, (**b**) permanent shear strains

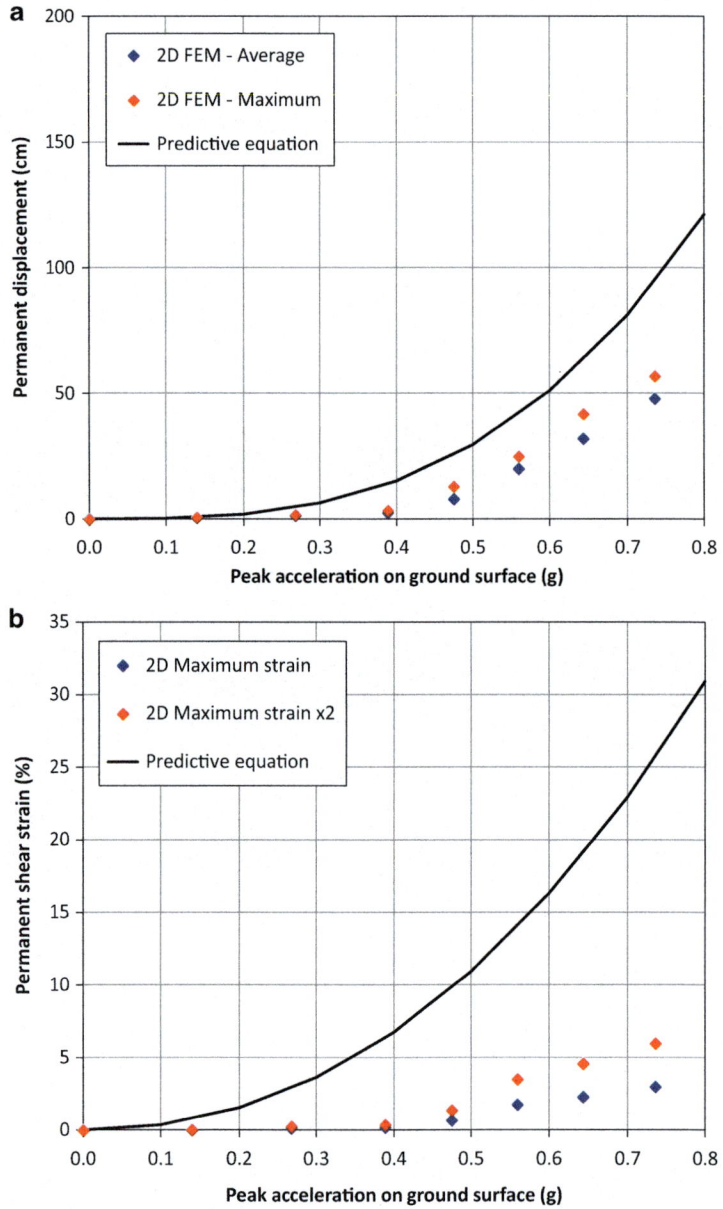

Fig. 18.21 Results from 2D FEM for FS = 1.5 versus best fit predictions, (**a**) permanent displacements, (**b**) permanent shear strains

Figure 18.21 presents similar comparisons for the case FS = 1.5. While comparison of the displacements by the FE model and predictive model is satisfactory, the predictive model for shear strain tends to overly estimate the shear strains. This is a typical effect of 2D geometry which represents a conservative situation compared to a 1D idealization. As the size of the slope increases, this effect is diminished, and the 2D results tend more to the 1D results as captured by the predictive models developed in this paper.

Open Access This chapter is distributed under the terms of the Creative Commons Attribution Noncommercial License, which permits any noncommercial use, distribution, and reproduction in any medium, provided the original author(s) and source are credited.

References

Abrahamson NA, Silva W (1997) Empirical response spectral attenuation relations for shallow crustal earthquakes. Seismol Res Lett 68:94–127

Andersen KH (2004) Cyclic clay data for foundation design of structures subjected to wave loading. In: Proceedings of the international conference of cyclic behaviour of soils and liquefaction phenomena. A. A. Balkema Publishers, Bochum, pp 371–387

Andersen KH (2009) Bearing capacity under cyclic loading — offshore, along the coast, and on land. The 21st Bjerrum Lecture presented in Oslo, 23 November 2007. Can Geotech J 46 (5):513–535

Biscontin G, Pestana JM, Nadim F (2004) Seismic triggering of submarine slides in soft cohesive soil deposits. Mar Geol 203(3 & 4):341–354

Boore DM, Atkinson GM (2008) Ground-motion prediction equations for the average horizontal component of PGA, PGV and 5 %-damped PSA at spectral periods between 0.01 s and 10.0 s. Earthq Spectra EERI 24(1):99–138

Bray JD, Travasarou T (2007) Simplified procedure for estimating earthquake-induced deviatoric slope displacements. J Geotech Geoenviron Eng ASCE 133(4):381–392

Bray JD, Rathje EM, Augello AJ, Merry SM (1998) Simplified seismic design procedure for lined solid-waste landfills. Geosynthet Int 5(1–2):203–235

Chopra AK (2001) Dynamics of structures. Theory and applications to earthquake engineering, 2nd edn. Prentice Hall, Englewood Cliffs

Jibson RW (2007) Regression models for estimating coseismic landslide displacement. Eng Geol 91:209–218

Johansson J, Lovolt F, Andersen KH, Madshus C, Aabøe R (2013) Impact of blast vibrations on the release of quick clay slides. In: Proceedings of 18th international conference on soil mechanics and geotechnical engineering, ICSMGE, Paris

Kaynia AM (2011) QUIVER_slope – numerical code for one-dimensional seismic response of slopes with strain softening behaviour. NGI report 20071851-00-79-R, 8 June 2012

Lunne T, Andersen KH (2007) Soft clay shear strength parameters for deepwater geotechnical design. In: Proceedings 6th international conference, society for underwater technology, offshore site investigation and geotechnics, London, pp 151–176

Nadim F, Kalsnes B (1997) Evaluation of clay strength for seismic slope stability analysis. In: Proceedings XIV ICSMFE, Hamburg, vol 1, pp 377–379, 6–12 Sept 1997

Newmark NM (1965) Effects of earthquakes on dams and embankments. Geotechnique 15 (2):139–160

PEER – PGMD (2011) Website – http://peer.berkeley.edu/peer_ground_motion_database/

Rathje EM, Bray JD (1999) An examination of simplified earthquake-induced displacement procedures for earth structures. Can Geotech J 36(1):72–87

Rathje EM, Bray JD (2000) Nonlinear coupled seismic sliding analysis of earth structures. J Geotech Geoenviron Eng 126(11):1002–1014

Rathje EM, Saygili G (2009) Probabilistic assessment of earthquake-induced sliding displacements of natural slopes. Bull N Z Soc Earthq Eng 42:18–27

Rathje EM, Faraj F, Russell S, Bray JD (2004) Empirical relationships for frequency content parameters of earthquake ground motions. Earthq Spectra 20(1):119–144

Saygili G, Rathje EM (2008) Empirical predictive models for earthquake-induced sliding displacements of slopes. J Geotech Geoenviron Eng ASCE 134(6):790–803

Travasarou T, Bray JD, Abrahamson NA (2003) Empirical attenuation relationship for Arias intensity. Earthq Eng Struct Dyn 32(7):1133–1155

Watson-Lamprey J, Abrahamson N (2006) Selection of ground motion time series and limits on scaling. Soil Dyn Earthq Eng 26(5):477–482

Chapter 19
Recent Advances in Seismic Soil Liquefaction Engineering

K. Önder Çetin and H. Tolga Bilge

Abstract The assessment of cyclic response of soils has been a major concern of geotechnical earthquake engineering since the very early days of the profession. The pioneering efforts were mostly focused on developing an understanding of the response of clean sands. These efforts were mostly confined to the assessment of the mechanisms of excess pore pressure buildup and corollary reduction in shear strength and stiffness, widely referred to as seismic soil liquefaction triggering. However, as the years passed, and earthquakes and laboratory testing programs continued to provide lessons and data, researchers and practitioners became increasingly aware of additional aspects, such as liquefaction susceptibility and cyclic degradation response of silt and clay mixtures. Inspired from the fact that these issues are still considered as the "soft" spots of the practice, the scope of this chapter is tailored to include a review of earlier efforts along with the introduction of new frameworks for the assessment of cyclic strength and straining performance of coarse- and fine-grained soils.

19.1 Introduction

The assessment of cyclic response of soils has been a major concern of geotechnical earthquake engineering since the very early days of the profession. Engineering treatment of liquefaction-induced problems evolved initially in the wake of the two devastating earthquakes of 1964 (Niigata, Japan and Great Alaska, USA), during

---

K. Önder Çetin (✉)
Department of Civil Engineering, Middle East Technical University, Ankara, Turkey
e-mail: kemalondercetin@gmail.com

H.T. Bilge
GeoDestek Geoengineering and Consultancy Services, Ankara, Turkey

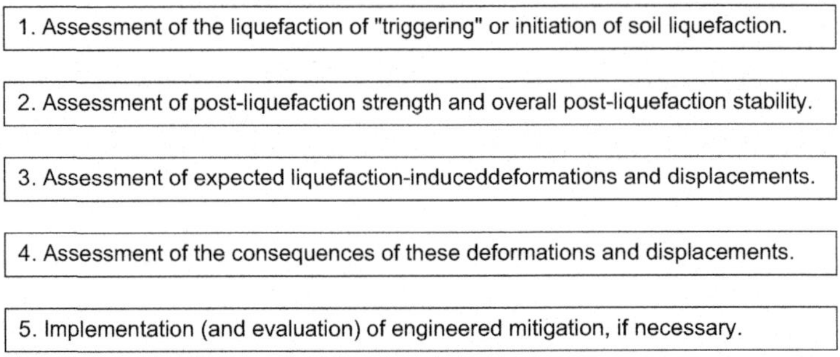

> 1. Assessment of the liquefaction of "triggering" or initiation of soil liquefaction.
>
> 2. Assessment of post-liquefaction strength and overall post-liquefaction stability.
>
> 3. Assessment of expected liquefaction-induceddeformations and displacements.
>
> 4. Assessment of the consequences of these deformations and displacements.
>
> 5. Implementation (and evaluation) of engineered mitigation, if necessary.

Fig. 19.1 Key elements of soil liquefaction engineering

which seismically-induced soil liquefaction was listed as one of the prime causes of structural failures. Pioneering efforts to resolve this problem have focused on developing an understanding on liquefaction triggering behavior of mostly clean sands. However, as earthquakes continued to provide lessons and data, researchers became increasingly aware of the problems associated with the cyclic response of silty and clayey soils.

Today, the profession of "soil liquefaction engineering" is emerging as a rapidly progressing field of practice. Within the scope of this chapter, in addition to the summary of the current state of practice, recent advances in this progressing field will be presented. As illustrated schematically in Fig. 19.1, consistent with the five major steps of seismic soil liquefaction engineering assessment, the discussion layout of the chapter is also structured to follow the same footprints.

Among these, the first step in seismic soil liquefaction engineering involves the assessment of soil liquefaction triggering and has drawn the highest level of research interest. Despite the level of current controversy, it can still be concluded as the most developed assessment stage in liquefaction engineering, and will be discussed next.

19.2 Assessment of Liquefaction Potential and Triggering

19.2.1 Potentially Liquefiable Soils

There has long been a consensus in the literature that "clean" sandy soils, with limited fines, are potentially vulnerable to seismically-induced liquefaction. There has, also been significant controversy and confusion regarding the liquefaction potential of silty soils (and silty/clayey soils), and also of coarser, gravelly soils and rockfills.

The cyclic behavior of coarse, gravelly soils is not very different than that of "sandy" soils. There are now a number of well-documented field cases of liquefaction of coarse, gravelly soils (e.g.: Ishihara 1985; Evans 1987; Harder 1988; Andrus et al. 1991). As discussed in Seed et al. (2001), these soils do differ in behavior from sandy soils in two ways: (1) they can be much more pervious, and so can often rapidly dissipate cyclically generated pore pressures, and (2) due to their larger particle masses, the coarse gravelly soils are seldom deposited "gently" and so they are not commonly encountered in loose state as compared with sandy soils. However, it should be noted that the apparent drainage advantages of coarse, gravelly soils can be eliminated (i) if they are surrounded and encapsulated by less pervious finer materials, (ii) if drainage is internally impeded by the presence of finer soils in the void spaces between the coarser particles, or (3) if the layer thickness is large, which in turn increase the distance over which drainage must occur (rapidly) during an earthquake. In these cases, the coarser soils should be considered to be potentially liquefiable and be assessed for liquefaction triggering hazard. This naturally requires the estimation of in-situ density state (or the penetration resistance), for which the Becker penetration test still continues to be the only practical tool, despite its major limitations.

Contrary to the consensus on liquefaction potential of clean sands, the susceptibility of silt and clay mixtures to liquefaction has been one of the controversial and widely discussed issues. As previously stated, in the early days of the profession, plastic silt and clay mixtures were considered to be resistant to cyclic loading, and consistently, most research was focused on cyclic response of saturated sandy soils mainly. This choice is also reinforced with liquefaction-induced ground failure case histories at coarse-grained (sandy) soil sites after the 1964 Alaska and Niigata earthquakes. However, in the following years, especially after fine-grained soil site failure case histories of 1975 Haicheng and 1979 Tangshan earthquakes from China (Wang 1979), increasing number of research studies focused on understanding their cyclic response.

On the basis of Wang's (1979) database and conclusions, a set of criteria to assess liquefaction potential of soils with fines (widely referred to as Chinese Criteria) was proposed by Seed and Idriss (1982). These criteria had been used widely with slight modifications (Finn et al. 1994; Perlea 2000; Andrews and Martin 2000). More recently, ground failure case histories compiled after 1989 Loma Prieta, 1994 Northridge, 1999 Adapazari and Chi-Chi earthquakes have refreshed research attention on assessing cyclic mobility response of clayey soils. Case histories from these earthquakes highlighted that low plasticity silt and clay mixtures may significantly strain soften, which may in turn cause significant damage to overlying structural systems. As an alternative to Chinese Criteria, Seed et al. (2003), Bray and Sancio (2006), Boulanger and Idriss (2006), and Bilge (2010) proposed new susceptibility criteria based on field observations and laboratory test results. Before the discussion of these methods, it is helpful to note that assessing susceptibility of soils to liquefaction, requires a potentially liquefiable soil definition, which ideally should be independent of the intensity and duration of earthquake loading. This is a difficult to achieve requirement and is listed as one of the common drawbacks of existing susceptibility criteria. Hence, in practice, for most cases, unfortunately liquefaction susceptibility (potential) assessments are combined with liquefaction triggering.

Fig. 19.2 Criteria for liquefaction susceptibility of fine-grained sediments proposed by Seed et al. (2003) (After Seed et al. 2003)

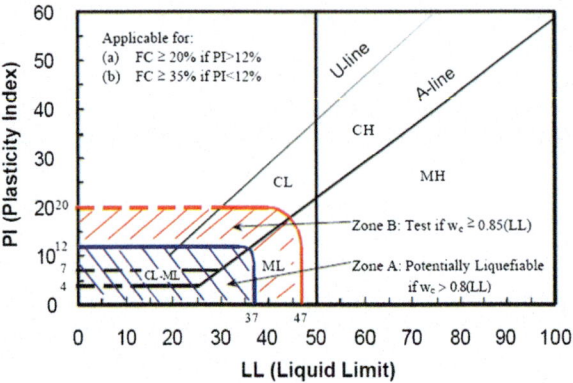

Fig. 19.3 Criteria for liquefaction susceptibility of fine-grained sediments proposed by Bray and Sancio (2006)

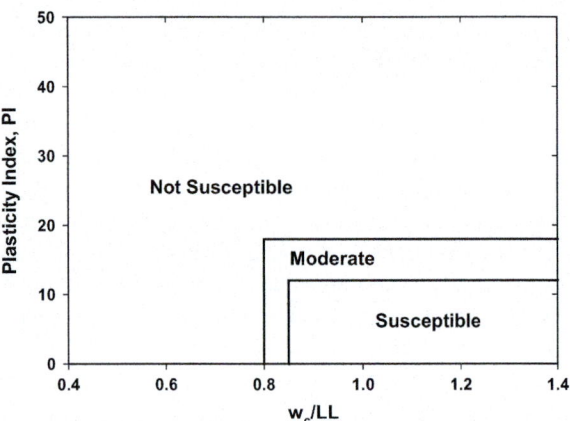

Seed et al. (2003) recommended a set of new susceptibility criteria inspired from the case histories and results of cyclic tests performed on "undisturbed" fine-grained soils documented after 1999 Adapazari and Chi-Chi earthquakes. As presented in Fig. 19.2, Seed et al. (2003) used liquid limit (LL), plasticity index (PI) and water content (w) to assess liquefaction susceptibility of soils. Fine grained soils with PI $\leq$ 12 and LL $\leq$ 37 are concluded to be potentially liquefiable, if the natural water content is wetter than 80 % of their liquid limit.

Bray and Sancio (2006) developed their liquefaction susceptibility criteria based on cyclic test results performed on undisturbed fine grained soil specimens retrieved from Adapazari province of Sakarya City, in Turkey. As summarized in Fig. 19.3, contrary to Seed et al. (2003), Bray and Sancio adopted the PI and w/LL ratio as the two input parameters of the problem. Fine grained soils with PI $\leq$ 12 are judged to be potentially liquefiable, if their natural water content is wetter than 85 % of their liquid limit. Also, it should be noted that unlike most of available methods to assess liquefaction susceptibility of fine grained soils, Bray and Sancio (2006) provided a complete documentation of their database (i.e. tested specimens and also

Fig. 19.4 Criteria for differentiating between sand-like and clay-like sediment behavior proposed by Boulanger and Idriss (2006) (After Boulanger and Idriss 2006)

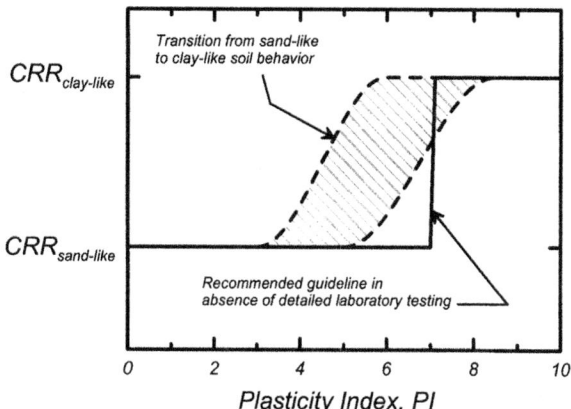

test conditions), which establish the basis of their recommendation. As clearly revealed by the adopted cyclic stress ratio levels and consolidation stress histories of soil samples, the intent of these criteria seems to assess liquefaction potential (better to refer to it as triggering) response of Adapazari soils, specifically subjected to 1999 Kocaeli Earthquake ($M_w = 7.5$) shaking. This limits the global validity of their findings.

Again, a relatively recent attempt was made by Boulanger and Idriss (2006) based on cyclic laboratory test results and on their extensive engineering judgment. As part of this new methodology, cyclic response of fine-grained soils are categorized as "sand-like" and "clay-like", where soils that behave "sand-like" are judged to be potentially liquefiable and have substantially lower values of cyclic resistance ratio (CRR) compared to those classified as to behave "clay-like". As presented in Fig. 19.4, the only input parameter was chosen as PI, and fine grained soils with PI>7 are judged to exhibit significantly "larger" cyclic resistance. The main drawback is that the y-axis of Fig. 19.4 is not to scale, thus the magnitude of larger CRR of "clay-like" soils as compared to "sand-like" ones cannot be clearly appreciated. Moreover, it should be noted that CRR definitions of the authors for "sand like" and "clay like" soils are quite different; hence a direct and a fair comparison between them is difficult.

As part of his Ph.D. studies under the supervision of Prof. Cetin, Bilge (2010) proposed a new liquefaction susceptibility criterion based on cyclic triaxial tests performed on a wide range of high quality "undisturbed" fine grained soil samples. As opposed to a r_u or γ_{max} threshold, occurrence of contraction – dilation cycles (i.e. banana loops), was used as the screening evidence for liquefaction triggering. Fine grained soils with PI values in excess of 30 are identified as "non-liquefiable" but with susceptibility to "cyclic mobility". Similarly, fine grained soils satisfying the following condition are classified as potentially "liquefiable"

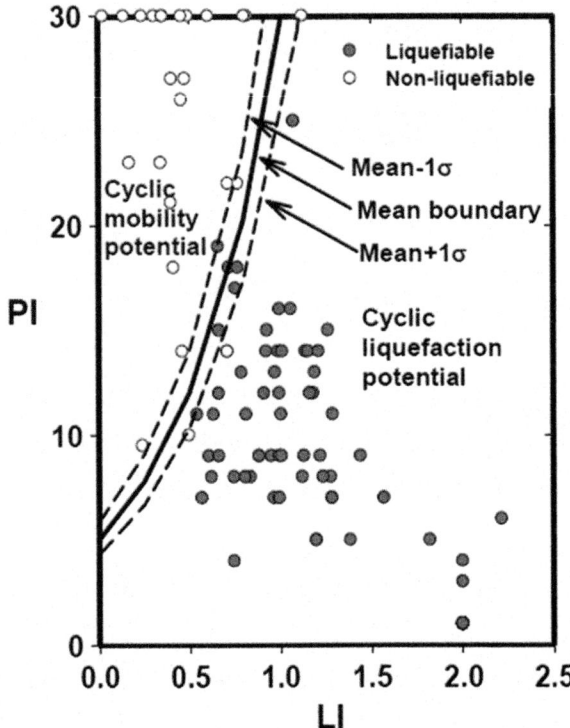

Fig. 19.5 Proposed liquefaction susceptibility criterion

$$LI \geq 0.578 \cdot \ln(PI) - 0.940 \qquad (19.1)$$

where LI is the liquidity index. The use of LI along with the occurrence of banana loops as a screening tool could be listed as the major contribution of this method. The proposed criterion along with the test database is presented in Fig. 19.5, and a complete documentation is available in Bilge (2010).

Although these studies were welcomed by practicing engineers as significant improvements over earlier efforts, they were observed to suffer from one or more of the following:

(i) ideally separate assessments of (a) identifying potentially liquefiable soils and (b) liquefaction triggering, were combined into a single assessment. When soil layers (in the field) or samples (in the laboratory) liquefied under a unique combination of CSR and number of equivalent loading cycle, N (or moment magnitude of the earthquake), they were erroneously labeled as "potentially liquefiable" rather than correctly labeled as "liquefied" at the selected CSR and N combination. These types of combined assessment procedures produce mostly biased classifications of potentially liquefiable soils.

(ii) judging liquefaction susceptibility of a soil layer or a sample through a unique combination of CSR and number of equivalent loading cycle (or moment magnitude of the earthquake) requires clear definition for liquefaction triggering. Unfortunately, there exist multiple and mostly conflicting strain, pore pressure or field performance based definitions, some of which are not even clearly documented.

(iii) liquefaction triggering manifestations and their extent are not unique in the field (sand boils, extensive settlements, lateral spreading etc.). There is no single liquefaction definition (exceedance of threshold r_u or γ_{max} levels) for laboratory-based evaluations either. The success rate of the existing assessment methodologies for identifying liquefiable soils depend strongly on the adopted threshold levels.

The authors of this chapter believe that either fine or coarse grained, every soil can be liquefied, and hence potentially liquefiable, if liquefaction triggering is defined by a threshold maximum shear strain, excess pore pressure ratio, or even the existence of banana loops. The dilemma, which is yet to be solved, is the identification of cyclic stress and number of loading cycle combinations to trigger liquefaction. Hence, with increasing popularity in performance based design practice, and available tools to assess cyclic straining and pore pressure responses of both fine and coarse grained soils, the elementary assessment steps of liquefaction susceptibility and triggering will be less popular and eventually eliminated. Alternatively, the assessments will directly start with the estimations of cyclically-induced strain or excess pore pressure levels. However, until this is achieved, existing liquefaction susceptibility and triggering methodologies will be used as initial screening tools.

19.2.2 Assessment of Liquefaction Triggering

Quantitative assessment of the likelihood of "triggering" or initiation of liquefaction is the necessary first step for most projects involving seismically-induced liquefaction problems. There exist two approaches for the purpose: the use of (1) laboratory testing of "undisturbed" samples, and (2) empirical relationships based on correlations with observed field behavior on the basis of various in-situ "index" tests.

The use of laboratory testing is complicated by difficulties associated with sample disturbance during both sampling and reconsolidation of cohesionless soils. It is also difficult and costly to perform high-quality cyclic simple shear testing, and additionally cyclic triaxial testing poorly represents the loading conditions of principal interest for most seismic problems. Both sets of problems can be ameliorated, to some extent, by use of appropriate "frozen" sampling techniques, and subsequent testing in a high quality cyclic simple or torsional shear apparatus. The difficulty and cost of these sophisticated techniques, however, places their use beyond the budget and scope of most engineering projects.

Accordingly, the use of in-situ "index" testing is the dominant approach in common engineering practice. As summarized in the recent state-of-the-art paper (Youd et al. 2001), four in-situ test methods have now reached to a level of sufficient maturity as to represent viable tools for this purpose. These are (1) the standard penetration test (SPT), (2) the cone penetration test (CPT), (3) measurement of in-situ shear wave velocity (V_s), and (4) the Becker penetration test (BPT). The oldest, and still the most widely used of these, is the SPT, and SPT-based methods will be the major focus of the following sections.

19.2.2.1 SPT-Based Triggering Assessment

The use of the SPT as a tool for the evaluation of liquefaction potential first began after the 1964 Great Alaskan Earthquake ($M = 8+$) and the 1964 Niigata Earthquake ($M \approx 7.5$), both of which produced significant number of liquefaction-induced failure case histories (e.g.: Kishida 1966; Seed and Idriss 1971). As discussed by the NCEER Working Group (NCEER 1997; Youd et al. 2001), one of the most widely accepted and widely used SPT-based correlations is the "deterministic" relationship proposed by Seed, et al. (1984, 1985). Figure 19.6 shows this relationship, with minor modification at low CSR (as recommended by the NCEER Working Group; NCEER 1997). This familiar relationship is based on comparison between SPT N-values, corrected for both effective overburden stress and energy, equipment and procedural factors affecting SPT testing (to $N_{1,60}$-values) vs. intensity of cyclic loading, expressed as magnitude-weighted equivalent uniform cyclic stress ratio (CSR_{eq}). As shown in Fig. 19.6, the relationship between corrected $N_{1,60}$-values and the intensity of cyclic loading required to trigger liquefaction is also a function of fines content. Although widely used in practice, this relationship is dated, and does not make use of an increasing body of field case history data from seismic events that have occurred since 1984. It is particularly lacking data from cases where peak ground shaking levels were high ($CSR > 0.25$), an increasingly common design range in regions of high seismicity. This correlation also has no formal probabilistic basis, and so provides no insight regarding either uncertainty or probability of liquefaction. Efforts at development of similar, but formally probabilistically-based, correlations have been published by a number of researchers, including Liao et al. (1988, 1998), and more recently Youd and Noble (1997), and Toprak et al. (1999). Cetin (2000) reassessed available case history data with improved understanding in geotechnical and earthquake engineering practice and recommended updated probabilistically-based liquefaction boundary curves for liquefaction triggering. Figure 19.6 comparatively presents these methods (boundaries corresponding to 5, 20, 50, 80 and 95 % probability of liquefaction) along with the "deterministic" boundaries given in the early work Seed et al. (1984). As revealed by this figure, Cetin et al. (2004) produces a more accurate and precise set of predictions.

Key elements in the development of Cetin et al. (2004) were: (1) accumulation of a significantly expanded database of field performance case histories, (2) use of

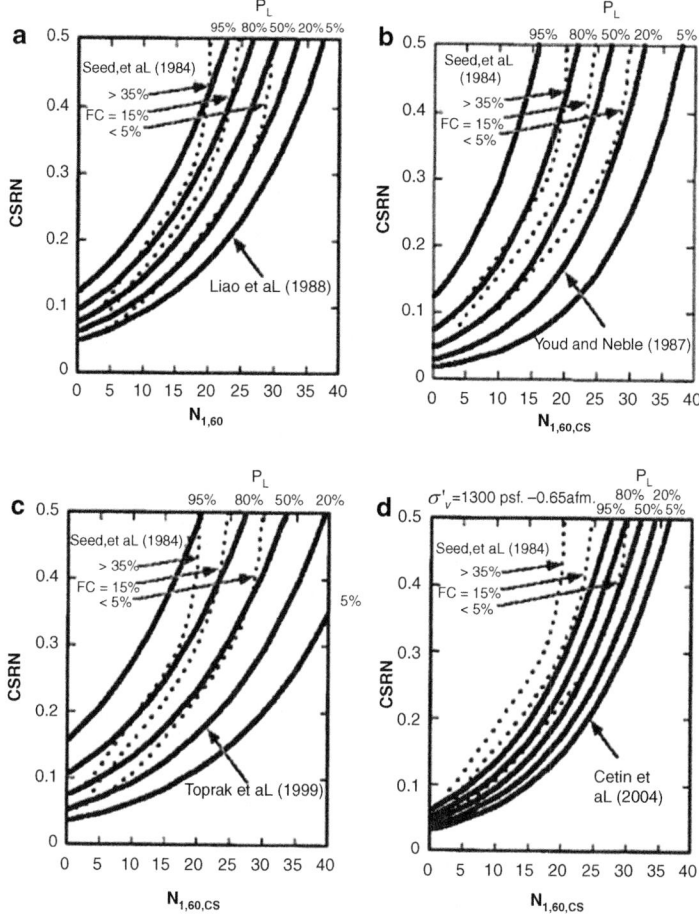

Fig. 19.6 Comparison of the existing methods for evaluation of liquefaction potential. (**a**) Liao et al. (1988). (**b**) Youd and Noble (1997). (**c**) Toprak et al. (1999). (**d**) Cetin et al. (2004)

improved knowledge and understanding of factors affecting interpretation of SPT data, (3) incorporation of improved understanding of factors affecting site-specific ground motions (including directivity effects, site-specific response, etc.), (4) use of improved methods for assessment of in-situ cyclic shear stress ratio (CSR), (5) screening of field data case histories on a quality/uncertainty basis, and (6) use of higher-order probabilistic tools (Bayesian Updating). Bayesian updating methodology (a) allowed for simultaneous use of more descriptive variables than most prior studies, and (b) allowed for appropriate treatment of various contributing sources of aleatory and epistemic uncertainty. The resulting relationships not only provide greatly reduced uncertainty, they also help to resolve a number of corollary issues that have long been difficult and controversial, including: (1) magnitude-correlated duration weighting factors, (2) adjustments for fines content, and

(3) corrections for effective overburden stress. Moreover, non-linear mass participation factor (r_d), which is a significant component of the "simplified procedure" of Seed and Idriss (1971) (Eq. 19.2), was re-evaluated based on the results of 2,153 seismic site response analyses. Cetin and Seed (2002) developed a relation in terms of depth (d), moment magnitude (M_w), peak horizontal ground surface acceleration (a_{max}) and stiffness of the site ($V^*_{s,12}$ in m/s) (Eq. 19.3).

$$CSR_{denk} = 0.65 \cdot \frac{a_{maks}}{g} \cdot \frac{\sigma_v}{\sigma'_v} \cdot r_d \qquad (19.2)$$

For d<20m

$$r_d(d, M_w, a_{max}, V^*_{s,12}) = \frac{\left[1 + \dfrac{-23.013 - 2.949 \cdot a_{max} + 0.999 \cdot M_w + 0.0525 \cdot V^*_{s,12}}{16.258 + 0.201 \cdot e^{0.341 \cdot (-d + 0.0785 \cdot V^*_{s,12} + 7.586)}}\right]}{\left[1 + \dfrac{-23.013 - 2.949 \cdot a_{max} + 0.999 \cdot M_w + 0.0525 \cdot V^*_{s,12}}{16.258 + 0.201 \cdot e^{0.341 \cdot (0.0785 \cdot V^*_{s,12} + 7.586)}}\right]} \pm \sigma_{\varepsilon r_d}$$

(19.3a)

For d≥20m

$$r_d(d, M_w, a_{max}, V^*_{s,12}) = \frac{\left[1 + \dfrac{-23.013 - 2.949 \cdot a_{max} + 0.999 \cdot M_w + 0.0525 \cdot V^*_{s,12}}{16.258 + 0.201 \cdot e^{0.341 \cdot (-20 + 0.0785 \cdot V^*_{s,12} + 7.586)}}\right]}{\left[1 + \dfrac{-23.013 - 2.949 \cdot a_{max} + 0.999 \cdot M_w + 0.0525 \cdot V^*_{s,12}}{16.258 + 0.201 \cdot e^{0.341 \cdot (0.0785 \cdot V^*_{s,12} + 7.586)}}\right]}$$
$$- 0.0046 \cdot (d - 20) \pm \sigma_{\varepsilon r_d}$$

(19.3b)

$$\text{For } d < 12 \text{ m } \sigma_{\varepsilon r_d}(d) = d^{0.850} \cdot 0.0198$$
$$\text{For } d \geq 12 \text{ m } \sigma_{\varepsilon r_d}(d) = 12^{0.850} \cdot 0.0198 \qquad (19.3c)$$

The close form solution of Cetin et al. (2004) for the assessment of the probability of liquefaction, which involves the corrections for the influence of fines content, duration and effective stress, is given in Eq. (19.4).

$$P_L\left(N_{1,60}, CSR, M_w, \sigma'_v, FC\right) = \Phi\left(-\frac{\begin{pmatrix} N_{1,60} \cdot (1 + 0.004 \cdot FC) - 13.32 \cdot \ln(CSR) \\ -29.53 \cdot \ln(M_w) - 3.70 \cdot \ln(\sigma'_v/P_a) \\ +0.05 \cdot FC + 16.85 \end{pmatrix}}{2.70}\right)$$

(19.4)

where P_L = probability of liquefaction in decimals (i.e. PL = 50 % is represented as 0.30); CSR_{eq} is not "adjusted" for magnitude (duration), overburden or fines effects (i.e.: corrections are executed within the equation itself); FC = percent fine content (by dry weight) expressed as an integer (e.g., 12 % fine is expressed as FC = 12) with the limit of $5 \leq FC \leq 35$; Pa = atmospheric pressure (=1 atm ~100 kPa~2,000 psf) in the same units as the in situ vertical effective stress; and Φ standard cumulative normal distribution. Also the cyclic resistance ratio for a given probability of liquefaction can be expressed as follows:

$$CRR\left(N_{1,60}, M_w, \sigma_v', FC, P_L\right) = \exp\left[\frac{\left(\begin{array}{c} N_{1,60} \cdot (1 + 0.004 \cdot FC) - 29.53 \cdot \ln(M_w) \\ -3.70 \cdot \ln\left(\sigma_v'/P_a\right) + 0.05 \cdot FC + \\ 16.85 + 2.70 \cdot \Phi^{-1}(P_L) \end{array}\right)}{13.32}\right]$$

(19.5)

where $\Phi^{-1}(P_L)$ = inverse of the standard cumulative normal distribution (i.e., mean = 0, and standard deviation = 1). For spreadsheet construction purposes, the command in *Microsoft Excel* for this specific function is "*NORMINV(P_L,0,1)*".

If a user prefers using this method to calculate factor of safety (i.e. for deterministic analysis), then CRR corresponding to $P_L = 50$ % (0.5) should be used as the capacity term. Note that a factor of safety in the range of 1.0–1.20 is typically used.

More recently, Idriss and Boulanger (2006) proposed a new semi-empirical approach for the evaluation of liquefaction triggering. The similarity of the proposed boundary curves with the ones proposed by Seed et al. (1985) is remarkable and should be noted. The presence of a number of alternative liquefaction triggering methodologies is a source of confusion for practicing engineers, and indicates the lack of consensus among researchers. For the purpose of clarifying the sources of this disagreement, integral components of liquefaction triggering assessments will be revisited, and the degree of consensus in these components will be discussed. For this purpose four sets of comparison charts were prepared. As shown in Fig. 19.7, the disagreement in the recommended r_d values is remarkable, and depending on the adopted r_d model, CSR values can be different by a factor of 1.1–1.2 at shallow depths. Similarly, the scatter in magnitude scaling (or duration weighting) factors, especially at smaller magnitude events is large and may produce CSR estimates different by a factor of 1.5–3. K_σ correction is another source of controversy and deserves further discussion. In 1984, Seed et al. presented their widely used relationship between procedure and overburden-corrected SPT blow counts, $N_{1,60}$ and CSR triggering liquefaction during a $M_w = 7.5$ event. Consistent with Seed (1983) and Seed et al. (1984), with the argument that K_σ corrections were not applied when assessing liquefaction triggering case histories (i.e.: back analysis), which establish the basis of liquefaction triggering relationship, consistently, it was recommended not to apply K_σ corrections for liquefaction engineering assessment

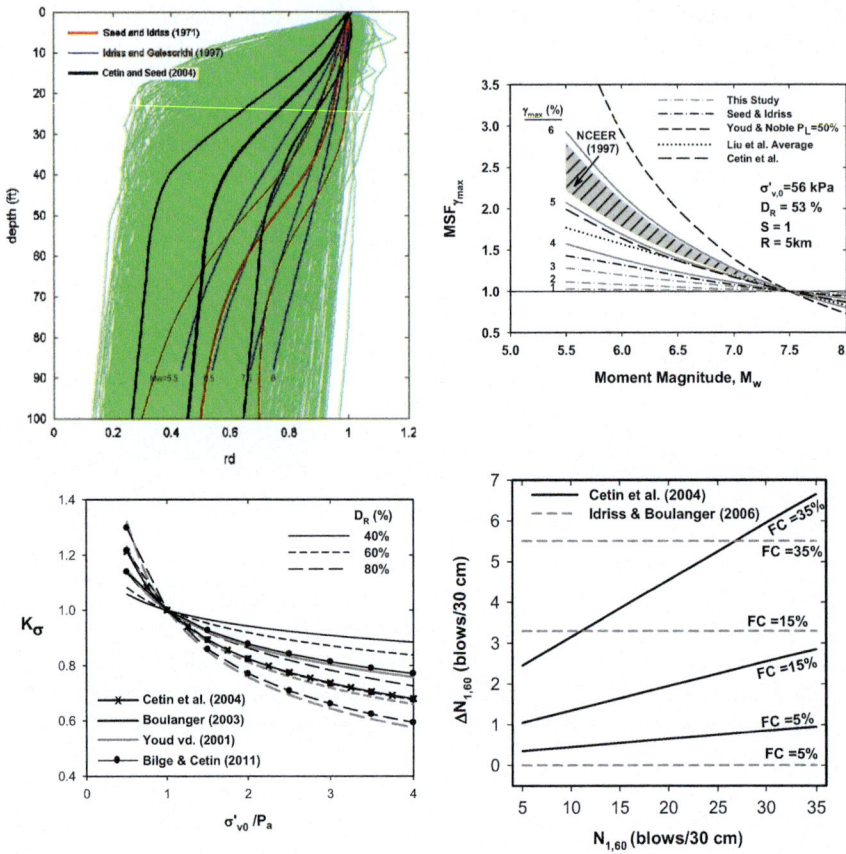

Fig. 19.7 Comparison of the existing methods for the evaluation of r_d, MSF, $K\sigma$ and fines corrections

of soil layers (i.e.: forward analysis) with a vertical effective stress less than 1 atm. Unfortunately, this -at first glance consistent and practical choice- produced unconservatively biased predictions for deep soil layers due to the fact that median vertical effective stress of liquefaction triggering case histories is 56 kPa (or 65 kPa if weighting applied, Cetin 2000) but not 100 kPa. Last but not least, due to asymptotic nature of triggering curves, fines corrections applied on $N_{1,60}$ can be extremely critical. In the literature, there exist contradicting arguments about if and how fines affect cyclic straining, pore pressure and stiffness degradation response of granular soils.

It is quite natural that the scattered correction factors produce a wide range of liquefaction triggering curves. However, it should be noted that practicing engineers may eliminate some of the uncertainty in liquefaction triggering predictions by consistently following the correction scheme of the original reference, since these corrections were consistently applied in the processing of case histories as well. Unfortunately, even consistency does not always guarantee the elimination of bias, if these models are used to predict the liquefaction performance of a site subjected to an earthquake shaking, which are different from "typical" (i.e.: median values) of the case history databases.

Within the confines of this chapter, due to page limitations and their wide use, only SPT-based methods were discussed. Regarding the CPT-based methods, readers are referred to the deterministic and probabilistic methods of Robertson and Wride (1998) and Moss et al. (2006), respectively. Shear wave velocity and Becker penetration test-based methods are relatively less frequently used; but readers are referred to Kayen et al. (2013) and Harder and Seed (1986), respectively, for a complete review of available literature.

It should be noted that all these methods are applicable to either clean sands or sands with limited amount of fines. As discussed earlier, silt and clay mixtures may also be susceptible to cyclic loading-induced strength loss and deformations. Unfortunately, research interest on their cyclic response picked up only recently, and hence, a comprehensive effort summarizing their cyclic performance is still missing. Yet, Boulanger and Idriss (2007) needs to be referred to as a practical tool, which is waiting to be tested via sufficient number of case histories.

Following sections are devoted to the discussion of seismic strength and deformation responses of soils, which allows a direct evaluation of seismic soil performance.

19.3 Assessment of Seismic Strength Response of Soils

There is a significant tendency towards the performance-based approaches in today's engineering profession. From seismic soil response point of view, this tendency puts forward the prediction of strength and deformation performances. Actually, they establish the basis of second and third level liquefaction engineering assessments, as outlined by Seed et al. (2001) (Fig. 19.1). For the sake of consistency, cyclic strength loss will be discussed before the discussion of cyclic straining.

19.3.1 Seismic Strength Performance of Clean Sands and Silt – Sand Mixtures

Most of the previous efforts have focused on saturated clean sands and non-plastic silt – sand mixtures. Shear strength of these soils solely rely on the effective stress state and inter granular friction. Thus, an increase in seismically-induced excess pore water pressure may cause a significant reduction in shear strength (most extreme case is liquefaction) of saturated cohesionless soils.

Consistent with liquefaction triggering methodologies, there exist two alternatives: (i) sampling and laboratory testing, and (ii) correlation of post-liquefaction strength with field case history data. The "steady-state" approach (e.g.: Poulos

Fig. 19.8 Recommended relationship between $s_{u,r}$ and $N_{1,60,CS}$ (After Seed and Harder 1990)

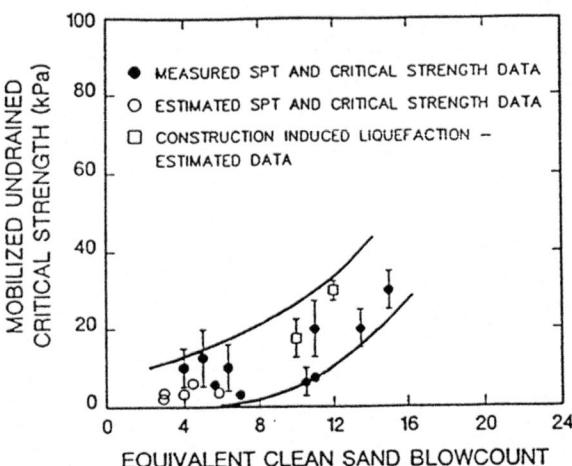

et al. 1985), has benefitted from laboratory testing of both reconstituted and high-quality "undisturbed" samples, and a systematic basis for correction has been proposed for post-liquefaction "steady-state" strengths due to inevitable disturbance and densification effects that occurred during sampling and re-consolidation phases of undrained shearing. The method was eventually claimed to produce post-liquefaction strengths that were much higher than those back-calculated from field failure case histories (e.g. Seed et al. 1989). Hence, most research has diverted to the latter approach.

After the pioneering work of Seed (1987), many researchers (e.g., Davis et al. 1988; Seed and Harder 1990; Robertson et al. 1992; Stark and Mesri 1992; Ishihara 1993; Wride et al. 1999) have performed extensive research to assess post-liquefaction shear strength of saturated sandy soils. Among these, Seed and Harder (1990) along with Stark and Mesri (1992) were widely accepted and used. Seed and Harder (1990) defined residual shear strength ($s_{u,r}$) in terms of procedure-, energy-, overburden stress- and fines- corrected SPT blow counts ($N_{1,60,CS}$) as presented in Fig. 19.8. Alternatively, Stark and Mesri (1992) normalized residual shear strength by initial vertical effective stress, and presented a chart solution as a function of $N_{1,60,CS}$ as shown in Fig. 19.9.

Recently, Olson and Stark (2002) revisited the available case history database and recommended the post-liquefaction shear strength relationships as a function of SPT blow counts and CPT tip resistance, as given in Eqs. (19.6) and (19.7), respectively.

$$\text{For } N_{1,60} \leq 12, \quad \frac{s_{u,S}}{\sigma'_{v0}} = 0.03 + 0.0075\left[(N_1)_{60}\right] \pm 0.03 \quad (19.6)$$

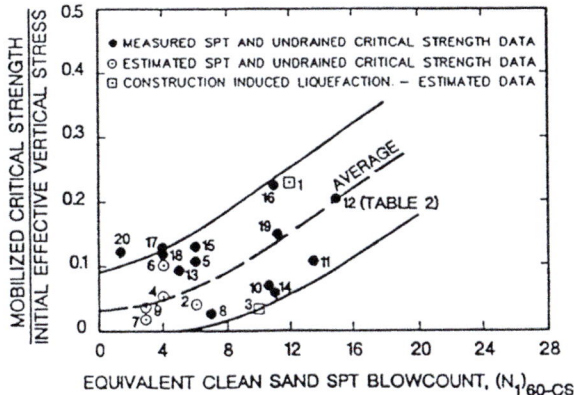

Fig. 19.9 Recommended relationship between $s_{u,r}/\sigma'_{v,0}$ and $N_{1,60,CS}$ (After Stark and Mesri 1992)

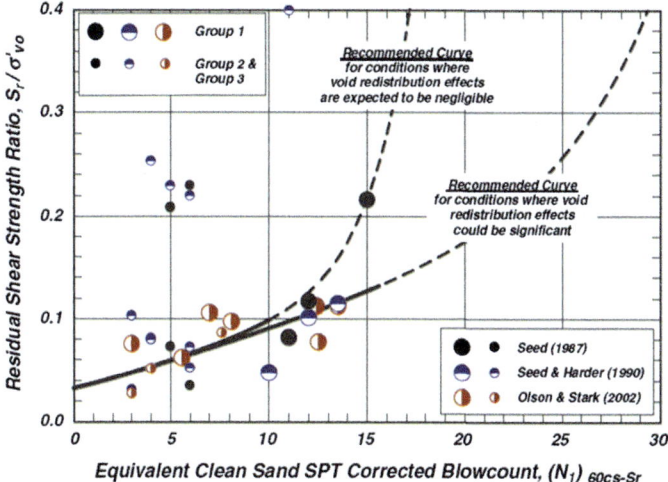

Fig. 19.10 Residual shear strength ratio, $S_r/\sigma'_{v,0}$, of liquefied soil versus equivalent clean-sand, SPT corrected blow count for $\sigma'_{v,0}$ less than 400 kPa (After Idriss and Boulanger 2007)

$$\text{For } q_{c1} \leq 6.5 MPa, \quad \frac{s_{u,S}}{\sigma'_{v0}} = 0.03 + 0.0143(q_{c1}) \pm 0.03 \quad (19.7)$$

More recently, Idriss and Boulanger (2007) re-assessed earlier efforts and existing case histories, and recommended two sets of solutions again for SPT and CPT data as presented in Figs. 19.10 and 19.11, respectively. Moreover, authors also developed the following close form solutions for the estimation of residual shear strength by taking into account void redistribution effects. If the influence of void redistribution is significant, residual shear can be estimated as follows:

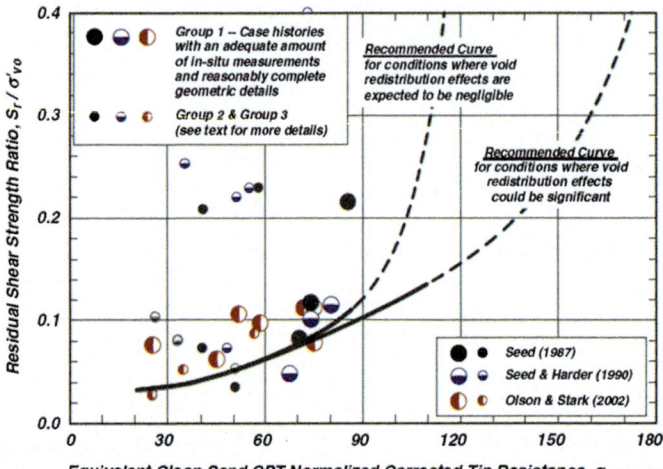

Fig. 19.11 Residual shear strength ratio, $S_r/\sigma'_{v,0}$, of liquefied soil versus CPT-qc1Ncs-Sr values for for/$\sigma'_{v,0}$ less than 400 kPa (After Idriss and Boulanger 2007)

$$\frac{S_{u,r}}{\sigma'_{v0}} = \exp\left(\frac{N_{1,60,CS}}{16} + \left(\frac{N_{1,60,CS} - 16}{21.2}\right)^3 - 3.0\right) \leq \tan\phi'$$

$$\frac{S_{u,r}}{\sigma'_{v0}} = \exp\left(\frac{q_{c,1N,CS}}{24.5} - \left(\frac{q_{c,1N,CS}}{61.7}\right)^2 + \left(\frac{q_{c,1N,CS}}{106}\right)^3 - 4.42\right) \leq \tan\phi' \quad (19.8)$$

Alternatively, if void redistribution effects are negligible,

$$\frac{S_{u,r}}{\sigma'_{v0}} = \exp\left(\frac{q_{c,1N,CS}}{24.5} - \left(\frac{q_{c,1N,CS}}{61.7}\right)^2 + \left(\frac{q_{c,1N,CS}}{106}\right)^3 - 4.42\right)$$

$$\times \left(1 + \exp\left(\frac{q_{c,1N,CS}}{11.1} - 9.82\right)\right) \leq \tan\phi'$$

$$\frac{S_{u,r}}{\sigma'_{v0}} = \exp\left(\frac{N_{1,60,CS}}{16} + \left(\frac{N_{1,60,CS} - 16}{21.2}\right)^3 - 3.0\right)$$

$$\times \left(1 + \exp\left(\frac{N_{1,60,CS}}{2.4} - 6.6\right)\right) \leq \tan\phi' \quad (19.9)$$

where ϕ' represents the effective stress based internal angle of friction

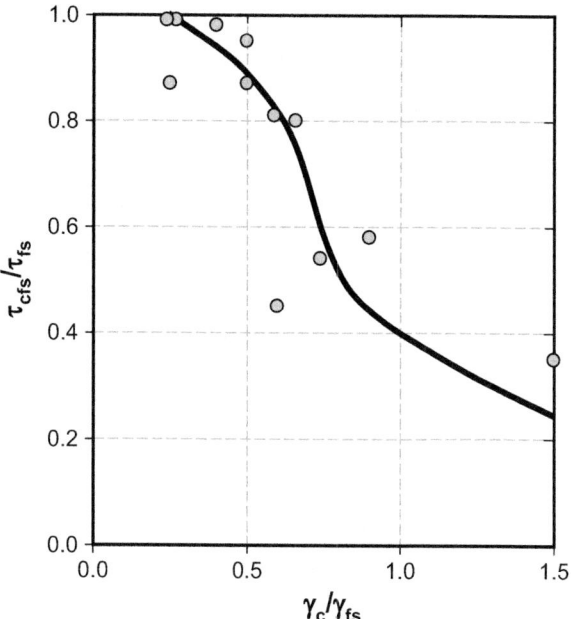

Fig. 19.12 Cyclic shear strain induced reduction in shear strength (After Castro and Christian 1976)

The recent effort of Idriss and Boulanger (2007) is considered to be an improvement over previous studies due to increased number of case history data points as well as consideration of the void redistribution effects. However, considering the scatter in case history data, the recommendation of a deterministic boundary curve (instead of upper and lower bounds or probabilistically based boundaries) is judged to be the limitation of the study.

19.3.2 Seismic Strength Performance of Silt – Clay Mixtures

Although post-cyclic strength loss is accepted to be more critical for saturated cohesionless soils, it could also be a serious threat for cohesive soils depending on intensity and duration of shaking and their undrained shear strength. Depending on the dilatancy properties of soil, the intensity of shaking and also post-cyclic stress path, post-cyclic shear strength may be significantly different than monotonic shear strength. Additionally, shear strength of most clays decreases due to remolding and excess pore pressure increase during cyclic loading.

In their pioneer study, Thiers and Seed (1969) proposed a chart solution given in Fig. 19.12, where the ratio of post-cyclic to monotonic shear strength was defined as a function of cyclic shear strain amplitude to shear strain at which monotonic failure takes place. Figure 19.12 reveals that strength loss may be a factor of five. However, as long as the amplitude of the cyclic shear strain (γ_c) is less than half of the strain level required for monotonic failure (γ_{fs}), the reduction in shear strength is less than

10 %. Later, Lee and Focht (1976), Koutsoftas (1978) and Sherif et al. (1977) provided experimental data supporting the findings of Thiers and Seed (1969). Additionally, Sangrey and France (1980) presented a supporting theoretical framework on the basis of critical state soil mechanics.

Castro and Christian (1976) also investigated post-cyclic shear strength of various types of soils. They addressed that $s_{u,pc}$ predictions using effective stress based Mohr Coulomb failure criterion might be misleading since this approach ignored the possible dilative nature of soil specimens. They have also stated that post-cyclic shear strength ($s_{u,pc}$) of clayey soils were very similar to their monotonic shear strength (s_u). The latter observation is based on the results of 4 cyclic tests performed on clayey soils having PI and LI values varying between 15–19 and 0.27–0.69, respectively. Thus, it is believed that the findings of the authors may not be valid for potentially liquefiable fine grained soils, and their statement on the similarity $s_{u,pc}$ and s_u values is, least to say, unconservative.

Van Eekelen and Potts (1978) proposed the following expression relating $s_{u,pc}$ and s_u of clayey soils.

$$\frac{s_{u,pc}}{s_u} = (1 - r_u)^{\chi/\lambda} \qquad (19.10)$$

where χ and λ are the critical state swell and compression coefficients, respectively, and the determination of them requires oedometer testing. Using consolidation theory as a theoretical basis, Yasuhara (1994) proposed a framework for estimating post-cyclic shear strength of cohesive soils considering both undrained and drained loading conditions. According to Yasuhara's observations the extent of the decrease in shear strength varies from 10 to 50 % of monotonic shear strength. Yasuhara (1994) proposed the close form solution presented in Eq. (19.11).

$$\frac{s_{u,cy}}{s_{u,NC}} = (OCR)_q \cdot \left(\frac{\Lambda_0}{1 - C_s/C_c} - 1 \right) \qquad (19.11)$$

where $s_{u,cy}$ and $s_{u,NC}$ are post-cyclic and monotonic shear strengths, respectively; C_s and C_c are swelling and compressibility indices, respectively; $(OCR)_q$ is the ratio of mean effective stresses before (p_i') and after (p_e') the application of cyclic shear stresses; and Λ_0 is a material constant determination of which requires additional consolidation testing. Based on the findings of Ue et al. (1991), Yasuhara proposed the following equation for the prediction of $\Lambda_0/(1 - C_s/C_c)$ term;

$$\Lambda_0/(1 - C_s/C_c) = 0.939 - 0.002 \cdot PI \qquad (19.12)$$

While this framework is arguably the most complete approach to assess post-cyclic shear strength of cohesive soils, it is judged to suffer from the following limitations: (i) applicability to post-liquefaction residual shear strength problems is still arguable, (ii) in the verification set, Yasuhara used clayey soils with high PI values reaching up to 320 and naturally none of the specimens experienced high r_u

Fig. 19.13 Database used for development of Eq. (19.12)

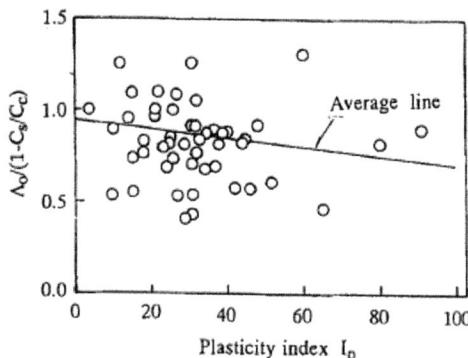

levels. (iii) there is no information on moisture content, so it is not possible to comment on liquefaction susceptibility of tested specimens. $(OCR)_q$ is another important component of this model; yet its estimation is not trivial. This term has been used by various researchers previously: Okamura (1971) referred to it as "disturbance ratio", Matsui et al. (1980) used the term "equivalent overconsolidation ratio" and Yasuhara et al. (1983) called it as "apparent" or "quasi-overconsolidation ratio". According to Yasuhara, its value depends on cyclically induced excess pore water pressure ratio. Following simplified expression was proposed for Yasuhara (1994) to predict $(OCR)_q$.

$$(OCR)_q = (OCR)^{1-C_s/C_c} \qquad (19.13)$$

where OCR is the overconsolidation ratio of the tested specimen. For the sake of producing a practical approach, Yasuhara once again adopted a relation given by Ue et al. (1991) for prediction of C_s/C_c which is given as follows:

$$C_s/C_c = 0.185 + 0.002 \cdot PI \qquad (19.14)$$

Expressing the parameters as a function of PI is a very practical approach; yet in turn, the success of Yasuhara's method strongly depends on Ue et al. (1991)'s correlations. Performance of these correlations is waiting to be tested since database of Ue et al. involves significant amount of data scatter as presented by Figs. 19.13 and 19.14 for Eqs. 19.12 and 19.14, respectively. Hence practicing engineers need to use it with caution due to the large uncertainty involved.

Although almost four decades have passed since the pioneer efforts on the evaluation of post-cyclic strength of silt and clay mixtures, current state of literature reveals that more needs to be done. This discussion revealed that these early efforts did not specifically focus on cyclic response of soils with significant straining and excess pore pressure generation potential. In these extremes, specimens may lose significant fraction of their initial shear strength. Inspired from this gap, a probabilistic-based semi-empirical model (Eq. 19.14) is developed to predict the ratio of the minimum shear strength during the course of cyclic loading to initial

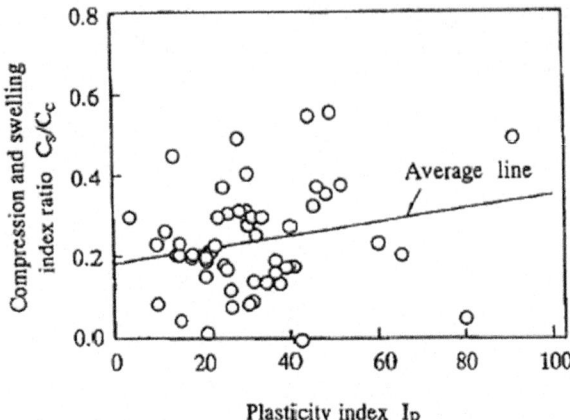

Fig. 19.14 Database used for development of Eq. (19.14)

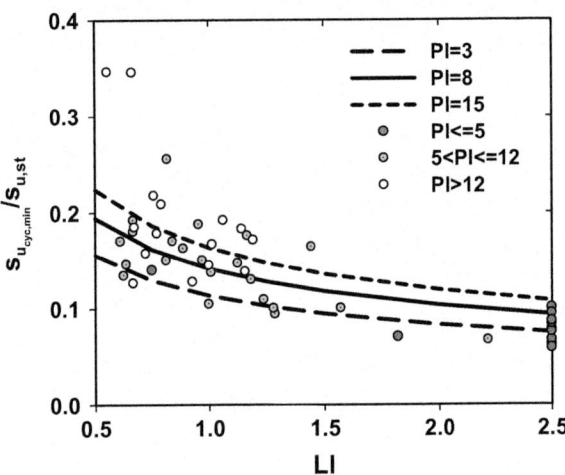

Fig. 19.15 Variation of strength ratio ($s_{u,liq}/s_{u,st}$) as a function of LI and PI

undrained shear strength as a function of Atterberg limits (PI and liquidity index, LI). Moreover, Fig. 19.15 presents the proposed model schematically. Readers are referred to Bilge (2010) for further details of model development.

$$\ln\left(\frac{s_{u_{cyc,min}}}{s_{u,st}}\right) = \ln\left(0.089 \cdot PI^{0.226} \cdot LI^{-0.455}\right) \pm 0.213 \qquad (19.15)$$

It should be noted that the proposed model is developed for the estimation of shear strength reduction due to cyclic-induced remolding and increase in excess pore pressure.

19.4 Assessment of Seismic Deformation Response of Soils

Despite major advances in soil liquefaction engineering, assessment of anticipated post-cyclic strain and deformations has remained a very "soft" area of practice. Within the confines of this chapter existing methods for assessment of cyclic-induced deformations will be discussed.

19.4.1 Seismic Deformation Response of Clean Sands and Silt – Sand Mixtures

Numerous researchers have tried to quantify cyclic (or sometimes liquefaction-induced) soil straining through use of deterministic techniques based on laboratory test results and/or correlations of in-situ "index" tests with observed field performance data. Seed and Idriss (1971) proposed "simplified procedure", a widely accepted and used methodology, where cyclic stress ratio (CSR), and overburden-, fines- and the procedure-corrected Standard Penetration Test (SPT) blow-counts ($N_{1,60,CS}$) were selected as the load and capacity terms, respectively, for the assessment of seismic soil liquefaction triggering. Using $N_{1,60,CS}$ and CSR terms, Tokimatsu and Seed (1984) recommended a set of chart solutions for the estimation of limiting shear and post-cyclic volumetric strains based on the results of cyclic triaxial and simple shear tests performed on clean sands, further calibrated with case history performance data. Similarly, based on the results of cyclic simple shear tests, Ishihara and Yoshimine (1992) proposed cyclically-induced maximum shear and post-cyclic volumetric strain correlations, where normalized demand term was chosen as factor of safety against liquefaction, and capacity term was defined as relative density (D_R), or cone tip resistance (q_c), or SPT blow count ($N_{1,72}$). Based on the results of cyclic torsional shear tests, Shamoto et al. (1998) recommended a semi-empirical constitutive model, as well as chart solutions, for the estimation of post-cyclic residual shear and volumetric strains. Recently, Wu et al. (2003) proposed cyclically-induced limiting shear and post-cyclic volumetric strain correlations based on the results of cyclic simple shear tests. Wu and Seed (2004) attempted to validate this volumetric strain relationship with ground settlement field case history data compiled from a number of earthquakes. Recommendations of all these four methods in the form of equi-shear or equi-volumetric strain contours are shown in Figs. 19.16, 19.17 and 19.18. However, direct comparisons are difficult and not fair due to different definitions of demand and capacity, as well as shear strain terms adopted.

All these deterministic methods have been regarded as the best of their kinds, and used in practical applications for many years. However, none of them considers the uncertainties associated with the nature of the problem. Recently, Cetin et al. (2009a) has introduced a new probabilistic-based framework based on the results of a comprehensive cyclic testing program. Semi-empirical models were

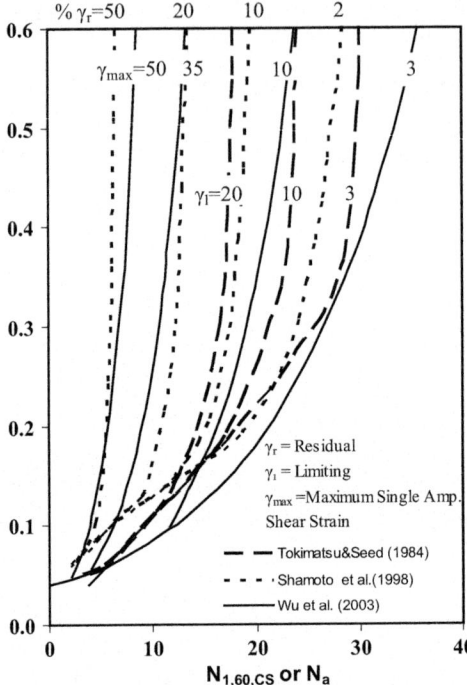

Fig. 19.16 Available methods for the estimation of cyclically-induced deviatoric strains

developed for estimation of maximum cyclic shear and post-cyclic reconsolidation volumetric strain potentials of saturated clean sands, as presented by Eqs. (19.16) and (19.17), respectively. Moreover, these models are presented schematically in Figs. 19.19 and 19.20, respectively.

$$\ln(\gamma_{max}) = \ln\left[\frac{-0.025 \cdot N_{1,60,CS} + \ln(CSR_{SS,20,1-D,1\,atm}) + 2.613}{0.004 \cdot N_{1,60,CS} + 0.001}\right] \pm 1.880$$

limit : $5 \leq N_{1,60,CS} \leq 40, 0.05 \leq CSR_{SS,20,1-D,1atm} \leq 0.60$ and $0\% \leq \gamma_{max} \leq 50\%$

(19.16)

$$\ln(\varepsilon_v) = \ln\left[1.879 \cdot \ln\left[\frac{780.416 \cdot \ln(CSR_{SS,20,1-D,1\,atm}) - N_{1,60,CS} + 2442.465}{636.613 \cdot N_{1,60,CS} + 306.732}\right]\right.$$
$$\left. + 5.583\right] \pm 0.689$$

limit : $5 \leq N_{1,60,CS} \leq 40, 0.05 \leq CSR_{SS,20,1-D,1atm} \leq 0.60$ and $0\% \leq \varepsilon_v \leq 5\%$

(19.17)

Fig. 19.17 Available methods for the estimation of post-cyclic volumetric strains

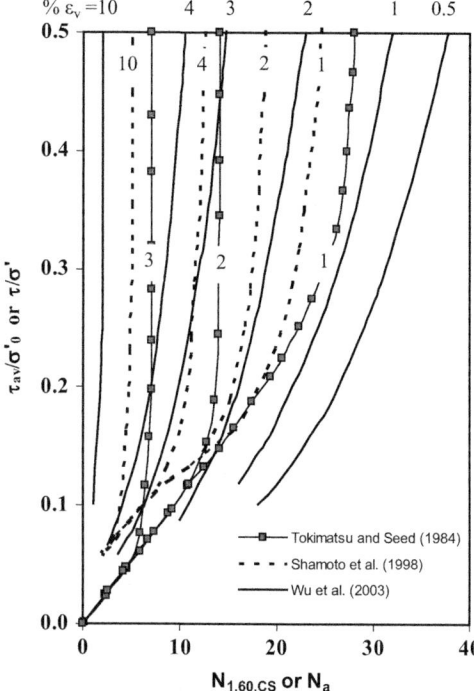

Fig. 19.18 Ishihara and Yoshimine (1992) method for determining the maximum shear and post-cyclic volumetric strains as a function of factor of safety against liquefaction

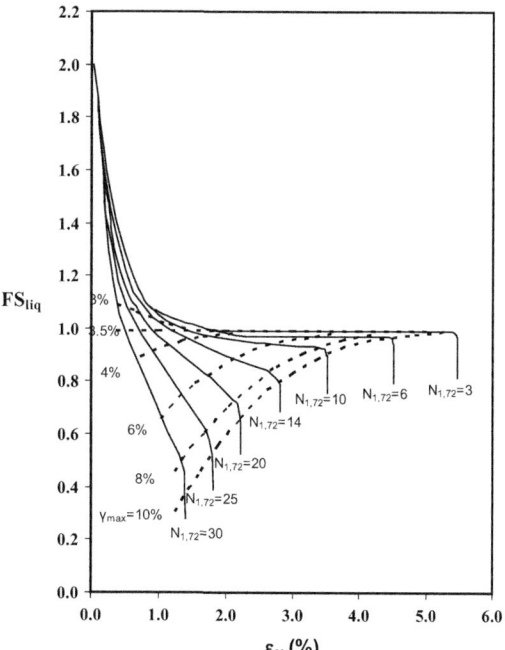

Fig. 19.19 Recommended maximum double amplitude shear strain boundary curves by Cetin et al. (2009a)

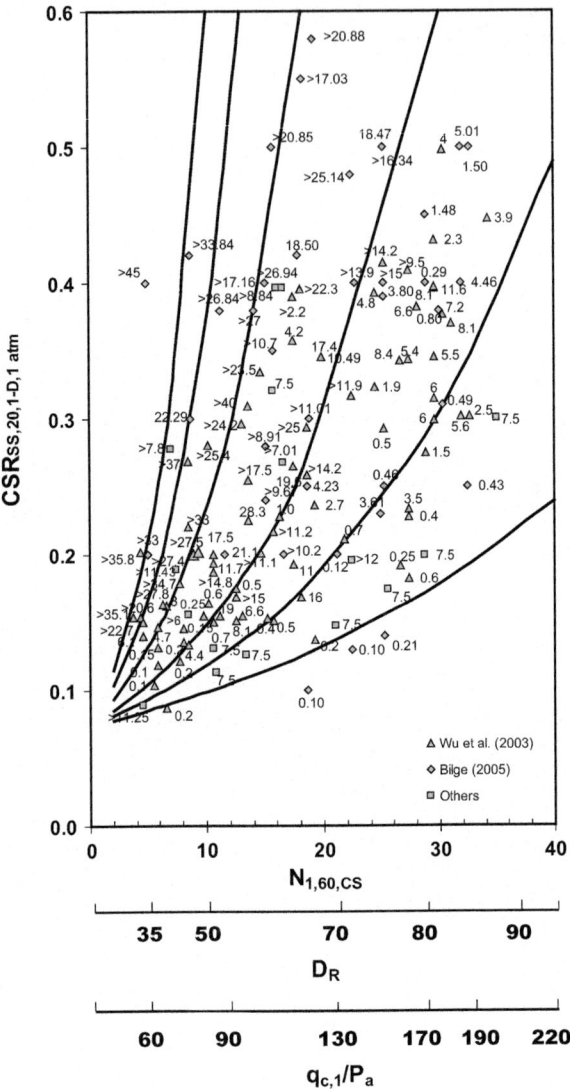

Proposed models were given in terms of $N_{1,60,CS}$ and $CSR_{SS,20,1-D,1\,atm}$ which is the CSR value corresponding to 1 dimensional, 20 uniform loading cycles simple shear test under a confining pressure of 100 kPa (=1 atm). Correction factors adopted to convert the CSR_{field} value to equivalent $CSR_{SS,20,1-D,1\,atm}$ are presented in Eq. (19.18).

$$CSR_{SS,20,1-D,1atm} = \frac{CSR_{field}}{K_{md} \cdot K_{M_W} \cdot K_\sigma} \qquad (19.18)$$

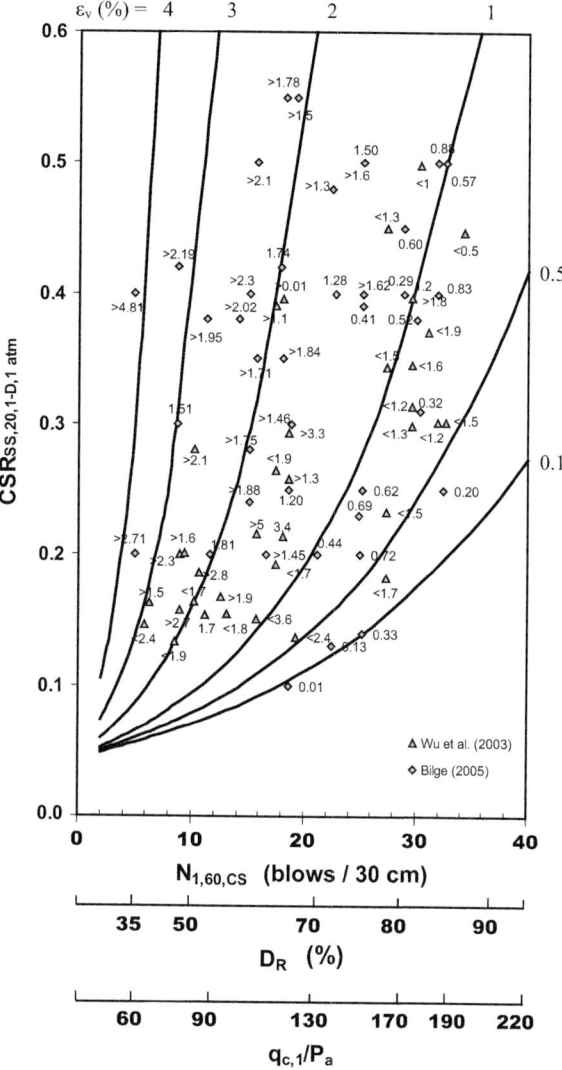

Fig. 19.20 Recommended post-cyclic volumetric strain boundary curves by Cetin et al. (2009a)

where K_{md}-correction is used to convert multi-directionally applied CSR_{field} value to the value of a uni-directionally applied laboratory CSR (Eq. 19.19), K_{Mw}-correction is used to take into account duration (magnitude) effects (Eq. 19.20) and K_σ is the correction factor for varying confining effective stress conditions (Eq. 19.21).

$$K_{md} = 0.361 \cdot \ln(D_R) - 0.579 \tag{19.19}$$

$$K_{M_W} = \frac{87.1}{M_w^{2.217}} \quad (19.20)$$

$$K_\sigma = \left(\frac{\sigma'_{v,0}}{P_a}\right)^{f-1} \text{ and } f = 1 - 0.005 \cdot D_R \quad (19.21)$$

The next step following the assessment of cyclic straining potential is prediction of soil deformations. In general, post-cyclic reconsolidation (volumetric) strains due to dissipation of excess pore water pressures are associated with settlements, whereas, cyclic shear strains are associated with lateral spreads. Following discussion will be devoted to the prediction of settlements and lateral spreads.

19.4.1.1 Assessment of Post-cyclic Settlements

Currently available approaches for predicting the magnitude of post-cyclic reconsolidation settlements are categorized as: (i) numerical analyses in the form of finite element and/or finite difference techniques (e.g., Martin et al. (1975), Seed et al. (1976), Booker et al. (1976), Finn et al. (1977), Liyanathirana and Poulos (2002)), and (ii) semi-empirical models developed based on laboratory, field test and performance data (e.g. Lee and Albeisa (1974), Tokimatsu and Seed (1984), Ishihara and Yoshimine (1992), Shamoto et al. (1998), Zhang et al. (2002), Wu and Seed (2004), Tsukamato et al. (2004), etc.). Due to difficulties in the determination of input model parameters necessary for numerical simulations, semi-empirical models continue to establish the state of practice for the assessment of cyclically-induced reconsolidation (volumetric) settlements. Even the best of their kind of these models cannot produce, at the moment, reasonably precise estimates of post-cyclic reconsolidation (volumetric) settlements.

Recently, Cetin et al. (2009b) has developed a new methodology based on their aforementioned semi-empirical post-cyclic volumetric strain estimation model. The proposed method was calibrated via 49 well-documented cyclically-induced ground settlement case histories from seven different earthquakes. Within the confines of that study, performance of the widely used methods of Tokimatsu and Seed (1984), Ishihara and Yoshimine (1992), Shamoto et al. (1998), Wu and Seed (2004) were comparatively evaluated. It was concluded that the proposed methodology, details of which will be given next, produced more accurate and precise settlement estimations compared to all other efforts.

Equation (19.16) constitutes the basis of the proposed method, and calculation of $N_{1,60,CS}$ and $CSR_{SS,20,1-D,1\ atm}$ is the necessary first step. Next, a weighting scheme, linearly decreasing with depth, inspired after the recommendations of Iwasaki et al. (1982), is implemented. Aside from the better model fit it produced, the rationale behind the use of a depth weighting factor, is based on (i) upward seepage, triggering void ratio redistribution, and resulting in unfavorably higher void ratios for the shallower sublayers of soil layers, (ii) reduced induced shear stresses and number of shear stress cycles transmitted to deeper soil layers due to initial liquefaction of surficial layers, and (iii) possible arching effects due to non-liquefied soil layers. All these may significantly reduce the contribution of volumetric settlement of deeper soil layers to the overall ground surface settlement. It is assumed that the contribution of layers to surface settlement diminishes as the depth of layer increases, and beyond a certain depth (z_{cr}) settlement of an individual layer cannot be traced at the ground surface. After statistical assessments, the optimum value of this threshold depth was found to be 18 m. The proposed depth weighting factor (DF_i) is defined in Eq. 19.22. Equivalent volumetric strain, $\varepsilon_{v,eqv.}$, of the soil profile is estimated by Eq. 19.23 and the estimated settlement, $s_{estimated}$, of the profile is simply calculated as the product of $\varepsilon_{v,eqv.}$ and the total thickness of the saturated cohesionless soil layers or sublayers, $\sum t_i$, as presented by Eq. 19.24. $s_{estimated}$ is further calibrated by θ for the estimation of field settlement values. In Eq. 19.25, σ_ε term designates the standard deviation of the calibration model. Further discussion of the σ_ε term is presented later in the manuscript.

$$DF_i = 1 - \frac{d_i}{z_{cr} = 18m}, \text{ where } d_i \text{ is the mid-depth of each saturated cohesionless soil layer from ground surface.} \quad (19.22)$$

$$\varepsilon_{v,eqv.} = \frac{\sum \varepsilon_{v,i} \cdot t_i \cdot DF_i}{\sum t_i \cdot DF_i} \quad (19.23)$$

$$s_{estimated} = \varepsilon_{v,eqv.} \cdot \sum t_i \quad (19.24)$$

$$\ln(s_{calibrated}) = \ln(\theta \cdot s_{estimated}) \pm \sigma_\varepsilon \quad (19.25)$$

In volumetric settlement assessment of the case histories, three cases were encountered regarding the application of DF: (i) a very dense cohesionless soil layer ($N_{1,60,CS} > 35$) or bedrock or a cohesive soil layer underlying the volumetric settlement vulnerable cohesionless soil layer, (ii) cohesionless soil layer continuing beyond the critical depth of 18 m with or without available SPT profile, and (iii) cohesionless soil site where the depth of boring is less than 18 m. For case (i), settlement calculations were performed till the depth to the top of the dense layer or bedrock or cohesive layer. For case (ii), potentially settlement vulnerable cohesionless layers beyond 18 m were simply ignored due to their limited contribution to the overall ground surface settlement. For case (iii), after confirming with the geological characteristics of soil site, for the soil sub-layers without an SPT value at a specific depth, SPT values were judgmentally extended beyond the maximum borehole depth to a depth of maximum 18 m., based on available SPT blow-counts. Whenever a cohesive soil layer was encountered, it was assumed that cyclically-induced volumetric strain due to this layer was negligible. In addition, thickness of this layer was not considered in the calculation of $\varepsilon_{v,eqv.}$.

For comparison purposes, each case history site (presented in detail in Bilge and Cetin 2007) was analyzed by using the methods of Tokimatsu and Seed (1984), Ishihara and Yoshimine (1992), Shamoto et al. (1998), Wu and Seed (2004) and finally the proposed method. The performance of the model predictions, expressed

Table 19.1 Comparison of the performance of existing models

Method	R^2	θ_1	σ_ε	$\sum likelihood\ fxn$
Çetin et al. (2009b)	0.64	1.15	0.61	−19.8
Tokimatsu and Seed (1984)	0.33	1.45	1.05	−31.1
Ishihara and Yoshimine (1992)	0.42	0.90	1.12	−32.7
Shamoto et al. (1998)	0.36	1.93	1.36	−36.7
Wu and Seed (2004)	0.33	0.98	0.71	−22.9

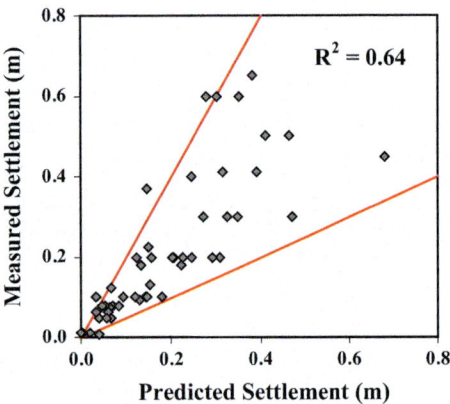

Fig. 19.21 Comparison between the measured and predicted ground settlements by Cetin et al. (2009b)

by Pearson product moment correlation coefficient, R^2, is summarized in Table 19.1. As a better alternative, which enabled the assessment of the model (calibration) error, predictions of each method were compared probabilistically by using the maximum likelihood analysis. Results of these analysis, a calibration coefficient (θ_1) which enables the model to produce unbiased predictions in the average is determined. These values are also presented in the same table along with the value of maximum likelihood and standard deviation of the random model correction term. It should be noted that higher values of maximum likelihood and lower values of standard deviation are also indicators of a better model. As the values of the calibration coefficient, θ, presented in Table 19.1 implies, existing methods of Shamoto et al. (1998), Tokimatsu and Seed (1984), and the proposed methodology under-predict the actual settlements by a factor of 1.91, 1.45 and 1.15, respectively. Similarly, Wu and Seed (2004), and Ishihara and Yoshimine (1992) over-predict settlements and need to be corrected by a factor of 0.98 and 0.90. Wu and Seed (2004) procedure produces the most unbiased settlement predictions (i.e.: the mean of the estimated settlements is about equal to the mean of the observed settlements). However, in terms of the uncertainty (or scatter) of the predictions, Wu and Seed (2004) methodology is ranked to be second to last with an R^2 value of 0.33. After scaling with the calibration coefficient, θ, the proposed model produces relatively the best predictions compared to the other four methods, also consistent with the R^2 trends presented in Table 19.1.

Performance of the proposed model is also highlighted by Fig. 19.21 in which predicted and observed settlements are paired and shown on figures along with the

1:2 and 1:0.5 boundary lines. Readers are referred to Cetin et al. (2009b) for the similar performance evaluation plots prepared for the other methods.

19.4.1.2 Assessment of Lateral Spreading

Lateral spreading is a liquefaction-induced deformation problem identified by surficial soil layers breaking into blocks that progressively slide downslope or toward a free face during and after earthquake shaking. As opposed to settlements, lateral ground deformations are generally more critical for the performance of overlying structures as well as of infrastructures due to their limited lateral resistance.

Currently available approaches for predicting the magnitude of lateral spreading ground deformations can be categorized as: (i) numerical analyses in the form of finite element and/or finite difference techniques (e.g., Finn et al. (1994), Arulanandan et al. (2000), and Liao et al. (2002)), (ii) soft computing techniques (e.g., Wang and Rahman (1999)), (iii) simplified analytical methods (e.g., Newmark (1965), Towhata et al. (1992), Kokusho and Fujita (2002), and Elgamal et al. (2003)), and (iv) empirical methods developed based on the assessment of either laboratory test data or statistical analyses of lateral spreading case histories (e.g., Hamada et al. (1986), Shamoto et al. (1998), and Youd et al. (2002)). Due to difficulties in the determination of input model parameters of currently existing numerical and analytical models, empirical and semi-empirical models continue to establish the state of practice for the assessment of liquefaction-induced lateral ground deformations.

Hamada et al.(1986), Youd and Perkins (1987), Rauch (1997), Shamoto et al. (1998), Bardet et al. (1999), and Youd et al. (2002), Kanibir (2003), Faris et al. (2006) introduced empirically-based models for the assessment of liquefaction-induced lateral spreading. With the exception of Shamoto et al. and Faris et al., these models were developed based on regression analyses of available lateral spreading case histories. The predictive approach of Shamoto et al. (1998) and Faris et al. (2006) employ laboratory-based estimates of liquefaction-induced limiting shear strains coupled with an empirical adjustment factor in order to relate these laboratory values to the observed field behavior. Among all of these models, in addition to the pioneering study of Hamada et al. (1986), widely accepted and used Youd et al. (2002), and laboratory-based and field- calibrated model of Faris et al. (2006) will be discussed in more detail next.

In 1986, Hamada et al. introduced a simple empirical equation for predicting liquefaction induced lateral ground deformations only in terms of ground slope and thickness of liquefied soil layer. This equation was based on the regression analysis of 60 earthquake case histories, mostly from Noshiro-Japan, and it was expressed as:

$$D_h = 0.75 \cdot H^{1/2} \cdot \theta^{1/3} \tag{19.26}$$

where: D_h is the predicted horizontal ground displacement (m), H is the thickness of liquefied zone (m), (when more than one sub-layer liquefies, H is measured as the distance from the top-most to the bottom-most liquefied sub-layers including all intermediate sub-layers), and θ is the larger slope of either ground surface or liquefied zone lower boundary (%). Despite its simplicity and ease of use, due to limited number of case histories which established the basis of the relationship, its use should be limited to only cases with similar conditions.

Starting in the early 1990s, Bartlett and Youd (1992, 1995) introduced empirical methods for predicting lateral spread displacements at liquefiable sites. The procedure of Youd et al. (2002) is a refinement of these early efforts and the new and improved predictive models for either (i) sloping ground conditions, or (ii) relatively level ground conditions with a "free face" towards which lateral displacements may occur, were developed through multi-linear regression of a case history database. The proposed predictive models for the sloping ground and "free face" conditions are given in Eqs. (19.27) and (19.28), respectively.

$$\log D_h = -16.213 + 1.532 \cdot M_w - 1.406 \cdot \log R^* - 0.012 \cdot R + 0.338 \cdot \log S + 0.54 \cdot \log T_{15} + 3.413 \cdot \log(100 - F_{15}) - 0.795 \cdot \log(D50_{15} + 0.1 \, mm)$$

(19.27)

$$\log D_h = -16.713 + 1.532 \cdot M_w - 1.406 \cdot \log R^* - 0.012 \cdot R + 0.592 \cdot \log W + 0.54 \cdot \log T_{15} + 3.413 \cdot \log(100 - F_{15}) - 0.795 \cdot \log(D50_{15} + 0.1 \, mm)$$

(19.28)

where; D_H is horizontal ground displacement in meters predicted by multiple linear regression model, M_w is earthquake magnitude, S is the gradient of surface topography or ground slope (%), W is the free-face ratio, defined as the height of the free-face divided by its distance to calculation point, T_{15} is the thickness of saturated layers with SPT- $N_{1,60} \leq 15$, F_{15} is the average fines content (particles < 0.075 mm) in T_{15} (%), $D50_{15}$ is the average D_{50} in T_{15}. R is the horizontal distance to the nearest seismic source or to nearest fault rupture (km), and R* is calculated according to following equation.

$$R^* = R + R_0 \text{ and } R_0 = 10^{0.89 \cdot M_w - 5.64}$$

(19.29)

The empirical model of Youd et al. (2002) is widely used in the engineering profession. The performance of the model was also evaluated by Youd et al., as presented in Fig. 19.22. Reported R^2 value of 83.6 % is concluded to be sufficiently high. However, it should be noted that (i) an attenuation-like intensity measure in terms of magnitude and distance is adopted as opposed to an independent peak soil ground acceleration term, which further brings along the uncertainties in the predictions of these attenuation-like formulations into the lateral spreading predictions, (ii) zero lateral displacement was produced for soil sites composed of sublayers with $(N_1)_{60}$ to be greater than 15 blows/30 cm. Moreover, the success rate at the displacement range of 0–3 m, which is believed to be more critical compared to large displacement range from performance point of view, is not satisfactorily high.

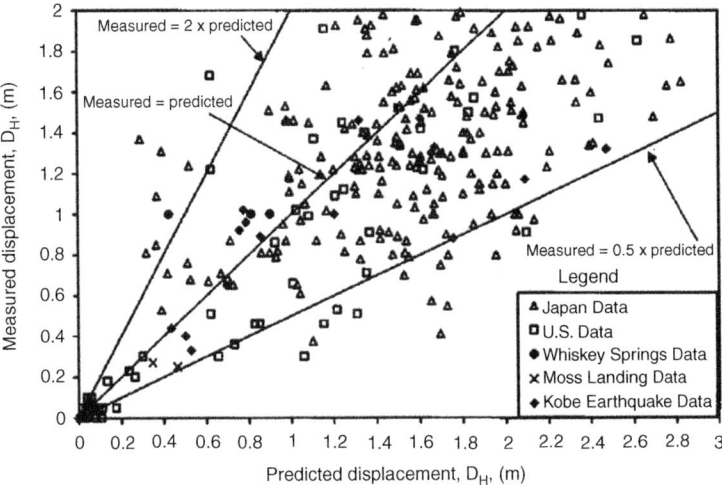

Fig. 19.22 Performance evaluation of Youd et al. (2002) lateral spreading prediction model

More recently, Faris et al. (2006) has presented the following semi-empirical model.

$$H_{\max} = \exp\left(1.0443 \cdot \ln(DPI_{\max}) + 0.0046 \cdot \ln\alpha + 0.0029 \cdot M_w\right) \quad (19.30)$$

where $H_{\max}$ is the lateral spreading in meters, $DPI_{\max}$ is the maximum cyclic shear strain potential (to be determined according to Wu et al. 2003; Fig. 19.16), α is the slope or free-face ratio, and M_w is the earthquake magnitude. Faris et al. has similarly performed a performance evaluation study results of which is presented in Fig. 19.23. Note that this framework takes into account the cyclic shear straining potential of soils, which is a physically meaningful term. However, similar to the method of Youd et al., the prediction success rate of this mode is not very high at the displacement range of 0–3 m.

Although these models are the best of their kind, due to large uncertainties associated with input parameters as well as model errors, more efforts are needed to achieve more precise models in the prediction of lateral spread-type soil deformations. Thus, practicing engineers are warned to be aware of the large uncertainty involved in the predictive models. A probabilistic approach addressing these sources of uncertainties could be a robust decision making approach and is strongly recommended.

19.4.2 Seismic Deformation Response of Silt and Clay Mixtures

Ohara and Matsuda (1988) presented one of the pioneering efforts, as part of which they expressed post-cyclic volumetric strain ($\varepsilon_{v,pc}$) as a function of excess pore

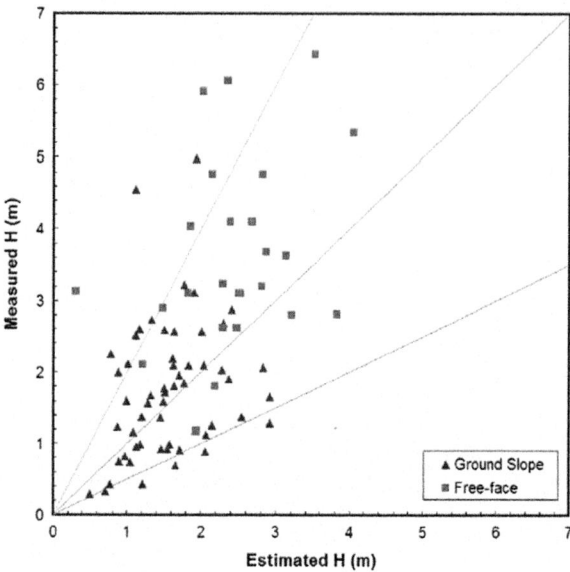

Fig. 19.23 Performance evaluation of Faris et al. (2006) lateral spreading prediction model

water pressure ratio (r_u), initial void ratio (e_0) and compression index induced by cyclic loading (C_{dyn}) as given by Eq. (19.31).

$$\varepsilon_{v,pc} = \frac{C_{dyn}}{1+e_0} \cdot \log\left(\frac{1}{1-r_u}\right) \qquad (19.31)$$

The relationship between C_{dyn} and OCR along with compression (C_c) and swelling (C_s) indices were given by Ohara and Matsuda as presented in Fig. 19.24. The authors also presented a model for prediction of cyclically-induced excess pore water pressure. However, this model is defined in terms of a large number of material coefficients which requires cyclic testing for each specific material. This limits the practical value of both r_u and also $\varepsilon_{v,pc}$ models significantly.

Yasuhara et al. (1992) has performed an experimental study and stated that the ratio of C_{dyn} to C_s was approximately equal to 1.5. Unfortunately, pore pressure generation response and corollary issues were not addressed by the researchers. Later, Yasuhara et al. (2001) proposed a design methodology for the assessment of post-cyclic volumetric settlements (i.e. strains) based on the early findings of Yasuhara's research teams (Yasuhara and Andersen 1991; Yasuhara et al. 1992; Yasuhara and Hyde 1997). As an input requirement of the methodology, the estimation of excess pore pressure is required, and authors recommended 2-D or 3-D dynamic numerical analysis for the determination of excess pore water pressure distribution within the soil media. The need of a 2-D or 3-D numerical analysis for the prediction of excess pore water pressure contradicts with authors' intention of producing a practical design procedure.

Recently, Hyde et al. (2007) studied post-cyclic recompression stiffness and cyclic strength of low plasticity silts. Based on cyclic tests results and 1-D

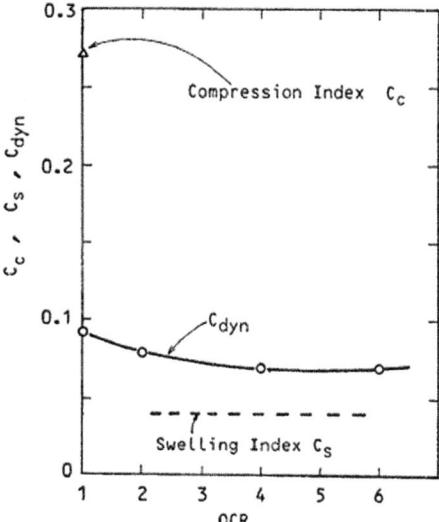

Fig. 19.24 Relationship between C_{dyn} and OCR (After Ohara and Matsuda 1988)

consolidation theory, authors proposed an expression in which $\varepsilon_{v,pc}$ was expressed as a function of initial sustained deviator stress ratio (q_s/p'_c), post-cyclic axial strain $(\varepsilon_{a,pc})$ and void ratio (e) of the tested material as follows:

$$\varepsilon_{v,pc} = \frac{1.74}{e^{1.71} \cdot (q_s/p'_c)} \cdot \varepsilon_{a,pc}^{0.461} \qquad (19.32)$$

Hyde et al. (2007) recommended an alternative approach by modeling $\varepsilon_{v,pc}$ as a function axial strain rather than excess pore water pressure. This approach has been used for saturated sandy soils by various researchers (e.g. Tatsuoka et al. 1984; Ishihara and Yoshimine 1992) but was not widely adopted for fine-grained soils, possibly due to absence of tools for predicting resulting axial strains. This fact also limits extensive use Hyde et al.'s model.

As presented so far, most of the attention has focused on the quantification of post-cyclic volumetric (reconsolidation) strains and cyclic shear straining response was not extensively studied. Except the theoretically-based attempts (e.g. Wilson and Greenwood 1974; Hyde and Brown 1976) proposed in the mid-1970s for the prediction of plastic deformation of plastic fine-grained subgrade soils under repeated loading, Hyodo et al. (1994) presented one of the few remarkable effort. Hyodo et al. (1994) attempted to correlate cyclically-induced shear strains with residual axial strains.

Considering the significant gap in the literature, the authors of this manuscript have performed a comprehensive experimental-based study. Using the results of cyclic and static triaxial test results on "undisturbed" silt and clay mixtures, following semi-empirical models are developed for the assessment cyclic maximum shear and residual strain potential of silt and clay mixtures.

$$\ln(\gamma_{max}) = \ln\left[\dfrac{9.939 \cdot \dfrac{26.163^{\left(0.995 \cdot \frac{w_c}{LL}\right)}}{\ln(PI)} \cdot \left(1 - 0.076 \cdot \ln\left(\dfrac{21.08}{PI}\right)\right)}{25.807 - \sqrt{\left(\dfrac{\tau_{st}}{s_u} - 5.870\right)^2 + \left(\dfrac{\tau_{cyc}}{s_u} - (-25.085)\right)^2}}\right] \pm 0.537$$

$$25.807 - 31.740$$

(19.33)

$$\ln(\gamma_{res}) = \ln\left[\gamma_{max} \cdot \left(\begin{array}{l}0.845 \cdot \gamma_{max}^{-0.332} + 0.404 \cdot SRR^{1.678} \\ +0.375 \cdot (\ln PI)^{0.446} + (7.564) \cdot \left(\dfrac{\tau_{st}}{s_u}\right)^{9.249} - 0.959\end{array}\right)^{1.438}\right]$$

$$\pm 0.586$$

(19.34)

where, τ_{st}/s_u and τ_{cyc}/s_u present the static and cyclic shear stress ratio for cohesive soils, respectively; whereas, SSR is the ratio of static to cyclic shear stresses (i.e. τ_{st}/τ_{cyc}).

The recommended framework requires index test results along with the undrained shear strength (s_u) of soils, which could be determined via laboratory or in-situ tests Ratio of τ_{cyc}/s_u presents the soil strength used by seismic loading nd it could be estimated by either the simplified procedure of Seed and Idriss (1971) or site response assessments; whereas, ratio of τ_{st}/s_u presents the soil strength used by available static shear stresses, if there exists any. This latter term could be estimated via simple analytical closed form elastic stress distribution solutions.

Assessment of the post-cyclic volumetric (reconsolidation) strains is the other issue which needs to be addressed. For the purpose, a consolidation-theory based approach is followed; however, unlike earlier efforts, C_{dyn} is defined as a function of over consolidation ratio (OCR), maximum cyclic shear strain potential under selected loading scenario and plasticity index of the soil, as presented in Eq. (19.35). As outlined before, estimation of excess pore water pressure constitutes the integral part of the problem, and by probabilistic assessment of the existing test data, a new cyclic-pore water pressure model was also developed for silt and clay mixtures as presented in Eq. (19.36).

$$C_{dinamik} = \left(1 + \dfrac{0.53 \cdot OCR^2 - 3.233 \cdot OCR + 5.927}{1 + 1.118 \cdot \gamma_{max}^{-0.404} + 0.829 \cdot \ln PI}\right) \cdot C_r \quad (19.35)$$

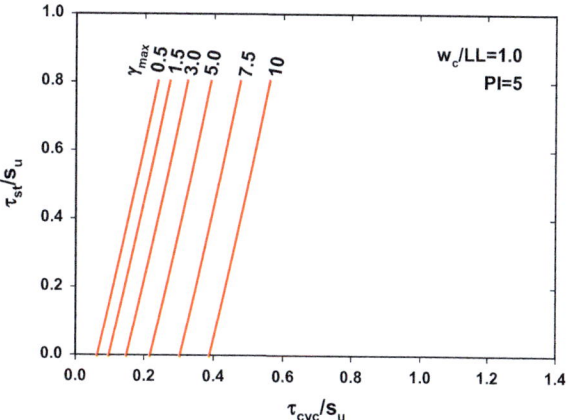

Fig. 19.25 Recommended maximum shear strain boundaries for wc/LL = 1.0 and PI = 10

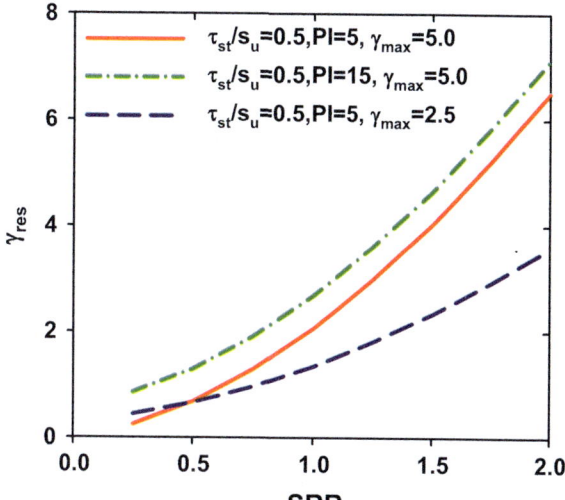

Fig. 19.26 Recommended residual shear strain boundaries

$$\ln(r_{u,N}) = \ln\left[1 - \exp\left(\frac{\gamma_{max,N}}{-1.991 \cdot \exp(0.02 \cdot PI - 0.05 \cdot LI) \cdot \left[\ln\left(\frac{FC}{0.01}\right)\right]^{0.328}}\right)\right]$$
$$\pm \left(\frac{1}{\gamma_{max,N}^{0.378} + 0.506}\right)$$

(19.36)

Although close form expressions are easier and more practical, the graphical solutions are also presented as given in Figs. 19.25, 19.26, and 19.27, to provide an

Fig. 19.27 Recommended values of C_{dyn} as a function of OCR and γ_{max}

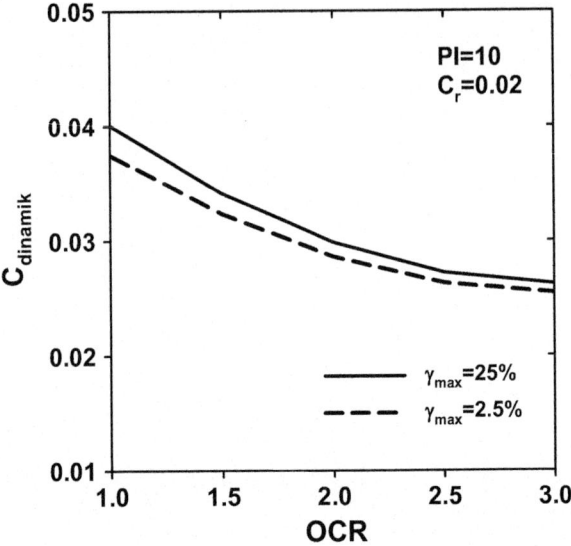

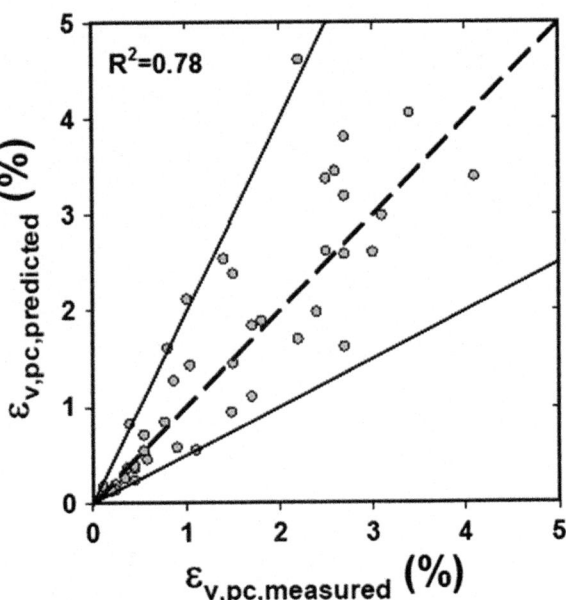

Fig. 19.28 Performance evaluation of the proposed post-cyclic volumetric strain prediction model

insight to the users. More detailed discussion on database compilation and model development phase are available in Bilge (2010).

The performance of these models was evaluated based on experimental measurements and presented in detail by Bilge (2010). Evaluation of the post-cyclic volumetric straining model is presented by Fig. 19.28, and it is concluded that the laboratory measurements could be estimated with a high success rate over a wide strain range. Yet case history based calibration is still needed.

19.5 Summary and Concluding Remarks

Within the confines of this chapter, a summary of current state of practice in seismic soil liquefaction engineering was presented. Since seismic soil liquefaction engineering problems involve a five step assessment framework including the assessment of (i) "triggering" or initiation of soil liquefaction, (ii) post-liquefaction strength and overall post-liquefaction stability, (iii) expected liquefaction-induced deformations and displacements, (iv) the consequences of these deformations and displacements, (v) mitigation alternatives, if necessary, the discussion scheme also followed the footprints of the first four steps of liquefaction engineering. Considering the increasing popularity of performance-based design trends, special emphasis was given on the assessment of cyclic strength and deformation performance of both cohesionless and cohesive soils. New frameworks were introduced and some recommendations listed for the practitioners. However, no conclusion can be complete without emphasizing the need for further research aiming to understand cyclic deformation response of soils.

Acknowledgements Some of the authors' findings and conclusions were compiled from their research performed with a number of collaborators including but not limited to Drs. R. B. Seed, R.E.S. Moss, A.M. Kammerer, J. Wu, J. M. Pestana, M. F. Riemer, R.E. Kayen, L.F. Harder Jr., A. Der Kiureaghian, K. Tokimatsu, J. Bray, R. Sancio, and T. L. Youd. Their contributions are strongly acknowledged and appreciated.

Open Access This chapter is distributed under the terms of the Creative Commons Attribution Noncommercial License, which permits any noncommercial use, distribution, and reproduction in any medium, provided the original author(s) and source are credited.

References

Andrews DC, Martin GR (2000) Criteria for liquefaction of silty sands. In: Proceedings of the 12th world conference on earthquake engineering, Auckland, New Zealand

Andrus RD, Stokoe KH, Roesset JM (1991) Liquefaction of gravelly soil at Pence Ranch during the 1983 Borah Peak, Idaho earthquake. In: First international conference on soil dynamics and earthquake engineering V, Karlsruhe, Germany

Arulanandan K, Li XS, Sivathasan K (2000) Numerical simulation of liquefaction-induced deformation. J Geotech Eng ASCE 126(7):657–666

Bardet JP, Mace N, Tobita T (1999) Liquefaction-induced ground deformation and failure. A report to PEER/PG&E, Task 4A – Phase 1, Civil Engineering Department, University of Southern California, Los Angeles

Bartlett SF, Youd TL (1992) Empirical analysis of horizontal ground displacement generated by liquefaction-induced lateral spreads. Technical report no. NCEER-92-0021, National Center for Earthquake Engineering Research, State University of New York, Buffalo, pp 5-14-15

Bartlett SF, Youd TL (1995) Empirical prediction of liquefaction-induced lateral spread. J Geotech Eng 121(4):316–329

Bilge HT (2010) Cyclic volumetric and shear strain responses of fine-grained soils. Ph.D. dissertation, Middle East Technical University, Ankara

Bilge HT, Cetin KO (2007) Field performance case histories for the assessment of cyclically-induced reconsolidation (volumetric) settlements. METU – EERC Report No. 07 – 01, Middle East Technical University Earthquake Engineering Research Center, Ankara

Booker JR, Rahman MS, Seed HB (1976) GADFLEA – A computer program for the analysis of pore pressure generation and dissipation during cyclic or earthquake loading. Report No. EERC 76-24, University of California at Berkeley, Berkeley, California

Boulanger RW, Idriss IM (2006) Liquefaction susceptibility criteria for silts and clays. J Geotech Geoenviron Eng ASCE 132(11):1413–1426

Boulanger RW, Idriss IM (2007) Evaluation of cyclic softening in silts and clays. J Geotech Geoenviron Eng ASCE 133(6):641–652

Bray JD, Sancio RB (2006) Assessment of liquefaction susceptibility of fine-grained soils. J Geotech Geoenviron Eng 132(9):1165–1177

Castro G, Christian JT (1976) Shear strength of soils and cyclic loading. J Geotech Eng 102(9):887–894

Cetin KO (2000) Reliability -based assessment of seismic soil liquefaction initiation hazard. Ph.D. dissertation, University of California, Berkeley, California

Cetin KO, Seed RB (2002) Nonlinear shear mass participation factor, rd for cyclic shear stress ratio evaluation. J Soil Dyn Earthquake Eng 24(2):103–113

Cetin KO, Seed RB, Der Kiureghian A, Tokimatsu K, Harder LF Jr, Kayen RE, Moss RES (2004) SPT-based probabilistic and deterministic assessment of seismic soil liquefaction potential. J Geotech Geoenviron Eng 130(12):1314–1340

Cetin KO, Bilge HT, Wu J, Kammerer A, Seed RB (2009a) Probabilistic models for cyclic straining of saturated clean sands. J Geotech Geoenviron Eng ASCE 135(3):371–386

Cetin KO, Bilge HT, Wu J, Kammerer A, Seed RB (2009b) Probabilistic model for the assessment of cyclically-induced reconsolidation (volumetric) settlements. J Geotech Geoenviron Eng ASCE 135(3):387–398

Davis AP, Poulos SJ, Castro G (1988) Strengths back figured from liquefaction case histories. In: Second international conference on case histories in geotechnical engineering, vol 3, St. Louis, pp 1693–1701

Evans MD (1987) Undrained cyclic triaxial testing of gravels: the effect of membrane compliance. Ph.D. dissertation, University of California, Berkeley

Elgamal A, Yang Z, Parra E, Ragheb A (2003) Modeling of cyclic mobility in saturated cohesionless soils. Int J Plast 19(6):883–905

Faris AT, Seed RB, Kayen RE, Wu J (2006) A semi-empirical model for the estimation of maximum horizontal displacement due to liquefaction-induced lateral spreading. In: Proceedings of the 8th US national conference on earthquake engineering, San Francisco, CA, USA, Paper No. 1323

Finn WDL, Lee KW, Martin GR (1977) An effective stress model for liquefaction. J Geotech Eng 103(6):517–533

Finn LW, Ledbetter RH, Guoxi WU (1994) Liquefaction in silty soils: design and analysis. Ground failures under seismic conditions. ASCE Geotechnical Special Publication No. 44, pp 51–79

Hamada M, Yasuda S, Isoyama R, Emoto K (1986) Study on liquefaction induced permanent ground displacement. Report for the Association for the Development of Earthquake Prediction, Japan

Harder LF Jr (1988) Use of penetration tests to determine the cyclic loading resistance of gravelly soils during earthquake shaking. Ph.D. dissertation, University of California, Berkeley

Harder LF Jr, Seed HB (1986) Determination of penetration resistance for coarse-grained soils using the Becker Hammer Drill. Earthquake Engineering Research Center, Report No. UCB/EERC-86/06, University of California, Berkeley

Hyde AFL, Brown SF (1976) The plastic deformation of a silty clay under creep and repeated loading. Géotechnique 26(1):173–184

Hyde AFL, Higuchi T, Yasuhara K (2007) Postcyclic recompression, stiffness, and consolidated cyclic strength of silt. J Geotech Geoenviron Eng ASCE 133(4):416–423

Hyodo M, Yamamoto Y, Sugiyama M (1994) Undrained cyclic shear behavior of normally consolidated clay subjected to initial static shear stress. Soils Found 34(4):1–11

Idriss IM, Boulanger R (2006) Semi-empirical procedures for evaluating liquefaction potential during earthquakes. Soil Dyn Earthquake Eng 26(2–4):115–130

Idriss IM, Boulanger RW (2007) SPT- and CPT-based relationships for the residual shear strength of liquefied soils. In: Pitilakis KD (ed) Earthquake geotechnical engineering, 4th international conference on earthquake geotechnical engineering – invited lectures. Springer, Dordrecht, pp 1–22

Ishihara K (1985) Stability of natural deposits during earthquakes. In: Proceedings of the eleventh international conference on soil mechanics and foundation engineering, San Francisco

Ishihara K (1993) Liquefaction and flow failure during earthquakes. Geotechnique 43(3):351–415

Ishihara K, Yoshimine M (1992) Evaluation of settlements in sand deposits following liquefaction during earthquakes. Soils Found 32(1):173–188

Iwasaki T, Arakawa T, Tokida K (1982) Standard penetration test and liquefaction potential evaluation. Proceedings, international conference of soil dynamics and earthquake engineering, vol 2, Southampton, pp 925–941

Kanibir A (2003) Investigation of the lateral spreading at Sapanca and suggestion of empirical relationships for predicting lateral spreading. M.Sc. thesis, Department of Geological Engineering, Hacettepe University, Ankara

Kayen R, Moss RES, Thompson EM, Seed RB, Cetin KO, Der Kiureghian A, Tanaka Y, Tokimatsu K (2013) Probabilistic and deterministic assessment of seismic soil liquefaction potential by shear wave velocity. ASCE J Geotech Geoenviron Eng 139(3):407–419

Kishida H (1966) Damage to reinforced concrete buildings in Niigata city with special reference to foundation engineering. Soils Found 2:38–44

Kokusho T, Fujita K (2002) Site investigations for involvement of water films in lateral flow in liquefied ground. J Geotech Geoenviron Eng 128(11):917–925

Koutsoftas DC (1978) Effect of cyclic loads on undrained shear strength of two marine clays. J Geotech Eng ASCE 104(5):609–620

Lee KL, Albeisa A (1974) Earthquake induced settlements in saturated sands. J Geotech Eng 100 (GT4):387–406

Lee KL, Focht JA (1976) Strength of clay subjected to cyclic loading. Marine Geotech 1(3):165–185

Liao SSC, Veneziano D, Whitman RV (1988) Regression models for evaluating liquefaction probability. J Geotech Eng 114(4):389–411

Liao T, McGillivray A, Mayne PW, Zavala G, Elhakim A (2002) Seismic ground deformation modeling. Fianl report for MAE HD-7a (year 1), Geosystems Engineering/School of Civil & Environmental Engineering, Georgia Institute of Technology, Atlanta, 12 Dec 2002

Liyanathirana DS, Poulos HG (2002) A numerical model for dynamic soil liquefaction analysis. Soil Dyn Earthquake Eng 22(9–12):1007–1015

Martin GR, Finn WDL, Seed HB (1975) Fundamentals of liquefaction under cyclic loading. J Geotech Eng Div 101(5):423–438

Matsui T, Ohara H, Itou N (1980) Cyclic stress – strain history and shear characteristics of clay. J Geotech Eng 106(10):1101–1120

Moss RES, Seed RB, Kayen RE, Stewart JP, Kiureghian AD, Cetin KO (2006) CPT-based probabilistic and deterministic assessment of in situ seismic soil liquefaction potential. J Geotech Geoenviron Eng 132(8):1032–1051

National Center for Earthquake Engineering Research (NCEER) (1997) In: Youd TL, Idriss IM (eds) Proceedings of the NCEER workshop on evaluation of liquefaction resistance of soils. Technical report no. NCEER-97-0022

Newmark NM (1965) Effects of earthquakes on embankments and dams. Géotechnique 15(2):139–160

Ohara S, Matsuda H (1988) Study on the settlement of saturated clay layer induced by cyclic shear. Soils Found 28(3):103–113

Okamura T (1971) The variation of mechanical properties of clay samples depending on its degree of disturbance: specialty session; quality in soil sampling. In: Proceedings of the 4th Asian regional conference on SMFE, vol 1, pp 73–81

Olson SM, Stark TD (2002) Liquefied strength ratio from liquefaction case histories. Can Geotech J 39:629–647

Perlea V (2000) Liquefaction of cohesive soils. In: Pak RYS, Yamamura J (eds) Liquefaction of cohesive soils. Soil dynamics and liquefaction 2000, Geotechnical Special Publications (GSP) GSP 107, pp 58–76

Poulos SJ, Castro G, France JW (1985) Liquefaction evaluation procedure. J Geotech Eng ASCE 111(6):772–792

Rauch AF (1997) An empirical method for predicting surface displacements due to liquefaction-induced lateral spreading in earthquakes. Ph.D. dissertation, Virginia Polytechnic Institute and State University, Blacksburg, Virginia

Robertson PK, Wride CE (1998) Evaluating cyclic liquefaction potential using the cone penetration test. Can Geotech J 35(3):442–459

Robertson PK, Woeller DJ, Finn WDL (1992) Seismic cone penetration test for evaluating liquefaction potential under cyclic loading. Can Geotech J 29:686–695

Sangrey DA, France JW (1980) Peak strength of clay soils after a repeated loading history. In: Proceedings of the international symposium soils under cyclic and transient loading, vol 1, pp 421–430

Seed HB (1983) Earthquake-resistant design of earth dams. In: Proceedings of the symposium on seismic design of embankments and caverns. ASCE, Philadelphia

Seed HB (1987) Design problems in soil liquefaction. J Geotech Eng ASCE 113(8):827–845

Seed RB, Harder LF Jr (1990) SPT-based analysis of cyclic pore pressure generation and undrained residual strength. In: Proceedings of the H. B. Seed memorial symposium, May 1990

Seed HB, Idriss IM (1971) Simplified procedure for evaluating soil liquefaction potential. J Soil Mech Found Div ASCE 97(SM9):1249–1273

Seed HB, Idriss IM (1982) Ground motions and soil liquefaction during earthquakes. Earthquake Engineering Research Institute, Berkeley, 134 pp

Seed HB, Martin GR, Lysmer J (1976) Pore-water pressure changes during soil liquefaction. J Geotech Eng 102(GT4):323–346

Seed HB, Tokimatsu K, Harder LF, Chung RM (1984) The influence of SPT procedures in soil liquefaction resistance evaluations. Earthquake Engineering Research Center Report No. UCB/EERC-84/15, University of California, Berkeley

Seed HB, Tokimatsu K, Harder LF, Chung RM (1985) The influence of SPT procedures in soil liquefaction resistance evaluations. J Geotech Eng 111(12):1425–1445

Seed HB, Seed RB, Harder LF, Jong H-L (1989) Re-evaluation of the Lower San Fernando Dam: Report 2, examination of the post-earthquake slide of February 9, 1971. US. Army Corps of Engineers Contract Report GL-89-2, US. Amy Corps of Engineers Waterways Experiment Station, Vicksburg, Mississippi

Seed RB, Cetin KO, Moss RES, Kammerer AM, Wu J, Pestana JM, Riemer MF (2001) Recent advances in soil liquefaction engineering and seismic site response evaluation. In: 4th international conference on recent advances in geotechnical earthquake engineering and soil dynamics, San Diego

Seed RB, Cetin KO, Moss RES, Kammerer AM, Wu J, Pestana JM, Riemer MF, Sancio RB, Bray RB, Kayen RE, Faris A (2003) Recent advances in soil liquefaction engineering: a unified and consistent framework. Report No. EERC 2003–06, Earthquake Engineering Research Center, University of California, Berkeley

Shamoto Y, Zhang JM, Tokimatsu K (1998) Methods for evaluating residual post-liquefaction ground settlement and horizontal displacement. Special Issue on the Geotechnical Aspects of the January 17 1995 Hyogoken-Nambu Earthquake, No. 2, pp 69–83

Sherif MA, Ishibashi I, Tsuchiya C (1977) Saturation effects on initial soil liquefaction. J Geotech Eng Div ASCE 103(8):914–917

Stark TD, Mesri G (1992) Undrained shear strength of liquefied sands for stability analysis. J Geotech Eng Div ASCE 118(11):1727–1747

Tatsuoka F, Sasaki T, Yamada S (1984) Settlement in saturated sand induced by cyclic undrained simple shear. In: 8th world conference on earthquake engineering, vol 3, San Francisco, CA, USA, pp 95–102

Thiers GR, Seed HB (1969) Strength and stress-strain characteristics of clays subjected to seismic loads. ASTM STP 450, symposium on vibration effects of earthquakes on soils and foundations, ASTM, pp 3–56

Tokimatsu K, Seed HB (1984) Simplified procedures of the evaluation of settlements in clean sands. Report No. UCB/GT-84/16, University of California, Berkeley, CA

Toprak S, Holzer TL, Bennett MJ, Tinsley JC (1999) CPT- and SPT-based probabilistic assessment of liquefaction potential. In: Proceedings of the 7th US-Japan workshop on earthquake resistant design of lifeline facilities and countermeasures against liquefaction, Seattle

Towhata I, Sasaki Y, Tokida K-I, Matsumoto H, Tamari Y, Yamada K (1992) Prediction of permanent displacement of liquefied ground by means of minimum energy principle. Soils Found JSSMFE 32(3):97–116

Tsukamato Y, Ishihara K, Sawada S (2004) Settlement of silty sand deposits following liquefaction during earthquakes. Soils Found 44(5):135–148

Ue S, Yasuhara K, Fujiwara H (1991) Influence of consolidation period on undrained strength of clays. Ground Construct 9(1):51–62

Van Eekelen HAM, Potts DM (1978) The behaviour of Drammen clay under cyclic loading. Géotechnique 28:173–196

Wang W (1979) Some findings in soil liquefaction. Report Water Conservancy and Hydro-electric Power Scientific Research Institute, Beijing, China, pp 1–17

Wang JG, Rahman MS (1999) A neural network model for liquefaction-induced horizontal ground displacement. Soil Dyn Earthquake Eng 18:555–568

Wilson NE, Greenwood JR (1974) Pore pressure and strains after repeated loading in saturated clay. Can Geotech J 11:269–277

Wride CE, Mcroberts EC, Robertson PK (1999) Reconsideration of case histories for estimating undrained shear strength in sandy soils. Can Geotech J 36(5):907–933

Wu J, Seed RB (2004) Estimating of liquefaction-induced ground settlement (case studies). In: Proceedings of the fifth international conference on case histories in geotechnical engineering, Paper 3.09, New York, NY, USA

Wu J, Seed RB, Pestana JM (2003) Liquefaction triggering and post liquefaction deformations of Monterey 0/30 sand under uni-directional cyclic simple shear loading. Geotechnical Engineering Research report no. UCB/GE-2003/01, University of California, Berkeley, CA

Yasuhara K (1994) Post-cyclic undrained strength for cohesive soils. J Geotech Eng ASCE 120(11):1961–1979

Yasuhara K, Andersen KH (1991) Post-cyclic recompression settlement of clay. Soils Found 31(1):83–94

Yasuhara K, Hyde AFL (1997) Method for estimating post-cyclic undrained secant modulus of clays. J Geotech Geoenviron Eng 123(3):204–211

Yasuhara K, Fujiwara H, Hirao K, Ue S (1983) Undrained shear behavior of quasi-overconsolidated clay induced by cyclic loading. In: Proceedings of the IUTAM symposium, Seabed Mechanics, pp 17–24

Yasuhara K, Hirao K, Hyde AFL (1992) Effects of cyclic loading on undrained strength and compressibility of clay. Soils Found 32(1):100–116

Yasuhara K, Murakami S, Toyota N, Hyde AFL (2001) Settlements in fine-grained soils under cyclic loading. Soils Found 41(6):25–36

Youd TL, Noble SK (1997) Liquefaction criteria based on statistical and probabilistic analyses. In: Proceedings of the NCEER workshop on evaluation of liquefaction resistance of soils, National Center for Earthquake Engineering Research, State University of New York at Buffalo, pp 201–215

Youd TL, Perkins JB (1987) Map showing liquefaction susceptibility of San Mateo County. U.S. Geological Survey, California, Map I-1257-G

Youd TL, Idriss IM, Andrus RD, Arango I, Castro G, Christian JT, Dobry R, Finn WDL, Harder LF Jr, Hynes ME, Ishihara K, Koester JP, Liao SC, Marcuson WF, Martin GR, Mitchell JK, Moriwaki Y, Power MS, Robertson PK, Seed RB, Stokoe KH (2001) Liquefaction resistance of soils: Summary report from the 1996 NCEER and 1998 NCEER/NSF workshops on evaluation of liquefaction resistance of soils. J Geotech Geoenviron Eng, ASCE 127(10):817–833

Youd TL, Hansen CM, Bartlett SF (2002) Revised multilinear regression equations for prediction of lateral spread displacement. J Geotech Geoenviron Eng ASCE 128:1007–1017

Zhang G, Robertson PK, Brachman RWI (2002) Estimating liquefaction-induced ground settlements from CPT for level ground. Canadian Geotech J 39:1168–1180

Chapter 20
Seismic Hazard and Seismic Design and Safety Aspects of Large Dam Projects

Martin Wieland

Abstract Earthquakes can affect large dam projects in many different ways. Usually, design engineers are focussing on ground shaking and neglect the other aspects. The May 12, 2008 Wenchuan earthquake has damaged 1803 dams and reservoirs. The widespread mass movements have caused substantial damage to dams and surface powerhouses in Sichuan province in China. The different features of the earthquake hazard are presented, the most important are ground shaking, faulting and mass movements. The basic requirement of any large dam is safety. Today, an integral dam safety concept is used, which includes (i) structural safety, (ii) dam safety monitoring, (iii) operational safety and maintenance, and (iv) emergency planning. The importance of these four safety elements is discussed. The long-term safety includes, first, the analysis of all hazards affecting the project, i.e. hazards from the natural environment, hazards from the man-made environment and project-specific and site-specific hazards. The role of the earthquake hazard on the seismic design and seismic safety of large dam projects are discussed as, today, the structural safety of large storage dams is often governed by the earthquake load case. The seismic design and performance criteria of dams and safety-relevant elements such as spillways and bottom outlets recommended by the seismic committee of the International Commission on Large Dams are presented. The conceptual and constructional requirements for the seismic design of concrete and embankment dams are given, which often are more important than the seismic design criteria that are used as a basis for dynamic analyses. Finally, the need and importance of periodic reviews of the seismic safety of existing dams is discussed.

M. Wieland (✉)
Committee on Seismic Aspects of Dam Design, International Commission on Large Dams, c/o Poyry Switzerland Ltd., Zurich, Switzerland
e-mail: martin.wieland@poyry.com

20.1 Introduction

Because strong earthquakes occur very seldom in Central Europe, hardly any dam engineer or dam owner has any experience with earthquakes. It is also very hard to find any dams which have been damaged during earthquakes, although the average age of dams in Europe is around 50 years, and the total number of years of exposure of large dams to seismic action has been very large.

However, as strong earthquakes may affect a large area, many dams may be subjected to strong ground shaking as in the case of the May 12, 2008 Wenchuan earthquake in China, where about 1,803 dams and reservoirs, most of them were small earth dams, and 403 hydropower plants were damaged, four dams had a height exceeding 100 m (Wieland and Chen 2009). Also, during the 2001 Bhuj earthquake in Gujarat, India, 245 dams – mainly small embankment dams – had to be rehabilitated or strengthened after the earthquake. The latest earthquake which affected many dams was the March 11, 2011 Tohoku earthquake in Japan where on 18 m high embankment dam failed and 8 people lost their live. Another 400 dams, subjected to earthquake shaking, had to be inspected.

These examples show that earthquake safety needs proper attention. Also, the field of seismic hazard analysis has developed very fast in the last years, and the estimated seismic hazard has been increasing steadily. In addition, the seismic design and performance criteria and methods of seismic analysis have developed but at a much slower pace than the seismic hazard analysis methods.

As most existing dams built before the 1990s were designed against earthquakes using either seismic design criteria and/or methods of dynamic analysis, which are considered obsolete or even wrong today, the earthquake safety of these dams is not known if modern criteria are applied. It has to be assumed that a few of them are structurally deficient. Consequently, there is a need for the systematic reassessment of the earthquake safety of large and also small dams (Wieland 2003, 2006).

The paper gives an overview on the current state of the seismic design and safety aspects of large dams and the role of the earthquake hazard within the comprehensive dam safety framework that should be used for large dams. The subjects presented were addressed by the Committee on Seismic Aspects of Dam Design of the International Commission on Large Dams (ICOLD) in recent years or are direct consequences of guidelines published by ICOLD's seismic committee. The paper also summarizes the main subjects of the papers given in the list of references and provides general guidelines for the seismic safety assessment and design of large storage dams.

20.2 Dam Safety

20.2.1 Integral Dam Safety Concept

The two main goals of every safety concept are the minimisation of all risks, and the mastering of the remaining risk in the best possible way. To reach these goals a comprehensive safety concept is used for large storage dams, which includes the following key elements (Wieland and Mueller 2009):

(i) structural safety (main elements: geologic, hydraulic and seismic design criteria; design criteria and methods of analysis may have to be updated when new data are available or new guidelines, regulations or codes are introduced);
(ii) dam safety monitoring (main elements: dam instrumentation, periodic safety assessments by dam experts, etc.);
(iii) operational safety (main elements: reliable rule curves for reservoir operation under normal and extraordinary (hydrological) conditions, training of personnel, dam maintenance, sediment flushing, engineering back-up. The most important element for a long service life is maintenance of all structures and components);
(iv) emergency planning (main elements: emergency action plans, inundation maps, water alarm systems, evacuation plans, etc.).

Therefore, as long as the proper implementation of these safety issues can be guaranteed according to this integral safety concept, a dam can be considered as safe.

Periodic safety assessments are indispensable as they will show what measures have to be taken to maintain or improve the safety and thus to even extend the lifespan. Deficiencies observed after commissioning must be rectified as early as possible.

20.2.2 Structural Safety

Structural safety is the main prerequisite for the safe operation of a storage facility and thus for its sustainability (Wieland 2012b). The basis for structural safety is laid mainly during design, as given by the design criteria. It is important that in the structural design all hazards, which can affect the dam are taken into account. The hazards are from the natural environment or are man-made. Furthermore, there are site and project related hazards (e.g. geology, hydro-geology, topography, vulnerability of dams to specific hazards etc.).

The design must be carried out assuming that the dam may become exposed to the worst possible scenario during a natural hazardous event, i.e. mainly floods and earthquakes.

Older dams are often not designed according to today's design standards or guidelines. One reason is that since their conception more data have become available, which allow a more realistic prediction of extreme events. Such storage projects may have inadequate spillway capacity or the dam structures were designed with inadequate seismic loading.

Inadequate spillway capacity can be handled by constructing an additional spillway, or where feasible, by rising the dam crest.

The most severe loading for dam structures originating from the natural environment is caused by strong earthquakes close to a dam. Since the 1930s concrete and embankment dams were generally designed against earthquakes in most parts of the world. The earthquake loading was represented by a seismic coefficient, which was used in a pseudo-static analysis. In general a seismic coefficient of 0.1 was assumed almost irrespective of the seismic hazard at the dam site. Using this concept, the earthquake load combination was usually not the governing one in dam design.

Field observations and seismic hazard analyses, however have shown that even in regions of moderate seismicity, such as Central Europe, earthquakes with magnitudes up to $M = 6.5$ are possible, although with a very low probability of occurrence. Such earthquakes can cause much higher peak ground acceleration than those assumed for the dam design.

Modern seismic design criteria were published by ICOLD in 1989 which were revised in 2010 (ICOLD 2014). These design criteria are different from those used for dams built before 1989. Therefore, dams designed with a pseudo-static analysis method and a seismic coefficient may not satisfy today's seismic safety criteria and it has to be assumed that some of these dams are structurally deficient. Only an earthquake analysis can show if an existing dam is safe. Of course, this also applies to dams, which have not been designed against earthquakes.

This change in seismic design concept shows clearly that a dam, which was safe at the time of completion and which has satisfied all safety criteria, does not necessarily remain safe forever even if it is kept in excellent condition.

20.2.3 Dam Safety Monitoring

Dam safety monitoring is a key activity in dam safety management and includes the following activities:

(i) Visual inspection of the entire dam and its appurtenances. It also includes checking the functioning of the flood control elements, i.e. spillway gates and the valves or gates for the bottom outlets, and the emergency power supply.
(ii) Measurements of physical quantities (mainly deformations, pressures, flow/seepage volumes, temperature, etc.) describing the status of the dam and its foundation. The measurements depend on the type of the dam and the local conditions.

Monitoring provides a rational insight into the safety of the dam-foundation system. With modern automatic data acquisition systems real-time monitoring becomes possible and rapidly changing conditions can be recorded.

Instrumental monitoring, if systematically performed, can detect a developing deficiency at an early stage, however, only at locations where corresponding instruments have been installed, e.g. piezometers, seepage weir, settlement point, etc. In other locations only visual inspection can detect whether something is wrong or unusual.

Unfortunately, there are still many older dams, which have inadequate monitoring facilities. Even today, some dam owners are reluctant to install instruments in their new dams if this is not required by the authorities.

Instrumental monitoring also requires a strict data management and a graphical display of the measurements to enable the rapid identification of irregularities caused by deficiencies or also by faulty measurements or deficient equipment. An important concept in monitoring is redundancy.

Dam safety monitoring is the main element of dam safety management, which includes the following (Swiss practice):

(i) dam safety monitoring and regular visual inspections by the dam owner,
(ii) annual dam safety inspection by a dam engineer, and
(iii) detailed dam safety inspection every 5 years by an independent dam engineer and a geologist. During the 5-year-inspection changes in the safety and design criteria, and new information on hazards affecting the dam are reviewed as well. If important changes have been observed a new safety check will be needed. In the past the safety checks included mainly the flood and earthquake safety.

During operation of the dam a dam safety authority should supervise the surveillance organisation of the owner, of the experienced engineer and of the experts.

If a dam does not comply with current dam safety standards or shows unusual behaviour, the most effective mean for reducing the risk is a reduction of the reservoir level.

20.2.4 Operational Safety

The importance of operational safety of dams is sometimes overlooked. In the case of hydropower plants it includes the following: Operational guidelines for the reservoir for usual, unusual and extreme conditions; training of personnel; experienced and technically qualified dam maintenance staff; dam maintenance procedures; engineering back-up to cope with unusual behaviour of the dam, etc.

Maintenance is the key issue as it is the prerequisite for long-term safety. If a dam designed for say 100 years is not maintained it can become unsafe within a very short period of time especially if the spillway gates and bottom outlets and the dam monitoring systems are no longer functioning properly.

20.2.5 Emergency Planning

The main risk for embankment dams is overtopping during large floods. Hence, upgrading of spillways with inadequate discharge capacity will reduce this risk in embankment dams.

In addition, storage dams should be provided with a bottom outlet, such that the reservoir can be drawn down to a safe level in an emergency situation, especially after a strong earthquake when parts of the dam may be damaged. This would require that the discharge capacity of the bottom outlet and other low level outlets must be larger than the average inflow into the reservoir. This safety requirement has been implemented in Switzerland where average inflows into reservoirs are moderate.

The need for lowering the reservoir was demonstrated during the Wenchuan earthquake in China, where the concrete face of the 156 m high Zipingpu concrete face rockfill dam was damaged and had to be repaired. Such repairs would be very difficult to perform under water.

Also if the power plant is shut down for long periods of time and due to limitations in the discharge capacity of low level outlets, the spillway may be the only way to control the reservoir level.

In the emergency planning concept it is assumed that every dam can fail or be destroyed. Therefore, the consequences of a dam failure, which is a flood wave caused by the uncontrolled release of the water from the reservoir, must be analysed.

Numerous dam failure scenarios could be considered, however, the main objective of emergency planning is to save lives, therefore, for alarming and evacuating people one has to focus on the worst scenarios with the largest consequences. No failure probabilities are considered for these scenarios. The worst scenario is the instantaneous failure of a dam with full reservoir, which may be due to military action. But also extreme flood events with overtopping of the dam and extreme water levels in the river downstream of the dam may be an extreme scenario as the water stored in the reservoir would be larger than the normal operation level to be considered in the case of instantaneous failure.

Emergency Action Plans (EAP) are intended to help the dam owner and operator, and the emergency officials to minimize the consequences of flooding caused by dam failure or the uncontrolled release of water from a reservoir. The EAP will guide the responsible personnel in identifying, monitoring, responding to, and mitigating emergency situations. It outlines "who does what, where, when, and how" in an emergency situation or unusual occurrence affecting the safety of the dam and the power plant. The EAP should be updated regularly and after important emergency events. Basically, the dam owner is responsible for maintaining a safe dam by means of safety monitoring, operations manual, maintenance, repair, and rehabilitation.

In an emergency situation, the dam owner is responsible for monitoring, determining appropriate alarm levels, making notifications, implementing emergency actions at the dam, determining when an emergency situation no longer exists, and documenting all activities. In the case of an emergency, the dam owner is responsible for immediate notification of the authorities, who are in charge of warning and evacuation of the affected population.

Warning is performed by special water alarm systems. The basis for evacuation planning is a dam breach flood wave analysis, which shows the inundated area for the worst-case failure scenario, i.e. the sudden failure of the dam. In addition, the arrival time of the flood wave, flow velocities and water depth are results obtained from such an analysis.

In Switzerland 65 large dams are equipped with a fully functional water alarm system. The first alarm systems were installed over 50 years ago as a consequence of the severe damage of two dams in Germany in 1943, which caused large numbers of casualties. Fortunately up to now these water alarm systems have never had to be used.

20.2.6 Consequences of Dam Failure and Risk Mitigating Measures

The consequences of dam failure are: loss of life and injuries (reduction of loss of life is the top priority of emergency planning); environmental damage; property damage in flood plain; damage of infrastructure; loss of power plant and electricity production; socio-economic impact; political impact, etc.

These consequences can be reduced by a number of structural and non-structural measures. The structural measures are mainly related to the safety of the dam, i.e. flood safety, earthquake safety, and site-specific and project-specific safety aspects. The non-structural measures include the following: safe operational guidelines for reservoir under normal and abnormal operational conditions; implementation of emergency action plans; implementation of water alarm systems; training of personnel; lowering of reservoir level in case of safety concerns; periodic safety checks; engineering back-up to cope effectively with abnormal and emergency situations; land use planning (political decision); insurance coverage, third party liability coverage (protection from economic losses), etc. The non-structural measures are often more effective than structural measures.

20.3 Hazards to Be Considered in Large Dam Projects

In the design of large dams all possible hazards affecting the project must be considered. A list of typical hazards is given in Table 20.1 (Wieland and Mueller 2009). A distinction can be made between hazards from the natural environment, structural or project-specific hazards, and man-made hazards. In the matrix shown

Table 20.1 Example of hazard matrix for hydropower plant showing hazards and required protective measures (Emergency classification: A: internal alert; B: developing situation; C: imminent situation)

Hazards	Protective measures				
	Rehabilitation	Partial reservoir drawdown	Full reservoir drawdown	Evacuation	Post-event evacuation
Natural hazards					
Floods	A	B		C	
Earthquake					C
Mass movements (landslides, rockfalls, avalanches, etc.)	A	B			
Extreme weather conditions (storm, rainfall, low temperatures, ice, etc.)	A				
Structural hazards					
Blockage of spillway gates (floating debris, pier or gate deformations, lack of maintenance), equipment failure, power supply, etc.			C	C	
Differential movement of structure (structural joints, interfaces between concrete structures and embankments etc.)	A	B	C	C	
Embankment piping or seepage		B	C		
Foundation seepage (damaged grout curtain, dissolution of minerals, etc.)	A				
Electrical or mechanical failure of equipment used for operation of vital gates and valves (gantry cranes on top of spillways etc.), failure of control units, faulty software, etc.	A				
Ageing, alkali-aggregate reactions etc.	A				
Man-made hazards					
Design errors, poor construction	A				
Faulty operation of equipment, inadequate rule curves for reservoir operation etc.	A				
Sabotage, terrorism, acts of war		B			C
Other hazards or unknown hazards (protective measures depend on the type of hazard)	(A)	(B)	(C)	(C)	(C)

also the possible protective measures are given if such hazards develop or events have happened. The protective measures include the following:

(i) Rehabilitation,
(ii) Partial reservoir drawdown,
(iii) Full reservoir drawdown,
(iv) Evacuation, and
(v) Post-event evacuation.

In the emergency classification a distinction is made between internal alert, developing situation, and imminent situation. If unusual behaviour of a dam is observed and if there is adequate time or if safety criteria have changed then rehabilitation of the dam may be required.

In case of a potentially dangerous situation a partial reservoir drawdown may be required.

Finally, in the case of an imminent situation when the hazard cannot be controlled and depending on the available time a full reservoir drawdown, evacuation or in the worst case post-event evacuation and rescue may be needed.

It is obvious from Table 20.1 that the most difficult hazards to handle are those where only post-event evacuation is possible as in the case of a dam failure caused by a strong earthquake or acts of war, terrorism or sabotage. As earthquake prediction is not an option for large dams, the dams must be structurally safe to resist the different features of the seismic hazard. Therefore, the earthquake hazard plays an important role in the design of large dams.

It should be added that in Switzerland the large storage dams had to be designed for specific scenarios of acts of war similar to those, which had led to the breach of the two dams in Germany in World War II. As a consequence the crest thickness of the largest concrete arch and gravity dams is generally larger than that of similar dams in countries, where such scenarios have not been taken into account and certain types of dams, whose reservoirs could not be lowered in a short period of time, such as buttress dams or hollow gravity dams were not permitted. Today this requirement is no longer needed. But a thick dam crest is certainly beneficial for the earthquake safety of both concrete and embankment dams.

20.4 Earthquakes Create Multiple Hazards in Large Dam Projects

We have to recognize that the earthquake hazard is a multi-hazard, which may affect large storage dams in different ways (Wieland and Chen 2009):

(i) ground shaking causing vibrations in dams, appurtenant structures and equipment, and their foundations (Fig. 20.1);
(ii) fault movements in the dam foundation or movements along discontinuities in dam foundation near major faults, which can be activated during strong earthquakes, causing structural distortions (Fig. 20.2);

Fig. 20.1 Crack at upstream face (*top*) and at the kink (*left bottom*) and crack showing sliding movement of wedge formed by cracks at the kink (*right bottom*) of the buttress at the downstream face of the Sefid Rud buttress dam caused by ground shaking during the 1990 Manjil earthquake in Iran

Fig. 20.2 Failure of two openings of the Shih-Kang weir caused by fault movements during the 1999 Chi-Chi earthquake in Taiwan

Fig. 20.3 Rockfalls in the Zipingpu reservoir area caused by the 2008 Wenchuan earthquake in China

Fig. 20.4 Infill wall and roof of powerhouse punctured by high-velocity rocks (*left*) and wall damage of building of Shapai power plant by large rock (*right*) caused by 2008 Wenchuan earthquake

(iii) fault movements in the reservoir causing water waves in the reservoir or loss of freeboard;
(iv) mass movements (rockfalls with large rocks) (Fig. 20.3), causing damage to surface powerhouses (Fig. 20.4), electro-mechanical equipment, gates, spillway piers (Fig. 20.5), retaining walls, penstocks, masts of transmission lines, etc.
(v) mass movements into the reservoir causing impulse waves in the reservoir (Fig. 20.3);

Fig. 20.5 Damaged pier of Futan weir looking downstream (*left*) and damage of sliding gate for power intake (indentation of steel leaf from rockfall) (*right*) caused by the 2008 Wenchuan earthquake in China

Fig. 20.6 Access roads to Sefid Rud dam site blocked by numerous rockfalls caused by the 1990 Manjil earthquake in Iran

- (vi) mass movements blocking rivers and forming landslide dams and lakes whose failure may lead to overtopping of run-of-river power plants or the inundation of powerhouses with equipment;
- (vii) mass movements blocking access roads to dam sites and appurtenant structures (Fig. 20.6);
- (viii) ground movements and settlements due to liquefaction and densification of soil, causing distortions in dams; and
- (ix) turbidity currents in reservoir blocking bottom outlets, power intakes and low level outlets.

Other seismic hazards such as surface water waves in reservoirs are of lesser importance for the earthquake safety of a dam as their dominant frequencies are much lower than the lowest eigenfrequencies of dams, i.e. the corresponding loads are of quasi static nature, and the maximum amplitude of surface water waves observed during strong ground shaking is less than 1 m.

Usually, the main hazard, which is addressed in codes and regulations, is the earthquake ground shaking. It causes stresses, deformations, cracking, sliding, overturning, etc.

An important hazard, which has generally been underestimated, is the rockfall hazard in mountainous regions.

During the 2008 Wenchuan earthquake, some 30 major landslide lakes were created. Tangjiashan landslide dam with a height of 124 m with a volume of about 20 Mm^3, created a reservoir with a volume of 320 Mm^3, threatening people living downstream of this natural dam.

Every time a strong earthquake occurs, the design guidelines have to be reviewed as new phenomena appear, which may have been overlooked. For example, during the Wenchuan earthquake, the problems of mass movements (mainly rockfalls in steep mountains) and landslide lakes have shown to be very important new features of strong earthquakes. In addition, an unprecedented large number of dams and run-of-river power plants have been affected by this earthquake. The Wenchuan earthquake has confirmed and demonstrated that dams, spillways and appurtenant structures must be able to withstand the multiple effects of strong earthquakes.

20.5 Seismic Design Criteria for Large Dams and Appurtenant Structures

The following design earthquakes are needed for the seismic design of the different structures and elements of a large dam project (ICOLD 2014; Wieland 2012a):

(i) Safety Evaluation Earthquake (SEE): The SEE is the earthquake ground motion a dam must be able to resist without uncontrolled release of the reservoir. The SEE is the governing earthquake ground motion for the safety assessment and seismic design of the dam and safety-relevant components, which have to be functioning after the SEE.

(ii) Design Basis Earthquake (DBE): The DBE with a return period of 475 years is the reference design earthquake for the appurtenant structures. The DBE ground motion parameters are estimated based on a probabilistic seismic hazard analysis (PSHA). The mean values of the ground motion parameters of the DBE can be taken. (Note: The return period of the DBE may be determined in accordance with the earthquake codes and regulations for buildings and bridges in the project region.)

(iii) Operating Basis Earthquake (OBE): The OBE may be expected to occur during the lifetime of the dam. No damage or loss of service must happen. It has a probability of occurrence of about 50 % during the service life of 100 years. The return period is taken as 145 years (ICOLD 2014). The OBE ground motion parameters are estimated based on a PSHA. The mean values of the ground motion parameters of the OBE can be taken.

(iv) Construction Earthquake (CE): The CE is to be used for the design of temporary structures such as coffer dams and takes into account the service life of the temporary structure. There are different methods to calculate this design earthquake. For the temporary diversion facilities a probability of exceedance of 10 % is assumed for the design life span of the diversion facilities. Alternatively the return period of the CE of the diversion facilities may be taken as that of the design flood of the river diversion

The SEE ground motion can be obtained from a probabilistic and/or a deterministic seismic hazard analysis, i.e.

- Maximum Credible Earthquake (MCE): The MCE is the event, which produces the largest ground motion expected at the dam site on the basis of the seismic history and the seismotectonic setup in the region. It is estimated based on deterministic earthquake scenarios. According to ICOLD (2014) the ground motion parameters of the MCE shall be taken as the 84 percentiles (mean plus one standard deviation).
- Maximum Design Earthquake (MDE): For large dams the return period of the MDE is taken as 10,000 years. For dams with small or limited damage potential shorter return periods can be specified. The MDE ground motion parameters are estimated based on a probabilistic seismic hazard analysis (PSHA). According to ICOLD (2014) the mean values of the ground motion parameters of the MDE shall be taken. In the case where a single seismic source (fault) contributes mainly to the seismic hazard, uniform hazard spectra can be used for the seismic design. Otherwise, based on the deaggregation of the seismic hazard (magnitude versus focal distance) different scenario earthquakes may be defined.

For major dams the SEE can be taken either as the MCE or MDE ground motions. Usually the most unfavourable ground motion parameters of these two earthquakes have to be taken. If it is not possible to make a realistic assessment of the MCE then the SEE shall be at least equal to the MDE.

MDE, DBE, OBE and CE ground motion parameters are usually determined by a probabilistic approach (mean values of ground motion parameters are recommended), while for the MCE ground motion deterministic earthquake scenarios are used (84 percentile values of ground motion parameters shall be used). However, for the MDE, DBE, OBE and CE also deterministic scenarios may be defined.

The different design earthquakes are characterized by the following seismic parameters:

- Peak ground acceleration (PGA) of horizontal and vertical earthquake components.
- Acceleration response spectra of horizontal and vertical earthquake components typically for 5 % damping, i.e. uniform hazard spectra for CE, OBE, DBE and MDE obtained from the probabilistic seismic hazard analysis (mean values) and 84 percentile values of acceleration spectra for MCE obtained from the deterministic analysis using different attenuation models.
- Spectrum-matched acceleration time histories for the horizontal and vertical components of the MCE ground motion determined either from a random process or by scaling of recorded earthquake ground motions. The artificially generated acceleration time histories of the horizontal and vertical earthquake components shall be stochastically independent. To account for aftershocks, it is recommended to increase the duration of strong ground shaking.

In case of fault movements, similar estimates are required as for the ground shaking. It appears that it is quite difficult for the dam designer to get quantitative estimates of fault movements for the different types of design earthquakes as the seismic hazard analyses are mainly concerned with ground shaking.

For underground structures where the effects of imposed deformations are more relevant than inertial effects, the displacement ground motion parameters or displacement time histories of the different design earthquakes are also needed.

The best description of the ground motion is by means of the acceleration time histories. They are needed for any nonlinear dynamic analysis of dams and components. It is also expected that inelastic deformations take place under the SEE ground motion. According to ICOLD (2014) the following aspects of the 'design acceleration time history' should be considered:

(i) The three components of the spectrum-matched acceleration time histories must be statistically independent.
(ii) The acceleration time histories of the horizontal earthquake components may be assumed to act in along river and across river directions. No modifications in the horizontal earthquake components are needed if they are applied to other directions.
(iii) The duration of strong ground shaking shall be selected in such a way that aftershocks are also covered, i.e. records with long duration of strong ground shaking shall be selected.
(iv) In the case of dams that are susceptible to damage processes, which are governed by the duration of strong ground shaking such as, e.g., the build-up of pore pressures, earthquake records with long duration of strong ground shaking shall be used.
(v) For the safety check of a dam at least three different earthquakes shall be considered for the SEE ground motion.

The spectrum-matched acceleration time histories with extended duration of strong ground shaking used for the seismic analysis and design of the dams may be quite different from real ones; however, their use will lead to a safe design, although this may difficult to understand or accept by seismologists and other experts, who are not familiar with the seismic design of dams.

In this connection it should be mentioned that in the design of any structures including large dams, the designer will use simplified load and analysis models that lead to a safe design, even if the load model does not comply with the real nature of the hazard and this also applies to the earthquake hazard and the earthquake ground motion.

20.5.1 Reservoir-Triggered Seismicity

For some dams an additional earthquake load case was defined for reservoir-triggered seismicity (RTS) or reservoir-induced seismicity (RIS), (Note: The term reservoir-induced seismicity, which in the past has often been used, is not correct as reservoirs cannot induce earthquakes, however, they can trigger earthquakes. Therefore the correct technical term, which also properly describes this phenomenon, is reservoir-triggered seismicity.). RTS has been observed in over 100 reservoir in general with a water depth of the reservoir of over 100 m. The largest magnitudes of RTS events reached 6.3, however, in most cases the magnitudes of these shallow-focus events were much smaller. If RTS is possible or expected in a large dam project then the DBE and OBE ground motion parameters should cover those from the assumed RTS scenarios as such events are expected to occur within a few years after the start of the impounding of the reservoir (ICOLD 2011).

20.6 Seismic Performance Criteria for Large Dams and Appurtenant Structures

The rather general performance criteria for the dam body and safety-relevant components and equipment given in ICOLD Bulletin 148 (2014) can be interpreted as follows:

- Performance of dam body during OBE: No structural damage (cracks, deformations, leakage etc.), which affect the operation of the dam and the reservoir, is permitted. Minor repairable damage is accepted. (Note: Crack width limitations do not have to be considered for OBE load combinations in reinforced concrete structures.)
- Performance of dam body during SEE: Structural damage (cracks, deformations, leakage etc.) is accepted as long as the stability of the dam is ensured and no large quantities of water are released from the reservoir causing flooding in the downstream region of the dam.

- Performance of safety-relevant components and equipment during and after OBE: These components and equipment shall be fully operable after the OBE and therefore should behave elastically during the OBE.
- Safety-relevant components and equipment during and after the SEE: These components and equipment must be fully operable after the SEE. Minor distortions and damage (e.g. leakage of seals of gates) are accepted as long as they have no impact on the proper functioning of the components and equipment.

More specific performance criteria may be given for the SEE, e.g. sliding stability safety factors of slopes of greater than 1.0 are required for an SEE with a return period of 2,500 years in Germany. Such requirements may be stricter than those given above as during strong ground shaking sliding movements of slopes can be accepted, i.e. sliding safety factors may temporarily drop to less than one during the earthquake. However, in this case the allowable sliding movements would have to be defined based on engineering judgement and the stability of the slope after the earthquake, which may be reduced due to the build-up of pore pressures, must be guaranteed. For that case the safety factors must be larger than 1 taking into account residual strength parameters (zero cohesion) and the effect of pore pressure. For the sliding stability of gravity dams or powerhouse complexes that retain the reservoir the same criteria apply. The dynamic sliding stability analyses can be done most easily using the Newmark sliding block method. In general the horizontal and vertical earthquake components should be taken into account in two-dimensional models of slopes or gravity structures. The sliding movements depend on (i) the so-called yield acceleration, which is obtained from a pseudo-static stability analysis of the slope or gravity structure and (ii) the duration of ground shaking. Therefore, if sliding movements are important then it is important to use earthquake records with long duration of strong ground shaking as discussed in the previous section.

In China it is also required that water stops in concrete arch dams shall not be damaged during the SEE with a return period of 5,000 years. This requirement is a criterion for specifying water stops in arch dams, which can cope with the maximum contraction joint opening during the SEE. Actually leakage of joints due to damaged water stops could be accepted, however, in dams with large reservoirs where lowering of the reservoir may be difficult, the repair of damaged water stops would have to be done under water.

The safety-relevant components and equipment are bottom outlets (low level outlets) and spillways and all related equipment (mainly gates), motors, hydraulic systems, control panels, power supply, software etc., as it must be possible to regulate and lower the reservoir after the SEE. As the repair of a damaged dam will need time, it is necessary that after an earthquake a moderate flood equal to about the river diversion flood used during dam construction can still be released safely. This may be a lesser problem for concrete dams or run-of-river power plants, where limited overtopping of the crest may be acceptable under extreme circumstances, however, in the case of embankment dams such overtopping cannot be accepted, thus after an earthquake the possibly damaged or partly inoperable spillway of an embankment dam must be able to release larger floods than that of a similar concrete dam. After the 2008 Wenchuan earthquake several run-of-river power plants were overtopped as the power plants were shut down mainly due to failure of the electric grid and the spillway gates could not be opened due to failure of the (emergency) power supply. No damage was caused to the overtopped concrete structures, however, mud was deposited, which required extensive cleaning of the equipment and inundated areas after the earthquake.

The main safety criteria for rockfill dams with impervious core for the SEE are as follows:

(i) loss of freeboard, i.e. after the earthquake the reservoir level shall be below the top of the impervious core of the dam,
(ii) internal erosion, i.e. after the earthquake at least 50 % of the initial thickness of the filter and transition zones must be available, and
(iii) the sliding safety factor of slopes (considering build-up of pore pressure and residual strength parameters of embankment materials) shall be larger than 1 after the earthquake.

The second criterion also applies for earth core rockfill dams located on faults or discontinuities in the dam foundation, which can be moving during a strong earthquake. Moreover, at such sites only conservatively designed earth core rockfill dams should be built.

For concrete dams the main seismic safety criteria are as follows:

(i) stability of dam foundation, i.e. stability of wedges in abutments of arch dams and sliding movements of gravity structures along potential sliding surfaces in the dam foundation, and
(ii) sliding and overturning stability of concrete blocks formed by contraction joints and cracks along lift elevations, i.e. concrete blocks close to the crest in the centre of dams experience the highest absolute acceleration response.

We can conclude that after strong earthquakes, the bottom outlet(s) and the spillway gates are operable, so a moderate flood can be released safely after the earthquake. It has to be assumed that the power plant will be shut down and water cannot be released through the power waterways. For controlling the water level in the reservoir after a strong earthquake it is not necessary that all openings of a spillway have to be functional. Therefore, it may be acceptable to focus on the gates that are essential and to strengthen them seismically. The other gates may remain blocked. However, this appears only feasible for concrete structures where limited overtopping may be accepted.

20.7 Conceptual and Constructional Requirements for the Seismic Design of Concrete and Embankment Dams

20.7.1 Concrete Dams

There are several design details that are regarded as contributing to a favourable seismic performance of concrete and in particular arch dams (ICOLD 2001):

- Design of a dam shape with symmetrical and anti-symmetrical mode shapes that are excited by along-river and cross-river components of ground shaking, respectively.
- Maintenance of continuous compressive loading along the foundation, by shaping of the foundation, by thickening of the arches towards the abutments (filets) or by a plinth structure to support the dam and transfer load to the foundation.
- Limiting the crest length to height ratio, to assure that the dam carries a substantial portion of the applied seismic forces by arch action, and that non-uniform ground motions excite higher modes and lead to undesired stress concentrations.
- Providing contraction joints with adequate interlocking (shear keys).
- Improving the dynamic resistance and consolidation of the foundation rock by appropriate excavation, grouting etc.
- Provision of well-prepared lift surfaces to maximize bond and tensile strength.
- Increasing the crest width to reduce high dynamic tensile stresses in arch direction in crest region.
- Minimizing unnecessary mass in the upper portion of the dam that does not contribute effectively to the stiffness of the crest.
- Maintenance of low concrete placing temperatures to minimize initial, heat-induced tensile stresses and shrinkage cracking.
- Development and maintenance of a good drainage system.

The structural features, which improve the seismic performance of gravity and buttress dams, are basically the same as that for arch dams. Earthquake observations have shown that a break in slope on the downstream faces of gravity and buttress dams should be avoided to eliminate local stress concentrations and cracking under moderate earthquakes. The webs of buttresses should be sufficiently massive to prevent damage from cross-river earthquake excitations.

The above criteria apply to conventional mass concrete dams. For RCC dams the same criteria apply. However, the high permeability of some RCC dams along the lifts with a typical vertical spacing of about 30 cm and the resulting pore pressures within the dam have a negative impact on the dynamic sliding stability of concrete blocks near the crest of the dam formed by the contraction joints and a horizontal crack along lift joints. The seismic sliding movements in downstream direction could be reduced by a watertight membrane or impermeable concrete face in the critical crest region of the dam. This also applies to conventional gravity dams at sites where strong ground shaking and significant amplification of the dynamic response (absolute acceleration response) in the central crest region is possible. The maximum amplification of the acceleration from the base to the crest during strong earthquakes can reach values of 4–6 for high gravity dams and 6–8 in high arch dams. For less intense ground motions these amplification factors can reach values up to 13 in very high arch dams, which indicates very low damping of these structures. When shear keys are provided in the contraction of gravity dams, the sliding movements of detached concrete blocks in the crest region is restrained. Therefore it would be favourable id some interlock is also provided at the contraction joints of RCC dams.

The main factor, which governs the dynamic response (stresses and deformations) of a concrete dam is damping. Structural damping ratios obtained from forced and ambient vibration tests are surprisingly low, i.e. damping ratios of the lowest modes of vibrations of large arch dams are of the order of 1 to 2 % of critical. In these field measurements the effect of radiation damping in the foundation and the reservoir are already included.

Linear-elastic dynamic interaction analyses of dam-foundation-reservoir systems would suggest damping ratios (structural and radiation damping) of about 10 % for the lowest modes of vibration and even higher values for the higher modes of large concrete dams. Accordingly, the maximum dynamic tensile stresses in an arch dam might be up to 2–3 times smaller when all dynamic interaction effects are considered than those obtained from an analysis with 5 % damping where the reservoir is assumed to be incompressible and the dynamic interaction effects with the foundation are represented by the foundation flexibility only (massless foundation). Unfortunately, there is still a lack of observational evidence, which would justify the use of large damping ratios in seismic analyses of concrete dams.

Moreover, in view of the fact that large concrete dams will exhibit nonlinear behaviour (joint opening and cracking) during the SEE, the linear dam-reservoir-foundation interaction models with analyses in the frequency domain are not applicable. Therefore, in view of the uncertainties in the estimation of the SEE ground motion, it is proposed to use damping ratios of maximum 5 % for large arch dams and not more than 7 % for gravity dams when no other information and data is available.

20.7.2 Embankment Dams

The seismic design of embankment dams is based on

(i) conceptual (empirical) criteria, which are mainly based on the observation of the behaviour of embankment dams during strong earthquakes and the behaviour of soils and rockfill under dynamic loadings, and

(ii) the results of seismic analysis of dams subjected to different types of design earthquakes, i.e. OBE and SEE. Usually several earthquakes must be analysed – at least three.

As a basis for the dynamic analysis, a static analysis that simulates the incremental construction of the dam body and the filling of the reservoir, and if applicable, a seepage analysis must be performed first before the earthquake ground motion can be applied.

The conceptual and constructional criteria for seismic-resistant fill dams are (ICOLD 2001):

- Foundations must be excavated to very dense materials or rock; alternatively the loose foundation materials must be densified, or removed and replaced with highly compacted materials, to guard against liquefaction or strength loss.
- Fill materials, which tend to build up significant pore water pressures during strong shaking must not be used.
- All zones of the embankment must be thoroughly compacted to prevent excessive settlements during an earthquake.
- All embankment dams, and especially homogeneous dams, must have high capacity internal drainage zones to intercept seepage from any transverse cracking caused by earthquakes, and to assure that embankment zones designed to be unsaturated remain so after any event that may have led to cracking.
- Filters must be provided on fractured foundation rock to preclude piping of embankment material into the foundation.
- Wide filter and drain zones must be used.
- The upstream and/or downstream transition zones should be 'self-healing', and of such gradation as to also heal cracking within the core.
- Sufficient freeboard should be provided in order to cover the settlement likely to occur during the earthquake and possible water waves in the reservoir due to mass movements etc.
- Since cracking of the crest is possible, the crest width should be wider than normal to produce longer seepage paths through any transverse cracks that may develop during earthquakes.

One of the most dangerous consequences of the dynamic loading of an embankment dam is the liquefaction of foundations or embankment zones that contain saturated fine-grained cohesionless and/or uncompacted materials.

The dynamic response of an embankment dam during strong ground shaking is governed by the deformational characteristics of the different soil materials. For large storage dams, the earthquake-induced permanent deformations must be calculated. The calculations of the permanent settlement of large rockfill dams based on dynamic analyses are still very approximate, as most of the dynamic soil tests are usually carried out with maximum aggregate size of less than 5 cm. This is a particular problem for rockfill dams and other dams with large rock aggregates and in dams, where the shell materials, containing coarse rock aggregates, have not been compacted at the time of construction. Poorly compacted rockfill may settle significantly during strong ground shaking but may well withstand strong earthquakes.

To get information on the dynamic material properties, dynamic direct shear or triaxial tests with large samples are needed. These tests are too costly for most rockfill dams. But as information on the dynamic behaviour of rockfill published in the literature is also scarce, the settlement prediction involves sensitivity analyses and engineering judgment.

At dam sites located on active or potentially active faults or discontinuities in the dam foundation, which can be moving during a strong earthquake, only conservatively designed earth core rockfill dams should be built. This means that in highly seismically active regions where there are doubts about possible movements along discontinuities in the dam foundation, earth core rockfill dams are the proper dam types (ICOLD 1998).

20.8 Exisiting Dams

The seismic safety aspects of existing dams is an important issue as most dam codes, regulations, recommendations and guidelines are primarily concerned with the design of new dams (Wieland 2006).

The design of a dam, which was considered as safe at the time it was commissioned may not be safe forever. This may be contradictory to the general opinion of owners and users of most structures. As earthquake engineering is still a relatively young discipline, design criteria, methods of analysis, design concepts etc. may be subject to changes especially when a large dam, designed according to the current state-of-practice, should be damaged during an earthquake. Thus there is a need for periodic checks of the seismic design criteria and the earthquake safety of large dams (and other structures as well), i.e. budgets for periodic seismic safety checks must be considered.

In general, dam owners and operators are reluctant to perform such checks unless there are laws and regulations and a dam safety organization, which has the authority and means to ensure that the rules are followed. In general, a thorough assessment of the design criteria is done when dam owners are applying for a new concession for their project. This may be adequate in the case of concession periods in the range of 30 years, but in some countries the concession periods are much longer such as, e.g. in Switzerland where the concession period for dam projects is 80 years. In this case reviews of the design criteria should be done as discussed in the previous section on Dam Safety Monitoring.

Again, the perception that what has been considered as safe once will remain safe forever is a dangerous misconception.

As a consequence during the long service life of a dam several seismic safety assessments will be needed.

In most European countries the economically feasible water resources have been developed. Although large dams belong to the first structures, which have been designed systematically against earthquakes since the 1930s, the seismic safety of these dams is unknown, as most of them have been designed using seismic design criteria and methods of dynamic analysis (pseudo-static analysis method) that are considered obsolete today.

The fact that no major dams have failed during earthquakes and that few lives have been lost may give the impression that well-designed dams are safe against earthquakes. We need to re-evaluate the seismic safety of existing dams based on current state-of-the-art practice and rehabilitate existing dams if necessary.

Additionally, there are a large number of smaller dams, especially earth structures, which were built either for irrigation or water supply by organisations or villagers with little experience in dam construction or they were built in previous centuries and subsequently abandoned. Earthquake effects on these dams have usually not been considered or in rather simplistic way.

As a prerequisite the seismic hazard at the dam sites must be reassessed to comply with the current seismic design criteria.

It must be pointed out that both new and existing large storage dams must satisfy today's safety criteria, which are equal for new and existing dams. Therefore a risk-based approach in which the remaining service life and the acceptable investment cost for saving additional lives is taken into account for existing dams, cannot be recommended.

20.9 Conclusions

In the seismic design and seismic safety assessment of the dams the following items are of main concern:

1. The seismic hazard is a multi-hazard for most dam projects. Ground shaking is the main hazard considered in all earthquake guidelines for dams. The other seismic hazards may even have been ignored.
2. Movements of active faults in the foot print of a dam or movements at discontinuities (faults, joints, bedding planes), which can be activated during strong nearby earthquakes, are the most critical seismic hazard for most dam types. If no other site can be selected then a conservatively designed earth core rockfill dam with wide filter and transition zones would be the right solution.
3. Dams are not inherently safe against earthquakes. However, the technology for designing and building dams and appurtenant structures that can safely resist the effects of strong ground shaking is available.
4. The concrete slab of concrete face rockfill dams is vulnerable to seismic settlements and seismic actions causing large inplane stresses if it acts as a monolithic structure. Open joints can almost completely eliminate these stresses resulting from the greatly different deformational behaviour and the great differences in the stiffness of the rockfill and the concrete.

5. As most dams built prior to 1989 when ICOLD has published its seismic design criteria of dams (ICOLD 2014), have not been checked for the SEE ground motion, the earthquake safety of these dams is not known and it must be assumed that a number of them do not satisfy today's seismic safety criteria. Therefore, owners of older dams shall start with the seismic safety checks of their dams.
6. The earthquake load case has evolved as the critical load case for most large dams even in regions of low to moderate seismicity.
7. Due to changes in the seismic design criteria and the design concepts it may be necessary to perform several seismic safety checks during the long economical life of a large dam.
8. Our knowledge on the behaviour of large dams during strong ground shaking is still very limited, therefore, each destructive earthquake affecting dams may reveal new features, which up to now have been overlooked or ignored.

Acknowledgements The author acknowledges the contributions of the members from the 32 countries participating in the works of the Committee on Seismic Aspects of Dam Design of the International Commission on Large Dams, which were published as ICOLD bulletins.

Open Access This chapter is distributed under the terms of the Creative Commons Attribution Noncommercial License, which permits any noncommercial use, distribution, and reproduction in any medium, provided the original author(s) and source are credited.

References

ICOLD (1998) Neotectonics and dams, bulletin 112, Committee on Seismic Aspects of Dam Design, International Commission on Large Dams (ICOLD), Paris
ICOLD (2001) Design features of dams to effectively resist seismic ground motion, bulletin 120, Committee on Seismic Aspects of Dam Design, International Commission on Large Dams (ICOLD), Paris
ICOLD (2011) Reservoirs and seismicity – state of knowledge, bulletin 137, Committee on Seismic Aspects of Dam Design, International Commission on Large Dams, International Commission on Large Dams (ICOLD), Paris
ICOLD (2014) Selecting seismic parameters for large dams, guidelines, bulletin 148, Committee on Seismic Aspects of Dam Design, International Commission on Large Dams (ICOLD), Paris
Wieland M (2003) Seismic aspects of dams, general report, Q.83 seismic aspects of dams. In: Proceedings of 21st international congress on large dams, ICOLD, Montreal
Wieland M (2006) Earthquake safety of existing dams, keynote lecture. In: Proceedings of first European conference on earthquake engineering and seismology, 1ECEES, (a joint event of the 13th European Conference on Earthquake Engineering & 30th General Assembly of the European Seismological Commission), Geneva, 3–8 Sept
Wieland M (2012a) Seismic design and performance criteria for large storage dams. In: Proceedings of 15th world conference on earthquake engineering, Lisbon, 24–28 Sep
Wieland M (2012b) Safety aspects of sustainable storage dams. In: Proceedings of 3rd International Symposium on Life-Cycle Civil Engineering (IALCCE 2012), Mini-symposium on sustainable dams and embankments, Vienna, 3–6 Oct
Wieland M, Chen H (2009) Lessons learnt from the Wenchuan earthquake. Int Water Pow Dam Constr 61(9):36–40
Wieland M, Mueller R (2009) Dam safety, emergency action plans, and water alarm systems. Int Water Pow Dam Constr 61:34–38

The manufacturer's authorised representative in the EU is Springer Nature Customer Service Centre GmbH, Europaplatz 3, 69115 Heidelberg, Germany. If you have any concerns regarding our products, please contact ProductSafety@springernature.com

Printed and bound by CPI Group (UK) Ltd, Croydon, CR0 4YY

23/03/2026

02076658-0007